General Airgap Field Modulation Theory for Electrical Machines

IEEE Press
445 Hoes Lane
Piscataway, NJ 08854

General Airgap Field Modulation Theory for Electrical Machines

Principles and Practice

Ming Cheng
Southeast University
Nanjing, China

Peng Han
Ansys, Inc.
Irvine, USA

Yi Du
Jiangsu University
Zhenjiang, China

Honghui Wen
Southeast University
Nanjing, China

IEEE PRESS
WILEY

Published by John Wiley & Sons, Inc., Hoboken, New Jersey.
Published simultaneously in Canada.

Library of Congress Cataloging-in-Publication Data applied for
Hardback ISBN: 9781119900344

Cover Design: Wiley
Cover Image: © Vectorig/Getty Images

Set in 9.5/12.5pt STIXTwoText by Straive, Chennai, India

Contents

Preface

Electrical machines are devices that convert mechanical energy into electrical energy or vice versa. They were invented in the 1800s and have a history of nearly 200 years. Other inventions of similar ages, such as the Watt steam engine, telegraph, incandescent light bulb, etc., have been outdated by emerging technologies. By contrast, the electrical machine shows great tenacity and vitality, becoming a living fossil of the Industrial Revolution.

Demand for high-performance electrical machines is increasing day by day with the rapid development of our social economy. Application areas of electrical machines have extended from conventional industrial drive to aerospace, transportation, numerical control machine tools, robots, and other high-tech fields, ranging from deep below the surface of the earth to deep space, from the furthest depths of the ocean to the surfaces of land and sea.

The diversity in performance requirements for different applications leads to the invention of novel electrical machine topologies with different performance advantages, especially those having multiple working spatial harmonics, such as the magnetically geared machine (MGM), permanent magnet vernier (PMV) machine, brushless doubly fed machines, just to name a few.

These new machine topologies show significant magnetic field modulation effects, posing great challenges to existing theories for the analysis of electrical machines. The operation of some emerging electrical machines, such as the PMV machine with dissimilar numbers of stator winding pole pairs and magnet polarities, can hardly be explained directly by the well-established theory for induction machines and synchronous machines. In addition, most theories and methods for the analysis of electrical machines were developed and therefore valid for only certain machine types.

Based on the extensive scientific and industrial research for high-performance electrical machines and drive technologies conducted by the Jiangsu Electrical Machines & Power Electronics League (JEMPEL), Southeast University, Nanjing, China, over the past decades, the authors noticed the generality of airgap magnetic field modulation phenomena in electrical machines and its instrumental role in improving performance of electrical machines. The discoveries were further examined against almost all the known electrical machine topologies, and then theorized to develop the general airgap field modulation theory.

The book is organized into seven chapters and three appendices, as outlined below:

- Chapter 1 reviews the historical development of electrical machines and their theories.
- Chapter 2 analyzes the airgap magnetic field modulation phenomena in common machine topologies, aiming to reveal the ubiquity of magnetic field modulation phenomena in electrical machines.
- Chapter 3 abstracts a unit machine with one stator, one rotor and one layer of airgap as a cascade of three key elements, based on which generalized mathematical models for the three elements are proposed, forming the general airgap field modulation theory framework for electrical machines.
- Chapter 4 analyzes the relationship between different modulation behaviors and their torque compositions.
- Chapter 5 applies the general airgap magnetic field modulation theory to multiple representative machine topologies to show its application in qualitative analysis and quantitative calculation of machine performance.
- Chapter 6 covers the innovation of machine topology with the guidance of the general airgap field modulation theory.
- Chapter 7 presents more application examples of the developed theory.
- Appendix A shows the mathematical derivation of the three typical modulation operators.
- Appendix B clarifies the relationship between electromagnetic forces/torques on conductors placed in the airgap and in-slot conductors.
- Appendix C presents the Maxwell Stress Tensor method and principle of virtual work, which are the basis of force/torque analysis using the general airgap field modulation theory.

The contents presented in this book are a selected collection of scientific and industrial research work conducted by the JEMPEL and its extended research groups during the past decades. The authors are very grateful to all the JEMPEL members, especially Prof. Xiaoyong Zhu, Prof. Yubin Wang, Dr. Le Sun, Dr. Xinkai Zhu, Dr. Jingxia Wang, Dr. Yu Zeng for their dedicated assistance in preparing the manuscript, and Dr. Gan Zhang, Mr. Zhengzhou Ma, Ms. Chenchen Zhao for their tremendous help in the preparation of main figures.

The authors would also like to express their sincere gratitude to Prof. Wei Hua, Prof. Jianzhong Zhang, Prof. Wenxiang Zhao, Prof. Xianglin Li, Dr. Qiang Sun, Dr. Hongyun Jia, Dr. Feng Yu, Dr. Feng Li, Dr. Lingyun Shao, Dr. Peng Su,

Dr. Minghao Tong, Dr. Xiaofeng Zhu and Dr. Peixin Wang for their excellent research outcomes, which are essential ingredients of this book.

We are deeply indebted to our colleagues and friends worldwide for their continuous support, encouragement, and discussion on the subject matter over the years, especially Prof. Ayman EL-Refaie, Prof. C.C. Chan, Prof. Dan M. Ionel, Prof. Hossein Torkaman, Prof. Ion Boldea, Prof. Iqbal Husain, Prof. James Kirtley, Prof. Jianguo Zhu, Prof. K.T. Chau, Prof. Longya Xu, Prof. W.N. Fu, Prof. Ying Fan, Prof. Z.Q. Zhu, Prof. Zhe Chen, Prof. Zheng Wang, Dr. Arijit Banerjee, Dr. Baoyun Ge, Dr. Hao Huang, Dr. Jianning Dong, Dr. Kaiyuan Lu, Dr. Ming Xu, Dr. Sa Zhu, Dr. Shengyi Liu, Dr. Ying Pang, Ms. Xiaoping Li, for their valuable feedback on this theory.

We also much appreciate the reviewers of this book for their thoughtful and constructive comments, and Editors of John Wiley & Sons for their professional and timely support in the review and production process.

Last but not least, we thank our families for nurturing and unconditionally supporting the writing of this book.

About the Authors

Ming Cheng received his B.Sc. and M.Sc. degrees from Southeast University, Nanjing, China, in 1982 and 1987, respectively, and his Ph.D. degree from the University of Hong Kong, Hong Kong, in 2001, all in electrical engineering. Since 1987, he has been with Southeast University, where he is currently an Endowed Chair Professor at the School of Electrical Engineering and the Director of the Research Center for Wind Power Generation.

He is a Fellow of the Institute of Electrical and Electronics Engineers (IEEE) and the Institution of Engineering and Technology (IET) and was a Distinguished Lecturer of the IEEE Industry Applications Society for 2015/2016. He has served as editor-in-chief, editor and editorial board member of various international journals, as well as chair and organizing committee member of many international conferences, especially in the area of Electrical Machines and Power Electronics.

His teaching and research interests include electrical machines, motor drives for EV, renewable energy generation, and servo motor & control. In these areas, he has published over 500 refereed technical papers and 7 books and holds over 150 invention patents.

He has received many awards, including Second Prize in the State Technological Invention Awards (given by the State Council of the People's Republic of China); First Prize in China's Ministry of Education's Natural Science Awards; First Prize in Jiangsu Provincial Government's Science and Technology Award; the IET Achievement Award; and the Environmental Excellence in Transportation Award for Education, Training, and Public Awareness by SAE International.

Peng Han received B.Sc. and Ph.D. degrees in electrical engineering from the School of Electrical Engineering, Southeast University, Nanjing, China, in 2012 and 2017, respectively.

From November 2014 to November 2015, he was a Guest Ph.D. student at the Department of Energy Technology, Aalborg University, Aalborg, Denmark, where he focused on brushless doubly fed machines' application in wind energy conversion and high-power drives. He is currently with Ansys, Inc., USA, as a Senior Application Engineer. Before joining Ansys, he was a Postdoctoral Researcher with the Center for High Performance Power Electronics (CHPPE) Department of Electrical and Computer Engineering, The Ohio State University, and later the SPARK Laboratory, Department of Electrical and Computer Engineering, University of Kentucky. His current research interests include electrical machines, machine drives, power electronics, and renewable energy.

He is an IEEE Senior Member and an Associate Editor for IEEE Transactions on Industrial Electronics, IEEE Transactions on Industry Applications and Journal of Power Electronics. He received two best paper/poster awards from IEEE conferences, and Third Prize in the IEEE IAS Student Thesis Contest in 2018.

Yi Du received B.Sc. and M.Sc. degrees in electrical engineering from Jiangsu University, Zhenjiang, China, in 2002 and 2007, respectively, and a Ph.D. degree in electrical engineering from Southeast University, Nanjing, China, in 2014.

He has been with Jiangsu University since 2002, where he is currently a Professor in the School of Electrical and Information Engineering. From 2018 to 2019, he was a Visiting Professor with the Department of Electronic & Electrical Engineering, The University of Sheffield, Sheffield, UK. His research interests include design and analysis of electrical machine systems with low-speed and high-torque outputs, and with wide-speed range.

Honghui Wen received B.Sc. and Ph.D. degrees in electrical engineering from the School of Electrical Engineering, Southeast University, Nanjing, China, in 2016 and 2021, respectively.

Since January 2022, he has been with the College of Electrical and Information Engineering, Hunan University, Changsha, China, where he is currently an Assistant Professor. His teaching and research interests include the design, analysis, and optimization of magnetic field electrical machines. In these areas, he has published one book and over 10 peer-reviewed technical papers.

About the Companion Website

This book is accompanied by a companion website:

www.wiley.com/go/genairgapfieldmodulationtheory

The website includes models and scripts.

1

Introduction

1.1 Review of Historical Development of Electrical Machines

Electrical machines are electromagnetic devices to achieve electromechanical energy conversion. Since Jacobi invented the DC machine (DCM) in 1834 and Tesla invented the induction machine (IM) in 1880s, electrical machine technology has developed rapidly, and its application has become more and more extensive. Electrical machines have become one of the most important energy and power technologies to support the national economy. Various types of machines with power levels ranging from several milliwatts to hundreds of megawatts, especially the three traditional machine topologies, namely the DCM, AC induction (asynchronous) machine, and AC synchronous machine (SM), have made great contributions to the development of human society [1, 2].

It is worth mentioning that the steam engine, telegraph, incandescent lamp, etc., which appeared at virtually the same time as electrical machines, have been replaced by emerging technologies and have gradually faded into disuse. By contrast, the electrical machine presents a tenacious vitality, and can be considered a surviving fossil of the modern industrial revolution, one which constantly renews its life. Correspondingly, the development and innovation of electrical machine theories and technologies have never stopped.

With the proliferation of electrification and automation technologies, the application of electrical machines has expanded from conventional industrial drive to aerospace, transportation, CNC equipment, robotics and other high-tech fields, and from ground to deep space, deep sea, and deep earth. The performance requirements of machines for different applications are constantly refined, and the traditional brushed DCMs, IMs, and SMs are difficult to meet the demanding requirements of new fields and applications. Meanwhile, the rapid development of material technology, processing and manufacturing technology, and control technology, combined with new application requirements, has given

General Airgap Field Modulation Theory for Electrical Machines: Principles and Practice,
First Edition. Ming Cheng, Peng Han, Yi Du, and Honghui Wen.

Companion website: www.wiley.com/go/genairgapfieldmodulationtheory

rise to various new electrical machines with different constructions, different working principles, and different performance advantages, such as: synchronous reluctance machines (SynRMs) [3–5], permanent magnet (PM) brushless machines [6, 7], vernier machines [8], brushless doubly-fed induction machines (BDFIMs) [9–11], brushless doubly-fed reluctance machines (BDFRMs) [12, 13], transverse flux machines [14], switched reluctance machines (SRMs) [15, 16], stator-PM brushless machines [17–20], PM vernier (PMV) machines [21, 22], magnetically-geared machines (MGMs) [23–26], dual-mechanical-port (DMP) machines [27, 28], etc.

In order to provide an overview of the relevant research on electrical machines, a bibliometric analysis was performed in IEEE Xplore digital library database at the end of 2021. The results summarized in Figure 1.1 show the diversity of machine topologies and soaring numbers of publications on these "unconventional" machines.

- As representative traditional electrical machines, IMs and wound-field salient-pole SMs have experienced the longest history of research. Considerable research work had been done before the field-oriented control (FOC) method was invented by F. Blaschke in 1971. Research on IMs grows steadily and peaked at 2014–2015. In contrast, research on wound-field salient-pole SMs was interrupted in the 1970s and 1980s and resumed later, but it is much less popular than that of IMs.

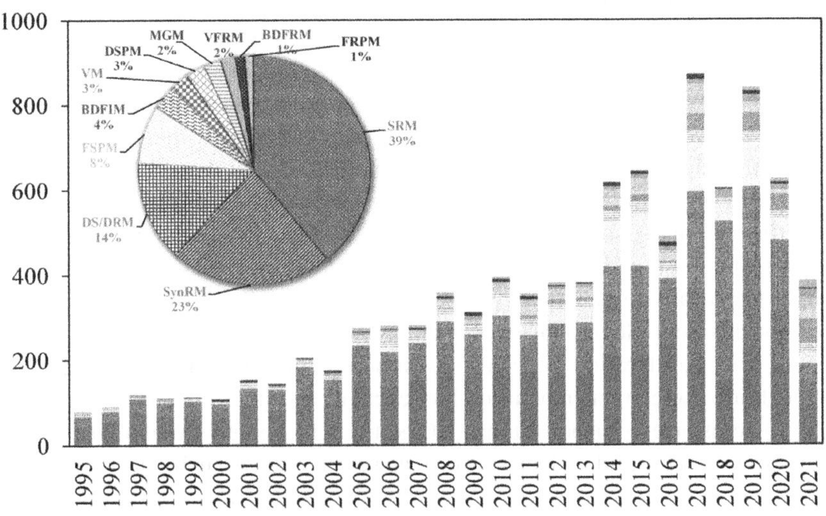

Figure 1.1 Number of publications on "unconventional machines" indexed by IEEE Xplore digital library database. DS/DRM – Dual-stator/dual-rotor machine, FSPM – Flux-switching PM, VM – Vernier machine, DSPM – Doubly-salient-pole PM, VFRM – Variable flux reluctance machine, FRPM – Flux-reversal PM.

- High-energy-product PM materials were introduced into the manufacturing and performance improvement of electrical machines in the late 1970s, after which the research on PM brushless machines began to grow steadily, with a dramatic increase in 2010. Afterwards, the research on PM machines decreased due to the large price fluctuation of PM materials, but still stayed at a high level.
- In addition to IMs and PMSMs, research on SRMs has continued to grow steadily since its revival in 1980. By contrast, research on SynRMs has been more stable, with a dramatic growth in 2014 and 2015. It was probably caused by the large price fluctuation of PM materials, which led to the concepts of "PM-free" and "less-PM" electrical machines.
- The so-called "stator-PM machines" invented in the 1990s which mainly include doubly-salient-pole PM machines, flux-switching PM machines (FSPMs) and flux-reversal PM machines started to gain their popularity after 2005. Compared to doubly-salient-pole PM machines and flux-reversal PM machines, FSPMs were more attractive to researchers in both academia and industry. Research on FSPMs witnessed a dramatic increase in 2014 and peaked in 2014–2015.
- Research on BDFIMs and BDFRMs dates back to early 1970s. It has been increasing since the 1990s and peaked in 2014–2015.
- Research on MGMs and vernier machines started in the early 1990s but did not receive much attention until 2010. It was hardly affected by the price fluctuation of PM materials.
- Dual-stator/dual-rotor machine topologies started to gain popularity after 2004 and keep increasing steadily afterwards, peaking in 2014–2015.

At the same time, theories for electrical machines have also been enriched and improved. In addition to the basic physical laws (such as the law of electromagnetic induction and the law of electromagnetic force) followed by electrical machines, a variety of theories and analysis methods have emerged accordingly to conduct the performance analysis and calculation of different machines.

For example, to address the difficulties caused by the unequal direct-axis (d-axis) and quadrature-axis (q-axis) reluctance of the salient-pole SM in armature reaction calculation, A. Blondel proposed the two-reaction theory [29, 30]. When the axis of the armature magnetomotive force (MMF) F_a coincides neither with the d-axis nor with the q-axis, F_a can be decomposed into the d-axis component F_{ad} and the q-axis component F_{aq}, as shown in Figure 1.2. Then the d-axis and q-axis armature reaction magnetic fields can be calculated separately, and finally they are superimposed. It has been proved that satisfactory results can be obtained by using the two-reaction theory when the magnetic circuit saturation is not taken into account. After that, R.H. Park further generalized and extended the two-reaction theory and proposed the famous Park transformation, thereby establishing general calculation formula for the current, voltage, power and torque of SMs under

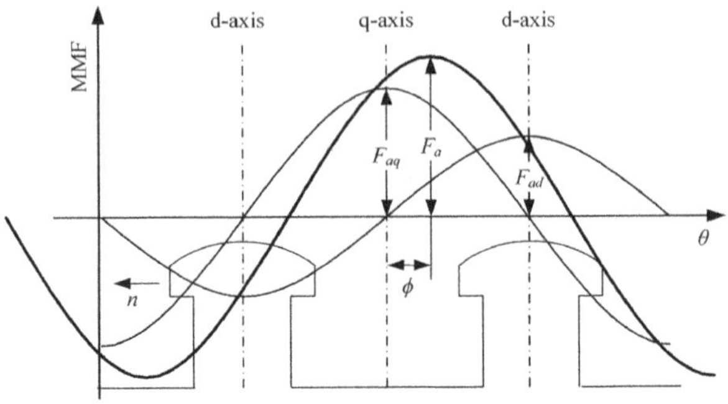

Figure 1.2 Decomposition of armature MMF of would-field salient-pole SM into d-axis and q-axis components.

steady and transient states [31], which brings great convenience to the analysis and calculation of SMs.

Another example is that G. Kron proposed a unified theory for analyzing electrical machines [32], which raised the theory of electrical machines to a new level. He analyzed machine characteristics from the perspective of energy for the first time. According to the law of energy conservation [33], there are four forms of energy in electrical machines: electrical energy, mechanical energy, electromagnetic field storage energy and thermal energy. According to the motor convention, the energy equation can be written as

$$\begin{pmatrix}\text{Input energy}\\ \text{from source}\end{pmatrix} = \begin{pmatrix}\text{Coupled field energy}\\ \text{storage increment}\end{pmatrix} + \begin{pmatrix}\text{Output mechanical}\\ \text{energy}\end{pmatrix} + \begin{pmatrix}\text{Energy loss}\\ \text{converted to heat}\end{pmatrix} \tag{1.1}$$

In addition, the concept of magnetic co-energy was proposed, and the relationship between magnetic co-energy and magnetic energy is:

$$W_m = \int_0^{\psi_1} i d\psi = i_1\psi_1 - \int_0^{i_1} \psi di = i_1\psi_1 - W'_m \tag{1.2}$$

where W_m is the magnetic energy, W'_m the magnetic co-energy, i_1 the winding current, and ψ_1 the magnetic flux linkage. When represented graphically, the magnetic co-energy is the area of the vertically shaded part in Figure 1.3. The rate of change of the magnetic co-energy to the mechanical displacement is the electromagnetic torque:

$$T_{em} = \frac{dW'_m}{d\theta} \tag{1.3}$$

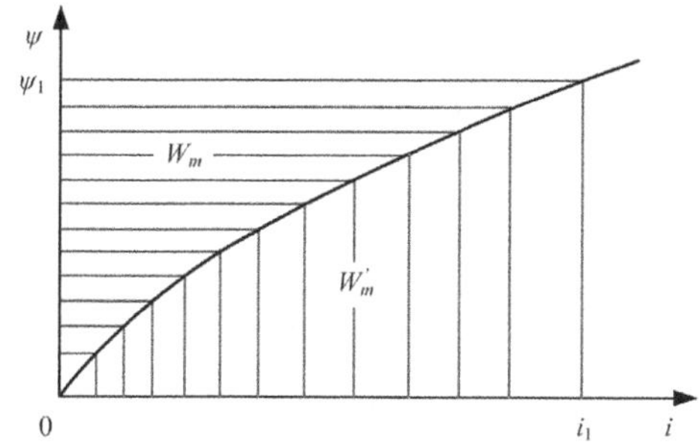

Figure 1.3 Magnetic energy W_m and magnetic co-energy $W_m^{'}$.

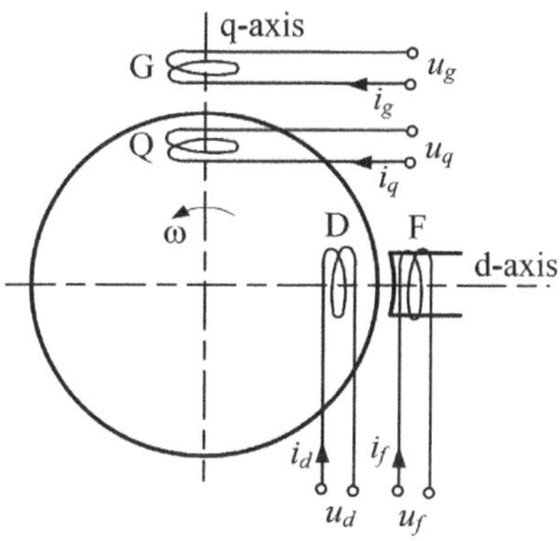

Figure 1.4 Two-axis primitive machine.

where T_{em} is the electromagnetic torque and θ the mechanical displacement angle of electrical machines.

Again, B. Adkins et al. proposed the general theory of AC machines. Any electrical machines can be equivalent to a primitive machine model by coordinate transformation, as shown in Figure 1.4. Based on this primitive machine model, a set of voltage equations representing the relationship between voltage and current of the primitive machine and torque equations representing the relationship between torque and current can be derived and then solved using a unified approach [34].

Cells with solid-line borders in Table 1.1 list typical electrical machine theories in the history of machine development as well as their applications, authors and other information, which together constitute the classical electrical machine theory and lay a solid theoretical foundation for the development of machine technology represented by DCMs, IMs, and SMs.

Table 1.1 Typical theories for electrical machines.

Year	Name of theory	Typical applications	Authors
1913	Two-reaction theory [29, 30]	Wound-field salient-pole SMs and SynRMs	A. Blondel (France)
1925	Rotating magnetic field theory [35]	AC electrical machines with sinusoidal back-electromotive force (back-EMF)	K.L. Hansen (United States)
1926	Cross-field theory [36]	AC electrical machines with sinusoidal back-EMF	H.R. West (United States)
1929	Park transformation [31]	AC electrical machines with sinusoidal back-EMF	R.H. Park (United States)
1930	Generalized theory of electrical machinery [32]	Electromechanical energy conversion devices	G. Kron (United States)
1954	Symmetrical components [37]	AC electrical machines with sinusoidal back-EMF	W.V. Lyon (United States)
1959	Space vector theory for transient analysis [38]	IMs and SMs	Pák K. Kovács (Hungary)
1973	General theory using equivalent magnetic circuit [39]	SMs and DCMs	J. Fienne (United Kingdom)
1975	The general theory of AC machines [34]	AC electrical machines	B. Adkins (United Kingdom)
1965	Winding function theory [40, 41]	IMs and SMs	N.L. Schmitz and D.W. Novotny (further improved by T.A. Lipo) (United States)
1992	Spiral vector theory [42]	AC circuits and machines	S. Yamamura (Japan)
1994	Unified theory of torque production [43]	All electrical machines	D.A. Staton, T.J.E. Miller, et al. (United Kingdom)
2017	General airgap field modulation theory [44]	All electrical machines	M. Cheng, P. Han, W. Hua (China)

1.2 Limitations of Classical Electrical Machine Theories

Although the classical electrical machine theory has brought great convenience to the analysis of traditional DC, induction and synchronous machines, it still seems to be inadequate in analyzing large numbers of new machine topologies. In summary, limitations of the classical electrical machine theory mainly lie in the following three aspects.

1.2.1 Fragmentation of Electrical Machine Theories

Among existing electrical machine theories, the two-reaction theory was primarily used to analyze SMs [29–31]. The rotating magnetic field theory [35] and cross-field theory [36] were suitable for AC machines with sinusoidal back-electromotive forces (EMFs). The unified theory of torque production based on magnetic flux linkage-current trajectories was employed to analyze all electrical machines by numerical methods [43]. The winding function theory was mainly for IMs and SMs [41]. The general theory of AC machines based on the two-axis primitive machine [34] and general theory using equivalent magnetic circuits [39] were mainly applied to IMs and SMs, though they can also be further extended for DCMs. It can be concluded that the existing theories are only valid for certain machine topologies and none of them are applicable to all machine types.

In addition, some theories can only be used as tools for quantitative analysis of machine performance, but fail to interpret the physics behind the machine operation, or vice versa. For instance, the general theory of AC electrical machines developed by B. Adkins et al. based on the two-axis idealized machine or primitive machine model [34] is only valid for conventional AC machines with explicit direct and quadrature axes and DCMs. The general theory using equivalent magnetic circuits developed by J. Fienne et al. [39] can be used for both AC machines and DCMs, but it can only be used for performance calculations and cannot reveal the internal mechanism and physical nature of electromechanical energy conversion.

On the other hand, machines with the same or similar structures can be treated as different machine types from different perspectives. For example, both machines shown in Figures 1.5 and 1.6 have three layers from the inside to the outside in the mechanical structure, namely, the innermost PM layer, the salient-pole rotor layer in the middle, and the outermost winding layer. However, the machine shown in Figure 1.5 was well known as a MGM [45], while the one shown in Figure 1.6 was considered as a partitioned-stator FSPM machine [46].

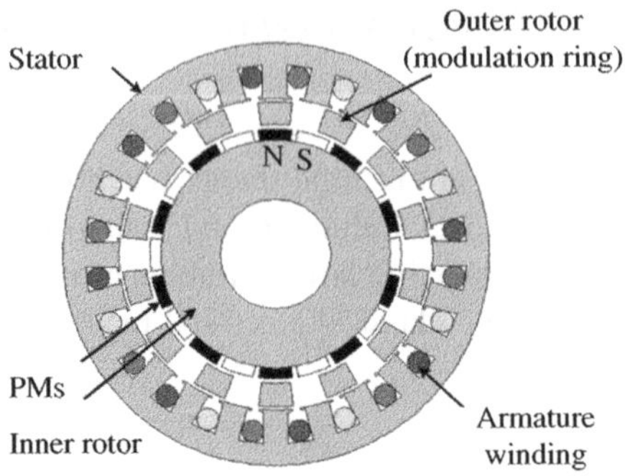

Figure 1.5 Dual-rotor MGM Adapted from [45].

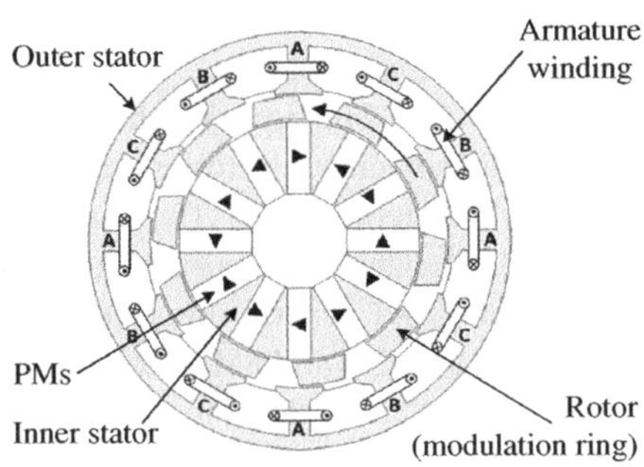

Figure 1.6 Partitioned-stator FSPM machine [46].

1.2.2 Limitations in Analysis of Operating Principles

For some new/special machines, the classical electrical machine theory is no longer fully applicable. In other words, it is difficult to directly use the classical theory to interpret the working principle and analyze the performance of these machines. For example, according to the classical electrical machine theory and the basic principles of electromechanical energy conversion [1, 2], the numbers of pole pairs of the stator winding and rotor field are required to be identical so as to achieve continuous electromechanical energy conversion. However, for the PMV machine [21] shown in Figure 1.7, the armature field created by the armature winding has one pole pair, but the rotor has 34 PM poles. Obviously, the numbers of stator and rotor pole pairs are unequal, so intuitively, the machine will not be able to operate. Even if the winding MMF harmonics are considered, the 17-pole-pair MMF harmonic is of a very low amplitude and the corresponding torque component is almost negligible. Therefore, this type of machine seems to be of no practical value according to the classical theory. However, research results show that this machine type can not only realize electromechanical energy

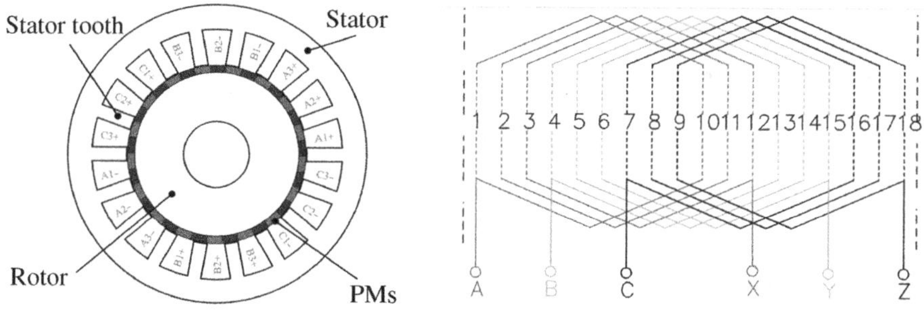

Figure 1.7 PMV machine, (a) cross-sectional view, (b) linear winding pattern.

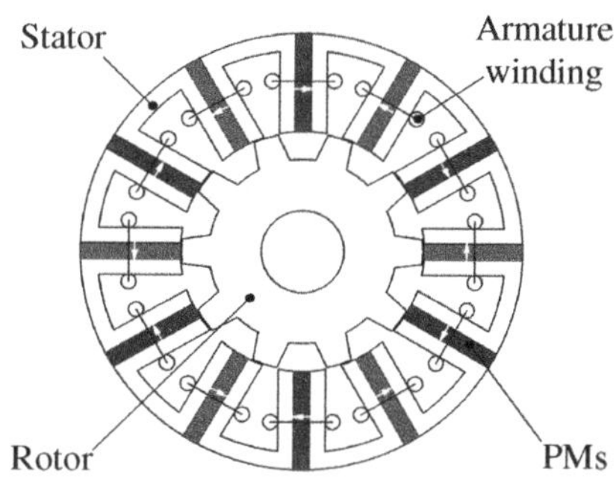

Figure 1.8 12/10 FSPM machine.

conversion, but also offers higher torque density than conventional PM machines [47], the reason of which will be explained later in this book.

FSPM machines are another example. Figure 1.8 shows a 12-slot stator and 10-pole rotor (12/10) FSPM machine [48, 49] with both the armature windings and PMs located in the stator and a simple salient-pole rotor. This machine has 12 PMs on the stator and adjacent magnets are magnetized in opposite directions, forming a stationary magnetic field with 6 pairs of poles, and 10 salient poles on the rotor. The equivalent number of main pole pairs obtained according to its speed and current frequency is 4, which neither equals to the number of pole pairs of the stator PM field nor the number of salient poles of the rotor. It is difficult to explain this mismatch by definitions of pole pairs in the classical machine theory.

1.2.3 Lack of Uniformity in Performance Analysis

The analysis of different types of machines is isolated from each other and lacks internal uniformity due to the fragmentation of electrical machine theories. Even for IMs and SMs, though both of them are AC machines, the analyses

Table 1.2 Comparison of analysis models for IMs and SMs.

<table>
<tr><th></th><th>IM</th><th>SM (taking the non-salient-pole type as an example)</th></tr>
<tr><td>Electrical equivalent circuit (per-phase)</td><td>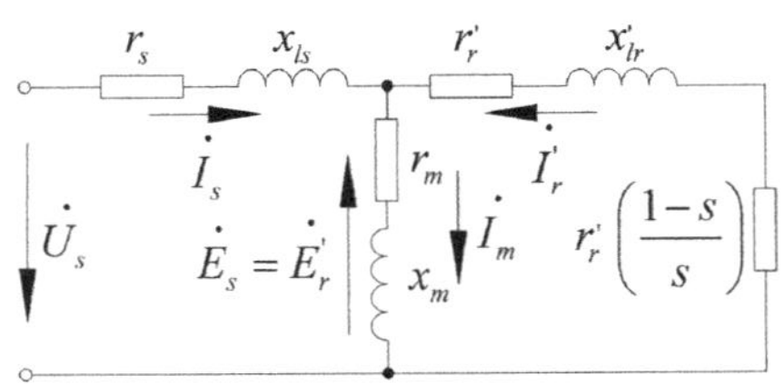</td><td>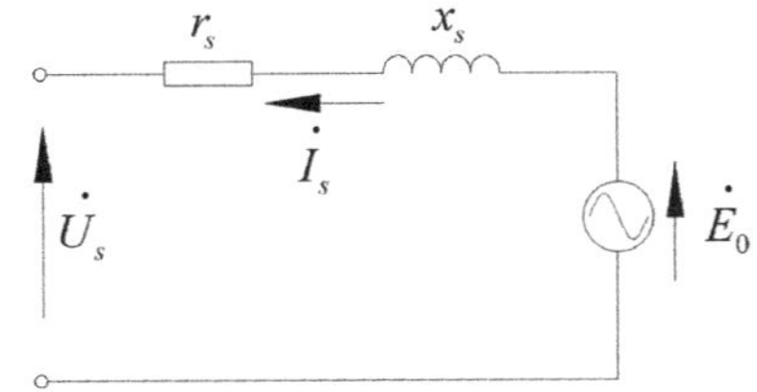</td></tr>
<tr><td>Phasor diagram (per-phase)</td><td>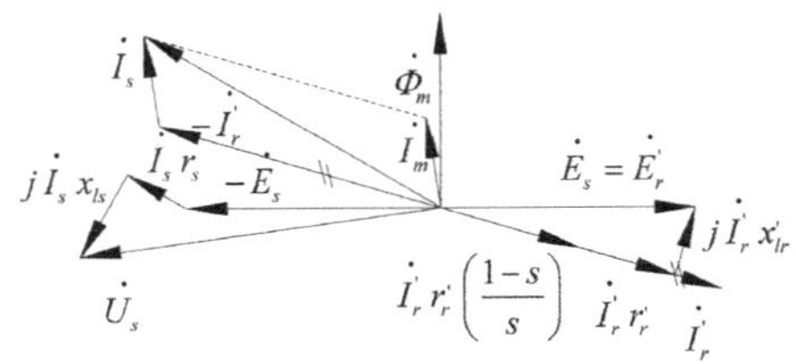</td><td>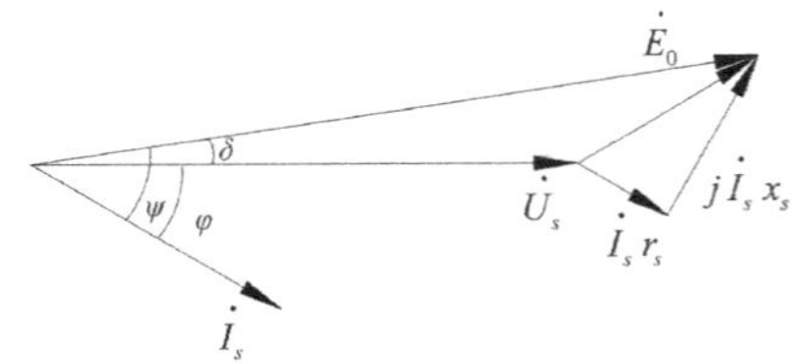</td></tr>
<tr><td>Torque equation</td><td>$T = \frac{p}{\omega} U_s^2 \frac{sr_r'}{\left(sr_s + c_1 r_r'\right)^2 + s^2\left(x_{ls} + c_1 x_{lr}'\right)^2}$</td><td>$T = \frac{p}{\omega} \frac{E_0 U_s}{x_s} \sin\delta$</td></tr>
<tr><td>Torque characteristic</td><td>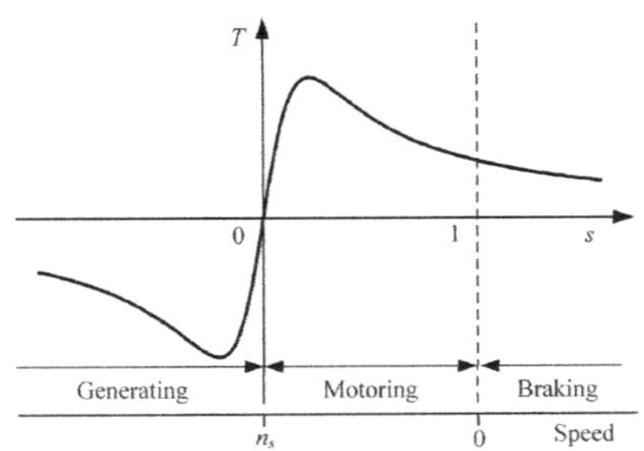
</td><td>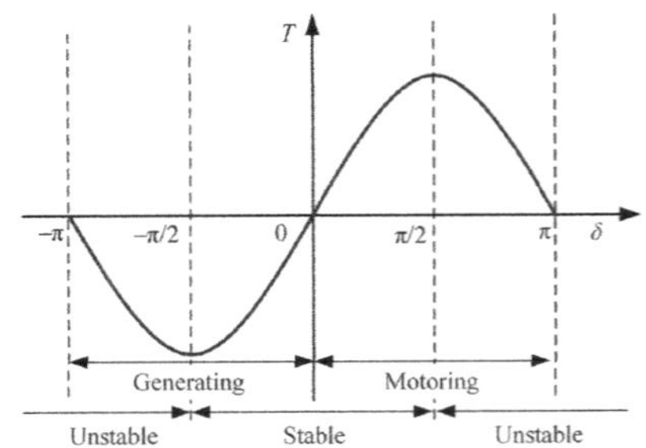
</td></tr>
</table>

are performed along two separate technical paths, and the results obtained, such as equivalent circuits, phasor diagrams, torque expressions, and torque characteristic curves, are incomparable, as shown in Table 1.2.

This dilemma may not be prominent in the past, because IMs were mainly used as electric motors and SMs mainly operated as generators, and there was little need to make direct comparisons between IMs and SMs. However, in recent years, SMs, especially PMSMs, have been widely used as motors. For a specific application, people are often faced with the question of how to choose between an induction motor and a synchronous motor. However, it is usually challenging to conduct a direct and fair performance comparison of the two motor types by existing electrical machine theories. The decision can only be made based on the field testing results of real motors, which not only brings inconvenience to engineering practice, but also slows down the development of electrical machine technologies.

1.3 Overview of Magnetic Field Modulation Machines and their Theories

The use of "magnetic field modulation" technology in electrical machines originates from the reluctance vernier motor proposed by C.H. Lee in the 1960s [8, 50], as shown in Figure 1.9. This motor has N_s stator teeth and p-pole-pair armature windings on the stator and N_r salient poles on the rotor. It relies on open slots in the stator and rotor to alter the airgap permeance and achieve speed reduction. C.H. Lee proposed the definition of the so-called "electric gear ratio" (hereinafter referred to as "magnetic gear ratio"). This type of motor uses salient poles in stator and rotor to reconstruct the airgap permeance profile and achieve speed change, which can be conceived as the origin of the term "magnetic field modulation."

To improve the torque density and operating efficiency of the reluctance vernier motor, A. Ishizaki et al. [51] proposed a PMV motor by embedding PMs in the stator and rotor slots. Subsequently, A. Toba et al. [21] further proposed the split-tooth, open-slot, and double-stator rotor-PM vernier motor topologies. E. Spooner [52] from Durham University proposed a hybrid vernier motor with PMs mounted on the stator tooth surfaces. Although both PM and hybrid vernier motors rely on magnets to generate excitation magnetic fields, the relationship among the number of stator teeth, the number of salient rotor poles, and the number of armature winding pole pairs is still consistent with that of reluctance vernier motors. They can still be analyzed using the airgap permeance reconstruction method.

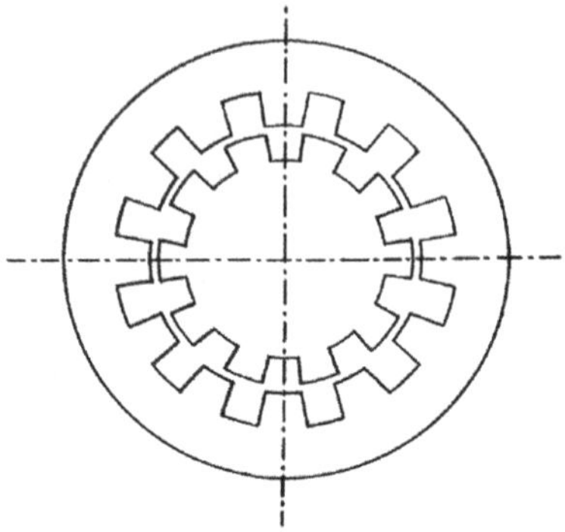

Figure 1.9 Reluctance vernier machine.

Brushless doubly-fed machines can be understood as vernier motors whose PMs are replaced by AC windings. References [53, 54] analyzed the airgap permeance, airgap magnetic field, and the relationship between them based on the principle of magnetic field modulation, and clarified the difference between "sum modulation" and "differential modulation," thus revealing the nature of "pole number converter" in doubly-fed reluctance machines. To the best of the authors' knowledge, reference [11] is probably the first use of the term "modulation" of the magnetic field.

In 2001, K. Atallah and D. Howe from the University of Sheffield, UK, proposed a magnetic field-modulated coaxial magnetic gear [23], which can be regarded as a doubly-fed machine whose two AC windings are replaced by two PM rotors, forming a dual-rotor structure. Based on the "magnetic field modulation" of the stationary reluctance ring in the middle, the torque and speed transmission between the inner and outer rotor is realized, as shown in Figure 1.10. The working principle of the magnetic gear was analyzed using the concept of magnetic field modulation in [55]. Therefore, the term "magnetic field modulation" was later also called "magnetic-gear effect," and the two terms were used together and considered interchangeable.

Recently, the "magnetic field modulation" phenomenon has received unprecedented attention. The so-called MGM was firstly proposed by K.T. Chau et al. from the University of Hong Kong by integrating a magnetic gear with a PMSM [25]. Since then, various types of MGMs based on the principle of "magnetic field modulation" [54, 56–59], PMV machines [55, 60], magnetic field modulated machines [61–64], partitioned-stator machines [65, 66], and other new machine topologies have emerged, enriching the research content of the electrical machine discipline.

At the same time, the phenomenon of "magnetic field modulation" in various types of machines has been revealed, providing a new perspective for the analysis of these machines. For example, Prof. Z.Q. Zhu from the University of Sheffield, UK [67], Prof. C.C. Chan et al. from the University of Hong Kong [68], and the

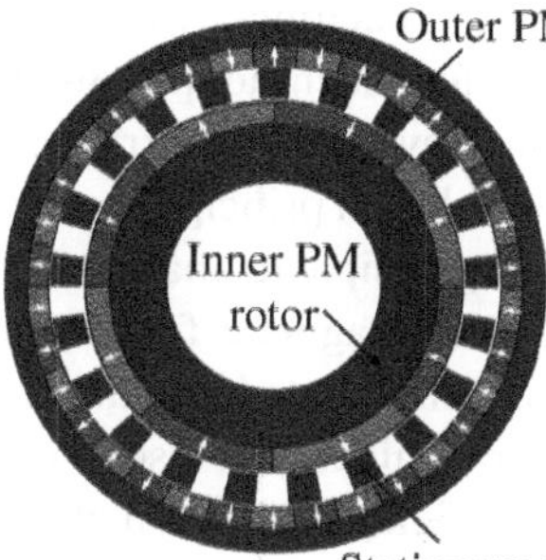

Figure 1.10 Coaxial magnetic gear.

authors' extended research group [69] have almost simultaneously discovered the phenomenon of "magnetic field modulation" in FSPM machines. In this phenomenon, the stationary permanent magnetic field generated by PMs located in the stator is "doubly modulated" by the stator teeth and rotor salient poles. Prof. R. Qu et al. from Huazhong University of Science and Technology, China, studied the relationship between magnetic gears and vernier motors [70], i.e. the principle of magnetic field modulation. Prof. W.N. Fu et al. from the Hong Kong Polytechnic University developed a magnetic field modulation analysis method based on the magnetic gear principle and conducted a comparative analysis of the basic magnetic field-modulated machine, PMV machine, doubly-fed reluctance machine and stator-rotor dual PM machine in a unified manner [71].

In summary, the discovery and understanding of the principle of magnetic field modulation play a critical role in interpreting the operation of electrical machines, as corroborated by detailed analysis of some machine topologies [28, 72, 73]. However, existing studies are limited to magnetic field-modulated machines in its narrow sense (including vernier machines, MGMs, etc.) and mainly focus on the influence of a few dominating magnetic field harmonics on the machine output generated by the alternating airgap permeance formed by salient reluctance poles (including stator teeth and rotor teeth).

The mechanism of the generation and variation of magnetic field harmonics caused by salient reluctance poles or short-circuited coils in various machine types has not been theoretically revealed and unified, nor has a comprehensive analysis and design of excitation sources and armature windings from the perspective of the magnetic field modulation phenomenon been conducted. The existing research is still at the stage of coincidental discovery of "modulation phenomena" and conceptual interpretation of "modulation principle." It has not yet reached a comprehensive understanding and utilization of magnetic field modulation, nor has its theorization been fully developed. It is thus hardly possible to analyze and design different types of electrical machines with different operating principles under a unified theoretical framework.

In the course of studying stator-PM machines [74–81], PMV machines [82], MGMs [83, 84], and brushless doubly-fed machines [85, 86] over the past decades, the authors' extended research group revealed the intrinsic unity of their operations based on the magnetic field modulation principle and further made attempts to use the magnetic field modulation principle to analyze traditional DCMs, IMs, and SMs, thus demonstrating that the magnetic field modulation principle is applicable not only to magnetic field-modulated machines in its narrow sense, such as magnetic gears and vernier machines, but also to conventional electrical machines. That is to say, the magnetic field modulation phenomenon is universal and exists in all types of electrical machines.

This ubiquity motivated us to think of the possibility of developing a new generalized theory for electrical machines since the magnetic field modulation phenomenon is universal. Inspired by this original idea, the authors' team, after years of continuous explorations and extensive trial-and-error, abstracted a unit machine as a cascade of three basic elements, i.e. an "excitation source-modulator-filter," for the first time, and constructed modulation operators to formulate the modulation phenomenon. The general airgap magnetic field modulation theory for electrical machines was then established [44, 87], and its validity has been verified by qualitative and quantitative analysis of numerous machine examples [88–97].

A number of special issues/sections focusing on the discussion and implementation of the developed theory have been organized and published during the past four years, including the IEEE TRANSACTIONS ON INDUSTRIAL ELECTRONICS Special Section on "Emerging Multiport Electrical Machines and Systems," the IEEE TRANSACTIONS ON ENERGY CONVERSION Special Section on "Emerging Electrical Machines and Drives for Smart Energy Conversion," the IEEE ACCESS Special Section on "Advanced Energy Conversion Systems Based on Multi-port Electrical Machines," and the Transactions of China Electrotechnical Society Special Issue on "Magnetic Field Modulation Machines and Theory," etc.

The general airgap field modulation theory has also attracted widespread attention and citations from numerous scholars all over the world since its publication. The affirmation and positive feedback from these colleagues greatly encouraged us and helped build our confidence to compile the relevant materials into this book.

1.4 Scope and Organization of the Book

The purpose of this book is to systematically introduce the general airgap field modulation theory for electrical machines to a broad audience, and to make an attempt to explain the working principle, analyze the characteristics, and guide the topological innovation of electrical machines using the developed theory, with the hope to attract more scholars' attention and participation to enrich the theory of electrical machines and advance the development of the discipline.

This book intends to establish the general airgap field modulation theory for electrical machines on the basis of revealing the phenomenon of magnetic field modulation in common electrical machines, including but not limited to, conventional DCMs, IMs, SMs, and then use the theory to qualitatively analyze the operating principles, quantitatively calculate their key parameters and performance, and guide the topology innovation of electrical machines. The general framework is as follows:

Chapter 1 reviews the historical development of electrical machines and their theories, points out limitations of the classical electrical machine theory, and elaborates the motivation and necessity of developing a general airgap field modulation theory for electrical machines.

Chapter 2 analyzes the airgap magnetic field modulation phenomena in common machine topologies, aiming to reveal the ubiquity of magnetic field modulation phenomena in electrical machines.

Chapter 3 is the core and focus of this book. A unit machine with one stator, one rotor and one layer of airgap is abstracted as a cascade of three elements, i.e. the excitation source, modulator, and filter. The excitation source and filter are characterized by winding functions, and the modulation operator function is further developed, based on which a generalized mathematical model for the three elements is proposed and the general airgap field modulation theory for electrical machines is established. For reasons of clarity, the unit machine is analyzed as a duality of a switching converter. Finally, several typical machines are analyzed as examples to show the applicability of the general theory.

Chapter 4 analyzes the relationship between different modulation behaviors and their torque components, compares the characteristics of three modulator types: salient reluctance poles, multilayer flux barriers and short-circuit coils. The equivalence and interchangeability of the three modulators are also analyzed.

Chapter 5 applies the general airgap magnetic field modulation theory to conduct qualitative analysis and quantitative calculation for a variety of typical electrical machines. The results are compared with the finite element numerical simulations and experimental measurements to validate the correctness, feasibility, and uniformity of the magnetic field modulation theory.

Chapter 6 introduces the innovation of machine topology under the guidance of the general airgap field modulation theory. After the general principles and methods of machine topology innovation are elaborated, several examples are presented to include some emerging machine topologies derived from the principle of magnetic field modulation in recent years, including stator-PM variable reluctance resolvers, dual-rotor magnetically-geared power split machines, etc.

Chapter 7 presents more possible applications of the general airgap field modulation theory in the analysis of radial electromagnetic forces, design of pole-changing windings, stray load loss calculation for IMs and fast core loss calculation of FSPM machines supplied by pulse width modulation (PWM) inverters, and assisting the design optimization of electrical machines.

It should be noted that, unless otherwise specified, "electrical machine" in this book refers to electromagnetic machines that use magnetic fields as the medium for electromechanical energy conversion.

References

1 Fitzgerald, A.E., Kingsley, C. Jr., and Umans, S.D. (2003). *Electric Machinery*, 6e. New York, USA: McGraw-Hill.

2 Seely, S. (1962). *Electromechanical Energy Conversion*, 1e. New York, USA: McGraw-Hill.

3 Talaat, M.E. (1951). Steady-state and transient synthesis of 3-phase reluctance motors (synchronous motors without field excitation). *Trans. AIEE* 70 (2): 1963–1970.

4 Zhou, E. and Gu, Z. (1983). Magnetic circuit analysis and parametric calculation of the new type of reluctance motors. *Sci. Sin. (Ser. A)* 26 (11): 1218–1230.

5 Zhou, E. and Cheng, M. (1989). Calculation of the steady-state performance of composite solid rotor reluctance motor. *J. Southeast Univ.* 19 (1): 24–31.

6 Brainard, M.W. (1952). Synchronous machines with rotating permanent-magnet fields. Part I. Characteristics and mechanical construction. *Trans. AIEE Part III: Power Apparat. Syst.* 71 (1): 670–676.

7 Chau, K.T., Chan, C.C., and Liu, C. (2008). Overview of permanent-magnet brushless drives for electric and hybrid electric vehicles. *IEEE Trans. Ind. Electron.* 55 (6): 2246–2257.

8 Lee, C.H. (1963). Vernier motor and its design. *IEEE Trans. Power Apparat. Syst.* 82 (66): 343–349.

9 Broadway, A.R.W. and Burbridge, L. (1970). Self-cascaded machine: a low speed motor or high frequency brushless alternator. *Proc. IEE* 117 (7): 1277–1290.

10 Han, P., Cheng, M., and Luo, R. (2016). Design and analysis of a brushless doubly-fed induction machine with dual-stator structure. *IEEE Trans. Energy Convers.* 31 (3): 1132–1141.

11 Broadway, A.R.W. (1971). Cageless induction machine. *Proc. IEE* 118 (11): 1593–1600.

12 Xu, L., Liang, F., and Lipo, T.A. (1991). Transient model of a doubly excited reluctance motor. *IEEE Trans. Energy Convers.* 6 (1): 126–133.

13 Han, P., Zhang, J., and Cheng, M. (2019). Analytical analysis and performance characterization of brushless doubly-fed machines with multi-barrier rotors. *IEEE Trans. Ind. Appl.* 55 (6): 5758–5767.

14 Laithwaite, E.R., Eastham, J.F., Bolton, H.R., and Fellows, T.G. (1971). Linear motors with transverse flux. *Proc. IEE* 118 (12): 1761–1767.

15 Lawrenson, P.J., Stephenson, J.M., Blenkinsop, P.T. et al. (1980). Variable-speed switched reluctance motors. *IEE Proc. Electr. Power Appl.* 127 (4): 253–265.

16 Jiang, Q. (1991). Study of basic theory of switched reluctance motors. Nanjing: Ph.D. Thesis of Southeast University.

17 Liao, Y., Liang, F., and Lipo, T.A. (1992). A novel permanent magnet motor with doubly salient structure. In: *Record IEEE Industry Applications Society Annual Meeting*, vol. 1, 308–314.

18 Cheng, M. (1997). Principle and linear analysis of static characteristics of a doubly salient variable speed permanent magnet motor. *Bull. Sci. Technol.* 13 (1): 16–20.

19 Cheng, M., Hua, W., Zhang, J., and Zhao, W. (2011). Overview of stator-permanent magnet brushless machines. *IEEE Trans. Ind. Electron.* 58 (11): 5087–5101.

20 Cheng, M., Zhang, G., and Hua, W. (2014). Overview of stator permanent magnet brushless machine systems and their key technologies. *Proc. CSEE* 34 (29): 5204–5220.

21 Toba, A. and Lipo, T.A. (1999). Novel dual-excitation permanent magnet vernier machine. In: *Record IEEE Industry Applications Society Annual Meeting*, vol. 4, 2539–2544.

22 Du, Y., Chau, K.T., Cheng, M. et al. (2011). Design and analysis of linear stator permanent magnet vernier machines. *IEEE Trans. Magn.* 47 (10): 4219–4222.

23 Atallah, K. and Howe, D. (2001). A novel high-performance magnetic gear. *IEEE Trans. Magn.* 37 (4): 2844–2846.

24 Atallah, K., Rens, J., Mezani, S., and Howe, D. (2008). A novel 'pseudo' direct-drive brushless permanent magnet machine. *IEEE Trans. Magn.* 44 (12): 4605–4617.

25 Chau, K.T., Zhang, D., Jiang, J.Z. et al. (2007). Design of a magnetic-geared outer-rotor permanent-magnet brushless motor for electric vehicles. *IEEE Trans. Magn.* 43 (6): 2504–2506.

26 Fan, Y., Jiang, H., Cheng, M., and Wang, Y. (2010). An improved magnetic-geared permanent magnet in-wheel motor for electric vehicles. In: *Proceeding of IEEE Vehicle Power and Propulsion Conference (VPPC)*, 1–5. Lille, France.

27 Xu, L. (2009). Dual-mechanical-port electric machines-concept and application of a new electric machine to hybrid electrical vehicles. *IEEE Ind. Appl. Mag.* 15 (4): 44–51.

28 Cheng, M., Han, P., Buja, G., and Jovanović, M.G. (2018). Emerging multi-port electrical machines and systems: past developments, current challenges and future prospects. *IEEE Trans. Ind. Electron.* 65 (7): 5422–5435.

29 Blondel, A.E., Mailloux, C.O., Adams, C.A. (1913). Synchronous Motors and Converters—Theory and Methods of Calculation and Testing. New York, USA: McGraw-Hill.

30 Doherty, R.E. and Nickle, C.A. (1926). Synchronous machines I-an extension of Blondel's two-reaction theory. *Trans. AIEE* XLV: 912–947.

31 Park, R.H. (1929). Two-reaction theory of synchronous machine generalized method of analysis – Part I. *Trans. AIEE* 48 (9): 716–727.

32 Kron, G. (1930). Generalized theory of electrical machinery. *Trans. AIEE* 49 (2): 666–683.

33 Gieras, J.F. (2017). Electrical Machines—Fundamentals of Electromechanical Energy Conversion. Boca Raton, FL, USA: CRC Press.

34 Adkins, B. and Harley, R.G. (1975). *The General Theory of Alternating Current Machines*. London, UK: Chapman and Hall Ltd.

35 Hansen, K.L. (1925). The rotating magnetic field theory of A–C motors. *Trans. AIEE* XLIV: 340–348.

36 West, H.R. (1926). The cross-field theory of alternating-current machines. *Trans. AIEE* XLV: 466–474.

37 Lyon, W.V. (1954). *Transient Analysis of Alternating-Current Machinery – An Application of the Method of Symmetrical Components*. New York, USA: Wiley.

38 Kovács, K.P. and Rácz, I. (1959). Transiente Vorgänge in Wechselstrom-Maschinen. Bd. I-II. (in German), Akadémiai Kiadó, Budapest, Hungary.

39 Fienne, J. (1973). New approach to general theory of electrical machines using magnetic equivalent circuit. *Proc. IEE* 120 (1): 94–104.

40 Schmitz, N.L. and Novotny, D.W. (1965). Introductory Electromagnetics, Hoboken, New Jersey, USA: John Wiley & Sons.

41 Lipo, T.A. (2012). Analysis of Synchronous Machine, Boca Raton, FL, USA: CRC Press.

42 Yamamura, S. (1992). *Spiral Vector Theory of AC Circuits and Machines*. Oxford, UK: Clearendon Press.

43 Staton, D.A., Soong, W.L., Deodhar, R.P. et al. (1994). Unified theory of torque production in ac, dc and reluctance motors. In: *Record IEEE IAS Annual Meeting*, 149–156. Denver, CO, USA.

44 Cheng, M., Han, P., and Hua, W. (2017). General airgap field modulation theory for electrical machines. *IEEE Trans. Ind. Electron.* 64 (8): 6063–6074.

45 Sun, L., Cheng, M., and Jia, H. (2015). Analysis of a novel magnetic-geared dual-rotor motor with complementary structure. *IEEE Trans. Ind. Electron.* 62 (11): 6737–6747.

46 Evans, D.J. and Zhu, Z.Q. (2015). Novel partitioned stator switched flux permanent magnet machines. *IEEE Trans. Magn.* 51 (1) Article# 8100114.

47 Li, X., Chau, K.T., and Cheng, M. (2015). Comparative analysis and experimental verification of an effective permanent-magnet vernier machine. *IEEE Trans. Magn.* 51 (7) Article# 8203009.

48 Hoang, E. and Ben-Ahmed, E.H. (1997). Switching flux permanent magnet polyphased synchronous machines. In: *Proceeding of European Conference on Power Electronics and Application*, 903–908. Trondheim, UK.

49 Zhu, Z.Q., Pang, Y., Howe, D. et al. (2005). Analysis of electromagnetic performance of flux-switching permanent magnet machines by non-linear adaptive lumped parameter magnetic circuit model. *IEEE Trans. Magn.* 41 (11): 4277–4287.

50 Mukherji, K.C. and Tustin, A. (1974). Vernier reluctance motor. *Proc. IEE* 121 (9): 965–974.

51 Ishizaki, A., Tanaka, T., Takasaki, K., and Nishikata, S. (1995). Theory and optimum design of PM vernier motor. In: *Proceeding of IEE International Conference on Electrical Machines Drives (ICEMD)*, 208–212. Durham, UK.

52 Spooner, E. and Haydock, L. (2003). Vernier hybrid machines. *IEE Proc. Electr. Power Appl.* 150 (6): 655–662.

53 Wang, F., Zhang, F., and Xu, L. (2002). Parameter and performance comparison of doubly fed brushless machine with cage and reluctance rotors. *IEEE Trans. Ind. Appl.* 38 (5): 1237–1243.

54 Wang, F. and Zhang, F. (2004). *Magnetic Field Modulation Brushless Doubly-Fed AC Machines (in Chinese)*. Changchun, China: Jilin University Press.

55 Li, J., Chau, K.T., Jiang, J.Z. et al. (2010). A new efficient permanent magnet vernier machine for wind power generation. *IEEE Trans. Magn.* 46 (6): 1475–1478.

56 Atallah, K., Calverley, S.D., and Howe, D. (2004). Design, analysis and realisation of a high-performance magnetic gear. *IEE Proc. Electr. Power Appl.* 151 (2): 135–143.

57 Jian, L., Chau, K.T., and Jiang, J.Z. (2009). A magnetic-geared outer-rotor permanent-magnet brushless machine for wind power generator. *IEEE Trans. Ind. Appl.* 45 (3): 954–962.

58 Du, Y., Chau, K.T., Cheng, M., and Wang, Y. (2010). A linear magnetic-geared permanent magnet machine for wave energy generation. In: *Proceeding of International Conference On Smart Environment Management And Solutions (ICEMS)*, 1538–1541. Incheon, Korea.

59 Zhao, X. and Niu, S. (2017). Design and optimization of a new magnetic-geared pole-changing hybrid excitation machine. *IEEE Trans. Ind. Electron.* 64 (12): 9943–9952.

60 Du, Y., Chau, K.T., Cheng, M. et al. (2011). Theory and comparison of the linear stator permanent magnet vernier machine. In: *Proceeding of International Conference On Smart Environment Management And Solutions (ICEMS)*, 1–4. Beijing, China.

61 Bai, J., Zheng, P., Cheng, L. et al. (2015). New magnetic-field-modulated brushless double-rotor machine. *IEEE Trans. Magn.* 51 (11) Article# 8112104.

62 Li, D., Qu, R., and Li, J. (2015). Topologies and analysis of flux-modulation machines. In: *IEEE Energy Conversion Congress and Exposition (ECCE)*, 2153–2160. Montreal, QC, Canada.

63 Li, X., Liu, S., and Wang, Y. (2017). Design and analysis of a stator HTS field-modulated machine for direct-drive applications. *IEEE Trans. Appl. Supercond.* 27 (4) Article# 5201005.

64 Cheng, M., Zhu, X., Wang, Y. et al. (2020). Effect and inhibition method of armature-reaction field on superconducting coil in field-modulation superconducting electrical machine. *IEEE Trans. Energy Convers.* 35 (1): 279–291.

65 Wu, Z.Z. and Zhu, Z.Q. (2016). Analysis of magnetic gearing effect in partitioned stator switched flux PM machines. *IEEE Trans. Energy Convers.* 31 (4): 1239–1249.

66 Zhu, X., Chen, Y., Xiang, Z. et al. (2016). Electromagnetic performance analysis of a new stator-partitioned flux memory machine capable of online flux control. *IEEE Trans. Magn.* 52 (7) Article# 8203704.

67 Wu, Z.Z. and Zhu, Z.Q. (2015). Analysis of air-gap field modulation and magnetic gearing effects in switched flux permanent magnet machines. *IEEE Trans. Magn.* 51 (5) Article# 8105012.

68 Shi, Y., Jian, L., Wei, J., Shao, Z., Li, W., and Chan, C.C. (2016). A new perspective on the operating principle of flux-switching permanent-magnet machines. *IEEE Trans. Ind. Electron.* 63 (3): 1425–1437.

69 Du, Y., Xiao, F., Hua, W. et al. (2016). Comparison of flux-switching PM motors with different winding configurations using magnetic gearing principle. *IEEE Trans. Magn.* 52 (5) Article# 8201908.

70 Qu, R., Li, D., and Wang, J. (2011). Relationshipe between magnetic gears and vernier machines. In: *Proceeding of IEEE International Conference On Smart Environment Management And Solutions (ICEMS)*, 1–6. Beijing, China.

71 Fu, W.N. and Liu, Y. (2016). A unified theory of flux-modulated electric machines. In: *Proceeding of International Symposium on Electrical and Electronics Engineering (ISEE)*, 1–13. Hong Kong.

72 Zhu, Z.Q. and Liu, Y. (2018). Analysis of air-gap field modulation and magnetic gearing effect in fractional slot concentrated winding permanent magnet synchronous machines. *IEEE Trans. Ind. Electron.* 65 (5): 3688–3698.

73 Chen, Y., Fu, W., and Weng, X. (2017). A concept of general flux-modulated electric machines based on a unified theory and its application to developing a novel doubly-fed dual-stator motor. *IEEE Trans. Ind. Electron.* 64 (12): 9914–9923.

74 Hua, W. (2007). Analysis, design and control of a novel flux-switching permanent magnet machine. Nanjing: Ph.D. Thesis of Southeast University

75 Zhu, X. (2008). Research on a hybrid excited doubly salient motor and drive control system. Nanjing: Ph.D. Thesis of Southeast University.

76 Zhang, J. (2008). Research on stator permanent magnet machine and its application for wind power generation. Nanjing: Ph.D. Thesis of Southeast University.

77 Zhao, W. (2010). Fault-tolerant control of stator-permanent-magnet motors with improved reliability. Nanjing: Ph.D. Thesis of Southeast University.

78 Yu, F. (2016). Research on control and fault tolerant controllability of nine-phase flux-switching permanent magnet machine. Nanjing: Ph.D. Thesis of Southeast University.

79 Zhang, G. (2016). Analysis and design of stator-excitation flux-switching brushless machines. Nanjing: Ph.D. Thesis of Southeast University.

80 Li, F. (2018). Design and analysis of a nine-phase flux-switching permanent magnet machine for wind power generation. Nanjing: Ph.D. Thesis of Southeast University.

81 Shao, L. (2018). Design and analysis of a twelve-phase flux-switching permanent magnet wind generator. Nanjing: Ph.D. Thesis of Southeast University.

82 Du, Y. (2013). Research on linear primary permanent magnet vernier generator and its control system for direct drive wave energy conversion. Nanjing: Ph.D. Thesis of Southeast University.

83 Li, X. (2015). Research on field modulated permanent magnet machine and its control system for direct drive wind power generation. Nanjing: Ph.D. Thesis of Southeast University.

84 Sun, L. (2016). Analysis, design and control of the magnetic-geared power split motor. Nanjing: Ph.D. Thesis of Southeast University.

85 Han, P. (2017). Design and drive control of dual-stator brushless doubly-fed machine. Nanjing: Ph.D. Thesis of Southeast University.

86 Wei, X. (2018). Research on control of dual stator-brushless doubly fed wind energy conversion system. Nanjing: Ph.D. Thesis of Southeast University

87 Cheng, M., Han, P., Du, Y. et al. (2022). A tutorial on general air-gap field modulation theory for electric machines. *IEEE J. Emerg. Sel. Topics Power Electron.* 10 (2): 1712–1732.

88 Cheng, M., Wen, H., and Han, P. (2019). Analysis of airgap field modulation principle of simple salient poles. *IEEE Trans. Ind. Electron.* 66 (4): 2628–2638.

89 Wen, H. and Cheng, M. (2019). Unified analysis of induction machine and synchronous machine based on the general airgap field modulation theory. *IEEE Trans. Ind. Electron.* 66 (12): 9205–9216.

90 Wen, H., Cheng, M., and Jiang, Y. (2018). A new perspective on the operating principle of direct current machine based on airgap field modulation theory. In: *Proceeding of International Conference On Smart Environment Management And Solutions (ICEMS)*, 620–627. Jeju, South Korea.

91 Han, P. and Cheng, M. (2019). Synthesis of airgap magnetic field modulation phenomena in electric machines. In: *IEEE Energy Conversion Congress and Exposition (ECCE)*, 283–290. Baltimore, MD, USA.

92 Cheng, M., Wen, H., Zeng, Y., and Jiang, Y. (2020). Analysis of airgap field modulation behavior and torque component in electric machines. *Trans. China Electrotech. Soc.* 35 (5): 921–930.

93 Su, P., Hua, W., Hu, M. et al. (2020). Analysis of PM eddy current loss in rotor-PM and stator-PM flux-switching machines by air-gap field modulation theory. *IEEE Trans. Ind. Electron.* 67 (3): 1824–1835.

94 Zhu, X., Hua, W., Wang, W., and Huang, W. (2019). Analysis of back-EMF in flux-reversal permanent magnet machines by air gap field modulation theory. *IEEE Trans. Ind. Electron.* 66 (5): 3344–3355.

95 Wen, H., Cheng, M., Jiang, Y. et al. (2020). Analysis of airgap field modulation principle of flux guides. *IEEE Trans. Ind. Appl.* 56 (5): 4758–4768.

96 Wen, H., Cheng, M., Wang, W., and Zhang, G. (2021). The modulation behaviors and interchangeability of modulators for electrical machines. *IET Electr. Power Appl.* 15 (5): 542–554.

97 Wen, H., Cheng, M., Zhou, J. et al. (2022). Optimization of rotor salient pole reluctance for typical field modulated electric machines. *IEEE J. Emerg. Sel. Top. Power Electron.* 10 (2): 1847–1859.

2

Airgap Magnetic Field Modulation Phenomena in Electrical Machines

The airgap magnetic field modulation phenomena have been widely observed in various electrical machines, from the slotting effect in traditional direct current machines (DCMs), induction machines (IMs), and synchronous machines (SMs), to the main torque production in field modulation magnetic gears (MGs), magnetically-geared machines (MGMs), permanent magnet vernier (PMV) machines, flux-switching permanent magnet (FSPM) machines, and brushless doubly-fed (induction and reluctance) machines.

The airgap magnetic field modulation phenomenon was also referred to as "magnetic gearing effect" or "magnetic flux modulation" in some references. In this book, the term "magnetic field modulation" is used not only for consistency but also because it means the modifications to the airgap magnetic field (flux density) distribution instead of magnetic flux linked by windings by special features including short-circuited coils, salient reluctance poles, and flux barriers. The "gearing effect" comes from MGs and only covers modulation behaviors of salient reluctance poles, lacking universality in describing this general phenomenon.

2.1 Traditional Electrical Machines

To better understand the magnetic field modulation phenomena in traditional machines, this section reviews the fundamentals of the DCM, IM, and SM, with a focus on the airgap magnetomotive force (MMF) distributions. In traditional electrical machines, the terminology "armature" is frequently used. An armature is the component of an electrical machine (motor or generator) which carries alternating current. The magnetizing MMF created by the field winding and armature MMF by the armature winding of brushed DCMs have a rectangular and triangular waveform, respectively. The airgap MMF distributions in IMs and non-salient-pole SMs are almost sinusoidal. The MMF distributions in

General Airgap Field Modulation Theory for Electrical Machines: Principles and Practice, First Edition. Ming Cheng, Peng Han, Yi Du, and Honghui Wen.

Companion website: www.wiley.com/go/genairgapfieldmodulationtheory

wound-field salient-pole SMs are different in the *d*-axis and *q*-axis, and typically have some third harmonics superimposed to the fundamental, as will be detailed below.

2.1.1 Brushed Direct Current Machines

The essential features of a brushed DCM are shown schematically in Figure 2.1a. The stator has salient reluctance poles and is excited by one or more field coils. The airgap flux density distribution created by the field windings is symmetric about the center line of the salient reluctance poles, which is typically called the field axis or direct axis.

The AC voltage generated in each rotating armature coil is converted to DC in the external armature terminals by a rotating commutator and stationary brushes to which the armature leads are connected. The commutator-brush combination forms a mechanical rectifier, resulting in a DC armature voltage as well as an armature MMF which is fixed in space.

The brushes are located so that commutation occurs when the coil sides are in the neutral zone, midway between the field poles. The axis of the armature MMF then is 90 electrical degrees from the axis of the field poles, i.e. in the quadrature axis. In the schematic representation shown in Figure 2.1a, the brushes are shown in the quadrature axis because this is the position of the coils to which they are connected. The armature MMF, then, is along the brush axis.

The outstanding advantages of DCMs arise from the wide variety of operating characteristics which can be obtained by employing different types of excitations of the field windings. Various connection diagrams are shown in Figure 2.1b. The method of excitation profoundly influences both the steady-state characteristics and the dynamic behavior of the machine in control systems.

Brushed DCMs are characterized by their versatility. They can be designed to have a wide variety of volt-ampere or torque-speed characteristics for both

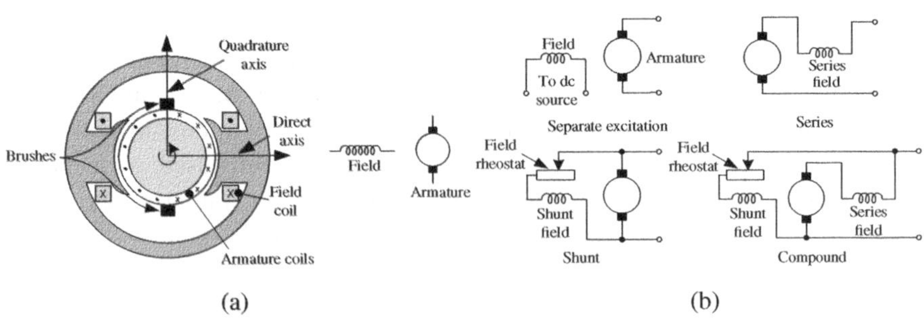

Figure 2.1 Brushed DCMs. (a) Schematic representation, (b) connections of field windings.

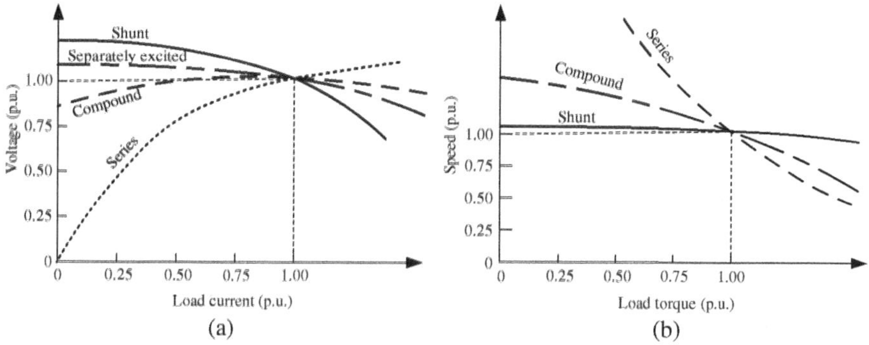

Figure 2.2 Volt–ampere characteristics and torque-speed characteristics of DCMs. (a) Generator, (b) motor.

dynamic and steady-state operation by means of various combinations of shunt-, series-, and separately-excited field windings, as illustrated in Figure 2.2.

The magnetizing MMF created by the stator salient reluctance poles with concentrated field windings is of the rectangular waveform. Armature MMF has definite effects on both the space distribution of the airgap flux density and the magnitude of the net flux per pole.

It is shown that the armature MMF can be closely approximated by a sawtooth wave, corresponding to the wave produced by a finely distributed armature winding or current sheet. For a machine with brushes in the neutral position, the idealized MMF wave is shown by the dashed sawtooth in Figure 2.3, in which a positive MMF ordinate denotes flux lines leaving the armature surface. Current directions in all windings other than the main field are indicated by black and cross-hatched bands. Because of the salient reluctance pole field structure found in almost all DCMs, the associated space distribution of flux density will not be triangular. The distribution of airgap flux density with only the armature excited is shown by the solid curve in Figure 2.3. As can be readily seen, it is appreciably decreased by the long air path in the interpole space.

When the armature and field windings are both excited, the resultant airgap flux density distribution takes the form of the solid curve in Figure 2.3b. Superimposed on this figure are the flux density distributions with only the armature excited (long-dash curve) and only the field excited (short-dash curve). The effect of cross-magnetizing armature reaction in decreasing the flux density under one pole tip and increasing it under the other can be seen by comparing the solid and short-dash curves.

In general, the solid curve is not the algebraic sum of the two dashed curves because of the nonlinearity of the iron magnetic circuit. Because of saturation of the iron, the flux density is decreased by a greater amount under one pole tip than it

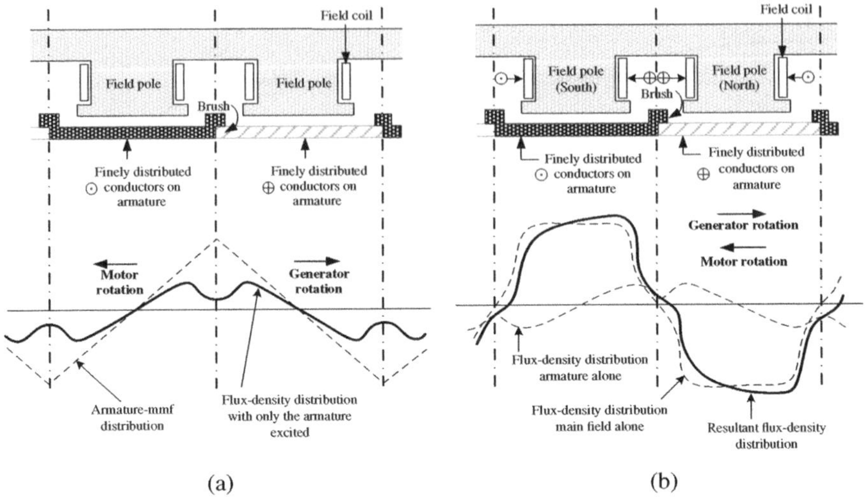

Figure 2.3 MMF in brushed DCMs. (a) Armature-MMF and flux density distribution with brushes on neutral and only the armature excited, (b) armature, main field, and resultant flux density distributions with brushes on neutral.

is increased under the other. Accordingly, the resultant flux per pole is lower than what would be produced by the field winding alone, a consequence known as the demagnetizing effect of cross-magnetizing armature reaction. Since it is caused by magnetic saturation, its magnitude is a nonlinear function of both the field current and the armature current. For normal machine operation at the flux densities used commercially, the effect is usually significant, especially at heavy loads, and must often be taken into account in analyses of performance. The salient reluctance poles with field windings can be replaced by PMs, yielding the brushed PM DCMs.

2.1.2 Induction Machines

An IM is one in which alternating current is supplied to the stator directly and to the rotor via induction or transformer action from the stator. The rotor of a polyphase IM may be of two types. A wound rotor is built with a polyphase winding similar to and wound with the same number of poles as the stator. The terminals of the rotor windings are connected to insulated slip rings mounted on the shaft. Carbon brushes bearing on these rings make the rotor terminals available external to the motor. Wound rotor IMs are relatively uncommon, being found only in a limited number of specialized applications, such as hoist and conveyor systems.

The polyphase IM more often has a squirrel-cage rotor with a winding consisting of conducting bars embedded in slots in the rotor core and short-circuited at each end by conducting end rings. The extreme simplicity and ruggedness of the

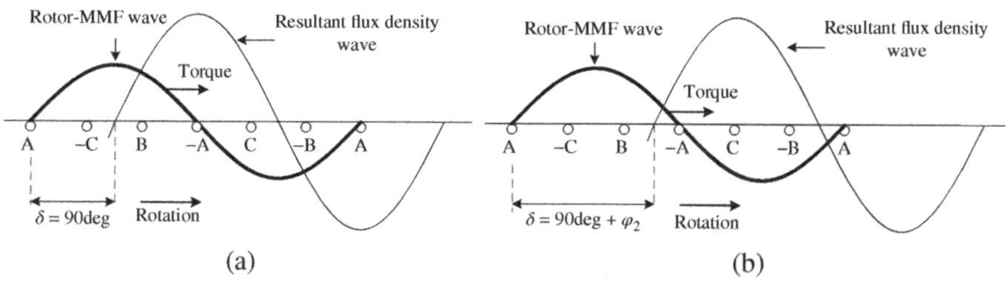

Figure 2.4 Developed rotor winding of an IM with its flux density and MMF waves in their relative positions for (a) zero and (b) nonzero rotor leakage reactance.

squirrel-cage construction are outstanding advantages of this type of IM and make it by far the most commonly used type of motor in sizes ranging from fractional horsepower on up.

For a wound rotor, the flux-MMF situation can be seen with the aid of Figure 2.4. This sketch shows the development of a simple two-pole, three-phase rotor winding in a two-pole field. The rotor flux density wave is moving to the right at the angular velocity ω_s and at the slip angular velocity $s\omega_s$ with respect to the rotor winding, which in tum is rotating to the right at the angular velocity $(1 - s)\omega_s$. It is shown in Figure 2.4 in the position of maximum instantaneous voltage in phase A.

If the rotor leakage reactance, equal to $s\omega_s$ times the rotor leakage inductance, is very small compared with the rotor resistance (which is typically the case at the small slips corresponding to normal operation), the phase-A current will also be a maximum. The rotor MMF wave will then be centered on phase A. It is so shown in Figure 2.4a. The displacement angle, or torque angle, δ under these conditions is at its optimum value of $-90°$.

If the rotor leakage reactance is appreciable however, the phase-A current lags the induced voltage by the power-factor angle φ_2 of the rotor leakage impedance. The phase-A current will not be at maximum until a correspondingly later time. The rotor-MMF wave will then not be centered on phase A until the flux wave has traveled φ_2 degrees farther down the gap, as shown in Figure 2.4b. The angle, δ is now $-(90\text{deg} + \varphi_2)$.

It departs from the optimum value of $-90°$ by the power factor angle of the rotor leakage impedance at slip frequency. The electromagnetic rotor torque is directed toward the right in Figure 2.4, or in the direction of the rotating flux density wave.

The comparable picture for a squirrel-cage rotor is given in Figure 2.5. A rotor having 16 bars placed in a two-pole field is shown in the developed form. To simplify the drawing, only a relatively small number of rotor bars have been chosen and the number is an integral multiple of the number of poles, a choice normally avoided in order to prevent harmful harmonic effects. In Figure 2.5a,

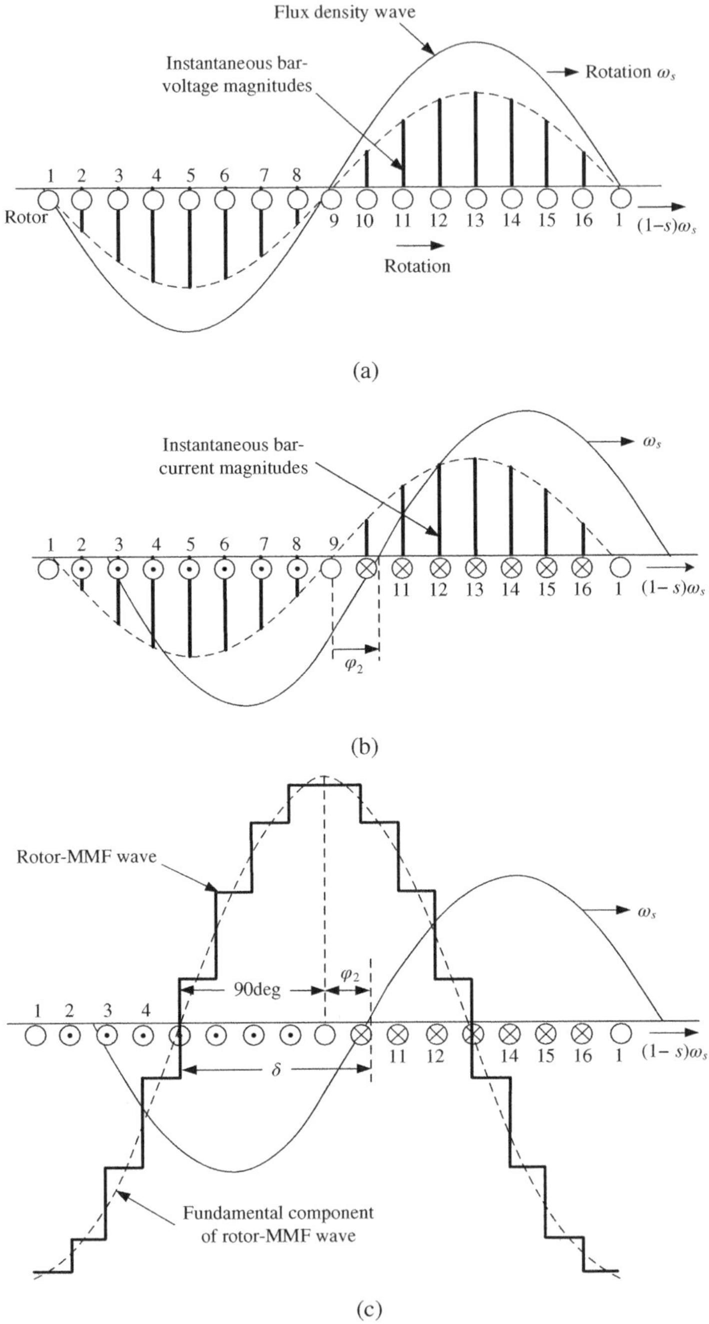

Figure 2.5 Reactions of a squirrel-cage rotor in a two-pole field. (a) Instantaneous bar voltages, (b) instantaneous bar currents, (c) rotor MMF.

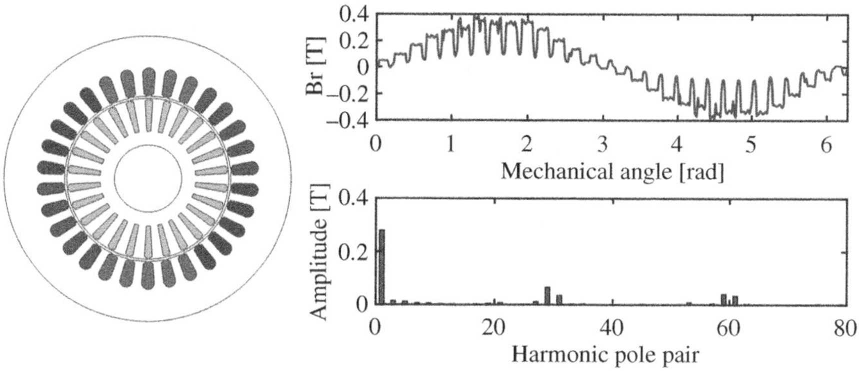

Figure 2.6 Radial airgap flux density waveform and its amplitude spectrum in a 2-pole squirrel-cage IM with 30 stator slots and 26 rotor bars from 2-dimensional (2D) finite element analysis (FEA).

the sinusoidal flux density wave induces a voltage in each bar which has an instantaneous value indicated by the solid vertical lines.

At a somewhat later instant of time, the bar currents assume the instantaneous values indicated by the solid vertical lines in Figure 2.5b, the time lag corresponding to the rotor power factor angle φ_2. In this time interval, the flux density wave has traveled in its direction of rotation with respect to the rotor through a space angle φ_2 and is then in the position shown in Figure 2.5b. The corresponding rotor-MMF wave is shown by the step wave of Figure 2.5c. The fundamental component is shown by the dashed sinusoid and the flux density wave by the solid sinusoid. Study of these figures confirms the general principle that the number of rotor poles in a squirrel-cage rotor is determined by the inducing flux density wave.

In addition to the winding harmonics introduced by the discrete distribution of winding conductors as shown in Figure 2.5, the slotting effect also causes spatial harmonics of high pole pairs in the airgap, as illustrated in Figure 2.6.

2.1.3 Synchronous Machines

A SM is an AC machine whose speed under steady-state conditions is proportional to the frequency of the current in its armature. The rotor, along with the magnetic field created by the DC field current on the rotor, rotates at the same speed as, or in synchronization with, the rotating magnetic field produced by the armature currents, and a steady torque results.

The armature winding is almost invariably on the stator and is usually a three-phase winding. The field winding is on the rotor. The cylindrical rotor construction shown in Figure 2.7 is used for two-pole turbine generators. The salient

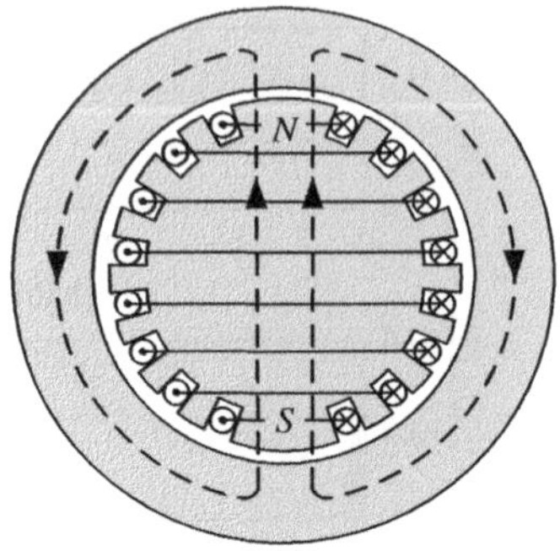

Figure 2.7 Wound-field non-salient-pole SM.

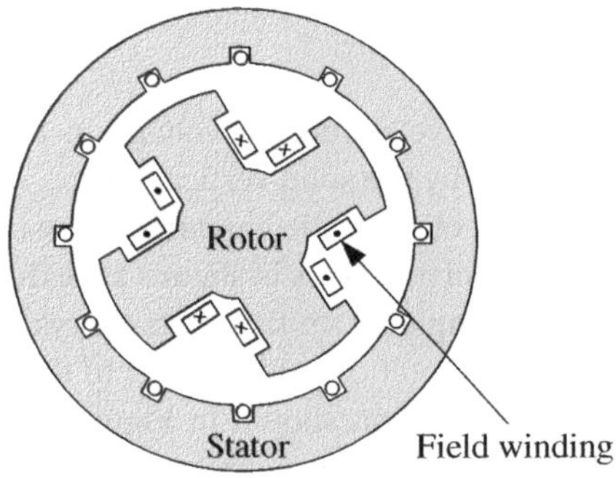

Figure 2.8 Wound-field salient-pole SM.

reluctance pole construction shown in Figure 2.8 is best adapted to multi-polar, slow-speed, hydroelectric generators and to most synchronous motors. The DC power required for excitation – approximately one to a few percent of the rating of the SM – is supplied by the excitation system.

Figure 2.9 shows the rotor of a typical 2-pole round rotor generator. Although the winding is symmetric with respect to the rotor axis, the number of turns per slot can be varied to control the various harmonics. In Figure 2.10, it can be seen that there are fewer turns in the slots nearest to the pole face. In addition, the designer can vary the spacing of the slots to optimize the airgap flux density distribution and thus the open-circuit back-electromotive force (back-EMF).

The open-circuit airgap flux density distribution of the wound-field salient-pole SM typically has a flat-topped waveform, containing significant fundamental and 3rd harmonic components, as shown in Figure 2.11.

The flux density produced by an MMF wave in a uniform airgap machine is independent of the spatial alignment of the wave with respect to the field poles.

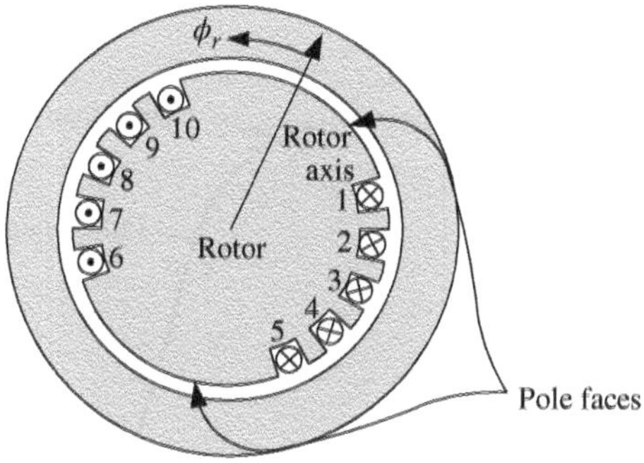

Figure 2.9 Cross-sectional view of a wound-field non-salient-pole SM.

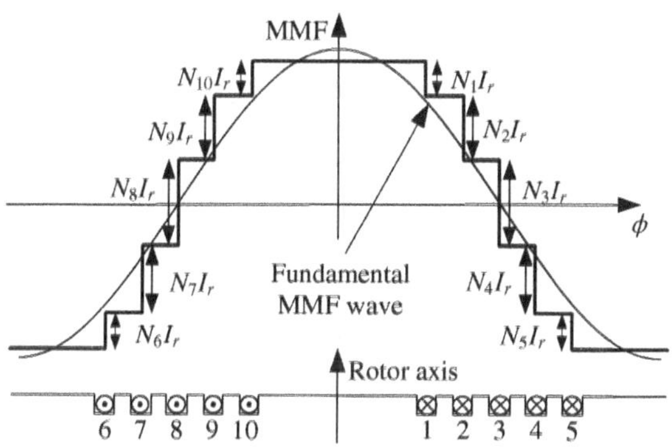

Figure 2.10 MMF distribution created by the field winding in the wound-field non-salient-pole SM.

In a wound-field salient-pole SM, such as that shown schematically in Figure 2.9, however, the preferred direction of magnetization is determined by the protruding field poles. The permeance along the polar axis, commonly referred to as the rotor direct axis, is appreciably greater than that along the interpolar axis, commonly referred to as the rotor quadrature axis.

Because of the longer airgap between the poles and the correspondingly larger reluctance, the fundamental armature-reaction flux when the armature reaction is along the quadrature axis is less than the fundamental armature reaction flux which would be created by the same armature current if the armature flux wave were directed along the direct axis. Hence, the quadrature-axis magnetizing reactance is less than that of the direct axis.

Focusing our attention on the fundamental components of the airgap flux density and MMF, the effects of salient poles can be taken into account by resolving

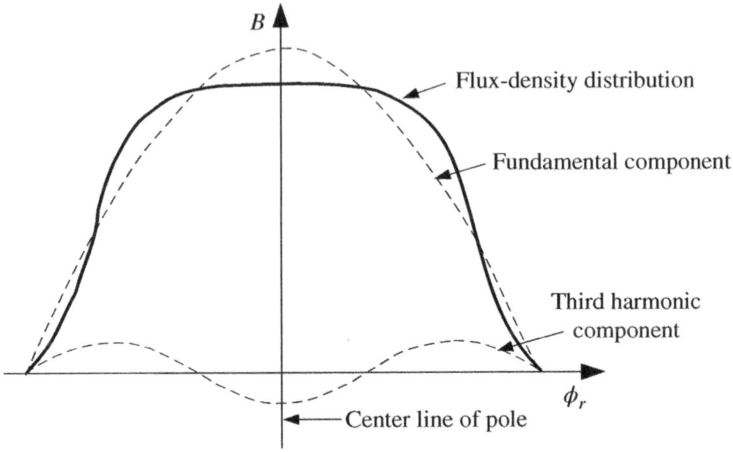

Figure 2.11 Airgap flux density waveform for a typical wound-field salient-pole SM under open-circuit condition.

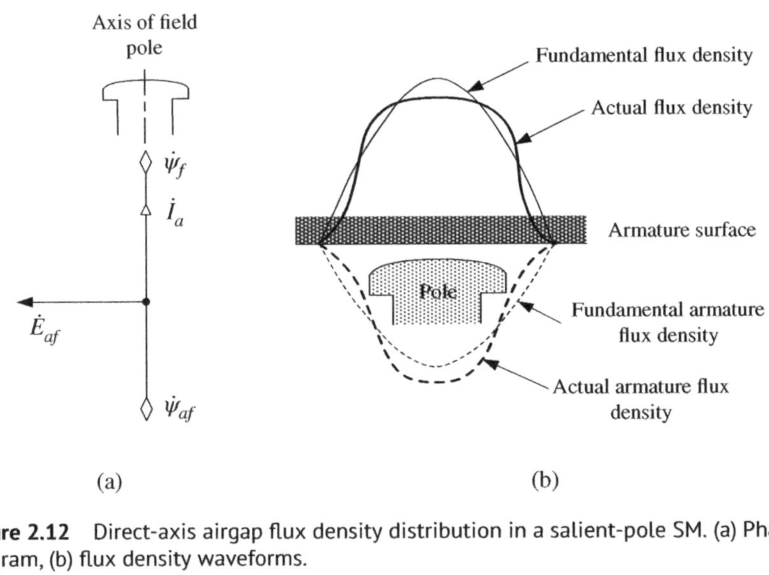

Figure 2.12 Direct-axis airgap flux density distribution in a salient-pole SM. (a) Phasor diagram, (b) flux density waveforms.

the armature current into two components, one along the direct axis and the other along the quadrature axis. The direct-axis and quadrature-axis airgap flux density distributions in a typical wound-field salient-pole SM are illustrated in Figures 2.12 and 2.13, respectively.

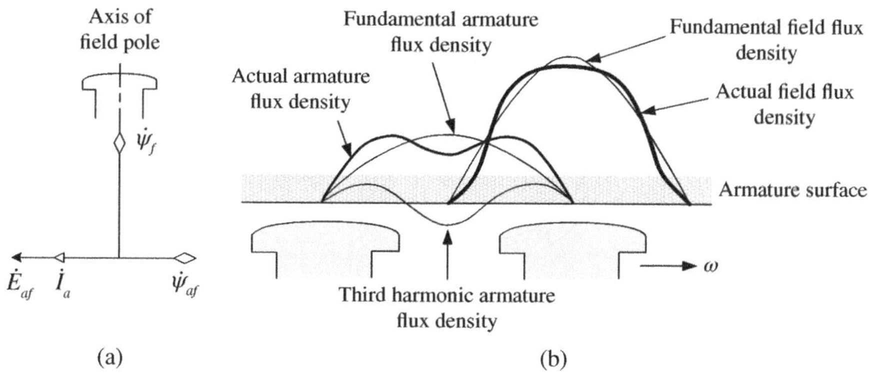

Figure 2.13 Quadrature-axis airgap flux density distribution in a salient-pole SM. (a) Phasor diagram, (b) flux density waveforms.

2.2 Field Modulation Magnetic Gears

Mechanical gears and gearboxes have been used in continuous variable transmission for a vast variety of industrial applications. Though mechanical gears have a high torque transmission capability, they also suffer from some inherent drawbacks, such as high operating noise, vibration and friction, high heat loss, high operating cost, need for lubrication and regular maintenance, irreversible damage risk in case of overload operation, and poor reliability. In variable speed transmission systems, mechanical gears are components often associated with a high rate of failure.

The concept of MG was proposed in 1916 based on electromagnets [1] to achieve the torque transmission based on magnetic field coupling instead of physical contact, showing numerous advantages over the mechanical counterpart, such as:

- Low vibration and acoustic noise
- No mechanical contact or friction loss
- No need for lubrication or maintenance
- Low operating cost
- High reliability with overload protection
- Physical isolation between input and output shafts.

In 1941, Faus further proposed an MG topology like a mechanical spur gear using ferrite magnets [2]. The MG has not attracted much attention until the high-performance neodymium iron born (NdFeB) PM material was introduced in the 1980s. Figure 2.14 shows a typical MG with parallel axes at that time. Though the structure is very simple, the torque density is poor due to the low utilization of PMs.

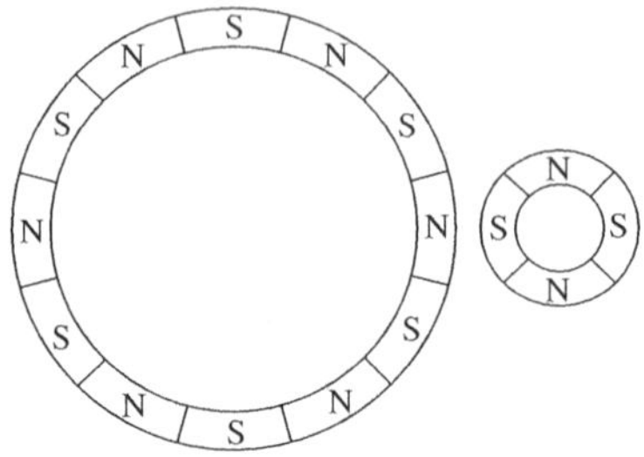

Figure 2.14 MG with parallel axes and radially-magnetized magnets.

2.2.1 Construction and Operating Principle

A high-performance coaxial MG working from the principle of magnetic field modulation was proposed in 2001 [3] and later thoroughly reported [4], showing a much higher torque density, because all the PMs can contribute to torque transmission simultaneously. Because its operation relies on the modulation behavior of ferromagnetic pole-pieces performed on the airgap magnetic fields created by PM rotors, the rotor or stator made of ferromagnetic pole-pieces is called the "modulating ring." A wide variety of new MG topologies has also been proposed following the studies of coaxial MGs [5–9].

A typical coaxial MG working by the principle of magnetic field modulation (also known as the field modulation MG) is shown in Figure 2.15. This consists of an outer rotor, an inner rotor, and a fixed modulating ring between the two rotors. The three parts are separated by the inner and outer airgaps, and the two rotors can rotate freely. The modulating ring is composed of ferromagnetic and non-magnetic blocks arranged alternately along the circumference; the ferromagnetic blocks are made of laminated silicon steel. The outer surface of the inner rotor core and the inner surface of the outer rotor core are mounted with PMs magnetized in the radial direction. Adjacent PMs are of the opposite magnetic polarity.

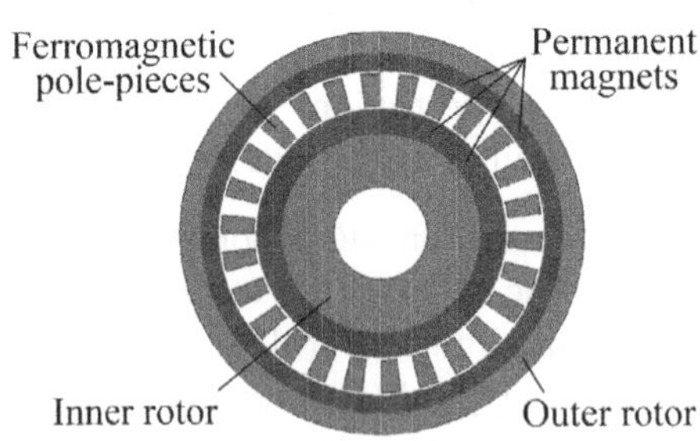

Figure 2.15 Field modulation co-axial MG with radially-magnetized magnets.

The PM magnetic field created by the inner (outer) rotor produces a series of magnetic field spatial harmonics in the outer (inner) airgap after the modulation of the modulating ring. The pole-pair number of the outer (inner) rotor is determined by the magnetic field spatial harmonic of the largest amplitude. When the number of ferromagnetic blocks in equal to the sum of the pole-pair numbers of the inner and outer rotors, a stable torque transmission can be achieved by the magnetic field coupling. The ratio between the inner and outer rotor speeds is equal to the pole ratio of outer and rotor PM rotors. Considering the conservation of energy, we can draw the conclusion that the torque transmitted by the inner and outer rotor is proportional to their PM pole pairs, showing the so-called "gearing effect" in variable speed transmission [3, 4].

Unlike traditional MGs derived directly from mechanical gears, all the magnetic field harmonics produced by the inner and outer PM rotors of the field modulation MG contribute to the torque. Numerical simulations and experimental tests show that the torque density of field modulation MGs with radially magnetized surface-mounted PMs can reach 100 kNm/m^3, and the torque transmission capability is comparable to mechanical gears. A quantitative comparison between the field modulation MG studied in [4] and various mechanical gears available on the market with similar speeds, transmission ratios and output torques is tabulated in Table 2.1 based on data from [10, 11], showing that the field modulation MG has significant edges over mechanical counterparts in terms of overall volume, mass and efficiency. In addition, field modulation MGs enjoy longer service life than mechanical counterparts due to its contactless and frictionless nature. The manufacturing cost of field modulation MGs is soaring due to the shortage of PM supplies and is typically much higher than that of mechanical gears, and the maintenance savings can offset the manufacturing cost to some extent.

A number of new topologies have been proposed to further improve the performance of field modulation MGs by changing the arrangement or magnetization mode of PMs, as shown in Figure 2.16.

The inner and outer rotor PMs shown in Figure 2.16a are magnetized in the Halbach mode, in which the arrangement of the Halbach PMs not only makes the airgap flux density waveform more sinusoidal, but also increases the fundamental component of the airgap flux density, thereby effectively improving the torque transmission capability of the MG [8]. The field modulation MG shown in Figure 2.16b employs a spoke-type PM arrangement in order to improve mechanical reliability and to prevent the surface-mounted PMs from falling off during the high-speed operation of the inner rotor. This structure also has the flux focusing effect, which can be used to improve the torque density [5].

To improve the mechanical integrity of the outer rotor, the outer rotor of the field modulation MG shown in Figure 2.16c adopts a homopolar surface-mounted

Table 2.1 Comparison between the radially-magnetized surface-mounted field modulation MG and mechanical gears.

Gear structure	Helical gear-worm gear reducer	Helical gear reducer	Helical gear-bevel gear reducer	Parallel shaft helical gear reducer	Field modulation MG [4]
Transmission ratio	7.81	6	5.7	5.54	5.75
Input speed (r/min)	1500				
Output speed (r/min)	192	250	263	271	261
Output torque (Nm)	54.74	57.32	54.49	52.88	55
Volume (cm^3)	3970	7410	4000	4680	769
Mass (kg)	10.1	9.8	11.2	12.9	4.6
Efficiency (%)	75	95			>97
Manufacturing cost ($)	236.9	221.2	308.1	260.7	472
Service life (year)	3~5				>10

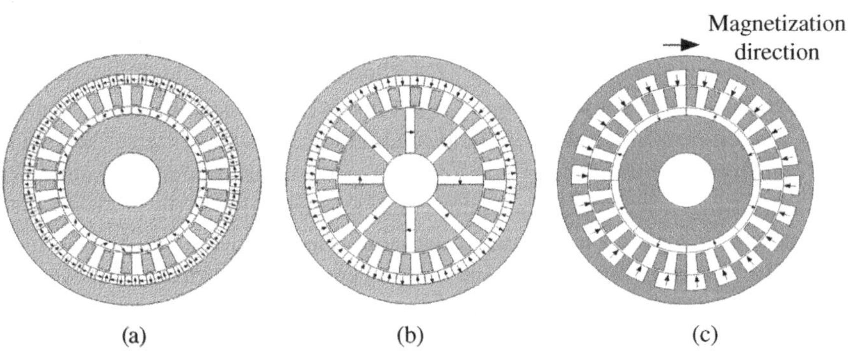

Figure 2.16 Field modulation MG variants with: (a) surface-mounted Halbach PM arrays, (b) spoke-type inner PM rotor, (c) surface-embedded homopolar outer PM rotor.

PM structure, which arranges PMs with the same magnetization direction uniformly along the inner surface of the outer rotor. This structure improves the PM utilization and reduces the manufacturing cost of the outer rotor [7].

2.2.2 Airgap Magnetic Field Modulation Behaviors

The key to the operation of a coaxial field modulation MG is the modulating ring between the inner and outer rotors, which provides a way to couple the magnetic fields created by the inner and outer rotor PM arrays of dissimilar pole pairs.

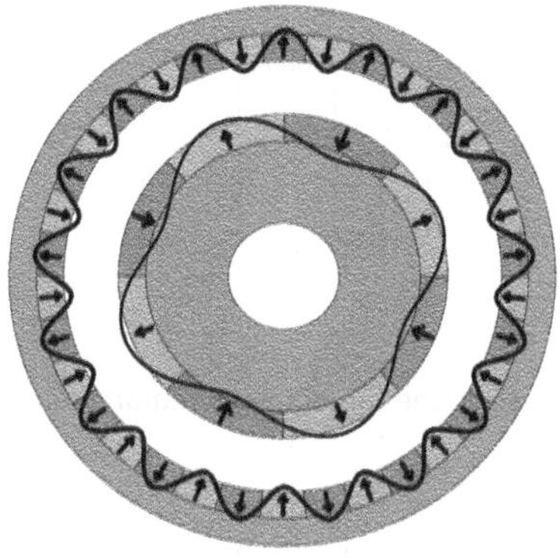

Figure 2.17 Fundamental magnetic field generated by the inner and outer rotor PMs without the modulation ring.

The working principle of the field modulation MG illustrated in Figure 2.17 can be explained as follows using some trigonometric equations [12, 13].

For simplicity, we assume that MMFs produced by the inner and outer rotor PM arrays are sinusoidally distributed along the circumferential direction, namely:

$$F_1 = F_{1m} \cos\left[p_1\left(\phi - \phi_{01} - c_1\omega_1 t\right)\right] \quad (2.1)$$

$$F_2 = F_{2m} \cos\left[p_2\left(\phi - \phi_{02} - c_2\omega_2 t\right)\right] \quad (2.2)$$

where F_{1m} and F_{2m} are amplitudes of MMFs produced by the inner and outer rotor PM arrays, respectively. ϕ_{01} and ϕ_{02} are initial positions of the inner and outer rotor PM arrays relative to the reference axis, respectively. ϕ is the mechanical angle along the circumferential direction. c_1 and c_2 are coefficients to denote the rotation direction of the inner and outer rotors, respectively, where $c = +1$ corresponds to the clockwise direction and $c = -1$ means the rotor rotates in the counterclockwise direction. ω_1, ω_2, and ω_{Fe} are the mechanical angular velocities of the inner rotor, outer rotor, and modulating ring, respectively. Similarly, p_1 and p_2 are the pole pair numbers of the inner and outer rotors, respectively, and N_{Fe}. is the number of ferromagnetic blocks in the modulating ring.

When there is no modulating ring, the airgap between the inner and outer rotors is uniform, and the airgap permeance at any position along the circumference is a constant. The flux density distributions produced by the rotor PMs are

$$B_1 = B_{1m} \cos\left[p_1\left(\phi - \phi_{01} - c_1\omega_1 t\right)\right] \quad (2.3)$$

$$B_2 = B_{2m} \cos\left[p_2\left(\phi - \phi_{02} - c_2\omega_2 t\right)\right] \quad (2.4)$$

where B_{1m} and B_{2m} are the amplitudes of flux densities produced by the inner and outer PMs. As shown in Figure 2.17, the magnetic field created by the inner rotor has eight poles, while the outer rotor produces a 34-pole magnetic field.

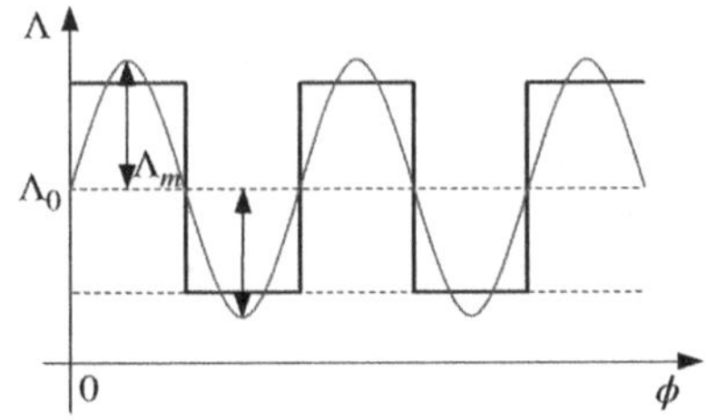

Figure 2.18 Permeance wave of the large airgap region including two physical airgaps and the modulating ring.

Apparently, since the two magnetic fields have dissimilar pole pairs, they cannot effectively couple with each other to produce a stable torque.

When the modulating ring is added between the two rotors, the permeance of the large airgap region —which includes the two physical airgaps and the modulating ring—can be approximated by a rectangular wave with a positive DC offset, as shown in Figure 2.18. Considering the fundamental component of the permeance waveform only, the airgap permeance can be written as

$$\Lambda = \Lambda_0 + \Lambda_m \cos\left[N_{Fe}\left(\phi - \phi_{0Fe} - c_{Fe}\omega_{Fe}t\right)\right] \tag{2.5}$$

where Λ_0 is the average value of the permeance in a full electrical cycle, Λ_m the amplitude of the fundamental alternating component. c_{Fe} is used to denote the rotation direction of the modulating ring. $c_{Fe} = 1$ means the clockwise direction and $c_{Fe} = -1$ the counterclockwise direction. ϕ_{0Fe} is the initial position of the modulating ring relative to the reference axis.

The airgap flux density generated by the inner rotor PM array can be obtained by (2.1) and (2.5), as

$$\begin{aligned} B_1 &= F_1\Lambda \\ &= F_{1m}\cos\left[p_1\left(\phi - \phi_{01} - c_1\omega_1 t\right)\right] \times \left\{\Lambda_0 + \Lambda_m \cos\left[N_{Fe}\left(\phi - \phi_{Fe} - c_{Fe}\omega_{Fe}t\right)\right]\right\} \\ &= \Lambda_0 F_{1m}\cos\left[p_1\left(\phi - \phi_{01} - c_1\omega_1 t\right)\right] + \Lambda_m F_{1m}\cos\left[p_1\left(\phi - \phi_{01} - c_1\omega_1 t\right)\right] \\ &\quad \cos\left[N_{Fe}\left(\phi - \phi_{Fe} - c_{Fe}\omega_{Fe}t\right)\right] \\ &= B_{11} + B_{12} + B_{13} \end{aligned} \tag{2.6}$$

where

$$B_{11} = \Lambda_0 F_{1m}\cos\left[p_1\left(\phi - \phi_{01} - c_1\omega_1 t\right)\right] \tag{2.7}$$

$$B_{12} = \frac{1}{2}\Lambda_m F_{1m}\cos\left[\left(N_{Fe} - p_1\right)\left(\phi - \frac{c_{Fe}N_{Fe}\omega_{Fe} - c_1 p_1\omega_1}{N_{Fe} - p_1}t\right) - \left(N_{Fe}\phi_{Fe} - p_1\phi_{01}\right)\right] \tag{2.8}$$

$$B_{13} = \frac{1}{2}\Lambda_m F_{1m} \cos\left[(N_{Fe} + p_1)\left(\phi - \frac{c_{Fe}N_{Fe}\omega_{Fe} + c_1 p_1 \omega_1}{N_{Fe} + p_1}t\right) - (N_{Fe}\phi_{Fe} + p_1\phi_{01})\right] \tag{2.9}$$

It can be seen that the airgap magnetic field generated by the inner rotor PM array has three components. The first component B_{11} is the fundamental component with the same pole-pair number and speed as those of the inner rotor, while the second and third components are introduced by the modulation behavior of the modulating ring.

Similarly, the airgap flux density created by the outer rotor PM array can be expressed as

$$B_2 = B_{21} + B_{22} + B_{23} \tag{2.10}$$

where

$$B_{21} = \Lambda_0 F_{2m} \cos\left[p_2\left(\phi - \phi_{02} - c_2\omega_2 t\right)\right] \tag{2.11}$$

$$B_{22} = \frac{1}{2}\Lambda_m F_{2m} \cos\left[(N_{Fe} - p_2)\left(\phi - \frac{c_{Fe}N_{Fe}\omega_{Fe} - c_2 p_2 \omega_2}{N_{Fe} - p_2}t\right) - (N_{Fe}\phi_{Fe} - p_2\phi_{02})\right] \tag{2.12}$$

$$B_{23} = \frac{1}{2}\Lambda_m F_{2m} \cos\left[(N_{Fe} + p_2)\left(\phi - \frac{c_{Fe}N_{Fe}\omega_{Fe} + c_2 p_2 \omega_2}{N_{Fe} + p_2}t\right) - (N_{Fe}\phi_{Fe} + p_2\phi_{02})\right] \tag{2.13}$$

The airgap magnetic field created by the outer rotor PM array again has the three components. For the MG shown in Figure 2.17, the modulating ring is stationary, i.e. $\omega_{Fe} = 0$. For simplicity, if we assume $\phi_{0Fe} = \phi_{01} = 0$, then (2.8) and (2.9) can be rewritten as

$$B_{12} = \frac{1}{2}\Lambda_m F_{1m} \cos\left[(N_{Fe} - p_1)\left(\phi + \frac{c_1 p_1 \omega_1}{N_{Fe} - p_1}t\right)\right] \tag{2.14}$$

$$B_{13} = \frac{1}{2}\Lambda_m F_{1m} \cos\left[(N_{Fe} + p_1)\left(\phi - \frac{c_1 p_1 \omega_1}{N_{Fe} + p_1}t\right)\right] \tag{2.15}$$

It can be seen that the pole-pair number of the magnetic field component B_{12} generated by the outer rotor PM array is $(N_{Fe} - p_1)$ and its rotating speed is $\frac{p_1}{N_{Fe}-p_1}\omega_1$ with a rotation direction opposite to that of the inner rotor. The pole-pair number and the rotatisng speed of the magnetic field component B_{13} are $(N_{Fe} + p_1)$ and $\frac{p_1}{N_{Fe}+p_1}\omega_1$ respectively, and the rotation direction is the same as that of the inner rotor.

If design parameters of the MG satisfy

$$N_{Fe} = p_1 + p_2, \tag{2.16}$$

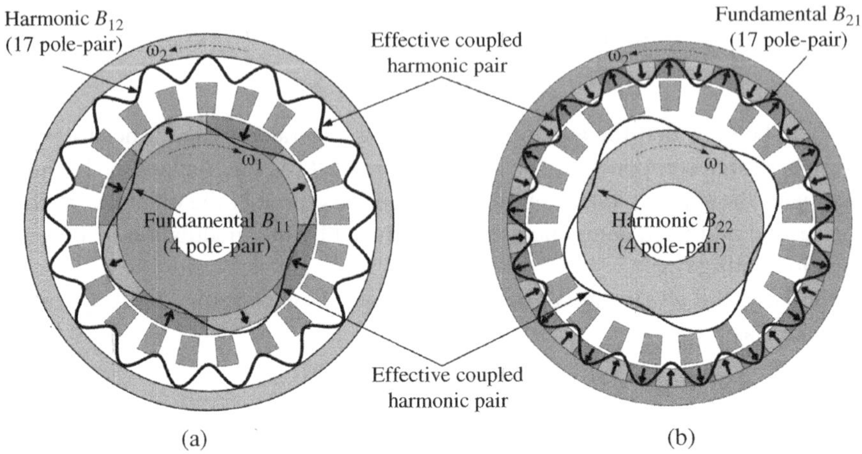

Figure 2.19 Magnetic field modulation behavior of the modulation ring performed on: (a) inner PM rotor only, (b) outer PM rotor only.

B_{12} will become

$$B_{12} = \frac{1}{2}\Lambda_m F_{1m} \cos\left[p_2\left(\theta - \frac{-c_1 p_1 \omega_1}{p_2}t\right)\right] \tag{2.17}$$

The pole-pair number of B_{12} is p_2, which is equal to that of the fundamental PM magnetic field of the outer rotor B_{21}. A stable torque can be produced by the interaction between B_{12} and B_{21} when the two magnetic field components share the same speed and rotation direction, as illustrated in Figure 2.19a. If we make $c_2 = -c_1$, the relationship of the rotating speed can be expressed as

$$\omega_2 = \frac{p_1}{p_2}\omega_1 \tag{2.18}$$

The speed ratio of MG is therefore defined as

$$G_r = \frac{\omega_1}{\omega_2} = \frac{p_2}{p_1} \tag{2.19}$$

Correspondingly, the flux density of the outer rotor can be obtained as

$$B_{22} = \frac{1}{2}\Lambda_m F_{2m} \cos\left[p_1\left(\phi + \frac{c_2 p_2 \omega_2}{p_1}t\right)\right] \tag{2.20}$$

The pole-pair number of B_{22} is p_1, which is equal to that of the fundamental PM magnetic field of the inner rotor B_{11}. When the rotating speeds of the inner and outer rotors satisfy equation (2.18), a stable torque can also be produced by the interaction between B_{22} and B_{11}, as shown in Figure 2.19b.

In summary, when design parameters of the field modulation MG satisfy equation (2.16), the inner rotor PM array will generate a fundamental magnetic

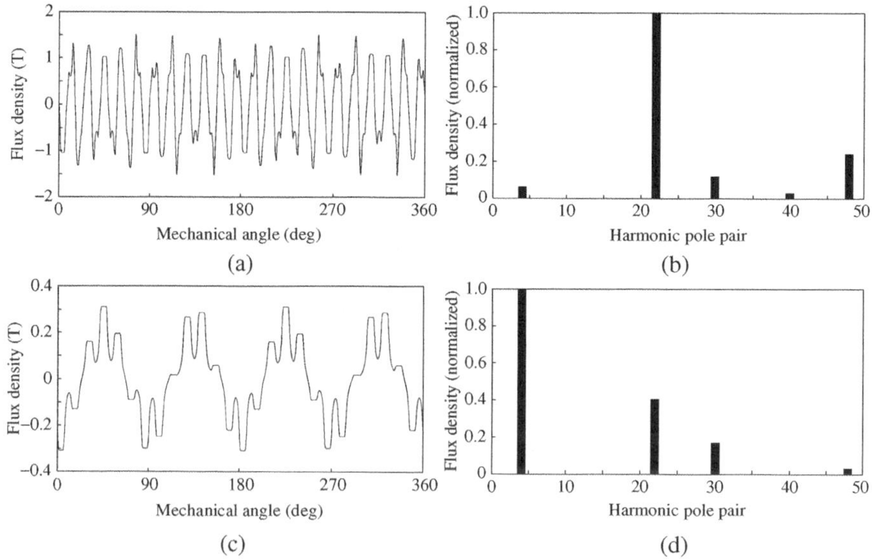

Figure 2.20 Radial airgap flux density produced by the outer rotor PMs. (a) Waveform in the outer airgap, (b) amplitude spectrum of (a), (c) waveform in the inner airgap, (d) amplitude spectrum of (c). In the studied MG, $p_1 = 4, p_2 = 22, N_{Fe} = 26$. Results are from 2D FEA without inner rotor PMs.

field with p_1 pole pairs in the inner airgap, and its rotating speed is the same as that of the inner rotor. With the modulation of the modulation ring, the inner rotor PM array also generates a heterodyned harmonic magnetic field with p_2 pole pairs in the outer airgap, whose rotating speed and rotation direction are the same as the outer rotor.

Likewise, the outer rotor PM array produces a fundamental magnetic field with p_2 pole pairs in the outer airgap, whose rotating speed is the same as that of the outer rotor. Together with the modulating ring, the outer rotor PM array generates a heterodyned harmonic magnetic field component with p_1 pole pairs in the inner airgap, whose speed and rotation direction are the same as the inner rotor.

Figure 2.20 shows waveforms of the radial airgap flux density distributions, their harmonic spectrums produced by the modulating ring and outer rotor PMs (it does not consider the inner rotor PMs). Results for the case with the inner rotor PMs and modulating ring (and not considering the outer rotor PMs) are illustrated in Figure 2.21. Both figures clearly show the modulation behaviors analyzed by (2.6) and (2.10).

It is with the interactions among these airgap magnetic field components that the torque transmission between the inner and outer rotors is achieved, where the inner and outer rotors rotate in opposite directions and their speeds are inversely

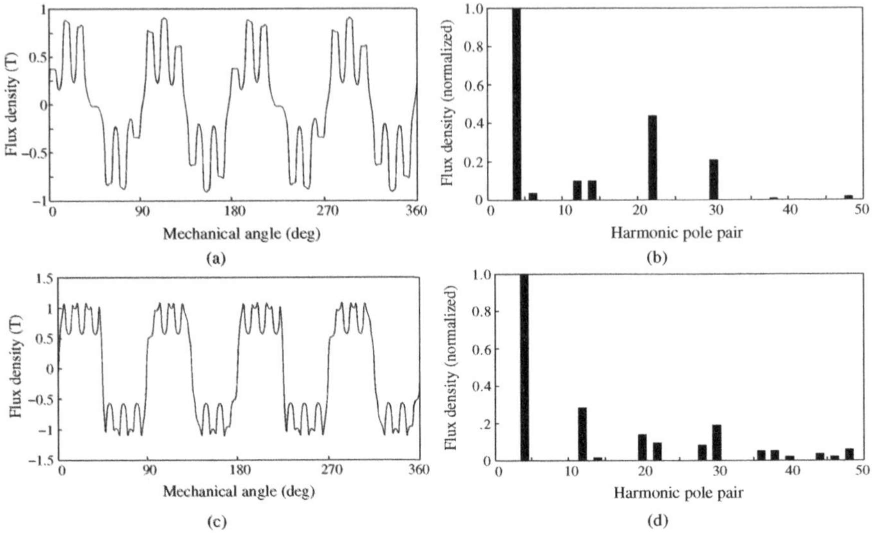

Figure 2.21 Radial airgap flux density produced by the inner rotor PMs. (a) Waveform in the outer airgap, (b) Amplitude spectrum of (a), (c) waveform in the inner airgap, (d) amplitude spectrum of (c). In the studied MG, $p_1 = 4$, $p_2 = 22$, $N_{Fe} = 26$. Results are from 2D FEA without outer rotor PMs.

proportional to the pole-pair numbers. The basic trigonometric relation described by (2.6) and (2.10) serve as the foundation for the so-called "gearing effect" in field modulation MGs. It will be revisited several times later in the following chapters when analyzing other field modulation machine types.

2.2.3 Other MG Types

Unlike MGs derived directly from mechanical gears, field modulation MGs show a higher torque transmission capability by taking advantage of the gearing effect, attracting more attention.

On the one hand, to meet the needs of different forms of torque/speed transmission, a variety of linear, axial-flux and transverse-flux field modulation MGs have been proposed. Figure 2.22 shows the tubular linear field modulation MG structure, whose force density is as high as 1.7 MN/m^3 [14]. When combined with the linear PM brushless motor, it can achieve the high thrust easily at low-speed operations, suitable for direct-drive applications, such as ocean wave energy harvesting and railway traction [15, 16]. Figure 2.23 shows two axial-flux MG examples. It has been reported that the torque density of the disk-type axial-flux field modulation MG shown in Figure 2.23a can reach 70 kNm/m^3, and the axial stress

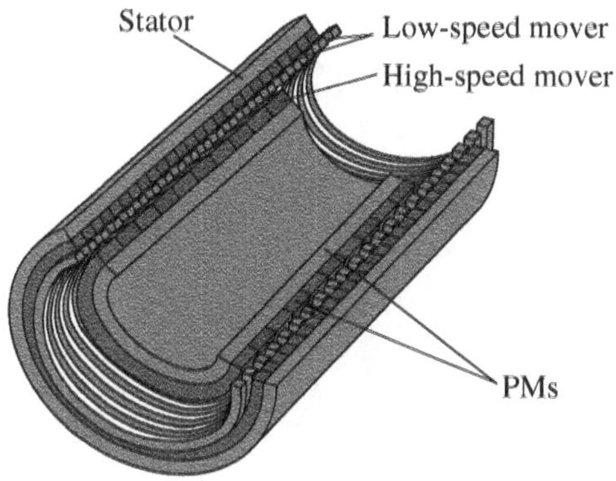

Figure 2.22 Tubular linear field modulation MG. The outer tube is stationary and mounted with PMs. The modulating ring is the low-speed mover and the inner tube with mounted PMs is the high-speed mover.

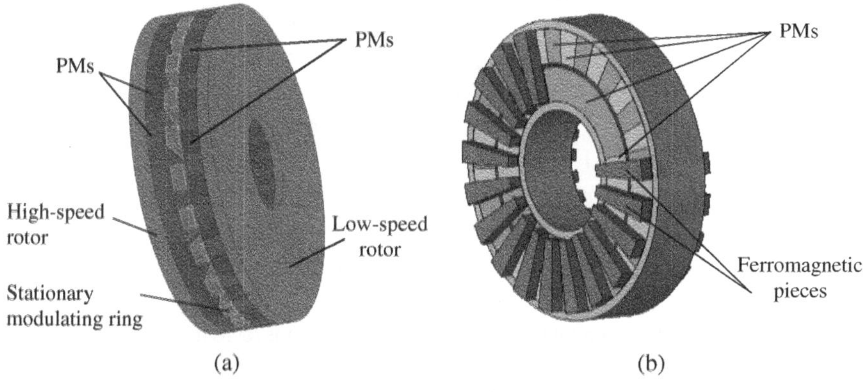

Figure 2.23 Axial-flux field modulation MGs. (a) Disk-type, (b) coaxial.

between the high-speed rotor, the low-speed rotor and the modulating ring is very low [17]. The coaxial axial-flux MG shown in Figure 2.23b effectively simplifies the mechanical structure by installing ferromagnetic blocks on end covers of the MG, and its torque density reaches 77 kNm/m^3, slightly higher than that of the disk-type axial-flux MG [18].

A transverse-flux MG has also been proposed [19], as shown in Figure 2.24, where PMs and iron core blocks are arranged alternately. PMs are magnetized tangentially, and adjacent PMs are magnetized in opposite directions. The most notable feature is that the two PM rotors are arranged in tandem and share the same shaft, while the modulating ring is located outside of the two rotors to form a transverse flux path. By introducing the T-shaped modulating ring, a new

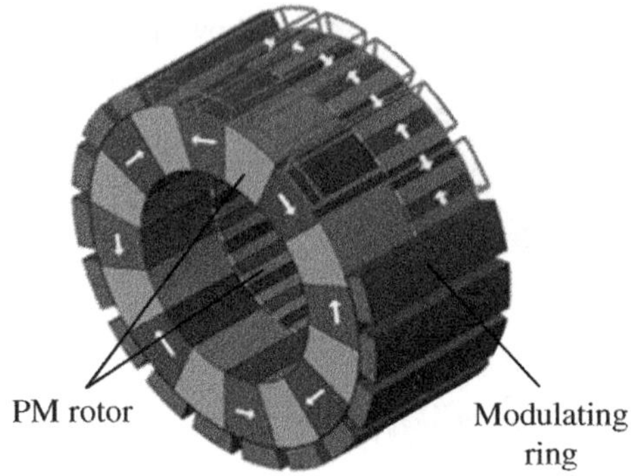

Figure 2.24 Transverse-flux MG.

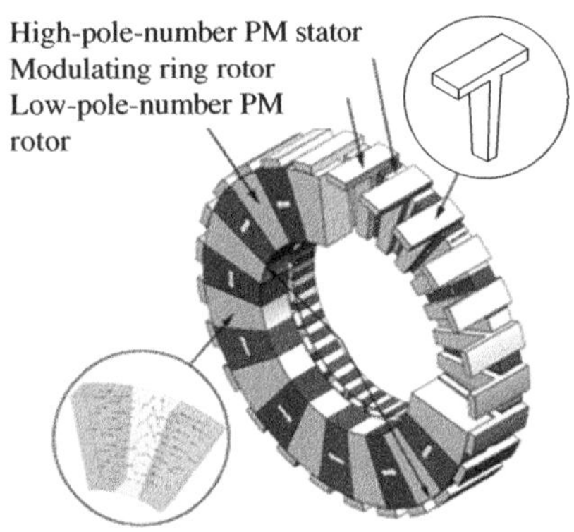

Figure 2.25 Hybrid field modulation MG with axial and transverse fluxes.

hybrid MG with both axial and transverse fluxes was proposed [20], as shown in Figure 2.25, whose torque density increases to 282.56 kNm/m^3.

On the other hand, to reduce the magnetic flux leakage between PMs, the low-speed rotor with high pole pairs is typically designed as an outer rotor with a larger diameter, and the high-speed rotor with low pole pairs is considered as an inner rotor. Although this arrangement can provide more space for the installation of outer rotor PMs, thereby improving the utilization of PMs, PMs on the high-speed rotor suffer from large centrifugal forces under high-speed operations, posing challenges for the mechanical design of MGs, especially those with surface-mounted PMs. Rotors with interior PMs can help improve the mechanical design to some extent [21]. Excessive eddy current losses in the high-speed PM rotor reduces the transmission efficiency significantly, which is another concern for field modulation MGs.

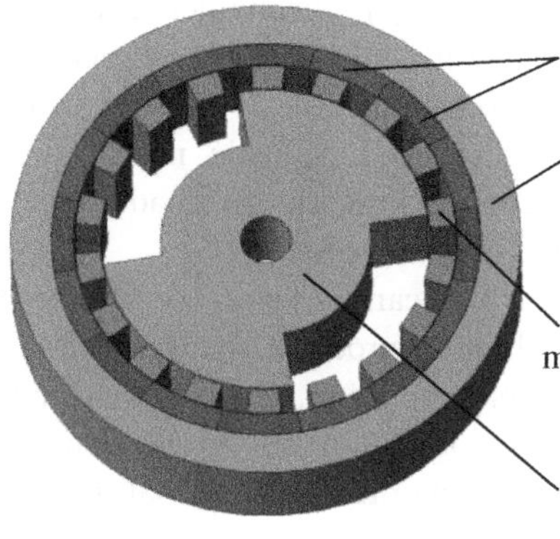

Figure 2.26 Reluctance type field modulation MG.

Considering the design challenges in mechanical robustness and loss suppression for the high-speed PM rotor, a field modulation MG with a high-speed reluctance rotor has been proposed [22], as shown in Figure 2.26. The low-speed PM rotor and the stationary modulating ring are the same as those shown in Figures 2.17 and 2.19, but the high-speed rotor adopts the salient reluctance pole rotor structure. This reluctance rotor field modulation MG is more suitable for high-speed and ultra-high-speed operations to increase the power density of the entire transmission system. Studies further show that when the high-speed rotor is made from amorphous alloy material, the transmission efficiency can be greatly improved. The torque density, however, is only about one-third of MGs with two PM rotors, and the torque ripple becomes larger.

In conclusion, it can be seen that the coaxial field modulation MG has a high torque density and can be easily integrated with existing PM brushless machines to achieve variable speed transmission, showing great promise in industrial applications. Overall, high-performance MGs in the near future should have the following characteristics to meet the application needs:

- High torque/force density comparable to their mechanical counterparts
- Simple structure and convenience in processing/production
- High reliability, stability, and robustness
- Low manufacture cost.

2.3 Magnetically Geared Machines

To improve the power density of electric machine drive systems, high-speed machines with high-ratio gearboxes have been adopted for many applications.

However, as described in Section 2.2, the mechanical gear typically has a high torque density yet suffers from some inherent issues such as contact friction, noise, vibration, and reliability. In contrast, the MG shows significant advantages in these aspects by providing reduced acoustic noise, minimum vibration, improved reliability, no maintenance, inherent overload protection, and physical isolation between the input and output shafts.

The coaxial field modulation MG can be favorably further integrated with a high-speed PM brushless machine to produce MGMs. MGMs are attractive for applications such as wind power generation, direct-drive electric vehicles, wave power harvesting, and robot joint actuation, due to the distinctive advantages of high torque density, low noise and maintenance-free nature.

2.3.1 Evolution of MGMs

As a novel solution to direct-drive applications, an MGM was proposed by Professor K. T. Chau in 2007 [23]. This MGM, as shown in Figure 2.27a, has an outer MG and an inner PM brushless DC motor sharing the same high-speed rotor. It has been reported that the three-airgap MGM can offer a high torque density when a field modulation MG is employed with a gear ratio between 5 : 1 and 10 : 1, and a high power factor at the winding terminal can be maintained.

When the high-speed inner PM rotor of the field modulation MG is replaced by a stator with a set of balanced AC windings, a rotating magnetic field can be created by the stator currents, which is similar to the rotating magnetic field generated by PMs mounted on the high-speed inner rotor. The three-airgap MGM can be converted into a new two-airgap MGM [24, 25]. As shown in Figure 2.27b, with a relatively simple structure, the two-airgap MGM simultaneously maintains a high frequency of the stator windings (SWs) and a low-speed direct-drive operation of the rotor.

The two-airgap MGM has reduced PM volumes, and the flux lines have to cross two airgaps for electromechanical energy transmission and conversion, so the torque density has been reduced. Considering that the ferromagnetic modulation ring is stationary relative to the stator, the inner airgap between them can be further removed, yielding a one-airgap MGM as shown in Figure 2.27c.

Compared to the two-airgap MGM, the one-airgap MGM seems to be closer to traditional PMSMs, but the working principle is essentially different. The one-airgap MGM shown in Figure 2.27c works as a PMV machine. When the number of ferromagnetic blocks and the number of stator teeth are the same, the one-airgap MGM becomes a PMV machine with a slotted stator [26]. When the number of ferromagnetic blocks is an integer multiple of the number of stator teeth, the one-airgap MGM becomes the stator-split-slot PMV machine [27].

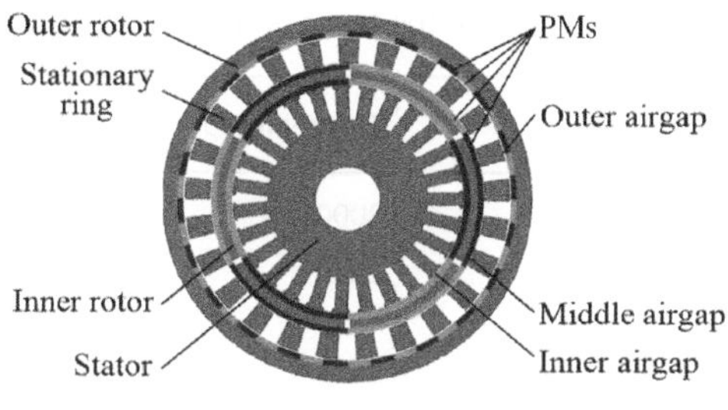

(a)

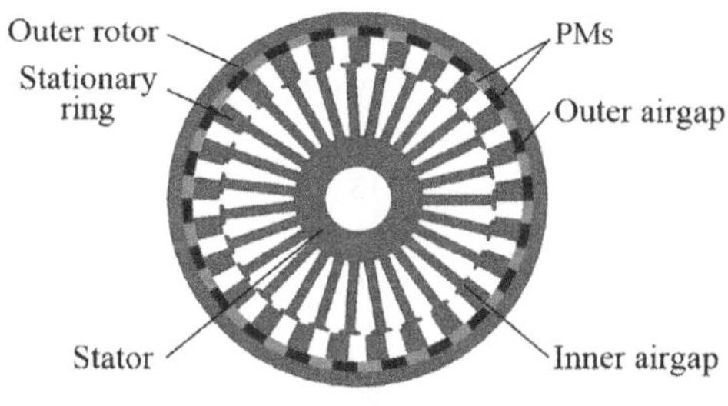

(b)

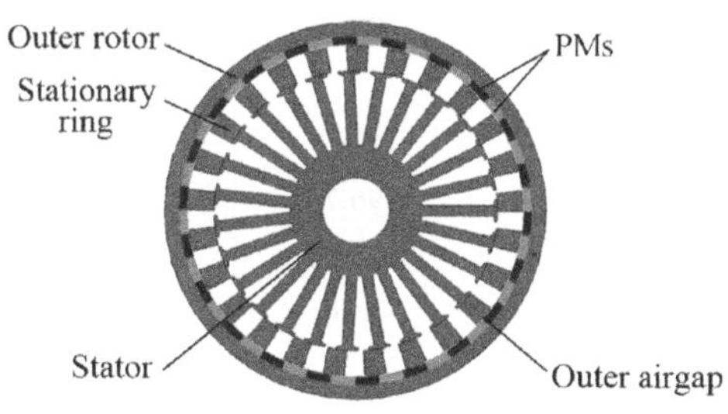

(c)

Figure 2.27 Cross-sectional view of MGMs, (a) with three airgaps, (b) with two airgaps, (c) with one airgap.

The electromagnetic performance of the three MGMs have been compared and tabulated in Table 2.2. The overall outer diameter, axial length and airgap length are kept the same as those of the three-airgap MGM designed in [28]. It is shown that the one-airgap topology provides the highest rated power for the same machine size and rotation speed. Although the three-airgap MGM also produces higher power than the two-airgap one, its structure is more complicated: three layers of PMs are needed, resulting in a higher PM consumption and a higher manufacturing cost. The PM consumption of the one-airgap MGM is about 20% less than that of the two-airgap one.

Table 2.2 Performance comparison of the three MGMs.

Type	Three-airgap	Two-airgap	One-airgap
Power (W)	3000	2500	3200
Number of pole pairs for outer rotor/stator winding	22/3		
Number of ferromagnetic blocks/stator slots	25/27		
Slot filling factor	0.46		
Copper current density (A/mm^2)	5		
Outer diameter/stator inner diameter (mm)	194/34		
Thickness of stationary modulating ring (mm)	13		
Stack length (mm)	40		
PM material		N38SH (NdFeB magnet)	
Number of turns in series per phase	27	90	72
Inner airgap length (mm)	0.6	0.6	N/A
Middle airgap length (mm)	0.6	N/A	N/A
Outer airgap length (mm)	1	0.6	0.6
Axial length (mm)	40		
Total volume of copper (cm^3)	96.7	278.9	265.4
Total volume of PM (cm^3)	132.5	109.3	88
Specific torque (active mass) (Nm/kg)	6.36	5.03	6.43
Ratio of torque to PM volume (kNm/m^3)	360.8	364.1	578.4

All the MGMs have three phases and a rated outer rotor speed of 600 r/min. Phase voltages are 36 Vrms.

Moreover, the specific torque of the one-airgap MGM is also shown to be almost the same as that of the three-airgap MGM, and about 25% higher than that of the two-airgap MGM. The one-airgap MGM has the highest ratio of torque to PM volume, the simplest structure, and the lowest price-performance ratio.

2.3.2 Airgap Magnetic Field Modulation Behaviors

No-load flux patterns of the three MGMs shown in Figure 2.28 indicate that although the pole-pair number of the outer rotor PM array is 22, the stator

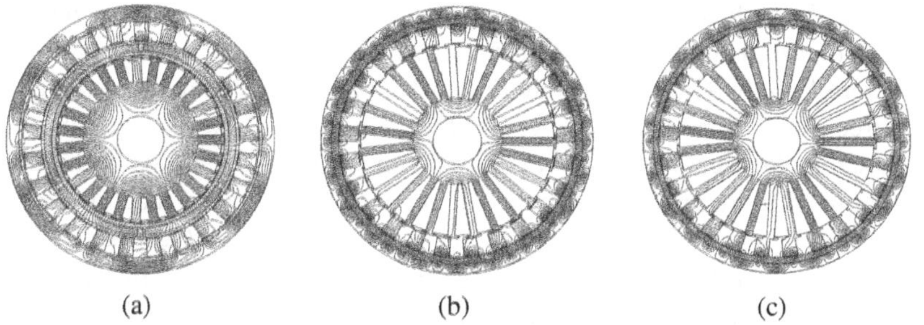

Figure 2.28 No-load magnetic field distribution of MGMs. (a) Three-airgap, (b) two-airgap, (c) one-airgap.

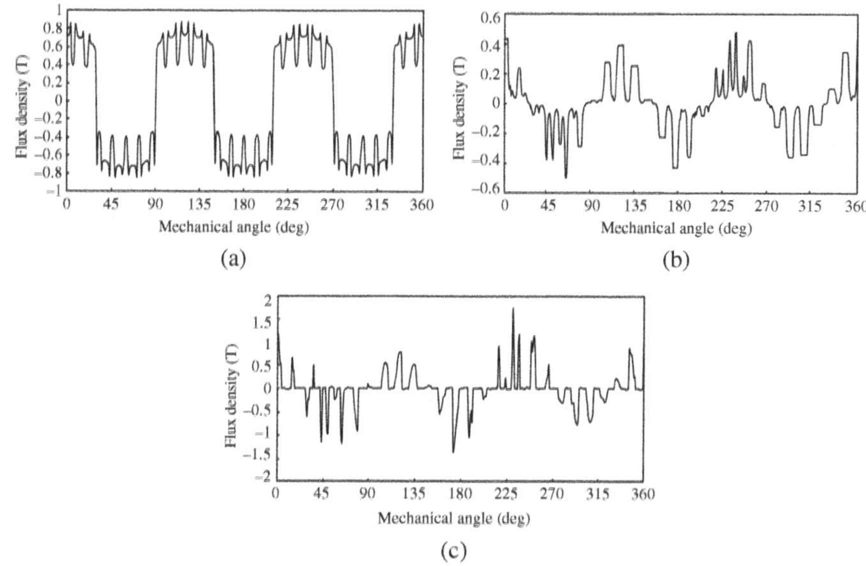

Figure 2.29 Radial airgap flux density waveforms of MGMs at the stator outer diameter under no-load condition. (a) Three-airgap, (b) two-airgap, (c) one-airgap.

magnetic fields of the two-airgap and one-airgap MGMs show a 6-pole pattern due to modulation behavior of ferromagnetic blocks. Therefore, the stator winding needs to be wound to have three pole pairs.

The corresponding radial airgap flux density waveforms at the stator outer diameter are shown in Figure 2.29. The radial flux density waveform in the inner airgap of the three-airgap MGM is similar to that in PMSMs, indicating that the magnetic field generated by the outer MG in this structure has little influence on the inner PM brushless machine.

Due to the reduced PM volume and magnetic field modulation behavior of the modulating ring, the amplitudes of radial flux densities at the stator outer diameter of the two-airgap and one-airgap MGMs are much smaller than that of the three-airgap one. Therefore, to produce the same phase voltage amplitude at the same electrical frequency, two-airgap and one-airgap MGMs both need to have more winding turns. Deeper slots are needed for these two machine types to accommodate more armature winding turns and meet the voltage requirements. On the other hand, more winding turns will not only increase phase winding inductance and reduce the power factor, but will also increase winding resistance and reduce efficiency. These aspects need to be considered carefully when designing MGMs.

2.4 PM Vernier Machine

The PMV machine as an emerging direct drive solution, which has dissimilar numbers of pole pairs for the stator winding and PM rotor. The stator winding can be designed based on the high-speed rotating magnetic field in the airgap while the rotor actually rotates at a lower speed to meet the direct drive requirements. Compared to PMSMs, the PMV machine can easily achieve the low-speed and high-torque transmission without changing mechanical structure or adding other mechanical components. This section presents the airgap magnetic field modulation phenomena in PMV machines based an outer-rotor PMV structure [11].

2.4.1 Machine Construction

The cross-sectional view of the three-phase PMV machine with 18 stator slots and 28 rotor poles is illustrated in Figure 2.30. It is designed to have an outer-rotor structure preferred in direct-drive applications such as wind turbines and in-wheel traction. The outer rotor consists of a laminated rotor core with tangentially magnetized PMs inserted. Adjacent PMs are magnetized in the opposite direction to achieve a flux-focusing or flux-concentration effect and improve the torque density.

The PMV machine may have open-slot or split-slot structures [29]. Compared with the split-slot stator structure, the open-slot type [27] has a higher utilization of PMs by suppressing PM flux leakage and is thus more favorable in designs.

2.4.2 Airgap Magnetic Field Modulation Behaviors

Figure 2.31 shows the radial airgap flux density waveform and its amplitude spectrum of an 18-slot 28-pole (18/28) PMV machine under no-load condition. It is

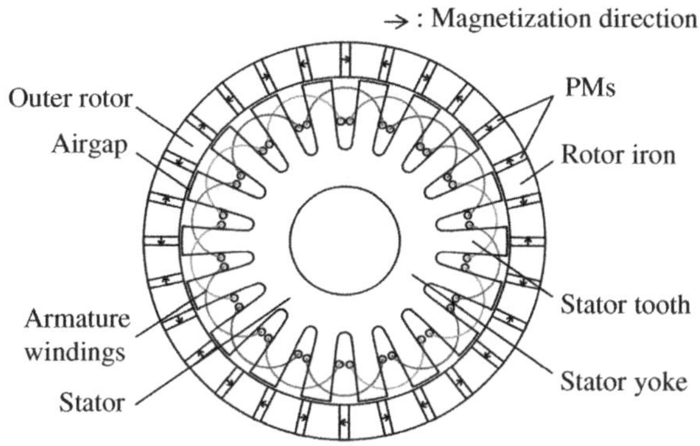

Figure 2.30 Cross-sectional view of the 18-slot 28-pole outer-rotor PMV machine.

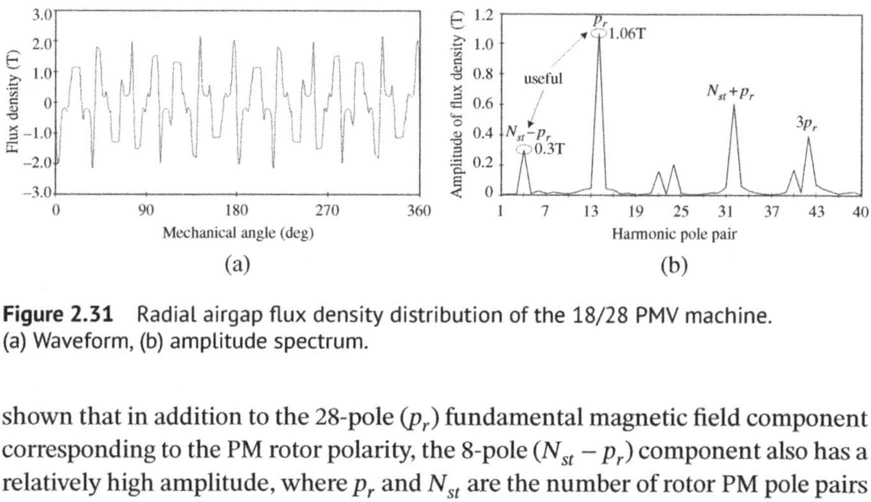

Figure 2.31 Radial airgap flux density distribution of the 18/28 PMV machine. (a) Waveform, (b) amplitude spectrum.

shown that in addition to the 28-pole (p_r) fundamental magnetic field component corresponding to the PM rotor polarity, the 8-pole ($N_{st} - p_r$) component also has a relatively high amplitude, where p_r and N_{st} are the number of rotor PM pole pairs and number of stator slots, respectively.

The number of stator slots is selected to be slightly larger than the pole-pair number of the rotor PMs to ensure that the ($N_{st} - p_r$)-pole-pair magnetic field harmonic in the airgap has a low polarity and a high rotating speed. Stator windings of PMV machines are designed based on the polarity of ($N_{st} - p_r$)-pole-pair magnetic field harmonic, instead of the actual rotor PM polarities, to reduce the number of stator slots and simplify the winding structure.

It is shown that when the above-mentioned design scheme is adopted, not only can the ($N_{st} - p_r$)-pole-pair harmonic in the airgap be utilized for electromechanical energy conversion; the fundamental harmonic with the p_r-pole-pair can also contribute to the torque production. Both the ($N_{st} - p_r$) and p_r-pole-pair harmonics will induce flux linkages of the same frequency in the stator winding.

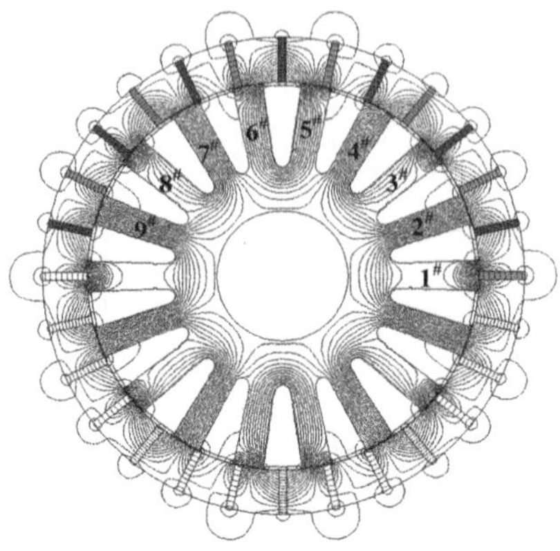

Figure 2.32 No-load magnetic field distribution of the 18/28 PMV machine.

To illustrate the operation of PMV machines in a more straightforward way, we plot the magnetic flux lines of the 18/28 PMV machine under no-load condition based on 2D FEA results in Figure 2.32. It can be seen that due to the airgap magnetic field modulation behavior performed by the stator open slots, the stator magnetic field distribution shows an 8-pole pattern, even though the outer rotor PM array has 14 pole pairs.

The electromagnetic performance of an 18/28 PMV machine and the 18-slot 8-pole (18/8) PMSM counterpart is compared and tabulated in Table 2.3. It is shown that the open-circuit back-EMF produced by the PMV machine is almost twice that of the 18/8 PMSM at the same rotor speed. It can be concluded that by utilizing more of the airgap field harmonics than the fundamental, the open-circuit phase back-EMF of the PMV machine can be greatly improved, contributing to more torque production.

2.5 Linear PMV Machine

In PMV machines discussed in Section 2.4, PMs are placed on the rotor, and naturally there is a relative motion between the magnetic field created by PMs and the armature winding placed on the stator. In the case of stator-PM machines, either the rotating or linear types, whose PMs and armature windings are located on the same stator, the operating principle shows great difference compared with machines with PMs on the rotor. Therefore, it is necessary to discuss the airgap magnetic field modulation phenomena in such machine types separately. This section takes a linear primary PMV machine as an example to illustrate the difference in their constructions, operating principles, and basic performance.

Table 2.3 Comparison of the PMV machine and PMSM.

Parameter	PMV machine	PMSM
Rotor speed (r/min)	214	
No-load phase back-EMF (Vrms)	228	123
Phase current (Arms)	8.33	
Number or rotor PM pole pairs	14	4
Number of stator teeth	18	
Number of pole pairs of stator winding	4	
Number of turns in series per phase	162	
Lamination material	DW360-50	
PM width (mm)	20.5	71.75
PM length in magnetization direction (rad)	0.039	
PM remanence (T)	1.15	
Relative permeability of PM	1.07	
Airgap diameter (mm)	178.5	
Airgap length/axial length (mm)	0.5/284	
Current density (A/mm^2)	4.15	
Average torque (Nm)	251.8 (BLAC)	129.8 (BLDC)/136.1 (BLAC)
Torque ripple (%)	6.6 (BLAC)	20.3 (BLDC)/18.6 (BLAC)

The airgap diameter, airgap length, stack length, PM volume, and specific electric loading are kept the same for the two machines. Both brushless DC (BLDC) and brushless AC (BLAC) operations apply to the PMSM due to the trapezoidal open-circuit back-EMF. The PMV machine has a more sinusoidal back-EMF so the BLAC control is used for the comparison.

2.5.1 Machine Construction

Figure 2.33 shows the cross-sectional view of a 6-slot 2-pole (6/2) linear PMV machine [30, 31] with a single-sided structure. The primary, i.e. stator, consists of six teeth and five magnet blocks mounted on the surface of each tooth. Adjacent PMs on a single tooth are magnetized in opposite directions. PMs on all the teeth are of the same pattern. Semi-closed slots are used to accommodate more magnets and slot openings share the same width as PMs. The secondary, i.e. mover, of the machine is a castellated magnetic core with salient reluctance poles, featuring a simple structure and high mechanical strength. It is very suitable for applications requiring high thrusts, such as the direct-drive ocean wave power generation.

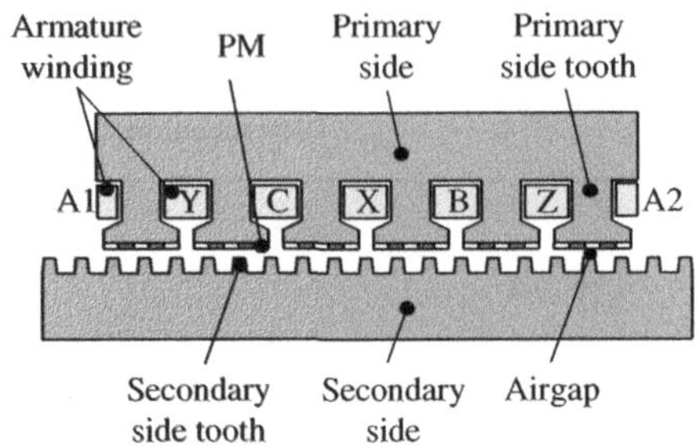

Figure 2.33 Cross-sectional view of the linear PMV machine.

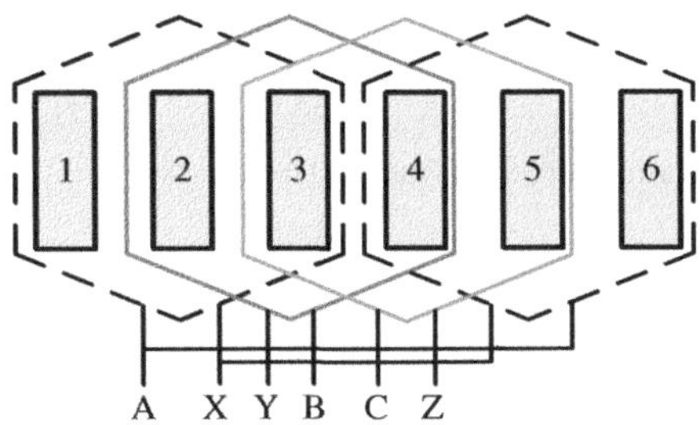

Figure 2.34 Armature winding layout of the 6/2 linear PMV machine.

Both the primary and secondary cores are manufactured by stacking multiple electrical steel sheets to suppress core losses and thus improve operating efficiency. Alternatively, they can be made from machined carbon steel blocks to simplify the manufacturing process at the cost of reduced efficiency.

The PM array has 18 pole pairs and the secondary has 17 salient reluctance poles. It will be shown in the following analysis that the pole-pair number of the working magnetic field harmonic is $18 - 1 = 1$. Therefore, the three-phase stator winding can be designed based on $Z = 6$ and $p = 1$, where Z is the number of stator slots and p is the number of winding pole pairs. Due to the linear structure, both phase B and phase C each has only one coil, while phase A consists of two coils in series placed at both ends of the primary to balance the magnetic circuit. Figure 2.34 shows the developed view of the three-phase armature winding.

2.5.2 Airgap Magnetic Field Modulation Behaviors

The operating principle of the linear PMV machine can be explained by visualizing the magnetic field distributions at four typical positions when the secondary moves, as shown in Figure 2.35. At the four positions, the phase-A flux linkage reaches the positive maximum, zero, the negative maximum, and zero

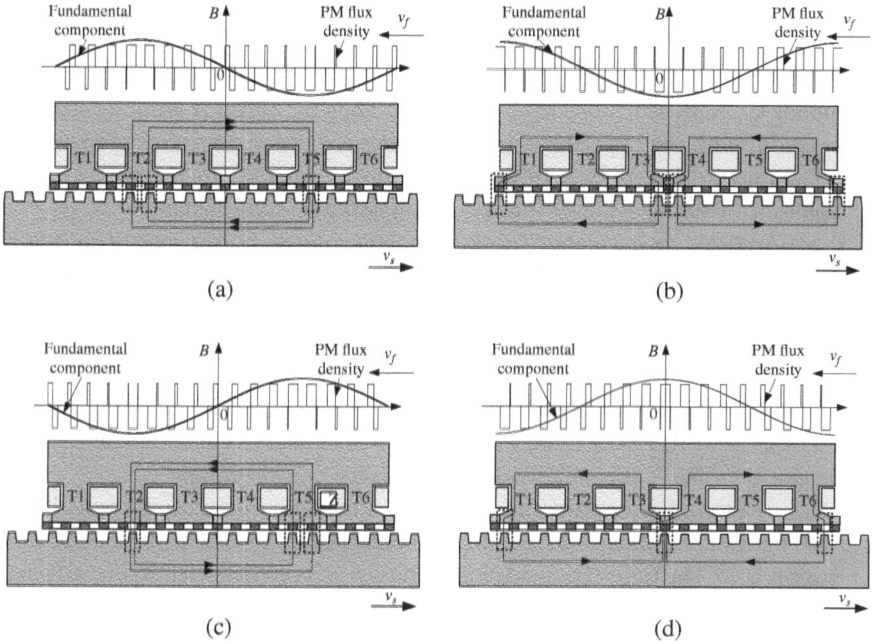

Figure 2.35 Four typical mover positions of the linear PMV machine. (a) Position A, (b) Position B, (c) Position C, (d) Position D.

successively. For the sake of clarity, slot openings between primary teeth are filled with fictitious magnets with the opposite magnetization of their adjacent PMs to neglect the slotting effect.

Position A: It is assumed that the magnetization direction of gray magnets in Figure 2.35 is upward and the white magnets is downward, and Position A is defined at $x_s = 0$. At this instant, the two gray magnets mounted on the primary tooth T2 have the largest overlapping area with the secondary salient reluctance poles. The white magnet in the middle of the primary tooth T5 is aligned to the centerline of a secondary salient reluctance pole, and any overlap between other magnets and other secondary salient reluctance poles is relatively small.

Since the airgap permeance at positions under secondary salient reluctance poles is much higher than that under the secondary slots, the magnetic flux generated by each PM is approximately proportional to the overlapping area between itself and the salient reluctance pole. According to the relative position of the PM and the secondary salient reluctance pole, the full waveform of the PM flux in the airgap shown in Figure 2.35a is obtained. Ignoring the high-order harmonic components, the waveform of the fundamental component can be obtained, where the positive maximum occurs at the middle of tooth T2.

At this instant, the main flux in the machine flows in a clockwise path: two gray PMs in the dotted box on tooth T2 → primary yoke → white PM in the dotted box on tooth T5 → airgap → secondary salient reluctance pole → secondary yoke → secondary salient reluctance poles → airgap. Since the phase-A winding consists of two coils with the opposite winding directions, and the directions of magnetic flux passing through the two coils are also opposite, the PM flux linkage linked by the phase-A winding achieves the positive maximum at this instant.

Position B: When the secondary of the machine moves by a quarter of one tooth pitch to the right, as shown in Figure 2.35b, the gray PMs on the primary teeth T1 and T6 are aligned with the secondary salient poles, as are the white PMs on T3 and T4, which is highlighted by the dashed box. Therefore, the magnetic circuit of the whole machine is divided into two parallel paths. The left-hand-side magnetic circuit follows the following path: gray PM on tooth T1 → tooth T1 → primary yoke → tooth T3 → white PM on tooth T3 → airgap → secondary salient reluctance poles → secondary yoke → secondary salient reluctance poles → airgap.

At this instant, the upward and downward magnetic flux passing through the phase-A coils are of the same magnitude but opposite directions, so the net flux linkage is zero. The right-hand-side magnetic circuit is symmetrical to the left part. As a result, the total flux linkage of the A-phase winding is zero at position B.

By examining the relative position of PMs and secondary salient reluctance poles, the detailed PM flux density waveform and its fundament component in the airgap can be obtained. It can be observed that the positive maximum is shifted by a half of one pole pitch to the left, compared with the case at position A. The positive maximum now appears at the left end of the primary core.

Position C: When the secondary continues to move rightward by a quarter of one tooth pitch, as shown in Position C of Figure 2.35c, the white PM at the middle of tooth T2 is aligned with one secondary salient reluctance pole, and the two gray PMs on tooth T5 overlap with the salient reluctance poles of the secondary with the largest overlapping area, as highlighted by the dashed box. The magnetic circuit at this instant mirrors the magnetic circuit at position A.

At this moment, the magnetic flux linkage linked by coil 1 of the A-phase winding reaches the negative maximum, while the magnetic flux linkage linked by coil 2 reaches the positive maximum. Since the two coils are wound in opposite directions, the net PM flux linkage of the A-phase winding is the negative maximum after the two coils are connected in series. Similarly, the waveforms of the PM flux density and its fundamental component in the airgap can be derived from the relative position of PMs and secondary salient reluctance poles. It can be observed that the peak fundamental flux linkage is shifted circularly to the left by a half of one pole pitch, locating at the middle of tooth T5.

Position D: When the secondary moves to the right further by a quarter of one tooth pitch, as denoted by Position D in Figure 2.35d, the white PMs on teeth T1 and T6 are aligned with the axis of the secondary salient reluctance poles, and the gray PM at the middle of teeth T3 and T4 are aligned with axis of another secondary salient reluctance pole. The magnetic circuit at this moment is similar to that at Position B, which also has two parallel paths.

Since the magnetization direction of PMs aligned with the secondary salient reluctance poles is opposite to that at Position B, the direction of the flux is reversed. The total PM flux linkage linked by the two phase-A coils is zero too. The peak airgap flux density moves to the left again, arriving at the middle position of teeth T3 and T4.

If the secondary is further moved rightward by a quarter tooth pitch, the relative position of PMs on the primary and salient reluctance poles of the secondary will repeat the case at position A, completing a full electric cycle.

The following conclusions can be drawn based on the analysis of the magnetic field distributions at the four representative positions:

1) PM flux density distributions similar to pulse width modulation (PWM) waveforms will be created in the airgap due to the modulation effect of secondary salient reluctance poles. The fundamental component produces the main flux.
2) The PM flux linkage linked by the armature winding will change by a full electric cycle if the secondary moves by a tooth pitch.
3) The fundamental component of the magnetic field will travel in the opposite direction of, but at the same speed as that of the secondary.
4) Since the pole pitch of the fundamental component is much larger than the secondary salient reluctance pole pitch, the mechanical speed of the fundamental field component is far greater than that of the secondary.
5) Linear PMV machines are suitable for low-speed high-thrust applications due to their high-polarity high-frequency nature.

The armature field can be analyzed similarly by looking at the four representative positions. The airgap magnetic field modulation phenomena can also be seen in the armature field when the magnetization of all the magnets is turned off in 2D FEA. The flux lines due to the armature field alone are shown in Figure 2.36. The corresponding airgap magnetic flux density waveform illustrated in Figure 2.37 shows a 2-pole pattern superimposed by high-polarity harmonic components introduced by the rotor salient reluctance poles, indicating the magnetic field modulation behavior of rotor poles.

2.6 Flux-switching PM Machine

Flux-switching machines have received extensive attention in recent years and a variety of new machine topologies have also been derived from them [32].

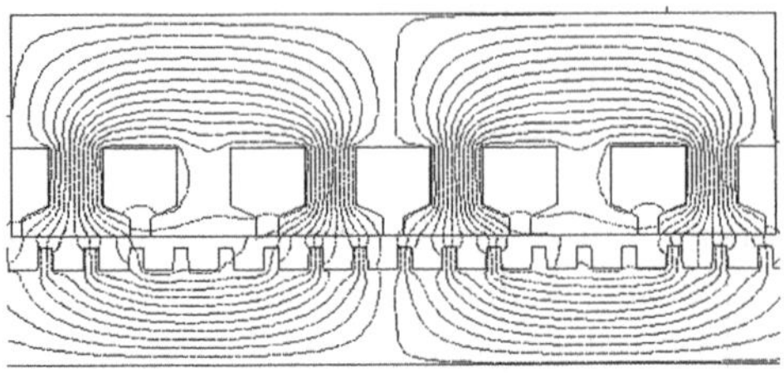

Figure 2.36 Armature field.

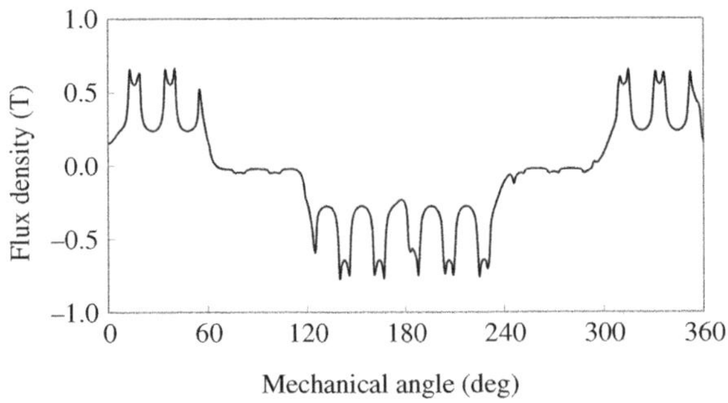

Figure 2.37 Radial airgap flux density created by three-phase armature currents only.

All these flux-switching machine variants show field modulation phenomena and share similar operating principles. This section takes the most representative flux-switching PM machine with 12 stator slots and 10 rotor salient reluctance poles as an example to illustrate the field modulation phenomenon in flux-switching machines using the simple MMF-permeance model presented in [33–35].

The three-dimensional (3D) view of the FSPM machine is shown in Figure 2.38. Both the stator and rotor have salient reluctance pole structures. The PM MMF and the armature MMF are modulated by the rotor salient reluctance poles, resulting in corresponding airgap flux density harmonics.

2.6.1 Magnetic Field Modulation Mechanism of PM Field

PMs in flux-switching PM machines are embedded between two adjacent U-shaped stator iron cores. The MMF created by PMs can be approximated by a rectangular wave along the airgap peripheral. Considering the influence of U-shaped stator cores, the PM MMF-permeance model is illustrated in

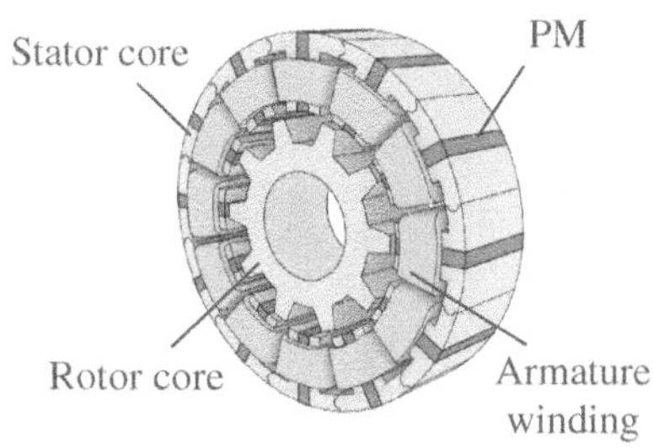

Figure 2.38 FSPM machine with 12 stator slots and 10 rotor salient reluctance poles.

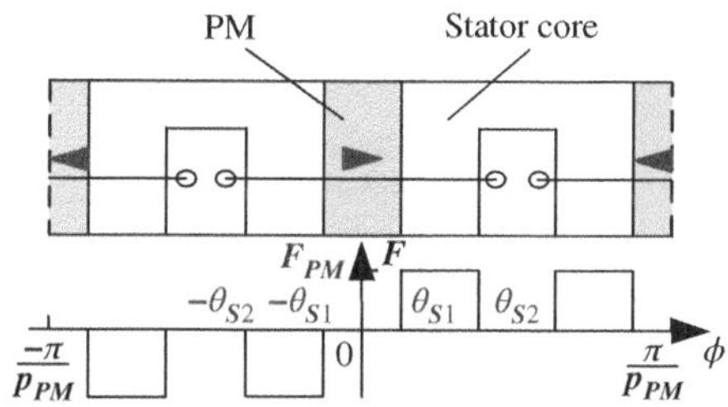

Figure 2.39 MMF distribution of the FSPM machine over a PM pole pitch.

Figure 2.39. By writing the PM MMF into its Fourier series form, we can get:

$$F_{PM}(\phi, t) = F_{PM0} + \sum_{n=1}^{\infty} F_{PMn} \sin\left[(2n-1)\, p_{PM}\phi\right] \tag{2.21}$$

where

$$F_{PM0} = -\frac{4F_{PM}}{\pi} \tag{2.22}$$

$$F_{PMn} = \frac{\cos\left[(2n-1)\, p_{PM}\theta_{S2}\right] - \cos\left[(2n-1)\, p_{PM}\theta_{S1}\right]}{2n-1} \tag{2.23}$$

where F_{PM} is the amplitude of the PM MMF. θ_{S1} and θ_{S2} are geometric variables. F_{PMn} is the Fourier coefficient, where n is a positive integer, and $n = 1, 2, 3, \ldots$. p_{PM} is the number of PM pole pairs, which is half of the number of stator slots p_s.

Equation (2.21) shows that the PM MMF only contains harmonics with a number of pole pairs of $(2n-1)p_{PM}$. In addition, PMs are located on the stator, so all the spatial harmonics are stationary too. Without considering the influence of rotor salient reluctance poles, the stator PM MMF spectrum of the 12-slot 10-pole (12/10) flux-switching PM machine mainly includes harmonics with pole pairs of 6, 18, 30, 42, etc., as shown in Figure 2.40. This is consistent with the analytical prediction from (2.21). The number of PM pole pairs is $p_{PM} = 6$, and the harmonic with six pole pairs has the largest amplitude.

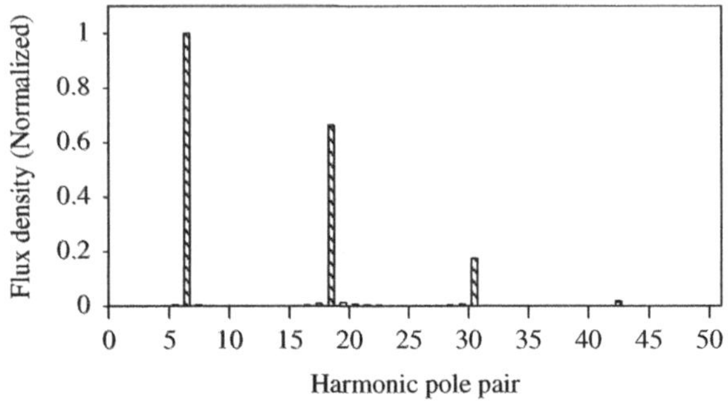

Figure 2.40 PM MMF distribution of the 12/10 flux-switching PM machine ignoring the influence of rotor salient reluctance poles.

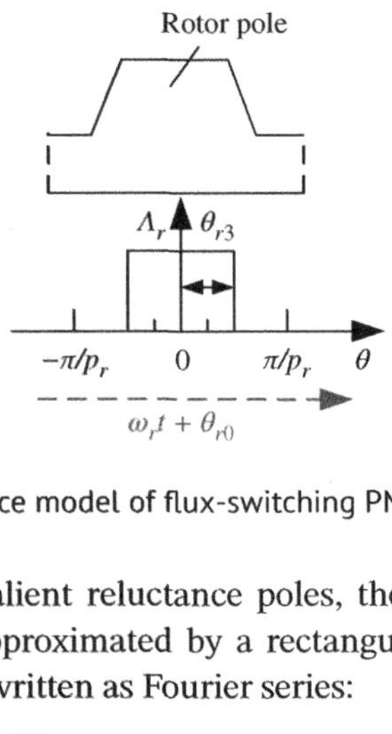

Figure 2.41 Airgap permeance model of flux-switching PM machines.

Considering the rotor salient reluctance poles, the airgap permeance over a rotor pole pitch can be approximated by a rectangular waveform as shown in Figure 2.41, which can be written as Fourier series:

$$\Lambda_{Sr}(\theta) = \Lambda_{Sr0} + \Lambda_{Srb}\sum_{k=1}^{\infty}\Lambda_{Srk}\cos\left(kp_r\theta\right), \theta = \phi - \left(\omega_r t + \theta_{r0}\right) \tag{2.24}$$

where Λ_{Sr0}, Λ_{Srb}, and Λ_{Srk} are the Fourier coefficients, and $k = 1, 2, 3, \ldots$.

$$\Lambda_{Sr0} = \frac{p_r\Lambda_r\theta_{r3}}{\pi} \tag{2.25}$$

$$\Lambda_{Srb} = \frac{2\Lambda_r}{\pi} \tag{2.26}$$

$$\Lambda_{Srk} = \frac{\sin\left(kp_r\theta_{r3}\right)}{k} \tag{2.27}$$

where Λ_{Sr} is the peak value of the airgap permeance at the rotor side, θ_{r3} is a half of the rotor tooth width, p_r is the number of rotor salient reluctance poles.

The open-circuit airgap flux density is determined by the PM MMF and the rotor airgap permeance together as:

$$B_g(\phi, t) = B_1(\phi, t) + B_2(\phi, t) + B_3(\phi, t) \tag{2.28}$$

where

$$B_1(\phi, t) = F_{PMb}\Lambda_{Sr0} + \sum_{n=1}^{\infty} F_{PMn}\Lambda_{Sr0} \sin\left[(2n-1)\,p_{PM}\phi\right] \tag{2.29}$$

$$B_2(\phi, t) = \frac{F_{PMb}\Lambda_{Srb}}{2}\sum_{n=1}^{\infty}\sum_{k=1}^{\infty} F_{PMn}\Lambda_{Srk} \sin\left\{\left[kp_r + (2n-1)\,p_{PM}\right]\phi - kp_r\left(\omega_r t + \theta_{r0}\right)\right\} \tag{2.30}$$

$$B_3(\phi, t) = -\frac{F_{PMb}\Lambda_{Srb}}{2}\sum_{n=1}^{\infty}\sum_{k=1}^{\infty} F_{PMn}\Lambda_{Srk} \sin\left\{\left[kp_r - (2n-1)\,p_{PM}\right]\phi - kp_r\left(\omega_r t + \theta_{r0}\right)\right\} \tag{2.31}$$

Based on (2.29)–(2.31), characteristics of open-circuit airgap flux density harmonics can be obtained, as summarized in Table 2.4. Since PMs are placed on the stator, the PM MMF is stationary and speeds of $(2n-1)p_{PM}$-pole-pair harmonics are zero. Due to the presence of rotor salient reluctance poles, new harmonics with $|(2n-1)p_{PM} \pm kp_r|$ pole pairs appear in the airgap, whose speeds are $kp_r\omega_r/[kp_r \pm (2n-1)p_{PM}]$.

The FEA-predicted open-circuit airgap flux density waveform and its harmonic spectrum of the 12/10 FSPM machine is shown in Figure 2.42, which mainly contains harmonic components with pole pairs of 4, 6, 8, 16, 18, and 28. Among them, harmonics with pole pairs of 4, 8, 16, and 28 are produced by the modulation effect of rotor salient reluctance poles on the PM MMF. Harmonics with pole pairs of 6 and 18 are directly from the PM MMF. According to the analytical equation derived from the PM MMF-permeance model, these pole-pair numbers can be expressed concisely as $|(2n-1)p_{PM} \pm kp_r|$, $k = 1$, $n = 1, 2$, and $(2n-1)p_{PM}$, $n = 1, 2$.

Table 2.4 Characteristics of no-load PM airgap flux density harmonics in flux-switching PM machines.

Harmonic pole pair	Speed of harmonics
$(2n-1)p_{PM}$	0
$kp_r + (2n-1)p_{PM}$	$kp_r\omega_r/[kp_r + (2n-1)p_{PM}]$
$\lvert kp_r - (2n-1)p_{PM}\rvert$	$kp_r\omega_r/[kp_r - (2n-1)p_{PM}]$

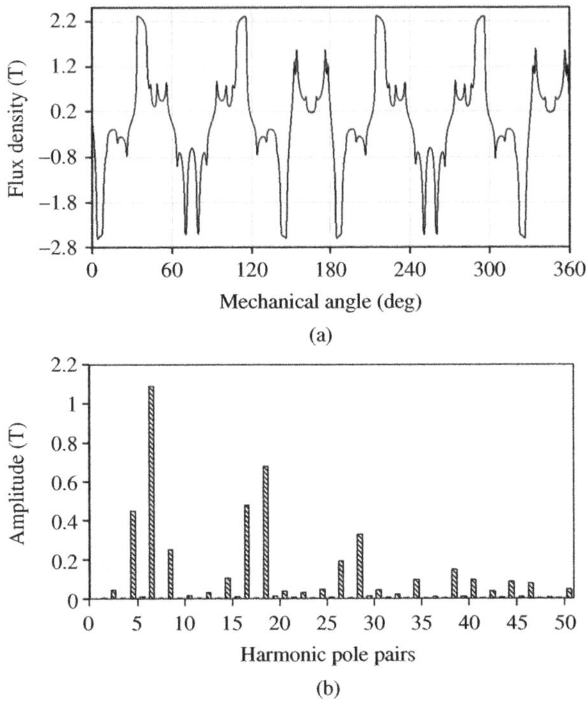

Figure 2.42 Open-circuit airgap flux density waveform and its harmonic spectrum of the 12/10 flux-switching PM machine. (a) Waveform, (b) amplitude spectrum of (a).

2.6.2 Magnetic Field Modulation Mechanism of Armature Field

Armature MMF distributions are shown in Figure 2.43. From the winding layout, it can be seen that the four coils of phase A are wound in the same direction, so they are all defined as positive values in the MMF distribution. Similarly, MMF distributions of phase B and phase C can be obtained, which are shifted from that of phase A by 30 and 60 mechanical angles, respectively.

The armature MMF distribution can be expressed by:

$$F_{Sw}(\phi, t) = \frac{2N_{Sc}}{\pi}\sum_{i=1}^{\infty} F_{Swi}\left\{ i_{SA}\sin[(4i)\phi] + i_{SB}\sin\left[(4i)\left(\phi - \frac{\pi}{6}\right)\right] \right.$$
$$\left. + i_{SC}\sin\left[(4i)\left(\phi + \frac{\pi}{6}\right)\right]\right\} \tag{2.32}$$

where N_{Sc} is the number of turns per coil, F_{Swi} is the Fourier coefficients, and $i = 1, 2, 3, \ldots$

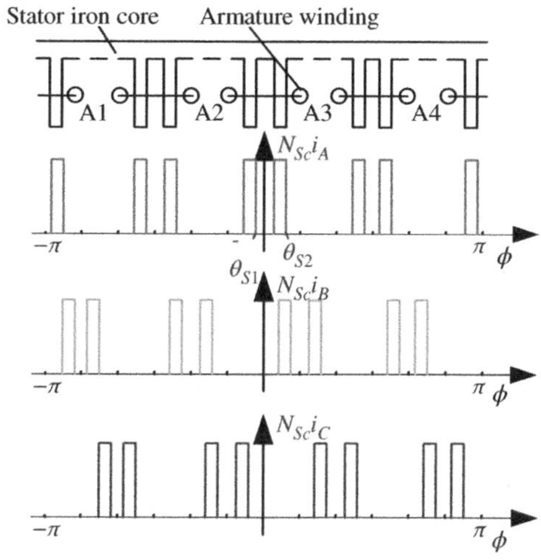

Figure 2.43 Armature MMF distribution of the 12/10 flux-switching PM machine.

When $i = 3r - 2$, where r is a positive integer, the armature MMF can be expressed as:

$$F_{Sw}(\phi, t) = \frac{3N_{Sc}I_{S\max}}{\pi}\sum_{i=1}^{\infty}F_{Swi}\sin\left(p_r\omega_r t - 4i\phi\right) \tag{2.33}$$

where I_{Smax} is the peak value of the armature current, and F_{Swi} can be written as:

$$F_{Swi} = \frac{\left[\sin\left(4i\theta_{S2}\right) - \sin\left(4i\theta_{S1}\right)\right]}{i} \tag{2.34}$$

When $i = 3r - 1$, (2.33) will become:

$$F_{Sw}(\phi, t) = \frac{3N_{Sc}I_{S\max}}{\pi}\sum_{i=1}^{\infty}F_{Swi}\sin\left(p_r\omega_r t + 4i\phi\right) \tag{2.35}$$

where

$$F_{Swi} = \frac{\left[\sin\left(4i\theta_{S2}\right) - \sin\left(4i\theta_{S1}\right)\right]}{i}. \tag{2.36}$$

When $i = 3r$, (2.35) will be:

$$F_{Sw}(\phi, t) = 0 \tag{2.37}$$

It can be concluded that the pole-pair number of the armature MMF harmonics can be expressed by $4i$, $i = 1, 2, \ldots$, excluding three and multiples of three. The harmonic spectrum of the armature MMF is shown in Figure 2.44, which mainly includes harmonics with pole pairs of 4, 8, 16, 20, 28, and 32. They correspond to

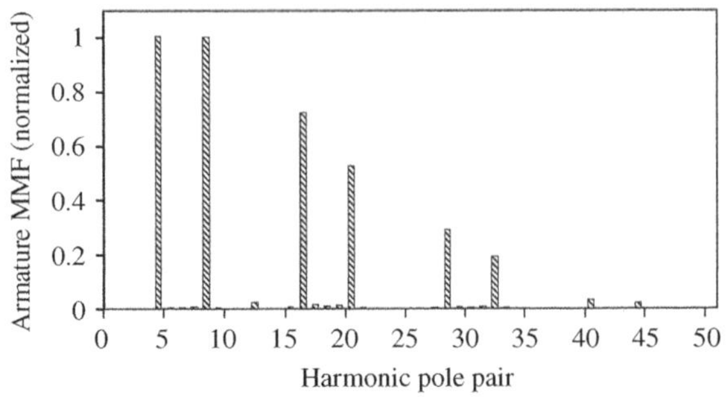

Figure 2.44 Amplitude spectrum of armature MMF in the 12/10 flux-switching PM machine based on (2.32).

the $4i$-pole-pair harmonics in the analytical armature MMF model with $i = 1, 2, 4, 5, 7$, and 8.

Using the airgap permeance expressed by (2.24), the airgap flux density produced by the armature MMF can be obtained as:

$$B_{Sw}(\phi, t) = F_{Sw}(\phi, t)\, \Lambda_{Sr}(\theta) \tag{2.38}$$

When $i = 3r - 2$, the airgap flux density can be written as:

$$\begin{aligned} B_{Sw}(\phi, t) = {} & \frac{3\Lambda_{Sr0} N_{Sc} I_{S\max}}{\pi} \sum_{i=1}^{\infty} F_{Swi} \cos\left(4i\phi - p_r\omega_r t + \frac{\pi}{2}\right) \\ & + \frac{3N_{Sc} I_{S\max}}{2\pi} \Lambda_{Srb} \sum_{i=1}^{\infty} \sum_{k=1}^{\infty} \Lambda_{Srk} F_{Swi} \left(\cos\beta_1 + \cos\beta_2\right) \end{aligned} \tag{2.39}$$

where

$$\begin{cases} \beta_1 = \left(kp_r - 4i\right)\phi - \left[(k-1)\,p_r\omega_r t + kp_r\theta_{r0} + \pi/2\right] \\ \beta_2 = \left(kp_r + 4i\right)\phi - \left[(k+1)\,p_r\omega_r t + kp_r\theta_0 - \pi/2\right]. \end{cases} \tag{2.40}$$

When $i = 3r - 1$, the airgap flux density can be written as:

$$\begin{aligned} B_{Sw}(\phi, t) = {} & \frac{3\Lambda_{Sr0} N_{Sc} I_{S\max}}{\pi} \sum_{i=1}^{\infty} F_{Swi} \sin\left(-4i\phi + p_r\omega_r t + \frac{\pi}{2}\right) \\ & + \frac{3N_{Sc} I_{S\max}}{2\pi} \Lambda_{Srb} \sum_{i=1}^{\infty} \sum_{k=1}^{\infty} \Lambda_{Srk} F_{swi} \left(\cos\beta_1 + \cos\beta_2\right) \end{aligned} \tag{2.41}$$

where

$$\begin{cases} \beta_1 = \left(kp_r + 4i\right)\phi - \left[(k-1)\,p_r\omega_r t + kp_r\theta_{r0} + \pi/2\right] \\ \beta_2 = \left(kp_r - 4i\right)\phi - \left[(k+1)\,p_r\omega_r t + kp_r\theta_{r0} - \pi/2\right]. \end{cases} \tag{2.42}$$

Main observations from (2.39) and (2.41) are summarized in Table 2.5. Main harmonics in the airgap flux density have the following two groups:

Table 2.5 Harmonic characteristic of armature field airgap flux density of FSPM machine.

$i = 3r - 2$		$i = 3r - 1$	
Pole-pair number of harmonic	**Speed of harmonic**	**Pole-pair number of harmonic**	**Speed of harmonic**
$4i$	$p_r\omega_r/4i$	$4i$	$-p_r/4i$
$\lvert kp_r - 4i\rvert$	$(k-1)p_r\omega_r/(kp_r - 4i)$	$kp_r + 4i$	$(k-1)p_r/(kp_r + 4i)$
$kp_r + 4i$	$(k+1)p_r\omega_r/(kp_r + 4i)$	$\lvert kp_r - 4i\rvert$	$(k+1)p_r/(kp_r - 4i)$

(1) Harmonics with $4i$ pole pairs, which are directly generated by the armature MMF and the average airgap, so they have the same pole-pair numbers as the armature MMF.
(2) Harmonics with $|p_r \pm 4i|$ pole pairs, which are generated by the armature field modulated by the rotor salient reluctance poles.

When $i = 3r - 2$, the speed of the $|p_r - 4i|$-pole-pair harmonics is zero, and the pole-pair number of harmonics can be rewritten as:

$$| p_r - 4i | = | 18 - 12r | = | [2(r-1) - 1] p_{PM} | \tag{2.43}$$

Obviously, these armature field harmonics and $(2n-1)p_{PM}$-pole-pair harmonics in the open-circuit airgap flux density shown in Table 2.4 are stationary.

When $i = 3r - 1$, the speed of the $(p_r + 4i)$-pole-pair armature field harmonics is also zero, and the pole-pair number of harmonics can be rewritten as:

$$| p_r + 4i | = | 12r + 6 | = | [2(r+1) - 1] p_{PM} | \tag{2.44}$$

Equation (2.44) shows these harmonics are also in synchronization with the $(2n-1)p_{PM}$-pole-pair harmonics in the open-circuit airgap flux density shown in Table 2.4. Both of them are stationary.

In addition to these stationary spatial harmonics, other harmonics are also synchronous with the corresponding harmonics of the open-circuit airgap flux density shown in Table 2.4. When $i = 3r - 2$, the pole-pair number of the armature field can be rewritten as:

$$4i = 12r - 8 = p_r + [2(r-1) - 1] p_{PM} \tag{2.45}$$

$$| kp_r - 4i | = | 10k - 12r + 8 | = | (k-1)p_r - [2(r-1) - 1] p_{PM} | \tag{2.46}$$

$$kp_r + 4i = 10k + 12r - 8 = (k+1)p_r + [2(r-1) - 1] p_{PM} \tag{2.47}$$

These armature field harmonics are all rotating in synchronization with the $(2n-1)p_{PM} \pm kP_r|$-pole-pair harmonics in the open-circuit airgap flux density

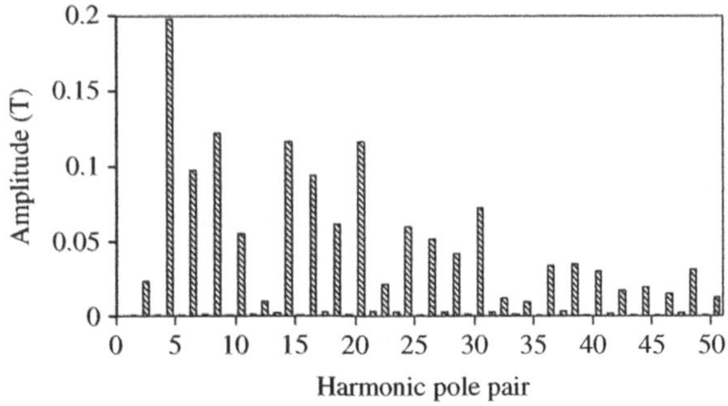

Figure 2.45 Harmonic spectrum of armature field in the 12/10 flux-switching PM machine based on 2D FEA.

shown in Table 2.4. Similarly, when $i = 3r - 1$, the pole-pair number of the armature field can be rewritten as:

$$4i = 12r - 4 = -\{p_r - [2(r+1) - 1]\,p_{PM}\} \tag{2.48}$$

$$|\,kp_r - 4i\,| = |\,10k - 12r + 4\,| = |\,(k+1)\,p_r - [2(r+1) - 1]\,p_{PM}\,| \tag{2.49}$$

$$kp_r + 4i = 10k + 12r - 4 = (k-1)\,p_r + [2(r+1) - 1]\,p_{PM} \tag{2.50}$$

Equations (2.48)–(2.50) show that such armature field harmonics are rotating and in synchronization with the $|(2n-1)p_{PM} \pm kp_r|$-pole-pair harmonics in the open-circuit airgap flux density shown in Table 2.4.

The amplitude spectrum of the armature field in the 12-slot 10-pole FSPM machine predicted from 2D FEA is shown in Figure 2.45. The armature field mainly includes harmonics with pole pairs of 4, 6, 8, 14, 16, and 20. Among them, the 4-, 8-, 16-, and 20-pole-pair harmonics are directly produced by the armature MMF, and the corresponding pole-pair number can be expressed as $4i$, $i = 1$, 2, 4, and 5. The 6- and 14-pole-pair harmonics are generated by the modulation behavior of the rotor salient reluctance poles performed on the armature MMF, whose pole-pair number can be expressed as $|kp_r \pm 4i|$, $k = 1$, $i = 1$.

2.7 Doubly-Fed Machines

Doubly-fed machines can be obtained by bringing out the rotor winding terminals of wound-rotor IMs through brushes and slip rings. Alternatively, they can also be specially designed for optimal electromagnetic performance. Since the rotor windings and rotor-side inverter only handle the slip power, the converter rating can be reduced significantly compared to traditional IMs and SMs in applications requiring only a narrow speed range, such as variable speed pumps, wind power harvesting, and ship shaft power generation [36].

How to remove brushes and slip rings used in double-fed machines has always been an important research topic in the field of electrical machines. Studies in recent years have shown that brushless doubly-fed machines inherit many advantages of the slip-ring doubly-fed machines, such as reduced converter rating and cost, flexible active and reactive power control, and low-voltage ride-through capability without crowbars, etc. They have become a strong competitor of slip-ring doubly-fed machines and have received wide attention from researchers all over the world.

Similar to the other machine types presented in previous sections of this chapter, the analysis of the airgap flux density waveform shows that there is obvious airgap field modulation phenomenon in both traditional slip-ring doubly-fed machines and various brushless counterparts. This means that in addition to the fundamental wave, there are multiple spatial harmonic components in the airgap magnetic field. The amplitude of these harmonic components can be large or small, depending on the electromagnetic design, and have an significant influence on the performance indicators, such as torque density/specific torque, power density/specific power, loss, efficiency, vibration and noise, etc.

Despite the large difference in rotor constructions, teeth and slots are commonly found in both brushed and brushless doubly-fed machines. They change the airgap permeance, resulting in so-called "tooth harmonics" in the airgap. In addition, there are a variety of working harmonics in different brushless doubly-fed machine topologies interacting with each other to produce electromagnetic torque and achieve electromechanical energy conversion.

2.7.1 Classification and Operating Principles

Doubly-fed machines can be classified into slip-ring doubly-fed machines and brushless doubly-fed machines. Taking a 4-pole slip-ring doubly-fed machine shown in Figure 2.46a as an example, the stator and rotor have 48 and 36 slots, respectively. Both the stator and rotor windings have two pole pairs. Ignoring the influence of spatial harmonics introduced by the winding layout and slotting effect, the MMFs created by stator and rotor windings are both sinusoidal traveling waves, and their speeds are the same.

When the machine is operating as a motor, the torque and power factor can be regulated by adjusting the amplitude and phase of the stator or rotor winding currents. The construction of a slip-ring doubly-fed machine is similar to that of a wound-rotor IM, except that the rotor winding is connected to the power electronic converter through brushes and slip rings. In contrast, the rotor winding of the wound rotor IM is directly short-circuited or connected to external resistors.

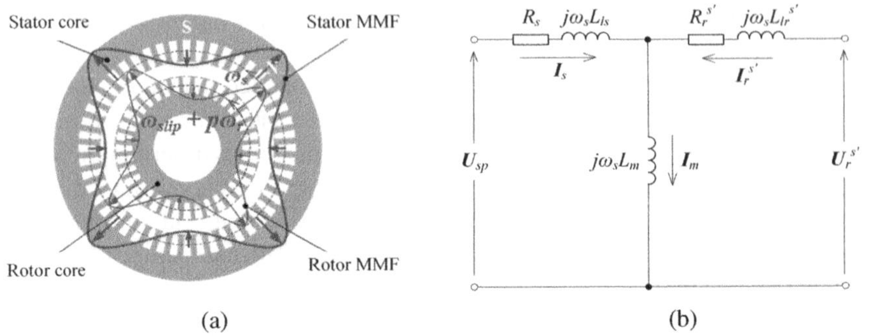

Figure 2.46 Slip-ring doubly-fed machine, (a) cross-sectional view, (b) steady-state equivalent circuit. ω_s is the electrical frequency of stator winding, ω_r the mechanical speed of rotor, ω_{slip} the slip frequency and p number of pole pairs.

The steady-state performance of the slip-ring doubly-fed machine can be obtained by analyzing the equivalent circuit illustrated in Figure 2.46b. It can be seen that, unlike traditional IMs and SMs, slip-ring doubly-fed machines have two electrical ports, namely the stator winding and the rotor winding, and the currents of the two windings can be controlled independently.

Existing brushless doubly-fed machine topologies can be regarded as the outcome of removing brushes from the slip-ring doubly-fed machines. In general, the elimination of brushes in electrical machines can be achieved in three ways, namely, PM excitation, cascading and modulation. The use of PMs will reduce the number of electrical ports, so it cannot be directly applied to brushless doubly-fed machines. Cascading means maintaining the basic doubly-fed IM configuration with an extra power device to transfer power onto the rotor winding in a contactless fashion, while modulation requires the use of short-circuited coils, salient reluctance poles or multilayer flux barriers to achieve indirect magnetic coupling between two sets of stator windings with dissimilar pole pairs. Either way, the two stator windings are indirectly coupled through the rotor action. Therefore, the operating principle of brushless doubly-fed machines can be described in a unified fashion by the schematic diagram shown in Figure 2.47 [37].

The brushless doubly-fed machine can be dated back to the cascaded operation of two multi-phase wound rotor IMs [38]. In cascaded construction, two wound rotor IMs share the same shaft. The rotor winding of one machine is connected to the stator winding of the second machine, and the rotor winding of the second machine is connected to external resistors through brushes and slip rings. This construction does not require a special winding design, and three different synchronous speeds can be obtained by switches, which met the urgent needs for variable speed in the early days of electrical machine development.

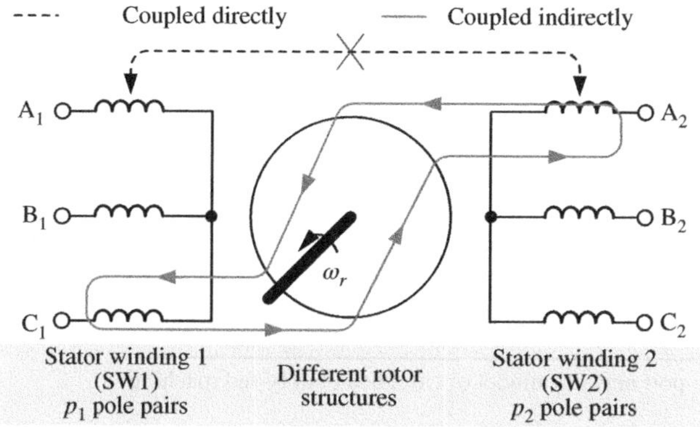

Figure 2.47 Unified description of the operating principle for brushless doubly-fed machines.

Then by improving the cascaded structure with the "rotor winding–stator winding" connection to the "rotor winding–rotor winding" connection, Hunt invented the first cascaded brushless double-fed induction machine. This change retains the advantages of three synchronous speeds, while effectively eliminating the brushes and slip rings, greatly improving the system reliability and reducing maintenance costs [39].

The idea of eliminating brushes has been further extended, and a variety of new brushless doubly-fed machine topologies have emerged one after another. Based on their method of eliminating brushes, they can be classified into two general types: the cascaded type and modulation type.

The cascaded type is the cascaded connection of the traditional slip-ring doubly-fed machine with other electromagnetic equipment to achieve bidirectional flow of slip power between the stationary power source and the rotor winding by the means of contactless power transfer. This type of brushless doubly-fed machine structure includes the cascaded brushless doubly-fed IM (CBDFIM) [40], the brushless doubly-fed IM with a rotary transformer (BDFIM-RT) [41] and the brushless doubly-fed IM with rotating power electronics (BDFIM-RPE) [42]. The operating principle of the cascaded brushless doubly-fed machines can be explained by the same steady-state equivalent circuit shown in Figure 2.48.

Unlike the cascade type, the modulation type uses a special rotor structure to generate two traveling magnetic field waves in the airgap with different pole pairs simultaneously. These two magnetic field components interact with two sets of stator windings respectively to produce effective torques. Modulation type brushless doubly-fed machines include the BDFIM [43], the brushless doubly-fed reluctance

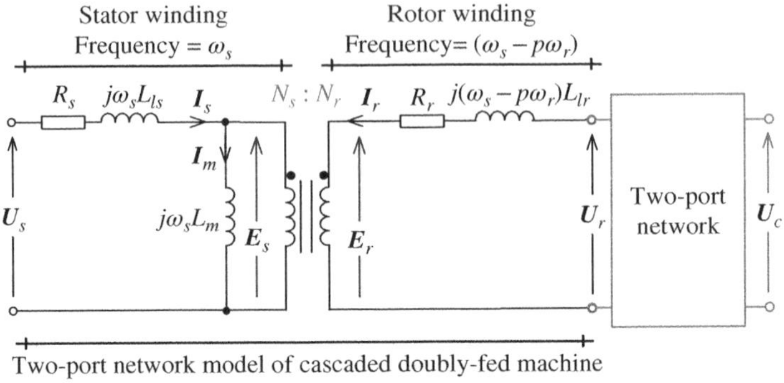

Figure 2.48 Unified steady-state equivalent circuit of cascaded brushless doubly-fed machines.

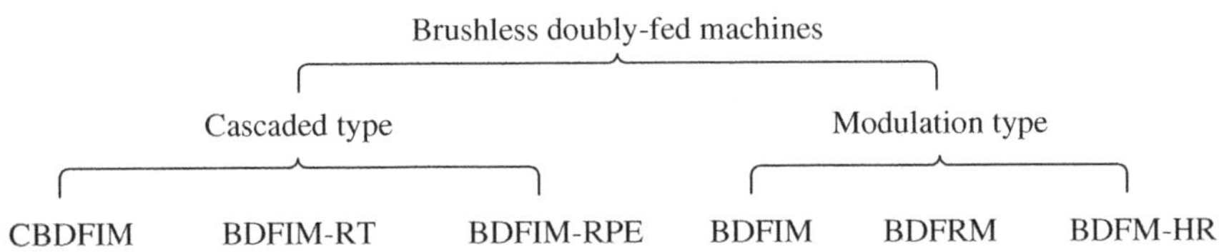

Figure 2.49 Classification of brushless doubly-fed machines.

machine (BDFRM) [44], and the brushless doubly-fed machine with a hybrid rotor (BDFM-HR) [45]. The detailed classification is shown in Figure 2.49.

2.7.2 Cascaded Type

The CBDFIM eliminates brushes through the cascading of two wound rotor IMs [38]. Figure 2.50 shows the evolution of the CBDFIM. As shown in Figure 2.50c, its basic construction is very similar to those shown in Figure 2.50a,b. The main differences are that structures shown in Figure 2.50a,b have only one electrical port and work in the asynchronous operation mode. The modern CBDFIM is produced by replacing external resistors with a power electronic converter. Its advantages are as follows:

- The construction is simple and easy to implement, and can be formed by cascading any two wound rotor IMs.
- The modeling is simple, and a mathematical model can be obtained by combining the dynamic and steady-state mathematical models of traditional IMs according to the rotor winding connection.
- The airgap flux density distribution is almost sinusoidal, and the electromagnetic torque is smooth.

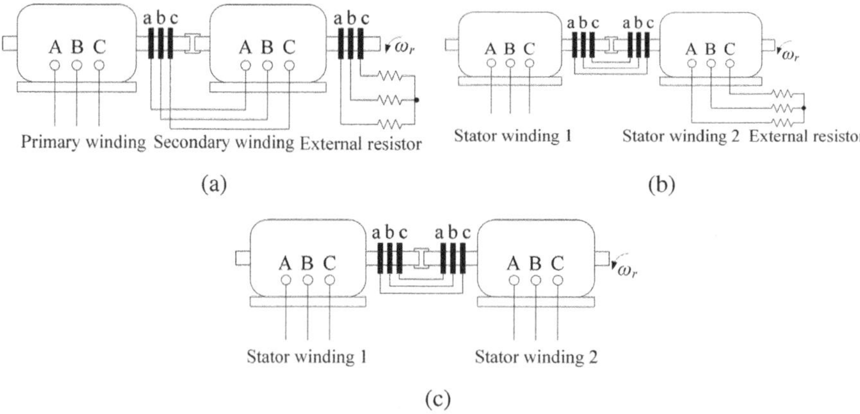

Figure 2.50 Evolution of CBDFIM. (a) Cascaded operation of two three-phase wound rotor IMs, (b) cascaded operation of two three-phase wound rotor IMs (the rotor windings of the two are connected), (c) CBDFIM.

From the perspectives of power flow, converter rating, system efficiency, and machine utilization, the optimal pole pair combination is $p_1 \geq p_2$ [45], where p_1 and p_2 are the pole-pair numbers of stator winding 1 (SW1) and stator winding 2 (SW2), respectively. One of the two sets of stator windings carries most of the power and is thus called the power winding; the other set carries the slip power and is called the control winding. This structure also has some drawbacks:

- The axial length is relatively long. Existing wound-rotor IMs are typically designed to have a low pole-pair number and a large ratio of axial length to diameter with long end winding extensions, reducing the power density.
- The wound rotor structure leads to a higher rotor resistance, which not only increases the copper loss, but also increases the proportion of the asynchornous torque component to the total torque, deteriorating the doubly-fed synchronous characteristic.

2.7.3 Modulation Type

1 BDFIM

The brushless doubly-fed machine with a nested-loop rotor is the most widely known BDFIM, and its name "brushless doubly-fed machine" originates from this structure. To simplify the rotor winding of the CBDFIM so as to obtain the robustness and manufacturability of rotors comparable to the squirrel-cage rotor of a traditional IM, Broadway and Burbridge proposed the nested-loop rotor structure, as shown in Figure 2.51a–c.

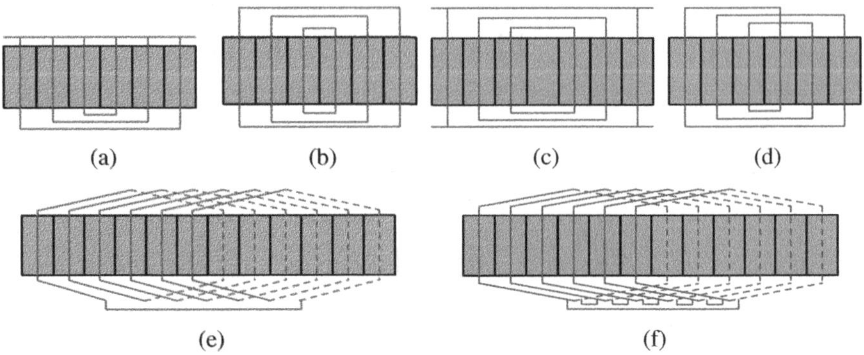

Figure 2.51 Possible rotor windings for BDFIMs. (a) Nested-loop winding with a common ring, (b) isolated nested-loop winding, (c) nested-loop winding with two common rings, (d) spiral-loop winding, (e) mutliphase double-layer winding with equal-turn coils, (f) multiphase double-layer winding with unequal-turn coils (six isolated coil in series).

The brushless doubly-fed machine with a nested-loop rotor has a single stator and a single rotor. Two sets of distributed windings with pole pairs of p_1 and p_2 are placed in the stator slots. The rotor winding includes $(p_1 + p_2)$ repeating units, each of which is called a nest, and each nest includes multiple concentric conductor loops. To avoid directly coupling between the two stator winding sets, $p_1 \neq p_2$ is required. Meanwhile, $|p_1 - p_2| > 1$ is required to avoid unbalanced magnetic pull (UMP).

Short-circuited coils on the rotor greatly increase the modeling complexity for this machine structure. Its fully-coupled circuit model has variable coefficients and shows high order and high nonlinearity [46]. The order of the fully-coupled circuit model must be reduced before it can be used in practical system simulations and controller design. Although the rotor structure was proposed mainly to simplify the manufacturing, studies have shown that the nested-loop rotor winding is not suitable for direct casting, and sufficient insulation must be added between in-slot bars and the core to obtain the ideal brushless doubly-fed characteristics [47]. Moreover, unlike the squirrel-cage rotor used in traditional IMs, the current density varies from bar to bar [48].

In addition to the two main magnetic field components in the airgap of the BDFIM, there are also a variety of ineffective magnetic field harmonic components [49], resulting in torque ripple and noise [50]. Figure 2.52a,b show the airgap flux density waveform and its harmonic spectrum under loaded condition, respectively. The two sets of stator windings have four- and two-pole pairs, respectively, and the rotor winding has six nests.

It can be seen that the radial and tangential airgap flux density waveforms contain the 8-pole and 4-pole dominating spatial harmonics at the same time,

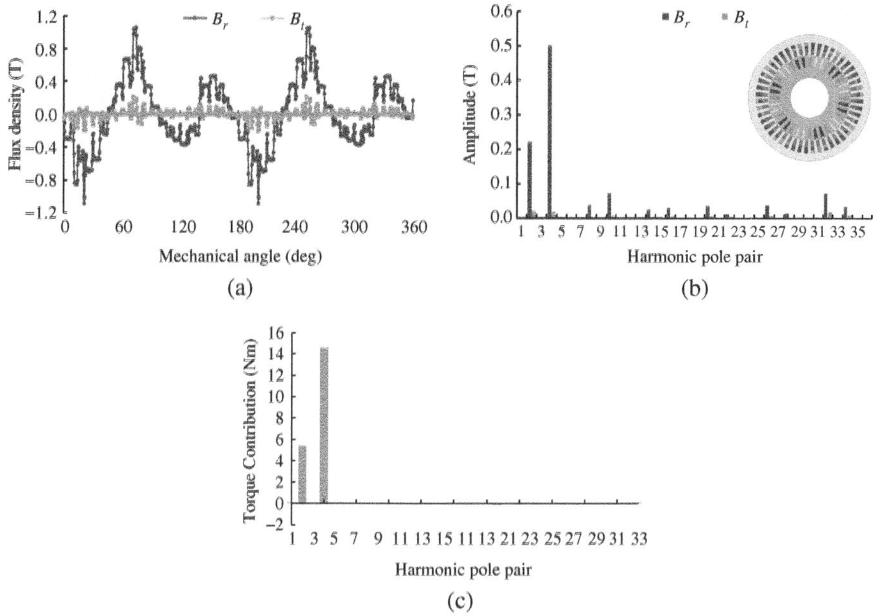

Figure 2.52 Airgap magnetic field modulation phenomenon in the BDFIM with a nested short-circuited loops winding based on 2D FEA results at a time moment under loaded operation. (a) Waveforms of the airgap flux density, (b) amplitude spectra of (a), (c) torque contribution of each harmonic.

as well as abundant other harmonics. Using the Maxwell stress tensor method to decompose the torque, we can see that the average electromagnetic torque is mainly produced by the 8-pole and 4-pole spatial harmonics. Moreover, the torque produced by the 8-pole harmonic is larger than that of the 4-pole harmonic, as shown in Figure 2.52c. The presence of these ineffective field harmonic components will increase the iron loss and copper loss, cause magnetic saturation, and reduce the power density and specific power.

Considering that the nested-loop rotor winding produces a large number of low-order spatial harmonics, a spiral-loop winding structure, as shown in Figure 2.51d, was proposed to improve the cross coupling between the two stator windings by optimizing coil spans [51]. A double sinusoidal winding structure was proposed based on the tooth harmonic principle, in order to enhance the two dominating field harmonics and suppress other harmonics by adjusting winding factors [52]. Figures 2.51e,f show one of the repeating segments of the short-circuited winding layout. Compared with the nested-loop rotor, the tooth harmonic wound rotor replaces the equal-turn concentric short-circuited coils with the equal-turn or unequal-turn distributed coil groups, so spatial harmonics other than the dominating components in the rotor winding MMF are minimized. The double sinusoidal winding structure greatly reduces the spatial harmonic content of the airgap magnetic field, but unavoidably increases the complexity of the rotor design and manufacturing. The rotor winding resistance increases as well.

2 BDFRM

While Broadway proposed the BDFIM with a nested-loop rotor, he noticed that salient reluctance poles can replace short-circuited coils to cause similar opposition to the alternating magnetic field. He thereafter proposed the brushless doubly-fed machine with a reluctance rotor, namely, the BDFRM [43]. Similar to the synchronous reluctance machine, the rotor can have a salient reluctance pole, axially-laminated anisotropic and multilayer flux barrier structures, as shown in Figure 2.53.

Studies have shown that the axially laminated anisotropic rotor has the strongest magnetic field conversion capability, followed by the multilayer flux barrier rotor, and the salient reluctance pole reluctance is the weakest [53, 54]. Since large eddy currents will be induced in axial laminations, the multilayer flux barrier rotor has been considered the most suitable rotor structure for BDFRMs.

Figure 2.54 shows the airgap flux density waveform and the corresponding harmonic spectrum of a BDFRM with a salient-pole reluctance rotor when only one stator winding set is energized. The stator windings have 8 and 4 poles, respectively, and the rotor has six salient reluctance poles. When the 8-pole winding is energized only, the airgap flux density waveform contains 4-, 8-, and 20-pole spatial harmonics at the same time, and the 4- and 20-pole spatial harmonics share the same amplitudes. Similarly, when only the 4-pole winding is energized, the airgap flux density waveform simultaneously contains 4-, 8-, and 16-pole spatial harmonics, and the 8- and 16-pole spatial harmonics share the same amplitudes.

A detailed analysis of the airgap magnetic field modulation phenomenon in a BDFRM with a multilayer flux barrier rotor has been presented in [55].

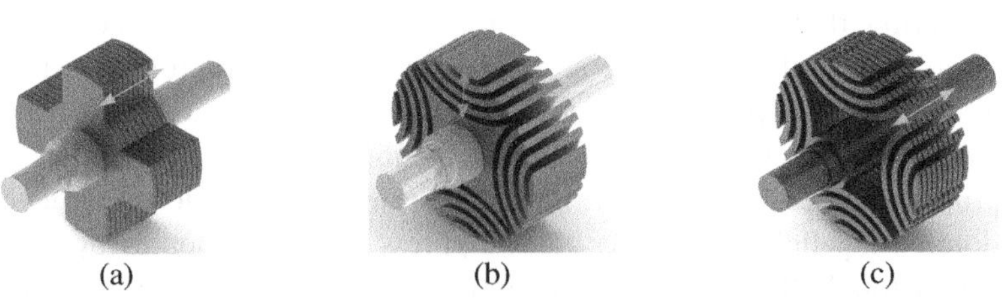

(a) (b) (c)

Figure 2.53 Available rotor structures for BDFRM. (a) Radially-laminated salient reluctance pole rotor, (b) axially-laminated anisotropic rotor, (c) radially-laminated multilayer flux barrier rotor. Arrows denote the stacking directions.

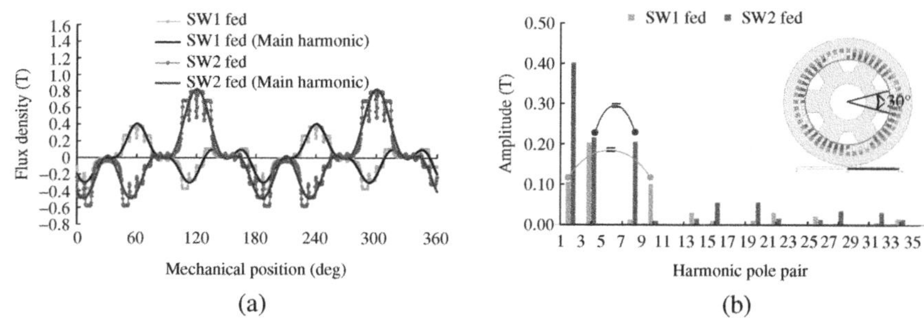

Figure 2.54 Radial airgap flux density waveforms and their harmonic spectra in a BDFRM with a radially-laminated salient pole reluctance rotor. (a) Waveforms, (b) amplitude spectra of (a). Results are from 2D FEA.

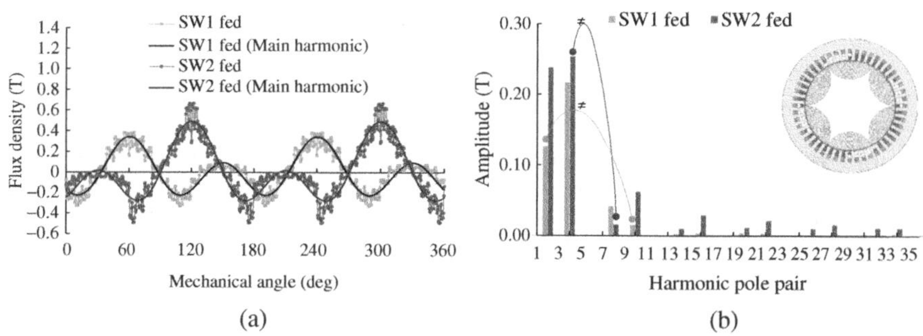

Figure 2.55 Radial airgap flux density waveforms and their harmonic spectra in a BDFRM with a radially-laminated multilayer flux barrier rotor. (a) Waveforms, (b) amplitude spectrums of (a). Results are from 2D FEA.

Figure 2.55 shows the airgap flux density waveform and the corresponding harmonic spectrum when one stator winding set is energized. The two sets of stator windings of the BDFRM with a multilayer flux barrier rotor have 8 and 4 pole-pairs, respectively, and the rotor has 6 multilayer flux barrier segments with a radially-laminated structure. Similar to the role of the salient pole reluctance rotor, when a stator winding set is energized, there will be three dominating spatial harmonics in the airgap flux density. Nevertheless, the amplitudes of the three dominating spatial harmonics are different from each other.

Further analysis of the airgap flux density waveform shows that amplitudes of the two spatial harmonics with the same pole-pair numbers as the stator windings remain basically unchanged during the normal operation. Figure 2.56 shows the magnetic flux patterns at six time moments of a full electric cycle under loaded condition. At any of the six time moments, the magnetic field distribution does not show 8 or 4 poles explicitly.

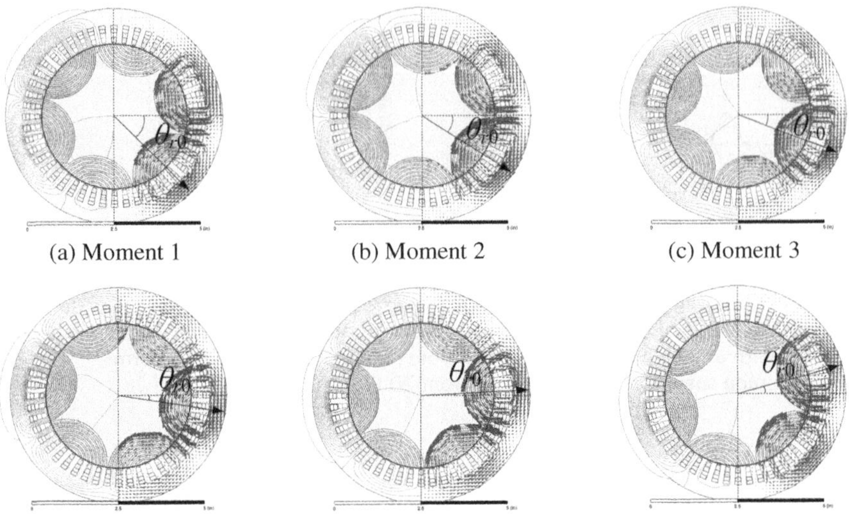

(a) Moment 1 (b) Moment 2 (c) Moment 3

(d) Moment 4 (e) Moment 5 (f) Moment 6

Figure 2.56 Magnetic flux patterns of the BDFRM at six time moments in the process of rotating the rotor by one flux barrier segment. (a) Moment 1, (b) moment 2, (c) moment 3, (d) moment 4, (e) moment 5, (f) moment 6. Results are from 2D FEA using a radially-laminated multilayer flux barrier rotor.

Figure 2.57 shows the airgap flux density waveforms and the corresponding harmonic spectrums at these six time moments. To visualize the dominating spatial harmonics more clearly, only the two spatial harmonics with the same pole-pair numbers as the stator windings are shown at all time moments. The waveform including all the harmonics is only shown for the case where the rotor position angle is 15°.

Using Maxwell's stress tensor method to decompose the average torque at each moment, we can see that only the 8- and 4-pole spatial harmonics can produce non-zero electromagnetic torque, and the torque generated by the 8-pole spatial harmonic is about twice that of the 4-pole spatial harmonic. This conclusion holds true for different rotor positions.

3 BDFM-HR

The BDFM-HR was proposed in [44]. The hybrid rotor shown in Figure 2.58a, is essentially a combination of a multilayer flux barrier rotor and a nested-loop rotor structure. The basic idea is to enhance the magnetic field conversion capability of the rotor by combining the modulation effects of the short-circuited rotor winding and rotor salient reluctance poles. How will the short-circuited coils and flux

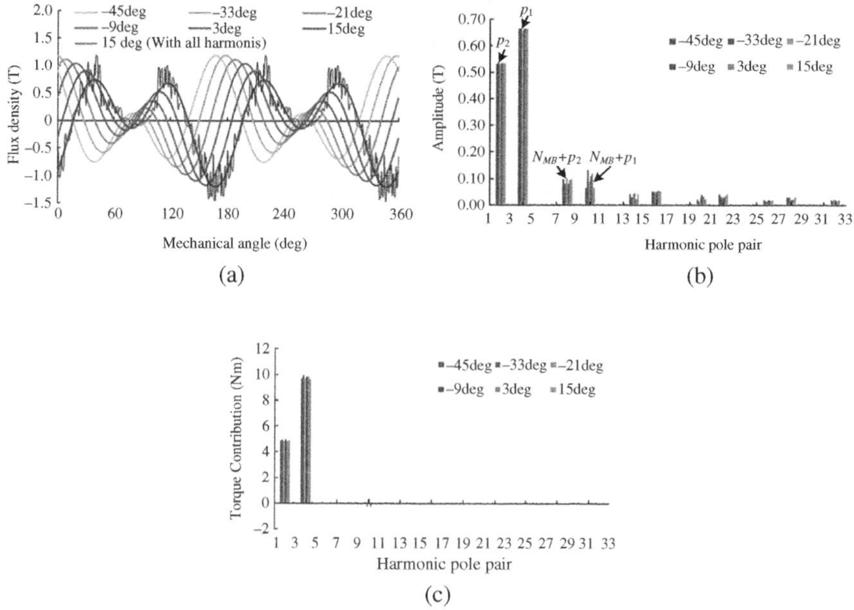

Figure 2.57 Airgap magnetic field modulation phenomenon in the BDFRM with a radially-laminated multilayer flux barrier rotor. (a) Waveforms of radial airgap flux density, (b) amplitude spectra of (a), (c) torque contribution of each harmonic.

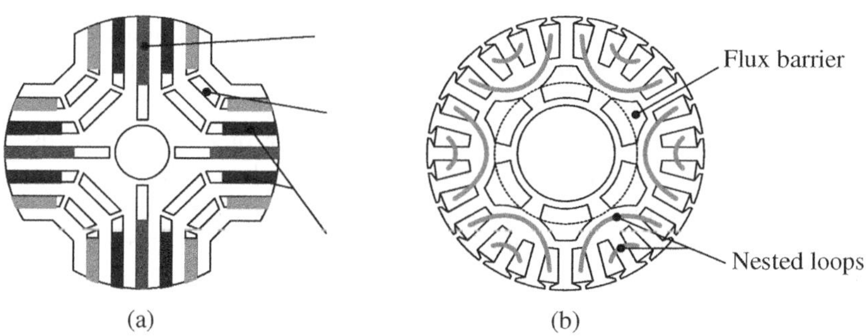

Figure 2.58 Hybrid rotors for BDFMs. (a) Combination of multilayer flux barriers with nested-loop short-circuited coils, (b) combination of simple flux barriers with nested-loop short-circuited coils.

barriers influence each other, and how to evaluate the magnetic field conversion capability of the hybrid rotor will be discussed in Chapter 6.

With the presence of magnetic saturation and eddy currents in bar conductors, the magnetic field distribution will be more complicated, so the analysis will rely more heavily on FEA tools. From the perspective of reducing the weight of the

BDFIM with a nested-loop rotor, it has been proposed to remove some areas of the rotor yoke to introduce magnetic barriers [56], as shown in Figure 2.58b. This structure can be intuitively considered as a combination of the radially-laminated simple magnetic barrier structure and nested-loop rotor windings.

2.7.4 Commonalities and Differences of Existing Brushless Doubly-fed Machines

In summary, brushless doubly-fed machines with different structures share some commonalities though some differences do exist [57]. The commonalities mainly lie in:

- There are no brush gears and PMs.
- There are two electrical ports (two sets of stator windings that are not electromagnetically coupled to each other directly) and one mechanical port (rotating shaft), and the energy of the two electrical ports can flow in both directions.
- The doubly-fed synchronous speed is:

$$\omega_{syn} = \frac{\omega_1 \pm \omega_2}{p_1 \pm p_2} \tag{2.51}$$

  where ω_1 and ω_2 are the current angular frequencies of the SW1 and SW2, respectively, p_1 and p_2 are pole-pair numbers of the SW1 and SW2. When the rotor speed meets $\omega_r = \omega_{syn}$, the machine will operate in the doubly-fed synchronous mode. In particular, when the frequency of one winding set is 0, the speed corresponding to (2.51) is called the natural synchronous speed.
- When one of the electrical ports is short-circuited, the electrical machine can work in the asynchronous operation mode.
- When the plus sign is used in the denominator of (2.51), the brushless doubly-fed machine has a large number of equivalent pole-pairs, which is suitable for medium- and low-speed applications; when the denominator of (2.51) is taken with the minus sign, the brushless doubly-fed machine has a small number of equivalent pole-pairs, which is suitable to design high-speed machines.

On the other hand, due to the different operating principles and implementations, there are differences among brushless doubly-fed machines, which mainly include:

- The pole-pair combination of the CBDFIM mainly depends on the airgap diameter ratio, and there is no certain relationship between the torque density and the pole-pair combination. However, for the modulation type, the pole-pair combination is mainly constrained by the electromagnetic relationship. From the perspective of improving the magnetic field coupling capability, the pole-pair numbers of the two sets of stator winding should be as close as possible, that is, the so-called near-pole combination.

- The cascaded type has a more sinusoidal airgap flux density waveform, but for the modulation type, the airgap magnetic field contains a variety of parasitic low-order spatial harmonics in addition to the two main sinusoidal spatial harmonic components.
- The vibration behavior of the cascaded type is the same as that of traditional AC machines. But for the modulation type, the two main magnetic field components with different pole-pair numbers in the airgap will produce additional vibration modes, and the vibration amplitude is closely related to the pole-pair combination. The far-pole combination helps to avoid low-order vibration modes.

2.8 Uniformity of Machine Operating Principles

From the representative examples presented this chapter, it can be seen that the airgap magnetic field modulation phenomenon generally exists in traditional DCMs, IMs, SMs, and new machine topologies, such as MGs, MGMs, PMV machines, FSPM machines, brushless doubly-fed machines, etc.

The phenomenon can be described as follows. The modulation behaviors performed by short-circuited coils, salient reluctance poles, and multilayer flux barriers on the magnetic field created by the excitation sources (PMs or field windings) not only change the amplitude and phase of the fundamental magnetic field, but also generate a variety of magnetic field harmonics. If the pole-pair number and speed of one or more of the magnetic field components match those of the armature magnetic field, coupled harmonic pairs are formed, which will interact with each other to achieve electromagnetic energy conversion. As a result, the pole-pair number and speed of the traveling magnetic field wave corresponding to the voltage and current outputs of the armature winding have a certain proportional relationship with the pole-pair number and speed of the excitation magnetic field, i.e. the gear ratio G_r.

Airgap magnetic field modulation phenomenon occurs in traditional DCMs, IMs, and SMs too, but the pole-pair number and speed of the main working magnetic field component before and after the modulation do not change, so they have attracted much less attention compared to field modulation machine topologies. In traditional machines, the gear ratio is $G_r = 1$, which is termed as "unit modulation" in this book, and can be regarded as a special case of the magnetic field modulation.

For non-salient-pole SMs, the distributed single-phase field winding on the rotor is supplied by a DC excitation current to create a sinusoidal magnetic field along the airgap peripheral. Once the slotting effect on the stator and rotor surfaces

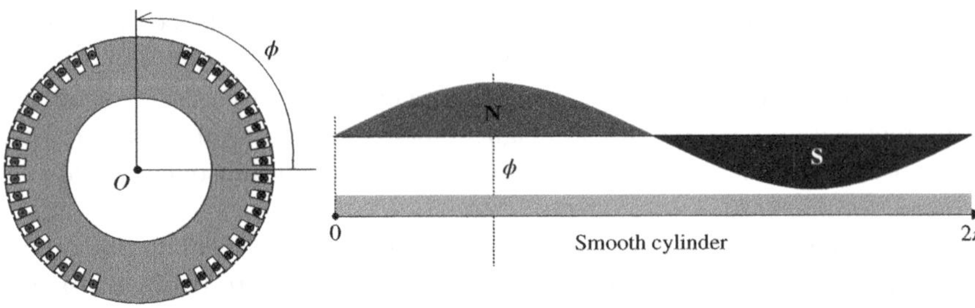

Figure 2.59 Magnetic field modulation behavior in the non-salient-pole SM.

are ignored, the airgap permeance will be a constant corresponding to the constant right-hand-side term in (2.5), so the modulated magnetic field has the fundamental wave only, as shown in Figure 2.59. To extract the fundamental magnetic field, the multi-phase distributed armature winding on the stator must have the same pole-pair number as the fundamental magnetic field. In other words, in non-salient-pole SMs, the modulation behavior does not change the pole-pair number of the magnetic field, which is a typical example of "unit modulation."

For squirrel-cage IMs, the airgap is uniform if the stator and rotor slotting effects are ignored. The stator winding serves as the field winding as well as the armature winding. For simplicity, we can assume that there are two sets of windings on the stator. One set is for the field winding, and the other is for the armature winding. The excitation MMF created by the field winding is sinusoidally distributed and travels at the synchronous speed. The squirrel-cage rotor winding is essentially a group of uniformly distributed short-circuit coils, which will induce voltage and current when there is a relative motion between itself and the excitation magnetic field, producing a secondary magnetic field with the same pole-pair number as the excitation MMF, as shown in Figure 2.60.

The modulated magnetic field is the superposition of the excitation magnetic field and the secondary magnetic field. To utilize the fundamental rotating magnetic field, the armature winding must have the same pole-pair number. Since the

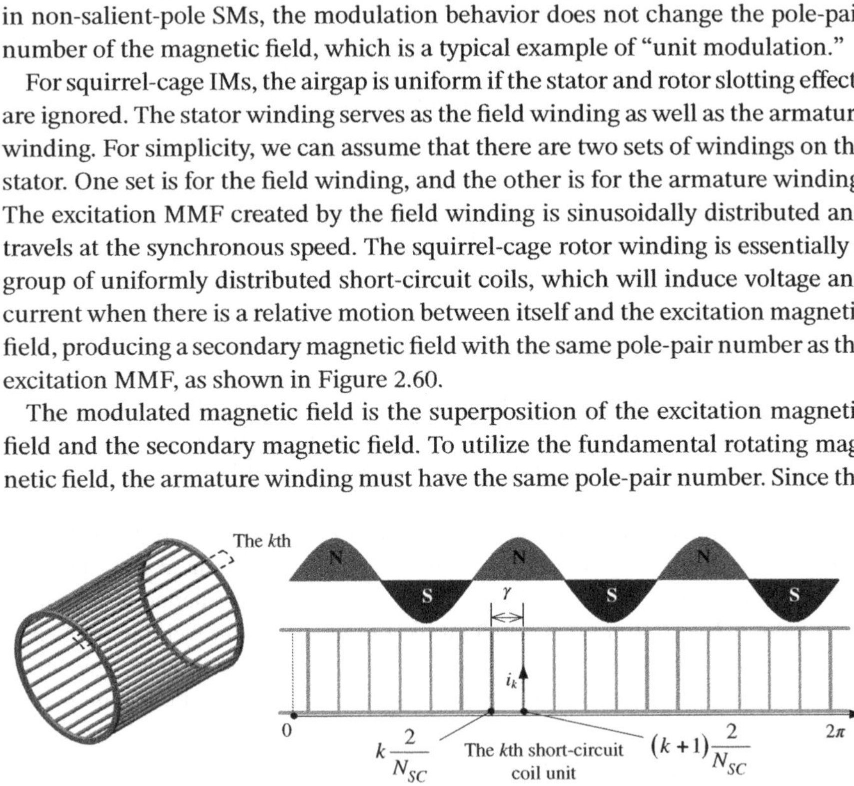

Figure 2.60 Magnetic field modualtion behavior in the squirrel-cage IM.

pole-pair number of the field and armature windings are the same, they can be combined into one. Therefore, there is only one set of multi-phase winding on the stator of real IMs, which has functionalities of both the field and armature windings.

In summary, the airgap magnetic field modulation phenomenon is widespread in electrical machines. Any type of electrical machine has three basic elements, i.e. the excitation source, modulator and armature winding. The excitation magnetic field generated by the excitation source is modulated by the modulator to produce a series of magnetic field harmonics. The armature winding extracts one or several magnetic field harmonics by matching the pole pairs, and filter out other useless magnetic field harmonics to accomplish the electromechanical energy conversion. In this sense, operating principles of electrical machines are unified.

The three basic elements of electrical machines may have different combinations, but at least one rotating (moving) element and one stationary element are required for a practical machine. For a specific electrical machine, one component may perform the functionalities of two basic elements at the same time. For example, the stator winding of an IM plays the roles of the excitation source and the armature winding simultaneously.

Mulitple components may also have the same basic element functionality in an electrical machine. For instance, in a flux-switching PM machine, both the stator teeth and rotor salient reluctance poles act as modulators, that is, there are two modulators in flux-switching PM machines. Another example comes from the hybrid excitation electrical machine, where both the magnets and DC field windings act as the excitation source.

Considering the generality of the airgap magnetic field modulation phenonenon, a new classification of electrical machines is proposed. Figure 2.61 shows the classification according to which of the three basic elements rotates. In the magnetically-geared double-rotor machine, both the PM rotor (excitation source) and the modulation ring rotate but at different speeds, while in salient-pole SMs, either the PM or wound-field type, the excitation source is fixed to (and thus rotates together with) the salient reluctance pole rotor.

It is possible to evolve from one machine type to another by simply changing the design parameters or implementations. Machines with a rotating excitation source are taken as an typical example here to show the evolution, as illustrated in Figure 2.62. When the MGM is designed to have the same number of ferromagnetic pole pieces of the modulating ring and stator teeth, the modulating ring and stator can be merged to form a new stator with fully open slots. Then the MGM evolves into a PMV machine. If the stator winding is further changed to have the same pole-pair number as the rotor PMs, the PMV machine will evolve into a PMSM. If the PM excitation on the rotor is replaced by a three-phase or multi-phase

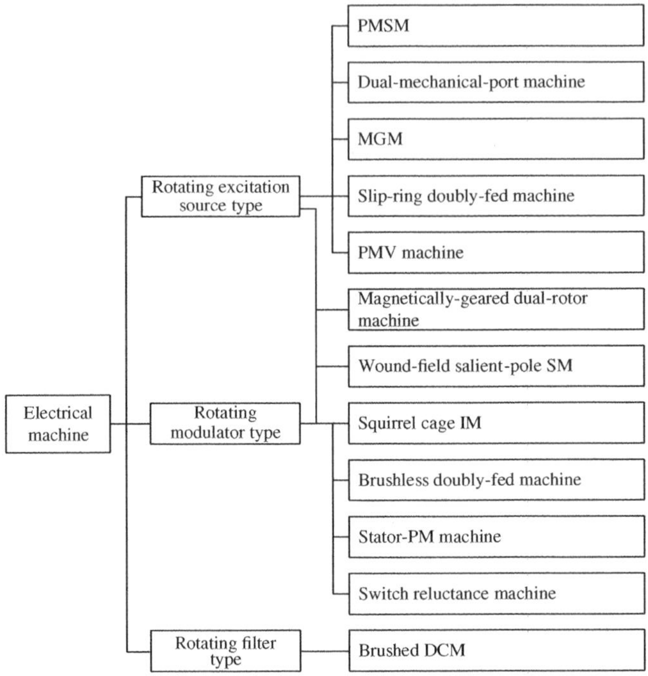

Figure 2.61 Classification of electrical machines from the perspective of airgap magnetic field modulation.

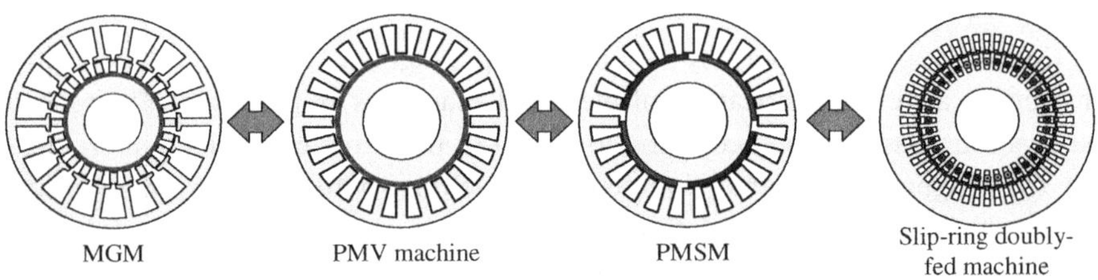

Figure 2.62 Evolution of electrical machines with rotating excitation sources.

winding connected to an external circuit using brushes and slip rings, the PMSM will evolve into a slip-ring doubly-fed machine.

References

1 Neuland, A.H. (1916). Apparatus for transmitting power. U.S. Patent 1 171 351.
2 Faus, H.T. (1941). Magnet gearing. U.S. Patent 2 243 555.

3 Atallah, K. and Howe, D. (2001). A novel high-performance magnetic gear. *IEEE Trans. Magn.* 37 (4): 2844–2846.
4 Atallah, K., Calverley, S.D., and Howe, D. (2004). Design, analysis and realization of a high-performance magnetic gear. *IEE Proc. Electr. Power Appl.* 151 (2): 135–143.
5 Rasmussen, P.O., Andersen, T.O., Jorgensen, F.T., and Nielsen, O. (2005). Development of a high-performance magnetic gear. *IEEE Trans. Ind. Appl.* 41 (3): 764–770.
6 Li, Y., Xing, J., Peng, K., and Lu, Y. (2008). Principle and simulation analysis of a novel structure magnetic gear. In: *Proceedings of the International Conference on Electrical Machines and Systems (ICEMS)*, Wuhan, China. 3845–3849.
7 Liu, X., Chau, K.T., Jiang, Z., and Yu, C. (2009). Design and analysis of interior-magnet outer-rotor concentric magnetic gears. *J. Appl. Phys.* 105: 1–3.
8 Jian, L. and Chau, K.T. (2010). A coaxial magnetic gear with Halbach permanent-magnet arrays. *IEEE Trans. Energy Convers.* 25 (2): 319–328.
9 Jian, L., Chau, K.T., Gong, Y. et al. (2009). Comparison of coaxial magnetic gears with different topologies. *IEEE Trans. Magn.* 45 (10): 4526–4529.
10 Li, X., Chau, K.T., Cheng, M., and Hua, W. (2013). Comparison of magnetic-geared permanent-magnet machines. *Prog. Electromag. Res. (PIER)* 133: 177–198.
11 Li, X. (2015). Research on field modulated permanent magnet machine and its control system for direct drive wind power generation. Nanjing: Ph.D. Thesis of Southeast University.
12 Lian, J. (2015). *Principle and Applications of Coaxial Magnetic Gear*, 1e. Beijing, China: Science Press.
13 Fu, W.N. and Liu, Y. (2016). A unified theory of flux-modulated electric machines. In: *Proceedings of the International Symposium of Electrical Engineering (ISEE)*, 1–13. Hong Kong.
14 Atallah, K., Wang, J., and Howe, D. (2005). A high-performance linear magnetic gear. *J. Appl. Phys.* 97: 1–3.
15 Li, W. and Chau, K.T. (2012). Analytical field calculation for linear tubular magnetic gears using equivalent anisotropic magnetic permeability. *Prog. Electromag. Res. (PIER)* 127: 155–171.
16 Li, W., Chau, K.T., and Jiang, J.Z. (2011). Application of linear magnetic gears for pseudo-direct-drive oceanic wave energy harvesting. *IEEE Trans. Magn.* 47 (10): 2624–2627.
17 Mezani, S., Atallah, K., and Howe, D. (2006). A high-performance axial-field magnetic gear. *J. Appl. Phys.* 99 (8) Article#08R303.
18 Zhu, D., Yang, F., Du, Y. et al. (2016). An axial-field flux-modulated magnetic gear. *IEEE Trans. Appl. Supercond.* 26 (4) Article#0604405.

19 Bomela, W., Bird, J.Z., and Acharya, V.M. (2014). The performance of a transverse flux magnetic gear. *IEEE Trans. Magn.* 50 (1) Article#4000104.

20 Yin, X., Pfister, P., and Fang, Y. (2015). A novel magnetic gear: toward a higher torque density. *IEEE Trans. Magn.* 51 (11) Article#8002804.

21 Frank, N.W. and Toliyat, H.A. (2011). Analysis of the concentric planetary magnetic gear with strengthened stator and interior permanent magnet inner rotor. *IEEE Trans. Ind. Appl.* 47 (4): 1652–1660.

22 Aiso, K., Akatsu, K., and Aoyama, Y. (2019). A novel reluctance magnetic gear for high-speed motor. *IEEE Trans. Ind. Appl.* 55 (3): 2690–2699.

23 Chau, K.T., Zhang, D., Jiang, J.Z. et al. (2007). Design of a magnetic-geared outer-rotor permanent-magnet brushless motor for electric vehicles. *IEEE Trans. Magn.* 43 (6): 2504–2506.

24 Wang, L.L., Shen, J.X., Luk, P.C.K. et al. (2009). Development of a magnetic-geared permanent magnet brushless motor. *IEEE Trans. Magn.* 45 (10): 4578–4581.

25 Fan, Y., Jiang, H., Cheng, M., and Wang, Y. (2010). An improved magnetic-geared permanent magnet in-wheel motor for electric vehicles. In: *Proceedings of the IEEE Vehicle Power and Propulsion Conference (VPPC)*, 1–5. Lille, France.

26 Toba, A. and Lipo, T.A. (2000). Generic torque-maximizing design methodology of surface permanent-magnet vernier machine. *IEEE Trans. Ind. Appl.* 36 (6): 1539–1546.

27 Li, J., Chau, K.T., Jiang, J.Z. et al. (2010). A new efficient permanent-magnet vernier machine for wind power generation. *IEEE Trans. Magn.* 46 (6): 1475–1478.

28 Jian, L., Chau, K.T., and Jiang, J.Z. (2009). A magnetic-geared outer-rotor permanent-magnet brushless machine for wind power generator. *IEEE Trans. Ind. Appl.* 45 (3): 954–962.

29 Toba, A. and Lipo, T.A. (1999). Novel dual-excitation permanent magnet vernier machine. In: *Record of IEEE IAS Annual Meeting*, 2539–2544, Phoenix, USA.

30 Du, Y. (2013). Research on linear primary permanent magnet vernier generator and its control system for direct drive wave energy conversion. Nanjing: Ph.D. Thesis of Southeast University.

31 Du, Y., Chau, K.T., Cheng, M. et al. (2011). Design and analysis of linear stator permanent magnet vernier machines. *IEEE Trans. Magn.* 47 (10): 4219–4222.

32 Cheng, M. and Hua, W. (2019). Theory, Design and Control of Stator Permanent Magnet Brushless Machine, 1e. Beijing, China: Science Press.

33 McFarland, J.D., Jahns, T.M., and El-Refaie, A.M. (2015). Analysis of the torque production mechanism for flux-switching permanent-magnet machines. *IEEE Trans. Ind. Appl.* 51 (4): 3041–3049.

34 Wu, Z.Z. and Zhu, Z.Q. (2015). Analysis of air-gap field modulation and magnetic gearing effects in switched flux permanent magnet machines. *IEEE Trans. Magn.* 51 (5) Article#8105012.

35 Du, Y., Xiao, F., Hua, W. et al. (2016). Comparison of flux-switching PM motors with different winding configurations using magnetic gearing principle. *IEEE Trans. Magn.* 52(5) Article#8201908.

36 Cheng, M., Han, P., Buja, G., and Jovanović, M.G. (2018). Emerging multiport electrical machines and systems: past developments, current challenges, and future prospects. *IEEE Trans. Ind. Electron.* 65 (7): 5422–5435.

37 Han, P., Cheng, M., Ademi, S., and Jovanović, M.G. (2018). Brushless doubly-fed machines: opportunities and challenges. *Chin. J. Electr. Eng.* 4 (2): 1–17.

38 Hunt, L. (1907). A new type of induction motor. *J. Inst. Electr. Eng.* 39 (186): 648–667.

39 Smith, B.H. (1967). Synchronous behavior of doubly fed twin stator induction machine. *IEEE Trans. Power Apparat. Syst.* PAS-86 (10): 1227–1236.

40 Ruviaro, M., Runcos, F., Sadowski, N. et al. (2012). Analysis and tests results of a brushless doubly-fed induction machine with rotary transformer. *IEEE Trans. Ind. Electron.* 59 (6): 2670–2677.

41 ur Rehman Malik, N. and Sadarangani, C. (2011). Brushless doubly-fed induction machine with rotating power electronic converter for power applications. In: *Proceedings of the International Conference on Electrical Machines and Systems (ICEMS)*, 1–6. Beijing, China.

42 Broadway, A.R.W. and Burbridge, L. (1970). Self-cascaded machine: a low speed motor or high frequency brushless alternator. *Proc. Inst. Electr. Eng.* 117 (7): 1277–1290.

43 Broadway, A.R.W. (1971). Cageless induction machine. *Proc. Inst. Electr. Eng.* 118 (11): 1593–1600.

44 Zhang, F., Li, Y., and Wang, X. (2009). The design and FEA of brushless doubly-fed machine with hybrid rotor. In: *Proceedings of the IEEE International Conference on Applied Superconductivity and Electromagnetic Devices*, Chengdu, 324–327.

45 Hopfensperger, B., Atkinson, D.J., and Lakin, R.A. (2002). Steady state of the cascaded doubly-fed induction machine. *Eur. Trans. Electr. Power* 12 (6): 427–437.

46 Robert, P.C., Long, T., McMahon, R.A. et al. (2013). Dynamic modeling of the brushless doubly fed machine. *IET Proc. Electr. Power Appl.* 7 (7): 544–556.

47 Williamson, S. and Boger, M.S. (1999). Impact of inter-bar currents on the performance of the brushless doubly fed motor. *IEEE Trans. Ind. Appl.* 35 (2): 453–460.

48 Kemp, A., Boger, M., Wiedenbrug, E. et al. (1996). Investigation of rotor current distribution in brushless doubly-fed machines. In: *Record of the IEEE IAS Annual Meeting*, vol. 1, 638–643. San Diego, CA.

49 Alexander, G.C. (1990). Characterization of the brushless doubly-fed machine by magnetic field analysis. In: *Record of the IEEE IAS Annual Meeting*, vol. 1, 67–74. Seattle, WA.

50 Logan, T., McMahon, R.A., and Seffen, K. (2014). Noise and vibration in brushless doubly fed machine and brushless doubly fed reluctance machine. *IET Electr. Power Appl.* 8 (2): 50–59.

51 Gorginpour, H., Jandaghi, B., and Oraee, H. (2013). A novel rotor configuration for brushless doubly-fed induction generators. *IET Electr. Power Appl.* 7 (2): 106–115.

52 Xiong, F. and Wang, X. (2014). Design of a low-harmonic-content wound rotor for the brushless doubly fed generator. *IEEE Trans. Energy Convers.* 29 (1): 158–168.

53 Xu, L., Tang, Y., and Ye, L. (1994). Comparative study of rotor structures of doubly excited brushless reluctance machine by finite element analysis. *IEEE Trans. Energy Convers.* 9 (1): 165–172.

54 Xu, L. and Wang, F. (1997). Comparative study of magnetic coupling for a doubly fed brushless machine with reluctance and cage rotors. In: *Record of the IEEE IAS Annual Meeting*, vol. 1, 326–332. New Orleans, LA.

55 Han, P., Zhang, J., and Cheng, M. (2019). Analytical analysis and performance characterization of brushless doubly fed machines with multibarrier rotors. *IEEE Trans. Ind. Appl.* 55 (6): 5758–5767.

56 Abdi, S., Abdi, E., Oraee, A., and McMahon, R.A. (2015). Optimization of magnetic circuit for brushless doubly fed machines. *IEEE Trans. Energy Convers.* 33 (4): 1611–1620.

57 Cheng, M., Han, P., and Wei, X. (2016). Design, analysis and control of brushless doubly-fed generators for wind power application. *Trans. Ch. Electrotech. Soc.* 31 (19): 36–52.

3

Three Key Elements Model for Electrical Machines

3.1 Introduction

The airgap magnetic field modulation phenomenon in electrical machines has been receiving increasing attention in recent years. Various new machine topologies, such as magnetic gears (MGs), magnetically geared machines (MGMs), permanent magnet (PM) vernier (PMV) machines, flux-switching PM (FSPM) machines, flux-reversal PM (FRPM) machines, and partitioned-stator electrical machines, just to name a few, are constantly emerging. The presence of these new machine topologies has not only enriched the scientific knowledge of electrical engineering but also greatly advanced the development of electric vehicles, wind power generation, ship power generation, aerospace, and other fields in this electrification era.

At the same time, the magnetic field modulation phenomenon has been continuously discovered in various new machine topologies, providing a new perspective for the design and analysis of electrical machines. The magnetic field modulation phenomenon and its significant role in the torque production have been preliminarily revealed in some machine types. Most of the existing research is, however, still limited to certain magnetic field modulation machine types, and mainly focuses on the effect of a few dominating airgap magnetic field harmonics generated by the varied airgap permeance from the salient reluctance poles/teeth on the torque output.

The mechanism of production and change of magnetic field harmonics caused by short-circuited coils, salient reluctance poles and multilayer flux barriers in various electrical machines have not been fully revealed and unified. Researchers are still struggling in the accidental discovery of "modulation phenomenon" and the theoretical explanation of "modulation principle." It is, therefore, hardly possible at this stage to analyze and design electrical machines of different types and operating principles within a unified theoretical framework.

General Airgap Field Modulation Theory for Electrical Machines: Principles and Practice,
First Edition. Ming Cheng, Peng Han, Yi Du, and Honghui Wen.

Companion website: www.wiley.com/go/genairgapfieldmodulationtheory

This chapter identifies the three key elements of electrical machines as the source, modulator, and filter by examining the analogy between electrical machines and switching converters. The simplest electrical machine with a stator, a rotor and a layer of airgap can be treated as the cascade of the source magnetomotive force (MMF), the modulator (short-circuited coils/salient reluctance poles/multilayer flux barriers), and the filter (armature windings) from the perspective of functionality, overcoming the limitations of specific machine structures. The three elements can be characterized by unified mathematical formulas. A generalized theory of airgap magnetic field modulation that is expected to be applied to all types of electrical machines is established in this chapter based on the magnetic coupling characterized by the three elements.

Windings are the basic component of electrical machines and the starting point for electromagnetic performance analysis. This chapter first reviews the basic methods for winding analysis and introduces related concepts, such as the conductor distribution function, linear current density distribution function, MMF distribution function, and magnetic flux density distribution function. The mathematical relationships among these distribution functions are also included. On this basis, the derivation of the classical winding function and its limitations are presented. Based on a large number of studies on the invention and evolution, structural and functional characteristics, and the analysis and calculation methods of existing machine topologies, three ways to alter the airgap MMF distribution are identified: short-circuited coils, salient reluctance poles, and multilayer flux barriers.

The introduction of modulation operators brings convenience to the analytical analysis of the airgap MMF and magnetic flux density distribution, but the airgap magnetic flux density distribution does not directly reflect the performance of electrical machines. The participation of the armature winding is required. The armature windings' frequency selection property for the airgap flux density waveform (i.e. armature windings and airgap flux density waveforms of the same number of pole pairs can interact with each other) motivates us to think about the similarity between electrical machines and switching converters. Based on such an analogy, the performance index of electrical machines, such as the no-load induced EMF, average torque, power, cogging torque, etc. are determined by the cascade of the three key elements, namely, the source MMF, modulator and armature winding. In addition, the three elements can be characterized by unified mathematical formulas. The general airgap magnetic field modulation theory developed in this chapter is expected to be valid for all types of electrical machines and serves as a new powerful tool for the topological analysis and innovation of electrical machines in subsequent chapters.

3.2 Classical Winding Function Theory and Its Limitations

3.2.1 Winding MMF

Windings are an essential component of electrical machines, which can be classified into field windings and armature windings according to their functionality. Currents flowing through field windings create the field MMF, while armature windings induce the electromotive force (EMF) by cutting the magnetic flux lines. Similarly, currents flowing through armature windings also create the armature MMF. In light of the analogy between the MMF and EMF, the analysis of machine windings can be simplified to the analysis of winding MMF, and then the conclusions obtained can be extended to the winding EMF.

The airgap is the place where electromechanical energy conversion occurs. The airgap magnetic flux density distribution is the basis for the calculation of flux linkage, induced EMF, force and torque. The purpose of winding MMF analysis is to determine the distribution of airgap flux density. In order to facilitate the understanding of related concepts in winding MMF analysis, it is assumed that the airgap is uniform and the magnetic core has infinite permeability.

The machine winding is composed of conductors discretely distributed along the airgap and connected together in a specific way. Its mathematical characteristics can be described by the conductor distribution function $C(\phi)$, where ϕ is the position angle along the circumference of the airgap in the stationary reference frame. Usually conductors are discretely distributed in the stator and rotor slots. The conductor distribution function $C(\phi)$ is an impulse train defined in the range of $0 \sim 2\pi$ mechanical angle, which can be expressed as:

$$C(\phi) = \sum_k C_k \delta(\phi - \phi_k) \tag{3.1}$$

where C_k is the number of conductors in series in the k-th stator slot, and it is positive when the current flows into the page and negative when the current flows out of the page. The conductor distribution function for a conductor in the k-th stator slot of is the unit impulse function shifted by the angular position of ϕ_k, which can be expressed as:

$$\left.\begin{array}{c} \delta(\phi - \phi_k) = 0, \quad \phi \neq \phi_k \\ \int_0^{2\pi} \delta(\phi) = 1 \end{array}\right\} \tag{3.2}$$

Figure 3.2 shows the conductor distribution function of Figure 3.1. The current flowing in the winding will produce a linear current distribution on the surface of the airgap, thereby creating an MMF and generating a magnetic field distribution. The distribution of the linear current along the airgap circumference at an instant

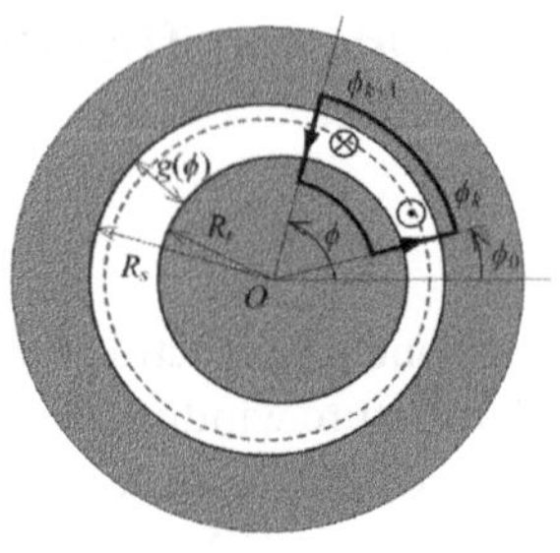

Figure 3.1 Derivation of classical winding function in a doubly cylindrical machine. The airgap is assumed non-uniform for generality.

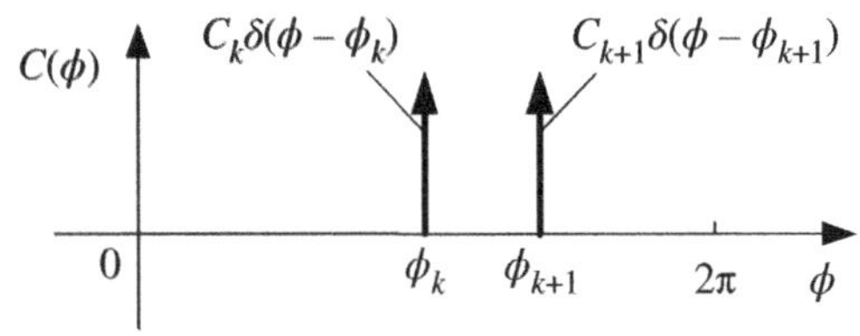

Figure 3.2 Conductor distribution function.

of time is described by the linear current density distribution function $A(\phi,t)$ [1]. Assuming that there are N_s conductors in a slot with a slot width of o (rad) (slot width/airgap circumference $\times\, 2\pi$), and at an instant of time t, the same current $i(t)$ in ampere flows in each conductor. Based on the assumption that the linear current layer is uniformly distributed on the slot opening, the linear current density over the slot opening is $A = N_s i/o$ (A/rad).

Usually in order to simplify the analysis, it is further considered that the overall current $N_s i$ in the slot is concentrated at the middle of the slot opening, then the linear current density produced by each slot becomes an impulse function (i.e. $A = \infty$ in the middle of the slot opening, and $A = 0$ at other points). The integral of this function along the circumference of the airgap is equal to the ampere-turn $N_s i$ in the slot. Thus, the linear current density distribution function $A(\phi,t)$ can be obtained by the product of the conductor distribution function $C(\phi)$ and the current in each conductor, namely:

$$A(\phi, t) = \sum C(\phi)\, i(t) \tag{3.3}$$

which is illustrated in Figure 3.3.

Now we take an arbitrary point on the circumference of the airgap as the starting point and mark it as A, and then take a closed loop through the starting point A and any point B with the coordinate of ϕ. According to Ampere's Law, the total

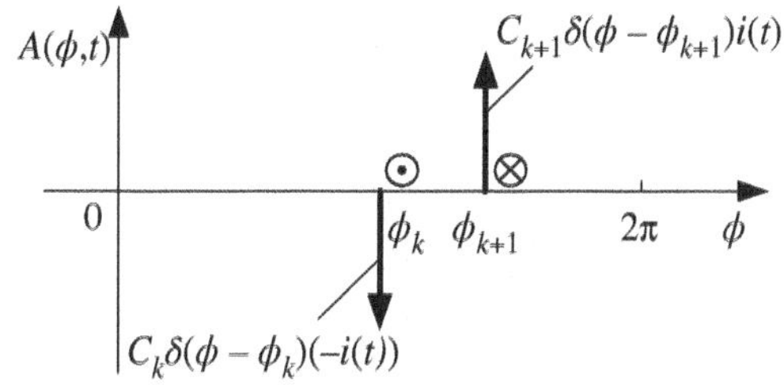

Figure 3.3 Linear current density distribution function.

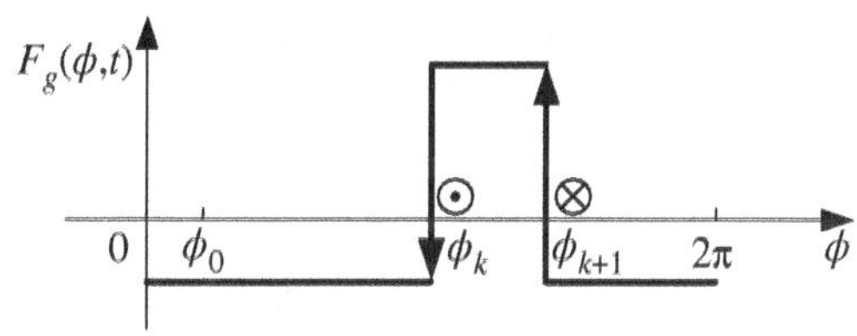

Figure 3.4 Airgap MMF distribution function.

MMF drop along the closed loop should be equal to the total current surrounded by the loop, that is, the integral of the linear current density distribution function $A(\phi, t)$ along AB. If the airgap MMF corresponding to point A is $F_g(\phi_0, t)$, then assuming that the relative permeability of the core is infinite, the airgap MMF at point B $F_g(\phi, t)$ is the sum of the total MMF of the closed loop and the airgap MMF at point A $F_g(\phi_0, t)$. Then the relationship between the airgap MMF distribution function $F_g(\phi, t)$ and the linear current density distribution function $A(\phi, t)$ can be obtained as

$$F_g(\phi, t) = \int_{\phi_0}^{\phi} A(\phi, t)\, d\phi + F_g(\phi_0, t) \tag{3.4}$$

Figure 3.4 shows the MMF distribution function of Figure 3.1.

The MMF distribution function describes the airgap MMF at each point along the airgap circumference. The magnetic field strength at each point along the airgap circumference can be obtained immediately by dividing the MMF distribution function by the airgap length of each point, and the corresponding airgap flux density distribution $B_g(\phi, t)$ is:

$$B_g(\phi, t) = \mu_0 \frac{F_g(\phi, t)}{g(\phi, \theta_r)} \tag{3.5}$$

The magnetic flux density distribution function in the case of a uniform airgap is shown in Figure 3.5.

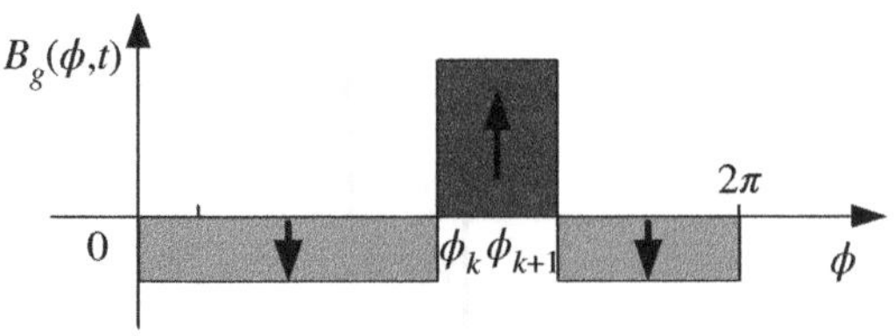

Figure 3.5 Airgap flux density distribution function.

3.2.2 Classical Winding Function Theory

The classical winding function method based on the aforementioned MMF analysis has been detailed in [2], and here is a brief review. Substituting (3.3) into the integral term in (3.4), we can get:

$$\int_{\phi_0}^{\phi} A(\phi,t)\,d\phi = i(t)\int_{\phi_0}^{\phi}\sum C(\phi,t)\,d\phi = T(\phi,\phi_0)\,i(t) \tag{3.6}$$

where $T(\phi,\phi_0)$ is the turn function, which is defined as the integral of the conductor distribution function over the position angle.

The MMF distribution functions on both sides of (3.4) are unknown, so the MMF at any position ϕ cannot be uniquely determined and additional constraints are required to obtain a unique solution. According to Gauss's theorem:

$$\int_0^{l_{stk}}\int_0^{2\pi} r_g\left[\mu_0\frac{F_g(\phi,t)}{g(\phi,\theta_r)}\right]d\phi\,dl = 0, \qquad r_g\in(R_r,R_s) \tag{3.7}$$

Integrating both sides of (3.4) on $0\sim 2\pi$, we can get:

$$\int_0^{2\pi}\frac{F_g(\phi,t)}{g(\phi,\theta_r)}d\phi = \int_0^{2\pi}\frac{T(\phi,\phi_0)\,i(t)}{g(\phi,\theta_r)}d\phi + \int_0^{2\pi}\frac{F_g(\phi_0,t)}{g(\phi,\theta_r)}d\phi \tag{3.8}$$

Simplifying (3.7) and substituting it into (3.8) yield:

$$F_g(\phi_0,t) = -\frac{\int_0^{2\pi} g^{-1}(\phi)\,T(\phi,\phi_0)\,d\phi}{\int_0^{2\pi} g^{-1}(\phi,\theta_r)\,d\phi}i(t) \tag{3.9}$$

Substituting (3.9) into (3.4) yields:

$$F_g(\phi,t) = \left[T(\phi,\phi_0) - \frac{\int_0^{2\pi} g^{-1}(\phi)\,T(\phi,\phi_0)\,d\phi}{\int_0^{2\pi} g^{-1}(\phi,\theta_r)\,d\phi}\right]i(t) = W(\phi)\,i(t) \tag{3.10}$$

where $W(\phi)$ is defined as the winding function, and $g^{-1}(\phi,\vartheta_r)$ is the reciprocal of the airgap function, which is called the inverse airgap function. For the case of a uniform airgap, the inverse airgap function is a constant value, and the winding function can be further simplified to:

$$W(\phi) = T(\phi,\phi_0) - \frac{1}{2}\int_0^{2\pi} T(\phi,\phi_0)d\phi \tag{3.11}$$

For the situation where one side of the airgap is uniform or nearly uniform and the other side has salient reluctance poles, such as wound-field salient-pole synchronous machines (SMs) and salient-pole SynRMs, the stator windings are all multiphase integer-slot distributed windings, and the corresponding MMF distribution has only odd harmonics. Considering that salient poles must appear in pairs, that is, the inverse airgap function has only even harmonics, the winding function can still be simplified to (3.11), so for salient-pole SMs and SynRMs, the winding functions can be easily calculated by (3.11).

The MMF distribution can be obtained by superimposing the products of winding functions of multiphase windings and the corresponding phase currents. Under the condition of a uniform or nearly uniform airgap, the airgap magnetic flux density distribution is consistent with the MMF distribution waveform, and the value is μ_0/g times of the MMF distribution. If a unit current is applied to the winding with the winding function $W_i(\phi)$, the flux linkage of any winding with the winding function $W_j(\phi)$ is the inductance between the i-th winding and the j-th winding, that is

$$L_{ij} = \mu_0 \int_0^{2\pi} [g^{-1}(\phi, \theta_r)\, W_i(\phi)]\, dS = \mu_0 r_g l_{\text{stk}} \int_0^{2\pi} [g^{-1}(\phi, \theta_r)\, W_i(\phi)]\, W_j(\phi)\, d\phi \tag{3.12}$$

There are many types of machine windings, which can be classified differently. For example, according to the number of phases, they can be divided into single-phase, two-phase, three-phase, four-phase, five-phase windings and above; according to the number of slots per pole per phase, they can be divided into integer-slot windings and fractional-slot windings; according to number of coils in each slot, they can be divided into single-layer winding, double-layer winding and multilayer winding; according to the coil span, they can be divided into full-pitch winding, short-pitch winding and long-pitch winding; according to the connection between phase windings, they can be divided into the star-connected and polygon-connected [3].

Reference [4] introduces the commonly used winding structures in electrical machines one by one and further enriches the winding forms, such as commutator windings in DC machines (DCMs), damping windings in salient-pole SMs and so on. In order to facilitate the application of the winding function method, the machine windings need to be reclassified from the perspective of winding structures, as shown in Figure 3.6.

Windings are first divided into asymmetric windings and symmetric windings, because the symmetry of the windings affects the spectrum of the phase and resultant winding MMF (or EMF). The asymmetric windings are further divided into single-phase and inter-phase asymmetric windings. The single-phase

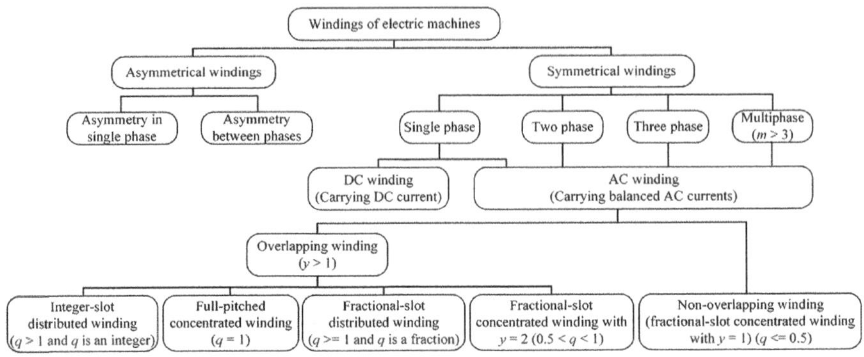

Figure 3.6 Classification of electrical machine windings.

asymmetric winding refers to phase windings that are distributed asymmetrically, such as nested-loop rotor windings, multiphase double-layer rotor windings, or spiral-loop rotor windings in brushless doubly-fed induction machines (BDFIMs). With the unit current applied, the positive and negative half waves of MMF distributions created by these windings are asymmetrical. In contrast, the inter-phase asymmetric winding refers to what is commonly referred to as asymmetric windings, such as the three-phase winding with a number of slots per pole per phase $q = Z/(2mp) = N/D$ ($m = 3$, D and N have no common divisor, and D is 3 and multiples of 3).

Symmetrical winding is the most commonly used form of machine windings, which can be divided into single-phase winding, two-phase winding, three-phase winding and multiphase winding according to the number of phases. (Single-phase winding and three-phase winding are more common.) When a DC current flows through a single-phase winding, it becomes a DC winding, such as the field winding in a DCM and a SM. In the following analysis, permanent magnets can be equivalent to a single-phase DC winding.

All symmetric windings flowing through AC currents become AC windings, which can be further divided into overlapping windings (coil span greater than 1) and non-overlapping windings (coil span equal to 1), which are also known as fractional-slot concentrated windings [3], according to whether the coils overlap or not.

Overlapping windings can be further divided into integer-slot distributed windings which require q to be an integer greater than 1, full-pitch concentrated windings whose q equals to 1, fractional-slot distributed windings which require q to be greater than or equal to 1, and fractional-slot concentrated windings with a span of 2, according to the number of slots per pole per phase. Fractional-slot concentrated windings with a coil span of 2 require the value of q to be between 0.5 and 1, so that the value of the pole pitch is between 1.5 and 3, and as a result, the pole pitch can only be 2.

3.2.3 Limitations of Classical Winding Function Theory

The classical winding function method requires that at least one of the two sides adjacent to the airgap is smooth or approximately smooth, and then the airgap flux density distribution function $B_g(\phi, t)$ can be obtained directly from the airgap MMF distribution function $F_g(\phi, t)$ by ignoring the MMF drop in the core. If both sides of the airgap are nonuniform, or the MMF drop in the core cannot be ignored due to the magnetic saturation, or the magnetic flux path is restricted by flux barriers, or short-circuited coils introduce additional MMFs, then the magnetic flux density distribution function $B_g(\phi, t)$ cannot be obtained directly from the MMF distribution function $F_g(\phi, t)$, because the magnetic flux density distribution also depends on specific magnetic flux paths. In all of these cases, the classical winding function method is no longer applicable.

The more common case in electrical machines is that short-circuited coils (used in induction machines (IMs) and BDFIMs), salient reluctance poles (used in double salient-pole machines) and multilayer flux barriers (used in SynRMs and brushless doubly-fed reluctance machines (BDFRMs) with multilayer flux barrier rotors) appear in the airgap. The existence of these structures will affect the airgap MMF distribution, and then affect the airgap magnetic flux density distribution, making the classical winding function method no longer applicable.

Detailed analysis reveals that these structures will not affect the linear current density distribution function generated by the current-carrying conductor in the slot on the surface of the airgap, and the linear current density distribution function is still discretely distributed at a point in the center of each slot opening. However, the existence of the three structures will change the distribution of MMF along the airgap, as shown in Figures 3.7–3.9.

1 Influence of Short-Circuited Coils

Assuming that the short-circuited coil is an ideal superconductive coil and the initial current is zero, at an instant of time, constant currents flowing in the stator windings will create the source MMF $F(\phi, t)$. At the same time, currents will be induced in short-circuited coils, creating an additional MMF distribution to oppose the source MMF [5], as shown in Figure 3.7. Taking into account the effect of short-circuited coils, the resultant airgap MMF becomes:

$$F_g(\phi, t) = F(\phi, t) + \sum_{j=1}^{N_{SC}} W_j(\phi) i_j(t) \tag{3.13}$$

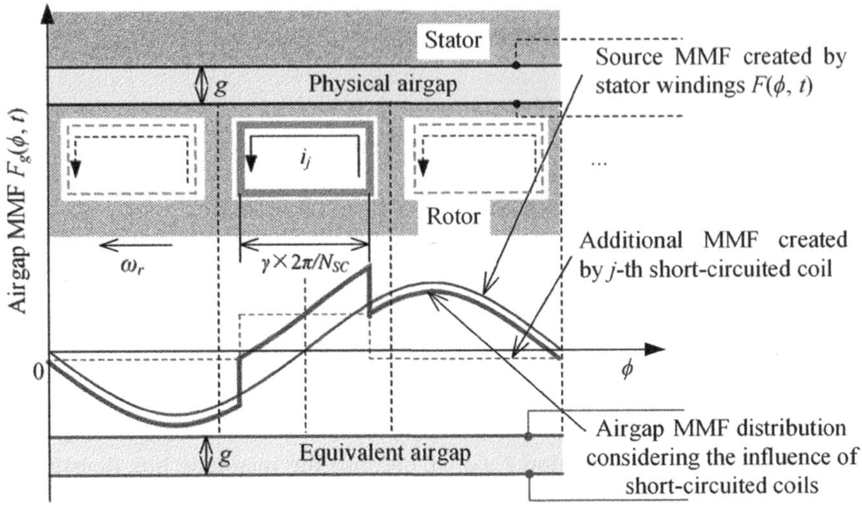

Figure 3.7 Influence of short-circuited coils on the airgap MMF distribution.

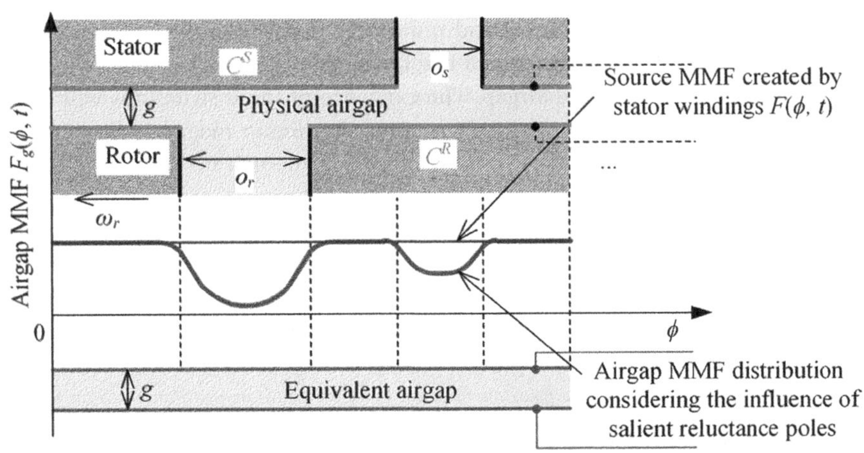

Figure 3.8 Influence of salient reluctance poles on the airgap MMF distribution.

where N_{SC} is the number of short-circuited coils, γ is the coil span factor, and i_j is the current in the j-th coil.

2 Influence of Salient Reluctance Poles

Assuming that there are salient-pole structures on both sides of the airgap, and there is a yoke connecting all salient poles to ensure that the integration path

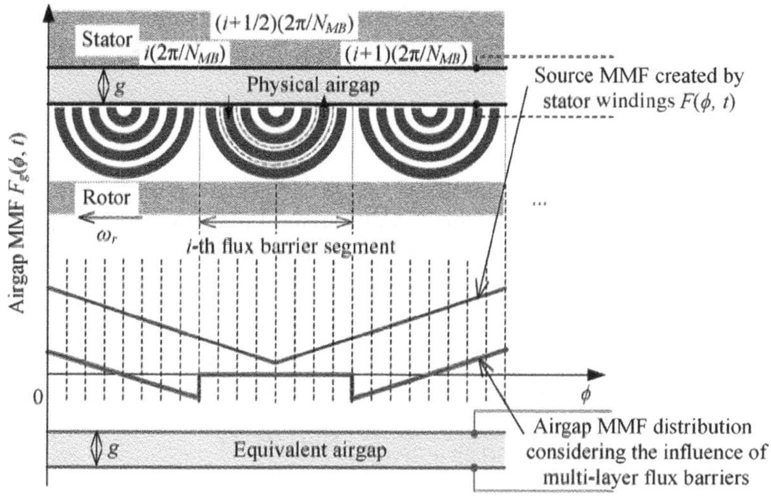

Figure 3.9 Influence of multilayer flux barriers on the airgap MMF distribution.

can be selected arbitrarily, the excitation MMF will concentrate in parts where the stator and rotor salient poles overlap. The MMF will be attenuated greatly in regions where the salient pole coincides with the slot and where the slot overlaps with the slot. The minimum MMF occurs at the center line of the slot, as shown in Figure 3.8. The resultant airgap MMF becomes:

$$F_g(\phi, t) = F(\phi, t)\lambda(\phi, \theta_r) \tag{3.14}$$

where $\lambda(\phi, \vartheta_r)$ is the normalized airgap permeability function, and one instance of the detailed empirical expression can be found in reference [6]:

$$\lambda(\phi, \theta_r) = \frac{\dfrac{1}{g(\phi, \theta_r)}}{\dfrac{1}{g}} = \frac{g}{g(\phi, \theta_r)} = \frac{g}{\dfrac{1}{g_s^{-1}(\phi, \theta_r)} + \dfrac{1}{g_r^{-1}(\phi, \theta_r)} - g} \tag{3.15}$$

where $g(\phi, \theta_r)$ is the actual airgap function, and $g_s^{-1}(\phi, \theta_r)$ the inverse airgap function when the rotor side of the airgap is assumed uniform and only the stator side is slotted. Similarly, $g_r^{-1}(\phi, \theta_r)$ is the inverse airgap function when the stator side is uniform and only the rotor side is slotted. The specific expression of the inverse airgap function depends on the slot width and tooth pitch. Both $g_s^{-1}(\phi, \theta_r)$ and $g_r^{-1}(\phi, \theta_r)$ can be written in the form of Fourier series, for example:

$$g_s^{-1}(\phi, \theta_r) = a_0 - \sum_{\nu=1}^{\infty} a_\nu \cos(\nu N_{SP1} \phi) \tag{3.16}$$

where N_{SP1} is the number of salient poles on the stator side. a_0 is a constant term and a_ν is the amplitude of the ν-th harmonic, which can be expressed by:

$$a_0 = \frac{t_d - 1.6\beta o}{t_d g} \tag{3.17}$$

$$a_\nu = \frac{\beta}{g} FC_\nu(\varepsilon), \quad \varepsilon = \frac{o}{t_d} \tag{3.18}$$

where t_d is the pole pitch (tooth pitch), and o is the slot opening width, which can be the stator slot opening width o_s or the rotor counterpart o_r. Both β and FC_ν are functions of the ratio of o to t_d:

$$\beta = \frac{1}{2} - \frac{1}{2}\left[1 + \left(\frac{o}{2g}\right)^2\right]^{-\frac{1}{2}} = \frac{1}{2} - \frac{1}{2}\left[1 + \left(\frac{o}{t_d}\frac{t_d}{2g}\right)^2\right]^{-\frac{1}{2}} \tag{3.19}$$

$$FC_\nu(\varepsilon) = \frac{4}{\nu\pi}\left[0.5 + \frac{(\nu\varepsilon)^2}{0.78125 - 2(\nu\varepsilon)^2}\right]\sin(1.6\pi\nu\varepsilon) \tag{3.20}$$

3 Influence of Multilayer Flux Barriers

When one side of the airgap is smooth or nearly smooth, and the other side has the flux barrier structure (multilayer flux barrier or axial lamination), as shown in Figure 3.9, the magnetic flux lines will be constrained within the magnetic guide layers, resulting in changes in the resultant airgap MMF distribution, that is

$$\begin{aligned} F_g(\underline{\phi}, t) &= \left[\frac{g(\underline{\phi}, \theta_r)}{g(\underline{\phi}, \theta_r) + g((2i+1)2\pi/N_{MB} - \underline{\phi})}\right] \\ &\quad \times \left[F(\underline{\phi}, t) - F\left((2i+1)\frac{2\pi}{N_{MB}} - \underline{\phi}, t\right)\right], \\ \underline{\phi} &= \phi - \theta_{MB} \in \left(i\frac{2\pi}{N_{MB}}, (i+1)\frac{2\pi}{N_{MB}}\right) \end{aligned} \tag{3.21}$$

where $\underline{\phi} = \phi - \theta_{MB}$ is the new coordinate of ϕ in the reference frame where the multilayer flux barrier is located, and N_{MB} is the number of multilayer flux barrier segments.

In the above three cases, the magnetic core on one side of the airgap with sources of MMF is continuous. Under the assumption of infinite permeability, the continuous magnetic core is an equipotential object. When the iron core on one side of the airgap with sources of MMF is discontinuous, it is no longer a continuous equipotential object, and the integration path of magnetic field strength is limited to several discrete areas. The change in the coincident surface directly affects the magnitude of the magnetic field strength, affecting the amplitude of the magnetic flux density in the airgap

When the magnetic core on one side of the airgap with sources of MMF is no longer continuous, Ampere's law and Gauss's law will not produce a simple relationship to bridge the resultant airgap MMF distribution function with the source MMF distribution function, although theoretically this relationship does exist. In this case, both the equivalent airgap length on the main magnetic circuit and the coincident surface in the actual airgap will influence the resultant airgap MMF distribution, greatly reducing the accuracy of the airgap flux density distribution directly obtained from the winding function method.

Typical machine topologies whose magnetic core on the airgap side with sources of MMF is discontinuous include double salient-pole PM (DSPM) machines, FSPM machines, and various variable-flux stator-PM machines. For this family of electrical machines, whether the influence on the airgap MMF distribution of three special structures can be determined from the perspective of winding functions depends on both the reluctance introduced by PMs on the main magnetic circuit and the actual airgap reluctance (or permeance).

In general, the reluctance introduced by PMs on the main magnetic circuit of well-designed DSPM machines and FSPM machines is smaller than the average value of the actual airgap reluctance, so the influence of the PM reluctance can be ignored, and the magnetic core is analyzed as if it is continuous. For example, although the stator cores of traditional FSPM machines are segmented and separated from each other by PM blocks, in general, due to the change of the magnetic circuit structure, when the rotor rotates by a rotor pitch, the airgap MMF distribution is the same as when the core is continuous. Reference [7] assumes that PMs are constant magnetic flux sources when analyzing a traditional FSPM machine. In fact, it assumes that the magnetic core on one side of the airgap with sources of MMF is continuous to make the winding function method applicable.

3.3 Three Key Elements

An electrical machine usually includes a stator, a rotor, and an airgap between the stator and rotor. The stator and rotor are usually made of soft magnetic materials with very high magnetic permeability. PMs or field windings located on the stator or rotor are used to create an airgap magnetic field, while the armature windings located on the stator or rotor generate induced EMF by cutting the magnetic flux lines.

A typical machine structure – a PM brushless DCM with six stator slots and four rotor poles – is shown in Figure 3.10. The above-mentioned machine with a stator, a rotor and a layer of airgap is called a "unit machine" here. For the sake of generality, any unit machine can be abstracted as a cascade of three key elements "excitation source-modulator-filter," as shown in Figure 3.11.

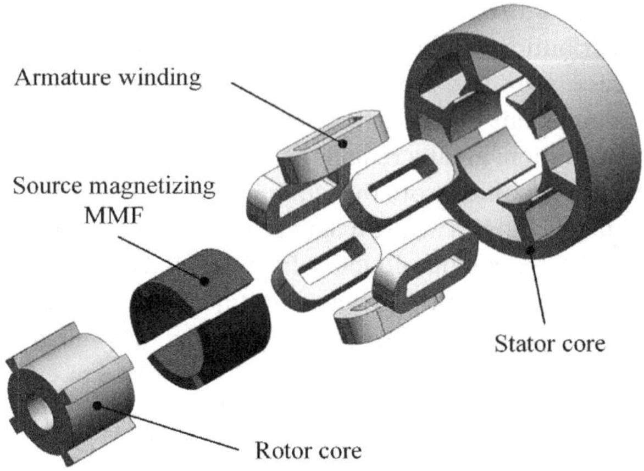

Figure 3.10 Exploded view of a brushless DC PM machine.

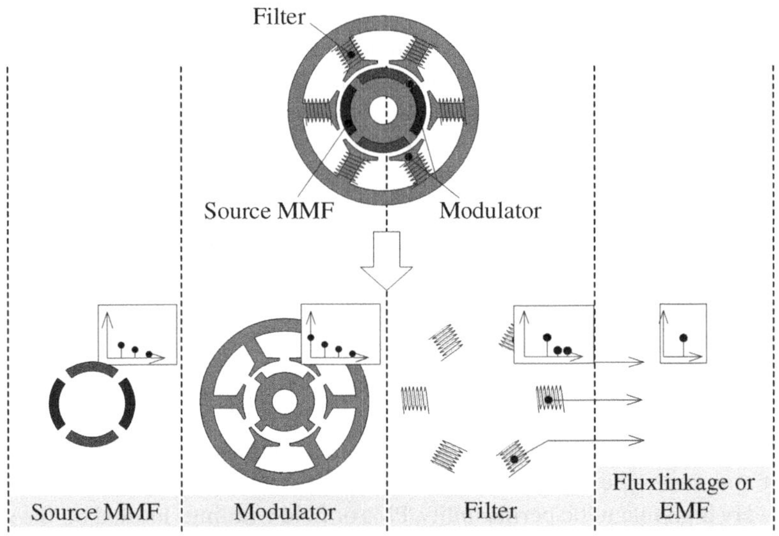

Figure 3.11 The cascade of three key elements "excitation source-modulator-filter."

The source of excitation (PMs in the figure) creates a source MMF distribution on the physical airgap. The modulator (salient poles in the figure) modulates the source MMF distribution to produce a series of MMF harmonic components and their corresponding magnetic flux density harmonic components. The filter (armature winding in the figure) is used as a spatial harmonic filter to extract effective airgap magnetic flux density harmonic components to generate magnetic flux linkage or EMF for energy conversion. In practice, each element may have multiple forms.

3.3.1 Source of Excitation

Usually, the source of excitation is provided by windings or PMs. In order to consider the problem in a unified way, PMs are replaced by a reluctance block surrounded by equivalent surface current. There are three main types of windings in real machines:

1 Single-Phase Winding (Typically with DC Current)
Single-phase windings supplied with DC currents are usually used as field windings to create an ideal rectangular-wave source excitation MMF that is stationary relative to the stator or rotor. A single-phase field winding fed with AC current will create a pulsating magnetic field, which can be further decomposed into two rotating magnetic fields with the same amplitude and speed but opposite directions, such as field windings in single-phase IMs [8] and universal motors [9].

2 Multi-phase Overlapping Winding (Typically with AC Current)
Multi-phase windings with balanced multiphase AC currents create ideal sinusoidally distributed source MMFs. Typical examples are the stator windings of three-phase IMs, the stator and rotor windings of DFIMs.

3 Multi-phase Non-overlapping Winding (Typically with AC Current)
Multi-phase fractional-slot concentrated winding supplied by balanced multiphase currents will generate many spatial harmonics at different speeds, so it is not as suitable a field winding as a distributed winding is. This also explains why IMs and SynRMs do adopt single-layer and double-layer fractional-slot concentrated windings that are commonly-used for PMSMs [10, 11].

3.3.2 Modulator

There are three basic forms of modulators, namely, short-circuited coils, salient reluctance poles and multilayer flux barriers.

1 Short-Circuited Coils
For alternating magnetic fields, the short-circuited coil is an ideal high impedance [5, 12]. Through careful arrangement, the short-circuited coil group can change the spectrum of the airgap MMF distribution by adding an induced additional MMF to the source MMF.

Table 3.1 Typical values of o/t_d for various electrical machine topologies.

Machine topologies	o_s/t_{ds}	o_r/t_{dr}
Squirrel-cage IM and BDFIM	0.28	0.18
PM brushless machines	0.22	—
DCM	0.33	0.25
Wound-field salient-pole SM	0.22	0.33
SynRM with a salient-pole rotor	0.26	0.38
BDFRM with a salient-pole rotor	0.26	0.38
MGM and PMV machine	0.2	0.5
FRPM machine	0.33	0.5
FSPM machine	0.5	0.71
Switched reluctance machine (SRM) and DSPM machine	0.5	0.56

2 Salient Reluctance Poles

Salient reluctance poles are the most common modulators. The basic principle is that a larger slot opening (the ratio of slot to tooth pitch $o/t_d \geq 0.3$) will significantly increase the influence of tooth harmonics [6]. Typical examples are the stator teeth and rotor poles in FRPM machines [13], [14], FSPM machines [15, 16], and the ferromagnetic poles in PMV machines [17]. The typical values of o/t_d for different machine topologies are shown in Table 3.1.

Specially designed pole shapes also belong to this category. A typical example is the rotor pole shoe of a wound-field salient-pole SM [2], which usually requires special design to make the airgap magnetic field distribution as sinusoidal as possible. Another example is that unequal airgap design is usually used in interior PMSMs to improve the back-EMF waveform [18].

3 Multilayer Flux Barriers

Multilayer flux barriers are usually equivalent to salient reluctance poles to facilitate understanding, but they are very different from salient reluctance poles in terms of modulation behavior. Multilayer flux barriers change the distribution of the airgap MMF by making full use of the difference in scalar magnetic potentials at different positions. Typical examples are SynRMs, BDFRMs, and the axially-laminated structure in FSPM machines [19–21].

3.3.3 Filter

Windings perform the function of a filter and are used to extract specific airgap spatial harmonics. Filters usually have two physical forms, namely, overlapping windings and non-overlapping windings.

1 Overlapping Windings

Multi-phase (or single-phase) overlapping windings (integer slots or fractional slots) are like a band-pass filter, with very strong frequency selection characteristics. It is typically the first choice for armature windings of almost all AC machines with sine-wave back-EMFs, such as the armature winding of a three-phase wound-field salient-pole SM, the stator winding of a three-phase squirrel-cage IMs and the stator winding of most rotor-PM brushless machines. Single-phase overlapping windings are used as spatial harmonic filters for AC machines that require single-phase sinusoidal output, such as the armature windings of single-phase IMs and single-phase BDFIMs [22].

2 Non-overlapping Windings

Some applications require the machine to have many pole pairs, short end windings or weak coupling between adjacent phases. In such cases, multiphase non-overlapping windings are more practical, such as the stator winding of rotor-PM machines with fractional-slot concentrated windings. The multiphase non-overlapping winding is also a band-pass filter, but the bandwidth is wider than the multiphase overlapping winding. Single-phase non-overlapping windings are only used in single-phase PM brushless machines that use fractional-slot concentrated windings.

3.4 Mathematical Representation of Three Key Elements

The similarity between electrical machines and switching converters provides a new perspective for the understanding of torque production mechanisms and electromagnetic energy conversion processes in electrical machines. This section focuses on how to use mathematical language to describe the behavior of excitation sources (source MMFs), modulators (short-circuited coils/salient reluctance poles/multilayer flux barriers) and filters (armature windings), as well as the relationship between machine performance and the three elements.

First, the following assumptions are made:

- PMs are replaced by reluctance blocks surrounded by equivalent surface currents;
- The magnetic core has infinite permeability;
- All the conductors in the slot are concentrated at the center of the slot opening;
- The magnetic flux lines produced by the airgap excitation MMF are perpendicular to the airgap surface;
- The influence of leakage flux and limited axial length is neglected;
- The magnetic saturation is accounted for by a constant MMF drop in the iron core.

3.4.1 Source MMF

The source MMF refers to the MMF distribution created by windings or PMs on a smooth and uniform airgap without considering the effects of short-circuited coils, salient reluctance poles and multilayer flux barriers. The source MMF can be analyzed using the slot star diagram [4] or the winding function introduced in Section 3.2.2 [2, 23]. The slot star diagram is convenient for hand calculation and is a very effective tool in the design and analysis of traditional AC machine windings. It is especially suitable for distributed windings (the number of slots per pole per phase q is an integer or a fraction greater than 1). In these machine types, the source MMF distribution created by the winding is almost sinusoidal, so the fundamental is the dominating component. The slot conductor vector has a clear physical meaning and it represents the space phasor of MMF. The main disadvantage of the slot star diagram is its frequency dependency.

In the MMF analysis, slot star diagrams of different MMF spatial harmonics are different. For example, the slot star diagram of the third harmonic will be different from the slot star diagram of the fundamental, so one has to redraw the slot star diagram when calculating the winding function of different harmonics. Moreover, when q is not an integer, the concept of unit winding has to be introduced to assist the analysis [3]. As the number of slots increases, the convenience of using slot star diagrams to analyze machine windings will no longer apply.

Compared with the slot star diagram, the winding function method can represent the waveform of the MMF distribution created by the winding more intuitively. The first step in the analysis of winding MMF using the winding function method is to draw the winding function. The winding function is a piecewise function, which can be obtained by scaling and shifting from the step function, which is suitable for programming. When using the winding function method to analyze the source MMF, the source MMF is obtained by multiplying the winding function and the current flowing through the winding. The winding configuration (the distribution of conductors and the way of connection between conductors) directly affects the distribution of the created MMF.

For a set of m-phase machine windings, the winding function $W(\phi)$ of one phase can be obtained from the winding layout, as expressed by (3.11). The frequency spectrum of the source MMF distribution created by a phase winding along the airgap completely depends on the winding configuration. Figure 3.12 shows the conductor distribution function, linear current density distribution function, turn function and winding function corresponding to a typical set of full-pitch distributed windings, where the number of slots Z is 12, the number of phases m is 3, the number of pole pairs p is 2, and the coil span y is 6. It can be seen that the winding function is independent of the initial position of the integral, while the turn function is different and is affected by the initial position of the integral.

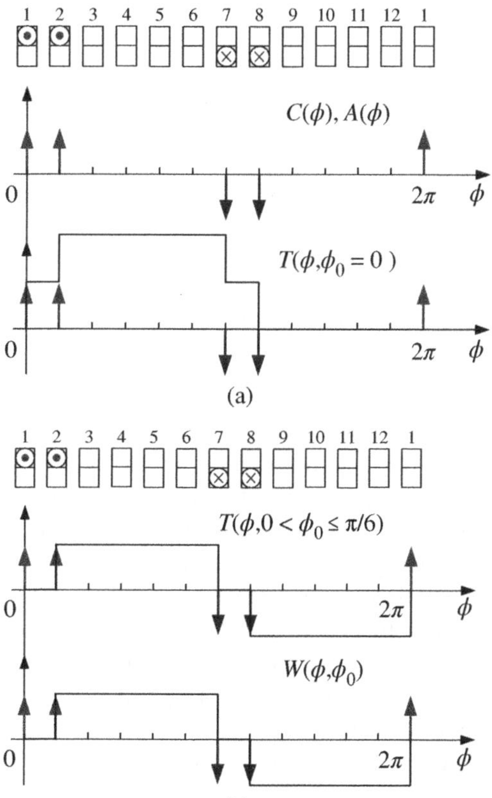

Figure 3.12 Conductor distribution function, turn function and winding function of a three-phase electrical machine with 12 slots and 2 poles. (a) $\phi_0 = 0$, (b) $0 < \phi_0 \leq \pi/6$.

Therefore, in the developed airgap field modulation theory, the winding function is use to characterize the winding configurations.

The most important features of windings are the principal pole pair number p and the winding factor $k_{w\nu}$ corresponding to each spatial harmonic. The number of principal pole pairs of a winding is defined here as the number of pole pairs corresponding to the harmonic component with the largest amplitude in the source MMF distribution created by the winding. This also means that the harmonic component with p pole pairs has the largest amplitude in the source MMF distribution created by the winding with a principal pole pair number of p. By writing winding functions in the form of Fourier series, the winding factor of each harmonic $k_{w\nu}$ can be easily obtained.

In addition to the chording and distribution of coil conductors that have been considered in winding factors, the connection (star connection, delta connection, etc.) between phase windings will also have an impact on the resultant MMF, which can be characterized by a coefficient $k_{c\nu}$. Taking these two factors into consideration, the resultant source MMF distribution created by a set of m-phase winding in a cross-section of radial-flux machines or a circular cut-plane of axial-flux machines can be obtained by:

$$F(\phi, t) = \underbrace{\sum_{i=1}^{m}}_{k_{c\nu}} \underbrace{W_i(\phi)\, i_i(t)}_{k_{w\nu}} = \sum_{\nu=1}^{\infty} F_\nu \cos(\nu\phi + \phi_\nu) \tag{3.22}$$

where the relationship between the amplitude of the ν-th spatial harmonic component of the source MMF and the number of harmonic pole pairs ν is:

$$|F_\nu| \propto k_{c\nu} \frac{k_{w\nu}}{\nu} \tag{3.23}$$

Traditional AC machines may have one or more sets of distributed windings, but the number of principal pole pairs of each set of windings is the same, so the resultant source MMF is a traveling wave with a certain number of pole pairs. Many new machine topologies have two or more sets of windings with different pole pairs, and their resultant source MMF is very different from traditional AC machines. Taking the brushless doubly-fed machine as an example, the stator has two sets of sinusoidally distributed three-phase windings. One of the two sets of windings or both of them can be supplied. If F_1 and F_2 denote the amplitudes of the fundamental MMF of the first and second set of stator windings, respectively, based on the assumptions that $F_1 + F_2$ is constant and the phase sequences are the same, changing the value of F_1/F_2 will alter the resultant MMF distribution, as shown in Figure 3.13.

When one of the windings is supplied, the resultant source MMF distribution is a traveling wave, and its amplitude shows a continuous parallel strip pattern in

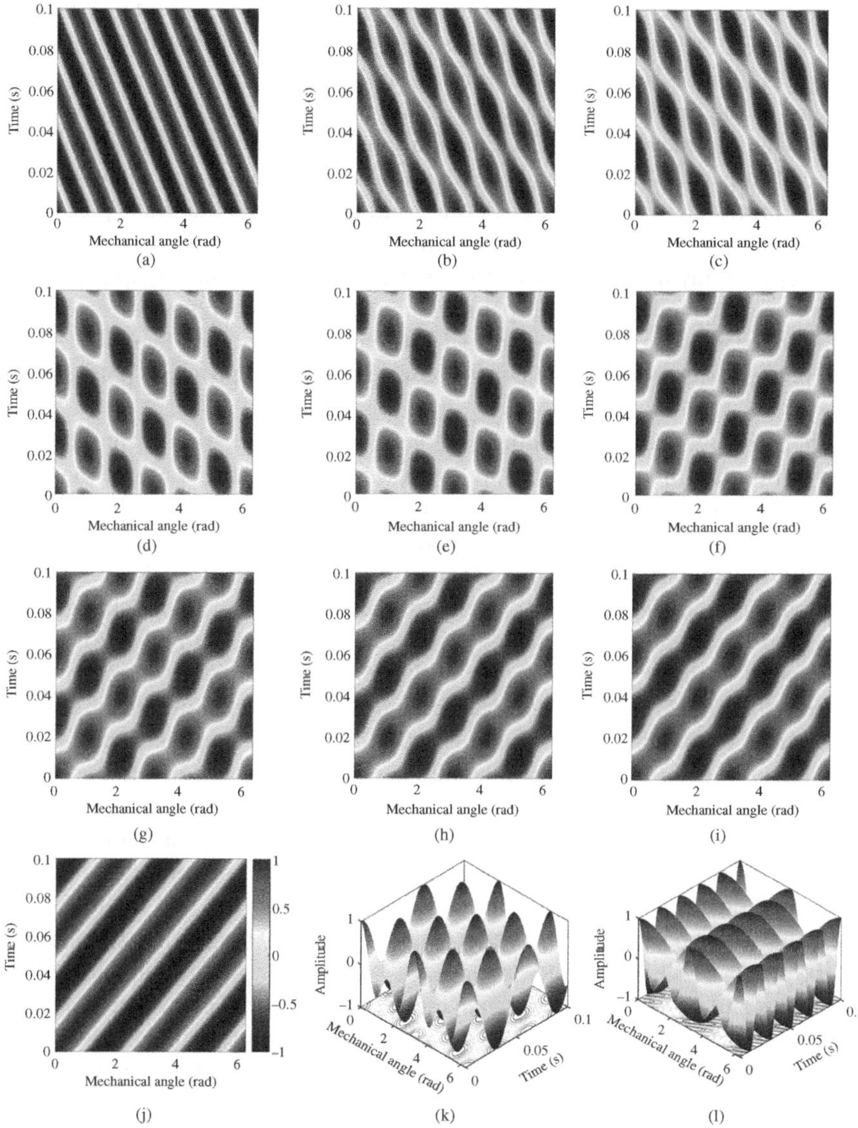

Figure 3.13 Resultant stator MMF distribution as a function of time and space in brushless doubly-fed machines. The principal pole pairs of the stator windings are $p_1 = 4$ and $p_2 = 2$. (a) $F_1/F_2 = 1/0$, (b) $F_1/F_2 = 3/1$, (c) $F_1/F_2 = 2/1$, (d) $F_1/F_2 = 1/1$, (e) $F_1/F_2 = 5/7$, (f) $F_1/F_2 = 1/2$, (g) $F_1/F_2 = 1/3$, (h) $F_1/F_2 = 1/4$, (i) $F_1/F_2 = 1/5$, (j) $F_1/F_2 = 0/1$, (k) 3-D view of (d) for sum modulation, (l) 3-D view of (d) for diff. modulation.

the two-dimensional plane of time and space, as shown in Figure 3.13a,j. As time goes by, the MMF shown in Figure 3.13a moves to the left, and the MMF shown in Figure 3.13j moves to the right. This is due to the analysis in the rotor coordinate system. In the brushless doubly-fed machine, MMFs created by the two sets of stator windings move in opposite directions relative to the rotor. In Figure 3.13a,j, the reddest part is the peak and the bluest part is the valley, and the peaks and valleys appear alternately. The peaks and valleys correspond to the N pole and S pole of the MMF respectively, and the number of peaks is equal to the number of valleys. At any moment, the distribution of MMF along the airgap peripheral shows an explicit number of pole pairs. Figure 3.13a has eight poles, corresponding to the stator winding with the principal pole pairs of $p_1 = 4$, and Figure 3.13b has four poles, corresponding to the stator winding with the principal pole pairs of $p_2 = 2$.

When the two sets of windings are supplied together, the resultant MMF distribution has the features of traveling waves and standing waves at the same time, which shows a discrete point pattern in the two-dimensional plane of time and space, as shown in Figure 3.13b–i, where the darkest part (the reddest part and the bluest part) is the antinode, and the middle position of the adjacent antinode is the node. The positions of antinodes and nodes do not change with time, and the number of antinodes is equal to the number of nodes. Antinodes correspond to the N pole and S pole of the MMF distribution. At any time, the distribution of MMF along the airgap peripheral has several antinodes, and the amplitudes of the antinodes are not exactly the same. When the rotor speed is taken as $\omega_r = (\omega_1 + \omega_2)/(p_1 + p_2)$, the number of antinodes is equal to the sum of the pole pairs of the two sets of windings, where ω_r is the rotor mechanical speed, ω_1 and ω_2 the angular frequencies of the two sets of stator winding currents, and p_1 and p_2 numbers of principal pole pairs of two sets of stator windings.

Cases shown in Figure 3.13b–i and k all have six antinodes and six nodes, and their equivalent pole pairs are the sum of the pole pairs of the two sets of windings, i.e. $p_1 + p_2 = 6$. When the rotor speed is $\omega_r = (\omega_1 + \omega_2)/|p_1 - p_2|$, the numbers of antinodes and nodes are both 2, which means the number of equivalent pole pairs becomes the difference between the two sets of winding pole pairs, i.e. $p_1 - p_2 = 2$. If the winding is a fractional-slot concentrated winding instead of a sinusoidal distributed winding, the distribution of the resultant MMF in space and time will become more complicated.

3.4.2 Modulation Operator

In Section 3.2.3, it has been pointed out that when there are short-circuited coils, salient reluctance poles or multilayer flux barriers on one or both sides of the airgap, the airgap MMF distribution will no longer be consistent with the source

MMF distribution, which means the airgap MMF distribution is affected by the modulation behavior of the three types of modulators. To describe this effect mathematically, the concept of the MMF modulation operator is introduced in this section. The MMF modulation operator is defined as a mapping from the source MMF distribution (function) to the actual airgap MMF distribution (function). Since a mathematical operator is a mapping from a set of functions to another set of functions, that is, the function of a function. The set of its independent variables is defined here as the collection of all the bivariate functions that are continuously differentiable and integrable in space and have a space period of 2π. $f(x,t)$ is an arbitrary element in the set of independent variables, where x and t are the space variable and the time variable, respectively. By definition, one can get $f(x,t) = f(x + 2\pi,t)$. The source MMF distribution function $F(\phi,t)$ is an instant of $f(x,t)$.

1 Short-Circuited Coils

The short-circuited coil is a passive component that relies on the speed difference between the source MMF (usually sinusoidal distribution) and the modulator to generate an alternating current in the short-circuited coil, thereby creating an additional source MMF on the airgap, as shown in Figure 3.7. The airgap MMF distribution considering the influence of the short-circuited coils becomes Eq. (3.13), so that the modulation operator of the short-circuited coils can be uniformly defined as:

$$M(N_{SC})[f(x,t)] = f(x,t) + \sum_{j=1}^{N_{SC}} W_j i_j(t), \quad x \in [0, 2\pi] \tag{3.24}$$

The detailed equation depends on the configuration of the short-circuited coil (conductor distribution and connection method). For example, there will be differences in the airgap MMF distributions when the same source MMF is modulated by a single-layer short-circuited coil, a multilayer nested-loop coil, and a spiral-loop coil.

2 Salient Reluctance Poles

The salient reluctance poles of the stator and rotor modulate the distribution of any source MMF applied to the airgap by the alternating changes of the permeance, as shown in Figure 3.8. Considering the influence of salient reluctance poles, the distribution of the airgap MMF becomes Eq. (3.14). Therefore, the modulation operator of salient reluctance poles can be defined as:

$$M(N_{SP})[f(x,t)] = \lambda(x, \theta_r) f(x,t) \tag{3.25}$$

Equation (3.25) takes into account the MMF distribution of slot openings on the stator side and rotor side [6], making the modulated airgap MMF distribution waveform more accurate. Salient-pole rotors of SynRMs and BDFRMs, ferromagnetic parts in the modulation rings of MGs and MGMs, stator poles of DCMs, stator and rotor salient poles of stator-PM machines, etc., are usually designed to have a large opening ($o/t_d \geq 0.3$) to produce a better modulation effect.

In most cases, stators and rotors with small openings ($o/t_d < 0.3$) can be considered slotless structures, unless the influence of slot openings is to be studied, such as in the study of harmonic losses in squirrel-cage IMs and research on the cogging torque in PMSMs. The slotless structure will not affect the distribution of the source MMF. It is defined as a unit modulation operator, and the corresponding modulation behavior is called unit modulation, which means the periodic function after modulation is the same as itself. In (3.25), when the slot openings on the stator and rotor sides of the airgap are both 0, $\lambda(x, \theta_r)$ becomes 1, and the salient reluctance pole modulation automatically turns into the unit modulation.

3 Multilayer Flux Barriers

The multilayer flux barrier structure modulates the non-uniformly distributed source MMF by pre-defining the paths of the magnetic flux lines, as shown in Figure 3.9. The ideal magnetic barrier is an axially-laminated structure, such as the axially-laminated anisotropic rotor of SynRMs, BDFRMs and FSPM machines. Considering the influence of the multilayer flux barriers, the distribution of the airgap MMF becomes (3.21). Therefore, the modulation operator of the multilayer flux barrier can be defined as:

$$M(N_{MB})[f(x,t)] = \left[\frac{g(x,\theta_r)}{g(x,\theta_r) + g((2i+1)2\pi/N_{MB} - x, \theta_r)}\right] \times \left[f(x,t) - f\left((2i+1)\frac{2\pi}{N_{MB}} - x, t\right)\right], \quad x \in \left(i\frac{2\pi}{N_{MB}}, (i+1)\frac{2\pi}{N_{MB}}\right) \tag{3.26}$$

In general, the two airgap lengths symmetric about $(2i+1)\pi/N_{MB}$ are equal, namely:

$$g(x,\theta_r) = g\left((2i+1)\frac{2\pi}{N_{MB}} - x, \theta_r\right) \tag{3.27}$$

Then, the MMF modulation operator can be simplified to:

$$M(N_{MB})[f(x,t)] = \frac{1}{2}\left[f(x,t) - f\left((2i+1)\frac{2\pi}{N_{MB}} - x, t\right)\right], \quad x \in \left(i\frac{2\pi}{N_{MB}}, (i+1)\frac{2\pi}{N_{MB}}\right) \tag{3.28}$$

In order to eliminate the influence of mechanical relative motion on the modulator and simplify the analysis, it is necessary to first convert the source MMF to the modulator reference frame, and then analyze the modulation behavior. For example, in order to analyze the modulation phenomenon in a flux-switching PM machine, we need to first analyze the modulation of the stationary PM field by the stator modulator in the stationary reference frame, and then analyze the complex airgap magnetic field modulated by the rotor modulator in the rotor-speed reference frame.

4 Mathematical Properties of Modulation Operators

As mathematical operators, modulation operators have some basic properties, as listed below:

(1) Linearity

The modulation operator is linear and satisfies the homogeneity and additivity, namely

$$M[\alpha f(x,t) + \beta g(x,t)] = \alpha M[f(x,t)] + \beta M[g(x,t)], \quad \alpha, \beta \in R \tag{3.29}$$

Linearity is the most basic mathematical property of modulation operators, which can be obtained according to their definitions. The derivation proof that the short-circuited coil modulator satisfies the linear law is given below. The derivation of the salient reluctance pole and the multilayer flux barrier modulators can be similarly obtained. In order to simplify the expressions, it is assumed that the N_{SC} short-circuited coils on the rotor do not overlap and are evenly distributed along the circumferential direction. The expression of the short-circuited coil modulation operator is shown in Eq. (3.24).

When the source MMF in the form of a unit cosine signal is applied to a uniform airgap, the resulting no-load airgap magnetic flux density distribution is:

$$B_g(\phi, t) = \frac{\mu_0}{g}\cos(p\phi - \omega t) \tag{3.30}$$

where μ_0 is the vacuum permeability, g is the airgap length, and ω is the angular frequency of the stator winding current. When the rotor rotates at a speed different from the synchronous speed, the slip frequency is $\omega_s = \omega - p\omega_r$ and the flux linkage of the j-th short-circuited coil is:

$$\begin{aligned}
\psi_j &= \int_{(1-\gamma)\pi/N_{SC}+(j-1)\times 2\pi/N_{SC}}^{(1+\gamma)\pi/N_{SC}+(j-1)\times 2\pi/N_{SC}} B_g(\phi, t) l_{\text{stk}} r_g d\phi \\
&= \frac{\mu_0 r_g l_{\text{stk}}}{g} \int_{(1-\gamma)\pi/N_{SC}+(j-1)\times 2\pi/N_{SC}}^{(1+\gamma)\pi/N_{SC}+(j-1)\times 2\pi/N_{SC}} \cos(p\phi - \omega_s t) d\phi \\
&= \frac{\mu_0 r_g l_{\text{stk}}}{g}\frac{2}{p}\cos\left[p(2j-1)\frac{\pi}{N_{SC}} - \omega_s t\right]\sin\left(\gamma\frac{p\pi}{N_{SC}}\right)
\end{aligned} \tag{3.31}$$

where r_g is the airgap radius, and l_{stk} is the effective stack length. Then the induced voltage in the j-th short-circuited coil is:

$$e_j(t) = -\frac{d\psi_j}{dt} = \omega_s \frac{\mu_0 r_g l_{stk}}{g} \frac{2}{p} \sin\left[p(2j-1)\frac{\pi}{N_{SC}} - \omega_s t\right] \sin\left(\gamma \frac{p\pi}{N_{SC}}\right) \tag{3.32}$$

The current in the j-th short-circuited coil is:

$$i_j(t) = \frac{\omega_s}{\sqrt{R^2 + (\omega_s L)^2}} \frac{\mu_0 r_g l_{stk}}{g} \frac{2}{p} \sin\left(\gamma \frac{p\pi}{N_{SC}}\right) \sin\left[p(2j-1)\frac{\pi}{N_{SC}} + \varphi - \omega_s t\right] \tag{3.33}$$

where R and L are the resistance and inductance of the j-th short-circuited coil respectively; φ is the impedance angle and $\varphi = \arctan(\omega_s L/R)$. In the rotor-speed coordinate system where the short-circuited coil is located, the modulated airgap MMF can be expressed as:

$$M(N_{SC})[F(\underline{\phi}, t)] = F(\underline{\phi}, t) + \begin{cases} W_1(\underline{\phi}) i_1(t), & \underline{\phi} \in \left[(1-\gamma)\frac{\pi}{N_{SC}} + 0 \times \frac{2\pi}{N_{SC}}, (1+\gamma)\frac{\pi}{N_{SC}} + 0 \times \frac{2\pi}{N_{SC}}\right], \\ \cdots \\ W_i(\underline{\phi}) i_i(t), & \underline{\phi} \in \left[(1-\gamma)\frac{\pi}{N_{SC}} + (i-1) \times \frac{2\pi}{N_{SC}}, (1+\gamma)\frac{\pi}{N_{SC}} + (i-1) \times \frac{2\pi}{N_{SC}}\right], \\ \cdots \\ W_{N_{SC}}(\underline{\phi}) i_{N_{SC}}(t), & \underline{\phi} \in \left[(1-\gamma)\frac{\pi}{N_{SC}} + (N_{SC}-1) \times \frac{2\pi}{N_{SC}}, (1+\gamma)\frac{\pi}{N_{SC}} + (N_{SC}-1) \times \frac{2\pi}{N_{SC}}\right], \\ 0, & \underline{\phi} \in [0, 2\pi] - \cup_{i=1}^{N_{SC}} \left[(1-\gamma)\frac{\pi}{N_{SC}} + (i-1) \times \frac{2\pi}{N_{SC}}, (1+\gamma)\frac{\pi}{N_{SC}} + (i-1) \times \frac{2\pi}{N_{SC}}\right] \end{cases} \tag{3.34}$$

where $\underline{\phi}$ is the coordinate position of ϕ in the coordinate system where the short-circuited coil is located, and $\underline{\phi} = \phi - \theta_r$. Expanding the modulated MMF of (3.34) into a Fourier series form can be obtained:

$$M(N_{SC})[F(\underline{\phi}, t)] = \frac{a_0}{2} + \sum_{k=1}^{\infty} a_k \cos(k\underline{\phi}) + \sum_{k=1}^{\infty} b_k \sin(k\underline{\phi}) \tag{3.35}$$

The coefficients are:

$$a_0 = \frac{1}{\pi} \int_0^{2\pi} M(N_{SC})[F(\underline{\phi}, t)] d\underline{\phi} \tag{3.36}$$

$$a_k = \frac{1}{\pi} \int_0^{2\pi} M(N_{SC})[F(\underline{\phi}, t)] \cos(k\underline{\phi}) d\underline{\phi} \tag{3.37}$$

$$b_k = \frac{1}{\pi} \int_0^{2\pi} M(N_{SC})[F(\underline{\phi}, t)] \sin(k\underline{\phi}) d\underline{\phi} \tag{3.38}$$

Therefore, the modulated airgap MMF is:

$$\begin{aligned} M(N_{SC})[F(\underline{\phi}, t)] &= C^d_{p,p}\cos(p\underline{\phi} - \omega_s t) + C^q_{p,p}\cos\left(p\underline{\phi} - \omega_s t + \frac{\pi}{2}\right) \\ &- \sum_{k=lN_{SC}-p}^{\infty} C_{\text{sum}}\sin\left[k\underline{\phi} + \omega_s t + (p+k)\frac{\pi}{N_{SC}} - \varphi\right] \\ &+ \sum_{k=lN_{SC}+p}^{\infty} C_{\text{dif}}\sin\left[k\underline{\phi} - \omega_s t + (k-p)\frac{\pi}{N_{SC}} + \varphi\right] \end{aligned} \tag{3.39}$$

In the stationary reference frame, the expression is:

$$\begin{aligned} M(N_{SC})[F(\phi, t)] &= C^d_{p,p}\cos(p\phi - \omega t) + C^q_{p,p}\cos\left(p\phi - \omega t + \frac{\pi}{2}\right) \\ &- \sum_{k=lN_{SC}-p}^{\infty} C_{\text{sum}}\sin\left[k\phi - (k+p)\omega_r t + \omega t + (p+k)\frac{\pi}{N_{SC}} - \varphi\right] \\ &+ \sum_{k=lN_{SC}+p}^{\infty} C_{\text{dif}}\sin\left[k\phi - (k-p)\omega_r t - \omega t + (k-p)\frac{\pi}{N_{SC}} + \varphi\right] \\ &= C^d_{p,p}\cos(p\phi - \omega t) + C^q_{p,p}\cos\left(p\phi - \omega t + \frac{\pi}{2}\right) \\ &- \sum_{l=1}^{\infty} C_{\text{sum}}\cos\left[(lN_{SC} - p)\phi + (\omega - lN_{SC}\omega_r)t + l\pi - \varphi - \frac{\pi}{2}\right] \\ &+ \sum_{l=1}^{\infty} C_{\text{dif}}\cos\left[(lN_{SC} + p)\phi - (\omega + lN_{SC}\omega_r)t + l\pi + \varphi - \frac{\pi}{2}\right] \end{aligned} \tag{3.40}$$

where the three coefficients can be expressed as:

$$\begin{cases} C^d_{p,p} = 1 - 2\pi N_{SC}\dfrac{\omega_s}{\sqrt{R^2 + (\omega_s L)^2}}\left(\dfrac{\mu_0 r_g L_{\text{stk}}}{g}\right)\left(\dfrac{\gamma}{N_{SC}}\right)^2\left[\dfrac{\sin\left(\gamma\frac{p\pi}{N_{SC}}\right)}{\left(\gamma\frac{p\pi}{N_{SC}}\right)}\right]^2\sin\varphi \\ C^q_{p,p} = 2\pi N_{SC}\dfrac{\omega_s}{\sqrt{R^2 + (\omega_s L)^2}}\left(\dfrac{\mu_0 r_g L_{\text{stk}}}{g}\right)\left(\dfrac{\gamma}{N_{SC}}\right)^2\left[\dfrac{\sin\left(\gamma\frac{p\pi}{N_{SC}}\right)}{\left(\gamma\frac{p\pi}{N_{SC}}\right)}\right]^2\cos\varphi \end{cases} \tag{3.41}$$

$$\begin{aligned} C_{\text{sum}} &= 2\pi N_{SC}\frac{\omega_s}{\sqrt{R^2 + (\omega_s L)^2}}\frac{\mu_0 r_g l_{\text{stk}}}{g}\left(\frac{\gamma}{N_{SC}}\right)^2\frac{\sin\left(\gamma\frac{p\pi}{N_{SC}}\right)}{\left(\gamma\frac{p\pi}{N_{SC}}\right)}\frac{\sin\left(\gamma\frac{k\pi}{N_{SC}}\right)}{\left(\gamma\frac{k\pi}{N_{SC}}\right)}, \\ k &= lN_{SC} - p \end{aligned} \tag{3.42}$$

$$C_{\text{dif}} = 2\pi N_{SC} \frac{\omega_s}{\sqrt{R^2 + (\omega_s L)^2}} \frac{\mu_0 r_g l_{\text{stk}}}{g} \left(\frac{\gamma}{N_{SC}}\right)^2 \frac{\sin\left(\gamma \frac{p\pi}{N_{SC}}\right)}{\left(\gamma \frac{p\pi}{N_{SC}}\right)} \frac{\sin\left(\gamma \frac{k\pi}{N_{SC}}\right)}{\left(\gamma \frac{k\pi}{N_{SC}}\right)},$$

$$k = lN_{SC} + p \tag{3.43}$$

where the subscript "sum" means sum modulation, and "dif" means difference modulation.

If the rotor resistance and leakage inductance are ignored, the short-circuited coil inductance can be expressed as $L = \mu_0 g^{-1} r_g l_{\text{stk}}(2\gamma\pi/N_{SC})$. The short-circuited coil becomes an ideal modulator, then the modulation coefficient can be simplified to:

$$\begin{cases} C_{p,p}^d\big|_{\text{ideal}} = 1 - \gamma \left[\dfrac{\sin\left(\gamma \frac{p\pi}{N_{SC}}\right)}{\left(\gamma \frac{p\pi}{N_{SC}}\right)}\right]^2 \\ C_{p,p}^q\big|_{\text{ideal}} = 0 \end{cases} \tag{3.44}$$

$$C_{\text{sum}}\big|_{\text{ideal}} = \gamma \frac{\sin\left(\gamma \frac{p\pi}{N_{SC}}\right)}{\left(\gamma \frac{p\pi}{N_{SC}}\right)} \frac{\sin\left(\gamma \frac{k\pi}{N_{SC}}\right)}{\left(\gamma \frac{k\pi}{N_{SC}}\right)}, \quad k = lN_{SC} - p \tag{3.45}$$

$$C_{\text{dif}}\big|_{\text{ideal}} = \gamma \frac{\sin\left(\gamma \frac{p\pi}{N_{SC}}\right)}{\left(\gamma \frac{p\pi}{N_{SC}}\right)} \frac{\sin\left(\gamma \frac{k\pi}{N_{SC}}\right)}{\left(\gamma \frac{k\pi}{N_{SC}}\right)}, \quad k = lN_{SC} + p \tag{3.46}$$

If the source MMF $\cos(p\phi - \omega t)$ is replaced with $\alpha\cos(p\phi - \omega t) + \beta\cos(q\phi - vt)$, following the derivation process of (3.30)–(3.46), the modulated airgap MMF can be expressed as:

$$\begin{aligned} &M(N_{SC})[\alpha\cos(p\phi - \omega t) + \beta\cos(q\phi - vt)] \\ &= \alpha\left[C_{p,p}^d \cos(p\phi - \omega t) + C_{p,p}^q \cos\left(p\phi - \omega t + \frac{\pi}{2}\right)\right] \\ &\quad - \sum_{k=lN_{SC}-p}^{\infty} \alpha C_{\text{sum},lN_{SC}-p} \sin\left[k\phi - (k+p)\omega_r t + \omega t + (p+k)\frac{\pi}{N_{SC}} - \varphi\right] \\ &\quad + \sum_{k=lN_{SC}+p}^{\infty} \alpha C_{\text{dif},lN_{SC}+p} \sin\left[k\phi - (k-p)\omega_r t - \omega t + (k-p)\frac{\pi}{N_{SC}} + \varphi\right] \\ &\quad + \beta\left[C_{q,q}^d \cos(q\phi - vt) + C_{q,q}^q \cos\left(q\phi - vt + \frac{\pi}{2}\right)\right] \end{aligned}$$

$$- \sum_{k=lN_{SC}-q}^{\infty} \beta C_{\mathrm{sum},lN_{SC}-q} \sin\left[k\phi - (k+p)\omega_r t + \omega t + (p+k)\frac{\pi}{N_{SC}} - \varphi\right]$$
$$+ \sum_{k=lN_{SC}+q}^{\infty} \beta C_{\mathrm{dif},lN_{SC}+q} \sin\left[k\phi - (k-p)\omega_r t - \omega t + (k-p)\frac{\pi}{N_{SC}} + \varphi\right] \tag{3.47}$$

The above formula is equal to $\alpha M(N_{SC})[\cos(p\phi - \omega t)] + \beta M(N_{SC})[\cos(q\phi - vt)]$, so the short-circuited coil modulation operator is linear and satisfies the superposition theorem.

(2) Commutative Property

The modulation operator satisfies the commutative law, namely

$$M_1[M_2[f(x,t)]] = M_2[M_1[f(x,t)]] \tag{3.48}$$

Taking the short-circuited coil modulation operator as an example, it is assumed that the unit cosine source MMF is modulated by $M(N_{SC1})[.]$ and then by $M(N_{SC2})[.]$. The mechanical speeds of the two modulators are the same, but the numbers and spans of the short-circuited coils are different ($N_{SC1} \neq N_{SC2}$, $\gamma_1 \neq \gamma_2$). Correspondingly, the impedance angles of the two modulators are also different ($\varphi_1 \neq \varphi_2$). Then the modulated MMF can be expressed as:

$$\begin{aligned}
&M(N_{SC2})[M(N_{SC1})[\cos(p\phi - \omega t)]] \\
&= M(N_{SC2})\left[C^d_{p,p}\cos(p\phi - \omega t) + C^q_{p,p}\cos\left(p\phi - \omega t + \frac{\pi}{2}\right)\right] \\
&+ M(N_{SC2})\left[\sum_{l=1}^{\infty} C_{\mathrm{sum},lN_{SC1}-p}\cos\left[(lN_{SC1}-p)\phi + (\omega - lN_{SC1}\omega_r)t + l\pi - \varphi_{SC1} - \frac{\pi}{2}\right]\right] \\
&+ M(N_{SC2})\left[\sum_{l=1}^{\infty} C_{\mathrm{dif},lN_{SC1}+p}\cos\left[(lN_{SC1}+p)\phi - (\omega + lN_{SC1}\omega_r)t + l\pi + \varphi_{SC1} - \frac{\pi}{2}\right]\right] \\
&= A + B + C
\end{aligned} \tag{3.49}$$

where

$$\begin{aligned}
A &= M(N_{SC2})\left[C^d_{p,p,N_{SC1}}\cos(p\phi - \omega t) + C^q_{p,p,N_{SC1}}\cos\left(p\phi - \omega t + \frac{\pi}{2}\right)\right] \\
&= C^d_{p,p,N_{SC1}}\left[C^d_{p,p,N_{SC2}}\cos(p\phi - \omega t) + C^q_{p,p,N_{SC2}}\cos\left(p\phi - \omega t + \frac{\pi}{2}\right)\right] \\
&+ C^d_{p,p,N_{SC1}}\sum_{l=1}^{\infty} C_{\mathrm{sum},lN_{SC2}-p}\cos\left[(lN_{SC2}-p)\phi + (\omega - lN_{SC2}\omega_r)t + l\pi - \varphi_{SC2} - \frac{\pi}{2}\right] \\
&+ C^d_{p,p,N_{SC1}}\sum_{l=1}^{\infty} C_{\mathrm{dif},lN_{SC2}+p}\cos\left[(lN_{SC2}+p)\phi - (\omega + lN_{SC2}\omega_r)t + l\pi + \varphi_{SC2} - \frac{\pi}{2}\right] \\
&+ C^q_{p,p,N_{SC1}}\left[C^d_{p,p,N_{SC2}}\cos\left(p\phi - \omega t + \frac{\pi}{2}\right) + C^q_{p,p,N_{SC2}}\cos(p\phi - \omega t + \pi)\right]
\end{aligned}$$

$$+ C^q_{p,p,N_{SC1}} \sum_{l=1}^{\infty} C_{\text{sum},lN_{SC2}-p} \cos[(lN_{SC2} - p)\phi + (\omega - lN_{SC2}\omega_r)t + l\pi - \varphi_{SC2} - \pi]$$

$$+ C^q_{p,p,N_{SC1}} \sum_{l=1}^{\infty} C_{\text{dif},lN_{SC2}+p} \cos[(lN_{SC2} + p)\phi - (\omega + lN_{SC2}\omega_r)t + l\pi + \varphi_{SC2}] \tag{3.50}$$

$$B = M(N_{SC2}) \left[\sum_{l=1}^{\infty} C_{\text{sum},lN_{SC1}-p} \cos\left[(lN_{SC1} - p)\phi + (\omega - lN_{SC1}\omega_r)t + l\pi - \varphi_{SC1} - \frac{\pi}{2}\right] \right]$$

$$= C_{p,N_{SC2}} \sum_{l=1}^{\infty} C_{\text{sum},lN_{SC1}-p} \cos\left[(lN_{SC1} - p)\phi + (\omega - lN_{SC1}\omega_r)t + l\pi - \varphi_{SC1} - \frac{\pi}{2}\right]$$

$$+ \sum_{n=1}^{\infty} C_{\text{sum},nN_{SC2}-p} \sum_{l=1}^{\infty} C_{\text{sum},lN_{SC1}-p} \cos \left\{ \begin{array}{l} nN_{SC2}(\phi - \omega_r t) + n\pi - \varphi_{SC2} - \frac{\pi}{2} \\ - \left[(lN_{SC1} - p)\phi + (\omega - lN_{SC1}\omega_r)t + l\pi - \varphi_{SC1} - \frac{\pi}{2}\right] \end{array} \right\}$$

$$+ \sum_{n=1}^{\infty} C_{\text{dif},nN_{SC2}+p} \sum_{l=1}^{\infty} C_{\text{sum},lN_{SC1}-p} \cos \left\{ \begin{array}{l} nN_{SC2}(\phi - \omega_r t) + n\pi - \varphi_{SC2} - \frac{\pi}{2} \\ - \left[(lN_{SC1} - p)\phi + (\omega - lN_{SC1}\omega_r)t + l\pi - \varphi_{SC1} - \frac{\pi}{2}\right] \end{array} \right\}$$

$$= \sum_{l=1}^{\infty} C_{p,N_{SC2}} C_{\text{sum},lN_{SC1}-p} \cos\left[lN_{SC1}(\phi - \omega_r t) - (p\phi - \omega_r t) + l\pi - \varphi_{SC1} - \frac{\pi}{2}\right]$$

$$+ \sum_{n=1}^{\infty} \sum_{l=1}^{\infty} C_{\text{sum},nN_{SC2}-p} C_{\text{sum},lN_{SC1}-p}$$

$$\cos\{(nN_{SC2} - lN_{SC1})(\phi - \omega_r t) - (p\phi - \omega t) + (n - l)\pi - \varphi_{SC1} - \varphi_{SC2} - \pi\}$$

$$+ \sum_{n=1}^{\infty} \sum_{l=1}^{\infty} C_{\text{dif},nN_{SC2}+p} C_{\text{sum},lN_{SC1}-p}$$

$$\cos\{(nN_{SC2} + lN_{SC1})(\phi - \omega_r t) + (p\phi - \omega t) + (n + l)\pi - \varphi_{SC1} - \varphi_{SC2} - \pi\} \tag{3.51}$$

$$C = M(N_{SC2}) \left[\sum_{l=1}^{\infty} C_{\text{dif},lN_{SC1}+p} \cos\left[(lN_{SC1} + p)\phi - (\omega + lN_{SC1}\omega_r)t + l\pi - \varphi_{SC1} - \frac{\pi}{2}\right] \right]$$

$$= \sum_{l=1}^{\infty} C_{p,N_{SC2}} C_{\text{dif},lN_{SC1}+p} \cos\left[lN_{SC1}(\phi - \omega_r t) + (p\phi - \omega_r t) + l\pi - \varphi_{SC1} - \frac{\pi}{2}\right]$$

$$+ \sum_{n=1}^{\infty} \sum_{l=1}^{\infty} C_{\text{sum},nN_{SC2}-p} C_{\text{dif},lN_{SC1}+p}$$

$$\cos\{(nN_{SC2} - lN_{SC1})(\phi - \omega_r t) + (p\phi - \omega t) + (n - l)\pi - \varphi_{SC1} - \varphi_{SC2} - \pi\}$$

$$+ \sum_{n=1}^{\infty} \sum_{l=1}^{\infty} C_{\text{dif},nN_{SC2}+p} C_{\text{dif},lN_{SC1}+p}$$

$$\cos\{(nN_{SC2} + lN_{SC1})(\phi - \omega_r t) - (p\phi - \omega t) + (n + l)\pi - \varphi_{SC1} - \varphi_{SC2} - \pi\} \tag{3.52}$$

Assuming that the source MMF is modulated by $M(N_{SC2},\gamma_2)[.]$ and then by $M(N_{SC1},\gamma_1)[.]$, the modulated MMF can be expressed as:

$$M(N_{SC1})[M(N_{SC2})[\cos(p\phi - \omega t)]]$$

$$= M(N_{SC1}) \left[C^d_{p,p,N_{SC2}} \cos(p\phi - \omega t) + C^q_{p,p,N_{SC2}} \cos\left(p\phi - \omega t + \frac{\pi}{2}\right) \right]$$

$$+ M(N_{SC1})\left[\sum_{l=1}^{\infty} C_{\text{sum},lN_{SC2}-p}\cos\left[(lN_{SC2}-p)\phi+(\omega-lN_{SC2}\omega_r)t+l\pi-\varphi_{SC2}-\frac{\pi}{2}\right]\right]$$
$$+ M(N_{SC1})\left[\sum_{l=1}^{\infty} C_{\text{dif},lN_{SC2}+p}\cos\left[(lN_{SC2}+p)\phi-(\omega+lN_{SC2}\omega_r)t+l\pi+\varphi_{SC2}-\frac{\pi}{2}\right]\right]$$
$$= D+E+F \tag{3.53}$$

where

$$\begin{aligned}
D &= M(N_{SC1})\left[C^d_{p,p,N_{SC2}}\cos(p\phi-\omega t)+C^q_{p,p,N_{SC2}}\cos\left(p\phi-\omega t+\frac{\pi}{2}\right)\right]\\
&= C^d_{p,p,N_{SC2}}\left[C^d_{p,p,N_{SC1}}\cos(p\phi-\omega t)+C^q_{p,p,N_{SC1}}\cos\left(p\phi-\omega t+\frac{\pi}{2}\right)\right]\\
&\quad + C^d_{p,p,N_{SC2}}\sum_{l=1}^{\infty}C_{\text{sum},lN_{SC1}-p}\cos\left[(lN_{SC1}-p)\phi+(\omega-lN_{SC1}\omega_r)t+l\pi-\varphi_{SC1}-\frac{\pi}{2}\right]\\
&\quad + C^d_{p,p,N_{SC2}}\sum_{l=1}^{\infty}C_{\text{dif},lN_{SC1}+p}\cos\left[(lN_{SC1}+p)\phi-(\omega+lN_{SC1}\omega_r)t+l\pi+\varphi_{SC1}-\frac{\pi}{2}\right]\\
&\quad + C^q_{p,p,N_{SC2}}\left[C^d_{p,p,N_{SC1}}\cos\left(p\phi-\omega t+\frac{\pi}{2}\right)+C^q_{p,p,N_{SC1}}\cos(p\phi-\omega t+\pi)\right]\\
&\quad + C^q_{p,p,N_{SC2}}\sum_{l=1}^{\infty}C_{\text{sum},lN_{SC1}-p}\cos[(lN_{SC1}-p)\phi+(\omega-lN_{SC1}\omega_r)t+l\pi-\varphi_{SC1}-\pi]\\
&\quad + C^q_{p,p,N_{SC2}}\sum_{l=1}^{\infty}C_{\text{dif},lN_{SC1}+p}\cos[(lN_{SC1}+p)\phi-(\omega+lN_{SC1}\omega_r)t+l\pi+\varphi_{SC1}]
\end{aligned} \tag{3.54}$$

$$\begin{aligned}
E &= M(N_{SC1})\left[\sum_{l=1}^{\infty}C_{\text{sum},lN_{SC2}-p}\cos\left[(lN_{SC2}-p)\phi+(\omega-lN_{SC2}\omega_r)t+l\pi-\varphi_{SC2}-\frac{\pi}{2}\right]\right]\\
&= \sum_{l=1}^{\infty}C_{p,N_{SC1}}C_{\text{sum},lN_{SC2}-p}\cos\left[(lN_{SC2}-p)\phi+(\omega-lN_{SC2}\omega_r)t+l\pi-\varphi_{SC2}-\frac{\pi}{2}\right]\\
&\quad + \sum_{n=1}^{\infty}C_{\text{sum},nN_{SC1}-p}\sum_{l=1}^{\infty}C_{\text{sum},lN_{SC2}-p}\\
&\quad \cos\left\{\begin{array}{l} nN_{SC1}(\phi-\omega_r t)+n\pi-\varphi_{SC1}-\frac{\pi}{2}\\ -\left[(lN_{SC2}-p)\phi+(\omega-lN_{SC2}\omega_r)t+l\pi-\varphi_{SC2}-\frac{\pi}{2}\right]\end{array}\right\}\\
&\quad + \sum_{n=1}^{\infty}C_{\text{dif},nN_{SC1}+p}\sum_{l=1}^{\infty}C_{\text{sum},lN_{SC2}-p}\\
&\quad \cos\left\{\begin{array}{l} nN_{SC1}(\phi-\omega_r t)+n\pi-\varphi_{SC1}-\frac{\pi}{2}\\ -\left[(lN_{SC2}-p)\phi+(\omega-lN_{SC2}\omega_r)t+l\pi-\varphi_{SC2}-\frac{\pi}{2}\right]\end{array}\right\}\\
&= \sum_{l=1}^{\infty}C_{p,N_{SC1}}C_{\text{sum},lN_{SC2}-p}\cos\left[lN_{SC2}(\phi-\omega_r t)-(p\phi-\omega_r t)+l\pi-\varphi_{SC2}-\frac{\pi}{2}\right]\\
&\quad + \sum_{n=1}^{\infty}\sum_{l=1}^{\infty}C_{\text{sum},nN_{SC1}-p}C_{\text{sum},lN_{SC2}-p}\\
&\quad \cos\{(nN_{SC1}-lN_{SC2})(\phi-\omega_r t)-(p\phi-\omega t)+(n-l)\pi-\varphi_{SC2}-\varphi_{SC1}-\pi\}
\end{aligned}$$

$$+\sum_{n=1}^{\infty}\sum_{l=1}^{\infty} C_{\mathrm{dif},nN_{SC1}+p} C_{\mathrm{sum},lN_{SC2}-p}$$
$$\cos\{(nN_{SC1}+lN_{SC2})(\phi-\omega_r t)+(p\phi-\omega t)+(n+l)\pi-\varphi_{SC2}-\varphi_{SC1}-\pi\} \tag{3.55}$$

$$F = M(N_{SC1})\left[\sum_{l=1}^{\infty} C_{\mathrm{dif},lN_{SC2}+p}\cos\left[(lN_{SC2}+p)\phi-(\omega+lN_{SC2}\omega_r)t+l\pi-\varphi_{SC2}-\frac{\pi}{2}\right]\right]$$
$$=\sum_{l=1}^{\infty} C_{p,N_{SC1}} C_{\mathrm{dif},lN_{SC2}+p}\cos\left[lN_{SC2}(\phi-\omega_r t)+(p\phi-\omega_r t)+l\pi-\varphi_{SC2}-\frac{\pi}{2}\right]$$
$$+\sum_{n=1}^{\infty}\sum_{l=1}^{\infty} C_{\mathrm{sum},nN_{SC1}-p} C_{\mathrm{dif},lN_{SC2}+p}$$
$$\cos\{(nN_{SC1}-lN_{SC2})(\phi-\omega_r t)+(p\phi-\omega t)+(n-l)\pi-\varphi_{SC2}-\varphi_{SC1}-\pi\}$$
$$+\sum_{n=1}^{\infty}\sum_{l=1}^{\infty} C_{\mathrm{dif},nN_{SC1}+p} C_{\mathrm{dif},lN_{SC2}+p}$$
$$\cos\{(nN_{SC1}+lN_{SC2})(\phi-\omega_r t)-(p\phi-\omega t)+(n+l)\pi-\varphi_{SC2}-\varphi_{SC1}-\pi\} \tag{3.56}$$

After simplification, it is obvious that $A + B + C = D + E + F$, so the short-circuited coil modulator satisfies the commutative law, and the derivation processes of the salient reluctance pole and the multilayer flux barrier modulator are similar.

(3) Differentiable Property

The modulated MMF can be regarded as the superposition of a series of trigonometric functions, and each trigonometric function is differentiable, so the source MMF is still differentiable after modulation.

When the differential operator directly acts on the airgap MMF modulated by short-circuited coils, the result can be expressed as:

$$\frac{d}{d\phi}M(N_{SC})[\cos(p\phi-\omega t)]$$
$$=\frac{d}{d\phi}\begin{bmatrix} C^d_{p,p}\cos(p\phi-\omega t)+C^q_{p,p}\cos\left(p\phi-\omega t+\frac{\pi}{2}\right) \\ +\sum_{l=1}^{\infty} C_{\mathrm{sum},lN_{SC}-p}\cos\left[(lN_{SC}-p)\phi+(\omega-lN_{SC}\omega_r)t+l\pi-\varphi-\frac{\pi}{2}\right] \\ +\sum_{l=1}^{\infty} C_{\mathrm{dif},lN_{SC}+p}\cos\left[(lN_{SC}+p)\phi-(\omega+lN_{SC}\omega_r)t+l\pi+\varphi-\frac{\pi}{2}\right] \end{bmatrix}$$
$$=C^d_{p,p}\left[\frac{d}{d\phi}\cos(p\phi-\omega t)\right]+C^q_{p,p}\left[\frac{d}{d\phi}\cos\left(p\phi-\omega t+\frac{\pi}{2}\right)\right]$$
$$+\sum_{l=1}^{\infty} C_{\mathrm{sum},lN_{SC}-p}\frac{d}{d\phi}\cos\left[(lN_{SC}-p)\phi+(\omega-lN_{SC}\omega_r)t+l\pi-\varphi-\frac{\pi}{2}\right]$$

$$
\begin{aligned}
&+\sum_{l=1}^{\infty} C_{\mathrm{dif},lN_{SC}+p} \frac{d}{d\phi} \cos\left[(lN_{SC}+p)\phi-(\omega+lN_{SC}\omega_r)t+l\pi+\varphi-\frac{\pi}{2}\right]\\
&= C_{p,p}^{d}[-p\sin(p\phi-\omega t)]+C_{p,p}^{q}\left[-p\sin\left(p\phi-\omega t+\frac{\pi}{2}\right)\right]\\
&+\sum_{l=1}^{\infty} C_{\mathrm{sum},lN_{SC}-p}(p-lN_{SC})\sin\left[(lN_{SC}-p)\phi+(\omega-lN_{SC}\omega_r)t+l\pi-\varphi-\frac{\pi}{2}\right]\\
&+\sum_{l=1}^{\infty} C_{\mathrm{dif},lN_{SC}+p}(-lN_{SC}-p)\sin\left[(lN_{SC}+p)\phi-(\omega+lN_{SC}\omega_r)t+l\pi+\varphi-\frac{\pi}{2}\right]
\end{aligned}
\tag{3.57}
$$

It can be concluded the source MMF is still differentiable after modulation, and the same conclusion is applicable to salient reluctance pole and multilayer flux barrier modulators.

(4) Power Operation

In addition, some modulators also satisfy the power operation property when the source MMF is stationary relative to the modulator and the modulation behavior does not change the number of principal pole pairs of the MMF distribution, which means the result of the same modulation operator acting k times is the same as that of acting once. Multilayer flux barrier and short-circuited coil modulators with zero resistance fall into this category. In salient reluctance pole modulators and short-circuited coil modulators with non-zero resistances, this property does not hold. When the modulation behavior changes the number of principal pole pairs of the MMF distribution, none of the three types of modulators satisfy the power law any more.

5 Modulation Factor

The airgap magnetic field modulation behavior changes the position and amplitude of the spectral lines of the periodic source MMF signal in the frequency domain, which in turn causes the power contained in the modulated airgap MMF signal to change. A modulation factor (MF) can be defined as the ratio of the power contained in the effective MMF component to the total power contained in the source MMF to quantify the modulation efficiency, namely:

$$
MF=\sqrt{\sum_{\mu\in E}|F_\mu|^2 \Big/ \left\{\frac{1}{2\pi}\int_{-\pi}^{\pi}[f(x,t)]^2\,dx\right\}}
\tag{3.58}
$$

where $\boldsymbol{E}$ is the set of harmonic orders corresponding to the effective components in the airgap MMF after modulation. It is obvious that the MF of the unit modulation operator is 1.

3.4.3 Filter

Based on the coordinate system shown in Figure 3.14, the induced EMF in any conductor in the phase winding can be obtained by:

$$\begin{aligned} E_c(\phi) &= \int_b^a (-\vec{v} \times \vec{B})\, d\vec{l} \\ &= r_g \omega_r l_{stk} B(-\phi) * S(-\phi + \phi_0) \\ &= \omega_r \mu_0 \frac{r_g l_{stk}}{g} M[F_g(-\phi)] * S(-\phi + \phi_0) \end{aligned} \tag{3.59}$$

where $S(\xi)$ is the relative inclination function. Its independent variable is ξ, and $\xi = \theta_m + \phi - \phi_0$. "*" stands for convolution operation. The relative inclination function is used to describe the influence of the relative deviation of the slot conductor and the airgap magnetic field distribution along the axial direction on the induced MMF in radial-flux machines. The expression of $S(\xi)$ is:

$$S(\xi) = \begin{cases} \dfrac{1}{l_{stk}} \dfrac{dl_z}{d\xi}, & \xi_b \le \xi \le \xi_a \\ 0, & \text{Otherwise} \end{cases} \tag{3.60}$$

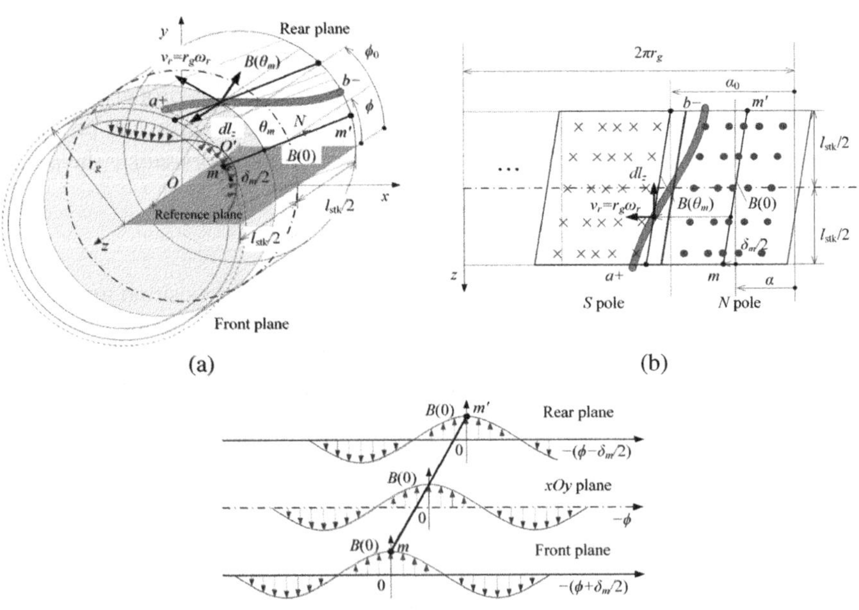

Figure 3.14 Derivation of induced EMF of a phase winding, (a) coordinate system, (b) unrolled view of airgap, (c) cut-planes in axial direction.

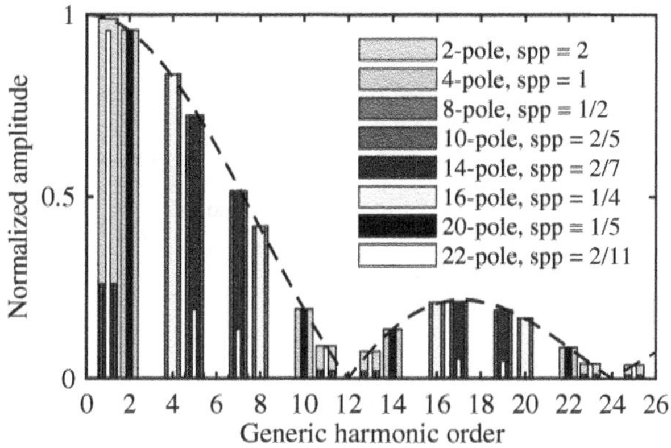

Figure 3.15 Amplitude spectrums of all the possibilities of three-phase 12-slot windings.

The induced EMF in the entire phase winding is:

$$E_w(\phi) = E_c(\phi) * C(\phi) \tag{3.61}$$

Equations (3.59) and (3.61) are spatial convolution operations, corresponding to the product in the spatial frequency domain. These two equations show that the generation of induced EMF is a filtering process. The winding can be regarded as a spatial harmonic filter described by a conductor distribution function, and performs filtering functionality through skewed slots (poles) or winding configurations (conductor distributions and connections).

The magnetic flux linkage linked by windings can be obtained by integrating the product of the winding function and airgap magnetic flux density distribution. The winding configuration determines the frequency selection characteristics of the filter. Figure 3.15 shows the amplitude spectrums of all available balanced 12-slot three-phase windings.

In Figure 3.15, q is the number of slots per pole per phase. It can be seen that the envelope is the absolute value of the sinc function. In digital signal processing, the sinc function is the Fourier transform of the symmetric periodic square wave signal, that is, the time domain expression of the frequency domain low-pass filter [24], indicating the filtering functionality of the winding.

3.4.4 Unified Airgap Model

By introducing modulation operators, physical airgaps with short-circuited coils, salient reluctance poles or multilayer flux barriers can be equivalent to a uniform airgap, as shown in Figure 3.16, where the modulated MMF $M[F_f(\phi,t)]$ and

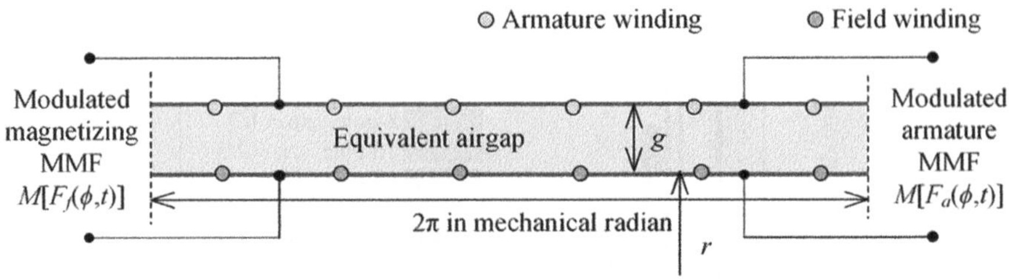

Figure 3.16 Equivalent airgap model.

$M[F_a(\phi,t)]$ can be expressed as:

$$M[F_f(\phi,t)] = M_1[M_2[\cdots[M_n[F_f(\phi,t)]]]] \tag{3.62}$$

$$M[F_a(\phi,t)] = M_1[M_2[\cdots[M_n[F_a(\phi,t)]]]] \tag{3.63}$$

where $F_f(\phi,t)$ is the source excitation MMF, $F_a(\phi,t)$ the source armature MMF, and:

$$F_f(\varphi,t) = \sum_{j=1}^{m}[W_{fj}(\varphi,t)\times i_{fj}(t)] \tag{3.64}$$

$$F_a(\varphi,t) = \sum_{j=1}^{m}[W_{aj}(\varphi,t)\times i_{aj}(t)] \tag{3.65}$$

where W_f and W_a are the winding functions of the field winding and armature winding, and i_f and i_a are the field current and armature current, respectively.

The airgap magnetic flux density distribution is:

$$B_g(\phi,t) = \frac{\mu_0}{g}\{M[F_f(\phi,t)] + M[F_a(\phi,t)]\} \tag{3.66}$$

The magnetic flux density distribution under no-load condition is:

$$B_f(\phi,t) = \frac{\mu_0}{g}M[F_f(\phi,t)] \tag{3.67}$$

The no-load flux linkage linked by each phase of the armature winding is:

$$\psi_{aj}(t) = \frac{\mu_0 r_g l_{\text{stk}}}{g}\int M[F_f(\phi,t)]\times W_{aj}(\phi,t)d\phi \tag{3.68}$$

The winding inductance is:

$$L_{ij} = k_{\text{sat}}\frac{\mu_0}{g}r_g l_{\text{stk}}\int_0^{2\pi} W_i(\phi)M[W_j(\phi)]\,d\phi \tag{3.69}$$

where k_{sat} is the saturation coefficient. When the core permeability is infinite, $k_{\text{sat}} = 1$. $W_i(\phi)$ and $W_j(\phi)$ are the winding functions of the i-th winding and the j-th winding (field winding and/or armature winding) respectively. When $i = j$, the inductance calculated according to (3.69) is the winding self-inductance; when $i \neq j$, it is the mutual inductance between the i-th winding and the j-th winding.

The no-load EMF of each phase generated in the armature winding can be obtained by:

$$E_{aj}(t) = -\frac{d}{dt}\left[\frac{\mu_0 r_g l_{\text{stk}}}{g}\int M[F_f(\phi,t)] \times W_{aj}(\phi,t)d\phi\right] \tag{3.70}$$

The expression of electromagnetic torque can be derived from the partial differential of magnetic energy and magnetic co-energy to the angle of rotation. The magnetic energy takes the flux linkage and the position angle of the rotor as the independent variables, while the magnetic co-energy takes the current and the rotor position angle as the independent variables [25]. In order to simplify the torque expression, the electromagnetic torque is derived here from the magnetic co-energy.

Using the airgap magnetic field as the analysis object, the magnetic co-energy $W_m{}'$ in the airgap is:

$$\begin{aligned} W'_m &= \int_V \frac{[B_g(\phi,t)]^2}{2\mu_0}dv \\ &= \int_V \left(\frac{\mu_0}{g}\right)^2 \frac{\{M[F_f(\phi,t)] + M[F_a(\phi,t)]\}^2}{2\mu_0}dv \\ &= \int_0^{2\pi} \left(\frac{\mu_0}{g}\right)^2 \frac{\{M[F_f(\phi,t)] + M[F_a(\phi,t)]\}^2}{2\mu_0} g l_{stk} r_g d\phi \\ &= \frac{\mu_0 r_g l_{\text{stk}}}{2g}\int_0^{2\pi} \{M[F_f(\phi,t)] + M[F_a(\phi,t)]\}^2 d\phi \end{aligned} \tag{3.71}$$

Using the basic relationship between torque and magnetic co-energy and keeping the winding current constant, that is, the magnitude of the MMF is constant, making a differential virtual displacement Δ to the rotor yields the electromagnetic torque expression:

$$T_{em}(t) = \frac{\mu_0 r_g l_{\text{stk}}}{2g}\frac{\partial}{\partial \Delta}\int_0^{2\pi} \{M[F_f(\phi,t)] + M[F_a(\phi,t)]\}^2 d\phi \tag{3.72}$$

If the motor winding is taken as the analysis object, for a machine with n windings, the magnetic co-energy of the machine is expressed by the current and the inductance calculated by Eq. (3.69), and the magnetic co-energy $W_m{}'$ is:

$$\begin{aligned} W'_m(i_1, i_2, \ldots, i_k, \ldots, i_n) = &\int_0^{i_1} \psi_1\left(i'_1, 0, \ldots, 0, \phi\right) di'_1 \\ &+ \int_0^{i_2} \psi_2\left(i_1, i'_2, 0, \ldots, 0, \phi\right) di'_2 + \ldots \\ &+ \int_0^{i_k} \psi_k\left(i_1, i_2, \ldots, i'_k, \ldots, 0, \phi\right) di'_k + \ldots \\ &+ \int_0^{i_n} \psi_n\left(i_1, i_2, \ldots, i'_n, \phi\right) di'_n \end{aligned} \tag{3.73}$$

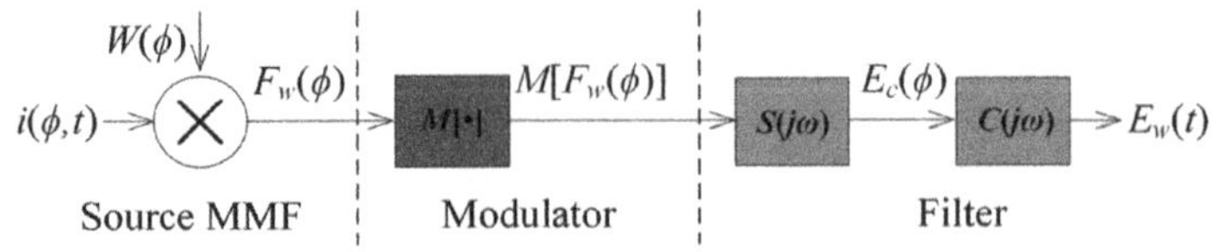

Figure 3.17 Energy flow of electrical machines.

Using the basic relationship between torque and magnetic co-energy, and keeping the winding current constant, making a differential virtual displacement Δ to the rotor can also yield the electromagnetic torque expression. The electromagnetic torque derived from the case where the machine winding is treated as the analysis object usually includes many items. It is necessary to make full use of the symmetry of winding inductances and currents to simplify the electromagnetic torque expression. Taking a three-phase 12/10 FSPM machine as an example, after replacing the PM array with a DC winding, the magnetic co-energy expression will have four integral terms, namely $k = 1, 2, 3, 4$, where the DC winding is the first winding, and the three-phase AC windings are the second, third, and fourth windings.

The energy flow diagram of electrical machines is shown in Figure 3.17. It is shown that the performance of an electrical machine mainly depends on three key elements: the excitation source (source MMF), modulator and filter (armature winding). $W(\phi)$ and $D(j\omega)$ can be for the same winding or different windings. If the MMF is produced by PM arrays instead of windings carrying currents, the flow diagram will start directly from $\mathcal{F}_w(\phi, t)$.

From the perspective of electromechanical energy conversion, at least one of the three parts is moving, so for a standard machine that includes an excitation source, a modulator and a filter, there will be $(2^3 - 1)$ possibilities in total. The modulator may be synchronized with the source excitation (or armature) MMF, or it may move relative to the source excitation (or armature) MMF. The former is called static modulation and the latter is called dynamic modulation. Static modulation can be regarded as a special case of dynamic modulation, that is, the relative velocity between the source MMF and the modulator is zero. These two modulation behaviors can be observed simultaneously in the traditional and new machine topologies.

3.4.5 Duality Between Electrical Machines and Switching Power Converters

From the derivation of those basic relations in spatial and spatial frequency domains, it can be concluded that the airgap magnetic field modulation theory applies to all electrical machines. Due to the nonlinearity of ferromagnetic materials, the accurate calculation of airgap flux density distribution from

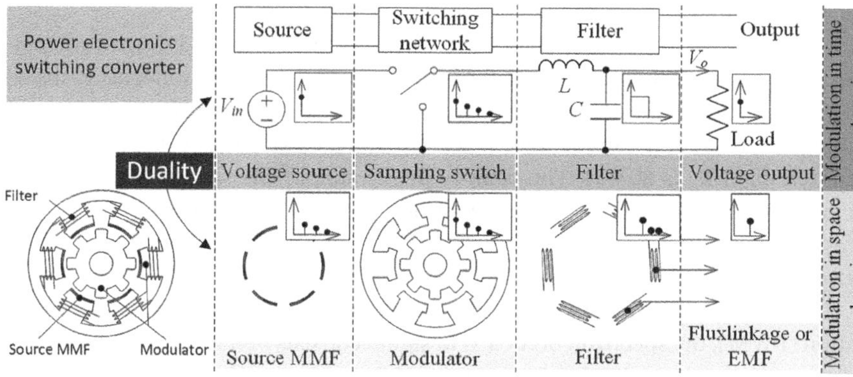

Figure 3.18 Duality of the electrical machine and switching converter.

the MMF produced by windings or PM arrays cannot always be simplified to one-dimensional problems. In such cases, the airgap flux density distribution should be solved with numerical methods, such as the magnetic equivalent circuit model [26], subdomain model [27], or FE method.

To better understand the chained three key elements, a duality between electrical machines and power electronics switching converters [28] is developed [29, 30], as illustrated in Figure 3.18. This similarity has solid mathematical foundations. References [31] and [32] deduced in detail the relationships between the induced EMF/torque ripple expressions and the arrangement of PMs/winding configurations in PMSMs from the spatial frequency domain, revealing the filtering effect of machine geometry on airgap flux density space harmonics. The difference is that the switching converter handles modulation of time-domain signals, while the electrical machine is related to modulation of spatial-domain signals.

Again we can see the source MMF provides the initial MMF distribution with possible space harmonics along the airgap peripheral. The short-circuited coils/salient reluctance poles/flux barriers modulate the initial MMF distribution to produce a spectrum of MMF harmonic components, and are therefore known as modulators. All the MMF components produce flux density components of the same pole numbers in the airgap. The armature winding plays the role of a spatial filter selecting certain airgap flux density harmonics to produce flux linkages and EMFs.

In generating mode, the armature winding serves as a voltage source and feeds the electric load. In motoring mode, the armature winding receives currents synchronous to the EMF to produce mechanical power output. The electromagnetic torque can then be derived from the airgap flux density using the virtual work principle, which will be shown in Section 3.5.

1 Voltage Mode Controlled Buck Converter

A 1 MHz voltage mode-controlled Buck converter modeled in [33] was studied to illustrate the modulation process in switching mode power converters, as shown in Figure 3.19. In this converter, the input voltage V_{in} is 12 V, the desired output voltage V_o is 1.2 V, the inductance L is 200 nH, the capacitance C is 1 mF, and the resistive load R_o is 80 mΩ. The peak-to-peak value of the sawtooth ramp is 1 V. A sinusoidal perturbation with a magnitude of 40 mV and frequency of 990 kHz was considered at the control voltage V_c, which was set to 0.1 V to achieve the 1.2 V output. The switching frequency is 1 MHz.

It has been shown that due to the sampling controlled by the pulse width modulation (PWM), the spectrum of the PWM signal consists of the DC component, the components at the switching frequency and its integer multiples (f_s, $2f_s$, $3f_s$, ...), and sideband frequencies ($f_s \pm f_p$, $2f_s \pm f_p$, $3f_s \pm f_p$, ...). The LC filter used in this example has a resonance frequency of 11.25 kHz and voltage components higher than this frequency are attenuated quickly. As a consequence, the output voltage V_o mainly contains the DC component and AC component whose frequency is $f_s - f_p = 1\,\text{MHz} - 990\,\text{kHz} = 10\,\text{kHz}$. The amplitudes of the DC component and AC component are approx. $12\,\text{V} \times 0.1 = 1.2\,\text{V}$ and $12\,\text{V} \times 0.04 = 0.48\,\text{V}$, respectively.

2 FRPM Machine

A FRPM machine with six stator teeth and eight rotor salient reluctance poles was analyzed here as an example to illustrate the modulation process in electric machines, as shown in Figure 3.20. The specifications of the FRPM machine are tabulated in Table 3.2. The source MMF created by the PM array with closed stator slots and a cylindrical rotor is a square wave with a cycle of $2\pi/6$. Its spectrum contains the fundamental component and odd harmonics.

By considering the wide stator slot opening, odd harmonics are eliminated and even harmonics emerge. When the rotor salient reluctance poles are further taken into account, sideband harmonics whose generic harmonic order, i.e. numbers of pole pairs, are $|6 \pm 8|$, 12 ± 8, 24 ± 8, ..., are introduced into the spectrum of the airgap MMF. The three-phase stator winding has a number of pole pairs of two and its number of slots per pole pair phase is 1/2. The winding factor for the four-pole field harmonic is 0.8660 and this magnetic field harmonic contributes to most of the flux linkage in the stator winding. Other field harmonics, for instance, the $6k$-pole field harmonics, are attenuated quickly due to the low winding factors, when k is a positive integer.

3 Dual Objects and Parameters

By comparing the modulation process in the voltage mode controlled buck converter and the FRPM machine, the dual objects and parameters can be identified, as tabulated in Table 3.3.

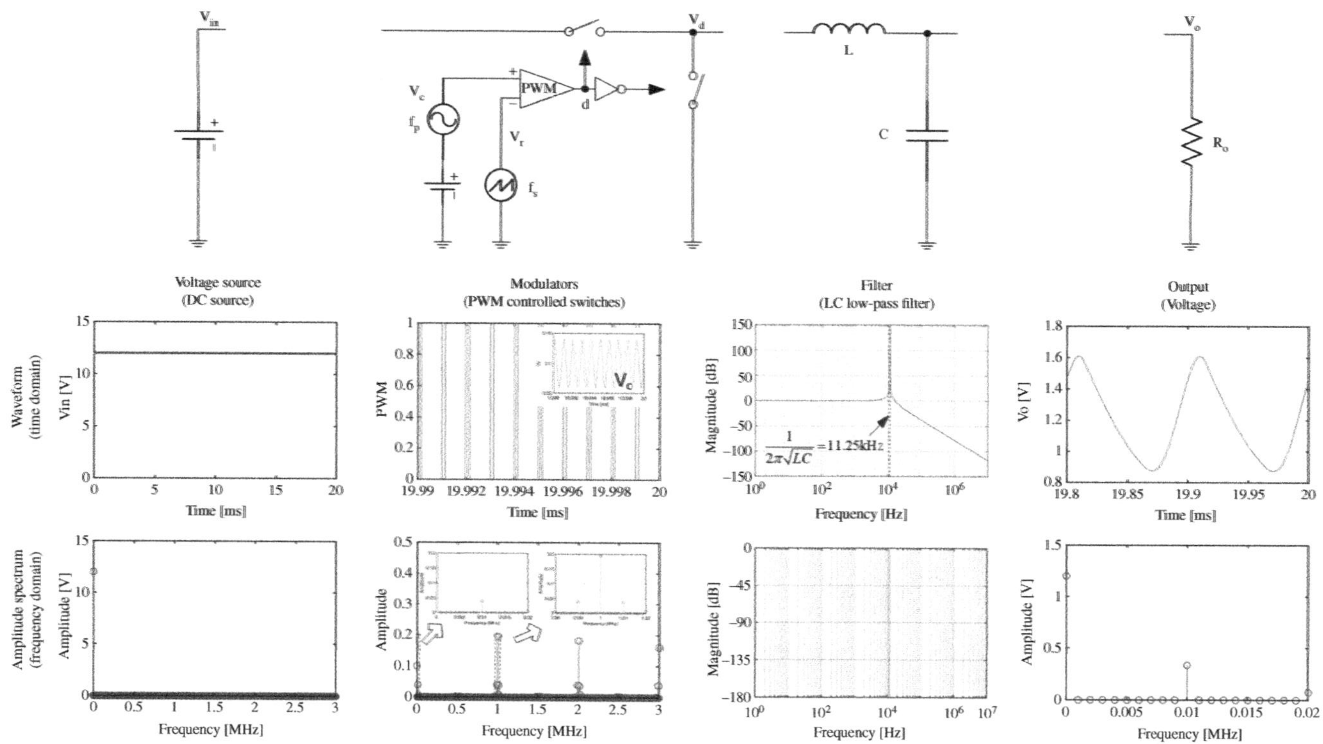

Figure 3.19 Modulation process in a 1 MHz voltage mode controlled Buck converter. The simulation was carried out in MATLAB/Simulink with a time step of 10 ns.

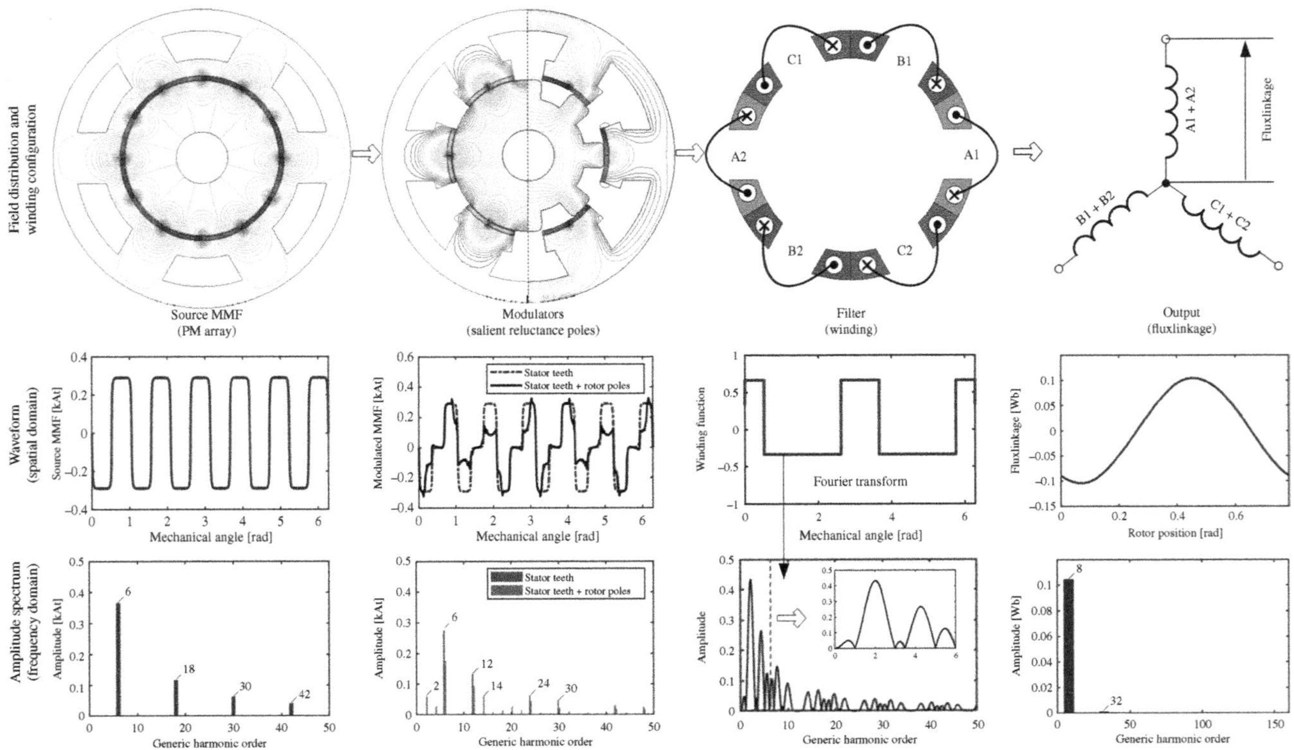

Figure 3.20 Modulation process in an FRPM machine with six stator teeth and eight rotor salient reluctance poles. The simulation was carried out in electromagnetic FE software.

Table 3.2 Specifications of the FRPM prototype.

Stator outer/inner diameter (mm)		128/70.4
Airgap/PM/stack length (mm)		0.35/1.84/75
Number of stator teeth/rotor salient poles		6/8
PM	Remanence at 20 °C (T)	1.2
	Relative permeability	1.05

Table 3.3 Dual objects and parameters between switching converters and electrical machines.

Dual parameters	Switching converters in voltage mode	Electrical machines using electromagnetic field
Source	Input voltage V_{in} in volts	Source MMF in ampere-turns
	Electrical frequency in Hz	Harmonic pole pair
Modulator	Perturbation frequency in Hz	Pole pairs of the source MMF
	Switching frequencies in Hz	N_{SC} (short-circuited coil)
		N_{SP} (Salient reluctance pole)
		N_{MB} (Multilayer barrier)
	Sideband frequencies	Sideband harmonic pole pairs
Filter	Filter circuit	Winding
	Filter gain	Winding factor
Output	Output voltage in volts	Flux linkage in Wb

3.5 Torque Decomposition

3.5.1 General Torque Equation

Starting from the torque Eq. (3.72), the more specific electromagnetic torque can be further derived as:

$$T_{em}(t) = T_{ff}(t) + T_{fa}(t) + T_{aa}(t) \tag{3.74}$$

where

$$T_{ff}(t) = \frac{\mu_0 r_g l_{\text{stk}}}{2g} \frac{\partial}{\partial \Delta} \int_0^{2\pi} M^2[F_f(\phi, t)]\, d\phi \tag{3.75}$$

$$T_{fa}(t) = \frac{\mu_0 r_g l_{\text{stk}}}{g} \frac{\partial}{\partial \Delta} \int_0^{2\pi} M[F_f(\phi, t)] M[F_a(\phi, t)]\, d\phi \tag{3.76}$$

$$T_{aa}(t) = \frac{\mu_0 r_g l_{\text{stk}}}{2g} \frac{\partial}{\partial \Delta} \int_0^{2\pi} M^2[F_a(\phi, t)] \, d\phi \tag{3.77}$$

T_{ff} is produced solely by the modulated field MMF. A typical example of this torque component is the cogging torque of PMSMs. This torque component does not contribute to the average electromagnetic torque and only produces periodic torque ripples. T_{fa} is produced by the interaction between the modulated field MMF and the modulated armature MMF. This component accounts for the majority of the electromagnetic torque. T_{aa} is produced solely by the modulated armature MMF. A typical example of this torque component is the effective torque in switched reluctance machines.

In most cases, the excitation source and armature winding are placed on the stator and rotor, respectively. Taking the brushed DC motor as an example, the DC field winding (excitation source) is located on the stator, and the armature winding is located on the rotor. The electromagnetic torque is the total torque acting on the rotor and it includes three components, i.e. T_{ff}, T_{fa}, and T_{aa}. T_{ff} is the torque produced by the stator MMF modulated by the stator poles and rotor teeth when the field winding is supplied and the armature winding is left open. It corresponds to the cogging torque. T_{fa} is the torque produced by the interaction between the modulated stator excitation MMF and the modulated rotor armature MMF when both the field winding and armature winding are supplied. T_{aa} is the torque produced by the modulated rotor armature MMF when the field winding is open and the armature winding is supplied.

There are also many other machine topologies whose excitation source and armature windings are placed on the same stator or rotor, such as IMs, SynRMs, and most newly-emerged machine topologies. When the excitation source and the armature winding are placed on the same stator or rotor, the electromagnetic torque also has all three components.

The stator winding of IMs provides both the field current and torque-producing current, so the field winding and armature winding are both located on the stator. The torque of an IM includes all three torque components. T_{ff} is the pulsating torque acting on the rotor when the IM has no load, and its average value in a time period is zero. T_{fa} is the torque produced by the interaction between the modulated field MMF and the modulated armature MMF when the IM is loaded. T_{aa} is the pulsating torque acting on the rotor when there is only the torque-producing current component in the stator winding.

There are two types of torque in the wound-field salient-pole SM, namely the torque component related to the rotor flux linkage and the reluctance torque component produced by the unequal d- and q-axis rotor reluctance.

The compositions of the two torque types are different. For the torque component related to the rotor flux linkage, the field winding and armature winding are placed on the rotor and stator, respectively. The electromagnetic torque is the total torque acting on the rotor, including T_{ff}, T_{af}, and T_{aa}. T_{ff} is the torque produced by the modulated rotor field MMF, that is, the cogging torque produced by applying a DC current to the rotor field winding. T_{fa} is the main torque generated by the interaction between the modulated rotor MMF and the modulated stator armature MMF. T_{aa} is the cogging torque produced solely by the modulated stator armature MMF.

For the reluctance torque component produced by the unequal reluctance of rotor d- and q-axes (the rotor field winding is open at this time), both the field current and the torque-producing current flow in the stator winding, and they correspond to the d-axis current i_d and q-axis current i_q. This also means that the stator winding plays the roles of both the field winding and armature winding. The SynRM only has the reluctance torque, and its torque composition is the same as that of the reluctance torque in wound-field salient-pole SMs.

There is no reluctance torque component in surface-mounted PMSMs, and its torque composition is similar to that of the torque component related to the rotor flux in wound-field salient-pole SMs, including all three components. Similar to this is the wound-field non-salient-pole SM. The interior PMSM is similar to the wound-field salient-pole SM, which includes not only the torque related to the rotor flux, but also the reluctance torque. Therefore, the electromagnetic torque or interior PM machines needs to be further decomposed according to the torque component related to the rotor flux and the reluctance torque component.

The BDFRM has two sets of stator windings, one of which is defined as the field winding, and the other the armature winding. The electromagnetic torque is the sum of the torque components acting on the rotor. T_{ff} is the torque produced by the modulated excitation MMF. T_{fu} is the torque produced by the interaction between the modulated excitation MMF and the modulated armature MMF. T_{aa} is the torque produced by the modulated armature MMF.

The torque composition of BDFIMs becomes very complicated due to the presence of short-circuited coils on the rotor. When the resistance of the short-circuited coil is negligible, its torque composition is similar to that of a BDFRM, including three torque components, of which T_{fa} produces a non-zero average electromagnetic torque. When the resistance of the short-circuited coil is taken into account, each set of stator windings and the rotor with short-circuited coils will form an additional IM that produces asynchronous torque. Each asynchronous torque includes all the three torque components.

Once the torque composition is clarified, the average torque can be easily obtained from the combination of equations (3.75)–(3.77). The torque equation derivation of several typical machines is given here, and equations for other machine types can be derived by similar approaches.

3.5.2 Wound-Field Salient-Pole SM

The average torque of the wound-field salient-pole SM is:

$$\begin{aligned} T_{avg\_SM} &= \frac{1}{T}\int_0^T T_{em\_SM}(t)dt \\ &= \frac{1}{T}\int_0^T T_{ff}(t)\,dt + \frac{1}{T}\int_0^T T_{fa}(t)\,dt + \frac{1}{T}\int_0^T [T_{aa}(t)]dt \end{aligned} \tag{3.78}$$

where the first term is the cogging torque produced by the field MMF modulated by the rotor salient poles and stator teeth, and the average value is zero in a full time period.

$$\frac{1}{T}\int_0^T T_{ff}(t)\,dt = \frac{1}{T}\int_0^T \left\{ \frac{\mu_0 r_g l_{\text{stk}}}{2g}\frac{\partial}{\partial\Delta}\int_0^{2\pi} M^2[F_f(\phi,t)]d\phi \right\} dt = 0 \tag{3.79}$$

Similarly, the average value of the third term is also zero. The second term is the main torque acting on the rotor produced by the interaction between the modulated field MMF and the modulated armature MMF:

$$\frac{1}{T}\int_0^T T_{fa}(t)\,dt = \frac{1}{T}\int_0^T \left\{ \frac{\mu_0 r_g l_{\text{stk}}}{g}\frac{\partial}{\partial\Delta}\int_0^{2\pi} M[F_f(\phi,t)]M[F_a(\phi,t)]\,d\phi \right\} dt \tag{3.80}$$

Considering the synchronous modulation behavior of the salient-pole rotor, and ignoring the influence of small stator slot openings, equation (3.80) can be further derived as:

$$\frac{1}{T}\int_0^T T_{fa}(t)dt = \frac{1}{T}\int_0^T \left\{ \frac{\mu_0 r_g l_{\text{stk}}}{g}\frac{\partial}{\partial\Delta}\int_0^{2\pi} F_f(\phi,t)F_a(\phi,t)\,d\phi \right\} dt \tag{3.81}$$

where the per-pole armature MMF is:

$$\begin{aligned} F_a(\phi,t) &= \left(\frac{m}{2}\right)\left(\frac{4}{\pi}\right)\left(\frac{N_{ph}k_{w1}\sqrt{2}I_{rms}}{2p}\right)\cos[p(\phi-\phi_0)-\omega t] \\ &= \frac{\sqrt{2}\mathrm{A}_a r_g}{p}\cos[p(\phi-\phi_0)-\omega t] \end{aligned} \tag{3.82}$$

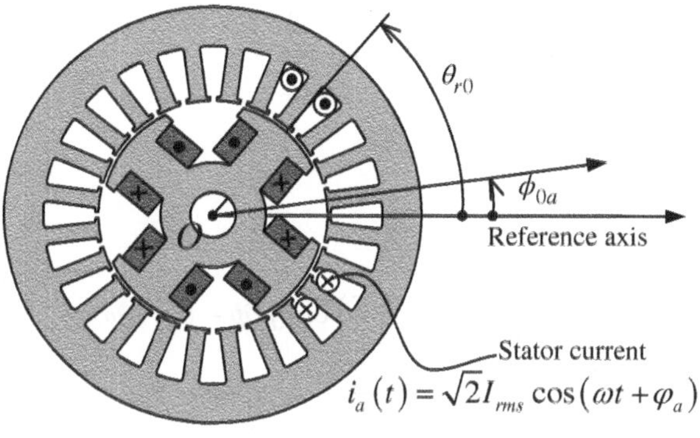

Figure 3.21 Coordinate system for wound-field salient-pole SMs.

where ϕ_{0a} is the angle from the reference axis to phase-A winding axis of the armature winding, as shown in Figure 3.21, and ω is the electrical angular frequency of the stator current.

The specific electric loading of the armature winding is:

$$\overline{A}_a = \frac{2mN_{ph}k_{w1}I_{rms}}{2\pi r_g} = \frac{mN_{ph}k_{w1}I_{rms}}{\pi r_g} \tag{3.83}$$

where m is the number of phases, N_{ph} turns in series per phase, k_{w1} the fundamental winding factor, and I_{rms} the phase current.

The no-load airgap magnetic flux density produced by the field winding is:

$$B_f(\phi, t) = \frac{\mu_0}{g}F_f(\phi, t) = B_{pk}\cos[p(\phi - \theta_{r0} - \omega_r t)] \tag{3.84}$$

where θ_{r0} is the angle from the reference axis to the center axis of the salient pole of the rotor, and ω_r is the mechanical angular velocity of the rotor. Substituting (3.82)–(3.84) into (3.81) yields:

$$\begin{aligned}
&\frac{1}{T}\int_0^T T_{fa}(t)dt \\
&= \frac{1}{T}\int_0^T \left\{ r_g l_{\text{stk}}\frac{\partial}{\partial\Delta}\int_0^{2\pi} B_f(\phi, t)\frac{\sqrt{2}\overline{A}_a r_g}{p}\cos[p(\phi - \phi_{0a}) - \omega t]\,d\phi \right\} dt \\
&= \frac{1}{T}\int_0^T \left\{ r_g^2 l_{\text{stk}}\frac{\partial}{\partial\Delta}\int_0^{2\pi} B_f(\phi, t)\frac{\sqrt{2}\overline{A}_a}{p}\cos[p(\phi - \phi_{0a}) - (\omega t + \varphi_a)]\,d\phi \right\} dt \\
&= \frac{1}{T}\int_0^T \left\{ r_g^2 l_{\text{stk}}\frac{\partial}{\partial\Delta}\int_0^{2\pi} B_{pk}\cos[p(\phi - \theta_{r0} - \omega_r t)]\frac{\sqrt{2}\overline{A}_a}{p}\cos[p(\phi - \phi_{0a}) - (\omega t + \varphi_a)]\,d\phi \right\} dt
\end{aligned} \tag{3.85}$$

Considering that the specific magnetic loading is:

$$\overline{B} = \frac{2}{\pi}B_{pk} \tag{3.86}$$

The relationship between the angular frequency of the armature winding current and the mechanical speed of the rotor is:

$$\omega = p\omega_r \tag{3.87}$$

The average torque becomes:

$$\begin{aligned}
&\frac{1}{T}\int_0^T T_{fa}(t)dt \\
&= \frac{1}{T}\int_0^T \left\{ r_g^2 l_{\text{stk}} \frac{\partial}{\partial \Delta} \int_0^{2\pi} \begin{pmatrix} \left(\frac{\pi}{2}\overline{\text{B}}\right)\cos[p(\phi-\theta_{r0}-\omega_r t)] \\ \times \frac{\sqrt{2}\overline{\text{A}}_a}{p}\cos[p(\phi-\phi_{0a})-(\omega t+\varphi_a)] \end{pmatrix} d\phi \right\} dt \\
&= \frac{1}{T}\int_0^T \left(\pi r_g^2 l_{\text{stk}}\right) \overline{\text{B}}\overline{\text{A}}_a \left(\frac{1}{p}\right) \\
&\quad \times \frac{\partial}{\partial \Delta}\left\{ \frac{1}{\sqrt{2}} \int_0^{2\pi} \begin{matrix} \cos[p(\phi-\theta_{r0}-\omega_r t)] \\ \times\cos[p(\phi-\phi_{0a})-(\omega t+\varphi_a)]d\phi \end{matrix} \right\} dt \\
&= \frac{1}{T}\int_0^T \left(\pi r_g^2 l_{\text{stk}}\right) \overline{\text{B}}\overline{\text{A}}_a \left(\frac{1}{p}\right) \frac{\partial}{\partial \Delta}\left\{ \frac{\pi}{\sqrt{2}}\cos[p(\phi_{0a}-\theta_{r0})+\varphi_a] \right\} dt \\
&= \left(\pi r_g^2 l_{\text{stk}}\right)\left(\frac{\pi}{\sqrt{2}}\overline{\text{B}}\right)\overline{\text{A}}_a\left(\frac{1}{p}\right)\frac{\partial}{\partial \Delta}\cos\Gamma
\end{aligned} \tag{3.88}$$

where Γ is the torque angle, and its expression is:

$$\Gamma = p\phi_{0a} + \varphi_a - p\theta_{r0} \tag{3.89}$$

Keeping the current constant and making a virtual displacement to the initial position of the rotor, the electromagnetic torque can be obtained:

$$\begin{aligned}
\frac{1}{T}\int_0^T T_{af}(t)dt &= \left(\pi r_g^2 l_{\text{stk}}\right)\left(\frac{\pi}{\sqrt{2}}\overline{\text{B}}\right)\overline{\text{A}}_a\left(\frac{1}{p}\right)\frac{d}{d\theta_{r0}}\cos\Gamma \\
&= \left(\pi r_g^2 l_{\text{stk}}\right)\left(\frac{\pi}{\sqrt{2}}\overline{\text{B}}\right)\overline{\text{A}}_a\sin\Gamma
\end{aligned} \tag{3.90}$$

It should be noted that the unit of the linear current density distribution function is A/rad, and the unit of the specific electric loading is A/m, and there is an airgap radius r_g between the two. Finally, when using $i_d = 0$ control, the torque per active rotor volume of SMs is:

$$TRV_{\text{SM}} = \frac{T_{avg\_SM}}{\pi r_g^2 l_{\text{stk}}} = \frac{\pi}{\sqrt{2}}\overline{\text{B}}\overline{\text{A}}_a \tag{3.91}$$

where $\overline{\text{B}}$ is the specific magnetic loading corresponding to the fundamental of the space harmonic (average value, $2/\pi$ times the peak of sine), and $\overline{\text{A}}_a$ is the specific electric loading corresponding to the fundamental space harmonic.

3.5.3 SynRM

Both the field winding and armature winding SynRMs are located on the stator, and the average torque is:

$$\begin{aligned} T_{avg\_SynRM} &= \frac{1}{T}\int_0^T T_{em\_SynRM}(t)dt \\ &= \frac{1}{T}\int_0^T T_{ff}(t)dt + \frac{1}{T}\int_0^T T_{fa}(t)dt + \frac{1}{T}\int_0^T T_{aa}(t)dt \end{aligned} \tag{3.92}$$

where the first term is the cogging torque produced by the modulated excitation MMF, and its average value in a full time period is zero. The third term is the cogging torque produced by the modulated armature MMF. The average torque is finally produced by the interaction between the modulated field MMF and the modulated armature MMF:

$$T_{avg\_SynRM} = \frac{1}{T}\int_0^T T_{fa}(t)\, dt \tag{3.93}$$

Similarly, when using $i_d = 0$ control, the torque per active rotor volume is:

$$TRV_{\text{SynRM}} = \frac{T_{avg\_SynRM}}{\pi r_g^2 l_{\text{stk}}} = \frac{\pi}{\sqrt{2}}\overline{\text{B}}\,\overline{\text{A}}_a \tag{3.94}$$

The torque density Eq. (3.94) is the same as (3.91), which is for wound-field salient-pole SMs. It is worth noting that the radial component of the airgap magnetic field of the SynRM is created by the field current component in the stator winding, while the tangential component is established by the armature current component. According to the definitions of specific magnetic loading and specific electric loading in traditional electric machinery, the specific magnetic and electric loadings of SynRMs are coupled.

In the developed general airgap field modulation theory, the specific magnetic loading corresponds to the field (or excitation) current component in the field winding and the specific electric loading corresponds to the armature current component in the armature winding. In addition, the specific magnetic and electric loadings are decoupled.

3.5.4 Squirrel-Cage IM

The average torque of the squirrel-cage IM is:

$$\begin{aligned} T_{avg\_IM} &= \frac{1}{T}\int_0^T T_{em\_IM}(t)dt \\ &= \frac{1}{T}\int_0^T T_{ff}(t)dt + \frac{1}{T}\int_0^T T_{fa}(t)dt + \frac{1}{T}\int_0^T T_{aa}(t)dt \end{aligned} \tag{3.95}$$

where the first term is the cogging torque produced by the modulated field MMF, and its average value in a full time period is zero. The third term is the cogging torque produced by the modulated armature MMF. The average torque is finally produced by the interaction between the modulated field MMF and the modulated armature MMF. Similarly, the torque per active rotor volume is:

$$TRV_{\mathrm{IM}} = \frac{T_{avg\_\mathrm{IM}}}{\pi r_g^2 l_{\mathrm{stk}}} = \frac{\pi}{\sqrt{2}} \overline{\mathrm{B}}\overline{\mathrm{A}}_a \tag{3.96}$$

The torque density equation for squirrel-cage IMs is the same as (3.91), which is for wound-field salient-pole SMs. Similarly, the specific magnetic and electric loadings are no longer coupled to each other. However, it is worth noting that the airgap magnetic field distribution of IMs is created by the excitation/field current component in the stator winding, so the specific magnetic and electric loadings defined in the traditional electric machinery are coupled.

By comparing equations (3.91), (3.94), and (3.96), it can be found that the torque density equations of wound-field salient-pole SMs, SynRMs and IMs are the same. In addition, the torque density is proportional to the magnetic field created by the excitation current component and the specific electric loading created by the armature current component, which is consistent with the traditional electric machinery.

It is also worth noting that the torque density discussed here is defined as the ratio of the average torque to the active rotor volume, and the influence of the specific magnetic and electric loadings on the stator volume is not taken into account, which is very common in traditional electric machinery. This conclusion is also applicable to other traditional AC machine topologies with only one working airgap magnetic field harmonic. When multiple magnetic field harmonics in the airgap participate in the production of the average torque, the torque density equation will change. Some typical examples are given below.

3.5.5 BDFRM

The average torque of the BDFRM is:

$$\begin{aligned} T_{avg\_\mathrm{BDFRM}} &= \frac{1}{T}\int_0^T T_{em\_\mathrm{BDFRM}}(t)dt \\ &= \frac{1}{T}\int_0^T T_{ff}(t)dt + \frac{1}{T}\int_0^T T_{fa}(t)dt + \frac{1}{T}\int_0^T T_{aa}(t)dt \end{aligned} \tag{3.97}$$

where the first and third terms are the cogging torques produced when the field winding and armature winding are supplied independently, and the average value in a full time period is zero. Therefore, the average torque can be further written as:

$$\frac{1}{T}\int_0^T T_{em\_BDFRM}(t)dt = \frac{1}{T}\int_0^T T_{fa}(t)dt$$
$$= \left(\pi r_g^2 l_{stk}\right)\frac{\pi}{\sqrt{2}}(\overline{B}_f\overline{A}_f\cos\varphi_f + \overline{B}_a\overline{A}_a\cos\varphi_a)\sin\Gamma \tag{3.98}$$

where Γ is the torque angle, and

$$\Gamma = (p_f\phi_{0f} + \varphi_f) + (p_a\phi_{0a} + \varphi_a) - (p_f + p_a)\theta_{r0} \tag{3.99}$$

where ϕ_{0f} is the mechanical angle from the reference axis to the phase-A winding axis of the field winding, and φ_f is the initial phase of the phase-A winding current of the field winding (cosine function form). ϕ_{0a} is the mechanical angle from the reference axis to the phase-A winding axis of the armature winding, and φ_a is the initial phase of the phase-A winding current of the armature winding (cosine function form). θ_{r0} is the initial position of the rotor, which is the mechanical angle from the reference axis to the centerline between adjacent magnetic barriers. The specific meaning of each variable is shown in Figure 3.22.

If we eliminate the specific magnetic loading from (3.98) and rewrite (3.98) with specific electric loadings of the two sets of windings, the torque expression will become:

$$\frac{1}{T}\int_0^T T_{em\_BDFRM}(t)dt = (p_f + p_a)\frac{2\mu_0 r_g}{g}C_{p_f,p_a}\left(\frac{\overline{A}_f}{p_f}\right)\left(\frac{\overline{A}_a}{p_a}\right)\sin\Gamma \tag{3.100}$$

The average torque of the BDFRM is produced by the interaction between the modulated field MMF and the modulated armature MMF. When Γ = 90deg and

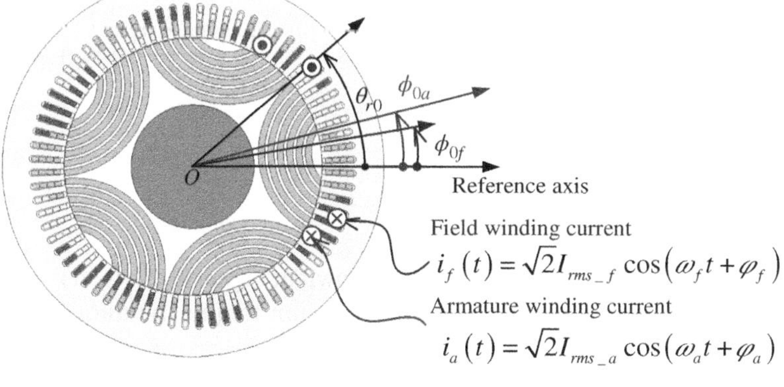

Figure 3.22 Coordinate system for BDFRMs.

the machine is operated to achieve the maximum torque, the torque per active rotor volume is:

$$TRV_{\text{BDFRM}} = \frac{T_{avg\_\text{BDFRM}}}{\pi r_g^2 l_{\text{stk}}} = \frac{\pi}{\sqrt{2}}(\overline{\text{B}}_f\overline{\text{A}}_f \cos \lambda_f + \overline{\text{B}}_a\overline{\text{A}}_a \cos \lambda_a) \tag{3.101}$$

It should be noted that the field winding and armature winding of brushless double-fed reluctance machines are dependent. The field winding with a principal pole pair number of p_f is not only the armature winding of the sub-machine with a pole pair number of p_f, but also the field winding of the sub-machine with a pole pair number of p_a. Similarly, the armature winding with a principal pole pair number of p_a is both the armature winding of the sub-machine with a pole pair number of p_a and the field winding of the sub-machine with a pole pair number of p_f. In this case, the specific electromagnetic loadings of the two sets of windings are coupled, and the coupling must be considered in order to obtain the electromagnetic torque expression.

The relationship between the specific electromagnetic loadings of the two sets of windings in BDFRMs can be written concisely as [20]:

$$\frac{\overline{\text{B}}_f\overline{\text{A}}_f \cos \lambda_f}{\overline{\text{B}}_a\overline{\text{A}}_a \cos \lambda_a} = \frac{p_f}{p_a} \tag{3.102}$$

where λ_f and λ_a represent the power factor angles of the field winding and armature winding, respectively. It is worth noting that since the phase angles of the winding terminal voltages are unknown, there is no direct relationships between λ_f and φ_f. Neither is λ_a and φ_a.

The torque per active rotor volume of BDFRMs can finally be expressed concisely as:

$$TRV_{\text{BDFRM}} = \frac{\pi}{\sqrt{2}}\left(\frac{p_f + p_a}{p_f}\right)\overline{\text{B}}_f\overline{\text{A}}_f \cos \lambda_f = \frac{\pi}{\sqrt{2}}\left(\frac{p_f + p_a}{p_a}\right)\overline{\text{B}}_a\overline{\text{A}}_a \cos \lambda_a \tag{3.103}$$

3.5.6 BDFIM

Similarly, based on the coordinate system shown in Figure 3.23, when $\Gamma = 90°$ and the machine is operated to achieve the maximum torque, the torque per active rotor volume of the BDFIM is:

$$TRV_{\text{BDFIM}} = \frac{T_{avg\_\text{BDFIM}}}{\pi r_g^2 l_{\text{stk}}} = \frac{\pi}{\sqrt{2}}(\overline{\text{B}}_f\overline{\text{A}}_f \cos \lambda_f + \overline{\text{B}}_a\overline{\text{A}}_a \cos \lambda_a) \tag{3.104}$$

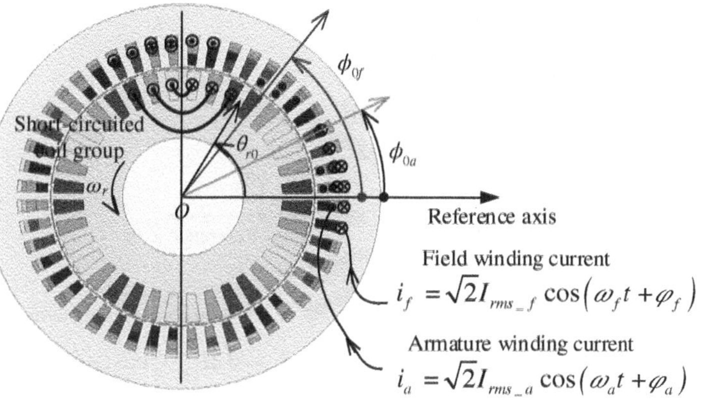

Figure 3.23 Coordinate system for BDFIMs.

Neglecting the resistance of the short-circuited rotor windings, the specific electromagnetic loadings of the two sets of windings are coupled as follows:

$$\frac{\overline{\mathrm{B}}_f \overline{\mathrm{A}}_f \cos \lambda_f}{\overline{\mathrm{B}}_a \overline{\mathrm{A}}_a \cos \lambda_a} = \frac{p_f}{p_a} \tag{3.105}$$

Therefore, the torque per active rotor volume of the BDFIM is:

$$\begin{aligned} TRV_{\mathrm{BDFIM}} &= \frac{T_{avg\_\mathrm{BDFIM}}}{\pi r_g^2 l_{\mathrm{stk}}} = \frac{\pi}{\sqrt{2}} \left(\frac{p_f + p_a}{p_f} \right) \overline{\mathrm{B}}_f \overline{\mathrm{A}}_f \cos \lambda_f \\ &= \frac{\pi}{\sqrt{2}} \left(\frac{p_f + p_a}{p_a} \right) \overline{\mathrm{B}}_a \overline{\mathrm{A}}_a \cos \lambda_a \end{aligned} \tag{3.106}$$

3.5.7 FSPM Machine

Similar to the brushless double-fed reluctance machine, both the field winding (or PMs) and armature winding of the FSPM machines are placed on the stator side. The difference is that the armature winding and the field winding equivalent to the PMs are both fractional-slot concentrated windings, resulting in rich spatial harmonics. In addition, multiple spatial harmonics contribute to the average torque.

$$\begin{aligned} T_{avg\_\mathrm{FSPM}} &= \frac{1}{T} \int_0^T T_{em\_\mathrm{FSPM}}(t) dt \\ &= \frac{1}{T} \int_0^T T_{ff}(t) dt + \frac{1}{T} \int_0^T T_{fa}(t) dt + \frac{1}{T} \int_0^T T_{aa}(t) dt \end{aligned} \tag{3.107}$$

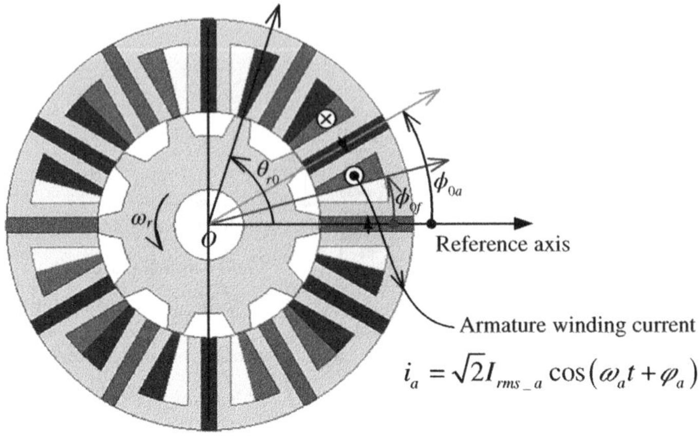

Figure 3.24 Coordinate system for the 12/10 FSPM machine.

where the first term is the cogging torque produced by the modulated field MMF, and its average value in a full time period is zero. The third term is the cogging torque produced by the modulated armature MMF. The second term is the electromagnetic torque produced by the interaction between the modulated field MMF and the modulated armature MMF. Based on the reference coordinate system shown in Figure 3.24, the average torque can be further written as [34]:

$$\begin{aligned} T_{avg\_FSPM} &= \frac{1}{T}\int_0^T T_{fa}(t)dt \\ &= \frac{\pi}{\sqrt{2}}\left(\pi r_g^2 l_{\mathrm{stk}}\right)\sum_n \left(\overline{\mathrm{B}}_f^n \overline{\mathrm{A}}_f^n \cos\lambda_f^n + \overline{\mathrm{B}}_a^n \overline{\mathrm{A}}_a^n \cos\lambda_a^n\right)\sin\Gamma, \\ & p_f^n + p_a^n = N_{RT}. \end{aligned} \tag{3.108}$$

where N_{RT} is the number of salient poles of the rotor, Γ is the torque angle, and $\Gamma = p_f\phi_{0f} + (p_a\phi_{0a} + \varphi_a) - N_{RT}\theta_{r0}$.

Similarly, when $\Gamma = 90°$ and the machine is operated to produce the maximum torque, the torque per active rotor volume is:

$$\begin{aligned} TRV_{\mathrm{FSPM}} = \frac{T_{avg\_\mathrm{FSPM}}}{\pi r_g^2 l_{\mathrm{stk}}} &= \frac{\pi}{\sqrt{2}}\sum_n \frac{N_{RT}}{p_f^n}\overline{\mathrm{B}}_f^n \overline{\mathrm{A}}_f^n \cos\lambda_f^n \\ &= \frac{\pi}{\sqrt{2}}\sum_n \frac{N_{RT}}{p_a^n}\overline{\mathrm{B}}_a^n \overline{\mathrm{A}}_a^n \cos\lambda_a^n \end{aligned} \tag{3.109}$$

In order to facilitate the understanding and verify the correctness of the torque decomposition, the waveforms of the three torque components of a FSPM machine based on finite element (FE) analysis and Maxwell stress tensor method are presented here.

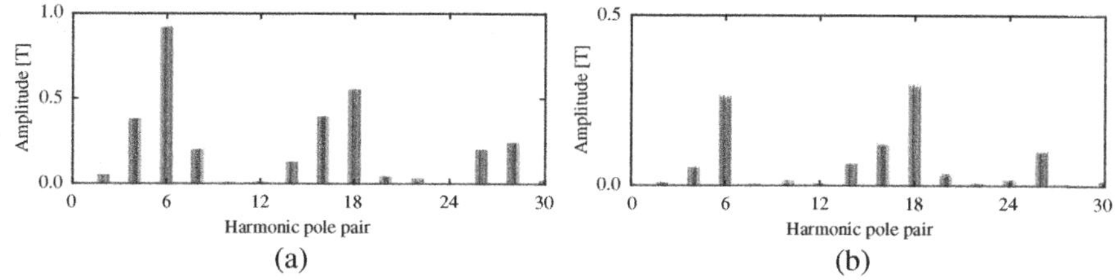

Figure 3.25 Amplitude spectrums of airgap flux density waveforms of a three-phase 12/10 FSPM machine over a full electric cycle under no-load condition, (a) Radial component, (b) Tangential component. Airgap flux density harmonics with pole pairs of 6, 10 ± 6, 18, $|10 \pm 18|$, 20 ± 6 have the largest amplitudes.

When the machine is operated with no load, i.e. the armature winding current is zero and the PM array is the only MMF source, there will be the torque component $T_{ff}(t)$ only. Figure 3.25 shows the variation of amplitudes of main airgap flux density harmonics with the rotor position under no-load condition of a three-phase 12/10 FSPM machine. It can be seen that under no-load condition, due to the presence of the salient-pole stator teeth and the salient rotor poles, the airgap magnetic flux density has both radial and tangential components. The radial component corresponds to the specific magnetic loading while the tangential component corresponds to the specific electric loading.

When both the stator bore and rotor are cylindrical, the airgap flux density distribution will only have radial components with pole pair numbers of $6\ (2l - 1)$, where $l = 1, 2, 3, \ldots$. Both the amplitude and phase of each harmonic are independent of the rotor position. With the presence of salient reluctance poles close to the airgap, the tangential component appears. In addition, due to the modulation behavior of salient reluctance poles, the radial and tangential components include not only harmonics whose numbers of pole pairs are $6(2l - 1)$ but also many other harmonic components, among which the main components have pole pairs of $|10k \pm 6(2l - 1)|$, $k = 1, 2$. Considering that the amplitudes of harmonic components with pole pairs greater than 30 are small and there are no odd harmonic components in the 12/10 FSPM machine, here we mainly study the even-order spatial harmonics with pole pairs no larger than 30, including their amplitudes and phases.

Figures 3.26–3.30 show the phase change of main airgap magnetic flux density harmonics with larger amplitudes and the resulting torque. It can be seen that the phase of components (both radial and tangential) with the number of pole pairs of $6(2l - 1)$ does not change with the rotor position (the 12-pole harmonic shown in Figure 3.27), that is, the harmonic component is stationary. On the contrary, the phase of components with the number of pole pairs of $|10k \pm 6(2l - 1)|$ changes linearly with the rotor position, that is, the harmonic component rotates in the

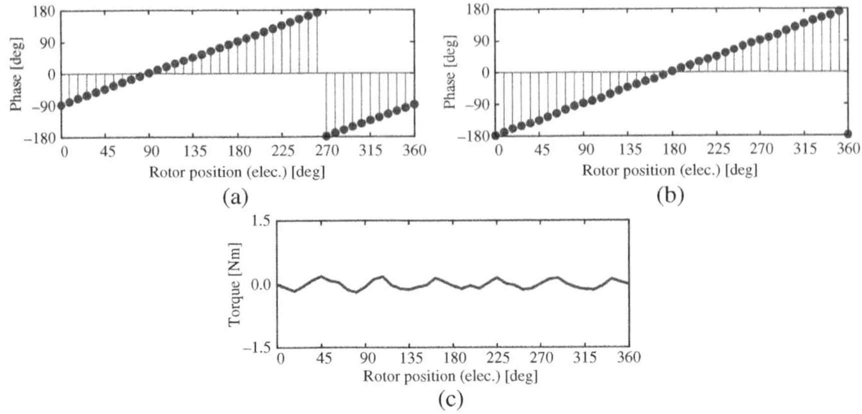

Figure 3.26 Phase spectrums of airgap flux density harmonics with pole pairs of four over a full electrical cycle and the corresponding torque variation under no-load condition. (a) Radial component, (b) Tangential component, (c) Torque produced by the interaction between the radial and tangential components.

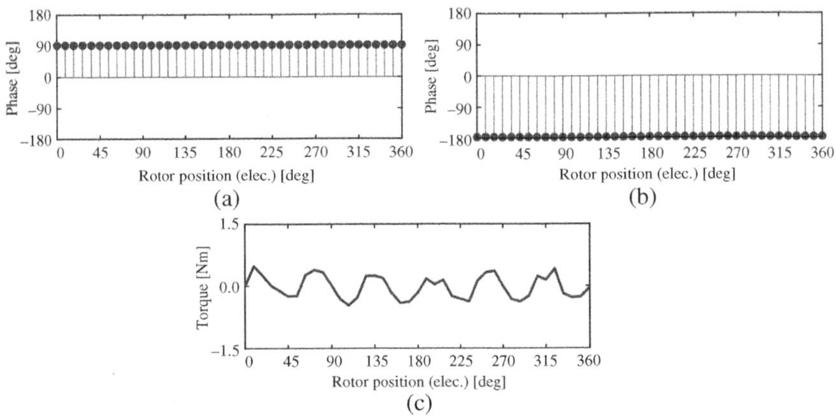

Figure 3.27 Phase spectrums of airgap flux density harmonics with pole pairs of six over a full electrical cycle and the corresponding torque variation under no-load condition. (a) Radial component, (b) Tangential component, (c) Torque produced by the interaction between the radial and tangential components.

airgap. Among them, only the harmonic component with a pole pair number of eight rotates in the reverse direction (clockwise), and other harmonic components rotate in the forward direction (counterclockwise). At the same time, it can be seen from the torque waveforms produced by various harmonics that all the main airgap flux density harmonics contribute to the no-load torque, i.e. the cogging torque caused by PMs.

When PMs are removed and the armature winding is energized alone, only the torque component of $T_{aa}(t)$ exists. When the stator bore and rotor are cylindrical, the airgap magnetic flux density distribution includes only the radial components, and their number of pole pairs depends on the winding configuration. In the example of the three-phase 12/10 FSPM machine mentioned here, the stator winding is a three-phase, 12-slot 8-pole double-layer fractional-slot concentrated winding, so the numbers of pole pairs of the winding harmonics are 4, 8, 16, 20, 28, 32, etc. The presence of salient-pole stator teeth and salient rotor poles introduces additional harmonic components into the airgap, whose numbers of pole pairs are mainly 2, 4, 6, 8, 14, 16, 18, 20, 24, 28, as shown in Figure 3.31.

Different from the no-load condition, when the armature winding is excited alone, the amplitude of each harmonic has a large fluctuation with the change of the rotor position. Among them, the airgap flux density harmonics with pole pairs of 2 and 14 are stationary, and other harmonic components all rotate in the airgap, which contribute to the cogging torque when only the armature winding is considered. Figures 3.32–3.35 show the phase and torque waveforms of several harmonics.

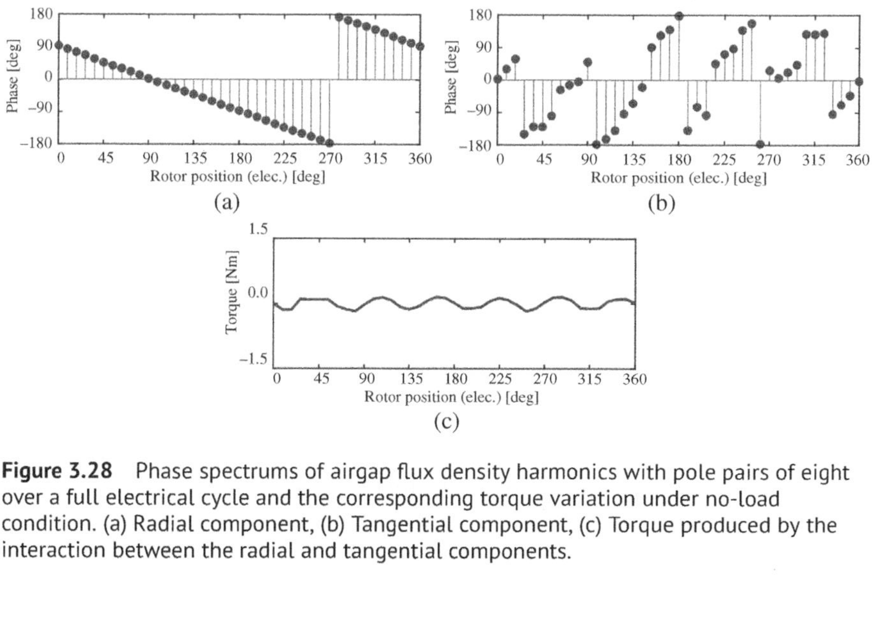

Figure 3.28 Phase spectrums of airgap flux density harmonics with pole pairs of eight over a full electrical cycle and the corresponding torque variation under no-load condition. (a) Radial component, (b) Tangential component, (c) Torque produced by the interaction between the radial and tangential components.

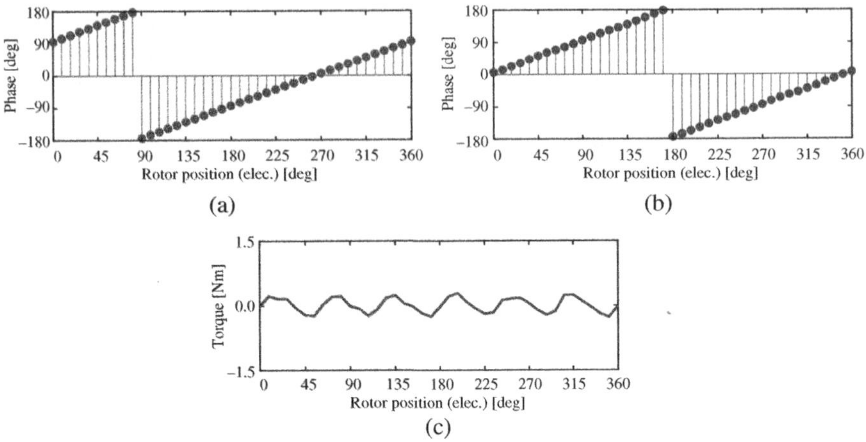

Figure 3.29 Phase spectrums of airgap flux density harmonics with pole pairs of 16 over a full electrical cycle and the corresponding torque variation under no-load condition. (a) Radial component, (b) Tangential component, (c) Torque produced by the interaction between the radial and tangential components.

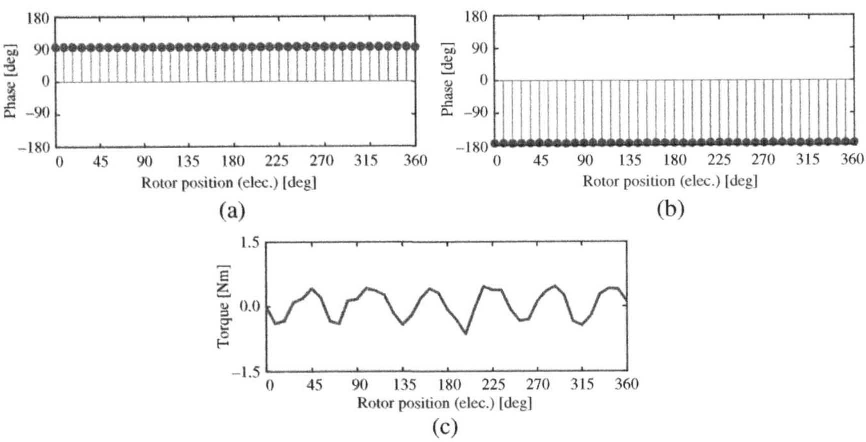

Figure 3.30 Phase spectrums of airgap flux density harmonics with pole pairs of 18 over a full electrical cycle and the corresponding torque variation under no-load condition. (a) Radial component, (b) Tangential component, (c) Torque produced by the interaction between the radial and tangential components.

When the machine is operated at rated load, the amplitude distribution of the radial and tangential components of the airgap flux density is very similar to that at no load, as shown in Figure 3.36. The reason is the three-phase 12/10 FSPM machine in this example uses neodymium iron boron (NdFeB) PMs, and the field

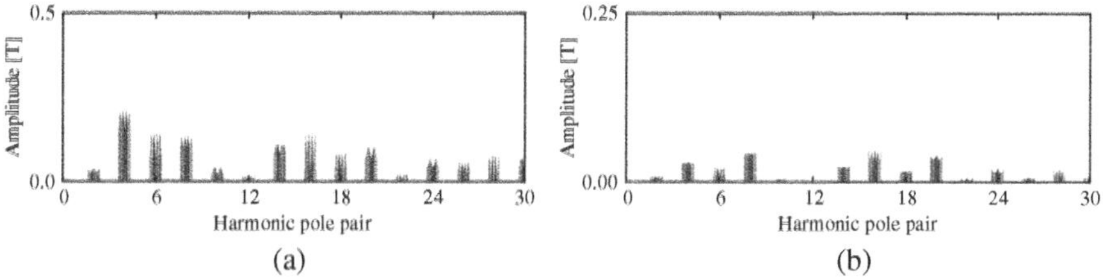

Figure 3.31 Amplitude spectrums of airgap flux density waveforms of a three-phase 12/10 FSPM machine over a full electrical cycle when PMs are removed and only the armature winding is supplied, (a) Radial component, (b) Tangential component. Airgap flux density harmonics with pole pairs of 4, 10 ± 4, 8, 10 ± 8, 20, 20 ± 4, 20 ± 8 have the largest amplitudes.

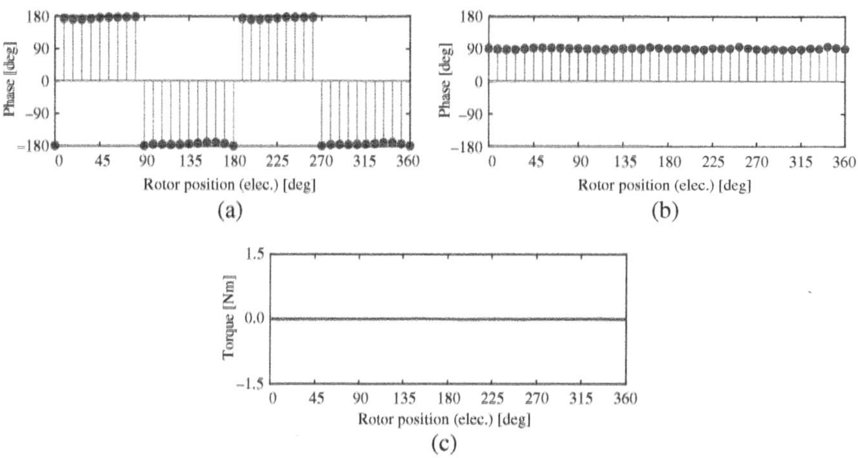

Figure 3.32 Phase spectrums of airgap flux density harmonics with pole pairs of two over a full electrical cycle and the corresponding torque variation when PMs are removed and only the armature winding is supplied. (a) Radial component, (b) Tangential component, (c) Torque produced by the interaction between the radial and tangential components.

MMF created by PMs is much larger than the armature MMF created by the armature winding. There are multiple flux density harmonic pairs in the airgap, such as harmonics with numbers of pole pairs of 6 and $|6 \pm 10|$, and harmonics with numbers of pole pairs of 18 and 18 ± 10. Among them, harmonics with pole pairs of 6 and 18 are stationary, while harmonics with other pole pairs rotate in the airgap. Figures 3.37–3.40 show the phase and torque waveforms of some harmonics. The contribution of each harmonic to the total torque is shown in Figure 3.41.

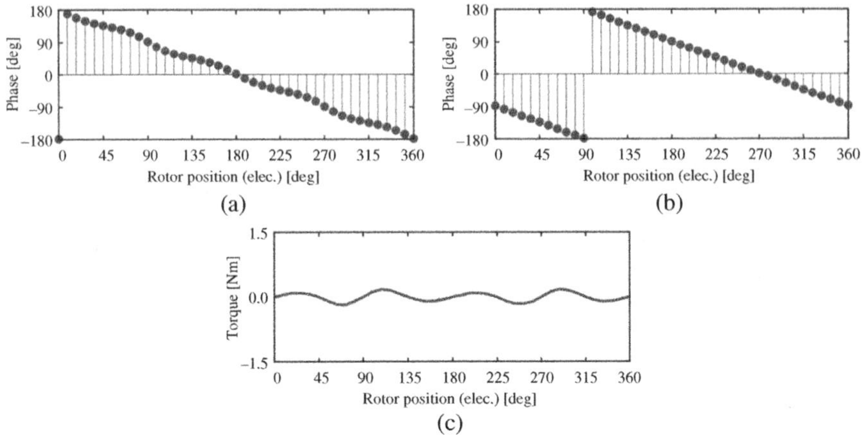

Figure 3.33 Phase spectrums of airgap flux density harmonics with pole pairs of four over a full electrical cycle and the corresponding torque variation when PMs are removed and only the armature winding is supplied. (a) Radial component, (b) Tangential component, (c) Torque produced by the interaction between the radial and tangential components.

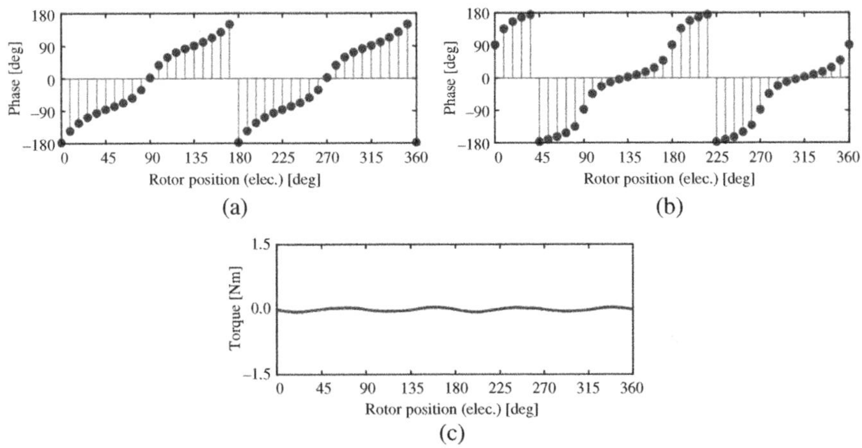

Figure 3.34 Phase spectrums of airgap flux density harmonics with pole pairs of six over a full electrical cycle and the corresponding torque variation when PMs are removed and only the armature winding is supplied. (a) Radial component, (b) Tangential component, (c) Torque produced by the interaction between the radial and tangential components.

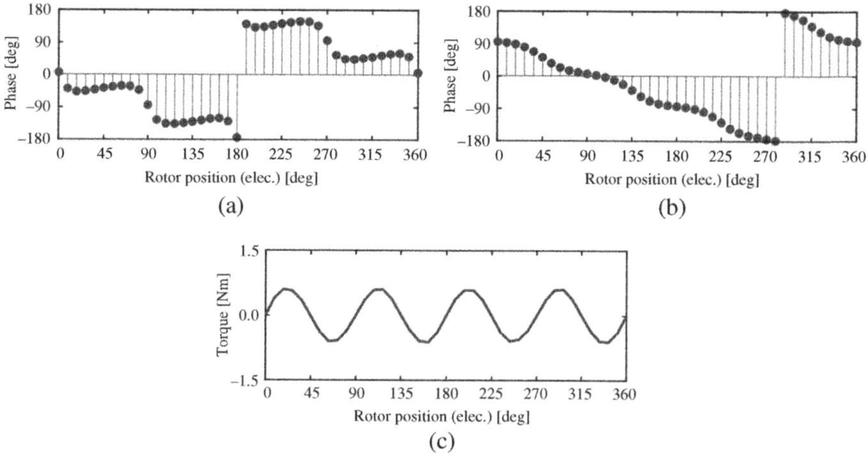

Figure 3.35 Phase spectrums of airgap flux density harmonics with pole pairs of 16 over a full electrical cycle and the corresponding torque variation when PMs are removed and only the armature winding is supplied. (a) Radial component, (b) Tangential component, (c) Torque produced by the interaction between the radial and tangential components.

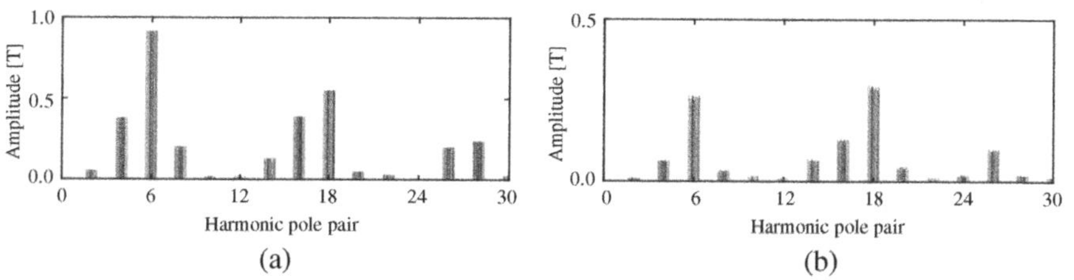

Figure 3.36 Amplitude spectrums of airgap flux density waveforms of a three-phase 12/10 FSPM machine over a full electrical cycle when operated with rated load, (a) Radial component, (b) Tangential component. Airgap flux density harmonics with pole pairs of 6, $|6 \pm 10|$, 18, 18 ± 10, 20 ± 6 have the largest amplitudes.

By plotting the cogging torque $T_{ff}(t)$ during no-load operation, the cogging torque $T_{aa}(t)$ when the armature winding is considered alone, and the total electromagnetic torque under the rated load in the same coordinate system, we can get Figure 3.42. Among them, the electromagnetic torque $T_{fa}(t)$ produced by the interaction between the modulated field MMF and the modulated armature MMF provides a non-zero average value, and its waveform can be further decomposed in accordance with harmonics that contribute to the torque. Here $T_{fa}(t)$ is decomposed into T_{fa} (PM) and T_{fa} (AW). T_{fa} (PM) is the sum of torques contributed by harmonics with pole pairs of 6, 18, 30,... produced by the PM array without the modulation of the salient-pole rotor. This can be considered

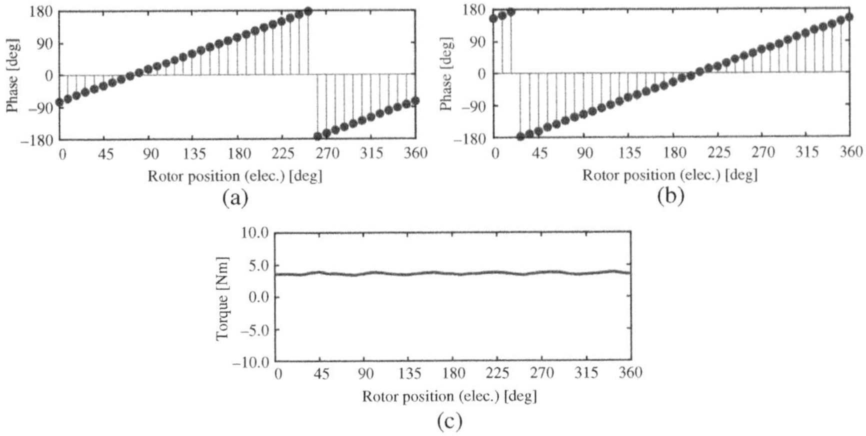

Figure 3.37 Phase spectrums of airgap flux density harmonics with pole pairs of four over a full electrical cycle and the corresponding torque variation when the machine is operated with rated load. (a) Radial component, (b) Tangential component, (c) Torque produced by the interaction between the radial and tangential components.

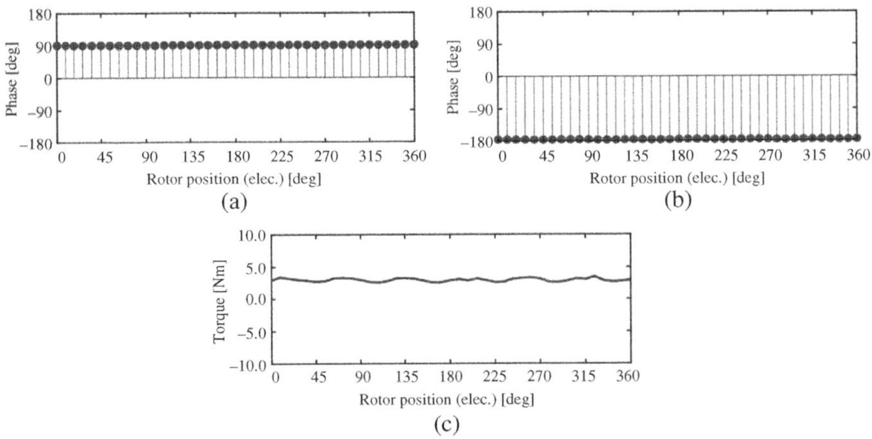

Figure 3.38 Phase spectrums of airgap flux density harmonics with pole pairs of 6 over a full electrical cycle and the corresponding torque variation when the machine is operated with rated load. (a) Radial component, (b) Tangential component, (c) Torque produced by the interaction between the radial and tangential components.

as the PM array provides the specific electric loading and the armature winding provides the specific magnetic loading. T_{fa} (AW) is the sum of torques contributed by harmonics with pole pairs of 4, 8, 16, 20,... produced by the armature winding without the modulation of the salient-pole rotor. This can be considered as the PM

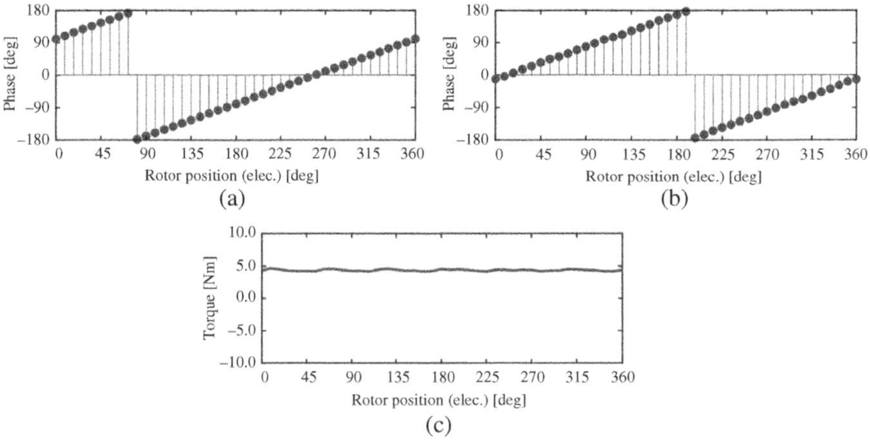

Figure 3.39 Phase spectrums of airgap flux density harmonics with pole pairs of 16 over a full electrical cycle and the corresponding torque variation when the machine is operated with rated load. (a) Radial component, (b) Tangential component, (c) Torque produced by the interaction between the radial and tangential components.

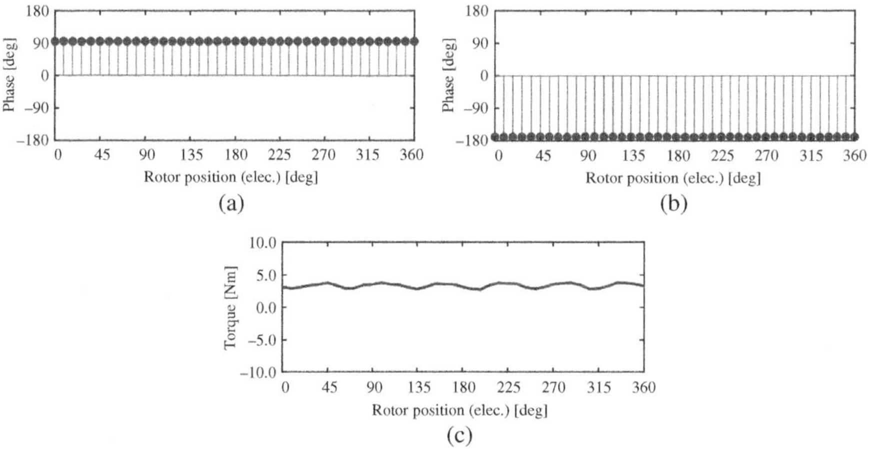

Figure 3.40 Phase spectrums of airgap flux density harmonics with pole pairs of 18 over a full electrical cycle and the corresponding torque variation when the machine is operated with rated load. (a) Radial component, (b) Tangential component, (c) Torque produced by the interaction between the radial and tangential components.

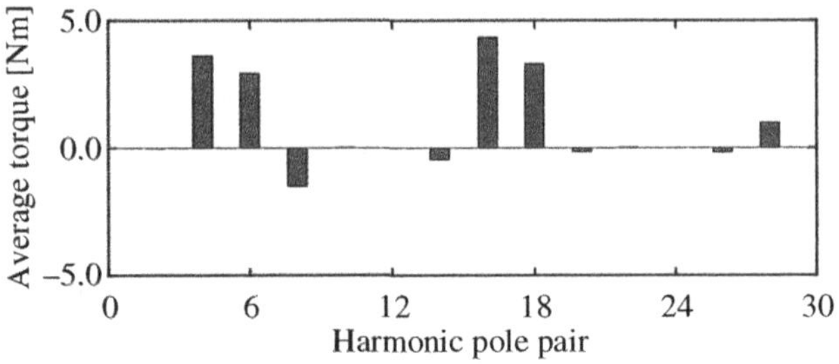

Figure 3.41 Average torques contributed by individual airgap flux density harmonic components of a three-phase 12/10 FSPM machine when operated with the rated load. The total average torque is 12.8 Nm.

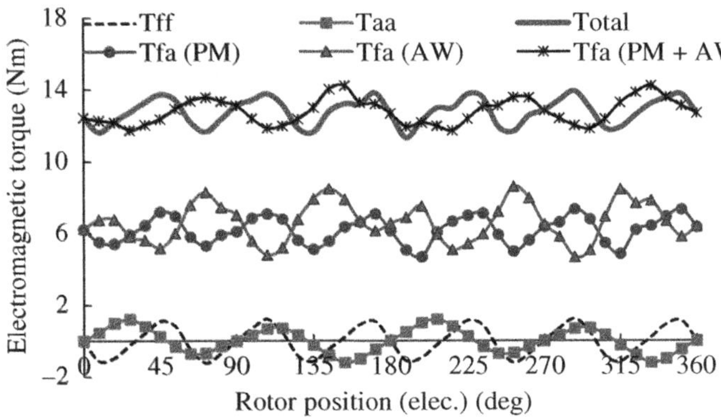

Figure 3.42 Torque decomposition of a three-phase 12/10 FSPM machine. T_{ff}, T_{aa}, and Total were all from FE analysis. T_{fa}(PM), T_{fa}(AW), and T_{fa}(PM + AW) were calculated from airgap flux density waveforms of a full electrical cycle by using the Maxwell stress tensor method. They can also be calculated using the magnetic co-energy based on the airgap flux density waveforms of a full electrical cycle. Results from these two methods will be close to each other.

array provides the specific magnetic loading and the armature winding provides the specific electric loading.

Through this torque decomposition, it has been demonstrated that there are three torque components in electrical machines, which is consistent with formulas (3.74)–(3.77). In addition, the average values of T_{fa} (PM) and T_{fa} (AW) are comparable, which proves that both the PM array and the armature winding in FSPM machines play both roles of the field winding and armature winding.

3.5.8 PMV Machine

In the PMV machine studied in [35], the armature winding and excitation source, i.e. PMs, are placed on the stator and rotor, respectively, and the electromagnetic torque has three components. The average torque is:

$$
\begin{aligned}
T_{avg\_PMV} &= \frac{1}{T}\int_0^T T_{em\_PMV}(t)dt \\
&= \frac{1}{T}\int_0^T T_{ff}(t)dt + \frac{1}{T}\int_0^T T_{fa}(t)dt + \frac{1}{T}\int_0^T T_{aa}(t)dt
\end{aligned} \tag{3.110}
$$

The first term is the torque produced by the modulated field MMF, which the cogging torque produced by the interaction between the PM array and the stator/rotor slots. Its average value is zero in a full time period. The average value of the third term in a full time period is also zero. The average torque is produced by the interaction between the modulated field MMF and the modulated armature MMF:

$$
\begin{aligned}
\frac{1}{T}\int_0^T T_{fa}(t)dt &= \frac{1}{T}\int_0^T \left\{ \frac{\mu_0 r_g l_{\text{stk}}}{g} \frac{\partial}{\partial\Delta} \int_0^{2\pi} M[F_f(\phi,t)]M[F_a(\phi,t)]\,d\phi \right\} dt \\
&= \frac{1}{T}\int_0^T r_g l_{\text{stk}} \frac{\partial}{\partial\Delta} \left\{ \int_0^{2\pi} \frac{\mu_0}{g} M[F_f(\phi,t)]M[F_a(\phi,t)]d\phi \right\} dt \\
&= \frac{1}{T}\int_0^T r_g l_{\text{stk}} \frac{\partial}{\partial\Delta} \left\{ \int_0^{2\pi} B_f(\phi,t)M[F_a(\phi,t)]d\phi \right\} dt
\end{aligned} \tag{3.111}
$$

For the sake of simplicity, the harmonic components in the source MMF distribution are ignored. Based on the coordinate system shown in Figure 3.43, the no-load airgap magnetic flux density distribution can be expressed as [36]:

$$
\begin{aligned}
B_f(\phi,t) &= C_{p_f}\left(\frac{\pi}{2}\overline{\text{B}}_f\right)\cos[p_f\phi - p_f\omega_r t - p_f(\theta_{r0} + \phi_{0f})] \\
&+ C_{p_f,N_{ST}-p_f}\left(\frac{\pi}{2}\overline{\text{B}}_f\right) \\
&\cos[(-p_f + N_{ST})\phi + p_f\omega_r t - \Gamma + (\varphi_a - p_a\phi_{0a})] \\
&+ C_{p_f,N_{ST}+p_f}\left(\frac{\pi}{2}\overline{\text{B}}_f\right) \\
&\cos[(p_f + N_{ST})\phi - p_f\omega_r t - \Gamma + (\varphi_a - p_a\phi_{0a}) - 2p_f(\theta_{r0} + \phi_{0f})]
\end{aligned} \tag{3.112}
$$

where C_{p_f}, $C_{p_f,N_{ST}-p_f}$, and $C_{p_f,N_{ST}+p_f}$ are the magnetic field conversion coefficients introduced after the source field MMF is modulated by the stator salient reluctance pole modulator. $\overline{\text{B}}_f$ is the airgap specific magnetic loading (average value) when the modulation behavior of stator salient poles is not considered, and it is $2/\pi$ times the amplitude of the sinusoidal no-load airgap magnetic flux density distribution created by the source field MMF. p_f is the number of principal pole pairs of the

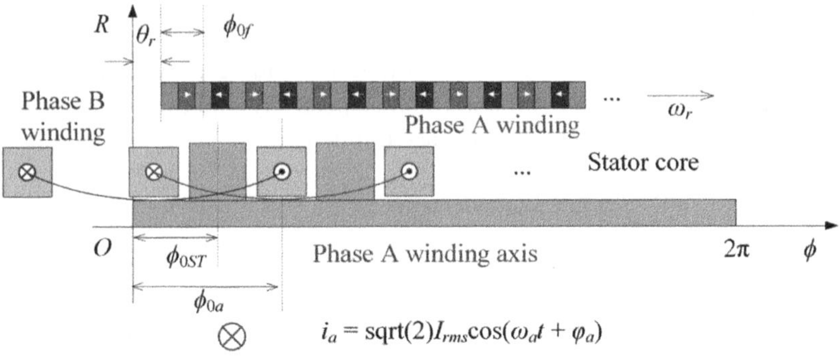

Figure 3.43 Coordinate system for the flux-concentrating field-modulated PM machine (alternatively called the PMV machine).

source field MMF created by the rotor PM array, ϕ is the mechanical position angle along the airgap peripheral in the stationary reference frame, t is the time, and ω_r is the mechanical angular velocity of the rotor. θ_{r0} is the initial position angle of the rotor, and ϕ_{0f} is the initial position angle of the N pole on the rotor. N_{ST} is the number of salient poles (teeth) of the stator. φ_a is the initial phase of the phase-A winding current of the armature winding, and the phase winding current adopts the cosine form. p_a is the number of principal pole pairs of the armature winding, and ϕ_{0a} is the mechanical position angle of the phase-A winding axis of the armature winding. Γ is the torque angle, and its expression is:

$$\Gamma = (\varphi_a - p_a\phi_{0a} + N_{ST}\phi_{0ST} - p_f\theta_{r0} - p_f\phi_{0f}) - \pi \tag{3.113}$$

The modulated armature MMF is:

$$\begin{aligned} M[F_a(\phi,t)] = {} & C_{p_a}F_a\cos[p_a\phi + \omega_a t + (\varphi_a - p_a\phi_{0a})] \\ & + C_{p_a,N_{ST}-p_a}F_a\cos[(-p_a + N_{ST})\phi - \omega_a t - \Gamma - p_f(\theta_{r0} + \phi_{0f})] \\ & + C_{p_a,N_{ST}+p_a}F_a \\ & \cos[(p_a + N_{ST})\phi + \omega_a t - \Gamma + 2(\varphi_a - p_a\phi_{0a}) - p_f(\theta_{r0} + \phi_{0f})] \end{aligned} \tag{3.114}$$

where C_{p_a}, $C_{p_a,N_{ST}-p_a}$, and $C_{p_a,N_{ST}+p_a}$ are the magnetic field conversion coefficients introduced after the source armature MMF is modulated by the stator salient pole modulator. $\overline{A}_a$ is the specific electric loading (A/m) corresponding to the fundamental wave with the principal pole pair number of p_a, whose expression is:

$$\overline{A}_a = \frac{2mN_{ph}k_{w1}I_{rms}}{2\pi r_g} \tag{3.115}$$

The relationship between $\overline{\mathrm{A}}_a$ and the magnitude of the MMF distribution with the principal pole pair number of p_a, i.e. F_a, is:

$$F_a = \left(\frac{m}{2}\right)\left(\frac{4}{\pi}\right)\frac{N_{ph}k_{w1}\sqrt{2}I_{rms}}{2p_a} = \frac{2\pi r_g\sqrt{2}\overline{\mathrm{A}}_a}{2\pi p_a} = \frac{\sqrt{2}r_g\overline{\mathrm{A}}_a}{p_a} \tag{3.116}$$

Rewriting (3.114) using the specific electric loading yields:

$$\begin{aligned} M[F_a(\phi,t)] &= \sqrt{2}r_g C_{p_a}\left(\frac{1}{p_a}\right)\overline{\mathrm{A}}_a \cos[p_a\phi + \omega_a t + (\varphi_a - p_a\phi_{0a})] \\ &+ \sqrt{2}r_g\left(\frac{1}{p_a}\right)C_{p_a,N_{ST}-p_a}\overline{\mathrm{A}}_a \\ &\cos[(-p_a + N_{ST})\phi - \omega_a t - \Gamma - p_f(\theta_{r0} + \phi_{0f})] \\ &+ \sqrt{2}r_g\left(\frac{1}{p_a}\right)C_{p_a,N_{ST}+p_a}\overline{\mathrm{A}}_a \\ &\cos[(p_a + N_{ST})\phi + \omega_a t - \Gamma + 2(\varphi_a - p_a\phi_{0a}) - p_f(\theta_{r0} + \phi_{0f})] \end{aligned} \tag{3.117}$$

In the machine topology adopting the sum modulation, equations (3.112) and (3.117) satisfy the following constraints on speed and number of pole pairs:

$$\omega_a = p_f\omega_r \tag{3.118}$$

$$p_a + p_f = N_{ST} \tag{3.119}$$

According to the orthogonality of trigonometric functions, equation (3.111) has two non-zero terms, and their expressions are:

$$\begin{aligned} &\int_0^{2\pi}\left\{\begin{array}{l} C_{p_f}\left(\frac{\pi}{2}\overline{\mathrm{B}}_f\right)\cos[p_f\phi - p_f\omega_r t - p_f(\theta_{r0} + \phi_{0f})]\left(\frac{1}{p_a}\right) \\ \times\sqrt{2}r_g C_{p_a,N_{ST}-p_a}\overline{\mathrm{A}}_a\cos[(-p_a + N_{ST})\phi - \omega_a t - \Gamma - p_f(\theta_{r0} + \phi_{0f})] \end{array}\right\} d\phi \\ &= \sqrt{2}r_g\left(\frac{1}{p_a}\right)C_{p_f}\left(\frac{\pi}{2}\overline{\mathrm{B}}_f\right)C_{p_a,N_{ST}-p_a}\overline{\mathrm{A}}_a \\ &\times\int_0^{2\pi}\cos[p_f\phi - p_f\omega_r t - p_f(\theta_{r0} + \phi_{0f})]\cos[p_f\phi - \omega_a t - \Gamma - p_f(\theta_{r0} + \phi_{0f})]\,d\phi \\ &= \sqrt{2}r_g\left(\frac{1}{p_a}\right)C_{p_f}\left(\frac{\pi}{2}\overline{\mathrm{B}}_f\right)C_{p_a,N_{ST}-p_a}\overline{\mathrm{A}}_a \\ &\times\left(\frac{1}{2}\right)\int_0^{2\pi}\{\cos\Gamma + \cos[2p_f\phi - 2p_f\omega_r t - 2p_f(\theta_{r0} + \phi_{0f}) - \Gamma]\}\,d\phi \\ &= \sqrt{2}\pi r_g\left(\frac{1}{p_a}\right)C_{p_f}\left(\frac{\pi}{2}\overline{\mathrm{B}}_f\right)C_{p_a,N_{ST}-p_a}\overline{\mathrm{A}}_a\cos\Gamma \\ &= \sqrt{2}\pi r_g\left(\frac{1}{p_a}\right)C_{p_f}\left(\frac{\pi}{2}\overline{\mathrm{B}}_f\right)C_{p_a,N_{ST}-p_a}\overline{\mathrm{A}}_a\cos\Gamma \end{aligned} \tag{3.120}$$

$$\int_0^{2\pi} \left\{ \begin{array}{l} C_{p_f,N_{ST}-p_f} \left(\frac{\pi}{2}\overline{\mathrm{B}}_f\right) \cos[(-p_f + N_{ST})\phi + p_f\omega_r t - \Gamma + (\varphi_a - p_a\phi_{0a})] \\ \times\sqrt{2} r_g C_{p_a} \left(\frac{1}{p_a}\right) \overline{\mathrm{A}}_a \cos[p_a\phi + \omega_a t + (\varphi_a - p_a\phi_{0a})] \end{array} \right\} d\phi$$

$$= \sqrt{2} r_g C_{p_f,N_{ST}-p_f} \left(\frac{1}{p_a}\right) \left(\frac{\pi}{2}\overline{\mathrm{B}}_f\right) C_{p_a}\overline{\mathrm{A}}_a$$

$$\times \int_0^{2\pi} \cos[p_a\phi + p_f\omega_r t - \Gamma + (\varphi_a - p_a\phi_{0a})] \cos[p_a\phi + \omega_a t + (\varphi_a - p_a\phi_{0a})]\, d\phi$$

$$= \sqrt{2} r_g C_{p_f,N_{ST}-p_f} \left(\frac{1}{p_a}\right) \left(\frac{\pi}{2}\overline{\mathrm{B}}_f\right) C_{p_a}\overline{\mathrm{A}}_a$$

$$\times \left(\frac{1}{2}\right) \int_0^{2\pi} \{\cos\Gamma + \cos[2p_a\phi + 2p_f\omega_r t - \Gamma + 2(\varphi_a - p_a\phi_{0a})]\}\, d\phi$$

$$= \sqrt{2}\pi r_g C_{p_f,N_{ST}-p_f} \left(\frac{1}{p_a}\right) \left(\frac{\pi}{2}\overline{\mathrm{B}}_f\right) C_{p_a}\overline{\mathrm{A}}_a \cos\Gamma$$

$$= -\sqrt{2}\pi r_g C_{p_f,N_{ST}-p_f} \left(\frac{1}{p_a}\right) \left(\frac{\pi}{2}\overline{\mathrm{B}}_f\right) C_{p_a}\overline{\mathrm{A}}_a \cos\Gamma \tag{3.121}$$

Therefore, the total torque is:

$$\frac{1}{T}\int_0^T T_{af}(t)dt$$

$$= \frac{1}{T}\int_0^T \frac{\partial}{\partial\Delta}\left[\sqrt{2}\pi r_g^2 l_{\mathrm{stk}} \left(\frac{1}{p_a}\right) (C_{p_f} C_{p_a,N_{ST}-p_a} - C_{p_f,N_{ST}-p_f} C_{p_a}) \left(\frac{\pi}{2}\overline{\mathrm{B}}_f\right) \overline{\mathrm{A}}_a \cos\Gamma\right] dt$$

$$= \frac{d}{d\theta_{r0}}\left[\pi r_g^2 l_{\mathrm{stk}} \left(\frac{\pi}{\sqrt{2}}\right) \left(\frac{1}{p_a}\right) (C_{p_f} C_{p_a,N_{ST}-p_a} - C_{p_f,N_{ST}-p_f} C_{p_a}) \overline{\mathrm{B}}_f \overline{\mathrm{A}}_a \cos\Gamma\right]$$

$$= \pi r_g^2 l_{\mathrm{stk}} \left(\frac{\pi}{\sqrt{2}}\right) \left(\frac{p_f}{p_a}\right) (C_{p_f} C_{p_a,N_{ST}-p_a} - C_{p_f,N_{ST}-p_f} C_{p_a}) \overline{\mathrm{B}}_f \overline{\mathrm{A}}_a \sin\Gamma \tag{3.122}$$

In the machine topology adopting the differential modulation, the constraints on number of pole pairs and speed are slightly different from (3.118)–(3.119). The torque equation can be derived similarly following steps shown in (3.120)–(3.122).

When $\Gamma = 90°$. and the machine is operated to produce the maximum torque, the torque per active rotor volume is:

$$TRV_{\mathrm{PMV}} = \frac{T_{avg\_\mathrm{PMV}}}{\pi r_g^2 l_{\mathrm{stk}}}$$

$$= \frac{\pi}{\sqrt{2}}\left(\frac{p_f}{p_a}\right) (C_{p_f}\overline{\mathrm{B}}_f)(C_{p_a,N_{ST}-p_a}\overline{\mathrm{A}}_a)$$

$$- \frac{\pi}{\sqrt{2}}\left(\frac{p_f}{p_a}\right) (C_{p_f,N_{ST}-p_f}\overline{\mathrm{B}}_f)(C_{p_a}\overline{\mathrm{A}}_a), \qquad p_a + p_f = N_{ST} \tag{3.123}$$

Equation (3.123) shows that the electromagnetic torque of this type of machine has two main components, the amplitudes of which are affected by the magnetic field conversion coefficient introduced in the modulation process. One of the torque components is produced by the airgap magnetic flux density distribution and linear current density distribution with the pole pair number of p_f, and the other is produced by the airgap magnetic flux density distribution and linear current density distribution with the pole pair number of p_a. Both torque components have their corresponding specific magnetic loading and specific electric loading. Whether the two torque components enhance or weaken each other depends on the combination of p_f, p_a, and N_{ST}. The average torque and torque density equations can be derived following the above process considering other spatial harmonics in the source MMF distribution.

Starting from the developed general airgap field modulation theory, the torque density equations of several typical electrical machines have been derived in this chapter. It can be seen that the torque of an electrical machine having multiple working magnetic flux density harmonic components is the superposition of torques produced by each working magnetic flux density harmonic component and the corresponding linear current density harmonic component. Each working magnetic flux density harmonic and linear current density harmonic pair corresponds to a specific magnetic loading and a specific electric loading.

3.5.9 Axial-Flux PMV Machine

Neglecting the end effects of windings and magnets, the axial-flux PMV machine shown in Figure 3.44 can be considered as consisting of a number of thin linear machine layers rolled up circumferentially and stacked in the radial direction.

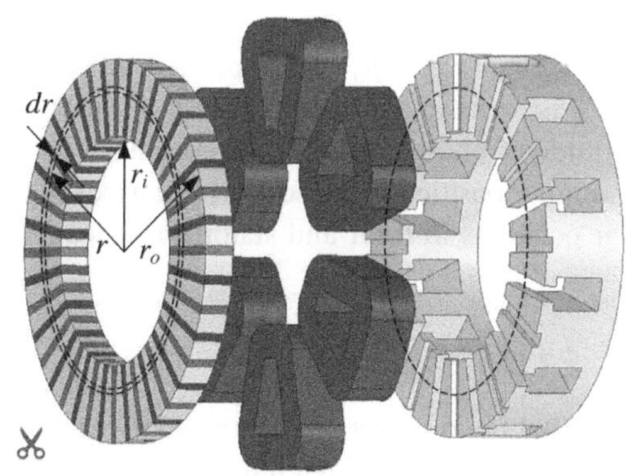

Figure 3.44 Exploded view of the axial-flux PMV machine with 16 stator slots and 56 rotor poles.

Each thin layer has a cross-sectional view illustrated in Figure 3.43, though the actual machine length depends on the diameter where the layer is located.

The analysis method, presented in Section 3.5.8, therefore, also applies to axial-flux PMV machines and the torque density equation can be easily derived from those for radial-flux counterparts. For simplicity, all the harmonic components in stator MMF produced by the stator winding are neglected. The airgap flux density distribution contributed by the rotor PM MMF and the stator armature MMF is:

$$\begin{aligned}
&[F_f(\underline{\phi}, t) + F_a(\phi, t)]\lambda(r, \phi)\\
&\quad = F_f\lambda_0(r)\cos[p_f\phi - p_f\omega_r t - p_f(\theta_{r0} + \phi_{0f})]\\
&\quad + \frac{1}{2}F_f\lambda_1(r)\cos[(-p_f + N_{ST})\phi + p_f\omega_r t - \Gamma + (\varphi_a - p_a\phi_{0a})]\\
&\quad + \frac{1}{2}F_f\lambda_1(r)\cos[(p_f + N_{ST})\phi - p_f\omega_r t - \Gamma + (\varphi_a - p_a\phi_{0a}) - 2p_f(\theta_{r0} + \phi_{0f})]\\
&\quad + F_a\lambda_0(r)\cos[p_a\phi + \omega_a t + (\varphi_a - p_a\phi_{0a})]\\
&\quad + \frac{1}{2}F_a\lambda_1(r)\cos[(-p_a + N_{ST})\phi - \omega_a t - \Gamma - p_f(\theta_{r0} + \phi_{0f})]\\
&\quad + \frac{1}{2}F_a\lambda_1(r)\cos[(p_a + N_{ST})\phi + \omega_a t - \Gamma + 2(\varphi_a - p_a\phi_{0a}) - p_f(\theta_{r0} + \phi_{0f})]
\end{aligned} \tag{3.124}$$

where $\underline{\phi}$ is the angular position in the rotor reference frame, F_f the amplitude of per-pole MMF created by the PM array, p_f the number of rotor pole pairs, ϕ_{0f} the position of the reference axis in the rotor reference frame, ϕ the angular position in the stator reference frame. N_{ST} is the number of auxiliary stator teeth and $N_{ST} = kN_s$, where k is a positive integer and N_s is the number of main teeth. ϕ_{0SP} is the position of the reference axis in the stator reference frame. λ_0 and λ_1 are the average airgap permeance and the amplitude of the pulsating component, respectively. $F_a = (4/\pi)[N_{ph}k_{w1}I_a/(2p_a)]$ is the per-pole MMF created by armature windings, where N_{ph} is turns in series per phase, k_{w1} the fundamental winding factor, I_a the amplitude of phase current, p_a the number of stator winding pole pairs, and ϕ_{0a} the angular position of phase-A winding axis in the stator reference frame. The angular positions in rotor and stator reference frames are related to each other by:

$$\phi = \underline{\phi} + \theta_r = \underline{\phi} + \omega_r t + \theta_{r0} \tag{3.125}$$

where θ_r is the angle that the rotor rotates relative to the stator, ω_r the rotor speed and θ_{r0} the initial position of rotor with reference to the stator. The axial airgap flux density waveform contains multiple spatial harmonics of dissimilar pole pairs, as shown in Figure 3.45.

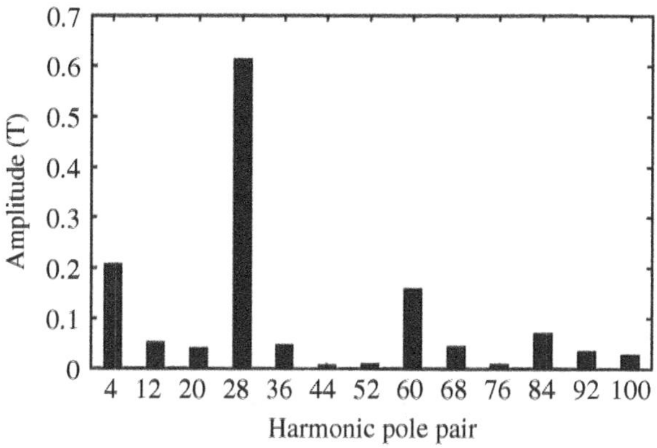

Figure 3.45 Spectrum of the axial airgap flux density waveform at the mean diameter of a 16-slot 56-pole axial-flux PMV machine under no-load condition. Results are from 3D FE analysis. The $2p_f$-pole field harmonic has the largest amplitude, followed by 2 $(N_{ST} \pm p_f)$-pole harmonics.

Considering the difference in radii of layers, the total PM flux linkage can be written as:

$$\psi_{PM} = \frac{\pi}{2}\left[\int_{r_i}^{r_o} \lambda_1(r) r dr\right] F_f \left[\frac{4}{\pi}\frac{N_{ph}k_{w1}}{2p_a}\right] e^{j(-\omega_a t + \Gamma - \varphi_a)} \tag{3.126}$$

where $\Gamma = (\varphi_a - p_a\phi_{0a} + N_{ST}\phi_{0ST} - p_f\vartheta_{r0} - p_f\phi_{0f}) - \pi$ is the current angle between the PM flux linkage vector and current vector. φ_a is the initial phase of phase-A current. ψ_{PM} is the space vector of PM flux linkage linked by the stator winding, i_s the space vector of stator current, r the radius of the circumferential cut-plane. r_i and r_o are stator inner radius and stator outer radius, respectively.

Equation (3.126) shows the PM flux linkage linked by the stator winding due to the flux modulation behavior of the teeth is related to the per-pole MMF of PM array F_f, effective turns per pole per phase $N_{ph}k_{w1}/(2p_a)$ and the pulsating component of the airgap permeance. The PM flux linkage can be increased by appropriately change the dimensions of the auxiliary teeth.

Other harmonic components in (3.124) may also contribute to the PM flux linkage if the number of pole pairs coincides with one of the stator winding harmonics. For the example with 16 main slots ($N_s = 6$) and 2 auxiliary teeth per main tooth ($k = 2$) and 56 rotor poles ($p_f = 28$), besides the 8-pole component ($p_a = 4$) produced by the modulation behavior, the 56-pole component also couples the seventh harmonic of the stator winding. The 120-pole component ($p_f + N_{ST} = 60$) produced by the modulation behavior also couples the 15th harmonic of the stator winding. The higher the harmonic order is, the smaller the contribution will be.

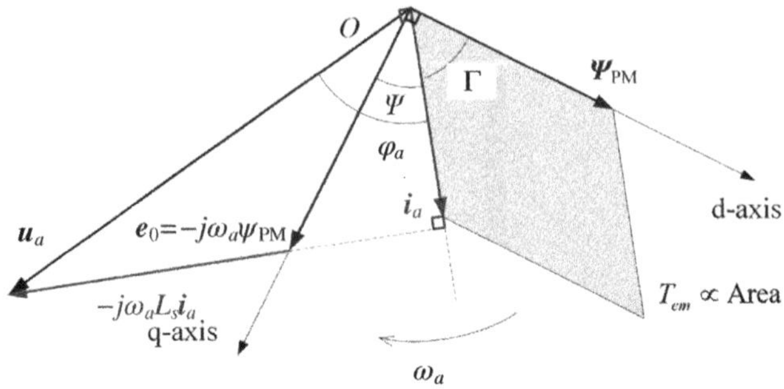

Figure 3.46 Vector diagram of the studied axial-flux PMV machine.

The electromagnetic torque contributed by the modulation behavior can be written as:

$$T_{em} = -p_f \Re\left[j\psi_{PM}(i_a)^*\right] = p_f \frac{\pi}{2}\left[\int_{r_i}^{r_o} \lambda_1(r) r dr\right] F_f F_a \sin\Gamma \tag{3.127}$$

Equation (3.127) shows the torque contributed by the modulation behavior is proportional to the number of rotor poles, PM MMF, stator MMF, and sinΓ. Compared to the radial-flux PMV machine, an integral term shows up in the torque equation for the axial-flux version. The integral term is related to the tooth geometry and may take different forms. The corresponding vector diagram is shown in Figure 3.46, $\boldsymbol{u}_a$ is the terminal voltage vector and $\boldsymbol{e}_0$ the no-load back-EMF vector. The torque contribution of each spatial harmonic in the airgap is calculated by the Maxwell Stress Tensor method. For each airgap, the flux density waveforms at a number of equally spaced circumferential paths are extracted from 3D FE analysis and then decomposed into z-component and ϕ-component for torque estimation. The 8-, 56-, 120- and 168-pole flux density harmonics contribute −8.3%, 99.8%, 14.1%, and −4.8%, respectively, to the total torque.

References

1 Xu, S. (1985). *Winding Theory of AC Machines*, 1e. Beijing, China: China Machinery Press.
2 Lipo, T.A. (2012). *Analysis of Synchronous Machine*, 2e. Boca Raton, FL, USA: CRC Press.
3 Tan, J. (2011). *Permanent Magnet Brushless DC Machine Technology*. Beijing, China: China Machinery Press.
4 Pyrhonen, J., Jokinen, T., and Hrabovcova, V. (2014). *Design of Rotating Electrical Machines*, 2e. Wiley.

5 Cheng, M., Qin, W., Zhu, X., and Wang, Z. (2022). Magnetic-inductance: concept, definition, and applications. *IEEE Trans. Power Electron.* 37 (10): 12406–12414.
6 Heller, V.H. (1977). *Harmonic Field Effects in Induction Machines.* Amsterdam, The Netherlands: Elsevier.
7 Chen, J.T., Zhu, Z.Q., and Iwasaki, S. (2011). A novel E-core switched-flux PM brushless AC machine. *IEEE Trans. Ind. Appl.* 47 (3): 1273–1282.
8 Langsdorf, A.S. (1909). The current locus of the single-phase induction motor. *Proc. AIEE* 28 (7): 731–742.
9 Creedy, P. (1909). A sketch of the theory of the adjustable-speed, single-phase, shunt induction motor. *Proc. AIEE* 28 (7): 831–866.
10 EL-Refaie, A.M. and Shah, M.R. (2009). Comparison of induction machine performance with distributed and fractional-slot concentrated windings. In: *Record of the IEEE Industry Applications Society Annual Meeting*, 1–8. Canada.
11 Spargo, C.M., Mecrow, B.C., Widmer, J.D. et al. (2015). Application of fractional slot-concentrated windings to synchronous reluctance motors. *IEEE Trans. Ind. Appl.* 51 (2): 1446–1455.
12 Broadway, A.R.W. and Burbridge, L. (1970). Self-cascaded machine: a low speed motor or high frequency brushless alternator. *Proc. IEE* 117 (7): 1277–1290.
13 Deodhar, R.P., Andersson, S., Boldea, I., and Miller, T.J.E. (1997). The flux-reversal machine: a new brushless doubly-salient permanent-magnet machine. *IEEE Trans. Ind. Appl.* 33 (4): 925–934.
14 More, D.S. and Fernandes, B.G. (2010). Analysis of flux-reversal machine based on fictitious electrical gear. *IEEE Trans. Energy Convers.* 25 (4): 940–947.
15 McFarland, J.D., Jahns, T.M., and EL-Refaie, A.M. (2015). Analysis of the torque production mechanism for flux-switching permanent-magnet machines. *IEEE Trans. Ind. Appl.* 51 (4): 3041–3045.
16 Wu, Z.Z. and Zhu, Z.Q. (2015). Analysis of airgap field modulation and magnetic gearing effects in switched flux permanent magnet machines. *IEEE Trans. Magn.* 51: 8105012–5.
17 Toba, A. and Lipo, T.A. (1999). Novel dual-excitation permanent magnet vernier machine. In: *Record of the IEEE Industry Applications Society Annual Meeting*, vol. 4, 2539–2544. Phoenix, AZ.
18 Chaithongsuk, S., N-Mobarakeh, B., Caron, J. et al. (2012). Optimal design of permanent magnet motors to improve field-weakening performances in variable speed drives. *IEEE Trans. Ind. Electron.* 59 (6): 2484–2494.
19 Boldea, I., Fu, Z.X., and Nasar, S.A. (1994). Performance evaluation of axially-laminated anisotropic (ALA) rotor reluctance synchronous motors. *IEEE Trans. Ind. Appl.* 30 (4): 977–985.

20 Han, P., Zhang, J., and Cheng, M. (2019). Analytical analysis and performance characterization of brushless doubly-fed machines with multi-barrier rotors. *IEEE Trans. Ind. Appl.* 55 (6): 5758–5767.

21 Xu, W., Zhu, J.G., Zhang, Y. et al. (2011). New axial laminated-structure flux-switching permanent magnet machine with 6/7 poles. *IEEE Trans. Magn.* 47 (10): 2823–2826.

22 Logan, T., Long, T., and McMahon, R.A. (2011). The single-phase brushless doubly-fed machine as a generator for wind turbines. In: *Proceedings of the IEEE International Electric Machines & Drives*, 795–800. Niagara Falls, Canada.

23 Raziee, S.M., Misir, O., and Ponick, B. (2017). Winding function approach for winding analysis. *IEEE Trans. Magn.* 53 (10): Article #8203809.

24 Oppenheim, A.V., Willsky, A.S., and Nawab, S.H. (1997). *Signal and Systems*, 2e. Hoboken, New York: Prentice Hall.

25 Tang, Y. (2011). *Electric Machinery*, 4e. Beijing, China: China Machinery Press.

26 Chen, J.T. and Zhu, Z.Q. (2010). Influence of the rotor pole number on optimal parameters in flux-switching PM brushless ac machines by the lumped-parameter magnetic circuit model. *IEEE Trans. Ind. Appl.* 46 (4): 1381–1388.

27 Boughrara, K., Lubin, T., and Ibtiouen, R. (2013). General subdomain model for predicting magnetic field in internal and external rotor multiphase flux-switching machines topologies. *IEEE Trans. Magn.* 49 (10): 5310–5325.

28 Erickson, R.W. and Maksimovic, D. (2004). *Fundamentals of Power Electronics*, 2e. Boulder, Colorado/New York, USA: University of Colorado/Kluwer Academic Publishers.

29 Cheng, M., Han, P., Du, Y. et al. (2022). A tutorial on general airgap field modulation theory for electrical machines. *IEEE Trans. Emerging Sel. Top. Power Electron.* 10 (2): 1712–1732.

30 Han, P. and Cheng, M. (2019). Synthesis of airgap magnetic field modulation phenomena in electric machines. In: *Proceedings of the IEEE Energy Conversion Congress and Exposition (ECCE)*, 283–290. Baltimore, MD, USA.

31 Simon-Sempere, V., Burgos-Payan, M., and Cerquides-Bueno, J.-R. (2012). Spatial filtering: a tool for selective harmonics elimination in the design of permanent–magnet synchronous motors. *IEEE Trans. Magn.* 48 (6): 2056–2067.

32 Simon-Sempere, V., Burgos-Payan, M., and Cerquides-Bueno, J.-R. (2013). Influence of manufacturing tolerances on the electromotive force in permanent-magnet motors. *IEEE Trans. Magn.* 49 (11): 5522–5532.

33 Qiu, Y., Xu, M., Yao, K. et al. (2006). Multifrequency small-signal model for buck and multiphase buck converters. *IEEE Trans. Power Electron.* 21 (5): 1185–1192.

34 Cheng, M., Han, P., and Hua, W. (2017). General airgap field modulation theory for electrical machines. *IEEE Trans. Ind. Electron.* 64 (8): 6063–6074.

35 Li, X., Chau, K.T., and Cheng, M. (2015). Comparative analysis and experimental verification of an effective permanent-magnet vernier machine. *IEEE Trans. Magn.* 51 (7): Article #8203009.

36 Rallabandi, V., Han, P., Kesgin, M. et al. (2019). Axial-field vernier-type flux modulation machines for low-speed direct-drive applications. In: *Proceedings of the IEEE Energy Conversion Congress and Exposition (ECCE)*, 3123–3128. Baltimore, MD, USA.

4

Analysis of Magnetic Field Modulation Behaviors

4.1 Introduction

Both the synchronous and asynchronous modulation behaviors introduced in Chapter 3 are very common in electrical machines [1, 2]. This chapter defines synchronous and asynchronous modulation behaviors, as well as the synchronous and asynchronous torque components based on the developed general airgap field modulation theory with detailed examples.

The relationship between the modulation behaviors and torque components has also been developed. Based on this relationship, this chapter discusses the equivalence and interchangeability of the three typical modulators, and the feasibility of combining multiple existing modulators to produce new modulators, laying a solid theoretical foundation for better understanding the modulation behavior of each individual modulator and optimizing modulator constructions.

4.2 Magnetic Field Modulation Behaviors and Torque Components

Magnetic field modulation behaviors and torque compositions vary from electrical machine to electrical machine, depending on the machine construction as well as the modes of operation. The brushless doubly-fed machine (BDFM) is a typical example with two AC electrical ports and a single mechanical port [3]. It has a variety of rotor structures, such as the nested-loop rotor, the salient reluctance pole rotor, the multi-layer flux barrier rotor, etc., and works in multiple modes of operation, including the singly-fed synchronous/asynchronous mode, doubly-fed synchronous/asynchronous mode, etc. The optimal design and operation of such machine topologies require a firm grasp of modulation behaviors, torque components, and their relationship.

General Airgap Field Modulation Theory for Electrical Machines: Principles and Practice,
First Edition. Ming Cheng, Peng Han, Yi Du, and Honghui Wen.

Companion website: www.wiley.com/go/genairgapfieldmodulationtheory

4.2.1 Asynchronous and Synchronous Modulation Behaviors

According to the relative motion state between the source MMF and the modulator, the magnetic field modulation behavior can be classified into the asynchronous modulation and synchronous modulation [1, 4]. If the modulator moves relative to the source MMF, the modulation behavior is asynchronous. Otherwise, it is a synchronous modulation behavior.

Typical examples of the asynchronous type are modulation behaviors in brushless doubly-fed induction machines (BDFIMs) with a short-circuited rotor winding, and flux-reversal PM (FRPM) machines with salient reluctance pole rotors, as depicted in Figure 4.1a. The modulation behavior of pole caps in V-shaped interior PM machines is synchronous, as illustrated in Figure 4.1b.

A detailed comparison between the asynchronous and synchronous modulation behaviors has been provided in Table 4.1, where the concepts of physical and functional dq axes are used. Specifically, in machines with a PM rotor, the physical d-axis is defined as the position where the centerline of the magnetic pole is located, and the physical q-axis is the position obtained by rotating the physical d-axis by 90 electrical degrees, i.e. the centerline between adjacent magnetic poles. While in machines with a salient reluctance pole rotor, the physical dq axes typically correspond to positions where the reluctance reaches the minimum and maximum, respectively.

The functional dq axes correspond to the positions where the winding phase self-inductance reaches the maximum and minimum, respectively. Therefore,

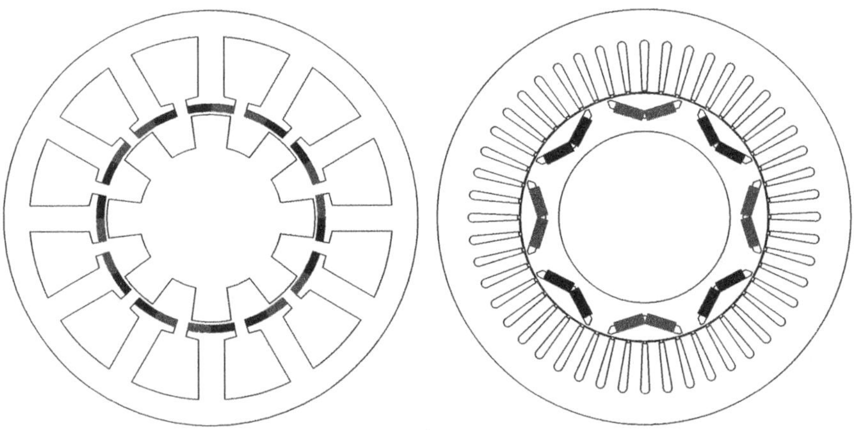

(a) (b)

Figure 4.1 Typical example of modulation behaviors, (a) Asynchronous modulation behavior of salient reluctance poles in FRPM machines, (b) Synchronous modulation behavior of pole caps in V-shaped interior PM machines.

Table 4.1 Comparison between asynchronous and synchronous modulation behaviors in electrical machine implementations.

Modulation type	Asynchronous modulation (dynamic modulation)	Synchronous modulation (static modulation)
Motion state of modulator relative to the source MMF	Moving	Stationary
Airgap magnetic field distribution	Irregular	Regular
Working harmonics	Multiple harmonics of different pole pairs	The fundamental harmonic or all harmonics
Relationship between magnetizing and armature fields	Close- or far-ratio pole combinations	Sharing the same pole pairs
Magnetic field harmonic utilization	Low	High
Power factor	Relatively low	High
Torque density	Relatively low	High
Role of salient reluctance poles	Producing constant reluctance torque component	Introducing magnetic field harmonics of different pole pairs
Relationship between physical and functional dq axes	Consistent	Inconsistent

the functional dq axes do not necessarily be in line with the physical dq axes. For example, in the 12-slot 10-pole FSPM machine, when the phase main self-inductance reaches the maximum or minimum, the rotor salient reluctance pole is not aligned with the winding axis but shifted slightly by a mechanical angle. In traditional DC machines (DCMs), induction machines (IMs) and synchronous machines (SMs), the stator and rotor have the same number of magnetic pole pairs. In addition, the physical dq axes are consistent with the functional dq axes.

Based on the comparison shown in Table 4.1, one can draw the following conclusions:

- The airgap magnetic field of synchronously modulated electrical machines typically contains only one working harmonic component, showing a sinusoidal airgap flux density distribution. A rectangular or triangular distribution containing full harmonics can also be seen in some electrical machine types.

By contrast, the airgap magnetic field of asynchronously modulated electrical machine typically has a large number of harmonic components of different pole pairs and speeds, featuring an irregular and time-varying flux density distribution;

- Electrical machines working by synchronous modulation behaviors typically utilize only the effective harmonic with a principal pole pair of p. There are exceptions, such as brushed DCMs with full-pitched or near-full-pitched armature windings. Those full-pitched coils located in the stator of DCMs allow almost all spatial harmonic components in the rectangularly distributed magnetizing field to pass through and produce flux linkages;
- Asynchronously modulated machines represented by the so-called stator-PM machines, can utilize a limited number of airgap magnetic field spatial harmonics and are typically dominated by two harmonic components with pole pairs of p and $(N_{RT} - p)$ or $(N_{RT} + p)$, where N_{RT} is the number of rotor salient reluctance poles;
- Electrical machines working by asynchronous modulation behaviors sacrifice part of the fundamental harmonic to produce a variety of additional spatial harmonic components. Some of them can contribute to the average torque while others are useless, leading to a low harmonic utilization. Traditional machines such as the DCM, IM, and SM have a sinusoidal magnetizing field and mainly use the fundamental component, so they inherently have a high harmonic utilization.

4.2.2 Asynchronous and Synchronous Torque Components

The concepts of synchronous and asynchronous torque components have been used in existing electric machinery. If a torque component is produced by the same magnetic field source and the rotor moves relative to the magnetic field source, it is called the asynchronous torque. The production mechanism of the asynchronous torque component is illustrated in Figure 4.2a, where only one set of stator windings creates the magnetizing magnetic field B_f, rotating at the synchronous speed ω_f/p_f. B_f' is the magnetic field created by the rotor winding with currents induced by B_f, and its mechanical angular speed is ω_r.

Typical asynchronous torque components include the main torque of squirrel-cage IMs and additional asynchronous torque components produced by high-order spatial harmonics, asynchronous torque components of BDFIMs in singly-fed and doubly-fed asynchronous modes, etc.

A torque component is synchronous if it is produced by the interaction of two magnetic field components from two independent magnetic field sources sharing the same principal pole pair number and speed. The production mechanism of the synchronous torque component is shown in Figure 4.2b, where the magnetizing field B_f and armature field B_a are from two independent magnetic field sources,

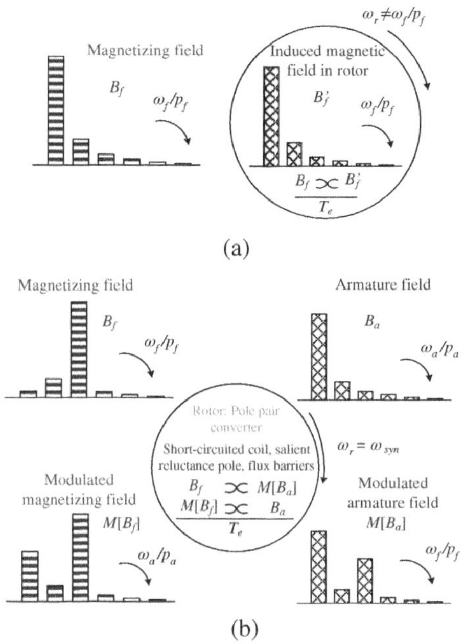

Figure 4.2 Production mechanisms of the asynchronous and synchronous torque components, (a) Asynchronous torque component, (b) Synchronous torque component.

whose mechanical angular velocity are ω_f/p_f and ω_a/p_a, respectively. $M[B_f]$ is the modulated magnetizing field having p_a pole pairs, and its speed has been changed to ω_a/p_a. Similarly, $M[B_a]$ is the modulated armature magnetic field having p_f pole pairs, and its speed is ω_f/p_f. ω_r is the mechanical rotor speed and it is determined by the electrical angular velocities and pole pairs of the magnetizing field and armature field together. In addition, the pole pairs of the magnetizing field and armature field – p_f and p_a can be equal or unequal. T_e is the synchronous torque component.

Typical synchronous torque components include the main torque of the interior PM synchronous machines, the main torque of magnetic field modulated electric machines featuring salient reluctance poles like the flux-switching PM machines, and the synchronous torque component of the BDFIMs in the doubly-fed synchronous mode.

4.3 Characterization of Modulation Behaviors in Typical Machine Topologies

This section presents the qualitative characterization of modulation behaviors in typical electrical machines common in existing literature. For clarity, N_{ST} and N_{RT}

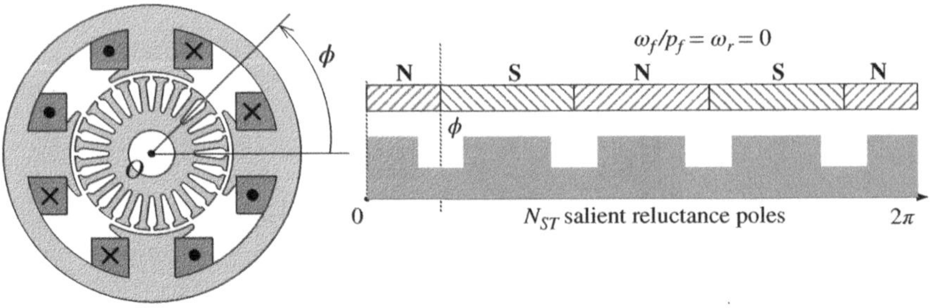

Figure 4.3 DCM with the synchronous modulation behavior of salient reluctance poles from the stator side.

represent the number of stator and rotor poles, respectively. p_f and p_a represent the principal pole pairs of the magnetizing and armature magnetic field, respectively. ω_f and ω_m denote the electrical angular frequency of the source magnetizing MMF and the mechanical rotor speed, respectively.

4.3.1 Brushed DCM

The brushed DCM is essentially an AC machine with a mechanical commutator. A typical DCM is shown in Figure 4.3, and the synchronous magnetic field modulation behavior of the stator salient reluctance pole is also given.

The stator of a DCM is usually constructed with salient reluctance poles to accommodate single-phase magnetizing windings generating a rectangular source magnetizing MMF on the airgap. The stator modulator is of the salient reluctance pole type. Since the source magnetizing MMF stays stationary relative to the stator, it is a synchronous modulation behavior.

After synchronously modulated by stator salient poles, the magnetizing MMF is further asynchronously modulated by the rotor salient poles. However, due to the fine slotted rotor, the rotor salient poles will have a negligible influence on the distribution of the airgap magnetic field. Therefore, the rotor modulation behavior is regarded as a unit modulation under most circumstances.

The armature winding located on the rotor is composed of full-pitched or close to full-pitched coils in series and/or parallel. The winding function of the full-pitched rotor coil has a wide spectrum, allowing almost all the spatial harmonic components produced by the rectangular modulated magnetizing MMF to pass through. Therefore, each coil located on the rotor links a triangular flux linkage and generates a rectangular induced electromotive force (EMF).

4.3.2 Wound-Field Salient-Pole SM

The wound-field salient-pole electrically excited SM having a salient-pole rotor with pole shoes and DC field windings is shown in Figure 4.4. The DC field

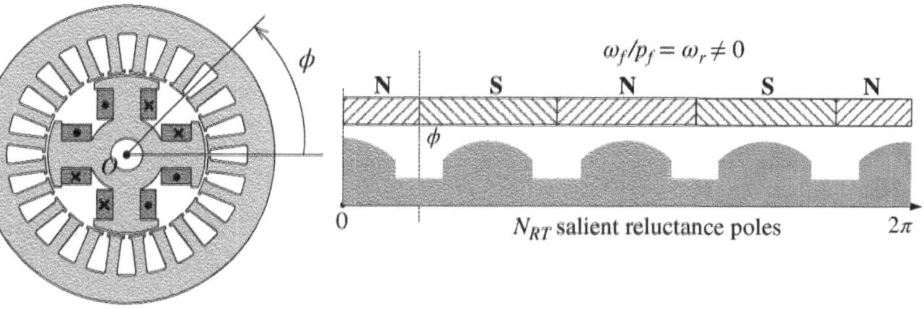

Figure 4.4 Wound-field salient-pole SM with the synchronous modulation behavior of salient reluctance poles from the rotor side.

windings create a rectangular source magnetizing MMF on the airgap which is then modulated synchronously by the salient rotor poles to produce a sinusoidal magnetizing MMF distribution. The fine slotted stator also performs negligible asynchronous modulation to the sinusoidal MMF distribution. The armature winding located on the stator is preferred to be a multi-phase overlapping winding which only allows the spatial harmonic component of a certain principal pole pair number to pass through. To maximize the flux linkage and induced EMF, the number of pole pairs of the armature winding should be equal to that of the modulated magnetizing MMF.

4.3.3 Wound-Field Non-Salient-Pole SM and Slip-Ring Doubly-Fed Induction Machine

The wound-field non-salient-pole SM as shown in Figure 4.5a has no distinct salient poles, but does have fine slots to accommodate sinusoidally distributed single-phase field windings carrying a direct current. The armature winding located on the fine slotted stator is multi-phase and non-overlapping, and primarily allows the spatial harmonic component of $p_f = 1$ pole pair to pass through.

The slip-ring doubly-fed induction machine (DFIM), sometimes referred to as an AC-excited SM, is nearly the same as the non-salient-pole wound-field SM except the replacement of the single-phase distributed DC field winding with a multi-phase distributed AC field winding, as shown in Figure 4.5b.

The difference lies in that the source magnetizing MMF created by the distributed single-phase field winding is sinusoidally distributed and stationary, while the MMF created by the multi-phase distributed AC field winding is a sinusoidal and rotating one. The stator and rotor modulation operators of these two machine types can be treated as unit due to fine slotting.

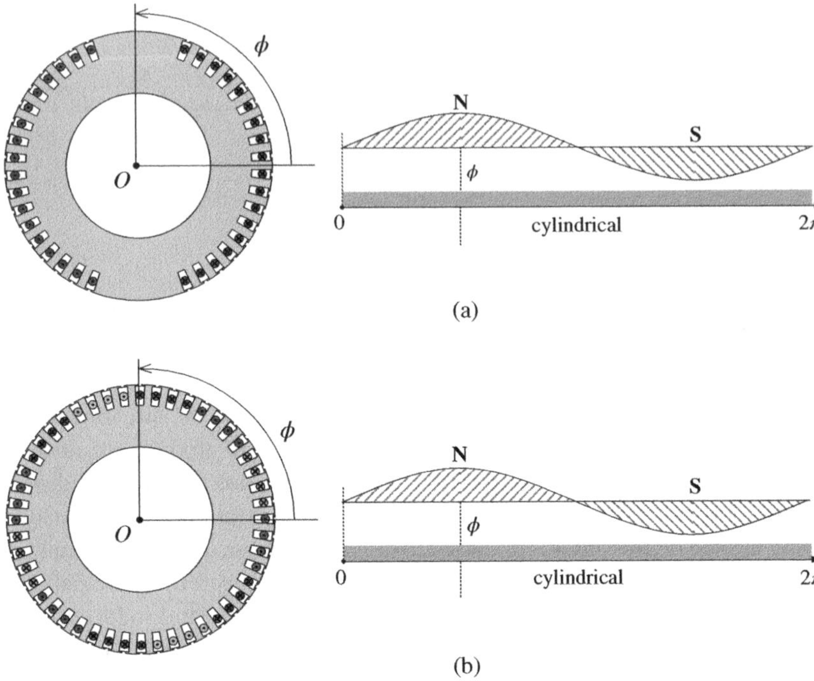

Figure 4.5 Wound-field non-salient-pole SM and slip-ring DFIM with the unit modulation behavior of cylindrical rotors. (a) Wound-field non-salient-pole SM, (b) Slip-ring DFIM.

4.3.4 Squirrel Cage IM and BDFIM

The squirrel-cage IM, as shown in Figure 4.6a, shares some common features with the BDFIM [5] shown in Figure 4.6b, as summarized below:

- Both the stator and rotor are fine slotted;
- The stator windings are sinusoidally distributed multi-phase ones;
- The rotors are equipped with short-circuited coils.

The source magnetizing MMF created by multi-phase distributed stator windings is sinusoidally distributed and travels along the airgap at the speed of ω_f/p_f. Once the rotor equipped with short-circuited coils rotates asynchronously with the source magnetizing MMF, alternating currents will flow in the rotor coils and further create an extra MMF on the airgap termed as the secondary magnetizing MMF. The modulated magnetizing MMF is superimposed by the source and the secondary ones.

The armature winding located on the stator extracts desired spatial harmonic components to produce flux linkage and induced EMF. For squirrel-cage IMs, the

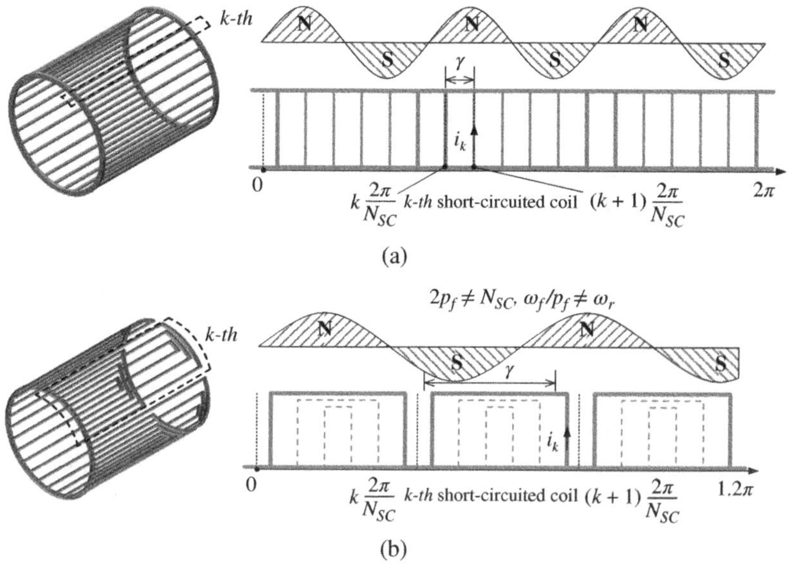

Figure 4.6 Squirrel-cage IM and BDFIM with the asynchronous modulation behavior of short-circuited rotor coils. (a) Squirrel-cage IM, (b) BDFIM. Only the rotor cages are shown for clarity.

fundamental component with $2p_f$ poles is to be extracted by the armature winding, so the pole pairs of the armature winding and the rotational speed of the armature MMF should be the same.

While for BDFIMs, another harmonic component with a different pole pair is to be extracted by the armature winding, so the number of pole pairs of the armature winding is dissimilar to that of the source magnetizing MMF, and so is the mechanical rotor speed. Therefore, typically there are two stator windings of dissimilar pole pairs in BDFIMs.

4.3.5 Synchronous Reluctance Machine and Brushless Doubly-Fed Reluctance Machine

The synchronous reluctance machine (SynRM) and brushless doubly-fed reluctance machine (BDFRM) [6], as shown in Figure 4.7a,b, respectively, share some common features as summarized below:

- Fine slotted stator whose modulator is unit and multi-phase distributed stator windings are adopted;
- Magnetic anisotropy can be achieved by salient reluctance poles and flux barriers (radial or axial lamination) for the rotor design requirement, whose modulation operator varies according to the rotor type.

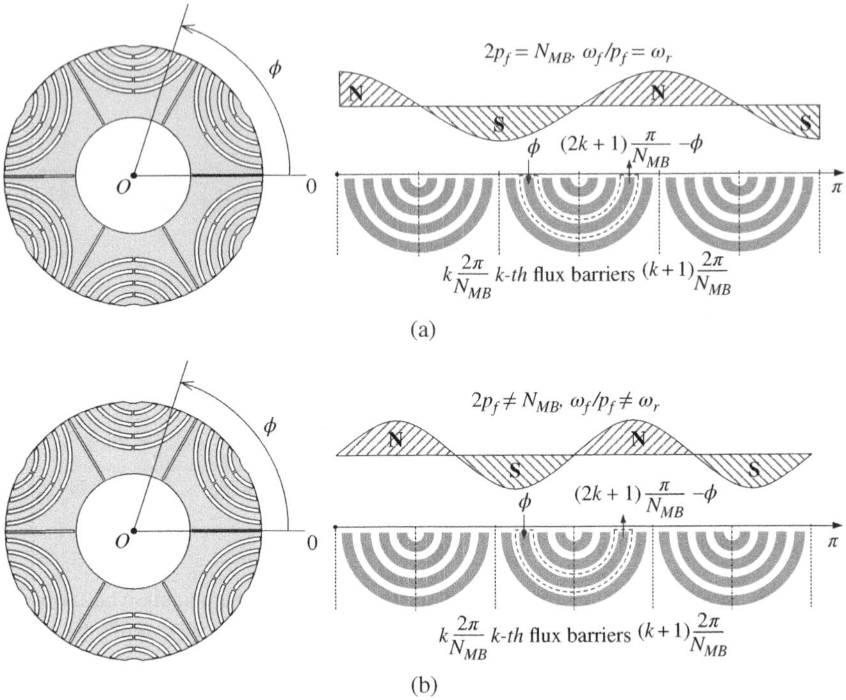

Figure 4.7 SynRM and BDFRM with the modulation behavior of flux barrier rotor, (a) SynRM, (b) BDFRM. Only the rotor cross-sectional views are shown for clarity.

Here the flux barrier rotor constructed from axial laminations is illustrated as an example. The source magnetizing MMF created by multi-phase distributed stator windings is sinusoidally distributed and travels along the airgap at the speed of ω_f/p_f. Once the rotor rotates synchronously with the source magnetizing MMF, the source magnetizing MMF will be synchronously modulated by the flux barriers to produce the modulated magnetizing MMF.

If the armature winding is to extract the source magnetizing MMF, then both the pole pair and mechanical rotating speed of armature MMF should be equal to those of the source magnetizing MMF, which leads to the SynRM.

If the armature winding is to extract another harmonic from the modulated magnetizing MMF, then both the pole pair and mechanical rotating speed of armature MMF are unequal to those of the source magnetizing MMF, which leads to the BDFRM.

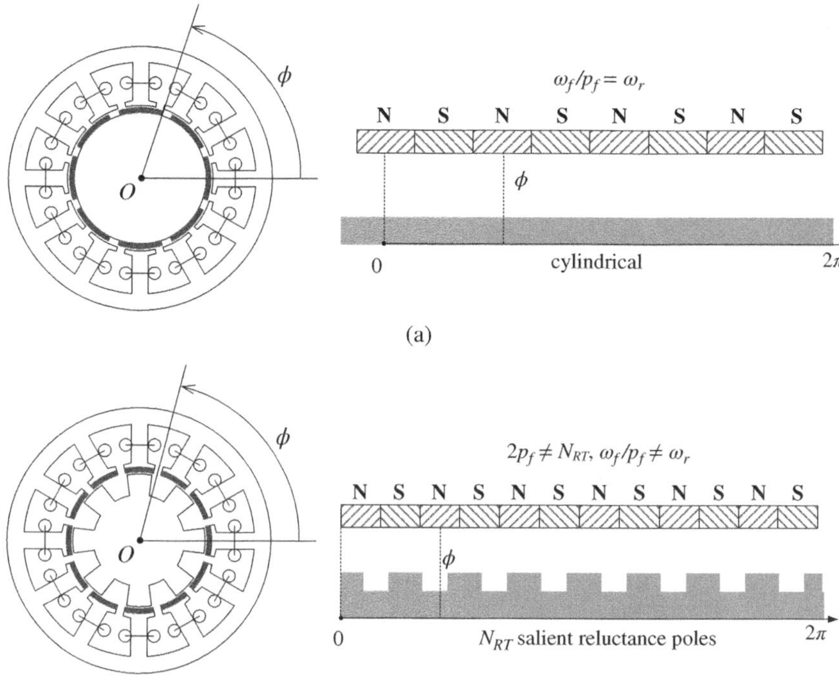

(a)

(b)

Figure 4.8 Surface-mounted PMSM and FRPM machine. (a) Surface-mounted PMSM with the unit modulation, (b) FRPM machine with the asynchronous modulation behavior of salient reluctance poles from the rotor side.

4.3.6 Surface-Mounted PMSM and FRPM Machine

The surface-mounted PMSM and FRPM machine [7] with small stator slot openings, as shown in Figure 4.8 share some common features as summarized below:

- PMs are placed in the main flux path and airgap;
- Stator windings can be configured as distributed or concentrated, and the machine can be driven in brushless DC or brushless AC mode.

In the surface-mounted PMSM, the source magnetizing MMF created by PMs is rectangularly distributed along the airgap, and modulated by the cylindrical rotor and fine slotted stator (both the modulators are units). The armature winding located on the stator extracts the fundamental component of $p_f = 4$ pole pair to produce flux linkage and EMF.

While in FRPM machines, the rectangularly distributed source magnetizing MMF is asynchronously modulated by the salient reluctance pole rotor to produce magnetic field harmonic components with $N_{RT} \pm \nu p_f = 8 \pm 6\nu = +2/(+14)/(-4)/(+20)/(-16)/(+32)\ldots$ $(\nu \neq 3k)$ principal pole pairs. The armature winding with two principal pole pairs located on the stator is used to primarily extract the traveling harmonic component of 2, 4, 14, 16, 20... pole pairs to produce flux linkage and EMF.

4.3.7 Interior PMSM and FSPM Machine

The interior PMSM and FSPM machine [7], as shown in Figure 4.9 share some common features as summarized below:

- PMs are placed in the main flux path and not in the airgap;
- Stator windings can be configured as distributed or concentrated, though concentrated winding is conventionally adopted in a flux-switching PM machine.

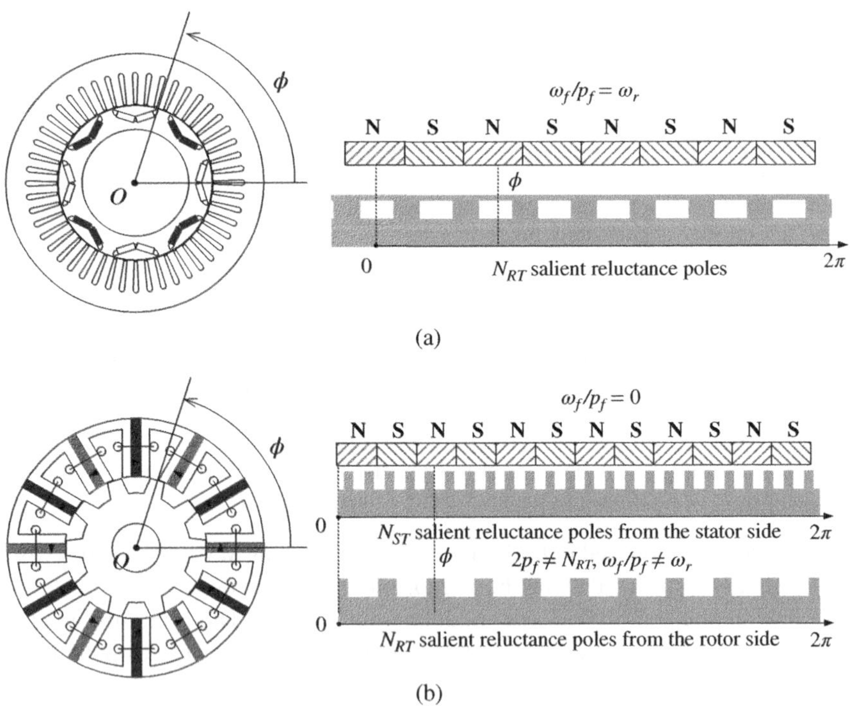

Figure 4.9 Interior PMSM and FSPM machine with the modulation behaviors of salient reluctance poles. (a) Interior PMSM, (b) FSPM machine.

In interior PMSMs, the source magnetizing MMF created by PMs is rectangularly distributed along the airgap, synchronously modulated by the salient-pole rotor and asynchronously modulated by the fine slotted stator. The armature winding located on the stator extracts the fundamental component of $p_f = 4$ pole pair to produce flux linkage and EMF.

While in FSPM machines, the rectangularly distributed source magnetizing MMF is first synchronously modulated by the salient stator poles to produce $(2t-1)p_f$ (t is an arbitrary positive integer) principal MMF components, which are then asynchronously modulated by the salient rotor poles to produce $N_{RT} \pm (2t-1)p_f = 10 \pm (2t-1)6 = +4/(+16)/(-8)/(+28)/(-20)/(+40)\ldots$ magnetic field harmonic components. The armature winding located on the stator with four principal pole pairs extracts the principal traveling harmonic components with 4, 8, 16, 20, 28… pole pairs to produce flux linkage and EMF.

4.3.8 Switched Reluctance Machine and Vernier Machine

As illustrated in Figure 4.10a, except for the double salient pole structure, the switched reluctance machine (SRM) has non-overlapping stator windings and is ideally fed by rectangular pulse currents to create the rectangular source magnetizing MMF, which moves forward step by step rather than continuously.

It is synchronously modulated by the $N_{ST} = 12$ salient stator poles to produce discrete distribution of modulated MMF. The asynchronously modulation behavior is accomplished by the $N_{RT} = 8$ salient rotor poles to produce principal MMF components of $(N_{RT} \pm p_f) = 8 \pm 1 = +7/(+9)$ pole pairs. Nevertheless, different from AC electrical machines, the SRM only utilizes a small spatial interval to complete asynchronously modulation behavior, which is from the position where stator tooth and rotor poles begin to overlap to the full aligned position.

If the concentrated windings are replaced by distributed ones and supplied with multi-phase balanced currents, then the SRM turns into the original vernier machine [8], as shown in Figure 4.10b.

Similar to the modulation process in SRMs, the asynchronous modulation behavior is accomplished by $N_{ST} = 12$ salient stator poles to produce $(N_{ST} \pm p_f) = 12 \pm 1 = +11/(+13)$ pole pair principal MMF components firstly, which is further asynchronously modulated by the $N_{RT} = 8$ salient rotor poles to produce $(N_{RT} \pm (N_{ST} \pm p_f)) = 10 \pm 11/(13) = -1/(+21)/(-3)/(+23)$ magnetic field harmonic components. An armature winding with one principal pole pair is used to extract the principal field component of one pole pair to produce flux linkage and EMF.

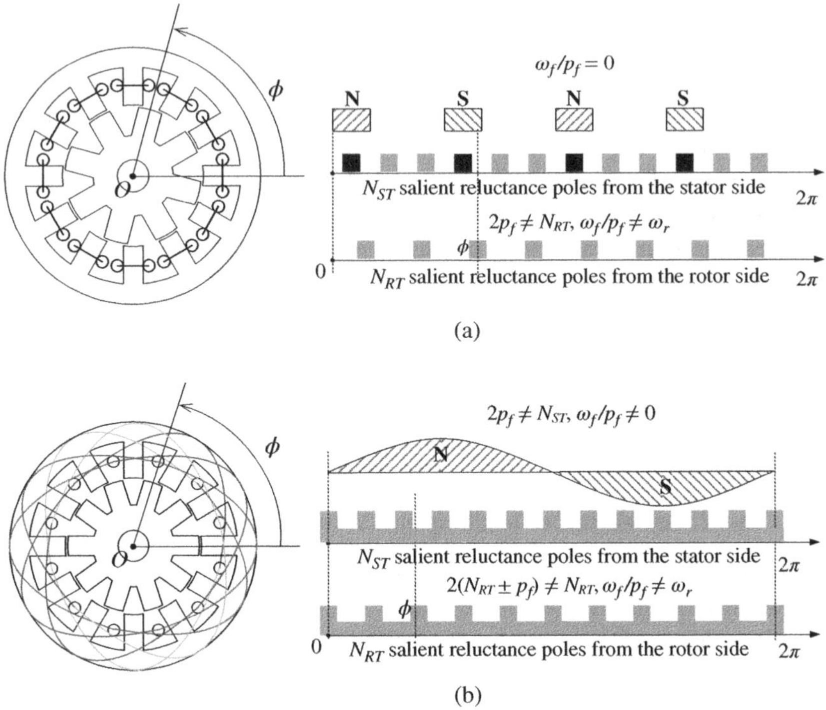

Figure 4.10 SRM and vernier machine with the modulation behaviors of salient reluctance poles from both the stator and rotor sides. (a) SRM, (b) vernier machine.

4.3.9 Magnetically-Geared Machine and PM Vernier Machine

The magnetically-geared machine (MGM) [9] is a simple mechanical combination of a magnetic gear [10] and a PMSM as illustrated in Figure 4.11a. There are two physical air gaps and a ferromagnetic pole-pieced stator.

The asynchronous modulation behavior is performed by the modulation ring which has $N_{Fe} = 25$ salient reluctance poles to produce $(N_{Fe} \pm p_{fo}) = 25 \pm 22 = +3/(+47)$-pole-pair principal magnetic field harmonic components. The principal harmonic components are extracted by the single-layer non-overlapping winding with a principal pole pair number of 3 to produce flux linkage. The unit modulation behavior occurring in the inner PMSM is neglected due to the fine slotted stator.

The stator tooth structure of the PM vernier (PMV) machine [11] performs the same magnetic field modulation effect as the ferromagnetic pieces do in an MGM. Additional working harmonics of the airgap flux density attributed to the so-called magnetic gear effect contribute to the increased torque density in PMV machines compared to traditional PMSMs.

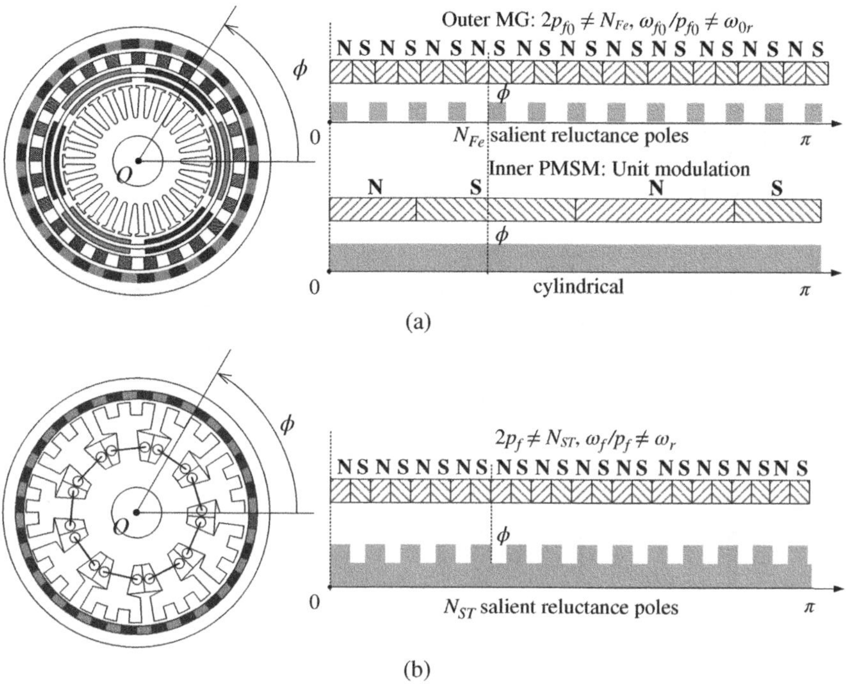

Figure 4.11 MGM and PMV machine with the modulation behaviors of salient reluctance poles from the stator and rotor sides.

As shown in Figure 4.11b, the asynchronous modulation behavior is accomplished in the PMV machine by $N_{ST} = 24$ salient reluctance poles of the inner stator to produce principal magnetic field components of $(N_{ST} \pm p_f) = 24 \pm 27 = -3/(+51)$ pole pairs, where the modulator (inner stator) moves relative to the source magnetizing (armature) MMF. The armature winding with three principal pole pairs is placed on the inner stator to extract the magnetic field component with the same pole pair.

4.4 Torque Composition of Typical Machine Topologies

Torque compositions of common electrical machines are shown in Table 4.2. Taking the salient-pole wound-field SM as an example, it includes the electromagnetic torque component related to the rotor magnetizing flux linkage and the reluctance torque component generated by the unequal reluctance in dq axes. It is worth noting that salient reluctance poles in traditional SMs are mainly used to produce the reluctance torque component, instead of changing the airgap permeance and

Table 4.2 Torque composition of typical electrical machine topologies. Only main modulation behaviors affecting the average torque are listed.

Electrical machine type				Main modulation type	Main torque component type	Number of main average torque components
DCM				Synchronous	Synchronous	1
IM				Asynchronous	Asynchronous	1
SM	Electrically excited	Salient-pole/Non-salient-pole SM		Synchronous	Synchronous	1
		SynRM		Synchronous	Synchronous	1
	PM excited	Surface-mounted/Interior PMSM		Synchronous	Synchronous	1
Magnetic field modulated machines	Salient reluctance pole	FSPM machine		Asynchronous	Synchronous	2
		FRPM machine		Asynchronous	Synchronous	2
		Doubly-salient-pole PM machine		Asynchronous	Synchronous	2
		PMV machine		Asynchronous	Synchronous	2
		Variable-flux reluctance machine		Asynchronous	Synchronous	2
	BDFM	BDFIM		Asynchronous	Synchronous + asynchronous	2 2
		BDFRM	Salient reluctance pole rotor	Asynchronous	Synchronous	2
			Flux barrier rotor	Asynchronous	Synchronous	2
SRM				Asynchronous	Synchronous	1

generate additional working spatial harmonics of different pole pairs and speeds. Therefore, the airgap magnetic field harmonic producing the main torque component is only related to the principal pole pairs of the magnetizing or armature magnetic field, and there is only one working magnetic field component. Both the PM torque and reluctance torque component in salient-pole wound-field SMs belong to the synchronous torque component.

The doubly-salient-pole PM machine is taken as another example, whose pole pairs of magnetizing and armature magnetic field are dissimilar. The asynchronous modulation behavior performed by the rotor salient reluctance poles produces additional working harmonics in the air gap. As a result, the average torque is mainly contributed by two airgap magnetic field harmonics related to the principal pole pairs of the magnetizing and the armature magnetic fields. The electromagnetic torque in the doubly-salient-pole PM machine is dominated by two synchronous torque components.

It is shown that newly-emerged magnetic field modulated electrical machines mainly utilize the asynchronous modulation behavior and rely heavily on additional working spatial harmonics introduced by the asynchronous modulation to produce the average torque. Since the principal pole pair number of the magnetizing and armature magnetic field are dissimilar, as are their electrical angular frequencies, the rotor has to rotate at the so-called equivalent synchronous speed determined by the magnetizing and armature magnetic fields together to produce non-zero average torques.

As can be seen from Table 4.2, whether the torque component is asynchronous or synchronous depends on the presence of short-circuited coil modulator. Synchronous torque components exist in electrical machines working by the synchronous modulation behavior, such as the interior PMSM shown in Figure 4.1b, or by the asynchronous modulation behavior, such as the FRPM machine illustrated in Figure 4.1a. By contrast, the asynchronous torque component only occurs in electrical machines working by the asynchronous modulation behavior.

4.4.1 Case Study I – BDFIM

BDFIMs can work in multiple modes of operation, as shown in Table 4.3 [3]. In the cascaded asynchronous mode, for example, only one of the two electrical ports is supplied by the alternating current and the other is short-circuited. There is no synchronous torque component in the cascade asynchronous mode.

If the power winding (PW) is short-circuited, the control winding (CW) is supplied with balanced alternating currents and the rotor moves relative to the source magnetic MMF created by the CW, a magnetic field with the same principal pole pairs as the CW will be created by the rotor short-circuited coils. This field interacts

Table 4.3 Modes of operation and torque composition of BDFIMs.

Operating mode	Condition of existence	Main torque component type	Number of main average torque components
Simple asynchronous	PW (CW) is supplied by alternating current, CW (PW) is open-circuited	Asynchronous	1
Cascaded asynchronous	PW (CW) is supplied by alternating current, CW (PW) is short-circuited	Asynchronous	2
Singly-fed synchronous	PW (CW) is supplied by alternating current, CW (PW) is supplied by direct current	Synchronous Asynchronous	2 2
Doubly-fed synchronous	PW and CW are both supplied by alternating currents, and (4.1) is satisfied	Synchronous Asynchronous	2 2
Doubly-fed asynchronous	PW and CW are both supplied by alternating currents, but (4.1) is not satisfied	Asynchronous	4

with the magnetic field created by the CW to produce the asynchronous torque component.

Similarly, although the PW is short-circuited, it is coupled with the rotor short-circuited coils and a magnetic field with the same principal pole pair and frequency as PW will be induced. Another asynchronous torque component will be produced by the interaction of the induced magnetic field and the magnetic field created by the PW.

Taking the doubly-fed synchronous operation mode as another example, both the PW and CW are supplied with alternating currents. The rotor rotates at the equivalent synchronous speed determined by the PW and CW magnetic fields, so the two separate frequency systems of PW and CW are unified into a single frequency system. The electrical frequencies and rotor speed satisfy:

$$\omega_p \pm \omega_c = (p_p + p_c)\omega_r \tag{4.1}$$

where the ω_p and ω_c are the electric angular frequencies of the rotating magnetic fields of PW and CW, respectively. ω_r is the rotor mechanical angular velocity, and p_p and p_c are the pole pair number of PW and CW, respectively.

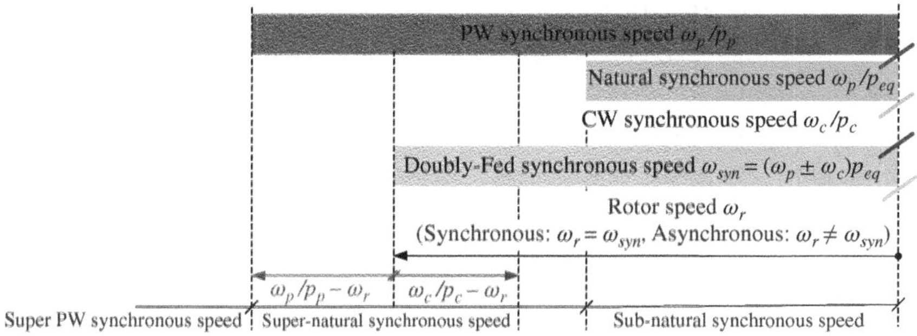

Figure 4.12 Definitions of different speeds in BDFIMs.

In the doubly-fed synchronous operation mode, the definitions of various speeds are shown in Figure 4.12. Both electrical ports are supplied with alternating currents and satisfy the constraint expressed by (4.1). Since the rotor rotates relative to the source MMF created by either PW or CW, the electromagnetic torque has two asynchronous torque components, from the PW and the CW side, respectively. They are produced by the interaction of the fundamental magnetic field created by the PW (CW) and the fundamental magnetic field induced by the PW (CW) in the rotor short-circuited coils.

On the other hand, the modulated magnetic field from the PW (CW) side by the rotor short-circuited coils has the same principal pole pairs and speeds as that from the CW (PW) side. These field components staying stationary relative to each other interact with each other to produce the synchronous torque components. Therefore, in the doubly-fed synchronous mode, the BDFIM has two asynchronous torque components and two synchronous torque components in total.

More specifically, the electromagnetic torque of the BDFIM can be expressed as [12]:

$$T = \underbrace{T_{pc} + T_{pp}}_{T_p} + \underbrace{T_{cp} + T_{cc}}_{T_c} \tag{4.2}$$

where the $T_{pc(cp)}$ is the synchronous torque component produced by the interaction between the magnetic fields in CW (PW) and PW (CW) side, $T_{pp(cc)}$ is the asynchronous torque component produced by the interaction between the magnetizing and the armature magnetic field created by the same PW (CW), and:

$$T_{pc} = -\frac{3p_p}{\omega_p}\frac{|E_{mp}||E_{mc}|}{|Z_{lr}|}\cos(\varphi_{lr} + \delta_{pc}) \tag{4.3}$$

$$T_{pp} = \frac{3p_p}{\omega_p}\frac{|E_{mp}|^2}{|Z_{lr}|}\frac{R_r}{s_r} \tag{4.4}$$

$$T_{cp} = -\frac{3p_c}{\omega_p}\frac{|E_{mp}||E_{mc}|}{|Z_{lr}|}\cos(\varphi_{lr} - \delta_{pc}) \tag{4.5}$$

$$T_{cc} = -\frac{3p_p}{\omega_p}\frac{|E_{mc}|^2}{|Z_{lr}|}\frac{R_r}{s_r} \tag{4.6}$$

where E_{mp} and E_{mc} are EMFs in the rotor circuit induced by the PW and CW magnetic fields, respectively. Z_{lr} is the rotor leakage impedance and $Z_{lr} = R_r/s_r + j\omega_p L_{lr} = Z_{lr} \angle \varphi_{lr}$, where R_r is the rotor resistance, S_r the rotor slip, L_{lr} the rotor leakage inductance, and φ_{lr} the leakage impedance angle. δ_{pc} is the phase difference between E_{mp} and E_{mc}.

Equations (4.2)–(4.6) indicate that for given PW and CW terminal voltages, the peak torque of BDFIMs is reversely proportional to the rotor leakage impedance when stator winding resistances and leakage inductances are neglected. The ratio of rotor leakage reactance to rotor resistance affects the proportion between the asynchronous and synchronous torque components. The greater the ratio is, the closer the rotor leakage reactance angle is to π/2, and the smaller the proportion of the asynchronous component will be. In the extreme case where the resistance of the rotor short-circuited coil is zero (e.g. the rotor short-circuited coils are made of superconductors), asynchronous torque components will disappear from (4.2).

BDFIMs in the single-fed synchronous mode shares some similarities with the salient-pole wound-field SM. The main similarities are:

- External characteristics: one of the two electrical ports is supplied by direct currents to create the magnetizing magnetic field, and the other electrical port is supplied by alternating currents to provide armature magnetic field;
- Magnetic field regulation: in grid-connected BDFIMs in single-fed synchronous mode, the power factor of PW can be changed like a wound-field SM by simply changing the DC field current amplitude;

The difference mainly lies in the following aspects:

- Magnetic field modulation behavior: the salient reluctance pole modulator is stationary relative to the source MMF created by the DC field winding in wound-field SMs, indicating that the wound-field SM is dominated by the synchronous modulation behavior. On the contrary, the short-circuited coil modulator in BDFIMs performs asynchronous modulation behavior;
- Torque composition: the wound-field SM has one synchronous torque component and no asynchronous torque component, as shown in Table 4.2, whereas the BDFIM has two synchronous and two asynchronous torque components in both the single-fed and doubly-fed synchronous modes.

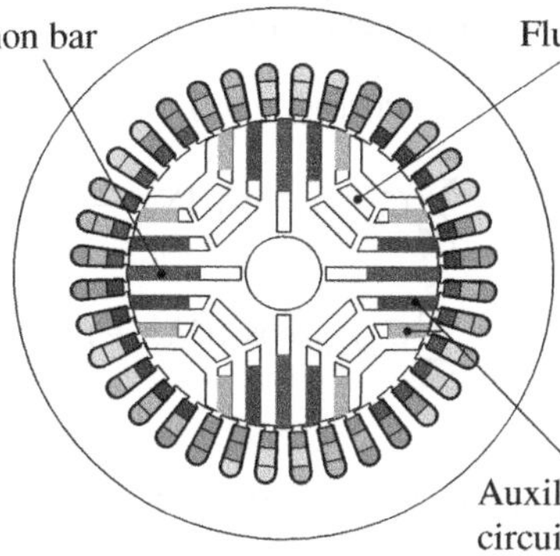

Figure 4.13 Cross-sectional view of the BDFM with a hybrid rotor. Common bars are connected by two end rings. Auxiliary short-circuited coils are nested loops.

4.4.2 Case Study II – BDFM with a Hybrid Rotor

By adding auxiliary short-circuited coils to a multi-layer flux barriers rotor, one can achieve a new rotor construction for BDFMs, as illustrated in Figure 4.13, which has been termed as the hybrid rotor [13]. The hybrid rotor was proposed with an attempt to enhance the magnetic field conversion by combining modulation behaviors of both the short-circuited coils and multi-layer flux barriers. Main magnetic field modulation behaviors in the BDFM with a hybrid rotor are asynchronous due to the relative motion between modulators and the source magnetizing MMF.

Assuming that the source magnetizing MMF is a unit cosine function expressed by:

$$f(\phi, t) = \cos(p\phi - \omega t) \tag{4.7}$$

Applying the multi-layer flux barriers modulator to (4.7) yields:

$$\begin{aligned} M(N_{MB})[f(\phi, t)] = {} & C_{p\_MB} \cos(p\phi - \omega t) \\ & + \sum_{l=1}^{\infty} C_{sum\_MB} \cos[(lN_{MB} - p)\phi + (\omega - lN_{MB}\omega_r)t - l\pi] \\ & + \sum_{l=1}^{\infty} C_{dif\_MB} \cos[(lN_{MB} + p)\phi + (\omega + lN_{MB}\omega_r)t + l\pi] \end{aligned} \tag{4.8}$$

where N_{MB} is the number of flux barrier segments, C_{p_MB}, C_{sum_MB}, and C_{dif_MB} are magnetic field conversion factors and represent interactions between harmonics with p, $(lN_{MB} - p)$ and $(lN_{MB} + p)$ pole pairs, respectively.

Applying the ideal short-circuited coil modulator to Eq. (4.7), one can get:

$$M_{SC}(N_{SC},\gamma)[f(\phi,t)] = C_{p\_SC}\cos(p\phi-\omega t) + \sum_{l=1}^{\infty} C_{sum\_SC}\cos[(lN_{SC}-p)\phi+(\omega-lN_{SC}\omega_r)t+l\pi] + \sum_{l=1}^{\infty} C_{dif\_SC}\cos[(lN_{SC}+p)\phi-(\omega+lN_{SC}\omega_r)t+l\pi] \tag{4.9}$$

where N_{SC} is the number of short-circuited coil groups. γ is the coil span to coil pitch ratio of a specific loop. C_{p_SC}, C_{sum_SC}, and C_{dif_SC} are magnetic field conversion factors introduced by the short-circuited coil modulator.

In the BDFM with a hybrid rotor, both the multi-layer flux barrier and short-circuited coil modulators show asynchronous modulation behaviors, and the modulated MMFs share similar harmonic spectrums considering $N_{MB} = N_{SC}$ and their positions. Assuming the two modulators are independent of each other and applying the linearity of modulation operators, the modulation behavior of the hybrid rotor can be obtained:

$$M(N_{MB})[M(N_{SC},\gamma)[f(\phi,t)]] = M(N_{MB})[f(\phi,t)] + M(N_{SC},\gamma)[f(\phi,t)] \tag{4.10}$$

It can be seen from Equations (4.9) and (4.10) that when the ideal short-circuited coil modulator is used, the amplitudes of harmonics with p and $(lN_{MB} - p)$ pole pairs will be enhanced, so are the corresponding magnetic field conversion factors C_p and $C_{p,NMB-p}$. How the amplitude of the harmonic with $(lN_{MB} + p)$ pole pairs will change depends on the equivalent pole pairs and rotor speed. A closer look at the third terms in (4.8) and (4.9) shows that harmonics with $(lN_{MB} + p)$ pole pairs introduced by the two modulators have different phase angles.

The detailed modulation process is illustrated in Figure 4.14, and the corresponding spectrum is shown in Figure 4.15. The modulation behavior of the hybrid rotor can be characterized as follows:

- The hybrid rotor can enhance the magnetic field conversion factors C_p and $C_{p,NMB-p}$;
- The magnetic field conversion capability of the hybrid rotor depends on the pole pair combination. For the same N_{MB} and N_{SC}, the close-ratio combination between CW and PW produces large mutual magnetic field conversion factors, and thus better cross-coupling between the two stator windings;
- For hybrid rotors with the same equivalent pole pairs, the magnetic field conversion factor is the largest when the pole pairs of PW and CW differ by 1. In practice, the pole pair difference is purposely changed to 2 to avoid the undesired unbalanced magnetic pull (UMP);

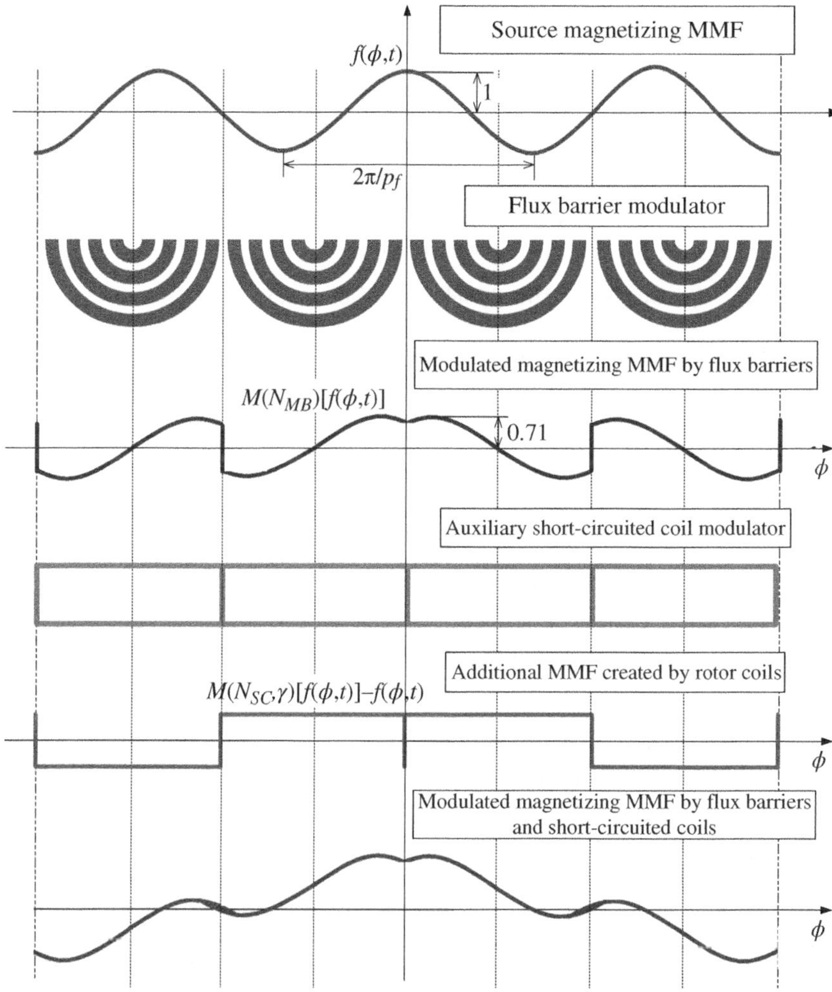

Figure 4.14 Modulation behavior of the hybrid rotor in BDFMs.

- The magnetic field conversion factors will be the same when N_{MB}, N_{SC} and the principal pole pair number p are all multiplied by the same integer number.

The main self-inductance of the CW for the BDFM with a hybrid rotor can be expressed by:

$$L_{msc}^{AA} = \frac{\mu_0 r_g l_{stk}}{g} \int_0^{2\pi} M[W_{scA}(\phi)] W_{scA}(\phi) d\phi \tag{4.11}$$

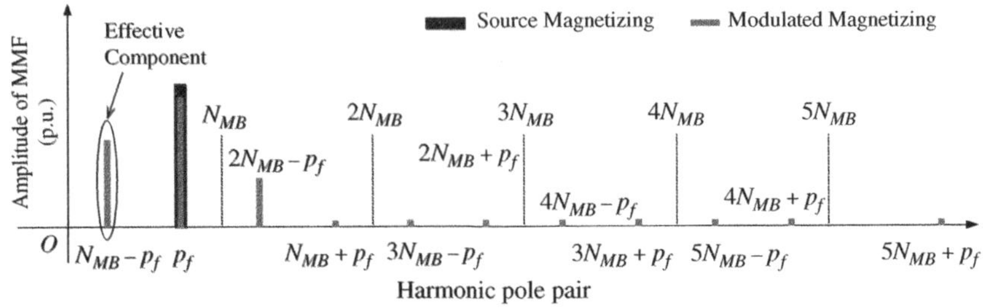

Figure 4.15 Spectrum shift of the BDFM with a hybrid rotor.

where $M[W(\phi)]$ and $W(\phi)$ are modulated and source winding functions, respectively. The subscript "scA" denote phase A of stator CW. Similarly, the inductance of PW can be obtained by replacing c with p.

The mutual inductance between the CW and PW can be expressed as:

$$L_{crp}^{AA}(\theta_r) = \frac{\mu_0 r_g l_{stk}}{g} \int_0^{2\pi} M[W_{scA}(\phi)]W_{spA}(\phi)d\phi \tag{4.12}$$

where θ_r is the mechanical angle along the airgap peripheral.

4.4.3 Case Study III – FSPM Machine

The FSPM machine can be considered a special case of the BDFM operating in singly-fed mode, where the field winding is replaced by PM arrays to create the magnetizing MMF distribution. The torque production mechanism of FSPM machines is illustrated in Figure 4.16.

The principal pole pair numbers of the magnetizing and armature magnetic fields are different, but the rotor rotates at the equivalent synchronous speed determined by the two magnetic fields so that harmonic components in the modulated magnetizing and armature magnetic fields with the same pole pairs interact with each other to produce synchronous torque components.

The magnetic field modulation behaviors and torque components of the magnetic field modulated electrical machines which have salient reluctance poles represented by FSPM and FRPM machines can be described uniformly as:

- The PM arrays create ideal square-wave source magnetizing MMF distribution.
- The source magnetizing MMF is synchronously modulated by the stator salient reluctance poles. The synchronous modulation behavior reduces the amplitude of the fundamental harmonic by approximately half and directly affects the source magnetizing MMF harmonic amplitudes and harmonic contents. The

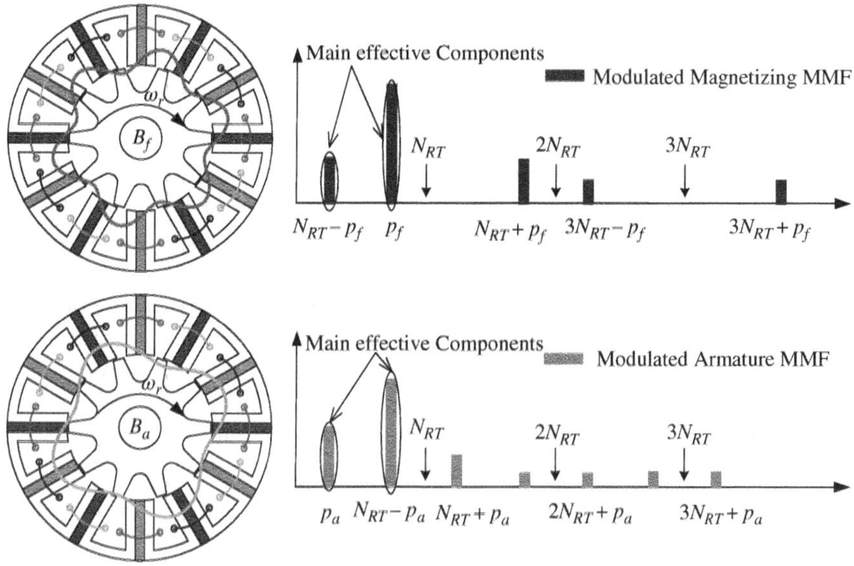

Figure 4.16 Torque production mechanism of FSPM machines.

modulated MMF has a series of spatial harmonics whose number of pole pairs is an odd multiple of the PM pole pair.

- The asynchronous modulation behavior performed by rotor salient reluctance poles modifies the locations of spectrum lines from ν to $lN_{RT} \pm \nu$. Therefore, the combination of principal pole pairs of the source magnetizing MMF and rotor salient reluctance poles has a significant influence on the armature winding layout. MMF harmonic components with $lN_{RT} \pm \nu$ pole pairs are produced, and their relative amplitudes can be characterized by three magnetic field conversion factors;
- The armature winding selectively interacts with the airgap magnetic field harmonic components to produce flux linkage and EMF, and the non-zero electromagnetic torque is produced when balanced multiphase currents with the same frequency as the EMF flows through the armature winding.
- Magnetic field modulated electrical machines with salient reluctance poles have both synchronous and asynchronous modulation behaviors, but the former only changes the amplitude of the source magnetizing MMF and does not affect the airgap magnetic field spectrum distribution. It is the asynchronous modulation behavior that allows the salient-pole rotor to operate at the equivalent synchronous speed determined by the magnetizing and armature magnetic fields to produce non-zero average torque.

4.5 Magnetic Field Modulation Behaviors of Various Modulators

4.5.1 Salient Reluctance Pole Modulator

Salient reluctance pole modulators are the most common in electrical machines, such as the stator and rotor salient pole/tooth structures in FSPM and FRPM machines, PMV machines, MGMs, and so on. Stator salient reluctance poles typically perform synchronous modulation behavior on the source magnetizing MMF, which keeps the positions of spectrum lines unchanged. Therefore, the stator modulator can be designed appropriately to improve mechanical robustness, reduce flux leakage, etc. By contrast, rotor salient poles usually perform asynchronous modulation behavior on an arbitrary source magnetizing MMF, as illustrated in Figure 4.17, where g is a constant representing the uniform airgap length neglecting the slots and d_s denotes the rotor slot depth.

Firstly, we assume that the salient reluctance pole modulator is an ideal modulator, i.e. the MMF distribution over the slot opening region is none. Then the modulation operator for salient reluctance poles presented in Chapter 3 can be simplified to:

$$M(N_{ST}, \varepsilon)[f(\phi)] = \begin{cases} f(\phi), & \phi \in C^S \\ 0, & \phi \in [0,2\pi] - C^S \end{cases} \tag{4.13}$$

$$M(N_{RT}, \varepsilon)[f(\phi)] = \begin{cases} f(\phi), & \phi \in C^R \\ 0, & \phi \in [0,2\pi] - C^R \end{cases} \tag{4.14}$$

where $C^{R(S)}$ is the set of discontinuous intervals occupied by rotor (stator) poles in mechanical radians. By considering the slot depth, we can define the modulation operator $M(N_{ST}, \varepsilon_{ST})[.]$ and $M(N_{RT}, \varepsilon_{RT})[.]$ as:

$$M(N_{ST}, \varepsilon_{ST})[f(\phi)] = \begin{cases} f(\phi), \phi \in C^S \\ \kappa(o_s/t_{ds})f(\phi), & \phi \in [0,2\pi] - C^S \end{cases} \tag{4.15}$$

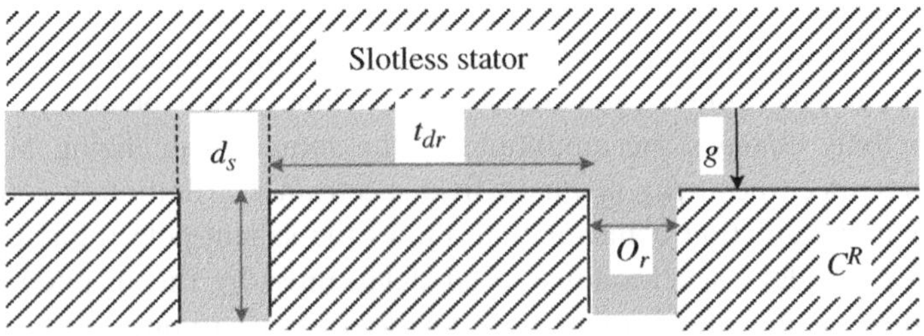

Figure 4.17 Visualization of rotor salient reluctance pole modulator.

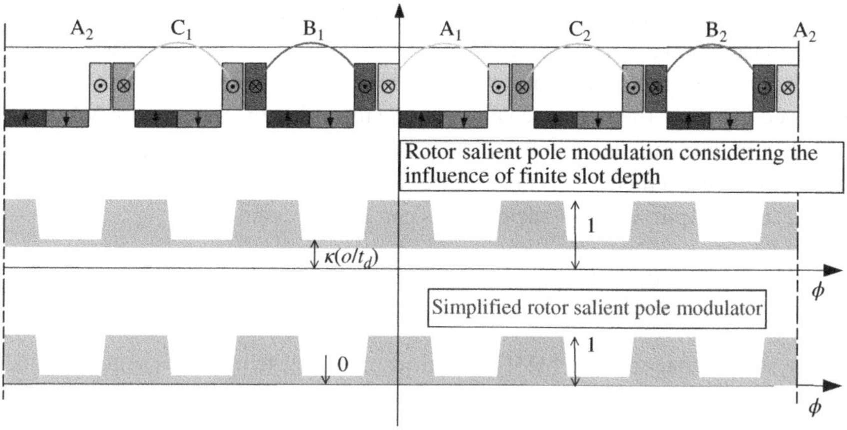

Figure 4.18 Modulation behaviors of salient reluctance poles in the 6/8 FRPM machine.

$$M(N_{RT}, \varepsilon_{RT})[f(\phi, t)] = \begin{cases} f_{RT}(\phi, t)f(\phi, t), \phi \in C \\ 0, \quad \phi \in [0,2\pi] - C \end{cases} \tag{4.16}$$

where C is intersection of the C^S and C^R, and,

$$f_{RT}(\phi, t) = \varepsilon_{RT} + \kappa(1 - \varepsilon_{RT}) + \sum_{l=1}^{\infty} \frac{2(1-\kappa)}{\pi} \frac{\sin(l\varepsilon_{RT}\pi)}{l} \cos[lN_{RT}(\phi - \omega_r t)] \tag{4.17}$$

$$\kappa = \frac{t_{dr} - \frac{8}{5}\beta o_r}{2t_{dr}\left(\frac{d_s}{t_{dr}}\right)} = \frac{t_{dr} - \frac{8}{5}\beta o_r}{2d_s} \tag{4.18}$$

The influence of rotor slot depth d_s on the airgap MMF distribution is taken into consideration by introducing a function κ, which reduces with the increase of d_s. The function κ can be set to zero under the ideal condition of an infinite slot depth for simplicity, as illustrated in Figure 4.18.

The 12/10 FSPM machine is taken as an example to analyze the magnetic field modulation behavior of salient reluctance pole modulators. Assuming that the stator teeth, slot openings and magnets are of the same width, the discontinuous intervals occupied by stator salient poles are:

$$C_{N_{ST}} = \cup_{i=1}^{N_{ST}} \left\{ \left[\left(i - \frac{7}{8}\right)\frac{2\pi}{N_{ST}}, \left(i - \frac{5}{8}\right)\frac{2\pi}{N_{ST}}\right] \cup \left[\left(i - \frac{3}{8}\right)\frac{2\pi}{N_{ST}}, \left(i - \frac{1}{8}\right)\frac{2\pi}{N_{ST}}\right] \right\} \tag{4.19}$$

The source magnetizing MMF distribution created by the PM array is an ideal square wave. For simplicity, we apply the unit cosine function as the input to obtain its response first. The response of each cosine MMF component from the Fourier series of the square-wave MMF created by the PM array can be readily achieved by referencing the cosine function response. All the input responses are then superimposed to obtain the response generated by inputting the unit square MMF.

The response can also be obtained by directly taking the unit square MMF as input. The second method is chosen here, with a unit square-wave MMF of amplitude 1 as the input, whose Fourier series can be expressed as:

$$f(\phi, t) = \sum_{\nu=2k+1}^{\infty} \left(\frac{4}{\pi}\right)\left(\frac{1}{\nu}\right)\sin(\nu p\phi) \tag{4.20}$$

The source magnetizing MMF with unit square distribution is synchronously modulated by the salient stator poles and then it becomes:

$$M(N_{ST}, 0.5)[f(\phi, t)] = \sum_{\nu=2k+1}^{\infty} \left(\frac{4}{\pi}\right)\left(\frac{2}{\nu}\right)\sin\left(\frac{\nu\pi}{8}\right)\cos\left(\frac{2\nu\pi}{8}\right)\sin\left(\frac{4\nu\pi}{8}\right)\sin(\nu p\phi) \tag{4.21}$$

The magnetic field modulation behavior of stator salient poles can be easily distinguished by comparing harmonic components of the modulated and source magnetizing MMF, as tabulated in Table 4.4. It is shown that for the 12/10 FSPM machine, the synchronous modulation behavior performed by stator salient poles almost halves the fundamental harmonic amplitude, while the amplitudes of the third and fifth harmonics are both significantly increased.

To visualize the modulation effect of stator salient poles alone on the source magnetizing MMF, one can change the rotor to a smooth cylinder, and then extract the radial flux density distributions along the airgap circumference from two-dimensional (2D) magnetostatic finite element analysis (FEA). If the rotor structure is not modified to a smooth cylinder, the radial airgap flux density distributions can be calculated for different mechanical rotor positions. By inspecting the envelope in the same coordinate system, the modulation effect of rotor salient reluctance poles can be eliminated and the modulation behavior of stator teeth can be obtained, as illustrated in Figure 4.19. It is shown that there are 24 explicit rectangular pulses in a full mechanical cycle from 0 to 2π, which is caused by the synchronous modulation behavior of the 24 stator teeth.

The magnetizing MMF distribution modulated further by rotor salient reluctance poles under no-load condition can be concisely expressed as:

$$M(N_{RT}, \varepsilon)[M(N_{ST}, 0.5)[f(\phi, t)]] = \begin{cases} M(N_{ST}, 0.5)[f(\phi, t)], & \phi \in C_{N_{ST}} \cap C_{N_{RT}} \\ 0, & \phi \in [0,2\pi] - C_{N_{ST}} \cap C_{N_{RT}}. \end{cases} \tag{4.22}$$

Table 4.4 Harmonic components of the source magnetizing MMF before and after the synchronous modulation performed by stator teeth for the 12/10 FSPM machine with $p = 6$. Results are from 2D FEA.

Harmonic order v	Harmonic pole pair vp	Amplitude of source magnetizing MMF (before synchronous modulation)	Amplitude of source magnetizing MMF (after synchronous modulation)
1	6	1.2732	0.6891
3	18	0.4244	0.5545
5	30	0.2546	0.3327
7	42	0.1819	0.0984
9	54	0.1415	0.0766
11	66	0.1157	0.1512
13	78	0.0979	0.1280
15	90	0.0849	0.0459
17	102	0.0749	0.0405

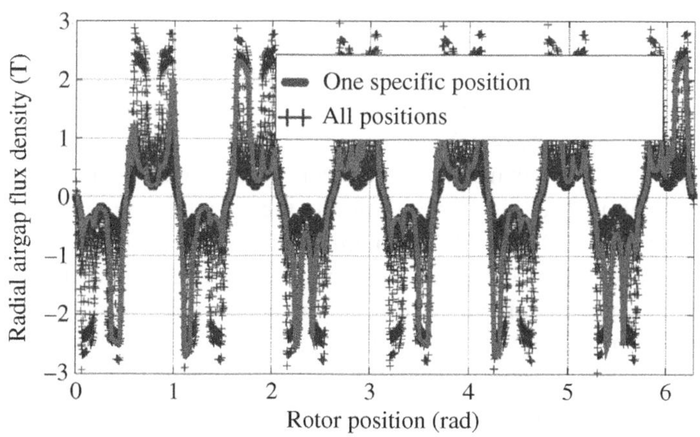

Figure 4.19 Visualization of the modulation effect of stator teeth on the magnetizing MMF in the 12/10 FSPM machine.

Writing (4.22) into its Fourier series, one can get:

$$M(N_{RT}, \varepsilon)[M(N_{ST}, 0.5)[f(\phi, t)]]$$
$$= \left\{ \varepsilon + \sum_{l=1}^{\infty} \left(\frac{2}{\pi}\right) \left[\frac{\sin(l\varepsilon\pi)}{l}\right] \cos[lN_{RT}(\phi - \omega t)] \right\}$$

$$\times \left[\sum_{\nu=2k+1}^{\infty} \left(\frac{4}{\pi}\right)\left(\frac{2}{\nu}\right)\sin\left(\frac{\nu\pi}{8}\right)\cos\left(\frac{2\nu\pi}{8}\right)\sin\left(\frac{4\nu\pi}{8}\right)\sin(\nu p\phi) \right]$$

$$= \sum_{\nu=2k+1}^{\infty} C_{\nu p,\nu p}\sin(\nu p\phi) + \sum_{l=1}^{\infty}\sum_{\nu=2k+1}^{\infty} C_{\nu p,\nu p-lN_{RT}}\sin[(\nu p - lN_{RT})\phi + lN_{RT}\omega t]$$

$$+ \sum_{l=1}^{\infty}\sum_{\nu=2k+1}^{\infty} C_{\nu p,\nu p+lN_{RT}}\sin[(\nu p + lN_{RT})\phi - lN_{RT}\omega t] \quad (4.23)$$

where the three magnetic field conversion factors characterizing the interactions between different pole pair of harmonics can be expressed as:

$$C_{\nu p,\nu p} = \varepsilon\left(\frac{4}{\pi}\right)\left(\frac{2}{\nu}\right)\sin\left(\frac{\nu\pi}{8}\right)\cos\left(\frac{2\nu\pi}{8}\right)\sin\left(\frac{4\nu\pi}{8}\right) \quad (4.24)$$

$$C_{\nu p,\nu p-lN_{RT}} = \left(\frac{1}{2}\right)\left(\frac{4}{\pi}\right)^2\left[\frac{\sin(l\varepsilon\pi)}{l}\right]\left(\frac{1}{\nu}\right)\sin\left(\frac{\nu\pi}{8}\right)\cos\left(\frac{2\nu\pi}{8}\right)\sin\left(\frac{4\nu\pi}{8}\right) \quad (4.25)$$

$$C_{\nu p,\nu p+lN_{RT}} = \left(\frac{1}{2}\right)\left(\frac{4}{\pi}\right)^2\left[\frac{\sin(l\varepsilon\pi)}{l}\right]\left(\frac{1}{\nu}\right)\sin\left(\frac{\nu\pi}{8}\right)\cos\left(\frac{2\nu\pi}{8}\right)\sin\left(\frac{4\nu\pi}{8}\right) \quad (4.26)$$

The stator teeth perform synchronous modulation behavior on the stationary rectangularly distributed source magnetizing MMF, modifying the relative amplitude of each harmonic component but keeping the positions of spectrum lines unchanged, as shown by (4.21).

The rotor salient reluctance poles perform asynchronous modulation behavior on the source magnetizing MMF to modify the harmonic amplitudes and shift the spectrum lines. The MMF harmonic components with $lN_{RT} - (2k-1)p$ and $lN_{RT} + (2k-1)p$ pole pairs are produced, and relative amplitudes of all the three harmonic components can be characterized by three magnetic field conversion factors $C_{\nu p,\nu p}$, $C_{\nu p,\nu p\text{-}lNRT}$, and $C_{\nu p,\nu p+lNRT}$, as indicated by (4.23)–(4.26). Further inspections show that harmonic components of $lN_{RT} - (2k-1)p$ and $lN_{RT} + (2k-1)p$ pole pairs have the same magnetic field conversion factors and thus the same amplitudes, as visualized in Figure 4.20, which is based on 2D FEA results. The detailed modulation process is illustrated in Figure 4.21 and changes of the amplitude spectrum are shown in Figure 4.22.

Another good example is the 6/8 FRPM machine. The Fourier series of the unit square-wave source magnetizing MMF created by the surface-mounted PMs can be expressed as:

$$F_{PM}(\phi, t) = \sum_{\substack{\nu=1 \\ \nu\neq 3k}}^{\infty} \left(\frac{1}{\nu}\right)\left(\frac{3}{\pi}\right)\sin(\nu p_f\phi) \quad (4.27)$$

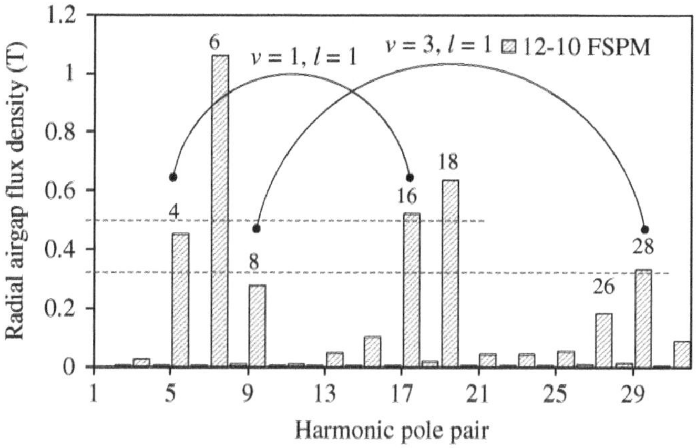

Figure 4.20 Amplitude spectrum of the radial airgap flux density waveform in the 12/10 FSPM machine obtained from 2D FEA.

The source magnetizing MMF expressed by (4.27) is synchronously modulated by stator salient reluctance poles and becomes:

$$M(N_{ST}, \varepsilon_{ST})[F_{PM}(\phi, t)] = \begin{cases} F_{PM}(\phi, t), & \phi \in C^S \\ 0, & \phi \in [0,2\pi] - C^S \end{cases} \tag{4.28}$$

Table 4.5 shows that in the 6/8 FRPM machine, the synchronous modulation behavior of stator salient reluctance poles hardly affects the harmonic amplitudes but eliminates $3k$-th spatial harmonics due to the specially designed pole-arc-to-pole-pitch ratio $\varepsilon_{ST} = 2/3$. It is interesting to see the synchronous modulation behavior performed by stator teeth in this FRPM machine affects the source magnetizing MMF differently compared to the 12/10 FSPM machine, which is mainly caused by the difference in locations of PMs. in FRPM machines, PMs are mounted on the stator tooth tops while in FSPM machines, PMs are sandwiched between stator teeth.

Figure 4.23 clearly shows 12 explicit rectangular pulses in a full mechanical cycle from 0 to 2π, which are caused by the source magnetizing MMF distribution and the synchronous modulation behavior of the 6 stator teeth.

The magnetizing MMF distribution further modulated by rotor salient reluctance poles can be concisely expressed as:

$$M(N_{RT}, \varepsilon_{RT})[M(N_{ST}, \varepsilon_{ST})[F_{PM}(\phi, t)]]$$
$$= \sum_{\substack{\nu=1 \\ \nu \neq 3k}}^{\infty} C_p \sin(\nu p_f \phi) + \sum_{l=1}^{\infty} \sum_{\substack{\nu=1 \\ \nu \neq 3k}}^{\infty} C_{sum} \sin[(\nu p_f - l N_{RT})\phi + l N_{RT} \omega_r t]$$

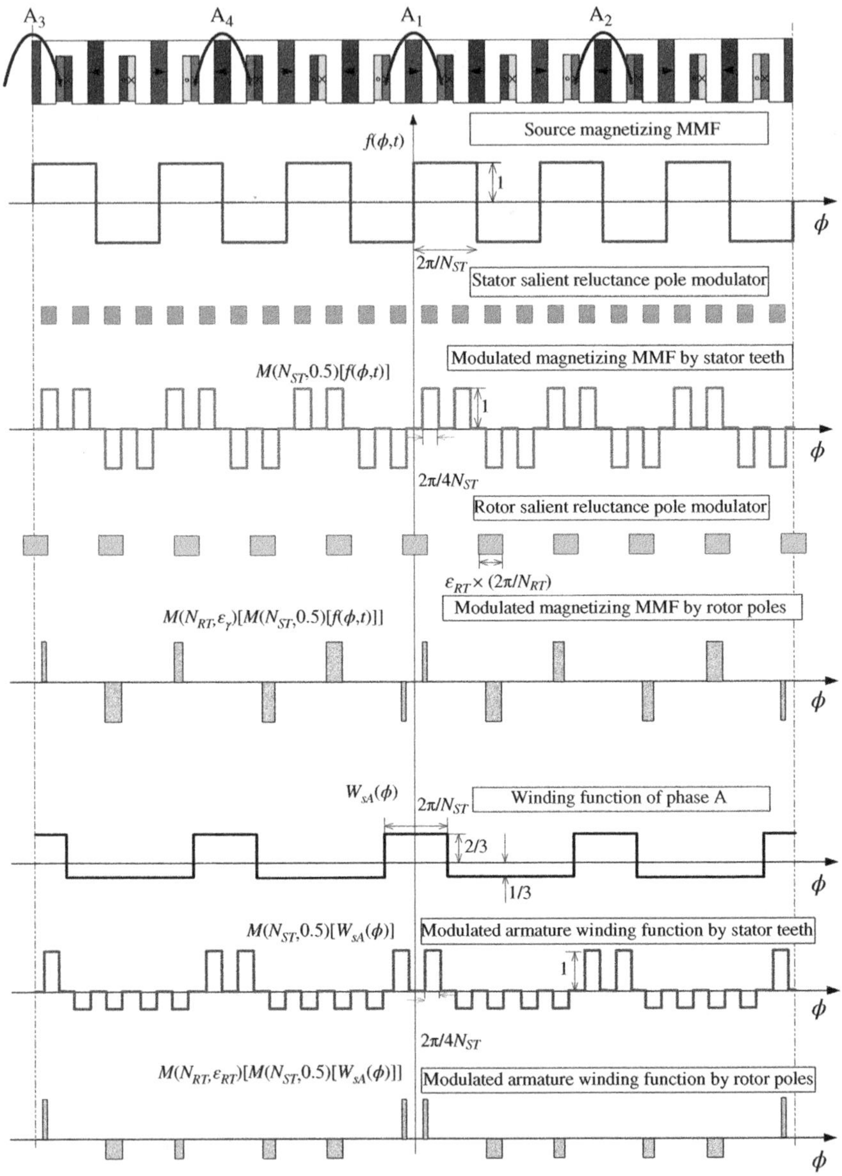

Figure 4.21 Modulation process in the 12/10 FSPM machine.

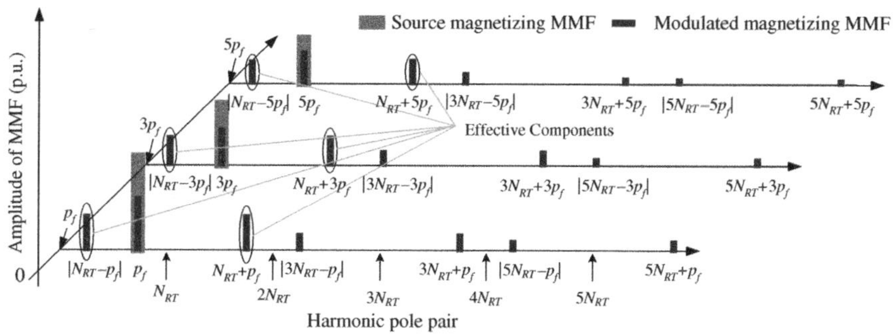

Figure 4.22 Spectrum shift in the 12/10 FSPM machine due to the modulation behaviors of stator teeth and rotor salient reluctance poles.

Table 4.5 Harmonic components of the source magnetizing MMF before and after the synchronous modulation performed by stator salient reluctance poles for the 6/8 FRPM machine with $p = 6$. Results are from 2D FEA.

Harmonic order v	Harmonic pole pair vp	Amplitude of source magnetizing MMF (before synchronous modulation)	Amplitude of source magnetizing MMF (after synchronous modulation)
1	6	1.0171	1.0187
2	12	0.4940	0.4913
4	24	0.2313	0.2312
5	30	0.1718	0.1712
7	42	0.1130	0.1128
8	48	0.0891	0.0893

$$+\sum_{l=1}^{\infty}\sum_{\substack{v=1\\v\neq 3k}}^{\infty} C_{dif}\sin[(vp_f + lN_{RT})\phi - lN_{RT}\omega_r t] \tag{4.29}$$

where the three magnetic field conversion factors characterizing interactions between harmonics of different pole pairs are:

$$C_{vp,vp} = \left(\frac{1}{v}\right)\left(\frac{3}{\pi}\right)[\varepsilon_{RT} + \kappa(1-\varepsilon_{RT})] \tag{4.30}$$

$$C_{vp,vp-lN_{RT}} = \left(\frac{3}{v}\right)\left(\frac{1}{\pi}\right)^2(1-\kappa)\frac{\sin(l\varepsilon_{RT}\pi)}{l} \tag{4.31}$$

$$C_{vp,vp+lN_{RT}} = \left(\frac{3}{v}\right)\left(\frac{1}{\pi}\right)^2(1-\kappa)\frac{\sin(l\varepsilon_{RT}\pi)}{l} \tag{4.32}$$

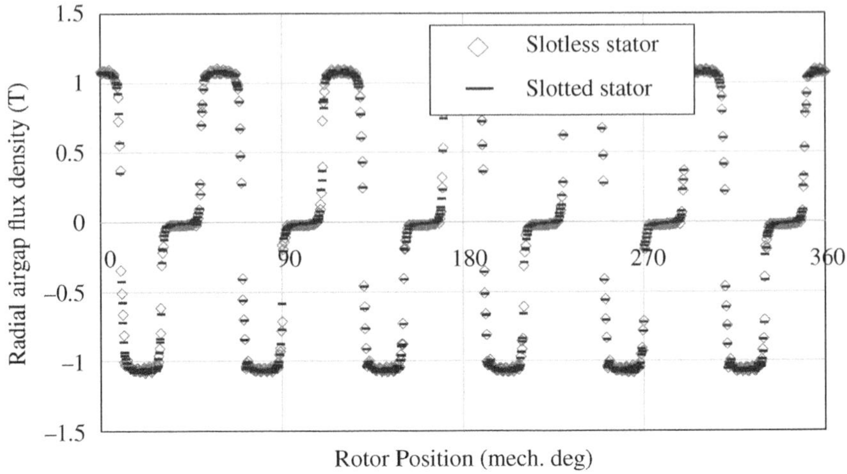

Figure 4.23 Visualization of the modulation effect of stator teeth on the magnetizing MMF in the 6/8 FRPM machine.

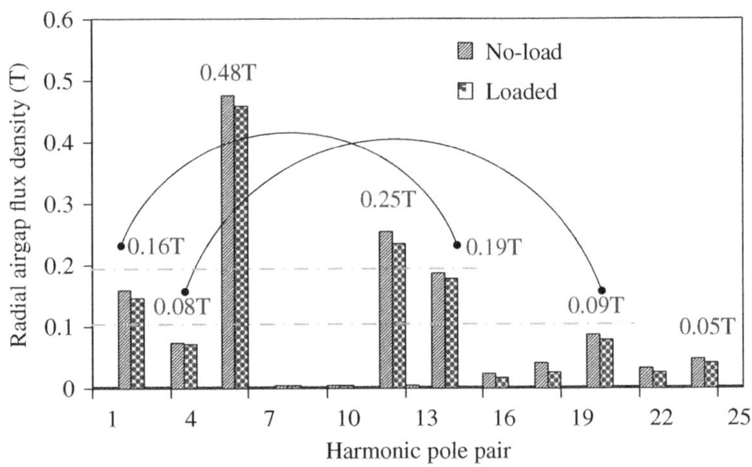

Figure 4.24 Amplitude spectrum of the radial airgap flux density waveform in the 6/8 FRPM machine obtained from 2D FEA.

Similarly, the magnetic field harmonics always appear in pairs and $C_{vp,vp-lNRT}$ equals $C_{vp,vp+lNRT}$ for the same v and l, as validated by 2D FEA results shown in Figure 4.24. Moreover, the vp-pole-pair harmonics are stationary but $(vp \pm lN_{RT})$-pole-pair harmonics are rotating, and the rotating direction depends only on which is larger between lN_{RT} and vp.

The detailed modulation process in the studied 6/8 FRPM machine is shown in Figure 4.25, and the corresponding spectrum change is illustrated in Figure 4.26.

Magnetic field conversion factors for salient reluctance pole modulators with $N_{RT} \leq 10$ have been calculated and tabulated in Table 4.6, where $\varepsilon_{RT} = 0.5$, $d_s = 0.5$ mm, and $g = 0.35$ mm. The following conclusions can be drawn for salient reluctance pole modulators based on the discussion presented in this section:

- The synchronous modulation behavior performed on the stationary rectangularly distributed source magnetizing MMF modifies the relative amplitude of each harmonic component, but leaves the positions of spectrum lines unchanged. On the contrary, the asynchronous modulation behavior not only modifies the harmonic amplitudes but also shift the spectrum lines.
- Magnetic field harmonics always appear in pairs, and there are three types of spatial harmonics corresponding to the source harmonic in the airgap, whose pole pair numbers are vp, $(vp + lN_{RT})$, and $(vp - lN_{RT})$, respectively.
- The magnetic field conversion factors and capability of salient reluctance pole modulators are independent of pole pair combination and determined by geometric parameters such as $\varepsilon_{S(R)T}$ and d_s only. It is suitable for MGMs with far-ratio pole pair combinations so relatively large magnetic field conversion factors can still be maintained.
- The best modulation effect or pole-changing capability can be achieved with $\varepsilon_{RT} = 0.5$. Magnetic field conversion factors $C_{p,NRT-p}$ and $C_{p,NRT+p}$ increase with the slot depth as opposed to $C_{p,p}$.
- The vp-pole-pair harmonics produced by the PM array are still stationary but $(vp \pm lN_{RT})$-pole-pair harmonics are rotating, and the rotating direction depends only on which is larger between lN_{RT} and vp.
- The so-called stator-PM machines represented by the FSPM and FRPM machines utilize multiple spatial harmonics to produce the average torque.

4.5.2 Multilayer Flux Barrier Modulator

Flux barrier rotors have been mainly used in SynRMs and BDFRMs, showing higher magnetic saliency and typically better modulation effect than rotors with salient reluctance poles. They may be stacked by radial or axial laminations. The radial lamination with flux guides and ducts can be considered a simplified version of the axial lamination, and it can effectively suppress the eddy current loss due to magnetic flux normal to the lamination plane.

The modulation behavior of multi-layer flux barriers is often explained by analogy to the salient reluctance poles. However, there are significant differences between the two in terms of modulation behaviors. Applying the flux barrier

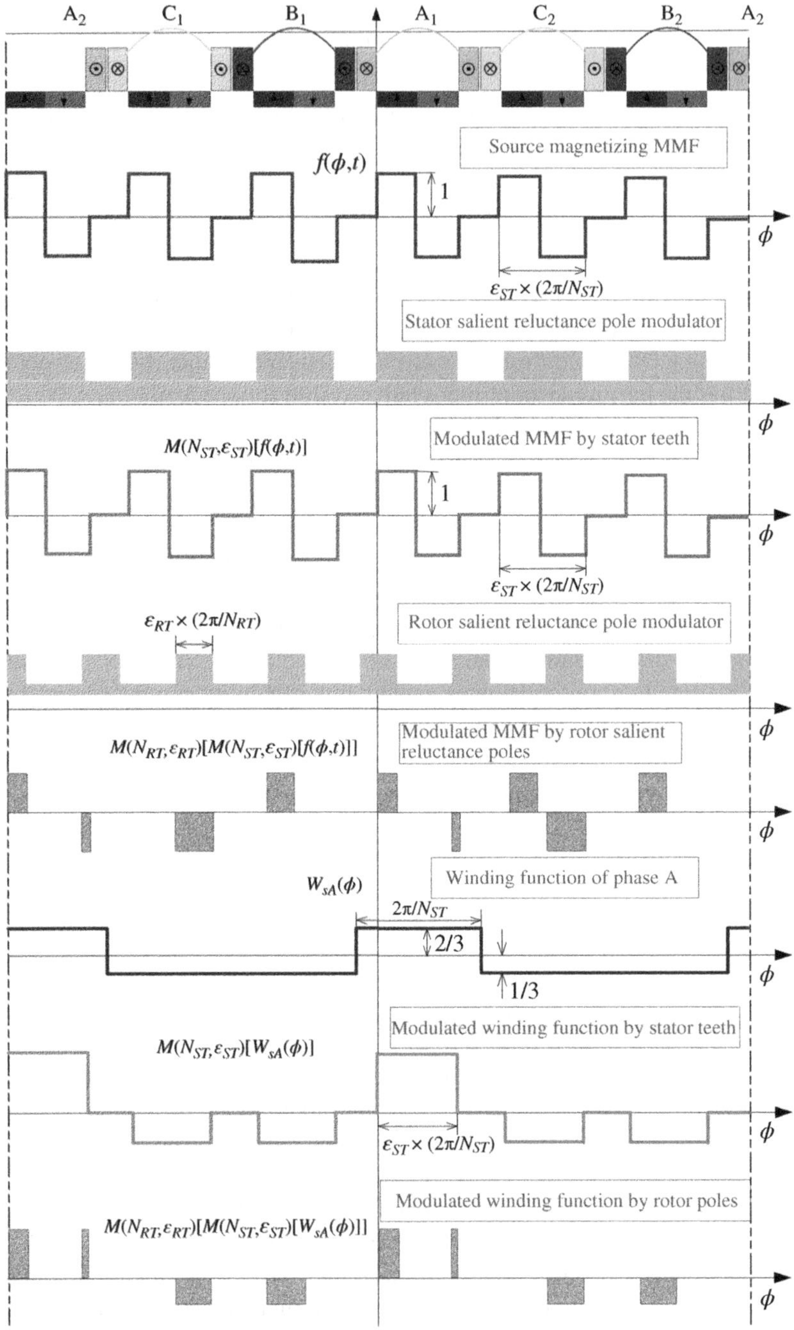

Figure 4.25 Modulation process in the 6/8 FRPM machine.

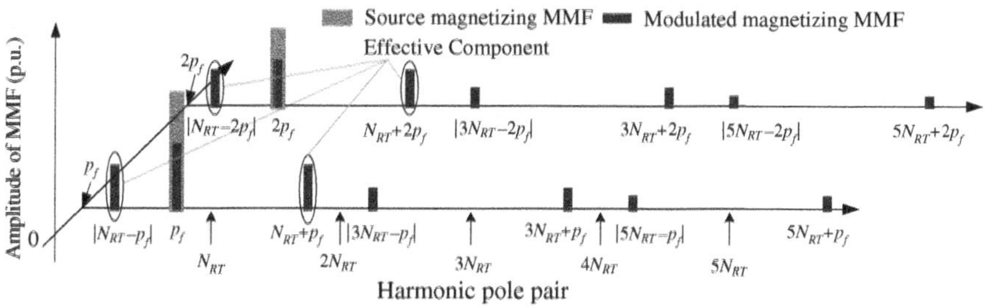

Figure 4.26 Spectrum shift in the 6/8 FRPM machine due to the modulation behaviors of stator teeth and rotor salient reluctance poles.

modulation operator to the unit cosine source magnetizing MMF, one can get [14, 15]:

$$M(N_{MB})[\cos(p\phi-\omega t)] = C_{p,p}\cos(p\phi-\omega t) + \sum_{k=lN_{MB}-p}^{\infty} C_{p,k}\cos\left[k\phi+\omega t-(k+p)\frac{\pi}{N_{MB}}\right] + \sum_{k=lN_{MB}+p}^{\infty} C_{p,k}\cos\left[k\phi+\omega t+(k-p)\frac{\pi}{N_{MB}}\right] \quad (4.33)$$

where the three magnetic field conversion factors are:

$$C_{p,p} = \left(\frac{1}{2}\right)\left[1-\frac{\sin\left(\frac{2p\pi}{N_{MB}}\right)}{\left(\frac{2p\pi}{N_{MB}}\right)}\right] \quad (4.34)$$

$$C_{p,lN_{MB}-p} = \frac{-\sin\left[\frac{(lN_{MB}-2p)\pi}{N_{MB}}\right]}{2\frac{(lN_{MB}-2p)\pi}{N_{MB}}}, \quad l=1,2,.. \quad (4.35)$$

$$C_{p,lN_{MB}+p} = \frac{-\sin\left[\frac{(lN_{MB}+2p)\pi}{N_{MB}}\right]}{2\frac{(lN_{MB}+2p)\pi}{N_{MB}}}, \quad l=1,2,.. \quad (4.36)$$

The MMF harmonic components with $|lN_{MB}-p|$ and $(lN_{MB}+p)$ pole pairs are produced, and relative amplitudes of all the three harmonic components can be characterized by magnetic field conversion factors: $C_{p,p}$, $C_{p,lNMB-p}$, and $C_{p,lNMB+p}$. The detailed modulation process in a BDFRM with two stator windings, one having six poles and the other two poles, is shown in Figure 4.27, and the corresponding spectrum shifting is visualized in Figure 4.28.

Magnetic field conversion factors for the studied BDFRM with $l \leq 5$ have been calculated and tabulated in Table 4.7. It is shown that the modulation behavior

Table 4.6 Magnetic field conversion factors of salient reluctance pole modulators for different pole combinations ($v = 1, l = 1$).

N_{RT}	p	$N_{RT}-p$	$N_{RT}+p$	Rotor salient reluctance pole with a finite slot depth			Rotor salient reluctance pole with an infinite slot depth		
				$C_{p,p}$	$C_{p,NRT-p}$	$C_{p,NRT+p}$	$C_{p,p}$	$C_{p,NRT-p}$	$C_{p,NRT+p}$
3	1	2	4						
4	1	3	5						
5	1	4	6						
5	2	3	7						
6	1	5	7						
6	2	4	8						
7	1	6	8						
7	2	5	9						
7	3	4	10						
8	1	7	9	0.626	0.210	0.210	0.477	0.304	0.304
8	2	6	10						
8	3	5	11						
9	1	8	10						
9	2	7	11						
9	3	6	12						
9	4	5	13						
10	1	9	11						
10	2	8	12						
10	3	7	13						
10	4	6	14						

of multi-layer flux barriers produces fewer magnetic field harmonics compared to the salient reluctance pole structure. The modulated MMF mainly contains a few low-order spatial harmonics with relatively high amplitudes. The useless spatial harmonic components can be further reduced by select the pole pair combination appropriately.

The magnetic field conversion capability of flux barriers modulator is closely related to the pole pair combination. For the same equivalent rotor pole pair

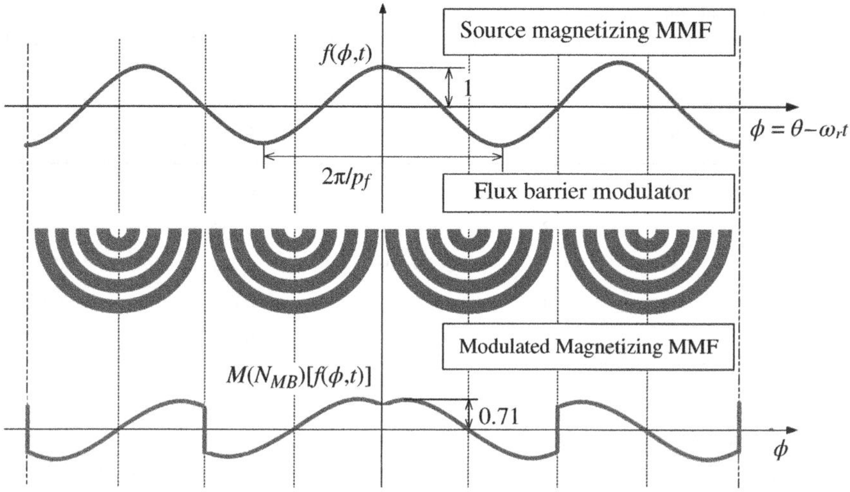

Figure 4.27 Modulation process of flux barriers on the unit cosine source magnetizing MMF.

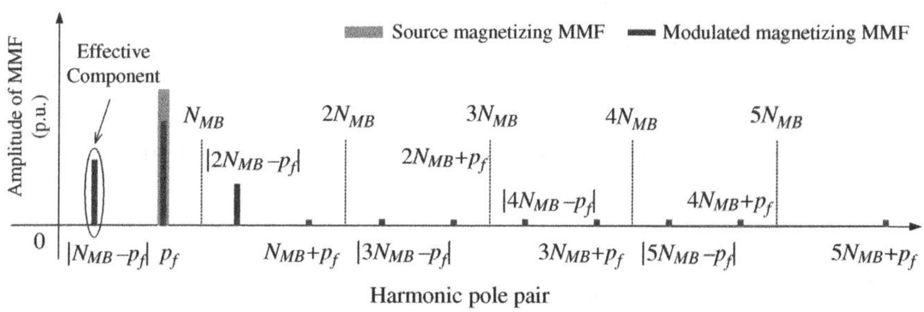

Figure 4.28 Spectrum shift of flux barriers on the unit cosine source magnetizing MMF.

number N_{MB}, the close-ratio pole pair combination for the two stator windings produces larger magnetic field conversion factors, showing better magnetic field conversion and cross-coupling capability between the two independent stator windings.

Magnetic field conversion factors for the BDFRM with a multi-barrier rotor with $N_{MB} \leq 10$ have been tabulated in Table 4.8. A few useful conclusions can be drawn:

- There are three types of spatial harmonics in response to the source magnetizing MMF, whose pole pair numbers are p, $|lN_{MB} - p|$ and $(lN_{MB} + p)$, respectively.
- Magnetic field conversion factors and capability of flux barrier modulators are closely related to the pole pair combination. For the same equivalent rotor pole

Table 4.7 Magnetic field conversion factors for a BDFRM with a multi-layer flux barrier rotor and two stator winding whose numbers of pole pairs are 3 and 1, respectively, with $l \leq 5$.

N_{MB}	p	l	$C_{p,p}$	$C_{p,lNMB-p}$	$C_{p,lNMB+p}$
		1		0.318	0.064
		2		0.318	0.045
4	3	3	0.606	0.106	0.035
		4		0.064	0.029
		5		0.045	0.024

pair number N_{MB}, the close-ratio pole pair combination produces larger magnetic field conversion factors, showing better magnetic field modulation and cross-coupling capability between the two stator windings.

- For the same N_{MB}, the largest magnetic field conversion factor is achieved when p and $(N_{MB} - p)$ differ by 1. In practice, numbers of pole pairs of the two windings are usually differed by 2 to avoid the undesired UMP.
- Magnetic field conversion factors will stay unchanged when N_{MB} and p are scaled by the same integer number.
- The flux barrier rotor structure can effectively suppress spatial harmonics of high pole pairs in asynchronous modulation by providing low magnetic field conversion factors.

4.5.3 Short-Circuited Coil Modulator

The short-circuited coil modulator is a passive device that relies on the slip frequency between the source MMF and the modulator to induce currents in coils. The induced alternating current will create an additional MMF to change the source MMF. The short-circuited coil modulator may take the form of the standard squirrel-cage winding with equally-spaced bar conductors, or more complicated short-circuited windings, such as the nest-loop, spiral-loop, multiphase double-layer windings, etc.

The modulation behavior of the short-circuited coil modulator can be described as:

$$M(N_{SC}, \gamma)[f(\phi, t)] = f(\phi, t) + \sum_{j=1}^{N_{SC}} W_j i_j, \quad \phi \in [0,2\pi] \tag{4.37}$$

The presence of short-circuited coils will change the MMF distribution along the airgap circumference and affect the airgap magnetic flux density distribution.

Table 4.8 Magnetic field conversion factors of flux barrier modulators with different pole combinations, $l = 1$.

				Flux barriers modulator		
N_{MB}	p	$N_{MB}-p$	$N_{MB}+p$	$C_{p,p}$	$C_{p,NMB-p}$	$C_{p,NMB+p}$
3	1	2	4	0.603	0.413	0.083
4	1	3	5	0.606	0.318	0.106
5	1	4	6	0.595	0.252	0.108
5	2	3	7	0.578	0.468	0.052
6	1	5	7	0.583	0.207	0.103
6	2	4	8	0.603	0.413	0.083
7	1	6	8	0.573	0.174	0.097
7	2	5	9	0.609	0.362	0.099
7	3	4	10	0.560	0.483	0.037
8	1	7	9	0.564	0.150	0.090
8	2	6	10	0.606	0.318	0.106
8	3	5	11	0.590	0.450	0.064
9	1	8	10	0.558	0.132	0.084
9	2	7	11	0.601	0.282	0.109
9	3	6	12	0.603	0.413	0.083
9	4	5	13	0.549	0.490	0.029
10	1	9	11	0.552	0.117	0.078
10	2	8	12	0.595	0.252	0.108
10	3	7	13	0.608	0.378	0.095
10	4	6	14	0.578	0.468	0.052

Balanced multiphase alternating currents flowing into stator windings create a source magnetizing MMF in the air gap. Currents are induced instantly in the rotor short-circuited coils and produce an additional MMF distribution to oppose the source magnetizing MMF, according to Lenz's law. Considering that the short-circuited coil modulator relies on induced currents to work, a speed difference between the source MMF and the modulator is required.

Considering the simplest rotor winding with N_{SC} independent loops equally spaced along the airgap circumference with a coil-span-to-coil-pitch ratio γ, as illustrated in Figure 4.29, the modulated MMF can be obtained as:

$$\begin{aligned} M(N_{SC},\gamma)[f(\phi,t)] &= C_{p,p}^{d}\cos(p\phi-\omega t)+C_{p,p}^{q}\cos\left(p\phi-\omega t+\frac{\pi}{2}\right)\\ &-\sum_{k=lN_{SC}-p}^{\infty}C_{p,k}\cos\left[k\phi+(\omega-lN_{SC}\omega_r)t+(p+k)\frac{\pi}{N_{SC}}-\phi-\frac{\pi}{2}\right]\\ &+\sum_{k=lN_{SC}+p}^{\infty}C_{p,k}\cos\left[k\phi-(\omega+lN_{SC}\omega_r)t+(k-p)\frac{\pi}{N_{SC}}+\phi-\frac{\pi}{2}\right] \end{aligned} \tag{4.38}$$

where $C_{p,p}$, $C_{p,lNSC-p}$, and $C_{p,lNSC+p}$ are magnetic field conversion factors, and can be expressed by:

$$\begin{cases} C_{p,p}^{d}=1-2\pi N_{SC}\dfrac{\omega_s}{\sqrt{R_r^2+(\omega_s L_r)^2}}\dfrac{\mu_0 r_g l_{stk}}{g}\left(\dfrac{\gamma}{N_{SC}}\right)^2\left[\dfrac{\sin\left(\gamma\frac{p\pi}{N_{SC}}\right)}{\gamma\frac{p\pi}{N_{SC}}}\right]^2\sin\phi\\ C_{p,p}^{q}=2\pi N_{SC}\dfrac{\omega_s}{\sqrt{R_r^2+(\omega_s L_r)^2}}\dfrac{\mu_0 r_g l_{stk}}{g}\left(\dfrac{\gamma}{N_{SC}}\right)^2\left[\dfrac{\sin\left(\gamma\frac{p\pi}{N_{SC}}\right)}{\gamma\frac{p\pi}{N_{SC}}}\right]^2\cos\phi \end{cases} \tag{4.39}$$

$$\begin{aligned} C_{p,lN_{SC}-p} &= 2\pi N_{SC}\frac{\omega_s}{\sqrt{R_r^2+(\omega_s L_r)^2}}\frac{\mu_0 r_g l_{stk}}{g}\left(\frac{\gamma}{N_{SC}}\right)^2\frac{\sin\left(\gamma\frac{p\pi}{N_{SC}}\right)}{\gamma\frac{p\pi}{N_{SC}}}\frac{\sin\left(\gamma\frac{k\pi}{N_{SC}}\right)}{\gamma\frac{k\pi}{N_{SC}}},\\ k &= lN_{SC}-p,\ l=1,2,3\ldots \end{aligned} \tag{4.40}$$

$$\begin{aligned} C_{p,lN_{SC}+p} &= 2\pi N_{SC}\frac{\omega_s}{\sqrt{R_r^2+(\omega_s L_r)^2}}\frac{\mu_0 r_g l_{stk}}{g}\left(\frac{\gamma}{N_{SC}}\right)^2\frac{\sin\left(\gamma\frac{p\pi}{N_{SC}}\right)}{\gamma\frac{p\pi}{N_{SC}}}\frac{\sin\left(\gamma\frac{k\pi}{N_{SC}}\right)}{\gamma\frac{k\pi}{N_{SC}}},\\ k &= lN_{SC}+p,\ l=1,2,3\ldots \end{aligned} \tag{4.41}$$

where $\omega_s=\omega-p\omega_r$ is the slip angular frequency. R_r and L_r are the loop resistance and inductance, respectively. $\varphi=\mathrm{atan}(\omega_s L_r/R_r)$ is the loop impedance angle, and when $R_r=0$, it will be 90°. When the end winding leakage inductance is neglected, the loop inductance can be expressed concisely as $L=\mu_0 r_g l_{stk} g^{-1}\gamma(2\pi/N_{SC})$.

The standard squirrel-cage winding is equivalent to a set of isolated loop windings with the bar resistance and leakage inductance scaled in steady state. In squirrel-cage IMs, N_{SC} is typically much larger than p, so (4.40) and (4.41)

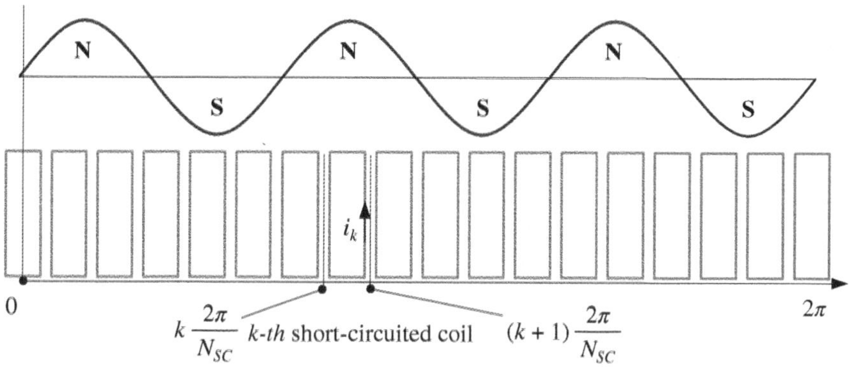

Figure 4.29 Modulation behavior of N_{SC} independent loops equally spaced along the air gap circumference. The coil-span-to-coil-pitch ratio γ is set to 0.7.

become close to zero, and (4.39) can be simplified to:

$$\begin{cases} C_{p,p}^{d} \approx 1 - 2\pi N_{SC}\dfrac{\omega_s}{\sqrt{R_r^2 + (\omega_s L_r)^2}}\dfrac{\mu_0 r_g l_{stk}}{g}\left(\dfrac{\gamma}{N_{SC}}\right)^2 \sin\varphi \\ C_{p,p}^{q} \approx 2\pi N_{SC}\dfrac{\omega_s}{\sqrt{R_r^2 + (\omega_s L_r)^2}}\dfrac{\mu_0 r_g l_{stk}}{g}\left(\dfrac{\gamma}{N_{SC}}\right)^2 \cos\varphi \end{cases} \tag{4.42}$$

Equation (4.42) indicates the asynchronous modulation behavior of the squirrel-cage winding decompose the airgap magnetic field into the d-axis and q-axis components. In addition, the modulation effect depends heavily on the loop resistance and inductance. The loop resistance cannot be neglected, otherwise both $C_{p,p}$ in d- and q-axis will be zero. The slip frequency also cannot be zero, otherwise the airgap magnetic field will have the magnetizing component only and the torque producing component will be zero, which is the ideal no-load operation of IMs.

In BDFIMs, the rotor resistance and loop leakage inductance can be neglected, and the magnetic field conversion factors can be simplified to:

$$C_{p,p}\Big|_{\text{ideal}} = 1 - \gamma\left[\frac{\sin\left(\gamma\frac{p\pi}{N_{SC}}\right)}{\gamma\frac{p\pi}{N_{SC}}}\right]^2 \tag{4.43}$$

$$C_{p,k}\Big|_{\text{ideal}} = \gamma\frac{\sin\left(\gamma\frac{p\pi}{N_{SC}}\right)}{\gamma\frac{p\pi}{N_{SC}}}\frac{\sin\left(\gamma\frac{k\pi}{N_{SC}}\right)}{\gamma\frac{k\pi}{N_{SC}}},\quad k = lN_{SC} - p \tag{4.44}$$

Table 4.9 Magnetic field conversion factors for a BDFRM with the simplest loop rotor winding and two stator winding whose numbers of pole pairs are 3 and 1, respectively, with $l \leq 5$.

N_{SC}	p	l	Conversion factors when $\gamma = 1.0$			Conversion factors when $\gamma = 0.7$		
			$C_{p,p}$	$C_{p,lNSC-p}$	$C_{p,lNSC+p}$	$C_{p,p}$	$C_{p,lNSC-p}$	$C_{p,lNSC+p}$
4	3	1	0.910	0.270	0.039	0.744	0.402	0.071
		2		0.054	0.025		0.059	0.016
		3		0.030	0.018		0.083	0.047
		4		0.021	0.014		0.045	0.035
		5		0.016	0.012		0.004	0.003
	1	1	0.189	0.270	0.162	0.368	0.402	0.093
		2		0.116	0.090		0.112	0.131
		3		0.074	0.062		0.026	0.071
		4		0.054	0.048		0.075	0.006
		5		0.043	0.039		0.054	0.049

$$C_{p,k}\Big|_{\text{ideal}} = \gamma \frac{\sin\left(\gamma \frac{p\pi}{N_{SC}}\right)}{\gamma \frac{p\pi}{N_{SC}}} \frac{\sin\left(\gamma \frac{k\pi}{N_{SC}}\right)}{\gamma \frac{k\pi}{N_{SC}}}, \quad k = lN_{SC} + p \tag{4.45}$$

It is shown that the source magnetizing MMF created by a set of multiphase winding supplied by balanced multiphase currents has only one main harmonic related to the pole pair number of the stator winding p, while there will be three different harmonic components whose number of pole pairs are p, $(lN_{SC} - p)$, and $(lN_{SC} + p)$ in the modulated MMF. All the three magnetic field conversion factors for the simplest loop winding depend on p, N_{SC}, and γ.

The BDFIM with four independent short-circuited loops equally spaced along the airgap circumference and two stator windings, whose number of pole pairs are 3 and 1, is taken as an example. The loop span is $2\pi/N_{SC}$, so $\gamma = 1$. Magnetic field conversion factors with $l \leq 5$ when the six-pole winding is excited have been calculated based on (4.43)–(4.45), and tabulated in Table 4.9 for better understanding.

Equations (4.43)–(4.45) indicate that the modulation behavior of short-circuited coil modulators is dependent on pole pair combinations. For the same N_{SC}, the close-ratio combination of $N_{SC} - p$ and p produces a larger magnetic field conversion factor $C_{p,NSC-p}$, showing better magnetic field modulation and cross-coupling capability between the two stator windings.

When p approaches $N_{SC}/2$, the two harmonic groups introduced by the modulation behavior of short-circuited coil modulators become more significant, among

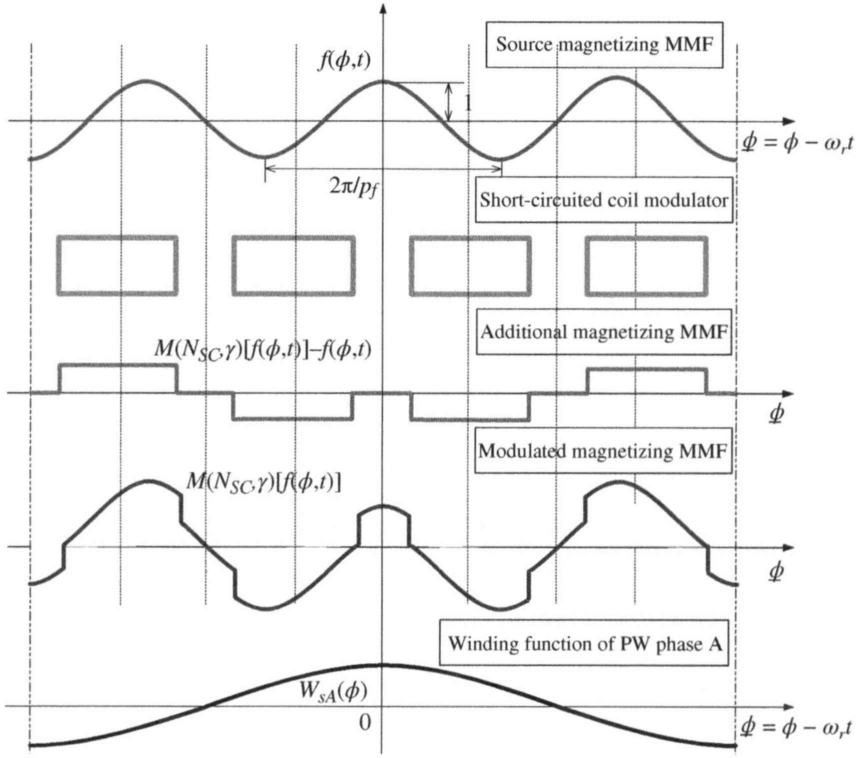

Figure 4.30 Modulation process in the BDFIM with a two-pole PW and six-pole CW. The modulation effect of the short-circuited coil modulator on the CW MMF is clearly shown. The PW is open-circuited. The underline denotes the position is in the rotor-speed reference frame.

which the two harmonics with $(N_{SC} \pm p)$ pole pairs have the largest amplitude. When the armature winding is designed to have $N_{SC} - p$ pole pairs, the IM turns into a BDFIM. The detailed modulation process in the studied BDFIM is illustrated in Figure 4.30, and the corresponding spectrum shifting is visualized in Figure 4.31.

In reality, the short-circuited rotor winding can be very complicated in terms of the conductor layout and connections between conductors, such as the nested-loop winding, spiral-loop winding, and multiphase double-layer winding with equal- or unequal-turn coils shown in Chapter 2. The key to derive magnetic field conversion factors for different rotor winding types lies in obtaining the winding functions and determining the current flowing in each short-circuited coil group.

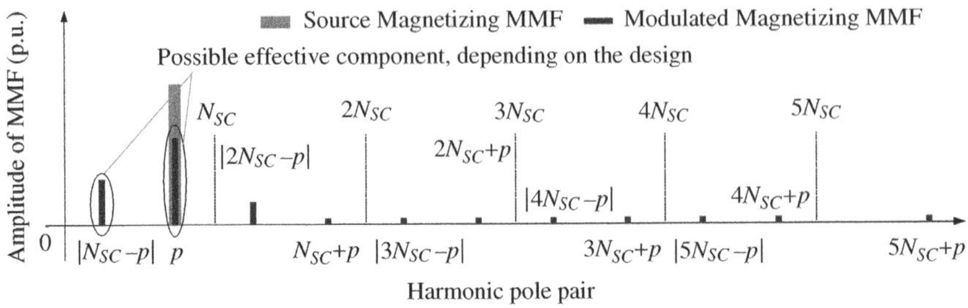

Figure 4.31 Spectrum shift in the BDFIM due to the modulation behaviors of short-circuited rotor windings.

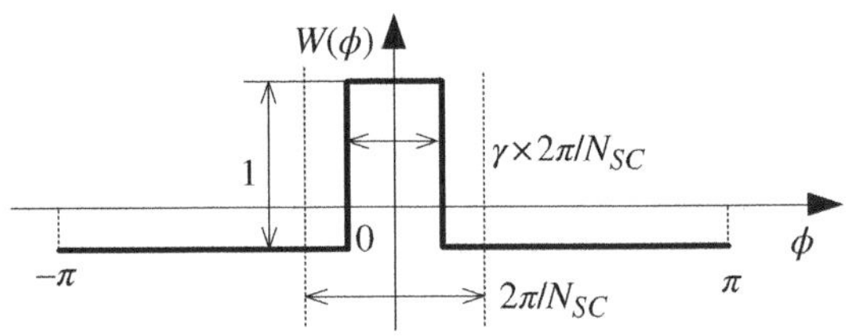

Figure 4.32 Winding function of an individual coil with a span of $\gamma \times 2\pi/N_{SC}$.

The winding function of a single short-circuited coil shown in Figure 4.32 can be calculated from the following definition:

$$W_j(\phi) = T_j(\phi, \phi_0) - \frac{1}{2\pi} \int_0^{2\pi} T_j(\phi, \phi_0) d\phi, \quad \phi \in [0, 2\pi], \tag{4.46}$$

where $T_j(\phi,\phi_0)$ is called the turn function and ϕ_0 is the lower bound of the integral, which does not appear in the winding function.

To determine the currents in the rotor short-circuited windings, one needs to look at the voltage equations of rotor short-circuited windings [16], which can be expressed in matrix form as

$$\mathbf{0} = [r_r]_{S\times S}[i_r]_{S\times 1} + \frac{d[\psi_r]_{S\times 1}}{dt}, \tag{4.47}$$

where $[r_r]_{S\times S}$ is the resistance matrix, $[i_r]_{S\times 1}$ is the rotor current vector, and $[\psi_r]_{S\times 1}$ is the rotor flux linkage vector.

According to the assumption that the rotor resistances are negligible in a doubly fed synchronous operation, the rotor flux linkage vector in the steady state should be zero, which means that

$$[\psi_r]_{S\times 1} = ([L_r]_{S\times S} - [L_{r-r}]_{S\times S})[i_r]_{S\times 1} + [\psi_{rs}]_{S\times 1} = \mathbf{0}, \tag{4.48}$$

where $[L_r]_{S\times S}$ is the group inductance matrix, which comprises the self- and mutual inductance of the loops within one winding group; $[L_{r-r}]_{S\times S}$ is the group–group mutual inductance matrix, which comprises the mutual inductance terms between loops of different groups; and $[\psi_{rs}]_{S\times 1}$ is the flux linkage linked by a rotor winding group resulting from the air-gap magnetic field created by the stator windings.

The flux linkage linked by the ith loop, whose coil span is $\tau_i \times 2\pi/N_{SC}$, can be calculated from

$$\begin{aligned}[\psi_{rs}]_i &= \int_{(1-\gamma_i)\pi/N_{SC}+(i-1)\times 2\pi/N_{SC}}^{(1+\gamma_i)\pi/N_{SC}+(i-1)\times 2\pi/N_{SC}} B_1(\phi,t) l_{\text{stk}} r_g d\phi, \\ &= \frac{\mu_0 r_g l_{\text{stk}}}{g_e}\left(\frac{2}{p_1}\right)\sin\left(\frac{\gamma_i p_1 \pi}{N_{SC}}\right) \\ &\quad \cos\left[\frac{(2i-1)p_1\pi}{N_{SC}} - (\omega_1 - p_1\omega_r)t + p_1\theta_{r0} - (p_1\phi_{01} + \varphi_1)\right]. \end{aligned} \tag{4.49}$$

For the simplest case, where there are N_{SC} isolated short-circuited loops equally spaced on the rotor surface, the closed-form equation for the rotor current can be obtained as

$$i_i(t) = -\frac{\sin\left(\frac{\gamma p_1 \pi}{N_{SC}}\right)}{\left(\frac{\gamma p_1 \pi}{N_{SC}}\right)} \cos\left[(2i-1)\frac{p_1\pi}{N_{SC}} - (\omega_1 - p_1\omega_r)t + p_1\theta_{r0} - (p_1\phi_{01} + \varphi_1)\right]. \tag{4.50}$$

It can be seen that the amplitude of the rotor current depends only on the coil span and the pole combination of the stator windings once the rotor resistance is neglected. When the number of loops per group, S, is greater than one or a single short-circuited winding group has multiple slot conductors, the closed-form analytical solutions for the rotor current will become very complicated. In such cases, the rotor current expressions can obtained by symbolic computations using the MATLAB Symbolic Math Toolbox or other similar tools.

By multiplying the rotor winding functions with rotor currents, the rotor MMF induced by a pure cosine stator MMF for a number of rotor winding configurations can be obtained, as shown in Figure 4.33, together with the winding configuration of a single rotor winding group.

It is worth noting that the presence of rotor resistance introduces an additional phase shift in the loop currents and the rotor MMF. Consequently, two asynchronous torque components produced by the interaction between the stator MMF and rotor MMF of the same pole pair numbers are introduced, which depend on the slip frequency and rotor resistance.

With more complex rotor winding configurations, the symbolic expressions for the conversion factors can be derived by symbolic computations. By substituting specific numbers of pole pairs and coil spans into these expressions, the numerical

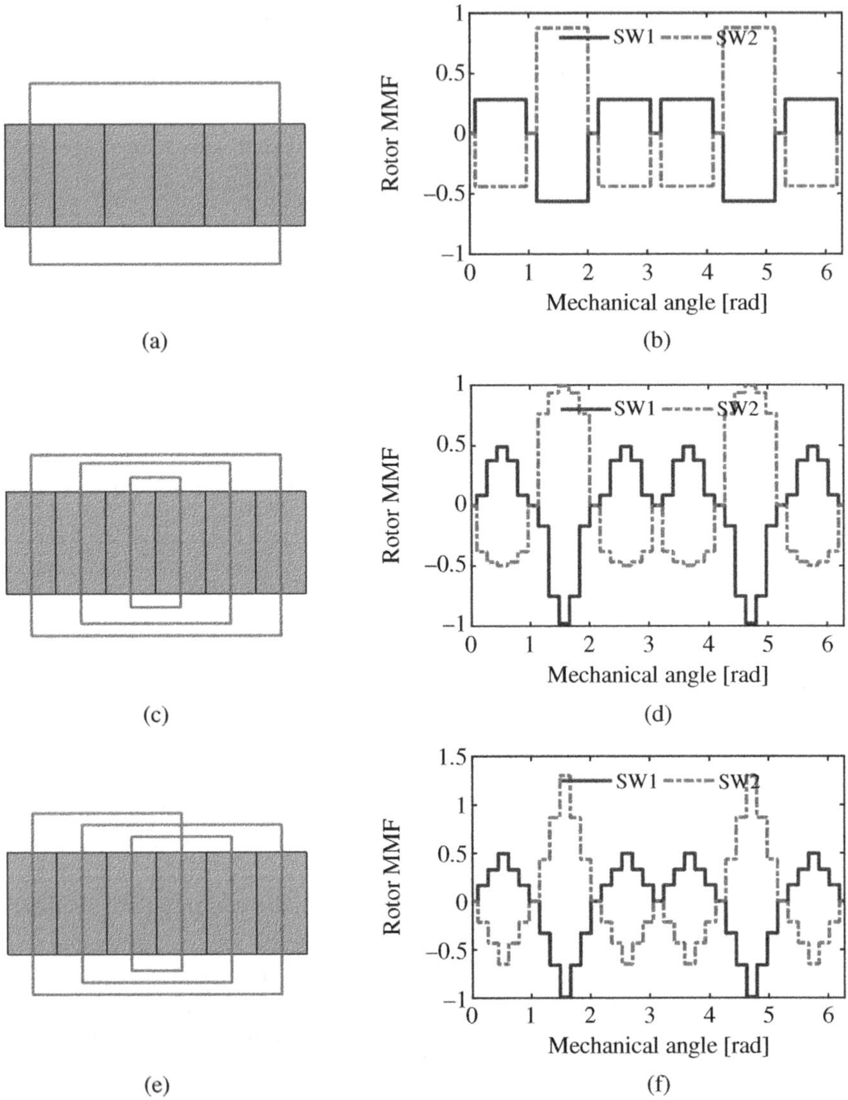

Figure 4.33 Normalized MMF waveforms of various rotor windings: (a) nested-loop winding ($N_{SC} = 6, Z_r = 36, S = 1, \tau = 5/6$), (b) MMF of (a), (c) nested-loop winding ($N_{SC} = 6, Z_r = 36, S = 3, \tau = 1/6$, 3/6, and 5/6) Adapted from [17], (d) MMF of (c), (e) spiral-loop winding ($N_{SC} = 6, Z_r = 36, S = 3, \tau = 1/6$, 3/6, and 5/6) Adapted from [18], (f) MMF of (e), (g) spiral-loop winding ($N_{SC} = 6, Z_r = 36, S = 3$, with an angle between adjacent conductors of 6.125°) Adapted from [18], (h) MMF of (g), (i) multiphase double-layer winding with unequal-turn coils ($N_{SC} = 6, Z_r = 54$, $N_1 : N_2 : N_3 : N_4 = 5:14:14:22$) Adapted from [19], (j) MMF of (i). Only 1/6 of the rotor windings are shown in (a), (c), (e), (g), and (i).

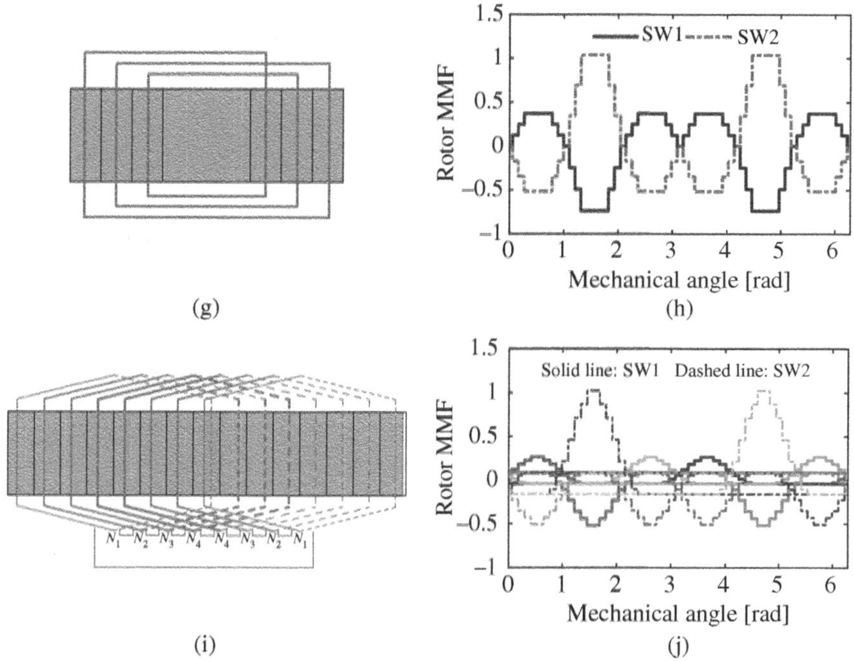

Figure 4.33 *(Continued)*

values can be obtained. Some examples are listed in Table 4.10, demonstrating the universality of the proposed approach.

In summary, the conversion factors are calculated in three steps using symbolic computational tools such as Wolfram Mathematica or MATLAB Symbolic Math Toolbox:

Step 1: Apply a unit cosine stator MMF of p pole pairs whose expression is $\cos(p\phi - \omega t)$ on the air gap and assume that the rotor is cylindrical without short-circuited windings.

Step 2: Calculate the total electromotive force induced in each short-circuited coil, which can be an individual loop in nested-loop windings or spiral loops as a whole, and express them in vector format.

Step 3: Calculate the winding function of each short-circuited coil, based on which the self- and mutual inductances related to rotor windings can be obtained and incorporated into the inductance matrix.

Step 4: Calculate the current in each short-circuited coil and develop the rotor MMF based on the electromotive force vector and inductance matrix.

Step 5: Add the rotor MMF to the stator MMF $\cos(p\phi - \omega t)$ to obtain the resultant air-gap MMF distribution.

Table 4.10 Magnetic field conversion factors for different short-circuit winding types.

Rotor topology	Winding pole pairs	Harmonic order													
		1	2	3	4	5	6	7	8	9	10	11	12	13	14
Nested-loop	1	**0.1894**	0	**0.2702**	0	0.1621	0	0.1158	0	0.0901	0	0.0737	0	0.0624	0
($N_{SC} = 4, S = 1, \tau = 1$)	3	**0.2702**	0	**0.9099**	0	0.0540	0	0.0386	0	0.0300	0	0.0246	0	0.0208	0
Nested-loop	2	0	**0.3579**	0	**0.4128**	0	0	0	0.0717	0	0.1575	0	0	0	0.0208
($N_{SC} = 6, S = 1, \tau = 5/6$)	4	0	**0.4128**	0	**0.7347**	0	0	0	0.0461	0	0.1013	0	0	0	0.0134
Nested-loop	2	0	**0.3500**	0	**0.4396**	0	0	0	0.0173	0	0.1115	0	0	0	0.0221
($N_{SC} = 6, S = 3$)	4	0	**0.4396**	0	**0.6426**	0	0	0	0.1413	0	0.0584	0	0	0	0.0142
Spiral-loop	2	0	**0.4031**	0	**0.4546**	0	0	0	0.1209	0	0.0271	0	0	0	0.0158
($N_{SC} = 6$, slot pitch 10°)	4	0	**0.4546**	0	**0.6538**	0	0	0	0.0921	0	0.0206	0	0	0	0.0120
Spiral-loop	2	0	**0.3529**	0	**0.4617**	0	0	0	0.0445	0	0.0610	0	0	0	0.0504
($N_{SC} = 6$, slot pitch 6.125°)	4	0	**0.4617**	0	**0.6706**	0	0	0	0.0317	0	0.0435	0	0	0	0.0359
Six-phase unequal-turn design	2	0	**0.2137**	0	**0.4024**	0	0	0	0.0047	0	0.0023	0	0	0	0.0003
($N_{SC} = 6$)	4	0	**0.4024**	0	**0.7941**	0	0	0	0.0024	0	0.0012	0	0	0	0.0002

Step 6: Expand the resultant airgap MMF into a Fourier series and extract the symbolic expressions for the coefficients of different spatial harmonic components.

The magnetic field conversion factors make it possible to compare different rotor winding configurations, including the reluctance rotor types [14]. For example, the spiral-loop rotor with six spiral loops and a slot pitch of 6.125deg from the optimization in [18] can achieve the highest conversion factor between the eight-pole and four-pole airgap field components. The six-phase unequal turn design in [19] had the lowest content of other spatial harmonics.

Magnetic field conversion factors with $N_{SC} \leq 10$ have been calculated for representative nested-loop and spiral-loop rotor windings, as tabulated in Table 4.11. A few useful conclusions can be drawn for short-circuited coil modulators:

- The standard squirrel-cage winding shows low magnetic field conversion factors $C_{p,lNSC\pm p}$ and thus low pole pair conversion capability.
- The magnetic field conversion capability of short-circuited coil modulators is closely related to the pole pair combination. For the same N_{SC}, the close-ratio combination of p and $(N_{SC} - p)$ pole pairs produces larger magnetic field conversion factors for the sum modulation. Similarly, larger magnetic field conversion factors can be achieved when p is close to N_{SC} for the difference modulation.
- Spatial harmonics introduced by the modulation behavior of short-circuited coils always appear in pairs. In addition, the magnetic field conversion factors for sum modulation are typically larger than those of difference modulation, indicating that the short-circuited coil modulators are not suitable for applications preferring difference modulations.
- For the same N_{SC}, the highest magnetic field conversion capability is achieved when the difference between p and $(N_{SC} - p)$ is 1. In practice, pole pair numbers of the two stator windings in BDFIMs are usually adjusted to 2 to avoid the undesired UMP.
- Magnetic field conversion factors for short-circuited coils also depend on the detailed layout and connection of conductors.
- Magnetic field conversion factors will remain unchanged when N_{SC} and the principal pole pair number p are scaled by the same integer.

4.6 Interchangeability of Modulators

4.6.1 Comparison of Three Basic Modulator Types

As presented in Section 4.5, all the three modulator types can perform asynchronous modulation behaviors and share some common features:

Table 4.11 Magnetic field conversion factors of short-circuited coil modulators with different pole combinations, $l = 1$. Rotor loop resistances have been neglected.

				Nested-loop ($S = 1, \gamma = 1$)			Nested-loop ($S = 3, \gamma = 1$, equally-spaced)		
N_{SC}	p	$N_{SC} - p$	$N_{SC} + p$	$C_{p,p}$	$C_{p,NSC-p}$	$C_{p,NSC+p}$	$C_{p,p}$	$C_{p,NSC-p}$	$C_{p,NSC+p}$
3	1	2	4	0.316	0.342	0.171	0.350	0.440	0.017
4	1	3	5	0.189	0.270	0.162	0.276	0.383	0
5	1	4	6	0.125	0.219	0.146	0.239	0.342	0.020
5	2	3	7	0.427	0.382	0.164	0.415	0.471	0.018
6	1	5	7	0.088	0.182	0.130	0.218	0.313	0.037
6	2	4	8	0.316	0.342	0.171	0.350	0.440	0.017
7	1	6	8	0.065	0.156	0.117	0.205	0.291	0.050
7	2	5	9	0.241	0.303	0.169	0.307	0.409	0.010
7	3	4	10	0.476	0.393	0.157	0.443	0.479	0.015
8	1	7	9	0.050	0.136	0.106	0.196	0.274	0.062
8	2	6	10	0.189	0.270	0.162	0.276	0.383	0.000
8	3	5	11	0.385	0.369	0.168	0.390	0.461	0.019
9	1	8	10	0.040	0.120	0.096	0.190	0.262	0.071
9	2	7	11	0.152	0.242	0.154	0.255	0.361	0.010
9	3	6	12	0.316	0.342	0.171	0.350	0.440	0.017
9	4	5	13	0.503	0.398	0.153	0.458	0.483	0.012
10	1	9	11	0.032	0.108	0.088	0.185	0.251	0.079
10	2	8	12	0.125	0.219	0.146	0.239	0.342	0.020
10	3	7	13	0.263	0.316	0.170	0.319	0.419	0.013
10	4	6	14	0.427	0.382	0.164	0.415	0.471	0.018

- The modulated MMF has three groups of spatial harmonic components, whose numbers of pole pairs are p and $N_r \pm p$, and their amplitudes can be fully characterized by magnetic field conversion factors, where p is the principal pole pair number of the source MMF and N_r can be N_{RT}, N_{MB} or N_{SC}.
- Electrical machines working by the asynchronous modulation behaviors of these three modulators inherently have irregular airgap magnetic field distributions and utilize at least two spatial harmonics of different pole pairs for torque production.
- The physical dq axes are not aligned with function dq axes.

It should be noted that magnetic field modulation behaviors discussed in this section refer to the dominant ones that affect harmonic spectrums of airgap field distributions and torque compositions, since most electrical machines have both synchronous and asynchronous modulation behaviors.

On the other hand, the three basic modulators also show some differences:

- Differences in modulation mechanisms – The short-circuited coil modulator relies on the relative motion between the source MMF and modulator to produce the extra MMF, so it only works for asynchronous modulation. In contrast, both the salient reluctance pole modulator and flux barrier modulation do not have this requirement. The salient reluctance pole modulator utilizes the variation of airgap reluctance/permeance to perform modulation behavior on the source MMF of an arbitrary distribution and motion state. The flux barrier modulator is similar to salient reluctance pole modulators in performing magnetic field modulation behaviors on both stationary and moving source MMFs of arbitrary distributions. The flux barrier modulation relies on the difference of magnetic potential at two positions shorted by the flux guide to change the airgap MMF distribution and flux barriers are crucial in constraining flux lines within narrow flux guides.
- Differences in magnetic field conversion capability and preferred pole pair combinations – The magnetic field conversion capability of flux barrier and short-circuited coil modulators depends heavily on the pole pair combination. For the same equivalent pole pair number, i.e. N_{MB} in flux barrier modulators and N_{SC} in short-circuited coil modulators, the close-ratio combination between p and $N_r - p$ can produce the largest magnetic field conversion factor $C_{p,Nr-p}$ in sum modulation. Mutual magnetic field conversion factors for the difference modulation are much lower than the sum modulation. By contrast, magnetic field conversion factors of salient reluctance pole modulators are independent of pole pair combination but closely related to geometric parameters $\varepsilon_{S(R)T}$ and d_s, suitable for applications using the far-ratio or close-ratio pole pair combinations, such as the MGM and FSPM machine, respectively.
- Differences in difference modulation – Mutual conversion factors for the sum and difference modulations are identical in electrical machines using salient reluctance pole modulators with the same v and l. In contrast, for short-circuited coil and flux barrier modulators, magnetic field conversion factors for the difference modulation case are much lower. Therefore, only the salient reluctance pole modulator is suitable for difference modulation.
- Differences in working harmonics – Both the heterodyne harmonics with $(vp \pm lN_r)$ pole pairs can contribute to the average torque in electrical machines using salient reluctance pole modulators, whereas only one component of the heterodyne harmonics is typically used in machines with flux barrier and/or short-circuited coil modulators.

Table 4.12 Comparison of three modulator types.

Modulator type	Modulation behavior	Relative motion between source MMF and modulator
Short-circuited coil	Asynchronous	Required
Salient reluctance pole	Synchronous or asynchronous	Not required
Multi-layer flux barrier	Synchronous or asynchronous	Not required

Table 4.13 Interchangeability between different modulators.

Modulation behavior	Relative motion between source MMF and modular	Interchangeability and feasibility of combination	Typical example
Synchronous	Stationary	Salient reluctance pole ↔ Flux barrier	Interior PMSM
Synchronous	Moving		SynRM
Asynchronous	Stationary		FSPM machine
Asynchronous	Moving	Salient reluctance pole ↔ Flux barrier Salient reluctance pole ↔ Short-circuited coil Flux barrier ↔ Short-circuited coil	BDFM

By considering the similarities and differences between different modulator types, it is clearly shown that under some circumstances, one modulator can be replaced or combined by another in electrical machines to achieve similar or even enhanced electromagnetic performance [20]. The interchangeability depends on the type of the intended modulation behavior to be performed by the modulator and motion state of the source MMF, as tabulated in Table 4.12. The flux barrier modulator can always be replaced or combined with the salient reluctance pole modulator, while to replace the short-circuited coil modulator with the flux barrier or salient reluctance pole modulator, or vice versa, the intended modulation behavior has to be asynchronous and there has to be a relative motion between the modulator and source MMF, as detailed in Table 4.13.

When an electrical machine works by asynchronous modulation, one can choose from a salient reluctance pole modulator, flux barrier modulator and short-circuited coil modulator freely, because they are interchangeable. The BDFRM is a good example. As shown in Figure 4.34, both the salient reluctance pole and flux barrier modulators work well for BDFRMs, and each modulator may take various forms, such as the axially-laminated and radially-laminated flux barrier rotors.

Based on the interchangeability, the salient reluctance pole and flux barrier modulators can be readily replaced by the short-circuited coil modulator, yielding the BDFIM, as illustrated in Figure 4.35. Similar electromagnetic performance can be achieved for those BDFRMs and BDFIMs, though they are very different in terms of mechanical robustness and manufacturability.

Similar rules also apply when combining multiple modulator types in a concerted manner. For example, when the intended modulation behaviors of the salient reluctance pole modulator, flux barrier modulator, and short-circuited coil modulator are asynchronous and there is a relative motion between the source MMF and modulators, they can be combined to change the electromagnetic performance, as illustrated in Figure 4.36.

Salient reluctance pole modulators and flux barrier modulators can be combined freely to achieve similar modulation behaviors, regardless of whether they are synchronous or asynchronous. It should be noticed that short-circuited coil modulators cannot be replaced or combined with salient reluctance pole or flux barrier modulators when there is no relative motion between the source MMF and modulators.

4.6.2 Influence of Modulators on Machine Performance

Though one of the three basic modulator types can be replaced by (or combined with) another based on the interchangeability requirement, the difference in electromagnetic performance can also be observed, originating from the difference in magnetic field conversion capabilities. BDFMs with four different rotor constructions are compared from the perspective of the airgap magnetic field modulation. The four rotor constructions under study are the salient reluctance pole rotor, the radially-laminated flux barrier rotor with iron bridges, the radially-laminated flux barrier rotor without iron bridges, and the nested-loop rotor. Cross-sectional views and magnetic field distributions when the CW is excited while the PW is open-circuited are illustrated in Figure 4.37, where the CW and PW have 2- and 4-pole pairs, respectively. The airgap magnetic flux density distribution, cross-coupling capability between the two stator windings, inductance characteristics, and torque capabilities are compared, as detailed below.

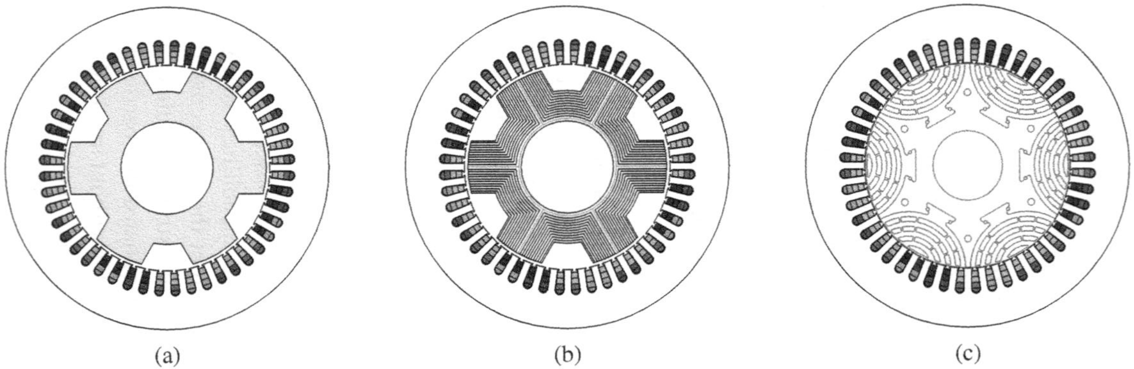

Figure 4.34 BDFM variations with (a) the salient reluctance pole rotor, (b) the axially-laminated flux barrier rotor, (c) the radially-laminated flux barrier rotor.

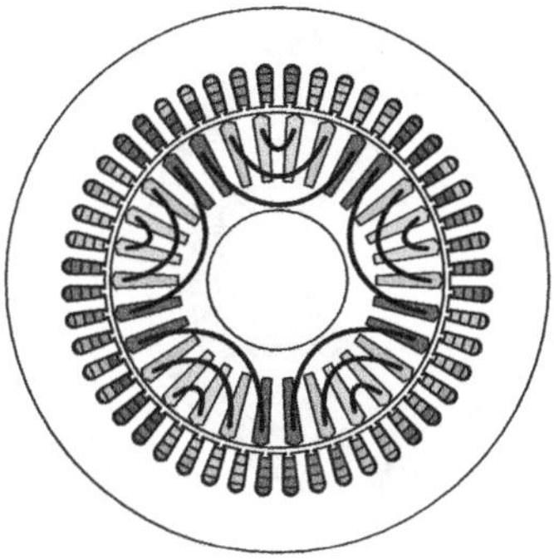

(a)

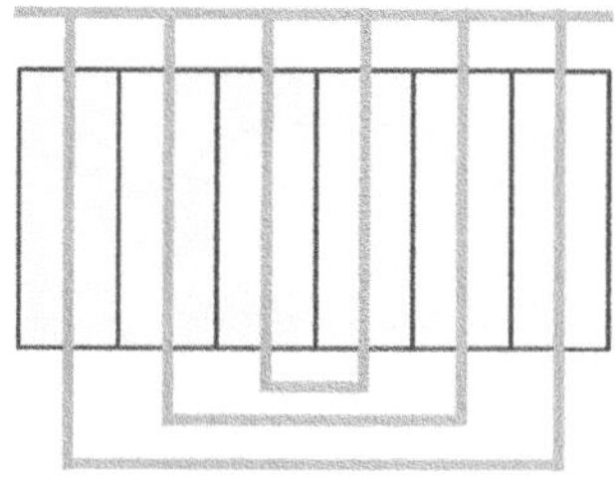

(b)

Figure 4.35 BDFM with a nested-loop rotor, (a) Cross-sectional view, (b) One nested-loop group.

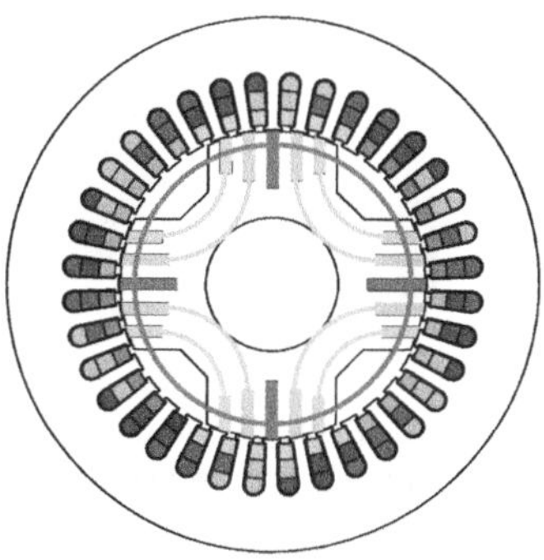

Figure 4.36 BDFM with a hybrid rotor. The rotor has a salient reluctance pole modulator and a short-circuited coil modulator working in the concerted manner.

1 Airgap Flux Density Distribution

Radial components of airgap magnetic flux density distributions in the four BDFMs have been extracted when the CW is excited with a 2.5 A_{rms} current and open-circuited PW. As shown in Figure 4.40, there are two dominating harmonics whose numbers of pole pairs are 2 and 4, respectively, due to the asynchronous modulation behavior performed on the 4-pole source magnetizing MMF. As the two spatial harmonics are of different pole pairs and rotating speeds, the airgap flux density distribution is irregular and time-varying.

Figure 4.38 also provides the corresponding harmonic spectrums. It is shown that the 4-pole harmonic of the flux barrier rotor with iron bridges is higher than that of the flux barrier rotor without bridges due to increased flux leakage via the bridges. The 8-pole harmonic is higher for the flux barrier rotor without bridges, indicating its better magnetic field conversion capability.

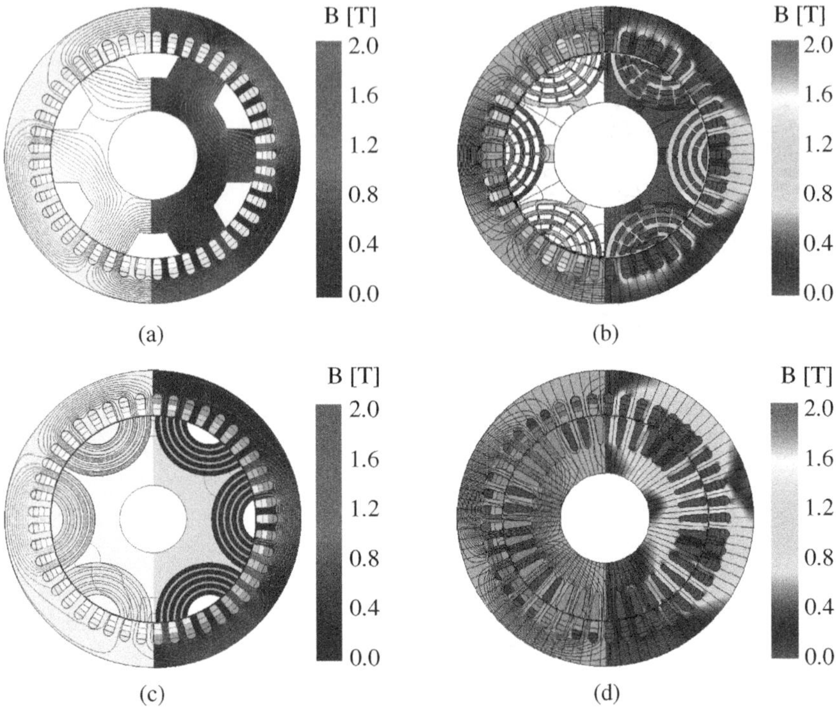

Figure 4.37 Flux lines and flux density distribution of BDFMs when the CW is excited and PW open-circuited. (a) BDFM with a salient reluctance pole rotor, (b) BDFM with a radially-laminated flux barrier rotor having iron bridges, (c) BDFM with a radially-laminated flux barrier rotor having no iron bridges, (d) BDFM with a nested-loop winding. Numbers of pole pairs of the CW and PW are 2 and 4, respectively.

It is also shown that amplitudes of the 4- and 8-pole-pair harmonics in BDFMs with the flux barrier and short-circuited coil modulators are very different, which is in line with the theoretical analysis that these two basic modulator types have different magnetic field conversion factors for sum and difference modulations. In addition, the two modulators can suppress spatial harmonics of high pole pairs more effectively than the salient reluctance pole modulator.

2 Cross-Coupling Capability

The cross-coupling capability of BDFMs characterizes the magnetic field conversion capability between magnetic fields created by the two stator windings. It can be measured in the voltage control mode and current control mode, as defined in [12]. Induced phase voltages in the voltage control mode almost vary linearly with the rotor speed and are symmetrical to the natural synchronous speed for both cases, where the CW and PW are excited with a 220 V_{rms} voltage at 50 Hz,

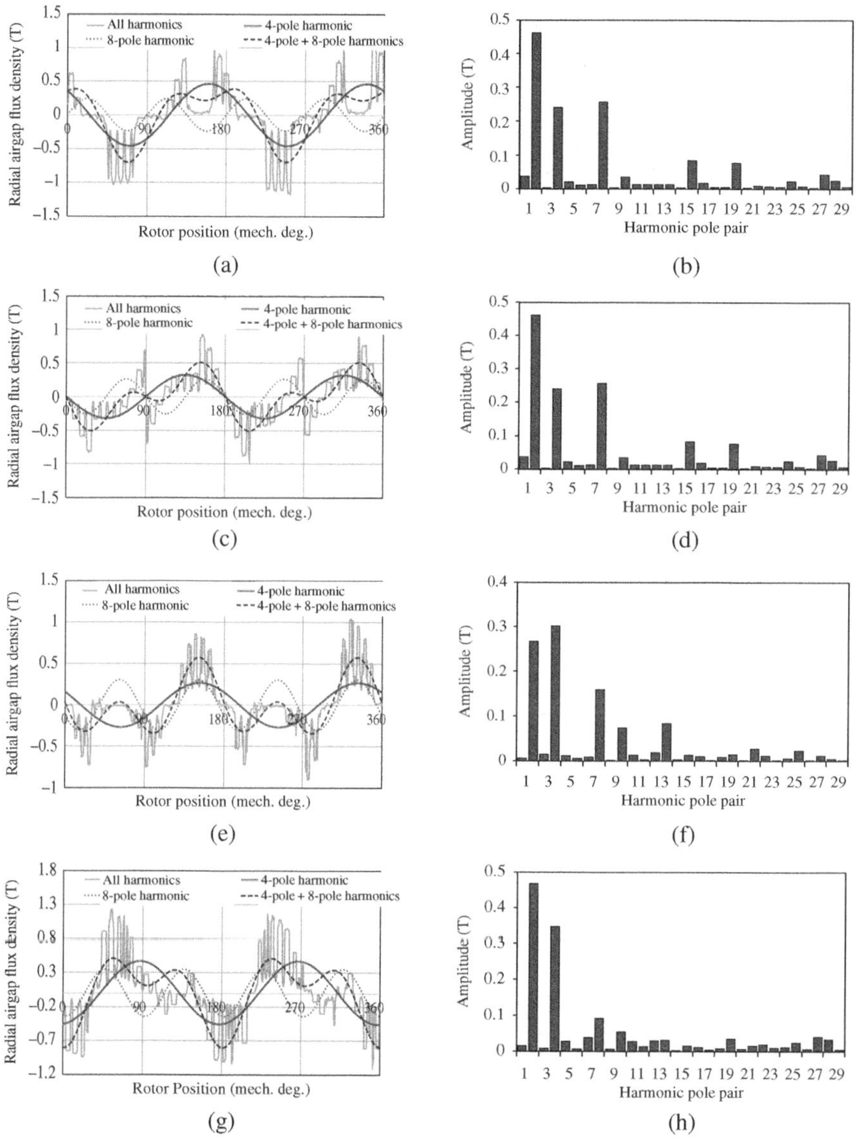

Figure 4.38 Radial airgap flux density waveforms and their amplitude spectrums when CW is excited and PW open-circuited, (a) Waveform for BDFM with a salient reluctance pole rotor, (b) Amplitude spectrum of (a), (c) Waveform for BDFM with a radially-laminated flux barrier rotor with iron bridges, (d) Amplitude spectrum of (c), (e) Waveform for BDFM with a radially-laminated flux barrier rotor without iron bridges, (f) Amplitude spectrum of (e), (g) Waveform for BDFM with a nested-loop rotor, (g) Amplitude spectrum of (f). Numbers of pole pairs of the CW and PW are 2 and 4, respectively.

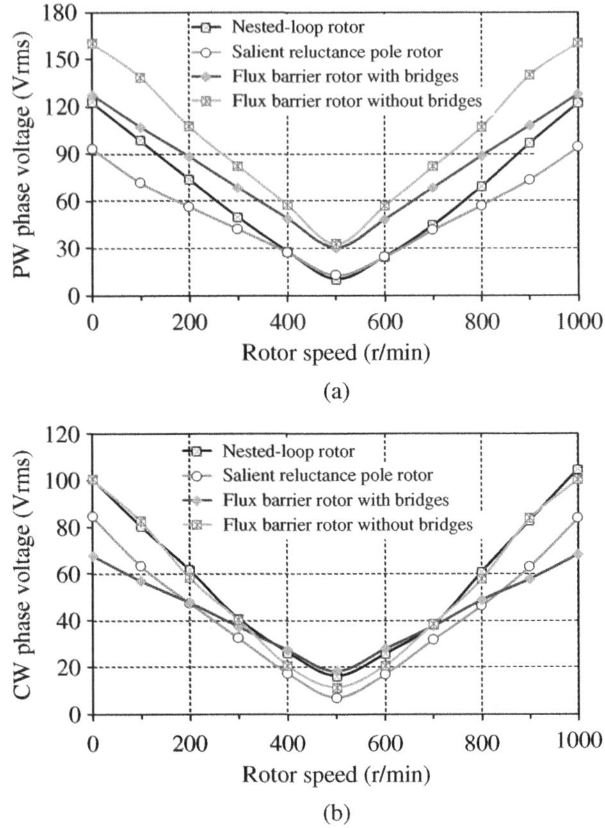

Figure 4.39 Cross-coupling characteristics in voltage control mode. (a) The PW voltage when CW is excited with a voltage of 220 V_{rms} and 50 Hz, (b) The CW voltage when PW is excited with a voltage of 220 V_{rms} and 50 Hz. Numbers of pole pairs of the CW and PW are 2 and 4, respectively.

respectively, as presented in Figure 4.39. The flux barrier rotor without iron bridges generally shows the highest induced voltages, confirming again its high magnetic field conversion capability.

3 Inductance Characteristics

Self-inductances and mutual inductances within the same stator winding set of BDFMs with four different rotor constructions are almost independent of the rotor position, as shown in Figure 4.40. For both the PW and CW, the self-inductance is approximately twice the mutual inductance between phases. The mutual inductance between two stator windings shows a sinusoidal waveform and repeats $N_r = 6$ time per rotor revolution. For clarity, the average self-inductances and

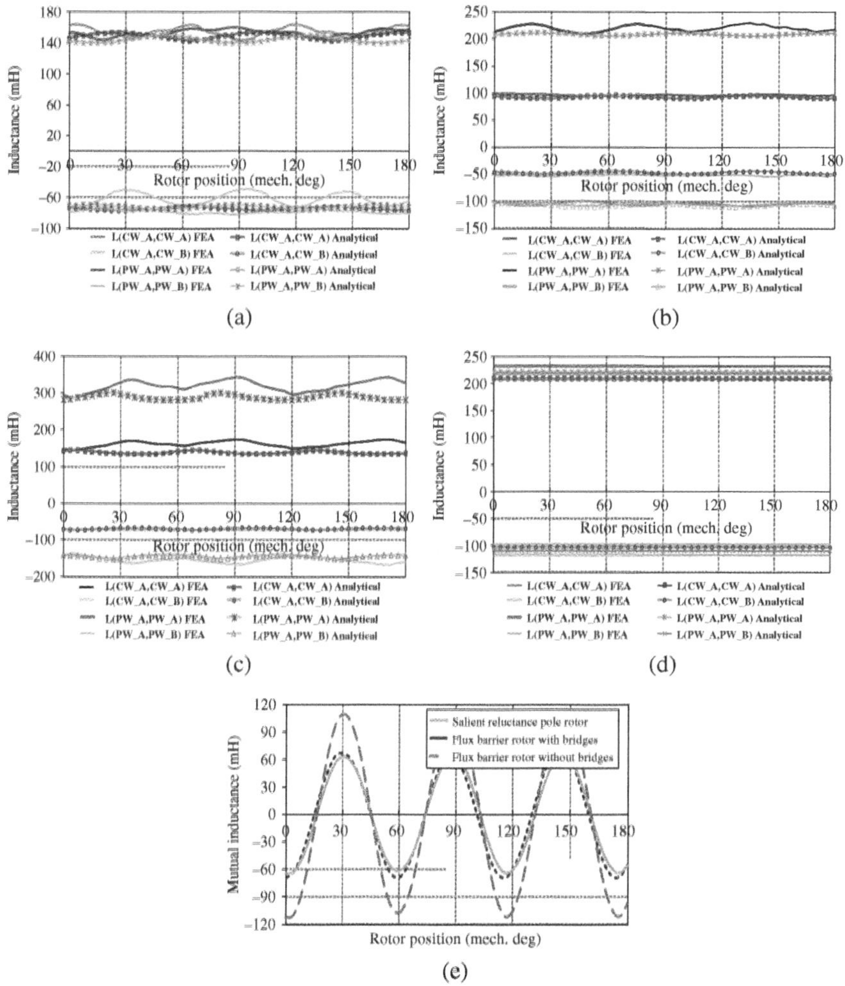

Figure 4.40 Inductance characteristics of BDFMs with different rotor types. (a) Inductances for the BDFM with a salient reluctance pole rotor, (b) Inductances of the BDFM with a flux barrier rotor having bridges, (c) Inductances of the BDFM with a flux barrier rotor having no bridges, (d) Inductances of the BDFM with a nested-loop rotor, (e) Mutual inductances between two stator windings from 2D FEA. Mutual inductances between two stator windings in the BDFM with a nested-loop rotor cannot be obtained from 2D FEA directly and thus excluded from (e).

Table 4.14 Comparison of inductances and magnetic field conversion factors. All the rotor types are radially-laminated. A leakage factor has to be introduced to match the ratios between inductances to account for the influence of iron bridges, magnetic saturation, etc., which will be studied in detail in Chapter 5.

Rotor types	Inductance (mH)			Conversion factors		
	L_{cc} (avg)	L_{prc} (peak)	L_{pp} (avg)	$C_{c,c}$	$C_{p,c}$	$C_{p,p}$
Salient reluctance pole rotor	162.682	81.118	153.341	0.477	0.304	0.477
Flux barrier rotor with bridges	98.888	68.975	213.702	0.293	0.414	0.603
Flux barrier rotor without bridges	160.281	112.397	317.312	0.293	0.414	0.603
Nested-loop rotor	218.120	78.870	232.399	0.350	0.440	0.643

peak mutual inductances between the two stator windings have been calculated and tabulated in Table 4.14, together with the conversion factors, which are proportional to the inductance values in theory.

4 Average Torque

Figure 4.41 shows that the average electromagnetic torque of BDFMs is almost proportional to the square of input currents when the current is below 2 A_{rms}, where the CW is excited and PW is connected to a three-phase resistive load. With larger currents, the torque increase becomes very slow, which is mainly due to the magnetic saturation.

The flux barrier rotor without iron bridges produces the largest torque when the phase current is relatively large (≥6 A_{rms} in this comparison), as expected. With a

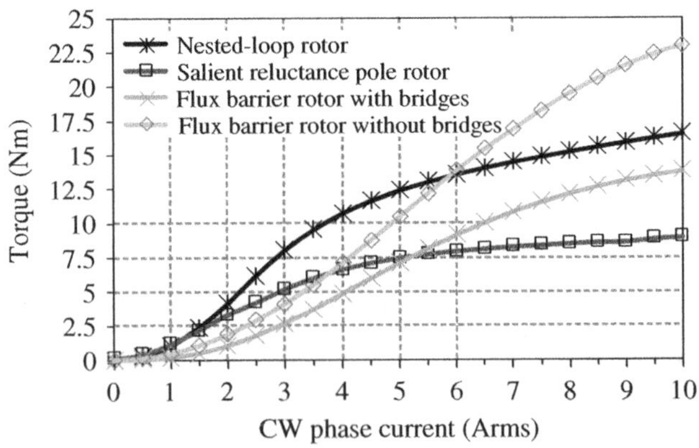

Figure 4.41 Average torque versus CW phase current when CW is excited, and PW connected to a 50 Ω resistive load.

relatively large CW current, the torque capability of the flux barrier rotor without any bridges is larger than the nested-loop rotor, followed by the flux barrier rotor with iron bridges. The salient reluctance pole rotor produces the lowest torque.

References

1 Cheng, M., Han, P., and Hua, W. (2017). General airgap field modulation theory for electrical machines. *IEEE Trans. Ind. Electron.* 64 (8): 6063–6074.

2 Cheng, M., Wen, H., Han, P., and Zhu, X. (2019). Analysis of airgap field modulation principle of simple salient poles. *IEEE Trans. Ind. Electron.* 66 (4): 2628–2638.

3 Han, P., Cheng, M., Ademi, S., and Jovanovic, M.G. (2018). Brushless doubly-fed machines: opportunities and challenges. *Chin. J. Electr. Eng.* 4 (2): 1–17.

4 Cheng, M., Wen, H., Zeng, Y., and Jiang, Y. (2020). Analysis of airgap field modulation behavior and torque component in electric machines. *Trans. China Electrotech. Soc.* 35 (5): 921–930.

5 Broadway, A.R.W. and Burbridge, L. (1970). Self-cascaded machine: a low speed motor or high frequency brushless alternator. *Proc. IEEE* 117 (7): 1277–1290.

6 Broadway, A.R.W. (1971). Cageless induction machine. *Proc. Inst. Electr. Eng.* 118 (11): 1593–1600.

7 Cheng, M., Hua, W., Zhang, J., and Zhao, W. (2011). Overview of stator-permanent magnet brushless machines. *IEEE Trans. Ind. Electron.* 58 (11): 5087–5101.

8 Lee, C.H. (1963). Vernier motor and its design. *IEEE Trans. Power Appl. Syst.* 82 (66): 343–349.

9 Chau, K.T., Zhang, D., Jiang, J.Z. et al. (2007). Design of a magnetic-geared outer-rotor permanent-magnet brushless motor for electric vehicles. *IEEE Trans. Magn.* 43 (6): 2504–2506.

10 Atallah, K., Calverly, S.D., and Howe, D. (2004). Design, analysis and realization of a high-performance magnetic gear. *Proc. IEE – Electr. Power Appl.* 151 (2): 135–143.

11 Toba, A. and Lipo, T.A. (1999). Novel dual-excitation permanent magnet vernier machine. *Rec. IEEE IAS Annu. Meet.* 4: 2539–2544.

12 Han, P., Cheng, M., Wei, X., and Jiang, Y. (2018). Steady-state characteristics of the dual-stator brushless doubly fed induction generator. *IEEE Trans. Ind. Electron.* 65 (1): 200–210.

13 Zhang, F., Yu, S., Wang, Y. et al. (2019). Design and performance comparisons of brushless doubly fed generators with different rotor structures. *IEEE Trans. Ind. Electron.* 66 (1): 631–640.

14 Han, P., Zhang, J., and Cheng, M. (2019). Analytical analysis and performance characterization of brushless doubly fed machines with multibarrier rotors. *IEEE Trans. Ind. Appl.* 55 (6): 5758–5767.

15 Wen, H., Cheng, M., Jiang, Y. et al. (2020). Analysis of airgap field modulation principle of flux guides. *IEEE Trans. Ind. Appl.* 56 (5): 4758–4768.

16 Han, P., Cheng, M., Zhu, X., and Chen, Z. (2021). Analytical analysis and performance characterization of brushless doubly fed induction machines based on general air-gap field modulation theory. *Chin. J. Electr. Eng.* 7 (3): 4–18.

17 Roberts, P.C., Long, T., McMahon, R.A. et al. (2013). Dynamic modelling of the brushless doubly fed machine. *IET Electr. Power Appl.* 7 (7): 544–556.

18 Gorginpour, H., Jandaghi, B., and Oraee, H. (2016). A novel rotor configuration for brushless doubly-fed induction generator. *IET Electr. Power Appl.* 5: 394–402.

19 Xiong, F. and Wang, X. (2014). Design of a low-harmonic-content wound rotor for the brushless doubly fed generator. *IEEE Trans. Energy Convers.* 29 (1): 158–168.

20 Wen, H., Cheng, M., Wang, W., and Zhang, G. (2021). Modulation behaviours and interchangeability of modulators for electrical machines. *IET Electr. Power Appl.* 15 (5): 542–554.

5

Performance Evaluation of Electrical Machines Based on General Airgap Field Modulation Theory

5.1 Introduction

The airgap is where electromechanical energy conversion occurs in electrical machines. Flux linkages linked by windings, winding inductances, normal and shear stresses in the airgap are all related to the airgap flux density distribution. Since the general airgap field modulation theory is developed based on the widely observed airgap magnetic field modulation phenomena, the theory is particularly well-suited for parameter calculations and basic performance evaluations, such as flux linkages, open-circuit back-EMFs, radial forces on the stator and rotor and electromagnetic torques.

This chapter takes a few representative machine topologies as examples to perform the parameter calculations and performance evaluations in a quantitative manner, including a squirrel-cage induction machine (IM) for traction applications, a brushless doubly-fed reluctance machine (BDFRM) with a multilayer flux barrier rotor and brushless doubly-fed induction machine (BDFIM) with a spiral-loop rotor for two-converter-based wide-speed-range operations, a synchronous reluctance machine (SynRM) with a multilayer flux barrier rotor for energy efficient industrial motor drives, and a flux-reversal permanent magnet (FRPM) machine. It is shown that various machine topologies can be quantitatively analyzed in a unified manner based on the developed general airgap field modulation theory.

5.2 Squirrel-Cage IM

Squirrel-cage IMs are one of the most widely-used electrical machine types and the theory for design and analysis of such machines have been well documented in textbooks and monographs, such as [1–3]. The cage rotor is robust and inherently

General Airgap Field Modulation Theory for Electrical Machines: Principles and Practice, First Edition. Ming Cheng, Peng Han, Yi Du, and Honghui Wen.

Companion website: www.wiley.com/go/genairgapfieldmodulationtheory

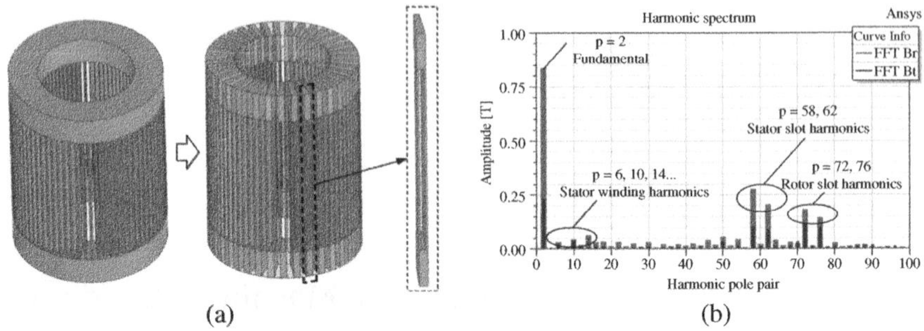

Figure 5.1 Rotor winding structures and harmonic spectrum. (a) Standard squirrel-cage winding and its loop winding equivalent, (b) amplitude spectrum of airgap flux density waveforms of the studied 4-pole IM, which has 60 semi-closed stator slots and 74 fully-closed rotor slots.

suitable for high-speed and high temperature operation, which is why it has attracted much attention from the automotive and aerospace industries.

There are multiple flux density spatial harmonics in the airgap of IMs, as discussed in [3]. The harmonic spectrum of the airgap flux density waveform for a representative traction IM using the standard copper squirrel-cage winding shown in Figure 5.1a is illustrated in Figure 5.1b. The stator and rotor winding harmonics are related to the winding layout and magnetic saturation of stator/rotor cores. Stator and rotor slot harmonics are introduced by slotting effects and related to the number of slots as well as slot openings. To consider these factors accurately in simulation models, two-dimensional (2D) or even three-dimenisonal (3D) transient electromagnetic finite element analysis (FEA) has to be employed. These spatial harmonics and their speed-dependent nature impose new challenges to FEA of IMs, such as the reduced time step size to capture the slotting effect, long steady-state simulation for harmonic force generation used in noise, vibration, and harshness (NVH) analysis, etc.

The closed cage rotor winding is the key for the operation of IMs and it was typically modeled by the coupled-circuit method, leading to a large number of voltage equations [4]. The complex vector notation which combines instantaneous phase variables into a complex variable simplifies the modeling and analysis of IMs significantly, providing additional insights into the circuit parameter extraction and referring [5]. In most textbooks, the squirrel-cage winding is equivalent to a balanced polyphase winding sharing the same number of phases as the stator winding (SW) for better understanding and phasor- or dq-variable-based performance analysis. The operating principle of the cage winding and its real role in the electromechanical energy conversion of IMs, however, have rarely been discussed. This section makes an attempt to analyze the squirrel-cage winding from the perspective of airgap magnetic field modulation, aiming to explain the operation of IMs based on the real cage winding instead of its three-phase equivalent and find a way to determine the minimum space–time periodicity of the flux pattern to benefit the transient FEA of IMs.

5.2.1 Airgap Magnetic Field Analysis

1 Stator Winding MMF

The basic relations in the spatial domain and spatial frequency domain offered by the general airgap field modulation theory provides a new perspective to analyze squirrel-cage IMs using closed-form analytical solutions following the signal flow starting from the stator magnetomotive force (MMF). Based on the coordinate system defined in Figure 5.2 and neglecting harmonics, the MMF created by the stator winding is:

$$F_s(\phi, t) = F_s \cos[p(\phi - \phi_{01}) - (\omega_s t + \varphi_s)] \tag{5.1}$$

where F_s is the amplitude of the fundamental MMF created by the stator winding and $F_s = (m_s/2)(4/\pi)(N_{ph} I_s k_{w1})/(2p)$, where m_s is the number of phases of the stator winding, N_{ph} number of turns in series per phase, k_{w1} the fundamental winding factor, I_s the peak value of winding current, and p number of pole pairs. ϕ is the angular position along the airgap periphery in the stationary reference frame, ϕ_{01} the angle from the reference axis to the stator winding axis, ω_s the electrical angular frequency of stator winding currents. t is time and φ_s is the initial phase of stator winding current.

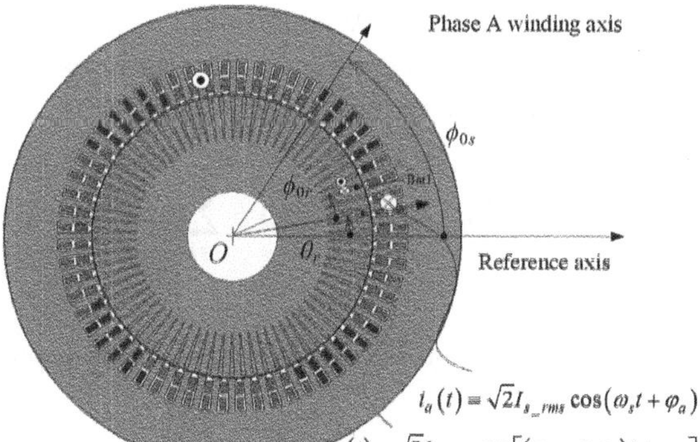

Figure 5.2 Coordinate system for analysis. The cross and dot denote the current flowing into the page and out of the page, respectively. The steady-state rotor current is shown explicitly but it is solved out from rotor voltage equations.

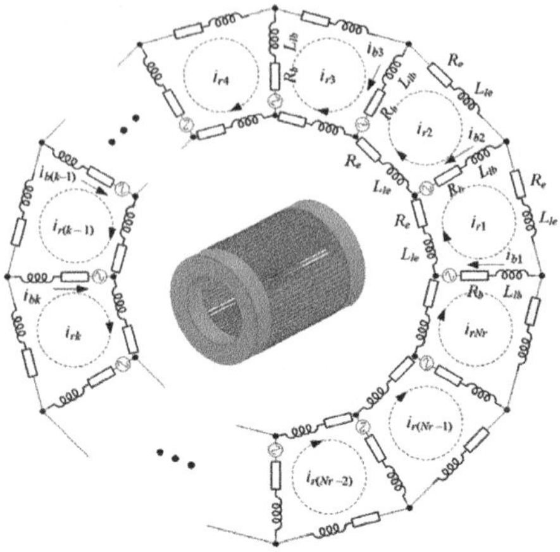

Figure 5.3 Equivalent circuit of the rotor cage. Each bar current denoted by the subscription "*b*" is the subtraction of end ring currents from the two adjacent loops denoted by the subscription "*r*." It is shown that adjacent loop currents are coupled by the shared copper bar.

2 Rotor Winding MMF

Considering the physical construction of the cage winding, its equivalent circuit can be drawn and show in Figure 5.3. The voltage–current relations in the steady state for the *k*-th loop can be expressed as [5]:

$$\begin{aligned}0 &= 2(R_b + R_e)i_{rk} - R_b(i_{r(k-1)} + i_{r(k+1)}) + d(\psi_{rks} + \psi_{rkr})/dt \\ &= R_r i_{rk} + d(\psi_{rks} + \psi_{rkr})/dt\end{aligned} \tag{5.2}$$

where k is an integer between 1 and N_r. When $k = N_r$, $i_{r(k+1)} = i_{r1}$. The flux linkage linked by k-th loop has two parts. One is due to stator currents as denoted by ψ_{rks} and the other is due to rotor currents as denoted by ψ_{rkr}:

$$\psi_{rkr} = 2(L_{lb} + L_{le})i_{rk} - L_{lb}(i_{r(k-1)} + i_{r(k+1)}) + \sum_{j=1}^{N_r} L_{kj} i_{rj} = L_{lr} i_{rk} + \sum_{j=1}^{N_r} L_{kj} i_{rj} \tag{5.3}$$

where R_r and L_{lr} are the equivalent loop resistance and loop leakage inductance, respectively.

$$R_r = 2R_e + 2R_b[1 - \cos(2\pi p/N_r)] \tag{5.4}$$

$$L_{lr} = 2L_{le} + 2L_{lb}[1 - \cos(2\pi p/N_r)] \tag{5.5}$$

The presence of short-circuited rotor windings introduces additional MMFs into the airgap. The rotor MMF depends on the rotor winding current and winding function. The rotor MMF can be expressed as:

$$F_r(\phi, t) = \sum_{k=1}^{N_r} W_k(\phi) i_{rk} \tag{5.6}$$

where $W_k(\phi)$ and i_{rk} are the winding function and current the k-th rotor loop, respectively.

5.2.2 Inductance and Torque Characteristics

Equations (5.2)–(5.5) show that in steady state, the squirrel-cage rotor winding is equivalent to a set of isolated loops as shown in Figure 5.1a, whose bar leakage inductance and bar resistance are scaled by a factor of $1 - \cos(2\pi p/N_r)$. A 2D FEA was conducted to verify this equivalence. As shown in Figure 5.4, the model with isolated loop windings shares with the standard squirrel-cage IM model a similar stator winding current, rotor bar current and electromagnetic torque. The difference in transients is in expectation because the equivalence is derived for the steady state. In addition, the inductance characteristics and rotor bar currents for the model with isolated loop windings are illustrated in Figure 5.5, showing similarities to the brushless doubly-fed IMs with short-circuited rotor windings [6].

The resultant airgap MMF can be expressed by:

$$\begin{aligned} M(N_r)[F_s(\phi, t)] &= C^d_{p,p} \cos(p\phi - \omega_s t) + C^q_{p,p} \cos\left(p\phi - \omega_s t + \frac{\pi}{2}\right) \\ &\quad - \sum_{l=1}^{\infty} C_{sum} \cos\left[(lN_r - p)\phi + (\omega_s - lN_r\omega_r)t + l\pi - \varphi - \frac{\pi}{2}\right] \\ &\quad + \sum_{l=1}^{\infty} C_{dif} \cos\left[(lN_r + p)\phi - (\omega_s + lN_r\omega_r)t + l\pi + \varphi - \frac{\pi}{2}\right] \end{aligned} \tag{5.7}$$

where $\varphi = \text{atan}[(\omega_s - p\omega_r)L_r/R_r]$ and when $R_r = 0$, φ will be 90°. The three magnetic field conversion factors are:

$$\begin{cases} C^d_{p,p} = 1 - 2\pi N_r \dfrac{(\omega_s - p\omega_r)}{\sqrt{R_r^2 + [(\omega_s - p\omega_r)L_r]^2}} \left(\dfrac{\mu_0 r_g l_{stk}}{g}\right)\left(\dfrac{1}{N_r}\right)^2 \left[\sin\left(\dfrac{p\pi}{N_r}\right) \Big/ \left(\dfrac{p\pi}{N_r}\right)\right]^2 \sin\varphi \\ C^q_{p,p} = 2\pi N_r \dfrac{(\omega_s - p\omega_r)}{\sqrt{R_r^2 + [(\omega_s - p\omega_r)L_r]^2}} \left(\dfrac{\mu_0 r_g l_{stk}}{g}\right)\left(\dfrac{1}{N_r}\right)^2 \left[\sin\left(\dfrac{p\pi}{N_r}\right) \Big/ \left(\dfrac{p\pi}{N_r}\right)\right]^2 \cos\varphi \end{cases} \tag{5.8}$$

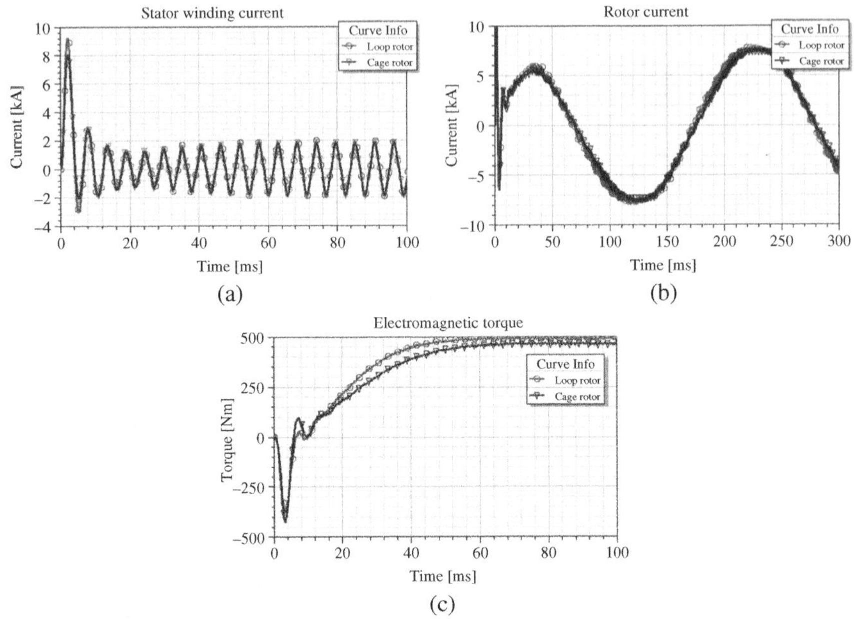

Figure 5.4 Performance comparison between IMs with the standard squirrel-cage winding and isolated loop windings. (a) Stator current, (b) bar 1 current, (c) electromagnetic torque. Results are from 2D FEA. In the IM with isolated loop windings, rotor bar conductivity has been scaled according to (5.4) but the bar leakage inductance has not due to implementation difficulties so there is a slight difference in the steady state.

$$C_{sum} = 2\pi N_r \frac{(\omega_s - p\omega_r)}{\sqrt{R_r^2 + [(\omega_s - p\omega_r)L_r]^2}} \left(\frac{\mu_0 r_g l_{stk}}{g}\right)\left(\frac{1}{N_r}\right)^2 \sin\left(\frac{p\pi}{N_r}\right)$$
$$\sin\left(\frac{k\pi}{N_r}\right) \Big/ \left[\left(\frac{p\pi}{N_r}\right)\left(\frac{k\pi}{N_r}\right)\right], k = lN_r - p \tag{5.9}$$

By replacing $k = lN_r - p$ with $k = lN_r + p$ in C_{sum}, C_{dif} can be readily obtained. It can be seen from (5.7) that the modulation behavior of the cage rotor introduces two additional spatial harmonic groups whose number of pole pairs are changed by the rotor slot number. When the rotor winding resistance and leakage inductance are neglected, C_p will be 0.0024, which means the 2-pole-pair main magnetic field cannot be created effectively by the stator MMF. On the contrary, when the rotor winding resistance is very large, C_p will be close to 1 and the 2-pole-pair air-gap magnetic field is almost created completely the stator winding MMF, which is consistent with the operation of IMs with the rotor windings open-circuit.

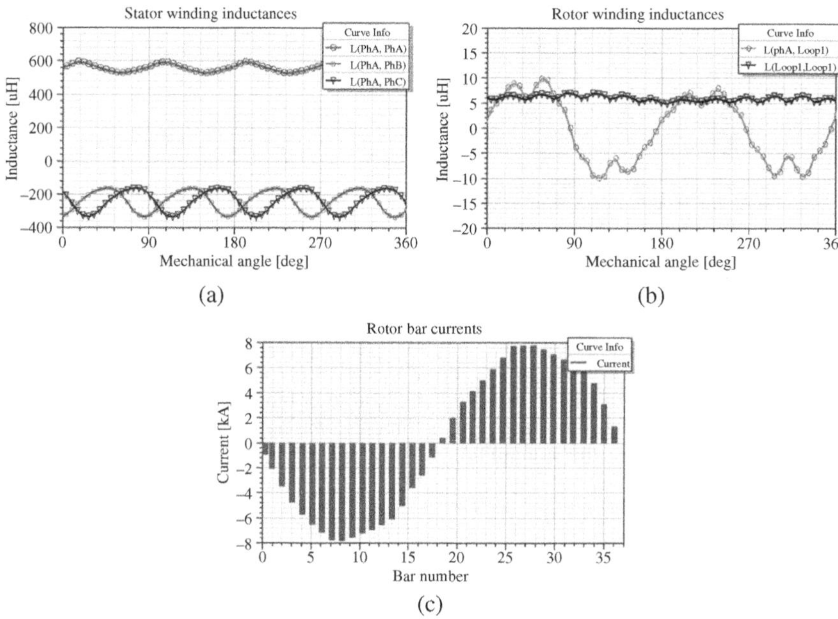

Figure 5.5 Inductances and rotor currents. (a) Self- and mutual inductances of stator windings, (b) loop self-inductance and mutual inductance between phase-A stator winding and rotor loop 1, (c) rotor current at an instant of time. The stator winding self- and mutual inductances are almost constant and pulsations are introduced by magnetic saturation. Loop self-inductance is almost constant and the mutual inductance between phase-A stator winding and rotor loop 1 is alternating.

By including more stator winding harmonics into (5.7), it is convenient to study the space–time periodicity of the airgap MMF distribution, which is a good indicator of the flux pattern in IMs. The time period ΔT needs to meet:

$$M(N_r)[\mathrm{F}_s(\phi + \omega_s \Delta T/p, t)] = M(N_r)[\mathrm{F}_s(\phi, t + \Delta T)] \tag{5.10}$$

Further derivation shows:

$$\Delta T = k\left(\frac{p}{sN_r}\right)\left(\frac{2\pi}{\omega_s}\right) = kT_0 = m\left(\frac{1}{2m_s}\right)\left(\frac{2\pi}{\omega_s}\right) = n\left[\frac{p}{(1-s)N_r}\right]\left(\frac{2\pi}{\omega_s}\right),\quad k, m, n \in Z^+. \tag{5.11}$$

Equation (5.11) indicates that the time period depends on the number of pole pairs, number of rotor slots, and the slip. For the studied traction IM operated at 5257 r/min with a slip of 0.02653, the combination of minimum (k, m, n) is (9, 55, 330). This is validated by the flux pattern plots from 2D FEA shown in Figure 5.6a,c. When k is set to another integer lower than 9, such as 6, the flux pattern will not repeat, as depicted in Figure 5.6b.

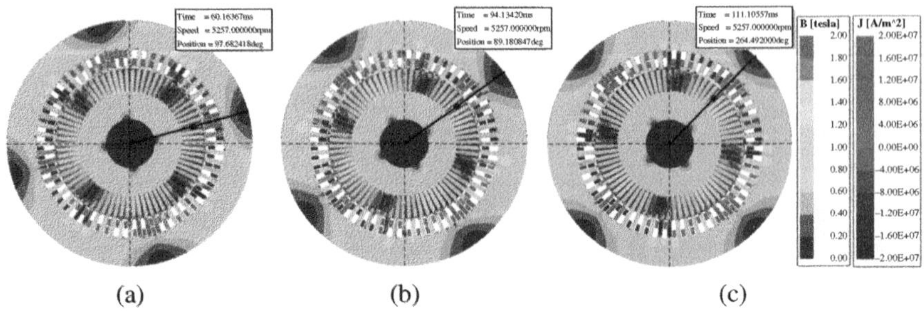

Figure 5.6 Visualization of periodicity in squirrel-cage IMs. (a) Flux pattern at $t_1 = 60.16367$ ms, (b) flux pattern at $t_2 = t_1 + 6T_0$, showing different pattern compared with (a), (c) flux pattern at $t_3 = t_1 + 9T_0$, showing similar pattern compared with (a). Based on (5.11), the flux pattern will be repeated after the rotor magnetic field rotates by nine rotor teeth relative to the rotor itself. The minimum combination of k, m, and n will be $k = 9$, $m = 55$, $n = 330$ for the studied IM with a slip of 0.02653 and rotor speed 5257 r/min.

5.3 Brushless Doubly-fed Machines

Unlike conventional single-fed machines, brushless double-fed machines are a family of machines with two AC electrical ports (power and control windings) and one mechanical port (rotor shaft). The two sets of stator windings are coupled indirectly through the rotor, and electromechanical energy conversion is accomplished during rotor rotation.

Brushless doubly-fed machines have been widely viewed as an ideal substitute for traditional slip-ring doubly-fed machines, especially in low to medium speed applications such as wind power generation [7–11]. Numerous brushless doubly-fed machine topologies have been proposed around the globe as mentioned in Chapter 2, such as the cascaded brushless doubly-fed induction machine (CBDFIM) [12], the BDFIM with a nested-loop rotor [8, 10] or multi-phase wound rotor [13], the BDFRM [14], the brushless doubly-fed machine with a hybrid rotor (BDFM-HR) [15], and the brushless doubly-fed induction machine with a rotary transformer (BDFIM-RT) [16], etc.

Rotors of brushless doubly-fed machines may include short-circuited coils, salient reluctance poles, or multi-layer flux barriers (either radially or axially laminated). The CBDFIM, BDFIM, and BDFRM are the most representative brushless doubly-fed machines. According to the general airgap field modulation theory, the modulation behavior in CBDFIMs is the unit modulation. Modulators in the BDFIM and BDFRM are short-circuited coils and salient reluctance poles/multilayer flux barriers, respectively.

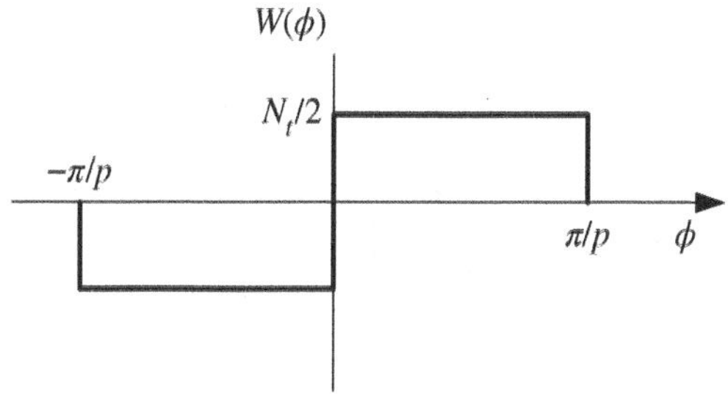

Figure 5.7 Winding function of a single-phase full-pitch winding. Only one pole pair is shown for clarity.

5.3.1 Stator Winding MMF

The power and control windings of brushless doubly-fed machines are typically designed as balanced multi-phase distributed windings, whose number of slots per pole per phase q is larger than 1. Each phase winding consists of a number of shifted full-pitch coils in series. Winding functions of the corresponding phase windings can be derived by superimposing those of individual full-pitch coils.

The winding function of a single-phase full-pitch winding having p pole pairs and N_t turns in series per pole per phase is illustrated in Figure 5.7, and it can be written as

$$W(\phi) = \frac{4}{\pi}\frac{N_t}{2}\sum_{\nu=2k+1}^{\infty}\frac{1}{\nu}\sin(\nu p\phi), k = 0, 1, 2, \dots, \infty \tag{5.12}$$

The winding function of an integer-slot double-layer short-pitch winding can be readily obtained by superimposing those of single-phase full-pitch windings located at the top and bottom layers shifted by the short-pitch angle β (mechanical angle), where the number of winding turns per pole per phase becomes $N_t/2$, as shown in Figure 5.8, namely:

$$\begin{aligned} W(\phi) &= \frac{1}{2}\left\{\frac{4}{\pi}\frac{N_t}{2}\sum_{\nu=2k+1}^{\infty}\frac{1}{\nu}\sin\left[\nu p\left(\phi-\frac{\beta}{2}\right)\right]\right. \\ &\quad \left.+\frac{4}{\pi}\frac{N_t}{2}\sum_{\nu=2k+1}^{\infty}\frac{1}{\nu}\sin\left[\nu p\left(\phi+\frac{\beta}{2}\right)\right]\right\}, k = 0, 1, 2, \dots, \infty \\ &= \frac{4}{\pi}\frac{N_t}{2}\sum_{\nu=2k+1}^{\infty}\frac{1}{\nu}\sin(\nu p\phi)\cos\left(\frac{\nu p\beta}{2}\right) \end{aligned} \tag{5.13}$$

The integer-slot double-layer short-pitch distributed winding can be regarded as the superposition of q double-layer short-pitch windings shifted by q slots, where q is the number of slots per pole per phase, and α is the slot pitch angle in electrical radians, as shown in Figure 5.9. The corresponding winding function can be

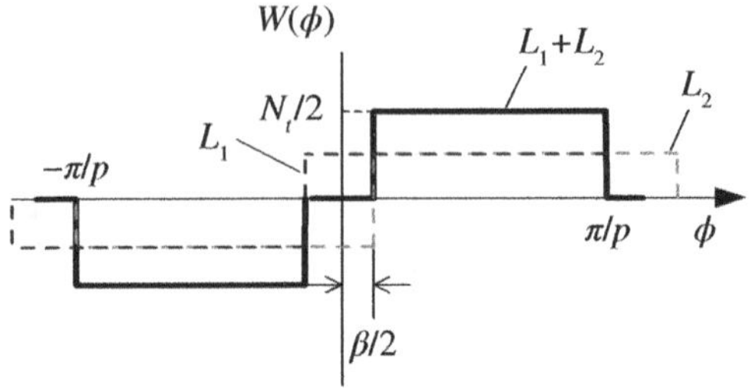

Figure 5.8 Winding function of a single-phase double-layer short-pitch winding. Only one pole pair is shown for clarity.

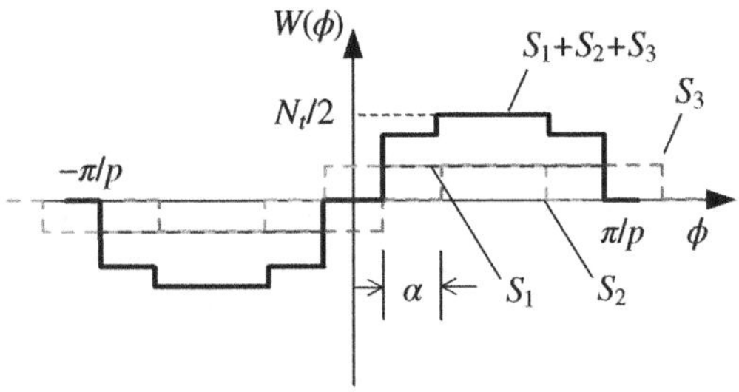

Figure 5.9 Winding function of a single-phase double-layer short-pitch distributed winding. Only one pole pair is shown for clarity.

expressed as:

$$W(\phi) = \frac{4}{\pi}\frac{N_t}{2}\sum_{v=2k+1}^{\infty}\frac{1}{v}\sum_{h=-\frac{q-1}{2}}^{\frac{q-1}{2}}\sin[vp(\phi - h\alpha)]\cos\left(\frac{vp\beta}{2}\right), q \in \text{odd},$$

$$= \frac{4}{\pi}\frac{N_t}{2}\sum_{v=2k+1}^{\infty}\frac{1}{v}\cos\left(\frac{vp\beta}{2}\right)\frac{1}{q}\sum_{h=-\frac{q-1}{2}}^{\frac{q-1}{2}}\sin[vp(\phi - h\alpha)] \quad (5.14)$$

Simplifying (5.14) yields:

$$W(\phi) = \frac{4}{\pi}\frac{N_t}{2}\sum_{v=2k+1}^{\infty}\frac{1}{v}\cos\left(\frac{vp\beta}{2}\right)\frac{\sin\left(qvp\frac{\alpha}{2}\right)}{q\sin\left(vp\frac{\alpha}{2}\right)}\sin(vp\phi)$$

$$= \frac{4}{\pi}\frac{N_t}{2}\sum_{v=2k+1}^{\infty}S_{vp}\sin(vp\phi), q \in \text{odd} \quad (5.15)$$

Equation (5.15) shows the Fourier series expansion of the winding function with the position angle ϕ as the variable. The Fourier coefficients and winding factors are linked by:

$$S_{vp} = \frac{1}{v} k_{p\_vp} k_{d\_vp} \quad (v = 2k + 1) \tag{5.16}$$

where the pitch factor, distribution factor and winding factor are:

$$k_{p\_vp} = \cos\left(\frac{vp\beta}{2}\right) \quad (v = 2k + 1) \tag{5.17}$$

$$k_{d\_vp} = \frac{\sin\left(qvp\frac{\alpha}{2}\right)}{q\sin\left(vp\frac{\alpha}{2}\right)} \quad (v = 2k + 1) \tag{5.18}$$

$$k_{w\_vp} = k_{d\_vp} k_{d\_vp} \quad (v = 2k + 1) \tag{5.19}$$

This feature provides a quick way to calculate winding factors for complex winding layouts by expanding the winding function into its Fourier series. Based on the linearity of modulation operators presented in Chapter 3, the modulated winding function can be obtained by summing up the modulated harmonic components of the original winding function.

For a three-phase integer-slot distributed winding, the $3i$-th harmonics will be cancelled out in the resultant MMF, where i is an arbitrary positive integer, so they can be neglected in analysis. The fundamental wave with $k = 0$ has the largest amplitude. The amplitude of the 5th harmonic, which is the closest to the fundamental wave and the main harmonic component, is less than 1/5 of the fundamental. Therefore, in most cases, the influence of harmonics can be ignored, and only the fundamental wave of the multi-phase integer-slot distributed winding is considered, namely:

$$W(\phi) = \frac{4}{\pi}\frac{N_t}{2} k_{w\_p} \sin(p\phi) + R_p \tag{5.20}$$

where the residual term R_p is:

$$R_p = \frac{4}{\pi}\frac{N_t}{2} \sum_{v=2k+1}^{\infty} S_{vp} \sin(vp\phi), q \in \text{odd}, k = 1, 2, \ldots, \infty \tag{5.21}$$

Unless otherwise indicated, the complete winding function with the residual term as shown (5.20) is used only when harmonic effects need to be considered. The power and control windings of brushless doubly-fed machines are typically multi-phase distributed windings, and by neglecting the residual term, the winding function of the three-phase control winding can be written concisely as:

$$\begin{cases} W_{scU}(\phi) = \dfrac{4}{\pi}\left(\dfrac{N_{ph\_sc}k_{w1\_sc}}{2p_c}\right)\sin(p_c\phi) \\ W_{scV}(\phi) = \dfrac{4}{\pi}\left(\dfrac{N_{ph\_sc}k_{w1\_sc}}{2p_c}\right)\sin\left(p_c\phi - \dfrac{2\pi}{3}\right) \\ W_{scW}(\phi) = \dfrac{4}{\pi}\left(\dfrac{N_{ph\_sc}k_{w1\_sc}}{2p_c}\right)\sin\left(p_c\phi + \dfrac{2\pi}{3}\right) \end{cases} \tag{5.22}$$

where the three phases are labeled by subscripts "U", "V", and "W", respectively, considering "c" is for the control winding. The winding function of the power winding can be readily obtained by replacing the subscript "c" with "p" in (5.22). The source MMF created by the control winding fed by balanced three-phase unit currents is:

$$\begin{aligned} F_c(\phi, t) &= W_{scU}(\phi)\sin(\omega t) + W_{scV}(\phi)\sin\left(\omega t - \frac{2\pi}{3}\right) + W_{scW}(\phi)\sin\left(\omega t + \frac{2\pi}{3}\right) \\ &= \frac{4}{\pi}\left(\frac{N_{ph\_sc}k_{w1\_sc}}{2p_c}\right)\left(\frac{3}{2}\right)\cos(p_c\phi - \omega t) \end{aligned} \tag{5.23}$$

Equation (5.23) shows that the resultant MMF is a travelling wave, and the amplitude is 3/2 times that of the single-phase counterpart. To facilitate the comparison between different brushless doubly-fed machines, the same source MMF for the control winding, i.e. a unit cosine function whose amplitude is 1, is used:

$$f(\phi, t) = \cos(p\phi - \omega t) \tag{5.24}$$

5.3.2 Airgap Magnetic Field and Inductances

1 CBDFIM

(1) Airgap Magnetic Field Distribution

Since both the stator and rotor in CBDFIMs are slotted, the modulation operator takes the form shown in (3.25). Furthermore, if we ignore the slotting effect of semi-closed slots, the modulation behavior turns into unit modulation. The modulated source MMF, therefore, is:

$$M[f(\phi, t)] = \cos(p_c\phi - \omega t) \tag{5.25}$$

The no-load airgap magnetic flux density distribution of the control machine can be obtained as:

$$B_f = \frac{\mu_0}{g}\cos(p_c\phi - \omega t) \tag{5.26}$$

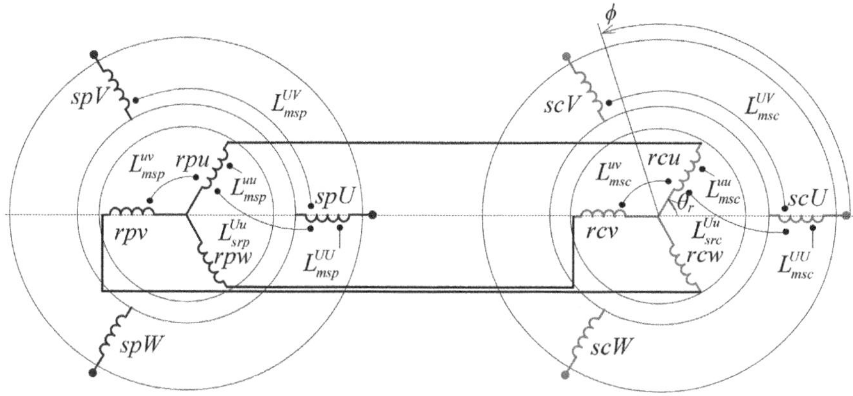

Figure 5.10 Inductances in the CBDFIM.

Equation (5.26) shows that, at any moment, the airgap magnetic field distribution in the control machine of the CBDFIM is sinusoidal. Similarly, by replacing the subscript "c" in (5.25) and (5.26) with "p," the sinusoidally distributed airgap magnetic field in the power machine can also be obtained.

(2) Analysis of Inductance Characteristics

Inductances in the CBDFIM include the phase main self-inductances and phase-to-phase mutual inductances of the power winding, phase main self-inductances and phase-to-phase mutual inductances of the control winding, mutual inductances between the power winding and rotor winding on the power machine side, and mutual inductances between the control winding and rotor winding on the control machine side, as visualized in Figure 5.10. Subscripts "U," "V," and "W" represent phase-U, -V, and -W of the stator winding. Similarly, subscripts "u," "v," and "w" represent phase-u, -v, and -w of the rotor winding. Subscripts "s" and "r" represent the stator and rotor, respectively. Subscripts "p" and "c" represent the power machine and control machine side, respectively.

The phase-U main self-inductance of the control winding is:

$$\begin{aligned} L_{msc\_CBDFIM}^{UU} &= \frac{\mu_0 r_g l_{stk}}{g} \int_0^{2\pi} M[W_{scU}(\phi)]W_{scU}(\phi)d\phi \\ &= \frac{\mu_0 r_g l_{stk}}{g} \int_0^{2\pi} \left[\frac{4}{\pi} \left(\frac{N_{ph\_sc}k_{w1\_sc}}{2p_c} \right) \sin(\phi) \right]^2 d\phi \\ &= \frac{\mu_0 \pi r_g l_{stk}}{g} \left[\frac{4}{\pi} \left(\frac{N_{ph\_sc}k_{w1\_sc}}{2p_c} \right) \right]^2 = L_{msc} \end{aligned} \tag{5.27}$$

The phase-V and phase-W main self-inductances are the same as the phase-U main self-inductance expressed by (5.27).

The mutual inductance between phase-U and phase-V of the control winding is:

$$\begin{aligned}
L^{UV}_{msc_CBDFIM} &= \frac{\mu_0 r_g l_{stk}}{g}\int_0^{2\pi} M[W_{scU}(\phi)]W_{scV}(\phi)d\phi \\
&= \frac{\mu_0 r_g l_{stk}}{g}\int_0^{2\pi}\left[\frac{4}{\pi}\left(\frac{N_{ph_sc}k_{w1_sc}}{2p_c}\right)\sin(\phi)\right] \\
&\quad \left[\frac{4}{\pi}\left(\frac{N_{ph_sc}k_{w1_sc}}{2p_c}\right)\sin\left(\phi-\frac{2\pi}{3}\right)\right]d\phi \\
&= \left(-\frac{1}{2}\right)\frac{\mu_0 \pi r_g l_{stk}}{g}\left[\frac{4}{\pi}\left(\frac{N_{ph_sc}k_{w1_sc}}{2p_c}\right)\right]^2 = \left(-\frac{1}{2}\right)L_{msc} \qquad (5.28)
\end{aligned}$$

Similarly, the mutual inductance between phase-V and phase-W, and the mutual inductance between phase-W and phase-U, are the same as that between the phase-U and phase-V.

The phase-u main self-inductance of rotor winding on the control machine side is:

$$\begin{aligned}
L^{uu}_{mrc_CBDFIM} &= \frac{\mu_0 r_g l_{stk}}{g}\int_0^{2\pi} M[W_{rcu}(\phi)]W_{rcu}(\phi)d\phi \\
&= \frac{\mu_0 r_g l_{stk}}{g}\int_0^{2\pi}\left[\frac{4}{\pi}\left(\frac{N_{ph_sc}k_{w1_sc}}{2p_c}\right)\sin(\phi)\right]^2 d\phi \\
&= \frac{\mu_0 \pi r_g l_{stk}}{g}\left[\frac{4}{\pi}\left(\frac{N_{ph_sc}k_{w1_sc}}{2p_c}\right)\right]^2 = L_{mrc} \qquad (5.29)
\end{aligned}$$

Similarly, the phase-v and phase-w main self-inductances on the control machine side are the same as the phase-u main self-inductance.

The mutual inductance between the phase-u and phase-v of the rotor winding on the control machine side is:

$$\begin{aligned}
L^{uv}_{mrc_CBDFIM} &= \frac{\mu_0 r_g l_{stk}}{g}\int_0^{2\pi} M[W_{rcu}(\phi)]W_{rcv}(\phi)d\phi \\
&= \frac{\mu_0 r_g l_{stk}}{g}\int_0^{2\pi}\left[\frac{4}{\pi}\left(\frac{N_{ph_rc}k_{w1_rc}}{2p_c}\right)\sin(\phi)\right] \\
&\quad \left[\frac{4}{\pi}\left(\frac{N_{ph_rc}k_{w1_rc}}{2p_c}\right)\sin\left(\phi-\frac{2\pi}{3}\right)\right]d\phi \\
&= \left(-\frac{1}{2}\right)\frac{\mu_0 \pi r_g l_{stk}}{g}\left[\frac{4}{\pi}\left(\frac{N_{ph_rc}k_{w1_rc}}{2p_c}\right)\right]^2 = \left(-\frac{1}{2}\right)L_{mrc} \qquad (5.30)
\end{aligned}$$

Similarly, the mutual inductance between the phase-v and phase-w, and the mutual inductance between the phase-w and phase-u, are the same as that between the phase-u and phase-v.

The mutual inductance between the control winding and the rotor winding on the control machine side is:

$$\begin{aligned}
L_{src_CBDFIM}^{Uu} &= \frac{\mu_0 r_g l_{stk}}{g}\int_0^{2\pi} M[W_{scU}(\phi)]W_{rcu}(\phi - p_c\theta_r)d\phi \\
&= \frac{\mu_0 r_g l_{stk}}{g}\int_0^{2\pi}\left[\frac{4}{\pi}\left(\frac{N_{ph_sc}k_{w1_sc}}{2p_c}\right)\sin(\phi)\right] \\
&\quad \left[\frac{4}{\pi}\left(\frac{N_{ph_rc}k_{w1_rc}}{2p_c}\right)\sin(\phi - p_c\theta_r)\right]d\phi \\
&= \frac{\mu_0 \pi r_g l_{stk}}{g}\left[\frac{4}{\pi}\left(\frac{N_{ph_sc}k_{w1_sc}}{2p_c}\right)\right]\left[\frac{4}{\pi}\left(\frac{N_{ph_rc}k_{w1_rc}}{2p_c}\right)\right] \\
&\quad \cos(p_c\theta_r) = L_{src}\cos(p_c\theta_r)
\end{aligned} \tag{5.31}$$

The mutual inductance between any stator phase winding and any rotor phase winding can be obtained in a similar manner. The main self-inductance of a single phase and the mutual inductance between any two phases of the stator or rotor in the power machine, and the mutual inductances between the stator and rotor windings can also be calculated in the same way. The main inductance matrix of the power machine can be obtained by replacing the subscript "c" in (5.27)–(5.31) with "p." Inductance formulas shown in (5.27)–(5.31) are the same as those in [17]. It is worth noting that these inductance formulas do not consider the magnetic saturation effect and need to be corrected by saturation factors when used to calculate saturated inductances in real machines.

2 BDFIM

(1) Airgap Magnetic Field Distribution

BDFIMs typically use nested-loop rotor windings to modulate the source MMF, and the modulation operator is of the form shown in (3.24). The simplest case with a single loop per nest ($S = 1$) is presented in this section to illustrate the analysis method.

It is assumed that the rotor winding has N_{SC} identical nests numbered by 1, 2, …, N_{SC} in order, and the coil-span-to-coil-pitch ratio is γ. The Fourier series expansion of the modulated MMF is expressed by (4.38). It is shown that the modulated MMF has three groups of field harmonics, and their numbers of pole pairs are p, $(lN_{SC} - p)$ and $(lN_{SC} + p)$, respectively. The magnetic field conversion factors are shown in (4.39)–(4.41).

Harmonic components with p-pole pairs maintain the same number of pole pairs as the source MMF while the $(lN_{SC} - p)$- and $(lN_{SC} + p)$-pole-pair harmonic components are introduced by the modulation behavior of the nested-loop winding. All magnetic field conversion factors are dependent on the number of short-circuited coils N_{SC}, number of the source MMF pole pairs p and coil-span-to-coil-pitch γ.

When N_{SC} approaches p, the content of harmonic components with $(lN_{SC} - p)$ and $(lN_{SC} + p)$ pole pairs increases, and the two harmonics with $l = 1$ whose numbers of pole pairs are $(N_{SC} - p)$ and $(N_{SC} + p)$, respectively, have the largest amplitudes. If the control winding is designed to have $(N_{SC} - p)$ or $(N_{SC} + p)$ pole pairs, the machine will operate as a BDFIM. Moreover, for the same number of nests N_{SC}, the closer the pole pair numbers of the power winding and control winding are, the larger the magnetic field conversion factors will be, which means the BDFIM prefers near-ratio combinations.

(2) Analysis of Inductance Characteristics

In BDFIMs, formulas for phase main self-inductances and mutual inductances between any two phases of the power winding and control winding are consistent with those for the CBDFIM, as illustrated in Figure 5.11. The difference mainly appears in the calculation of inductances related to the nested-loop rotor winding, including main self-inductances and mutual inductances of nests and mutual inductances between each loop and the power winding/control winding. Besides the nested-loop winding, there are many other rotor winding structures available for BDFIMs. Here we take the nested-loop rotor winding as an example to illustrate the analysis.

Assuming that each nest is composed of S concentric nested conductor loops, inductances of the rotor winding itself include two types: the main self-inductances and mutual inductances within the same nest, and mutual

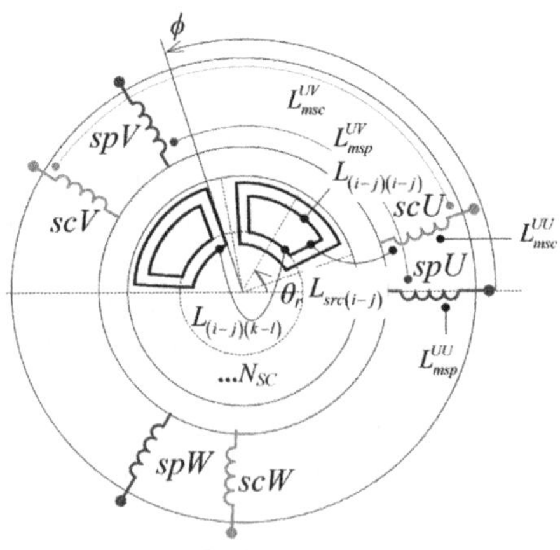

Figure 5.11 Inductances in the BDFIM with a nested-loop rotor.

inductances between different nests. A single conductor loop can be treated as the single-phase asymmetric winding based on the winding classification shown in Chapter 2. When a DC current flows in the conductor loop, the positive and negative half cycles of the MMF created by the loop are of different shapes. For clarity, we analyze the winding function of a single conductor loop first.

The winding function of a single conductor loop is shown in Figure 4.32, whose Fourier series expansion is expressed by (4.38). The inductance between the j-th loop in the i-th nest and the l-th loop in the k-th nest is:

$$
\begin{aligned}
L_{(i-j)(k-l)} &= \frac{\mu_0 r_g l_{stk}}{g} \int_0^{2\pi} W_j^i(\phi) W_l^k(\phi) d\phi \\
&= \frac{\mu_0 \pi r_g l_{stk}}{g} \sum_{n=1}^{\infty} \left(\frac{2\gamma_j}{N_{SC}}\right) \left[\frac{\sin\left(\gamma_j \frac{n\pi}{N_{SC}}\right)}{\left(\gamma_j \frac{n\pi}{N_{SC}}\right)}\right] \left(\frac{2\gamma_l}{N_{SC}}\right) \left[\frac{\sin\left(\gamma_l \frac{n\pi}{N_{SC}}\right)}{\left(\gamma_l \frac{n\pi}{N_{SC}}\right)}\right] \\
&\quad \cos n \left[(k-i)\frac{2\pi}{N_{SC}}\right]
\end{aligned}
\tag{5.32}
$$

The simplification of (5.32) depends on the relative position of the two loops. Letting k and l be equal to i and j, respectively, we can get the main self-inductance of the j-th loop in the i-th nest:

$$
\begin{aligned}
L_{loop(j,j)} &= \frac{\mu_0 r_g l_{stk}}{g} \int_0^{2\pi} W_{ij}^2(\phi) d\phi \\
&= \frac{\mu_0 r_g l_{stk}}{g} \left\{ \int_{\left[-\gamma_j\left(\frac{\pi}{N_{SC}}\right), \gamma_j\left(\frac{\pi}{N_{SC}}\right)\right]} \left[1 - \gamma_j \left(\frac{2\pi}{N_{SC}}\right) \frac{1}{2\pi}\right]^2 d\phi \right. \\
&\quad \left. + \int_{[0,2\pi]-\left[-\gamma_j\left(\frac{\pi}{N_{SC}}\right), \gamma_j\left(\frac{\pi}{N_{SC}}\right)\right]} \left[-\gamma_j \left(\frac{2\pi}{N_{SC}}\right) \frac{1}{2\pi}\right]^2 d\phi \right\} \\
&= \frac{\mu_0 r_g l_{stk}}{g} \left[\gamma_j \left(\frac{2\pi}{N_{SC}}\right)\right] \left[1 - \gamma_j \left(\frac{1}{N_{SC}}\right)\right]
\end{aligned}
\tag{5.33}
$$

Letting l be equal to j, we can get the mutual inductance between any two j-th loops in different nests:

$$
L_{loops(j-j)} = \frac{\mu_0 r_g l_{stk}}{g} \int_0^{2\pi} W_{ij}(\phi) W_{ij}\left(\phi - h\frac{2\pi}{N_{SC}}\right) d\phi
$$

$$= \frac{\mu_0 r_g l_{stk}}{g}\left\{2\int_{\left[-\gamma_j\left(\frac{\pi}{N_{SC}}\right),\gamma_j\left(\frac{\pi}{N_{SC}}\right)\right]}\left[1-\gamma_j\left(\frac{2\pi}{N_{SC}}\right)\frac{1}{2\pi}\right]\left[-\gamma_j\left(\frac{2\pi}{N_{SC}}\right)\frac{1}{2\pi}\right]d\phi\right.$$

$$\left.+\int_{[0,2\pi]-2\left[-\gamma_j\left(\frac{\pi}{N_{SC}}\right),\gamma_j\left(\frac{\pi}{N_{SC}}\right)\right]}\left[-\gamma_j\left(\frac{2\pi}{N_{SC}}\right)\frac{1}{2\pi}\right]^2 d\phi\right\}$$

$$= \frac{\mu_0 r_g l_{stk}}{g}\left\{\left[\gamma_j\left(\frac{2\pi}{N_{SC}}\right)\frac{1}{2\pi}\right]\left[-\gamma_j\left(\frac{2\pi}{N_{SC}}\right)\right]\right\} \tag{5.34}$$

where h is a positive integer, and $h \in [1, 2, \ldots, N_{SC}]$.

The mutual inductance between the control winding and the j-th loop of the i-th nest is:

$$L_{src(j)} = \frac{\mu_0 r_g l_{stk}}{g}\int_0^{2\pi} W_{scU}(\phi)W_j^i(\phi)d\phi$$

$$= \frac{\mu_0 r_g l_{stk}}{g}\int_0^{2\pi}\frac{4}{\pi}\left(\frac{N_{ph\_sc}k_{w1\_sc}}{2p_c}\right)\sin(p_c\phi)\sum_{n=1}^{\infty}\left(\frac{2\gamma_j}{N_{SC}}\right)\left[\frac{\sin\left(\gamma_j\frac{n\pi}{N_{SC}}\right)}{\left(\gamma_j\frac{n\pi}{N_{SC}}\right)}\right]$$

$$\cos n\left[\phi-(i-1)\frac{2\pi}{N_{SC}}-\theta_r\right]d\phi$$

$$= \frac{\mu_0 \pi r_g l_{stk}}{g}\frac{4}{\pi}\left(\frac{N_{ph\_sc}k_{w1\_sc}}{2p_c}\right)\left(1-\frac{\gamma_j}{N_{SC}}\right)\sin\left\{p_c\left[(i-1)\frac{2\pi}{N_{SC}}+\theta_r\right]\right\} \tag{5.35}$$

3 BDFRM with a Salient Reluctance Pole Rotor

The BDFRM may have three rotor types, i.e. the salient reluctance pole rotor, the axially-laminated flux barrier rotor, and the radially-laminated flux barrier rotor. The salient reluctance pole rotor uses the variation of airgap permeance to change the distribution of airgap MMF, while flux barrier rotors short airgaps at different locations through the flux guides. The radially-laminated structure is a simplified version of the axially-laminated anisotropic rotor and it can suppress the eddy current loss in the rotor core effectively.

(1) Airgap Magnetic Field Distribution

The modulation operator of BDFRMs with salient reluctance pole rotors takes the form shown in (3.25). Due to the fine slot structure on the stator side, the modulation behavior of stator teeth can be seen as a unit modulation. Therefore, only the modulation behavior of rotor salient reluctance poles is considered. For simplicity, it is assumed that the relative airgap permeance under rotor pole faces is 1.0 and 0 between adjacent teeth. The discontinuous interval C^R occupied by rotor salient reluctance poles can be expressed as:

$$C^R = \cup_{i=1}^{N_{SP}} \left[(i-1) \times \frac{2\pi}{N_{SP}} + (1-\varepsilon)\frac{\pi}{N_{SP}}, (i-1) \times \frac{2\pi}{N_{SP}} + (1+\varepsilon)\frac{\pi}{N_{SP}} \right], \varepsilon \in (0,1) \tag{5.36}$$

where N_{SP} is the number of rotor salient reluctance poles, and ε is the ratio of rotor pole arc width to the rotor pole pitch.

Applying the rotor salient reluctance pole modulation operator to the unit cosine source MMF, we can get:

$$M(N_{SP}, \varepsilon)[\cos(p\phi - \omega_s t)] = \begin{cases} \cos(p\phi - \omega_s t), & \phi \in C^R, \\ 0, & \phi \in [0,2\pi] - C^R. \end{cases} \tag{5.37}$$

Expanding (5.37) into its Fourier series yields:

$$\begin{aligned} M(N_{SP}, \varepsilon)[\cos(p\phi - \omega_s t)] = {} & C_{p,p} \cos(p\phi - \omega_s t) \\ & + \sum_{k=lN_{SP}-p}^{\infty} C_{p,k} \cos\left[k\phi + \omega_s t + \frac{(p+k)\pi}{N_{SP}}\right] \\ & + \sum_{k=lN_{SP}+p}^{\infty} C_{p,k} \cos\left[k\phi - \omega_s t + \frac{(k-p)\pi}{N_{SP}}\right], l \in Z^+ \end{aligned} \tag{5.38}$$

where the three magnetic field conversion factors are:

$$C_{p,p} = \varepsilon \tag{5.39}$$

$$C_{p,lN_{SP}-p} = \frac{1}{\pi}\left[\frac{\sin(\varepsilon l \pi)}{l}\right], l = 1, 2, \dots, \infty \tag{5.40}$$

$$C_{p,lN_{SP}+p} = \frac{1}{\pi}\left[\frac{\sin(\varepsilon l \pi)}{l}\right], l = 1, 2, \dots, \infty \tag{5.41}$$

Equation (5.38) shows that the modulated source MMF includes three groups of harmonic components, whose numbers of pole pairs are p, $lN_{SP} - p$, and $lN_{SP} + p$, respectively. The complete modulation process is illustrated in Figure 5.12.

Unlike BDFIMs, magnetic field conversion factors of the BDFRM with a salient reluctance pole rotor are independent of the pole pair combination and are only affected by the pole-arc-to-pole-pitch ratio ε. If we only consider the modulation behavior on the fundamental component of the source MMF modulation, magnetic field conversion factors shown in (5.40) and (5.41) will reach the maximum when $\varepsilon = 0.5$, which is consistent with the conclusion obtained by FEA in [18].

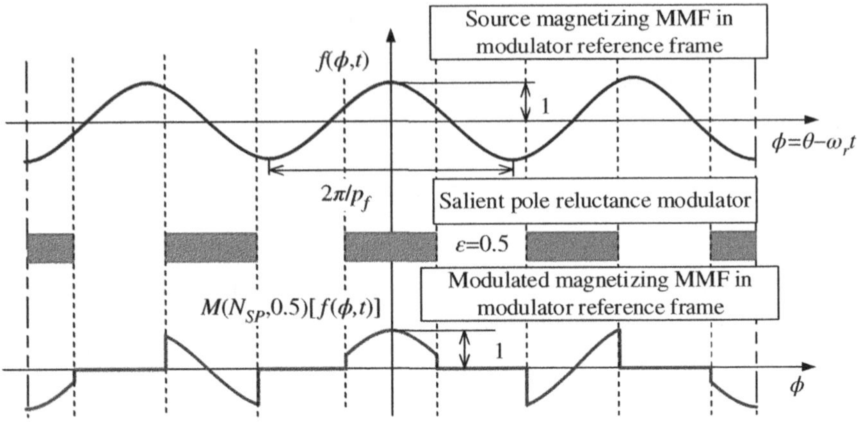

Figure 5.12 Modulation process of the BDFRM with a salient reluctance pole rotor.

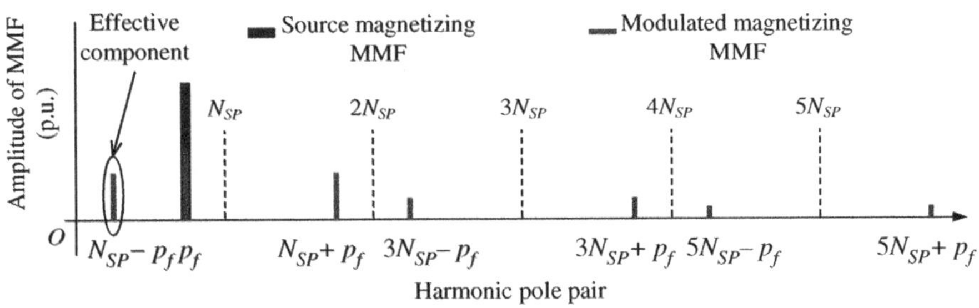

Figure 5.13 Spectrum shift of the BDFRM with a salient reluctance pole rotor.

Let us take the BDFRM with a salient reluctance pole rotor and two stator windings whose pole pairs are 3 and 1 as an example. The number of rotor salient reluctance poles N_{SP} is 4 and ε is 0.5. Magnetic field conversion factors for each harmonic with $l \leq 5$ can be obtained by (5.39)–(5.41), and the corresponding spectrum shift is shown in Figure 5.13. It can be seen that after the source MMF is modulated by rotor salient reluctance poles in the BDFRM, new MMF harmonics with $(lN_{SP} + p)$ or $(lN_{SP} - p)$ pole pairs will have only approximately $1/\pi$ of the source MMF to interact with the other winding, so the utilization of the source MMF is relatively low.

(2) Analysis of Inductance Characteristics

In BDFRMs, both the power and control windings are typically designed as multi-phase distributed windings, and the MMF by each stator winding is dominated by the fundamental wave. Winding functions of the power and control windings are the same as those of the CBDFIM. Inductances in the

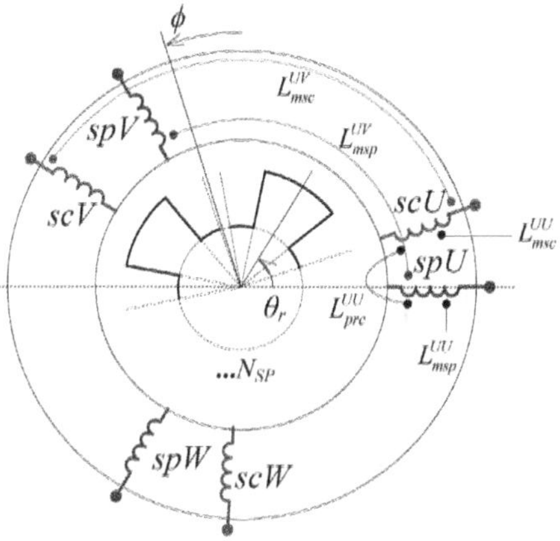

Figure 5.14 Inductances in the BDFRM with a salient reluctance pole rotor.

BDFRM with a salient reluctance pole rotor include phase main self-inductances and phase-to-phase mutual inductances of the power and control windings, as illustrated in Figure 5.14.

The modulated phase-U winding function of the control winding is:

$$M(N_{SP},\varepsilon)[W_{scU}(\phi)] = \begin{cases} W_{scU}(\phi), & \phi \in \cup_{l=1}^{N_{SP}} \left[(i-1)\times\frac{2\pi}{N_{SP}}+(1-\varepsilon)\frac{\pi}{N_{SP}},(i-1)\times\frac{2\pi}{N_{SP}}+(1+\varepsilon)\frac{\pi}{N_{SP}}\right], \\ 0, & \phi \in [0,2\pi]-\cup_{l=1}^{N_{SP}}\left[(i-1)\times\frac{2\pi}{N_{SP}}+(1-\varepsilon)\frac{\pi}{N_{SP}},(i-1) \right. \\ & \left. \times\frac{2\pi}{N_{SP}}+(1+\varepsilon)\frac{\pi}{N_{SP}}\right]. \end{cases} \quad (5.42)$$

Expanding (5.42) into its Fourier series yields:

$$M(N_{SP},\varepsilon)[W_{scU}(\phi)] = \frac{4}{\pi}\left(\frac{N_{ph\_sc}k_{w1\_sc}}{2p_c}\right)\left\{\begin{array}{l} C_{p_c,p_c}\cos(p_c\phi-\omega_s t) \\ +\sum_{k=lN_{SP}-p_c}^{\infty} C_{p_c,k}\cos\left[k\phi+\omega_s t+(p_c+k)\frac{\pi}{N_{SP}}\right] \\ +\sum_{k=lN_{SP}+p_c}^{\infty} C_{p_c,k}\cos\left[k\phi-\omega_s t+(k-p_c)\frac{\pi}{N_{SP}}\right] \end{array}\right\}, l\in Z^+ \quad (5.43)$$

The phase-U main self-inductance of the control winding is:

$$\begin{aligned} L_{msc}^{UU} &= \frac{\mu_0 r_g l_{stk}}{g} \int_0^{2\pi} M[W_{scU}(\phi)] W_{scU}(\phi) d\phi \\ &= \frac{\mu_0 r_g l_{stk}}{g} \int_0^{2\pi} \left\{ \frac{4}{\pi} \left(\frac{N_{ph\_sc} k_{w1\_sc}}{2p_c} \right) C_{p_c,p_c} \sin(\phi + p_c \theta_r) \right\} \\ & \frac{4}{\pi} \left(\frac{N_{ph\_sc} k_{w1\_sc}}{2p_c} \right) \sin(\phi + p_c \theta_r) d\theta \\ &= C_{p_c,p_c} \frac{\mu_0 \pi r_g l_{stk}}{g} \left[\frac{4}{\pi} \left(\frac{N_{ph\_sc} k_{w1\_sc}}{2p_c} \right) \right]^2 \end{aligned} \tag{5.44}$$

where,

$$C_{p_c,p_c} = \varepsilon \tag{5.45}$$

The phase-U main self-inductance of the power winding can be obtained by replacing the subscript "c" with "p" in (5.44). Mutual inductances between the power winding and control winding can be derived similarly.

4 BDFRM with a Multilayer Flux Barrier Rotor

(1) Airgap Magnetic Field Distribution

The modulation operator in the BDFRM with a multilayer flux barrier rotor, the modulation operator takes the form shown in (3.28). The magnetic field modulation effect of the multilayer flux barrier rotor on the unit cosine source MMF is shown in (4.33), and the three magnetic field conversion factors are (4.34)–(4.36). The modulated source MMF has three groups of harmonic components, whose numbers of pole pairs are p, $lN_{MB} - p$, and $lN_{MB} + p$, respectively. Similar to BDFIMs, the BDFRM with a multilayer flux barrier rotor prefers near-ratio pole pair combinations. The closer the pole pair numbers of power and control windings are, the larger the magnetic field conversion factors will be. This conclusion is consistent with that obtained by 2D FEA in [19, 20].

Taking the BDFRM with a multilayer flux barrier rotor, a 6-pole power winding and a 2-pole control winding as an example, the detailed magnetic field modulation process is shown in Figure 4.27. The corresponding spectrum changes are shown in Figure 4.28. It is shown that the content of field harmonics introduced by the multilayer barrier modulator is much lower than that by the salient reluctance pole modulator.

(2) Analysis of Inductance Characteristics

Inductances in the BDFRM with a multi-layer flux barrier rotor are similar to those in the BDFRM with a salient reluctance pole rotor, as shown in Figure 5.15.

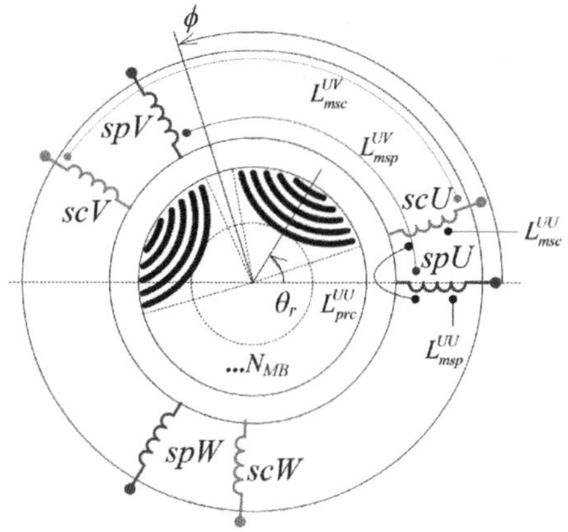

Figure 5.15 Inductance of BDFRM with a multilayer flux barrier rotor.

The modulated phase-U winding function of the control winding is:

$$\begin{aligned}
&M(N_{MB})[N_{scU}(\phi)] \\
&= \frac{4}{\pi}\left(\frac{N_{ph_sc}k_{w1_sc}}{2p_c}\right)\left(\frac{1}{2}\right) \\
&\qquad \sum\left[\sin(p_c\phi + p_c\theta_r) - \sin\left((2i+1)\frac{2\pi}{N_{SC}} - (p_c\phi + p_c\theta_r)\right)\right] \\
&= \frac{4}{\pi}\left(\frac{N_{ph_sc}k_{w1_sc}}{2p_c}\right)\left\{\begin{array}{l}
C_{p_c,p_c}\cos(p_c\phi - p_c\theta_r) \\
-\sum\limits_{k=lN_{MB}-p_c}^{\infty} C^{sum}_{p_c,k}\cos\left[k\phi - p_c\theta_r - \left(\frac{p_c+k}{2}\right)\frac{2\pi}{N_{MB}}\right] \\
+\sum\limits_{k=lN_{MB}+p_c}^{\infty} C^{dif}_{p_c,k}\cos\left[k\phi - p_c\theta_r - \left(\frac{p_c-k}{2}\right)\frac{2\pi}{N_{MB}}\right]
\end{array}\right\}
\end{aligned} \tag{5.46}$$

The phase-U main self-inductance of the control winding is:

$$\begin{aligned}
L^{UU}_{msc} &= \frac{\mu_0 r_g l_{stk}}{g}\int_0^{2\pi} M[W_{scU}(\phi)]W_{scU}(\phi)d\phi \\
&= \frac{\mu_0 r_g l_{stk}}{g}\int_0^{2\pi} \frac{4}{\pi}\left(\frac{N_{ph_sc}k_{w1_sc}}{2p_c}\right) C_{p_c,p_c}\sin(p_c\phi + p_c\theta_r) \\
&\quad \frac{4}{\pi}\left(\frac{N_{ph_sc}k_{w1_sc}}{2p_c}\right)\sin(p_c\phi + p_c\theta_r)d\phi
\end{aligned}$$

$$= C_{p_c,p_c} \frac{\mu_0 \pi r_g l_{stk}}{g} \left[\frac{4}{\pi} \left(\frac{N_{ph\_sc} k_{w1\_sc}}{2p_c} \right) \right]^2 \tag{5.47}$$

where

$$C_{p_c,p_c} = \left(\frac{1}{2}\right) \left[1 - \frac{\sin\left(p_c \frac{2\pi}{N_{MB}}\right)}{\left(p_c \frac{2\pi}{N_{MB}}\right)} \right] \tag{5.48}$$

The phase-U main self-inductance of the power winding can be obtained by replacing the subscript "c" with "p" in (5.48). Mutual inductances between the power winding and control winding can be derived similarly.

5.3.3 Quantitative Analysis of 4/2 BDFRM

1 Magnetic Field Modulation Behavior

The studied BDFRM shown in Figure 5.16 has two 3-phase stator windings with the number of pole pairs of $p_1 = 2$ and $p_2 = 4$, respectively. They are referred to as the stator winding 1 (SW1) and stator winding 2 (SW2) in this section. The stator has 45 slots. Both stator windings are fractional-slot distributed windings, i.e. the number of slots per pole per phase $q > 1$ and q is a fraction. The rotor has six identical lamination segments. A holder made of stainless steel 304 is used to

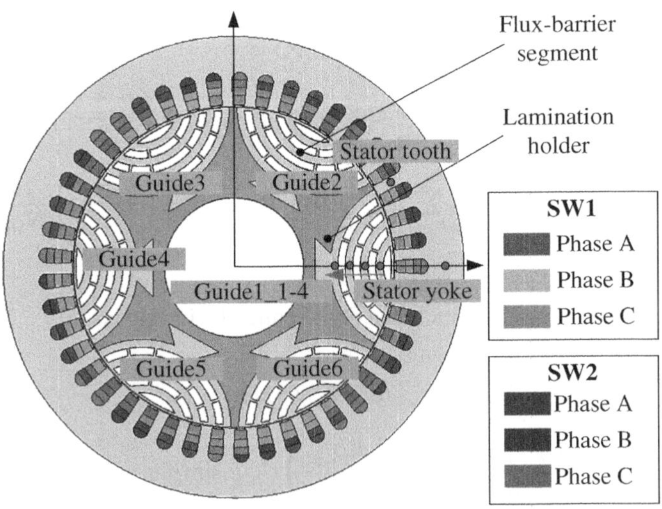

Figure 5.16 Cross-sectional view of the studied BDFRM.

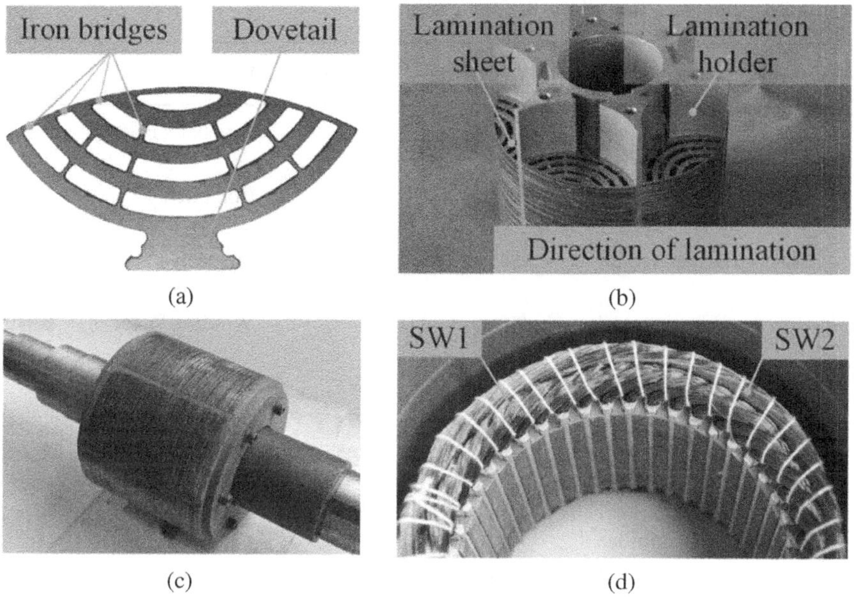

Figure 5.17 The BDFRM prototype with a modular multiple-barrier rotor. (a) Lamination sheet of one flux-barrier segment, (b) assembling of the rotor, (c) rotor assembly, (d) stator windings.

accommodate the segments, as illustrated in Figure 5.17. The specifications are tabulated in Table 5.1.

The winding functions and associated spatial harmonic spectrums of the stator winding are shown in Figure 5.18. The MMFs created by SW1 and SW2 are stair-case waves. In addition to the fundamental component, some low-order winding harmonics also exist. Therefore, the multilayer barrier rotor modulates not only the fundamental but also other main harmonics. With the modulation operator, analyzing the modulation phenomenon involving a complicated source MMF waveform can be greatly simplified by decomposing the applied MMF into a series of harmonics and then synthesizing the modulation result of each component. In this section, only the fundamental components are considered to reveal the production mechanism of the average torque. The harmonic components can then be analyzed in a similar way to investigate the torque ripple.

It is assumed that SW2 with p_2 pole pairs is open-circuited and the amplitude of the fundamental component of the resultant MMF created by the p_1-pole-pair SW1 is F_1. Applying the multilayer barrier modulation operator to the fundamental component, the modulated airgap MMF distribution can be concisely expressed as:

Table 5.1 Specifications of the BDFRM prototype.

Symbol	Name	Value	Windings	Name	Value
R_{s_out}	Stator outer diameter	91 mm	SW_1	Pole pairs p_1	2
R_{s_in}	Stator inner diameter	65 mm		Coil pitch	10
g	Airgap length	0.35 mm		Turns in series per phase	255
l_{stk}	Effective stack length	90 mm		Winding factor	0.9406
N_{MB}	Rotor module number	6	SW_2	Pole pairs p_2	4
N_{slot}	Stator slot number	45		Coil pitch	5
N_{duct}	Flux barrier layer number	4		Turns in series per phase	510
n_s	Fundamental speed	1000 r/min		Winding factor	0.9406

$$\begin{aligned} M(N_{MB})[F_1(\phi,t)] = F_1 \Big\{ & C_{p_1,p_1}\cos(p_1\phi - \omega_1 t - p_1\phi_{01}) \\ & + \sum_{l=1}^{\infty} C_{p_1,lN_{MB}-p_1}\cos[(lN_{MB}-p_1)\phi - lN_{MB}\theta_{r0} + (\omega_1 - lN_{MB}\omega_r)t + p_1\phi_{01} + l\pi] \\ & + \sum_{l=1}^{\infty} C_{p_1,lN_{MB}+p_1}\cos[(lN_{MB}+p_1)\phi - lN_{MB}\theta_{r0} - (\omega_1 + lN_{MB}\omega_r)t - p_1\phi_{01} + l\pi] \Big\} \end{aligned} \tag{5.49}$$

$$C_{p_1,p_1} = \left(\frac{1}{2}\right)\left[1 - \frac{\sin\left(p_1\frac{2\pi}{N_{MB}}\right)}{\left(p_1\frac{2\pi}{N_{MB}}\right)}\right] \tag{5.50}$$

$$C_{p_1,lN_{MB}\pm p_1} = (1-\sigma_l)\left(-\frac{1}{2}\right)\frac{\sin\left[\frac{(lN_{MB}\pm 2p_1)\pi}{N_{MB}}\right]}{\left[\frac{(lN_{MB}\pm 2p_1)\pi}{N_{MB}}\right]} \tag{5.51}$$

where $F_1(\phi,t)$ is the source MMF distribution created by SW1. F_1 and ω_1 are the amplitude and the electrical angular frequency of SW1, and ω_r is the mechanical angular velocity of the rotor.

Compared to (4.35) and (4.36) in Chapter 4, the flux leakage factor σ_l is included in (5.51) to account for the influence of leakage flux on the magnetic coupling capability caused by iron bridges between flux guides in practical multilayer

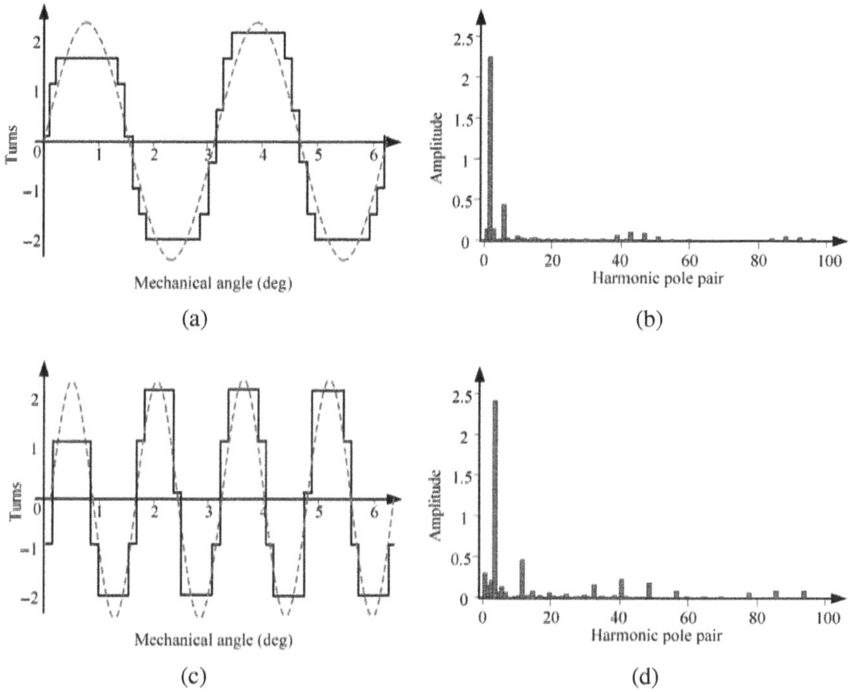

Figure 5.18 Winding function of stator winding. (a) Waveforms of winding function of SW1. (b) Harmonics of winding function of SW1. (c) Waveforms of winding function of SW2. (d) Harmonics of winding function of SW2.

barrier rotors. In ideal cases, σ_l is set to 0, which means all the flux leakages are neglected.

Equation (5.51) indicates that the specially designed rotor with N_{MB} multiple-barrier segments can produce three groups of airgap flux density harmonics in response to the p_1-pole-pair MMF created by SW1 and their pole pairs are p_1, $lN_{MB} - p_1$, and $lN_{MB} + p_1$, respectively. Amplitudes of different harmonic components can be measured by three magnetic field conversion factors, i.e. $C_{p1,p1}$, $C_{p1,lNMB-p1}$, and $C_{p1,lNMB+p1}$, which only depend on the rotor structure.

By replacing p_1 with p_2, the airgap field distribution in response to p_2-pole-pair MMF created by SW2 can be readily obtained. Due to the multi-harmonic nature, the flux pattern of the studied BDFRM is quite different from those in ordinary AC machines. Figure 5.19 illustrates the flux contours at two different time moments T0 and T1 when SW1 is excited and SW2 is left open. The radial airgap flux density distribution and the corresponding amplitude spectrum are shown in Figure 5.20. The harmonic orders coincide well with the those predicted by (5.49).

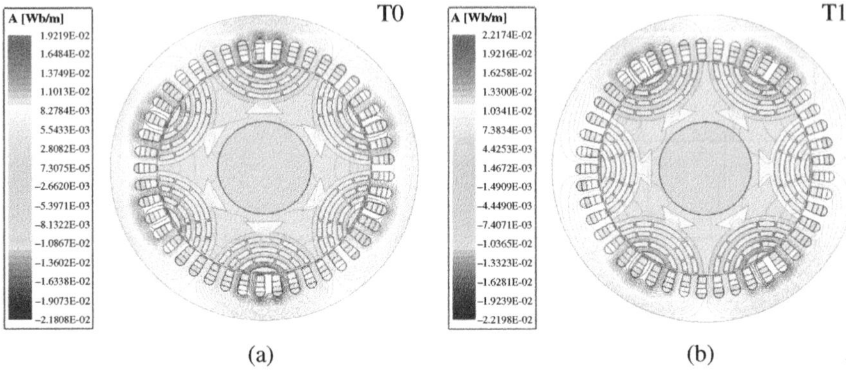

Figure 5.19 Flux patterns at two instants of time T0 and T1 when SW1 is excited and SW2 open-circuited. (a) Time T0. (b) Time T1.

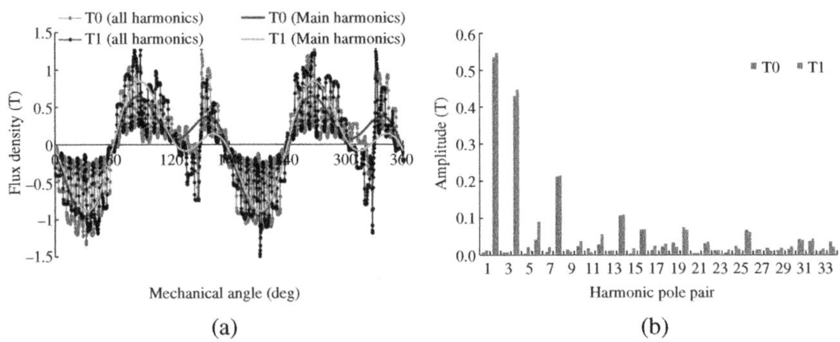

Figure 5.20 Airgap flux density at two instants of time T0 and T1 when SW1 is excited and SW2 open-circuited. (a) Flux density waveforms, (b) harmonic analysis.

The flux density distribution at an instant of time when both SW1 and SW2 are excited are shown in Figure 5.21. The corresponding radial airgap flux densities and associated amplitude spectrums are illustrated in Figure 5.22. The steady-state torque is decomposed by the Maxwell stress tensor method, as shown in Figure 5.23. Torque ripple can be clearly observed as the rotor rotates. The torque contributed by the p_1-pole-pair harmonic is about half that contributed by the p_2-pole-pair harmonic.

In addition, to study the variation of flux densities in the laminated rotor core, several probing points marked in Figure 5.16 are chosen (Stator tooth, Stator yoke, Guidex_y, where $x = 1, 2, \ldots, 6$, and $y = 1, 2, 3$). A few useful conclusions have been reached:

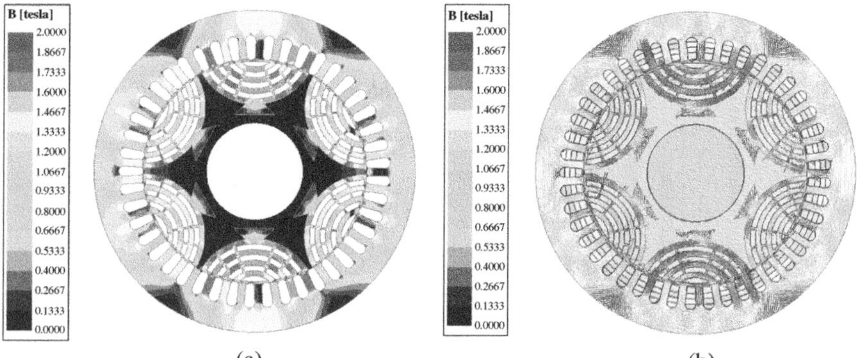

Figure 5.21 Flux density distribution of the BDFRM with both SW1 and SW2 excited. (a) Flux density map, (b) vector plot.

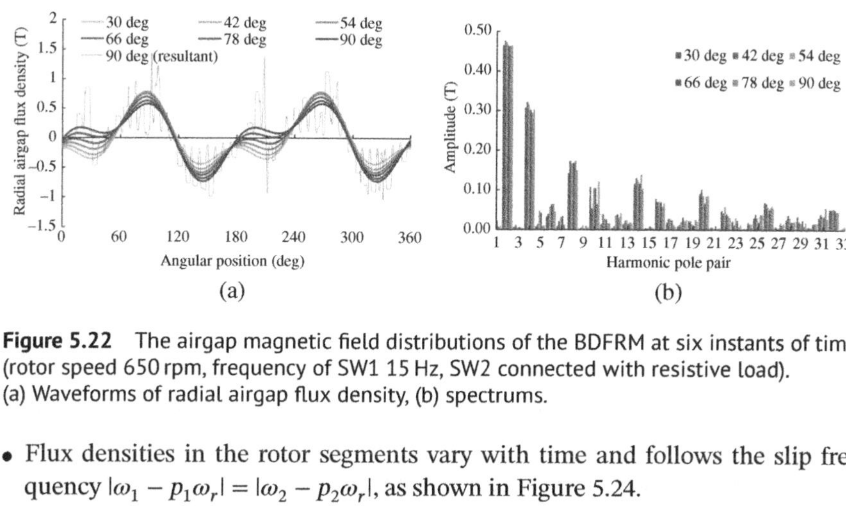

Figure 5.22 The airgap magnetic field distributions of the BDFRM at six instants of time (rotor speed 650 rpm, frequency of SW1 15 Hz, SW2 connected with resistive load). (a) Waveforms of radial airgap flux density, (b) spectrums.

- Flux densities in the rotor segments vary with time and follows the slip frequency $|\omega_1 - p_1\omega_r| = |\omega_2 - p_2\omega_r|$, as shown in Figure 5.24.
- There is a constant phase shift between flux densities at the same locations in two adjacent rotor segments, as illustrated in Figure 5.25.
- Flux densities in the stator core also vary with time, but the frequency is the greatest common factor of frequencies of SW1 and SW2.
- Magnetizing in stator yoke and stator tooth is rotational and alternating, respectively.

2 Cross-coupling Capability

The cross-coupling capability denotes the magnetic coupling between SW1 and SW2 through the intermediate action of the rotor. From the point of view of electric circuit, the brushless doubly-fed machine can be regarded as a set of rotor-position

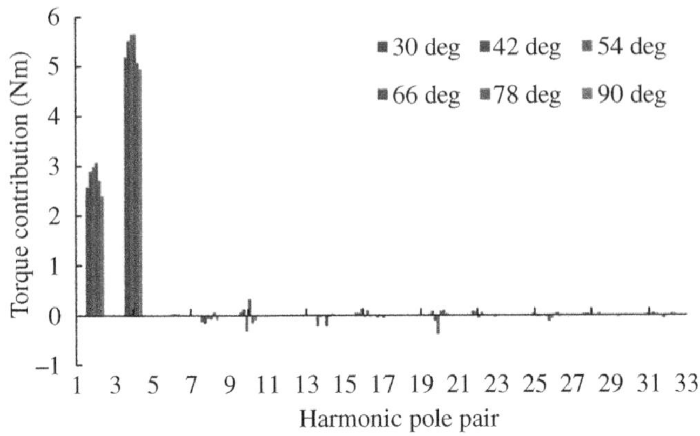

Figure 5.23 The torque compositions of the BDFRM at six instants of time.

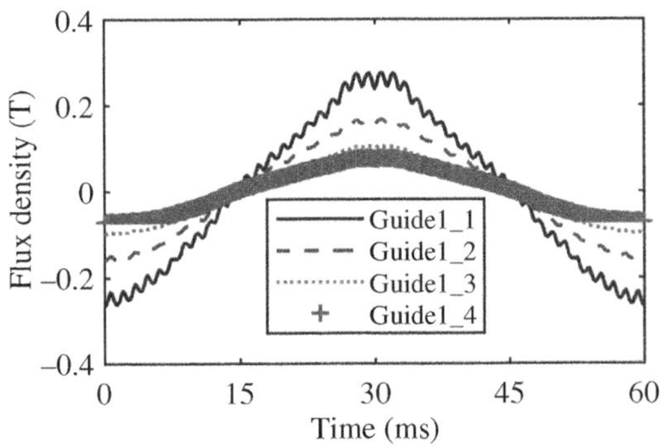

Figure 5.24 Waveforms of the flux density in different flux guides of the same segment.

dependent inductances. The concept of coupling factor (CF) is thus introduced to characterize the cross-coupling capability of brushless doubly-fed machines [21]. It is defined as:

$$\mathrm{CF} = \frac{|C_{p_1,p_2}|}{\sqrt{C_{p_1,p_1} C_{p_2,p_2}}} \tag{5.52}$$

Magnetic field conversion factors and coupling factors of the BDFRM with a multilayer barrier rotor using different pole pair combinations are tabulated in Tables 5.2–5.4. Generally, coupling factors and conversion factors of the difference modulation from $2p_1$-pole excitation to $2p_2$-pole MMF component are much smaller than the sum modulation. This explains why sum modulation is more preferable for BDFRMs with a multilayer barrier rotor.

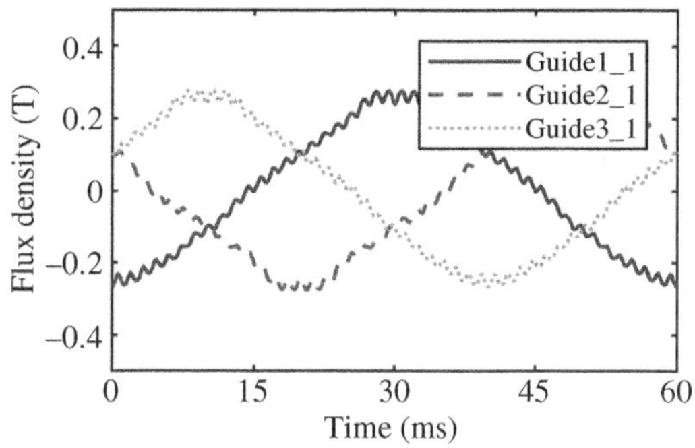

Figure 5.25 Waveforms of the flux density in the same flux guide in different segments.

Table 5.2 Magnetic field conversion factors and coupling factors of the multilayer barrier rotor (sum modulation).

N_{MB}	p_1	$p_2 = N_{MB} - p_1$	$C_{p1,p1}$	$C_{p2,p2}$	$C_{p1,p2}$	CF
3	1	2	0.293	0.603	0.414	0.983
4	1	3	0.182	0.606	0.318	0.959
5	1	4	0.122	0.595	0.252	0.938
5	2	3	0.383	0.578	0.468	0.994
6	1	5	0.087	0.583	0.207	0.921
6	2	4	0.293	0.603	0.414	0.983
7	1	6	0.064	0.573	0.174	0.907
7	2	5	0.228	0.609	0.362	0.971
7	3	4	0.419	0.560	0.483	0.997
8	1	7	0.050	0.564	0.150	0.895
8	2	6	0.182	0.606	0.318	0.959
8	3	5	0.350	0.590	0.450	0.991

Unlike in sum modulation, where the near-ratio combination is preferred, the far-ratio combination is more attractive in difference modulation. Aside from the commonly used sum modulation $p_1 + p_2 = N_{MB}$ and difference modulation $|p_1 - p_2| = N_{MB}$, one more pole pair combination from the sum modulation family $p_1 + p_2 = 2N_{MB}$ is also included because in some cases, the conversion factor of the pole pair combination $p_1 + p_2 = 2N_{MB}$ is even slightly larger than that of the combination $|p_1 - p_2| = N_{MB}$.

Table 5.3 Magnetic field conversion factors and coupling factors of the multilayer barrier rotor (difference modulation).

N_{MB}	p_1	$p_2 = N_{MB} + p_1$	$C_{p1,p1}$	$C_{p2,p2}$	$C_{p1,p2}$	CF
1	1	2	0.5	0.5	0	0
2	1	3	0.5	0.5	0	0
3	1	4	0.293	0.448	0.083	0.228
3	2	5	0.603	0.541	0.059	0.103
4	1	5	0.182	0.436	0.106	0.377
4	2	6	0.5	0.5	0	0
4	3	7	0.606	0.545	0.064	0.111
5	1	6	0.122	0.437	0.108	0.469
5	2	7	0.383	0.467	0.052	0.123
5	3	8	0.578	0.529	0.043	0.077
5	4	9	0.595	0.542	0.058	0.103
6	1	7	0.087	0.441	0.103	0.529
6	2	8	0.293	0.448	0.083	0.228
6	3	9	0.5	0.5	0	0
6	4	10	0603	0.541	0.059	0.103
6	5	11	0.583	0.538	0.052	0.092
7	1	8	0.064	0.446	0.097	0.571
7	2	9	0.228	0.440	0.098	0.312
7	3	10	0.419	0.476	0.037	0.083
7	4	11	0.560	0.522	0.032	0.060
7	5	12	0.609	0.545	0.064	0.111
7	6	13	0.573	0.534	0.046	0.083
8	1	9	0.050	0.450	0.090	0.601
8	2	10	0.182	0.436	0.106	0.377
8	3	11	0.350	0.459	0.064	0.160
8	4	12	0.5	0.5	0	0
8	5	13	0.590	0.535	0.050	0.089
8	6	14	0.606	0.545	0.064	0.111
8	7	15	0.564	0.530	0.041	0.075

Table 5.4 Magnetic field conversion factors and coupling factors of the multilayer barrier rotor ($p_1 + p_2 = 2N_{MB}$ special sum modulation).

N_{MB}	p_1	$p_2 = 2N_{MB} - p_1$	$C_{p1,p1}$	$C_{p2,p2}$	$C_{p1,p2}$	CF
3	1	5	0.293	0.541	0.103	0.259
4	1	7	0.182	0.545	0.106	0.337
5	1	9	0.122	0.542	0.095	0.369
5	2	8	0.383	0.529	0.078	0.173
6	1	11	0.087	0.538	0.083	0.383
6	2	10	0.293	0.541	0.103	0.259
7	1	13	0.064	0.534	0.073	0.391
7	2	12	0.228	0.545	0.109	0.308
7	3	11	0.419	0.522	0.060	0.129
8	1	15	0.050	0.530	0.064	0.396
8	2	14	0.182	0.545	0.106	0.337
8	3	13	0.350	0.535	0.090	0.208

3 Torque Production and Circuit Parameters Calculation

This section makes an attempt to analyze the torque production mechanism of BDFRMs with flux barrier rotors using the analytical expression of stator MMF to clearly show the direct magnetic field interaction between the resultant stator MMF and rotor. Little work has been done in this aspect perhaps due to the complexity of the rotor geometry. The resultant MMF created by the two stator windings ($F_1 \neq 0$ and $F_2 \neq 0$) in the rotor reference frame under doubly-fed synchronous operation can be expressed as:

$$F_1^r(\underline{\phi}, t) + F_2^r(\underline{\phi}, t) = F_s \cos\left\{(\omega_1 - p_1\omega_r)t + \left(\frac{\alpha - \beta}{2}\right) - \vartheta\right\} \tag{5.53}$$

$$F_s = \sqrt{F_1^2 + F_2^2 + 2F_1F_2\cos(\alpha + \beta)} \tag{5.54}$$

$$\vartheta = \arctan 2\left[(F_1 + F_2)\cos\left(\frac{\alpha + \beta}{2}\right), -(F_1 - F_2)\sin\left(\frac{\alpha + \beta}{2}\right)\right] \tag{5.55}$$

$$\alpha = (\varphi_1 + p_1\theta_{01}) - p_1(\underline{\phi} + \theta_{r0}) \tag{5.56}$$

$$\beta = (\varphi_2 + p_2\theta_{02}) - p_2(\underline{\phi} + \theta_{r0}) \tag{5.57}$$

where the superscript "r" of MMFs indicates that the variable is in the rotor-speed reference frame. φ_1 and φ_2 are the initial phases of SW1 and SW2, respectively.

θ_{01} and θ_{02} are the positions of Phase A winding axis of SW1 and SW2, respectively. θ_{r0} is the initial position of the rotor.

It can be concluded from (5.53) that the amplitude of the resultant MMF is position-dependent and has a spatial period of $2\pi/(p_1 + p_2)$. Though the stator core is continuous and distributed stator windings are used, the stator MMF is split into $(p_1 + p_2)$ independent segments equally spaced along the rotor circumference. As a result, the rotor is inherently modularized and can be manufactured to have $(p_1 + p_2)$ repetitive segments.

When seen from the rotor, the resultant MMF established by SW1 and SW2 is a proximate standing wave that has $(p_1 + p_2)$ nodes if the rotor rotates at the speed of $(\omega_1 + \omega_2)/(p_1 + p_2)$. The pulsating frequency of the proximate standing wave is $|\omega_1 - p_1\omega_r|$. In this respect, the difference between operating principles of the SynRM and the BDFRM with a multilayer barrier rotor only lies in the resultant MMF created by stator windings.

As illustrated in Figure 5.26, the resultant stator MMF of a BDFRM with two distributed stator windings, one 2-pole and the other 6-pole, has four pole pairs with an ideal cylindrical rotor, as denoted by the four flux loops. At an arbitrary instant of time, only at some positions the adjacent flux lines have the same polarity. The magnetic polarity along the rotor peripheral is "NN-SN-SS-NS," which is totally different from the case in SynRM whose polarity distribution is "NN-SS-NN-SS." Once the multilayer barrier rotor is placed in the stator bore, the flux lines passing through the rotor will be confined in flux guides, but the positions of the same flux loops on the stator part remain unchanged. As a result, the flux lines are bent when crossing the airgap. The MMF variation along the airgap within each segment tries to pull the adjacent multilayer barrier group into an appropriate position to form the shortest path for flux lines, producing a notable torque to rotate the rotor clockwise.

If one of the stator winding MMFs is too small (or zero) compared with the other, the segmented feature of the resultant MMF will be significantly reduced, and consequently, the resultant MMF turns out to have a spatial period of p_1 or p_2, and the synchronization between the stator MMF segments and rotor multilayer barrier groups tend to be lost.

Inductance parameters can also be determined analytically based on the developed theory, which not only allows more accurate parameter estimation from the machine geometry in the first-cut design but also provides better guidelines to adjust the inductances to meet the design requirements in the optimization phase. The main inductances in the BDFRM with a multilayer barrier rotor is shown in Figure 5.15. The self-inductance of a single phase, for instance, phase-A, and mutual inductance between two stator windings can be derived in conjugation

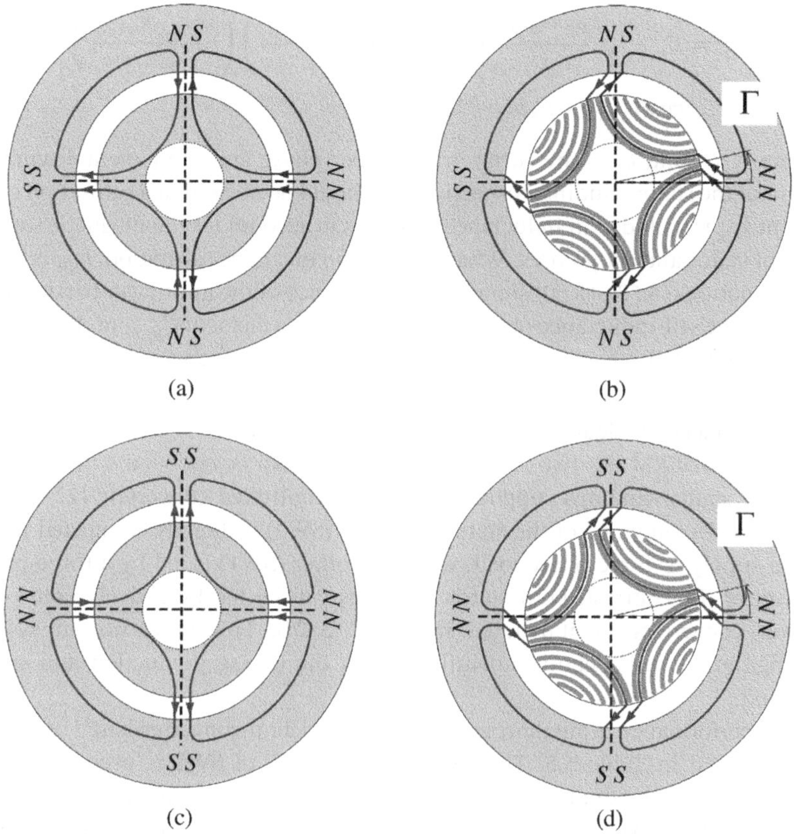

Figure 5.26 Torque production mechanism of the BDFRM and SynRM with a multilayer barrier rotor (the airgap has been widen for better readability). (a) Resultant MMF of a 6-pole and a 2-pole stator winding acting on a cylindrical rotor; (b) resultant MMF of a 6-pole and a 2-pole stator winding acting on a multilayer barrier rotor; (c) MMF created by a 4-pole stator winding acting on a cylindrical rotor; (d) MMF created by a 4-pole stator winding acting on a multilayer barrier rotor.

with the classical winding function theory and expressed as:

$$L_{ms1}^{AA} = C_{p_1,p_1} k_{\text{sat}} \frac{\mu_0 \pi r_g l_{stk}}{g} \left[\frac{4}{\pi} \left(\frac{N_{ph\_s1} k_{w1\_s1}}{2p_1} \right) \right]^2 \tag{5.58}$$

$$L_{ms2}^{AA} = C_{p_2,p_2} k_{\text{sat}} \frac{\mu_0 \pi r_g l_{stk}}{g} \left[\frac{4}{\pi} \left(\frac{N_{ph\_s2} k_{w1\_s2}}{2p_2} \right) \right]^2 \tag{5.59}$$

$$L_{1r2}^{AA} = C_{p_1,p_2} k_{\text{sat}} \frac{\mu_0 \pi r_g l_{\text{stk}}}{g} \left[\left(\frac{4}{\pi}\right)^2 \left(\frac{N_{ph\_s1} k_{w1\_s1}}{2p_1}\right) \left(\frac{N_{ph\_s2} k_{w1\_s2}}{2p_2}\right) \right]$$
$$\cos(p_1 \phi_{01} + p_2 \phi_{02} - N_{\text{MB}} \theta_r) \tag{5.60}$$

where N_{ph_s} and k_{w1_s} are turns in series per phase and fundamental winding factor of the stator winding, respectively. It should be noted that a saturation coefficient k_{sat} is included in inductance formulas to account for the magnetic saturation of stator and rotor cores. When there is no magnetic saturation, k_{sat} is set to 1.0. Equations (5.58)–(5.60) show that the presence of the multilayer barrier rotor reduces the self-inductance of the SW1 and SW2 by a factor of $C_{p1,p1}$ and $C_{p2,p2}$ and mutual inductance between SW1 and SW2 by $C_{p1,p2}$, respectively.

4 Experimental Validation

Since the BDFRM has two AC electrical ports, it can be operated with only one or both stator windings supplied by actively controlled converters [22–25]. In Figure 5.27a, only one of the stator windings (SW2) is directly connected to the main supply or three-phase load, while the other (SW1) is fed by a three-phase converter. By contrast, both stator windings can be fed by actively controlled power converters, as illustrated in Figure 5.27b. This difference in system configuration has a significant influence on the torque production and sizing method.

The performance in single-converter-based and dual-converter-based operation is compared in Table 5.5. The same overall MMF of SW1 is used in the FE simulations. In the single-converter-based operation, the power factor of grid-side stator winding (SW1 in Figure 5.27a) is usually required to be constant or vary in a narrow range, e.g. 0.85–1.0. This requirement on grid-side power factor disables

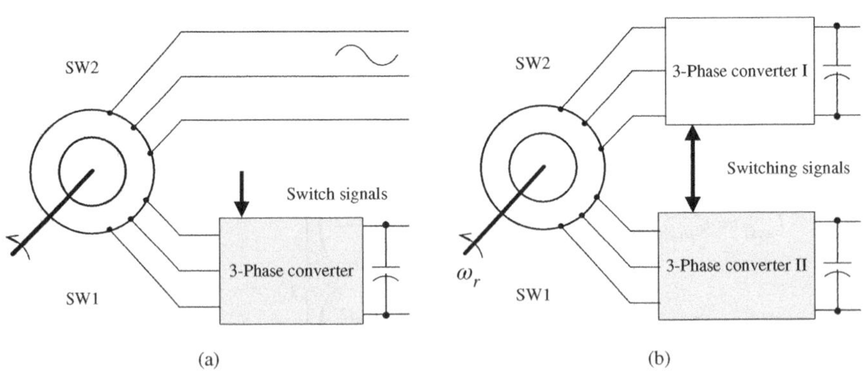

Figure 5.27 Available system configurations of the BDFRM. (a) The single converter-based, (b) the two-converter-based.

Table 5.5 Performance comparison of BDFRMs based on a single and dual power converters.

Parameters		Single power converter	Dual power converter
SW1	Active power (W)	309.2	1368.5
	Power factor	1.0	0.56
	Rated voltage (Vrms)	50	162.7
	Rated current (Arms)	5	5
	Line load (A/m)	18 000	18 000
	Frequency (Hz)	15	50
SW2	Active power (W)	−371.5	1073.8
	Power factor	0.412	0.79
	Rated voltage (Vrms)	79.5	181.4
	Rated current (Arms)	1.59	2.5
	Line load (A/m)	11 500	18 000
	Frequency (Hz)	50	50
Copper loss (W)		468.75	558.3
speed (r/min)		650	1 000
Electromagnetic torque (Nm)		−8.0	18.0
Iron loss (W)		12.5	20
Efficiency (%)		43.6	76.3
Torque/rotor volume (kNm/m^3)		18.0	63.2
Volume torque density (kNm/m^3)		8.66	30.4
Mass torque density (Nm/kg)		0.38	1.34

Results are from 2D FEA.

the maximum torque per ampere (MTPA) operation of the BDFRM where the synchronous angle is supposed to be 90deg (without saturation) or slightly larger than 90deg (with saturation). Consequently, the torque density is unavoidably reduced.

To validate the presented theoretical analysis, experimental tests were conducted on a 1.5-kW BDFRM prototype with a maximum speed of 4000 r/min. The inductance was measured following these steps:

- **Step 1**: The rotor is locked into a given position by an external locking device.

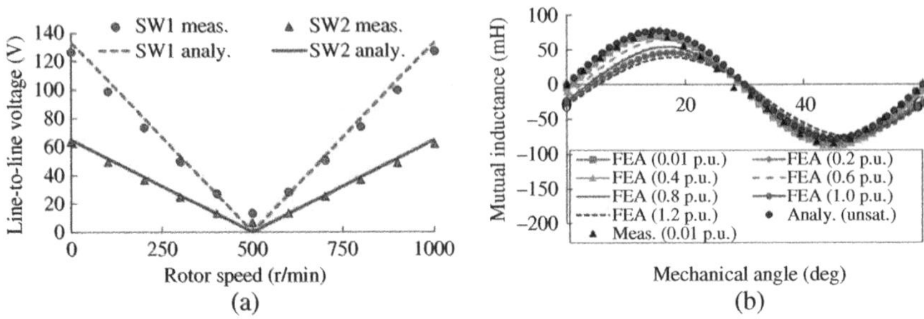

Figure 5.28 Experimental test results. (a) Phase voltage vs. speed. (b) Mutual inductance.

- **Step 2**: The three-phase SW1 is connected in such a way to form a single-phase winding where two phases are connected in parallel and then in series with the third phase. It is then supplied by a single-phase AC voltage source to produce a pulsating magnetic field in the Phase A winding axis of SW1.
- **Step 3**: SW2 is also connected in a similar way to form a single-phase winding.
- **Step 4**: The current flowing in SW1 and terminal voltages of both stator windings are recorded to calculate the self and mutual inductances.

The tested cross coupling characteristic in voltage control mode is shown in Figure 5.28a. In the vicinity of the natural synchronous speed 500 r/min, the induced phase voltage is close to zero.

The mutual inductance between SW1 and SW2 is an AC quantity with a zero average value, as shown in Figure 5.28b. It varies with the rotor position and changes six times in one rotor revolution, which corresponds to $N_{MB} = 6$. The experimental measurement is in good agreement with the analytical and 2D FEA results when the magnetic core is not saturated. With a higher current, the mutual inductance will reduce, and a phase shift also appears.

The measured self-inductances of SW1 and SW2 are shown in Figure 5.29a. There are some ripples superimposed on the constant value predicted by (5.58) and (5.59). The static torque measured at current angle $\Gamma = 90$deg is shown in Figure 5.29b. The torque vs. phase current curve shows a quadratic growth when the magnetic core is unsaturated because we increase both stator winding currents. With larger currents, the torque increase slows down, showing noticeable influence of magnetic saturation.

5.3.4 Quantitative Analysis of 4/2 BDFIM

1 Inductance Characteristics

The studied BDFIM shown in Figure 5.30 shares the same stator core and windings with the BDFRM analyzed in Section 5.3.3. It has two 3-phase stator windings with

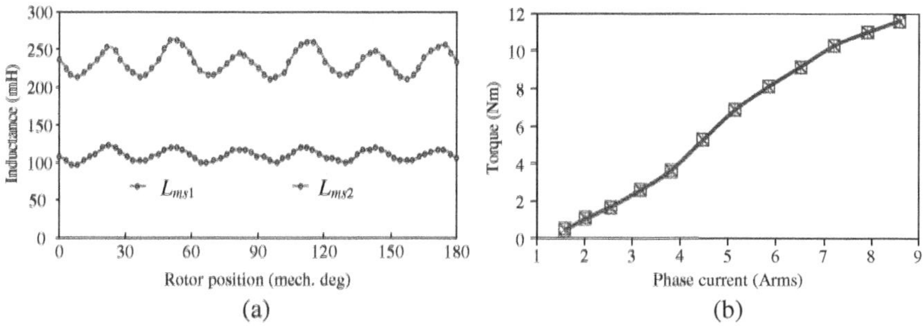

Figure 5.29 Experimental results. (a) Unsaturated inductance waveforms, (b) torque vs. current.

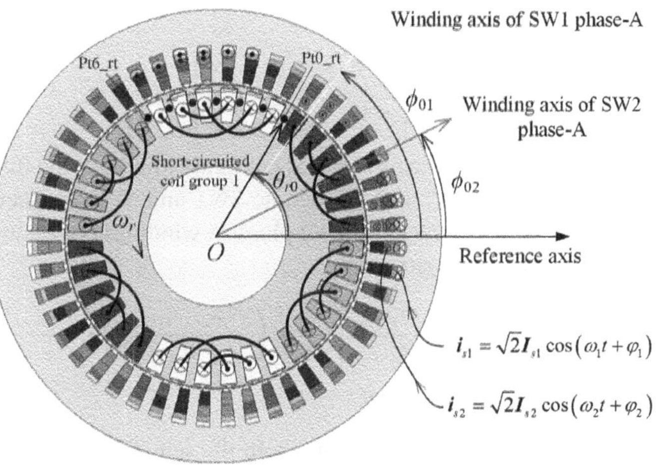

Figure 5.30 Cross-sectional view of the studied BDFIM and the coordinate system.

the number of pole pairs of $p_1 = 2$ and $p_2 = 4$, respectively. They are referred to as the stator winding 1 (SW1) and stator winding 2 (SW2) in this section as well.

Details of the motor components are shown in Figure 5.31. The rotor has 36 slots and 6 separate spiral loops wound by 0.95 mm enamel copper magnet wires. Each spiral loop was wound loosely to have a bare copper slot fill factor of 0.335 to increase the rotor resistance on purpose and reveal the influence of rotor resistance. The 4-pole winding, i.e. SW2, was placed at the slot bottom due to its long end turns and the 8-pole winding (SW1) is closer to the airgap.

Based on the developed theory, the inductance between stator winding i and stator winding j can be obtained by:

$$L_{ij} = k_{\text{sat}} \frac{\mu_0}{g_e} r_g l_{\text{stk}} \int_0^{2\pi} W_i(\phi) M(N_{SC}) [W_j(\phi)] d\phi, \tag{5.61}$$

(a) (b) (c)

Figure 5.31 Prototype of the BDFIM with spiral-loop rotor windings. (a) Rotor assembly, (b) stator, (c) schematic diagram of one spiral loop.

where $W_i(\phi)$ and $W_j(\phi)$ are winding functions of stator winding i and stator winding j, respectively. The self-inductance of phase winding is independent of the rotor position and can be written as:

$$L_{msk}^{AA} = k_{sat} C_{p_k, p_k} \frac{\mu_0 \pi r_g l_{stk}}{g_e} \left[\frac{4}{\pi} \left(\frac{N_{ph\_sk} k_{w1\_sk}}{2p_k} \right) \right]^2, k = 1{,}2. \tag{5.62}$$

where N_{ph_s} is the number of turns in series per phase and k_{w1_s} the fundamental winding factor. Subscripts "1" and "2" denote SW1 and SW2, respectively. The mutual inductance between SW1 and SW2 phase-A windings is a cosine function of rotor position and can be written as:

$$L_{1r2}^{AA} = L_{2r1}^{AA} = -k_{sat} C_{p_1, p_2} \frac{\mu_0 \pi r_g l_{stk}}{g_e} \left[\left(\frac{4}{\pi} \right)^2 \left(\frac{N_{ph\_s1} k_{w1\_s1}}{2p_1} \right) \left(\frac{N_{ph\_s2} k_{w1\_s2}}{2p_2} \right) \right]$$
$$\cos(p_1 \phi_{01} + p_2 \phi_{02} - N_{SC} \theta_r) \tag{5.63}$$

Equations (5.62) and (5.63) indicate that both the self-inductances and mutual inductance are reduced by magnetic field conversion factors due to the presence of the short-circuited rotor windings.

2 Torque Production

The developed theory can be used to derive the torque equation and explain the torque production mechanism of BDFIMs. Since the MMF drop or flux density distribution across the uniform airgap is known by applying the modulation operator to the resultant stator MMF, the magnetic co-energy W_m' in the airgap can be obtained by:

$$W_m' = \int_V \left(\frac{\mu_0}{g_e} \right)^2 \frac{\left\{ M(N_{SC}) \left[F_{1+2}^r(\phi, t) \right] \right\}^2}{2\mu_0} dV \tag{5.64}$$

where μ_0 is the permeability of vacuum, g_e the effective airgap length corrected by Carter's factor, and V the airgap volume.

The electromagnetic torque can be derived from the magnetic co-energy by the virtual work principle. Keeping the current unchanged and making a virtual displacement for initial rotor position, the electromagnetic torque can be written as:

$$T_{em} = \left.\frac{\partial W'_m}{\partial \theta_{r0}}\right|_{\text{current=const.}} = -(p_1 + p_2)\frac{2\mu_0 \pi r_g^3 l_{\text{stk}}}{g_e}\left(\frac{\bar{A}_1}{p_1}\right)\left(\frac{\bar{A}_2}{p_2}\right)$$
$$(C_{p_1,p_1} + C_{p_2,p_2})C_{p_1,p_2}\sin\Gamma \tag{5.65}$$

where r_g is the airgap radius, l_{stk} the stack length. $\bar{A}_1$ and $\bar{A}_2$ are the specific electric loadings of SW1 and SW2, respectively. The current angle Γ takes the same form as (3.99) and is defined as:

$$\Gamma = (p_1\phi_{01} + \varphi_1) + (p_2\phi_{02} + \varphi_2) - N_{SC}\theta_{r0} \tag{5.66}$$

Equation (5.72) shows the synchronous electromagnetic torque in BDFIMs under two-converter-based operation is closely related to the pole pair combination and specific electric loadings of both stator windings. For a given overall specific electric loading, the highest torque density will be achieved by designs with a pole pair combination whose numbers of poles are differed by 1 and equal allocation of specific electric loadings between two stator windings with an optimal $\Gamma = \pi/2$.

Starting from (5.72), the torque density equation in terms of rotor volume, which has been commonly used for initial sizing of electrical machines, can be obtained:

$$\frac{T_{em}}{\pi r_g^2 l_{\text{stk}}} = (p_1 + p_2)\frac{2\mu_0 r_g}{g_e}k_{\text{sat}}\left(\frac{\bar{A}_1}{p_1}\right)\left(\frac{\bar{A}_2}{p_2}\right)(C_{p_1,p_1} + C_{p_2,p_2})C_{p_1,p_2}\sin\Gamma, \tag{5.67}$$

where the specific electric loading of stator winding j is:

$$\bar{A}_j = \frac{2m_j(N_{ph\_sj}k_{w1\_sj}I_{sj})}{2\pi r_g} = \frac{p_j F_j}{\sqrt{2}r_g}, \quad j - 1{,}2. \tag{5.68}$$

where m is the number of phases and I_s the rms value of phase current. The specific magnetic loadings depend on the specific electric loadings, as presented by:

$$\bar{B}_{ij} = k_{\text{sat}}\frac{\sqrt{2}\mu_0}{g_e}\left(\frac{2}{\pi}\right)r_g\left(\frac{C_{p_i,p_j}}{p_j}\right)\bar{A}_j, \quad i,j = 1{,}2. \tag{5.69}$$

Based on design Equations (5.67)–(5.69), a prototype was designed whose specifications and key design parameters are tabulated in Table 5.6.

3 Two-converters-based Operation

To show the performance characteristics of two-converter-based operation of BDFIMs, the 2D transient FEA was conducted for a full range of current

Table 5.6 Specifications and key design parameters of the 1.5 kW BDFIM prototype for two-converter-based operation.

Stator outer/inner diameter (mm)	182/127
DC bus voltage (V)	600
Airgap/stack length (mm)	0.35/90
Number of stator/rotor slots	45/36
Rated torque (Nm)	14.0 (design value)/12.6 (achieved)
Rated speed (r/min)	1000 (freq. of SW1/2 = 50 Hz)
Pole pairs of SW1/2	2/4
Rotor winding configuration	6 equally-spaced spiral loops
Turns in series per phase of SW1/SW2	510/255
Turns in series per spiral loop	15
Specific electric loading $\bar{A}_1, \bar{A}_2$ (kA/m)	18.13, 18.13
Power factor of SW1/SW2	0.53/0.56
Coefficient introduced by saturation, flux leakage and rotor resistance	0.265
Torque per rotor volume (kNm/m^3)	44.70
Torque density (kNm/m^3)	21.53
Specific torque (Nm/kg)	0.89
Efficiency (%)/coolant	76.0/Air

angles and various frequency combinations following the derived current angle Equation (5.66). The results are summarized in Figure 5.32. In order to reach the steady state, each operating point was calculated to have five full electric cycles of the low stator winding frequency and in total 200 time steps [6].

With linear magnetic material, the torque curve at 1000 r/min is a sine wave with a positive DC offset, as shown in Figure 5.32a. The maximum motoring torque occurs when the current angle is $\pi/2$, as indicated by (5.65). The breakdown of the total electromagnetic torque is also included. The DC offsets denote the asynchronous torque components introduced by the rotor winding resistance. After the magnetic saturation is considered, the optimal current angle for maximum motoring torque deviates from (but still stays very close to) $\pi/2$, which can significantly reduce the searching range for the optimal current angle in large-scale design optimizations, as shown in Figure 5.32d.

Under the linear condition, the active and reactive powers of each stator winding are sine wave and cosine wave with positive offsets, respectively, at 1000 r/min. The total active and reactive powers show similar trends. The total active power peaks at a current angle of $\pi/2$ and the maximum total reactive power occurs at a current angle of zero. When the saturation effect is taken into account, the active power

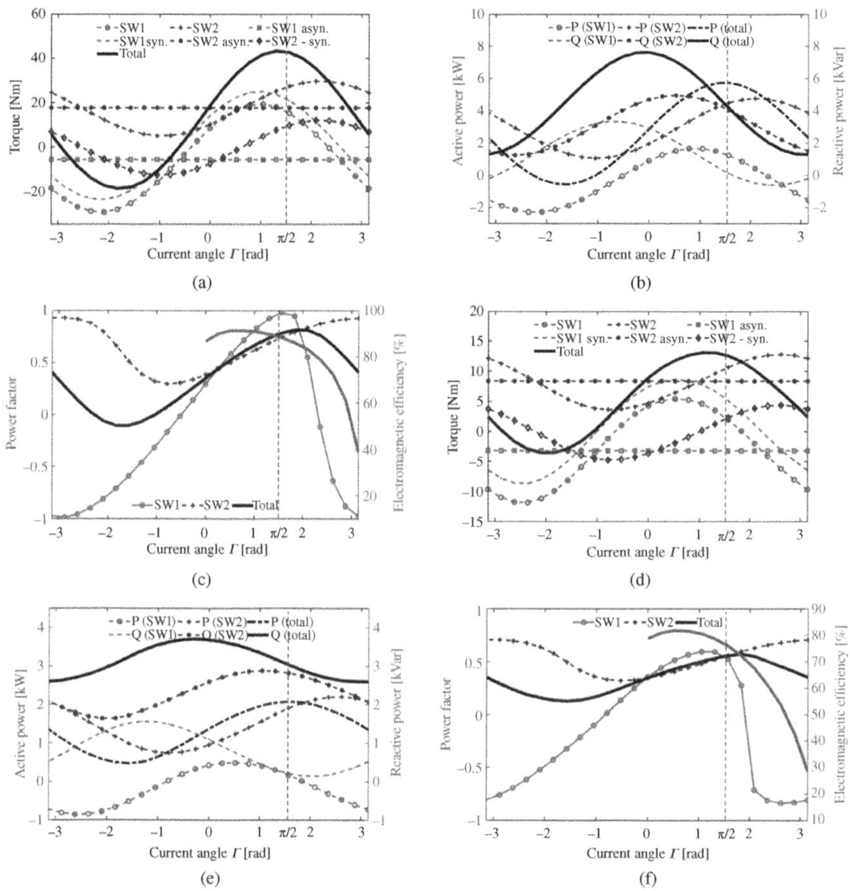

Figure 5.32 Performance of a 1.5-kW prototype at 1000 r/min with linear and nonlinear lamination materials predicted from analytical analysis assisted 2-D FEA (SW1: 50 Hz, SW2: 50 Hz). (a) Torque vs. current angle with linear material, (b) power vs. current angle with linear material, (c) power factor and electromagnetic efficiency vs. current angle with linear material, (d) torque vs. current angle with 50W470, (e) power vs. current angle with 50W470, (f) power factor and electromagnetic efficiency vs. current angle with 50W470. Appling the derived optimal Γ into FEA simulation settings makes it possible to avoid the time-consuming searching process for the maximum torque and to predict the peak torque by one transient FEA simulation.

curves are distorted and the maximum values occur at a current angle slightly larger than $\pi/2$ at 1000 r/min.

One common observation from Figures 5.32b,e is the active and reactive powers of SW2, which is the less-pole-pair stator winding, are much larger than those of the more-pole-pair winding, i.e. SW1.

High power factors and electromagnetic efficiency can be achieved in a narrow current angle interval from 0 to 2 rad (approx. 115deg) regardless of the magnetic material type, indicating the machine is very suitable for motoring operation in this range of current angle under the two-converter-based operation.

It should be noted that in Figure 5.32c,f, a positive power factor means a lagging power factor, and the machine is absorbing active power from the supply. A negative value denotes a leading power factor. Though under the linear condition, with some current angles (approx. from −110deg to −60deg), the machine as a whole can deliver some active power, the amount is very low. The magnetic saturation effect almost eliminates the generating range under this new two-converter-based operation.

The flux pattern at an instant of time when both SW1 and SW2 are excited are shown in Figure 5.33. It is shown that when the per-pole MMFs created by SW1 and SW2 are equal, it shows an explicit 6-pole pattern. With more per-pole MMF from SW2 which has two pole pairs, the flux pattern looks more like a 4-pole one.

In addition, to study the variation of flux densities in the laminated rotor core, several probing points marked in Figure 5.30 are chosen (Ptx_rt, where $x = 0, 1, \ldots, 6$). It is shown that flux densities in the rotor teeth vary sinusoidally with time when the magnetic core is not saturated and follow the slip frequency $|\omega_1 - p_1\omega_r| = |\omega_2 - p_2\omega_r|$. There is a constant phase shift between flux densities at the same location every six rotor teeth, as illustrated in Figure 5.34a. When magnetic saturation is considered, the flux density variation is not sinusoidal anymore. Instead, the waveform shows a flat top.

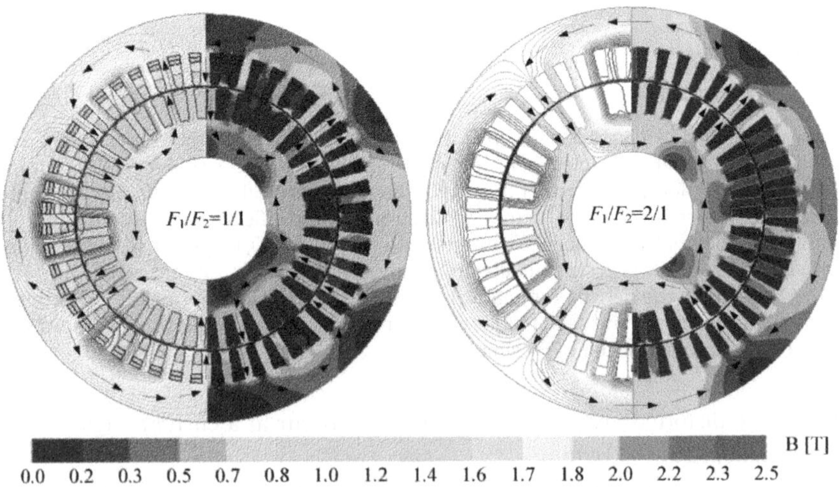

Figure 5.33 Flux patterns with different combinations of SW1 and SW2 MMFs, where F_1 and F_2 are per-pole MMF created by SW1 and SW2, respectively.

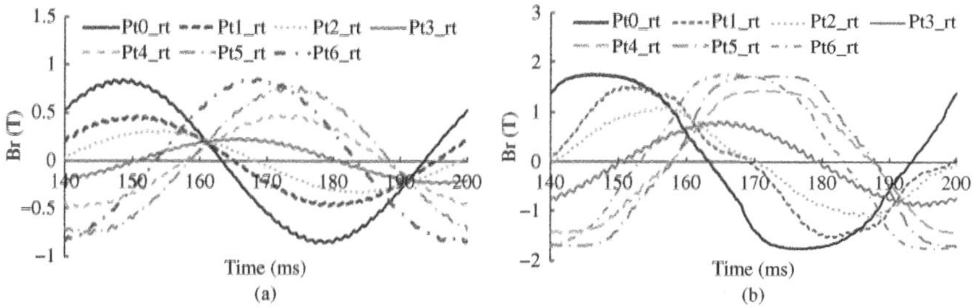

Figure 5.34 Flux density in rotor core under (a) unsaturated condition, (b) saturated condition.

4 Experimental Validation

To validate the presented theoretical analysis, the 1.5-kW BDFIM with a spiral-loop rotor was tested. The experimental setup for the inductance measurement and open-circuit testing is shown in Figure 5.35. It was tailored from a more general test rig for various brushless doubly-fed machine types and various modes of operation shown in [26]. The two DC–AC converters were of the same rating and used to supply SW1 and SW2 independently with variable voltage and frequency. A 5.5 kW permanent magnet SM (PMSM) was used as the load machine when the BDFIM under test was operated as a motor fed by two independent converters.

To test the BDFIM in both the motoring and generating operations, braking choppers are necessary which need to be connected in parallel with the two independent DC buses.

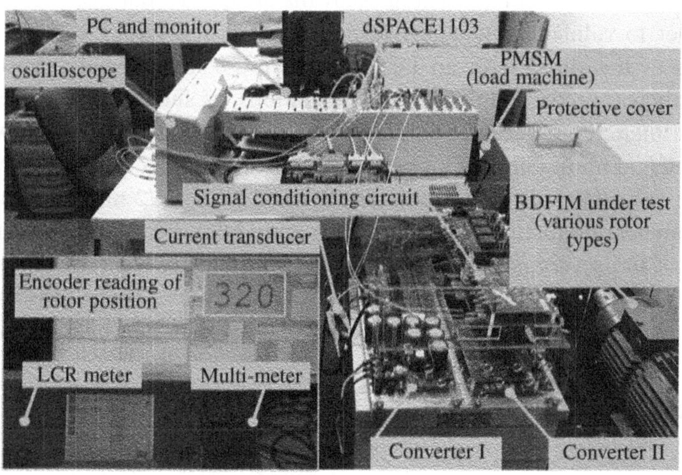

Figure 5.35 Experimental setup for the testing of the 1.5 kW BDFIM prototype with a spiral loop rotor.

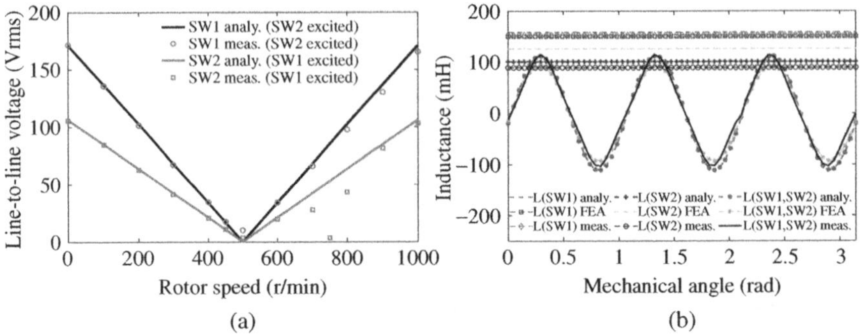

Figure 5.36 Experimental results. (a) Phase voltage vs. speed, (b) unsaturated inductances. The phase voltage vs. speed plot shows the cross-coupling characteristic in voltage model (excitation voltage [line-to-line]: 190 Vrms, 50 Hz).

In order to validate the presented derivation for magnetic field conversion factors and those specific values for the rotor winding with six equally-spaced spiral loops, the cross-coupling characteristic was measured in voltage mode, as shown in Figure 5.36a, which generally agree well with the analytical predictions. In this test, one of the stator windings was left open to measure the open-circuit voltage while the other stator winding was supplied by a 190 Vrms (line-to-line value) and 50 Hz three-phase voltage source in a wide speed range from 0 to 1000 r/min. When SW1 is excited, the synchronous speed of the 8-pole field is 750 r/min, so there will be no current induced in the rotor winding and no voltage at the terminals of SW2. This explains the groove in the measured voltage curve in the vicinity of 750 r/min, which cannot be captured by the analytical analysis.

In order to validate inductance formulas (5.62) and (5.63), the unsaturated inductances measured with an LCR meter were compared with those results from analytical predictions and FEA calculations, as shown in Figure 5.36b. In the experimental measurements, the influence of rotor short-circuited windings on the self and mutual inductances related to SW1 and SW2 has been considered. It can be seen that self-inductances of SW1 and SW1–SW2 mutual inductances from analytical prediction based on (5.62) and (5.63), FE calculation and measurements are in very good agreement. Though there are some variations in the SW2 self-inductances, the differences are still within 25% range. The measured self-inductances have been averaged to remove the ripples for a better comparison.

5.4 SynRM

5.4.1 Inductances

SynRMs are a type of synchronous machines with a reluctance rotor and operates by sinusoidal currents [27]. They share similar rotor constructions with the

BDFRM. A SynRM with the same stator core and multilayer barrier rotor as the studied BDFRM in Section 5.3.3 is analyzed in this section. The stator was rewound to have a 6-pole 3-phase winding.

It has been known that flux barriers modulate the source MMF by defining the path of flux lines and flux barriers usually have multiple layers magnetically insolated by non-magnetic materials. Therefore, each layer behaves like a flux tube that only allows flux lines to enter or exit from the two end points symmetrical around the middle line of each flux barrier segment at $(i + 1/2)2\pi/N_{MB}$.

The MMF drop on the closed path through the two points will be equally shared by the airgaps at the two end points if the airgap is assumed uniform. The modulation behavior of flux barriers can be characterized by:

$$M(N_{MB})[f(x,t)] = \rho(x,\theta_r) \times [f(x,t) - f((2i+1)2\pi/N_{MB} - x, t)],$$
$$x \in (2\pi i/N_{MB}, 2\pi(i+1)/N_{MB}). \tag{5.70}$$

where N_{MB} is the number of multi-layer barrier segments, $\rho(x, \theta_r)$ is a ratio for airgap length defined by:

$$\rho(x,\theta_r) = \frac{g(x,\theta_r)}{g(x,\theta_r) + g[(2i+1)2\pi/N_{MB} - x, \theta_r]} \tag{5.71}$$

The pictorial derivation of the modulator for multi-layer barriers is illustrated in Figure 5.37.

By applying the multilayer barrier modulation operator to the winding function of SynRM, the closed-form equation for main self-inductance of phase-A winding

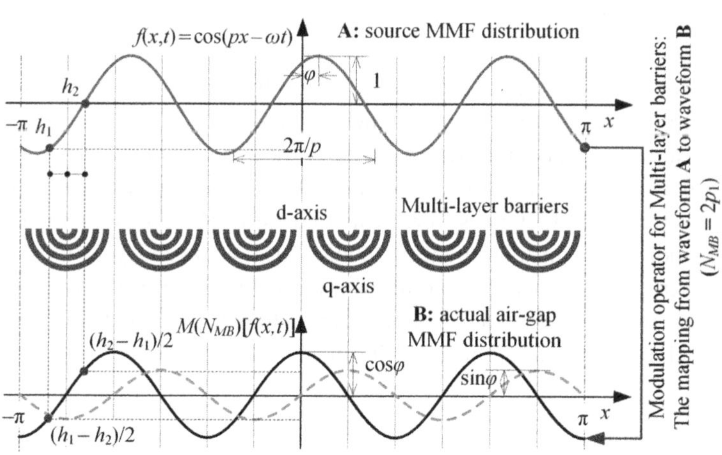

Figure 5.37 Modulation process in a SynRM with a multi-layer barrier rotor, where the number of flux barrier segments is twice of the stator winding pole pairs. In this case, the modulation decomposes the source MMF distribution into *d*- and *q*-axis components. This also holds true for the wound-field salient-pole SM.

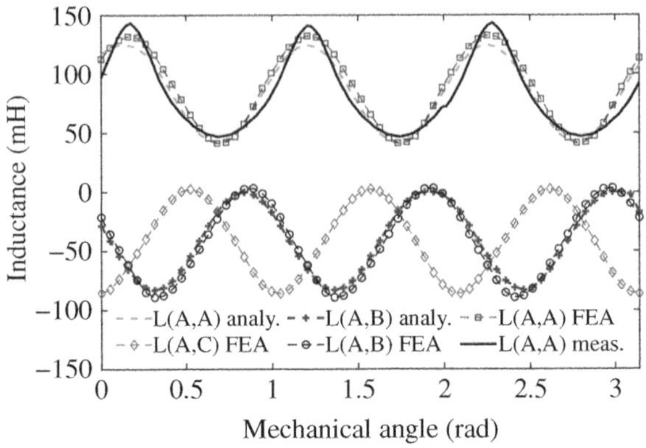

Figure 5.38 Inductances in a 6-pole SynRM prototype with a multilayer barrier rotor.

in a 6-pole SynRM prototype with a multi-layer barrier rotor can be written as:

$$
\begin{aligned}
L_{ms\_SynRM}^{AA}(\theta_r) &= k_{sat}\frac{\mu_0 r_g l_{stk}}{g_e}\int_0^{2\pi} W_{sA}(\phi)M[W_{sA}(\phi)]d\phi \\
&= k_{sat}\frac{\mu_0 \pi r_g l_{stk}}{g_e}\left(\frac{4}{\pi}\right)^2\left(\frac{N_{ph\_s}k_{w1\_s}}{2p}\right)^2 \\
&\quad \left\{\left(C_{p,p}^d + C_{p,p}^q\right) - \left(C_{p,p}^d - C_{p,p}^q\right)\cos[2p(\phi_0 - \theta_r)]\right\}
\end{aligned} \tag{5.72}
$$

where $C_{p,p}^d$ and $C_{p,p}^q$ are magnetic field conversion factors between the source MMF and the airgap MMF drop after modulation in d-axis and q-axis positions, respectively. N_{ph_s} and k_{w1_s} are turns in series per phase and fundamental winding factors of stator windings, respectively. ϕ_0 is the mechanical angle from the reference axis to the phase winding axis, and θ_r the rotor position angle.

The unsaturated inductances from (5.72) for the SynRM is shown in Figure 5.38, which are in good agreement with the 2D FEA results and measured waveforms.

The unsaturated inductances were measured with a Hioki LCR meter at various rotor positions with an increment of 360deg/1024 = 2.8125deg. An incremental encoder with 1024 pulses per revolution was used to monitor the rotor position. To facilitate the measurement, the phase-A winding was connected in series with the parallel of phase-B and phase-C windings.

5.4.2 Torque Characteristic

Since the MMF drop or flux density distribution across the equivalent uniform airgap is known by applying the developed airgap field modulation theory, the magnetic co-energy W_m' in the airgap can be obtained by:

$$W_m' = \int_V \frac{[B_g(\phi,t)]^2}{2\mu_0} dV = \frac{\mu_0 r_g l_{\text{stk}}}{2g_e} \int_0^{2\pi} \{M[F_1(\phi,t)] + M[F_2(\phi,t)]\}^2 d\phi \tag{5.73}$$

The electromagnetic torque is:

$$T_{em} = \frac{\partial W_m'}{\partial \theta_r} \tag{5.74}$$

Starting from (5.73) and (5.74), the average torque equations for the SynRM prototype is:

$$T_{\text{avg\_SynRM}} = (\pi r_g^2 l_{\text{stk}}) \left(\frac{\pi}{\sqrt{2}} \bar{\text{B}}_d \right) \bar{\text{A}}_q \sin \Gamma \tag{5.75}$$

where $\bar{\text{B}}_d$ is the specific magnetic loading produced by the d-axis current, $\bar{\text{A}}_q$ the specific electric loading produced by the q-axis current, and $\Gamma = (p\phi_0 + \varphi_0) - p\theta_{r0}$ the torque angle, where φ_0 is the initial phase angle of phase-A current when taking the form of cosine functions.

The torque vs. torque angle under light-load condition calculated from (5.75) is compared with the 2D FEA results in Figure 5.39. Both the analytical calculation and FEA results are in line with the measured data points. The steady-state torque with different combinations of d-axis and q-axis currents were measured by the torque transducer at a constant speed due to the lack of a locking mechanism. The 3ϕ RLC load was connected to the PMSM to provide retarding torque to counteract the torque produced by the machine under test.

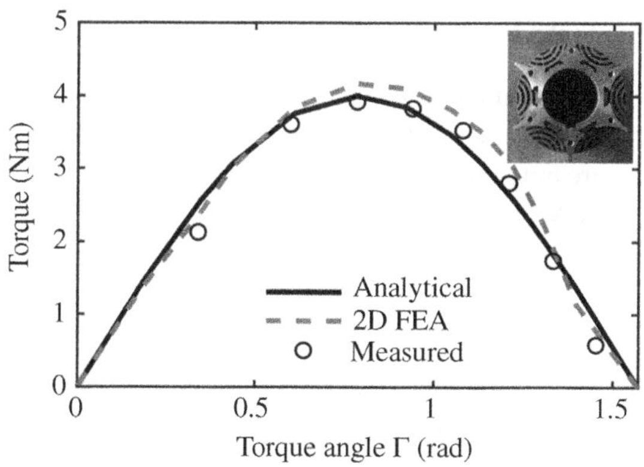

Figure 5.39 Torque vs. torque angle Γ of a 6-pole SynRM prototype with a multilayer barrier rotor.

5.5 FRPM Machine

Electrical machines with salient reluctance pole structures constitute the majority of traditional and emerging machine topologies, and the salient reluctance pole modulator is the most common of the three typical modulator types presented in Chapter 3.

The FEA, nonlinear lumped parameter magnetic circuit models [28–30], and linear subdomain models [31] have been widely used in analyzing electrical machines with salient reluctance pole structures. These methods are accurate in predicting the electromagnetic performance but show difficulties in revealing the operating principle of such machines. Some attempts have also been made to study the operating principle and performance comparison of field modulation machines with salient reluctance poles in a unified manner. For instance, the torque production mechanism was investigated in [32, 33] based on the simple "MMF-permeance" model. The airgap magnetic field harmonic orders and rotating speeds were predicted analytically. The relationship between flux-reversal PM machines and PM vernier machines was explored using a graphical approach in [34]. In addition, a number of field modulation machine topologies were reviewed to form a preliminary unified theory based on the "MMF-permeance" model [35].

A FRPM machine is analyzed as a representative example of electrical machines with salient reluctance pole structures based on the developed theory. The influence of slot opening width and airgap length on the modulation effect are also discussed. An attempt is made to bridge the electromagnetic performance with machine topological parameters and provide generalized design guidelines to electrical machine designers.

5.5.1 Magnetic Field Modulation Behavior

The studied FRPM machine has six stator slots and eight rotor salient reluctance poles, as illustrated in Figure 5.40. PMs are mounted on the tooth surfaces and adjacent magnets are magnetized in the opposite direction to create a rectangularly distributed source MMF:

$$F_{PM}(\phi, t) = \left(\frac{3}{\pi}\right) F_{PM} \sum_{\nu=1,\nu\neq 3k}^{\infty} \left(\frac{1}{\nu}\right) \sin[\nu N_{ST}(\phi - \phi_0)] \tag{5.76}$$

The stator pole-arc-to-pole-pitch ratio has been set to 2/3 to eliminate $3k$-th harmonics in the modulated source MMF considering the synchronous modulation behavior of stator teeth, where k is a positive integer. To consider the modulation behaviors of stator and rotor salient reluctance poles more accurately, we apply the modulation operator shown in (3.14) to the source MMF created by PMs without

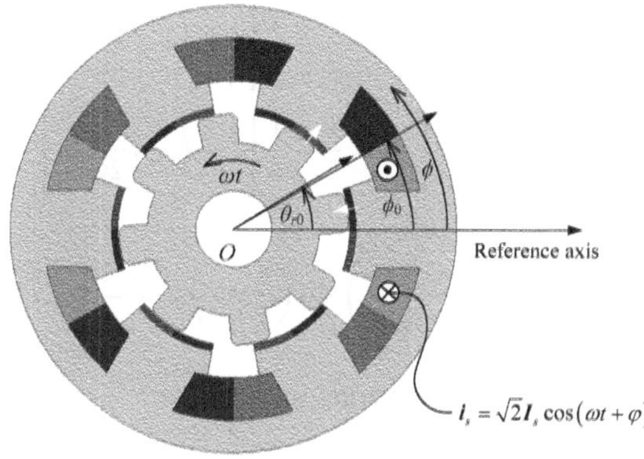

Figure 5.40 Cross-sectional view of the studied FRPM machine and the coordinate system.

any simplifications:

$$\begin{aligned} M(N_{RT}, \varepsilon_{RT})[M(N_{ST}, \varepsilon_{ST})[F_{PM}(\phi, t)]] &= F_{PM}(\phi, t)\lambda(\phi, t) \\ &= \frac{g}{\frac{1}{g_s^{-1}(\phi,\theta_r)} + \frac{1}{g_r^{-1}(\phi,\theta_r)} - g} F_{PM}(\phi, t) \\ &= \frac{1}{\left[\frac{1}{g \cdot g_s^{-1}(\phi,\theta_r)} + \frac{1}{g \cdot g_r^{-1}(\phi,\theta_r)} - 1\right]} F_{PM}(\phi, t) \end{aligned} \tag{5.77}$$

where $g.g_s^{-1}(\phi, t)$ is the normalized inverse airgap function when the stator is slotted and the rotor surface is smooth. Similarly, $g.g_r^{-1}(\phi, \theta_r)$ is the normalized inverse airgap function when the rotor is slotted and the stator surface is smooth.

$$\begin{cases} g \cdot g_s^{-1}(\phi, \theta_r) = \dfrac{t_{ds} - \frac{8}{5}\beta_s o_s}{t_{ds}} - \displaystyle\sum_{\nu=1}^{\infty} \beta_s FC_\nu(\varepsilon_{ST}) \cos[\nu N_{ST}(\phi - \phi_0)] \\ g \cdot g_r^{-1}(\phi, \theta_r) = \dfrac{t_{dr} - \frac{8}{5}\beta_r o_r}{t_{dr}} - \displaystyle\sum_{\nu=1}^{\infty} \beta_r FC_\nu(\varepsilon_{RT}) \cos[\nu N_{RT}(\phi - \omega_r t - \theta_{r0})] \end{cases} \tag{5.78}$$

where

$$\begin{cases} t_{ds} = \dfrac{2\pi R_{si}}{N_{ST}} \\ t_{dr} = \dfrac{2\pi R_{ro}}{N_{RT}} \end{cases} \tag{5.79}$$

$$\begin{cases} \beta_s = \dfrac{1}{2} - \dfrac{1}{2}\left[1 + \left(\dfrac{o_s}{2g}\right)^2\right]^{-\frac{1}{2}} = \dfrac{1}{2} - \dfrac{1}{2}\left[1 + \left(\varepsilon_{ST}\dfrac{t_{ds}}{2g}\right)^2\right]^{-\frac{1}{2}} \\ \beta_r = \dfrac{1}{2} - \dfrac{1}{2}\left[1 + \left(\dfrac{o_r}{2g}\right)^2\right]^{-\frac{1}{2}} = \dfrac{1}{2} - \dfrac{1}{2}\left[1 + \left(\varepsilon_{RT}\dfrac{t_{dr}}{2g}\right)^2\right]^{-\frac{1}{2}} \end{cases} \tag{5.80}$$

$$FC_\nu(\varepsilon) = \frac{4}{\nu\pi}\left(\frac{1}{2} + \frac{(\nu\varepsilon)^2}{\frac{25}{32} - 2(\nu\varepsilon)^2}\right)\sin\left(\frac{8}{5}\pi\nu\varepsilon\right), \quad \varepsilon = \varepsilon_{ST}, \varepsilon_{RT} \tag{5.81}$$

Expanding (5.77) into its Fourier series, one can get:

$$\begin{aligned} M(N_{RT}, \varepsilon_{RT})[M(N_{ST}, \varepsilon_{ST})[F_{PM}(\phi, t)]] &= \sum_{\substack{\nu=1 \\ \nu\neq 3k}}^{\infty} C_{\nu N_{ST},\nu N_{ST}} \sin[\nu N_{ST}(\phi - \phi_0)] \\ &+ \sum_{l=1}^{\infty}\sum_{\substack{\nu=1 \\ \nu\neq 3k}}^{\infty} C_{\nu N_{ST},\nu N_{ST}-lN_{RT}} \sin[(\nu N_{ST} - lN_{RT})\phi - (\nu N_{ST}\phi_0 - lN_{RT}\theta_{r0}) + lN_{RT}\omega_r t] \\ &+ \sum_{l=1}^{\infty}\sum_{\substack{\nu=1 \\ \nu\neq 3k}}^{\infty} C_{\nu N_{ST},\nu N_{ST}+lN_{RT}} \sin[(\nu N_{ST} + lN_{RT})\phi - (\nu N_{ST}\phi_0 + lN_{RT}\theta_{r0}) - lN_{RT}\omega_r t] \end{aligned} \tag{5.82}$$

where $C_{\nu NST,\nu NST}$, $C_{\nu NST,\nu NST-lNRT}$, and $C_{\nu NST,\nu NST+lNRT}$ are the three magnetic field conversion factors, which includes the information of stator slot and rotor salient reluctance pole geometries.

It has been shown in Chapter 4 that by assuming the stator and rotor airgap permeances are rectangular waves, we can get simplified modulation operators for the stator and rotor separately, with which we can further derive closed-form solutions for the three magnetic field conversion factors. For better accuracy of the airgap MMF and flux density waveforms, we need to consider the smooth permeance change in the slot opening region more carefully, so the modulation operation shown in (3.14) is preferable.

The source MMF created by surface-mounted PMs is modulated by stator teeth first, yielding one group of harmonics whose number of pole pairs is νN_{ST}. After the modulation behavior of rotor salient reluctance poles is considered, there are three groups of harmonics components, whose numbers of pole pairs are νN_{ST}, $(\nu N_{ST} + lN_{RT})$, and $(\nu N_{ST} - lN_{RT})$. Amplitudes of the three harmonic groups can be characterized by the three magnetic field conversion factors in (5.82).

The combination of the principal pole pair number of the PM array and the number of rotor salient reluctance poles has a significant influence on the armature winding layout. Furthermore, harmonics in the modulated source MMF always appear in pairs and the heterodyned harmonics share the same magnetic field conversion factors for the same v and l, as shown in Figure 4.24. The airgap flux density waveform includes six main harmonic components grouped into four pairs whose numbers of pole pairs are (6, 2), (6, 14), (12, −4), and (12, 20). The pole pair sum or difference of the two harmonic components in each pair is $N_{RT} = 8$. The negative sign indicates the corresponding harmonic travels in the opposite direction of the rotor.

The airgap flux density distribution can be calculated by (5.77) or 2D FEA directly, showing similar waveforms under the no-load and loaded condition. Moreover, the vN_{ST}-pole-pair harmonics are still stationary, while those with $(vN_{ST} \pm lN_{RT})$ pole pairs harmonics are rotating, and the rotation direction depends on which is larger between lN_{RT} and vN_{ST}. The detailed modulation process of the studied flux-reversal PM machine is shown in Figure 4.25, and the corresponding spectrum shift is illustrated in Figure 4.26.

5.5.2 Influence of Key Topological Parameters

1 Ratio of Slot Opening Width to Slot Pitch

Values of $FC_v(\varepsilon)$ for various ε are plotted in Figure 5.41. It is shown that $FC_v(\varepsilon)$ increases and then decreases with the increase of ε, causing noticeable changes of the airgap permeance and further influencing the airgap MMF distribution [3]. The following conclusions can be obtained from Figure 5.41:

- When ε is lower than 0.25, $FC_2(\varepsilon)$ is comparable to $FC_1(\varepsilon)$, and both of them play an important role in the airgap permeance function.
- When ε is larger than 0.25, $FC_2(\varepsilon)$ starts to decrease while $FC_1(\varepsilon)$ keeps increasing until ε reaches 0.5. In this range, $FC_2(\varepsilon)$ is always smaller than $FC_1(\varepsilon)$.
- Higher-order harmonics with $v \geq 3$ can be neglected due to the relatively low magnitudes.

It has been recommended to have $\varepsilon \geq 0.3$ [3] so the modulation effect of salient reluctance poles on the source MMF is noticeable, such as the flux-reversal PM machine, PM vernier machine, and switched reluctance machine (SRM). Slots with $\varepsilon < 0.3$ has been typically considered as fine slots and the corresponding modulation behavior is treated as the unit modulation. If only the fundamental component of the relative permeance function is considered, the best modulation effect can be achieved when ε is 0.5, where $FC_1(\varepsilon)$ reaches its maximum.

2 Airgap Length

The airgap length has a profound influence on the modulation effect, too. When the effective airgap is sufficiently small, flux lines are almost perpendicular to the stator and rotor surfaces. The airgap MMF drop and excitation magnetic field distribution are almost one-dimensional functions of the angular position along the

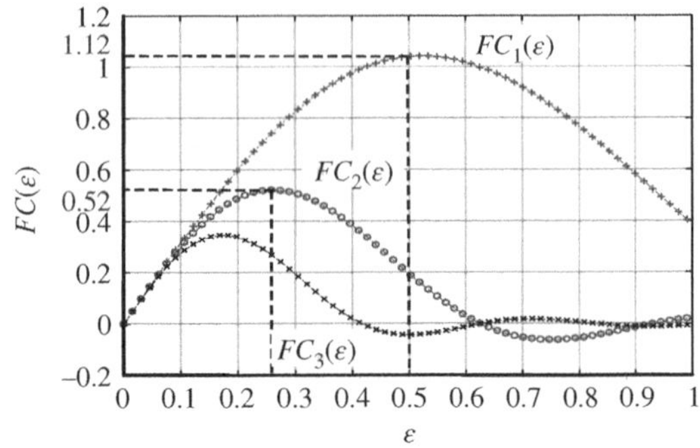

Figure 5.41 FC_v vs. ε.

airgap peripheral, so the MMF-permeance model can be used. It should be noted that the radial and tangential components of the airgap flux density waveform correspond to the magnetizing and armature fields, respectively.

In the case of a relatively large airgap between the stator and rotor, the magnetic modulation may still occur, but flux lines cannot be considered as perpendicular to the stator and rotor surfaces anymore due to the leakage flux and flux density changes along the radial and/or circumferential directions. The airgap MMF drop and excitation magnetic field distribution are two-dimensional functions of the angular position along the airgap peripheral and the position in radial direction. The modulation effect of salient reluctance poles will be reduced significantly when the airgap length is large. In this case, it becomes more important to get the accurate airgap MMF modulated by salient reluctance poles, with which the developed theory can then be applied to analyze the open-circuit back-EMF, inductance characteristics, average torque, and cogging torque.

5.5.3 Experimental Validation

The studied flux-reversal PM machine has been prototyped and tested, as shown in Figure 5.42, to validate the presented theoretical analysis. Main design parameters are tabulated in Table 5.7.

1 Open-circuit Back-EMF

The airgap flux density distribution is the basis to calculate flux linkages linked by armature windings and the induced voltages. It can be readily obtained from the modulated airgap MMF distribution expressed by (5.83):

$$B_g(\phi, t) = \sum_{\substack{v=1 \\ v\neq 3k}}^{\infty} B_{vN_{ST}} \sin[vN_{ST}(\phi - \phi_0)] + \sum_{l=1}^{\infty} \sum_{\substack{v=1 \\ v\neq 3k}}^{\infty}$$

(a) (b)

Figure 5.42 The studied flux-reversal PM machine prototype with six stator slots and eight rotor salient reluctance poles. (a) Stator and rotor, (b) whole assembly.

Table 5.7 Main design parameters of the flux-reversal PM machine prototype.

Parameter	Value
Stator outer/inner diameter (mm)	128/70.4
Airgap length (mm)	0.35
Stack length (mm)	75
PM remanence (T)	1.26
PM relative permeability	1.2
PM length (mm)	1.84
Rated speed (r/min)	1500

$$B_{vN_{ST}-lN_{RT}}\sin[(vN_{ST}-lN_{RT})\phi-(vN_{ST}\phi_0-lN_{RT}\theta_{r0})+lN_{RT}\omega_r t]$$
$$+\sum_{l=1}^{\infty}\sum_{\substack{v=1\\ v\neq 3k}}^{\infty}$$
$$B_{vN_{ST}+lN_{RT}}\sin[(vN_{ST}+lN_{RT})\phi-(vN_{ST}\phi_0+lN_{RT}\theta_{r0})-lN_{RT}\omega_r t] \quad (5.83)$$

where $B_g(\phi,t)$ is the airgap flux density distribution. B_{vNST} and $B_{vNST\pm lNRT}$ are amplitudes of flux density harmonics with vN_{ST} and $(vN_{ST}\pm lN_{RT})$ pole pairs.

The PM flux linkage linked by phase-A armature winding is:

$$\psi_{A\_PM}(t)=2r_g l_{stk}N_a\int_0^{\frac{2\pi}{N_{ST}}}B_g(\phi,t)d\phi$$
$$=2r_g l_{stk}N_a\sin(lN_{RT}\omega_r t)\sum_{\substack{l=1\\ v=1,v\neq 3k}}^{\infty}\left(\frac{2B_{vN_{ST}-lN_{RT}}}{vN_{ST}-lN_{RT}}\right)$$

$$\sin\left[(\nu N_{ST} - lN_{RT})\frac{\pi}{N_{ST}}\right]$$
$$- 2r_g l_{stk} N_a \sin(lN_{RT}\omega_r t) \sum_{\substack{l=1\\ \nu=1,\nu\neq 3k}}^{\infty} \left(\frac{2B_{\nu N_{ST}+lN_{RT}}}{\nu N_{ST} + lN_{RT}}\right)$$
$$\sin\left[(\nu N_{ST} + lN_{RT})\frac{\pi}{N_{ST}}\right] \tag{5.84}$$

where N_a is the number of turns per coil. The no-load back-EMF can be obtained by the time derivative of the flux linkage:

$$e_A(t) = -\frac{d\psi_{A\_PM}}{dt} = -2r_g l_{stk} N_a \cos(lN_{RT}\omega_r t)$$
$$\sum_{\substack{l=1\\ \nu=1,\nu\neq 3k}}^{\infty} \left(\frac{2lN_{RT}\omega_r B_{\nu N_{ST}-lN_{RT}}}{\nu N_{ST} - lN_{RT}}\right) \sin\left[(\nu N_{ST} - lN_{RT})\frac{\pi}{N_{ST}}\right]$$
$$+ 2r_g l_{stk} N_a \cos(lN_{RT}\omega_r t)$$
$$\sum_{\substack{l=1\\ \nu=1,\nu\neq 3k}}^{\infty} \left(\frac{2lN_{RT}\omega_r B_{\nu N_{ST}+lN_{RT}}}{\nu N_{ST} + lN_{RT}}\right) \sin\left[(\nu N_{ST} + lN_{RT})\frac{\pi}{N_{ST}}\right] \tag{5.85}$$

Equation (5.85) shows that both the heterodyned harmonics whose numbers of pole pairs are $(\nu N_{ST} \pm lN_{RT})$ can contribute to the PM flux linkage and the open-circuit back-EMF for the fractional-slot concentrated winding structure. Compared to the BDFIM and BDFRM where only one of the heterodyned harmonics is used, the utilization of field harmonics in flux-reversal PM machines is higher.

Figure 5.43 illustrates the open-circuit back-EMF waveforms and the associated amplitude spectrums at 1500 r/min. The result predicted by the developed theory, or more specifically, Equation (5.85), agrees well with the 2D FEA and measured results in terms of the amplitude and frequency. It can be noted that the phase open-circuit back-EMF of the studied flux-reversal PM machine with six stator slots and eight rotor salient reluctance poles is almost sinusoidal and superimposed by even-order harmonics. Among them, the 4th harmonic is noticeable. Since there is little saliency in FRPM machines and the average torque is proportional to the PM flux linkage and q-axis current, it is in expectation to see the good match between the torque predicted by the developed theory and that from 2D FEA and measurements, as shown in Figure 5.44.

2 Cogging Torque

Flux-reversal PM machines typically have higher cogging torques than traditional PMSMs. The cogging torque is a load-independent torque component and contributes nothing to the average torque. It is produced by the interaction between

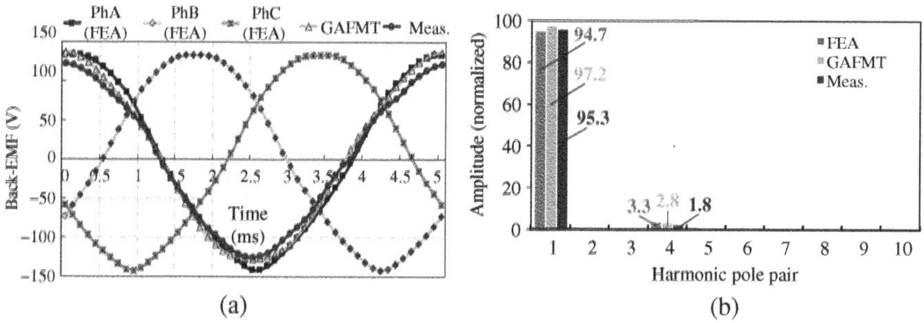

Figure 5.43 No-load back EMF. (a) Waveforms. (b) Spectrum.

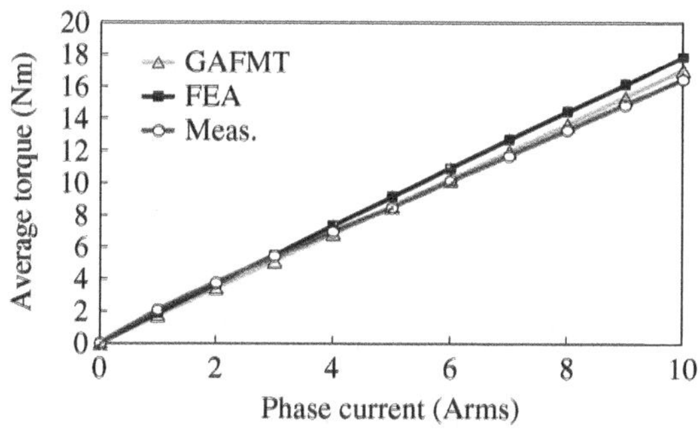

Figure 5.44 Electromagnetic torque with the phase current rms value.

the PM MMF and salient reluctance poles on the stator and rotor. It can be derived from the partial derivation of the magnetic co-energy in the airgap with respect to θ_r:

$$
\begin{aligned}
T_{cog}(t) &= \frac{\partial W'_m}{\partial \theta_r} \approx \frac{\partial}{\partial \theta_r}\left[\frac{1}{2\mu_0}\int_V B_g^2(\phi, t)dV\right] \\
&= \frac{\partial}{\partial \theta_r}\left[\frac{\left(R_{si}^2 - R_{ro}^2\right) l_{stk}}{2\mu_0}\int_0^{2\pi} B_g^2(\phi, t)d\phi\right] \\
&= \frac{\pi\left(R_{ro}^2 - R_{si}^2\right) N_{RT} l_{stk}}{2\mu_0} \\
&\quad \cdot \sum_{n=1}^{\infty}(l \pm z)\sin(nN_L\omega_r t)(B_{vN_{ST}-lN_{RT}}B_{vN_{ST}-zN_{RT}} + 2B_{vN_{ST}}B_{vN_{ST}-lN_{RT}} \\
&\quad + 2B_{vN_{ST}-lN_{RT}}B_{vN_{ST}+zN_{RT}}), \quad l \pm z = \frac{nN_L}{N_{RT}}
\end{aligned}
\tag{5.86}
$$

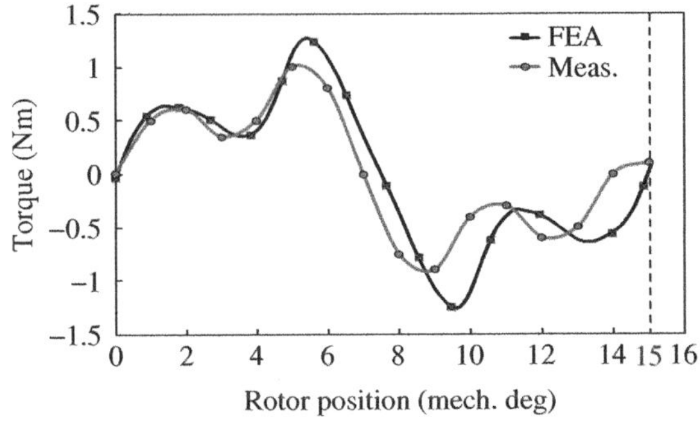

Figure 5.45 Cogging torque waveform.

where W_m' is the magnetic co-energy. N_L is the least common multiplier (LCM) of N_{ST} and N_{RT}, and n is a positive integer. In addition, the following relationship exists between the number of stator/rotor pole pairs and N_L:

$$vN_{ST} = lN_{RT} = nLCM(N_{ST}, N_{RT}) = nN_L \tag{5.87}$$

It can be seen that the periodicity of the cogging torque waveform is related to N_L, which is 24 for the studied flux-reversal PM machine with six stator slots and eight rotor salient reluctance poles. Figure 5.45 compares the cogging torque waveforms from 2D FEA and experimental measurements, both showing a cycle of 360deg/15 = 24deg. It coincides well with the theoretical analysis. The peak-to-peak value of cogging torque depends on ε_{ST}, ε_{RT}, and N_L. A larger N_L will reduce B_{vNST} and $B_{vNST \pm lNRT}$ and is thus recommended to suppress the cogging torque, which can be achieved by fine tuning the slot opening width and the combination of stator slots and rotor salient reluctance poles.

5.6 Comparison of Representative Machine Topologies

This section summarizes the qualitative analysis of representative machine topologies and provides a preliminary comparison among them in terms of design and operational features, advantages, disadvantages and applicability in practice, as tabulated in Table 5.8. The source MMF, modulator, and filter for each typical topology are also included, where DW and FSCW are short for "distributed winding" and "fractional-slot concentrated winding", respectively.

It is shown that "unconventional" PM machine topologies using salient reluctance pole modulators, such as PM vernier (PMV) machines, doubly salient-pole PM (DSPM) machines, flux-switching PM (FSPM) machines, flux-reversal PM (FRPM) machines, and magnetic gears (MGs), typically have low utilization of source MMF created by PMs than PMSMs. According to the simplified

Table 5.8 Qualitative comparison of typical "unconventional" machine topologies. "VM" stands for "vernier machine", "DSPM" for "double salient-pole PM machine", "FSPM" for "flux-switching PM machine", "FRPM" for "flux-reversal PM machine", "MG" for "magnetic gear", and "DS/DRM" for "dual-stator/dual-rotor machine".

Typical topology	Source MMF	Modulator	Filter	Design and operational features	Advantages	Disadvantages	Applicability
SynRM [27, 36]	Sinusoidal MMF by stator DW	Salient reluctance pole and/or multi-layer barrier	The same stator DW that creates source MMF	• High saliency ratio (L_d/L_q) preferred	• Comparable torque density to IMs • Relatively high efficiency	• Complex rotor construction • Relatively low power factor	Suitable for a wide range of applications, with a focus on industrial motor drives
SRM [37, 38]	Pulsed MMF by tooth-wound stator windings	Salient reluctance pole	The same tooth-wound stator windings	• Simple and robust stator and rotor constructions • Small airgap preferred	• Inherent fault-tolerance • Harsh environment withstanding capability • Wide speed range	• High torque ripple w/o sophisticated current profiling • Relatively low power factor • Vibration and acoustic noise	Suitable for a wide range of applications, especially those requiring high-temperature, high-speed operation and high reliability
VM [39–41]	Square wave MMF by PM array and/or field windings	Salient reluctance pole	Stator armature windings	• High magnetic polarity and high fundamental supply frequency	• High torque density and low torque ripple (appropriate pole combinations)	• Low utilization of source MMF created by PMs • Relatively low power factor	Low-speed direct-drive applications, such as in-wheel motors, etc.

(continued)

Table 5.8 (Continued)

Typical topology	Source MMF	Modulator	Filter	Design and operational features	Advantages	Disadvantages	Applicability
BDFIM [42, 43]	Sinusoidal MMF by one of the stator DWs	Short-circuited coil	The other stator DW	• Two electrical ports and a common mechanical port • Variable-speed constant frequency capability	• More degrees of freedom for speed and power control • Inherent fault-tolerance • High reliability	• Relatively lower torque density • Inferior efficiency • Harmonic-related issues (vibration and acoustic noise, output voltage quality, etc.)	Suitable for limited-speed-range applications for reduced converter rating and high reliability, such as wind /ship shaft power generation
BDFRM [14, 21, 44]		Salient reluctance pole and/or multi-layer barrier					
DSPM [45, 46]	Square wave MMF by PM array	Salient reluctance pole	Stator FSCW	• Simple and robust rotor construction • Modularized stator • PMs in the main flux paths • Low saliency ratio (L_d/L_q)	• Relatively high torque density and low torque ripple (appropriate pole combinations) • Relatively high efficiency due to PMs • Inherent fault-tolerance	• Relatively low utilization of source MMF created by PMs	Suitable for a wide range of applications as a brushless PM machine, especially those requiring high-speed operation

FSPM [32, 33, 46, 47]							
FRPM [46, 48]							
MG [49, 50]	Square wave MMF by one PM array	Salient reluctance pole (ferro-magnetic pole pieces)	Equivalent winding of the other PM array	• Non-contact torque transmission	• Reduced acoustic noise and vibration • High reliability • Overload protection • High torque density and low torque ripple (appropriate pole combinations)	• Relatively low utilization of source MMF created by PMs • Core losses in ferromagnetic pole pieces	Applications requiring non-contact torque transmission and/or scenarios where low maintenance, low acoustic noise and vibration are preferred
DS/DRM with field modulation [51–53]	Square wave MMF by one PM array	Salient reluctance pole (ferro-magnetic pole pieces)	Equivalent winding of the other PM array or armature windings	• Compact construction • Multiple electrical/ mechanical ports	• Flexible power conversion and/or transmission • Relatively high torque density due to PMs	• Complex assembly • Difficult to design and optimization due to increased number of variables	Applications requiring flexible power flows, such as power split for hybrid electric vehicles, etc.

analysis presented in [32, 33, 54] based on the square-wave source MMF and square-wave relative permeance function approximations, the amplitude of the useful MMF harmonic component after the modulation is no larger than half of the fundamental of the square-wave MMF created by PMs. Most of the other MMF harmonics after the modulation are not involved in the torque production but introduce additional core losses and PM eddy current loss. Therefore, in this sense, the utilization of source MMFs created by PMs in these machines are lower than PMSMs.

It is worth noting that the low utilization of source MMF created by PMs does not necessarily mean a lower torque to PM weight or volume ratio. For example, the magnetically geared machine (MGM) studied in [55] has substantially higher torque to PM weight and volume ratios than the direct-drive PMSM counterpart thanks to the selected large gear ratio (10.6) and high torque density of the integrated MG despite the low utilization of source MMFs created by magnets of the MG.

It has to be noted that to further have a fair and comprehensive comparison among different topologies, large-scale design optimization studies, such as those presented in [56, 57], are recommended.

References

1 Boldea, I. (2020). *Induction Machine Handbook*, 3e. CRC Press.

2 Lipo, T.A. (2017). *Introduction to AC Machine Design*, 1e. Wiley-IEEE Press.

3 Heller, B. and Hamata, V. (1977). *Harmonic Field Effects in Induction Machines*. Amsterdam, The Netherlands: Elsevier.

4 Wallace, A.K. and Wright, A. (1974). Novel simulation of cage windings based on mesh circuit model. *IEEE Trans. Power Apparat.* PAS-93 (1): 377–382.

5 Munoz, A.R. and Lipo, T.A. (1999). Complex vector model of the squirrel-cage induction machine including instantaneous rotor bar currents. *IEEE Trans. Ind. Appl.* 35 (6): 1332–1340.

6 Han, P., Cheng, M., Zhu, X., and Chen, Z. (2021). Analytical analysis and performance characterization of brushless doubly fed induction machines based on general air-gap field modulation theory. *Chin. J. Electr. Eng.* 7 (3): 4–19.

7 Cheng, M. and Zhu, Y. (2014). The state of the art of wind energy conversion systems and technologies: a review. *Energy Convers. Manag.* 88: 332–347.

8 Roberts, P.C. (2005). A study of brushless doubly-fed (induction) machines. Cambridge, U.K.: Ph.D. Thesis of University of Cambridge.

9 Guan, B. (2014). Design and control of a high-efficiency doubly-fed brushless machine for power generation applications. Columbus, USA: Ph.D. Thesis of The Ohio State University.

10 Strous, T.D. (2016). Brushless doubly-fed induction machines for wind turbine drive-train applications. Delft, Netherlands: Ph.D. Thesis of Delft University of Technology.

11 Duke, A. (2015). Brushless doubly-fed reluctance machines for aerospace electrical power generation systems. Sheffield, U.K.: Ph.D. Thesis of The University of Sheffield.

12 Hopfensperger, B., Atkinson, D.J., and Lakin, R.A. (2002). Steady state of the cascaded doubly-fed induction machine. *Eur. Trans. Electr. Power* 12 (6): 427–437.

13 Xiong, F. and Wang, X. (2014). Design of a low-harmonic-content wound rotor for the brushless doubly fed generator. *IEEE Trans. Energy Convers.* 29 (1): 158–168.

14 Broadway, A.R.W. (1971). Cageless induction machine. *Proc. Inst. Electr. Eng.* 118 (11): 1593–1600.

15 Zhang, F., Jia, G., Zhao, Y. et al. (2015). Simulation and experimental analysis of a brushless electrically excited synchronous machine with hybrid rotor. *IEEE Trans. Magn.* 51 (12), Article# 8115007.

16 Ruviaro, M., Runcos, F., Sadowski, N. et al. (2012). Analysis and test results of a brushless doubly-fed induction machine with rotary transformer. *IEEE Trans. Ind. Electron.* 59 (6): 2670–2677.

17 Lipo, T.A. (2012). *Analysis of Synchronous Machine*, 2e. Boca Raton, FL, USA: CRC Press.

18 Schulz, E.M. and Betz, R.E. (2003). Optimal rotor design for brushless doubly fed reluctance machines. In: *Record of IEEE IAS Annual Meeting*, Vol. 1, 256–261, Salt Lake City, USA.

19 Liu, H. (2009). Research on brushless doubly-fed machines with radially-laminated flux barrier rotors. Beijing, China: Ph.D. Thesis of Beijing Jiaotong University.

20 Knight, A.M., Betz, R.E., and Dorrell, D.G. (2013). Design and analysis of brushless doubly fed reluctance machines. *IEEE Trans. Ind. Appl.* 49 (1): 50–58.

21 Han, P., Zhang, J., and Cheng, M. (2019). Analytical analysis and performance characterization of brushless doubly fed machines with multibarrier rotors. *IEEE Trans. Ind. Appl.* 55 (6): 5758–5767.

22 Xu, L., Liu, Y., and Wen, X. (2011). Comparison study of singly-fed electric machine with doubly-fed machine for EV/HEV applications. In: *Proceedings of the International Conference on Electrical Machines and Systems (ICEMS)*, Beijing, China.

23 Han, P., Cheng, M., Ademi, S., and Jovanovic, M.G. (2018). Brushless doubly-fed machines: opportunities and challenges. *Chin. J. Electr. Eng.* 4 (2): 1–17.

24 Han, P., Cheng, M., and Chen, Z. (2017). Dual-electrical-port control of cascaded doubly-fed induction machine for EV/HEV applications. *IEEE Trans. Ind. Appl.* 53 (2): 1390–1398.

25 Rebeiro, R.S. and Knight, A.M. (2018). Two-converters-based synchronous operation and control of a brushless doubly fed reluctance machine. *IEEE Trans. Magn.* 54 (11), Article# 8107205.

26 Cheng, M., Han, P., Du, Y. et al. (2022). A tutorial on general air-gap field modulation theory for electrical machines. *IEEE J. Emerg. Sel. Top. Power Electron.* 10 (2): 1712–1732.

27 Moghaddam, R.R., Magnussen, F., and Sadarangani, C. (2010). Theoretical and experimental reevaluation of synchronous reluctance machine. *IEEE Trans. Ind. Electron.* 57 (1): 6–13.

28 Cheng, M., Chau, K.T., Chan, C.C. et al. (2000). Nonlinear varying-network magnetic circuit analysis for doubly salient permanent magnet motors. *IEEE Trans. Magn.* 36 (1): 339–348.

29 Zhu, Z.Q., Pang, Y., Howe, D. et al. (2005). Analysis of electromagnetic performance of flux-switching permanent-magnet machines by nonlinear adaptive lumped parameter magnetic circuit model. *IEEE Trans. Magn.* 41 (11): 4277–4287.

30 Zhang, G., Hua, W., and Cheng, M. (2016). Nonlinear magnetic network models for flux-switching permanent magnet machines. *Sci. China Technol. Sci.* 59 (3): 494–505.

31 Boughrara, K., Lubin, T., and Ibtiouen, R. (2013). General subdomain model for predicting magnetic field in internal and external rotor multiphase flux-switching machines topologies. *IEEE Trans. Magn.* 49 (10): 5310–5325.

32 McFarland, J.D., Johns, T.M., and El-Refaie, A. (2015). Analysis of the torque production mechanism for flux-switching permanent-magnet machines. *IEEE Trans. Ind. Appl.* 51 (4): 3041–3049.

33 Wu, Z.Z. and Zhu, Z.Q. (2015). Analysis of air-gap field modulation and magnetic gearing effects in switched flux permanent magnet machines. *IEEE Trans. Magn.* 51 (5), Article # 8105012.

34 Li, D., Qu, R., and Li, J. (2015). Topologies and analysis of flux-modulation machines. In: *Proceedings of the IEEE Energy Conversion Congress and Exposition (ECCE)*, 2153–2160. Montreal, Canada.

35 Fu, W.N. and Liu, Y. (2016). A unified theory of flux-modulated electric machines. In: *Proceedings of the International Symposium on Electrical and Electronics Engineering (ISEE)*, 1–13. Dalian, China.

36 Talaat, M.E. (1951). Steady-state and transient synthesis of 3-phase reluctance motors (synchronous motors without field excitation). *Trans. Am. Inst. Electr. Eng.* 70 (2): 1963–1970.

37 Lawrenson, P.J., Stephenson, J.M., Blenkinsop, P.T. et al. (1980). Variable-speed switched reluctance motors. *Inst. Elect. Eng. Elect. Power Appl.* 127 (4): 253–265.

38 Hua, H., Hua, W., Zhao, G., and Cheng, M. (2020). Torque production mechanism of switched reluctance machines with air-gap field modulation principle. *IEEE Trans. Energy Convers.* 35 (3): 1617–1627.

39 Toba, A. and Lipo, T.A. (1999). Novel dual-excitation permanent magnet vernier machine. In: *Record of IEEE IAS Annual Meeting*, Phoenix, USA. vol. 4, 2539–2544.

40 Wu, F. and El-Refaie, A. (2019). Permanent magnet vernier machine: a review. *IET Electr. Power Appl.* 13 (2): 127–137.

41 Li, D., Qu, R., Liu, J. et al. (2016). Analysis of torque capability and quality in vernier permanent-magnet machines. *IEEE Trans. Ind. Appl.* 52 (1): 125–135.

42 Broadway, A.R.W. and Burbridge, L. (1970). Self-cascaded machine: a low speed motor or high-frequency brushless alternator. *Proc. IEE* 117 (7): 1277–1290.

43 Long, T., Shao, S., Malliband, P. et al. (2013). Crowbarless fault ride-through of the brushless doubly fed induction generator in a wind turbine under symmetrical voltage dips. *IEEE Trans. Ind. Electron.* 60 (7): 2833–2841.

44 Agrawal, S., Province, A., and Banerjee, A. (2020). An approach to maximize torque density in a brushless doubly fed reluctance machine. *IEEE Trans. Ind. Appl.* 56 (5): 4829–4838.

45 Liao, Y., Liang, F., and Lipo, T.A. (1992). A novel permanent magnet motor with doubly salient structure. In: *Record of IEEE IAS Annual Meeting*, Houston, USA. vol. 1, 308–314.

46 Cheng, M., Hua, W., Zhang, J., and Zhao, W. (2011). Overview of stator-permanent magnet brushless machines. *IEEE Trans. Ind. Electron.* 58 (11): 5087–5101.

47 Chen, H., El-Refaie, A.M., and Demerdash, N.A.O. (2020). Flux-switching permanent magnet machines: a review of opportunities and challenges – Part I: fundamentals and topologies. *IEEE Trans. Energy Convers.* 35 (2): 684–698.

48 More, D.S. and Fernandes, B.G. (2010). Analysis of flux-reversal machine based on fictitious electrical gear. *IEEE Trans. Energy Convers.* 25 (4): 940–947.

49 Martin, T.B. (1968). Magnetic transmission, US Patent, no. 3378710A.

50 Atallah, K., Calverley, S.D., and Howe, D. (2004). Design, analysis and realization of a high-performance magnetic gear. *Inst. Elect. Eng. Elect. Power Appl.* 151 (2): 135–143.

51 Chau, K.T., Zhang, D., Jiang, J.Z. et al. (2007). Design of a magnetic-geared outer-rotor permanent-magnet brushless motor for electric vehicles. *IEEE Trans. Magn.* 43 (6): 2504–2506.

52 Li, X., Chau, K.T., Cheng, M., and Hua, W. (2013). Comparison of magnetic-geared permanent-magnet machines. *Prog. Electromagn. Res.* 133: 177–178.

53 Sun, X., Cheng, M., Hua, W., and Xu, L. (2009). Optimal design of double-layer permanent magnet dual mechanical port machine for wind power application. *IEEE Trans. Magn.* 45 (10): 4613–4616.

54 Du, Y., Xiao, F., Hua, W. et al. (2016). Comparison of flux-switching PM motors with different winding configurations using magnetic gearing principle. *IEEE Trans. Magn.* 52 (5), Article# 8201908.

55 Kjaer, A.B., Korsgaard, S., Nielsen, S.S. et al. (2020). Design, fabrication, test, and benchmark of a magnetically geared permanent magnet generator for wind power generation. *IEEE Trans. Energy Convers.* 35 (1): 24–32.

56 Zhang, P., Sizov, G.Y., Ionel, D.M., and Demerdash, N.A.O. (2015). Establishing the relative merits of interior and spoke-type permanent-magnet machines with ferrite or NdFeB through systematic design optimization. *IEEE Trans. Ind. Appl.* 51 (4): 2940–2948.

57 Rallabandi, V., Han, P., Wu, J. et al. (2021). Design optimization and comparison of direct-drive outer-rotor SRMs based on fast current profile estimation and transient FEA. *IEEE Trans. Ind. Appl.* 57 (1): 236–245.

6

Innovation of Electrical Machine Topologies

6.1 Innovation Methods

An electrical machine can be abstracted into three key elements based on the general airgap field modulation theory (GAFMT) developed in Chapter 3, which not only contributes to the analysis and calculation of electromagnetic performance, but also promotes the topology innovation of electrical machines to satisfy diverse application needs. Research in the electrical machine field has been greatly enriched by changing one or all of the three basic elements. This chapter aims to illustrate how to make topology innovation under the guidance of the GAFMT by taking flux-switching permanent-magnet (FSPM) machines and brushless doubly-fed machines (BDFMs) as two typical examples. Operating principles and basic electromagnetic performance of a few new machine topologies derived from the conventional FSPM machine and BDFM machine are presented, with comprehensive simulation and experimental studies for validations.

According to the GAFMT, the operating principle of the conventional 12-slot 10-pole (12/10) FSPM machine as shown in Figure 6.1 can be explained as:

Step 1: Permanent-magnets (PMs) on the stator create the source excitation magnetomotive force (MMF) with rectangular-wave distribution along the airgap circumference, whose principal pole-pair number is 6.

Step 2: The source excitation MMF is firstly affected by the synchronous modulation of stator teeth, and then by the asynchronous modulation of rotor salient reluctance poles. The excitation MMF drop across the actual airgap, of which the principal pole-pair number is 4, produces the corresponding flux density distribution.

Step 3: The three-phase armature winding, which is a set of fractional-slot concentrated windings with a principal pole-pair number of 4, extracts effective harmonic components of the airgap flux density distribution and induces electromotive force (EMF) in the winding.

General Airgap Field Modulation Theory for Electrical Machines: Principles and Practice,
First Edition. Ming Cheng, Peng Han, Yi Du, and Honghui Wen.

Companion website: www.wiley.com/go/genairgapfieldmodulationtheory

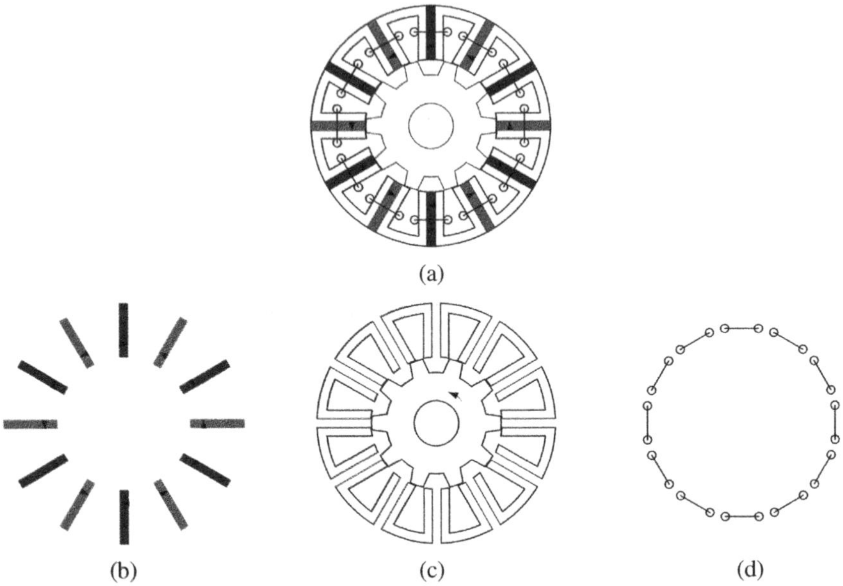

Figure 6.1 Conventional 12/10 FSPM machine and the three basic elements. (a) Cross-sectional view of the conventional 12/10 FSPM machine, (b) Source excitation MMF, (c) Salient reluctance pole modulators, (d) Armature winding.

Step 4: When AC current is fed into the armature winding, the source armature MMF is created and it interacts with the excitation MMF in the airgap to produce the electromagnetic torque.

It can be seen from the description of the operating principle that the performance of FSPM machines is only determined by the cascade of the PM array (source excitation MMF), stator teeth and rotor salient reluctance poles (modulators) and stator winding (filter). Its operating principle will not be affected by simply changing the shape or relative position of the three elements. The difference in electromagnetic performance, however, does exist due to the difference in the shape and size of PMs, the geometry of stator/rotor salient poles, and the layout of armature windings, producing a large design space for comparative studies and systematic performance optimization.

6.1.1 Change of Source MMF

By moving the PM array of a conventional 12/10 FSPM machine from the interior of stator teeth to the stator yoke [1], to the stator tooth tops [2] or rotating the PMs clockwise by 90° around the shaft [3], three stator-PM machine variations

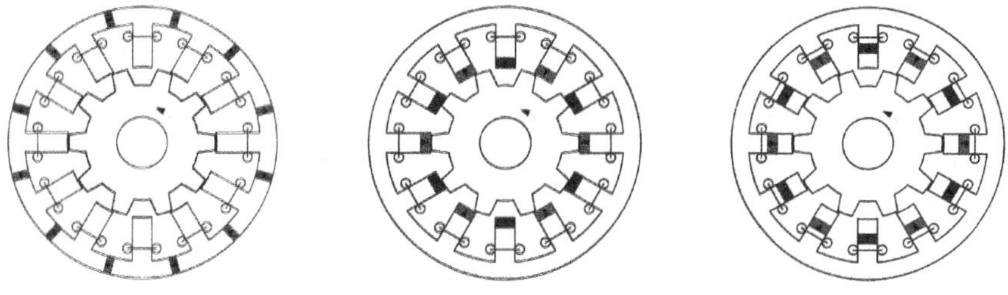

(a) (b) (c)

Figure 6.2 New stator-PM machines derived by changing the source excitation MMF. (a) PMs located in the stator yoke (magnetized in the circumferential direction), (b) PMs located on the stator tooth tops (magnetized in the radial direction), (c) PMs located in the middle of stator teeth (magnetized in the radial direction).

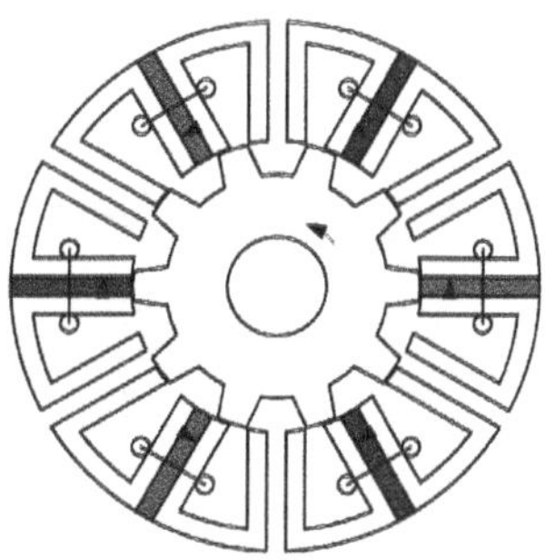

Figure 6.3 A new stator-PM machine derived by changing the source excitation MMF and the corresponding armature winding.

can be obtained, as shown in Figure 6.2. In these machine topologies, the rotor and armature windings remain unchanged, and the stator core is adjusted accordingly to accommodate PMs.

Figure 6.2a shows the so-called doubly salient permanent magnet (DSPM) machine; see Sections 6.2 and 6.3 for more details. By widening PMs located on the stator tooth surface and rotating them around the shaft by a half stator tooth pitch, the machine shown in Figure 6.2b will turn into a flux-reversal permanent magnet (FRPM) machine, which will be discussed in Section 6.4 in detail.

If the principal pole-pair number also changes when the position of the PM array is changed, then the number of salient reluctance poles of the modulator and/or the principal pole-pair number of the armature winding should be adjusted accordingly. Another stator-PM machine variant is shown in Figure 6.3. The PM array and armature winding change simultaneously to satisfy the modulation principle introduced in Chapter 3.

6.1.2 Change of Modulator

By changing the structure of the stator or rotor modulator, such as the number of stator teeth [4], the shape of stator slot opening [5] and the stacking direction of stator and rotor laminations [6], other stator-PM machine variations will be obtained as shown in Figure 6.4.

6.1.3 Change of Filter

By using the single-layer fractional-slot concentrated winding instead of the double-layer one, a new stator-PM machine variant can be obtained [7] with the unchanged excitation MMF and modulator, as shown in Figure 6.5a.

By using the single-layer full-pitch distributed winding and adjusting the number of rotor salient reluctance poles accordingly, another new stator-PM machine

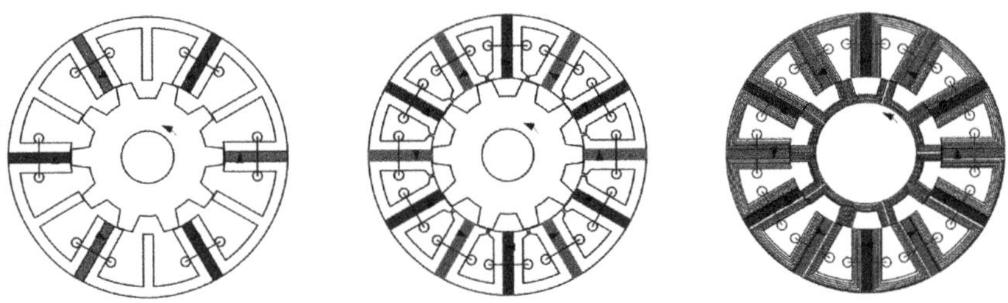

(a) (b) (c)

Figure 6.4 Other new stator-PM machine variants derived by changing the structure of the modulator. (a) E-shaped stator core, (b) Stator core with semi-closed slots, (c) Axially-laminated stator and rotor cores.

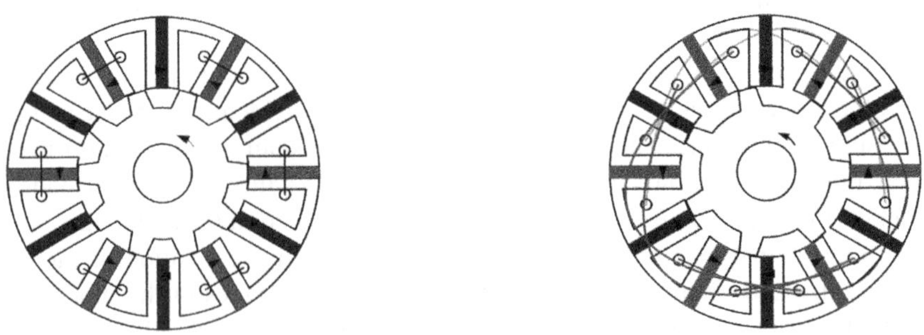

(a) (b)

Figure 6.5 Other new stator-PM machine variants derived by changing armature windings. (a) Single-layer fractional-slot concentrated winding, (b) Single-layer full-pitch distributed winding.

can be obtained, and it is called the "full pitched winding FSPM" in reference [8], as shown in Figure 6.5b. See Section 6.5 for more details.

6.1.4 Change of Relative Position of Three Key Elements

By moving the PM array from stator to rotor, a novel type of PM machine can be obtained, as shown in Figure 6.6a, which is called the rotor-PM FSPM machine in reference [9]. See Section 6.6 for more details. It has similar electromagnetic characteristics to the conventional FSPM machine. Besides the fundamental component of the airgap magnetic field, some other harmonic components generated by the asynchronous modulation of stator teeth can also be used to produce electromagnetic torque.

By housing the PMs in the inner stator and retaining the basic geometry of the outer stator and salient reluctance pole rotor, the so-called partitioned-stator FSPM machine [10] or dual-stator FSPM machine can be obtained, as shown in Figure 6.6b.

6.1.5 Change of Relative Motion of Three Key Elements

When the inner stator of the dual-stator machine shown in Figure 6.6b is allowed to rotate, it can evolve into a magnetically-geared dual-rotor machine [11], as shown in Figure 6.7. These two rotors operate independently and the rotational speed can be different. It belongs to the dual-mechanical-port machine [12]. See Section 6.7 for more details.

Some or all the PMs in aforementioned machines can be replaced by excitation windings, so that new machines with the electrical excitation or hybrid excitation

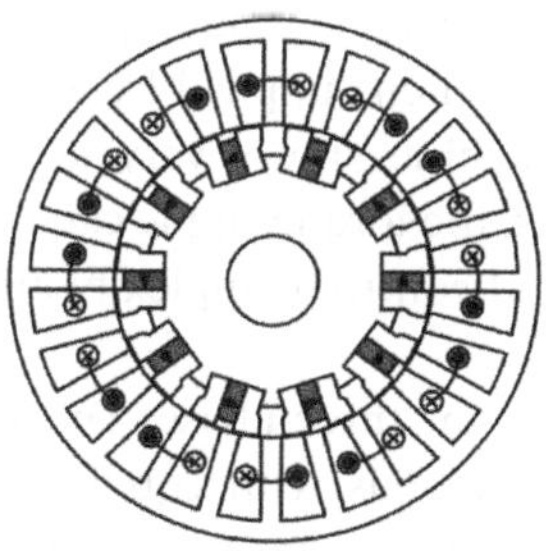

(a)

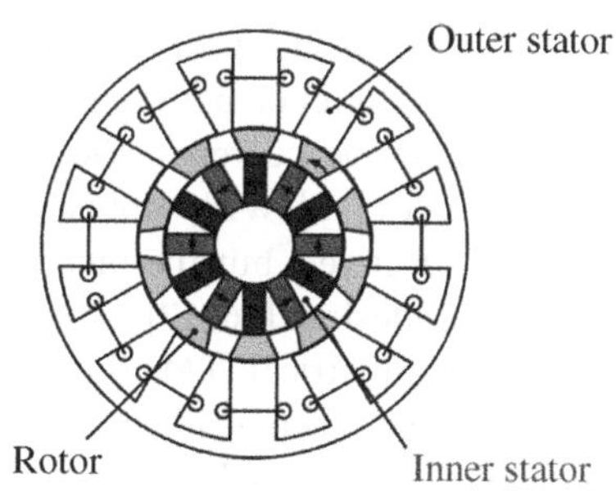

(b)

Figure 6.6 New FSPM machine variations derived by changing the position of source excitation MMF. (a) Rotor-PM FSPM machine, (b) Partitioned-stator FSPM machine.

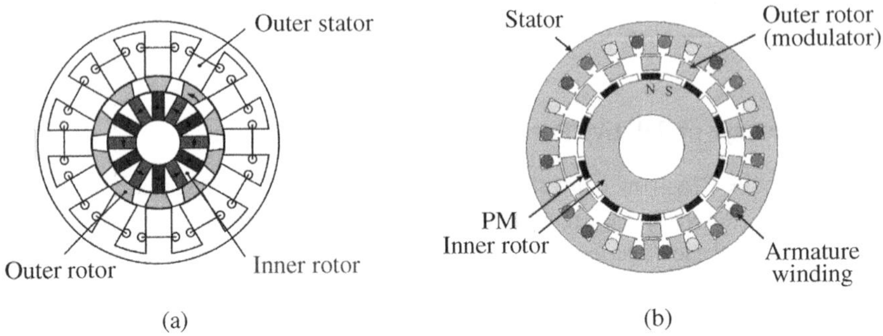

Figure 6.7 Dual-rotor PM machine derived by changing the relative motion among the three basic elements. (a) Spoke-type PM rotor, (b) Surface-mounted PM rotor.

can be obtained [13]. For example, PMs on the inner stator of the machine shown in Figure 6.6b can be replaced by high-temperature superconducting (HTS) coils for excitation, leading to a superconducting dual-stator field modulation machine with a stationary seal [14]. See Section 6.8 for more details.

In BDFMs, the modulator may also be the short-circuited coil type or the multi-layer barrier type apart from the most commonly-used salient reluctance poles in stator-PM machines. By combining multiple types of modulators and/or changing the relative position/motion of these modulators, new machines can also be obtained. The BDFM with an asymmetrical composite rotor is presented as an example in Section 6.9.

6.2 DSPM Machine with Π-Shaped Stator Core

The three-phase DSPM machine usually suffers from the asymmetrical PM flux linkage between the positive and negative half cycles as well as among phases due to the difference between phase magnetic circuits. When supplied by ideal trapezoidal-wave currents, the three-phase DSPM machine shows very large torque ripples [15]. With sophisticated control strategies, the torque ripple can be reduced effectively but the increased hardware cost and control complexity are typically undesired [16, 17].

The asymmetrical PM flux linkage phenomenon in three-phase DSPM machines has been studied in [18] and two design methods with equal and unequal slot widths have been proposed to overcome the issue and reduce the torque ripple [19]. Based on the design method with unequal slot widths, a modular complementary linear DSPM machine has been proposed in [20]. The primary side consists of two modules and by adjusting the width of the non-magnetic barrier, the distance between the two modules is set to be 180 electrical degrees. As a result, the complementary electromagnetic performance between the two primary modules can be achieved. The asymmetry in PM flux linkage and the cogging force of whole linear machine can be reduced greatly. It has also been shown that more balanced magnetic circuits can be achieved by replacing PMs with field windings [21].

Starting from the GAFMT, the authors addressed the asymmetrical PM flux linkage issue by changing the stator modulator. A new three-phase DSPM machine with Π-shaped stator core, referred as the Π-DSPM machine for short, has been proposed [22].

6.2.1 Machine Construction and Operating Principle

1 Machine Construction

Figure 6.8 illustrates a typical three-phase 12-slot 8-pole (12/8) DSPM machine with E-shaped stator cores, showing asymmetrical PM flux linkages. For short, it will be referred to as the E-DSPM machine in this section.

The stator is composed of four E-shaped core modules, between which tangentially magnetized PMs are sandwiched. The magnetization directions of adjacent PMs are opposite, and each stator tooth is wound with a concentrated winding. It can be seen that there are three stator teeth between two adjacent PMs, and the three stator teeth belong to phase A, phase B, and phase C, respectively, thus forming an "A ~ B ~ C ~ (—A) ~ (—B) ~ (—C)" armature winding pattern. The distances between the PMs and phase windings are different, resulting in the unbalance between phase magnetic circuits.

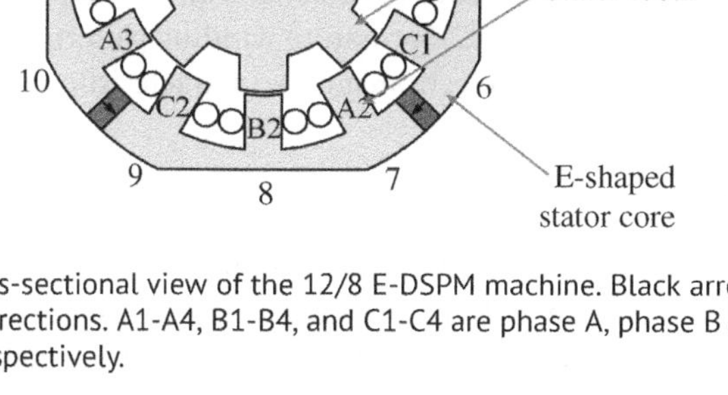

Figure 6.8 Cross-sectional view of the 12/8 E-DSPM machine. Black arrows denote the magnetization directions. A1-A4, B1-B4, and C1-C4 are phase A, phase B and phase C winding coils, respectively.

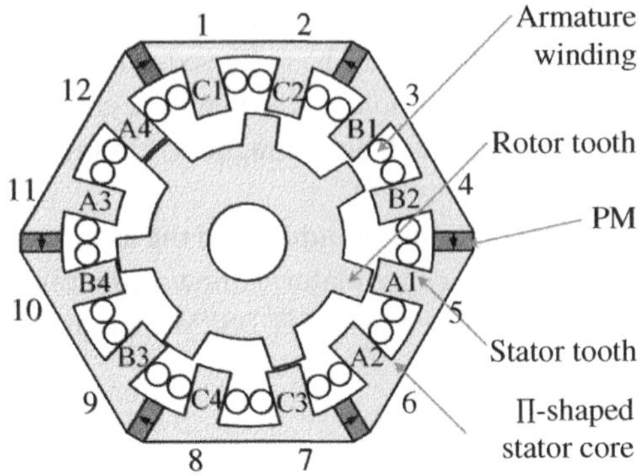

Figure 6.9 Cross-sectional view of the new 12/7 Π-DSPM machine. Black arrows denote the magnetization directions. A1-A4, B1-B4, and C1-C4 are phase A, phase B and phase C winding coils, respectively.

The new three-phase 12-slot 7-pole (12/7) Π-DSPM machine is shown in Figure 6.9. Compared to the E-DSPM machine, the main difference lies in the shape of the stator core module. There are two stator teeth between two adjacent PMs, which is different from the number of phases. Because the PMs have to be in pairs and each Π-shaped stator core has two teeth, a three phase Π-DSPM machine has to have at least 12 stator teeth to form a unit machine, i.e. the minimum geometric and magnetic periodicity. The number of rotor salient poles in such Π-DSPM machines can be any integer except multiples of the phase number, as pointed out by researchers for the variable flux reluctance machine with DC field windings [21].

For the unit machine with 12 stator teeth, the typical number of rotor salient reluctance poles can be 7, 8, 10, 13, etc. Here, the number of rotor salient reluctance poles is selected as 7 to achieve a high and symmetrical no-load back-EMF. Because of the adoption of the Π-shaped stator core, six PM blocks are needed to have 12 stator teeth, which divide the stator circumference into six equal parts. There are four equal parts in the 12/8 E-DSPM machine. As shown in Figure 6.9, both stator teeth between adjacent PMs belong to the same phase. Therefore, an "A ~ (—A) ~ B ~ (—B) ~ C ~ (—C)" armature winding pattern is achieved, which is the key to produce balanced electromagnetic characteristics between phases.

2 Flux Linkage and Back-EMF

DSPM machines are excited by PMs on the stator so the magnetic field created by PMs is stationary relative to armature windings. When the rotor rotates, the excitation field will change with the variation of the airgap magnetic permeance, achieving expected change of the PM flux linkage linked by armature windings, as

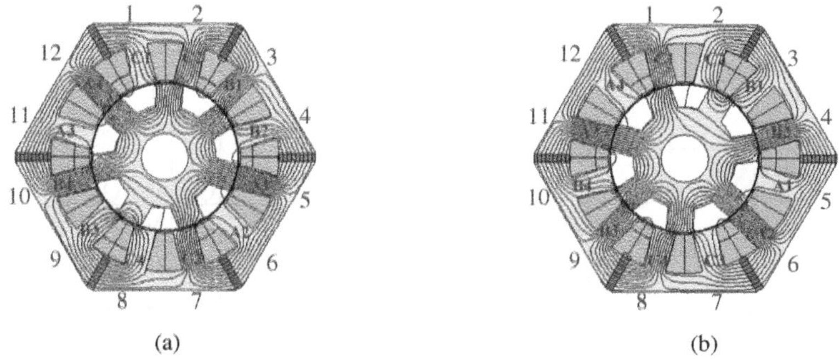

Figure 6.10 Magnetic field distribution of the new 12/7 Π-DSPM machine. (a) Position A where flux linkages linked by Coil A1 and Coil A4 reaches the maximum, and flux linkages linked by Coil A2 and Coil A4 are the minimum, (b) Position B where flux linkages linked by Coil A1 and Coil A4 reaches the minimum, and flux linkages linked by Coil A2 and A4 are the maximum.

well as inducing no-load back-EMF. Consequently, power output can be realized when suitable three-phase currents are fed into armature windings. Therefore, the operating principle of the 12/7 Π-DSPM machine can be explained by examining the airgap magnetic field distributions at different rotor positions, as illustrated in Figure 6.10.

As shown in Figure 6.10a, the 5th and 12th stator teeth are almost aligned with two rotor salient reluctance poles and the 6th and 11th stator teeth are nearly in line with two rotor slots. At this moment, the PM flux linkages linked by Coil A1 and Coil A2 reach the maximum and minimum, respectively, of which the magnetic field directions are toward the rotor side from the stator side.

Similarly, the PM flux linkages linked by Coil A4 and Coil A3 reach the maximum and minimum, respectively, but their magnetic field directions are both toward the stator side from the rotor side. When the rotor rotates by one half tooth pitch, as shown in Figure 6.10b, the PM flux linkage linked by Coil A1 and Coil A4 reaches the minimum. The PM flux linkage linked by Coil A2 and Coil A3 reaches the maximum. The flux directions of these four coils are the same as those shown in Figure 6.10a.

It can be concluded that the amplitude of the PM flux linkage linked by the four coils will change with the rotor position, but the direction remains unchanged, namely the PM flux linkage in each coil is unipolar. When the four coils are connected in series to form the phase A winding (A1 − A2 + A3 − A4), a bipolar flux linkage can be obtained, as shown in Figure 6.11a.

In addition, the PM flux linkage and no-load back-EMF of the 12/7 Π-DSPM machine are sinusoidal. Therefore, to reduce the torque ripple, brushless AC operation is recommended. In contrast, the flux linkages linked by each coil and each

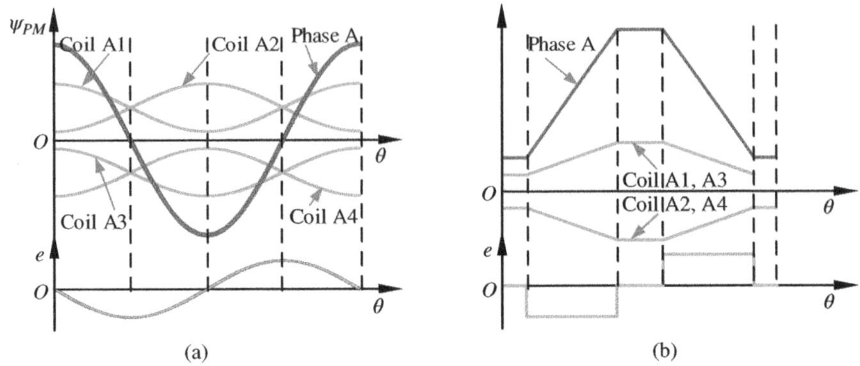

Figure 6.11 PM flux linkage and no-load back-EMF waveforms of DSPM machines. (a) The 12/7 Π-DSPM machine, (b) The 12/8 E-DSPM machine.

phase armature winding of the E-DSPM machine are all unipolar, and the phase no-load back-EMF is trapezoidal, as shown in Figure 6.11b; brushless DC operation is more suitable here.

3 Operating Principle Analysis Based on GAFMT

Based on the developed GAFMT, the airgap flux density distribution created by PMs in the DSPM machine can be expressed as

$$\begin{aligned} B_g(\theta, t) &= F_{PM}(\theta)\Lambda(\theta) \\ &= \frac{\Lambda_0}{2}\sum_{n=1}^{\infty} F_{PMn} \sin\left(\frac{nN_s}{2n_s}\theta\right) \\ &+ \sum_{n=1}^{\infty}\sum_{m=1}^{\infty} \frac{F_{PMn}\Lambda_m}{2} \sin\left[\left(\frac{nN_s}{2n_s} + mN_r\right)\left(\theta - \frac{mN_r\theta_0 + mN_r\omega_r t}{\frac{nN_s}{2n_s} + mN_r}\right)\right] \\ &+ \sum_{n=1}^{\infty}\sum_{m=1}^{\infty} \frac{F_{PMn}\Lambda_m}{2} \sin\left[\left(\frac{nN_s}{2n_s} - mN_r\right)\left(\theta + \frac{mN_r\theta_0 + mN_r\omega_r t}{\frac{nN_s}{2n_s} - mN_r}\right)\right] \end{aligned} \tag{6.1}$$

in which $F_{PM}(\theta)$ is the PM MMF drop across the effective airgap in the radial direction, $\Lambda_r(\theta)$ is the effective airgap magnetic permeance caused by the physical airgap taking into consideration the influence of stator and rotor slots, $\Lambda_0/2$ is the DC component of the magnetic permeance per unit area, F_{PMn}, and Λ_m are Fourier coefficients, which can be determined based on the discussion in [23]. N_s and N_r

Table 6.1 Speed and initial phase of airgap flux density harmonics.

Group	Pole-pair number	Rotation speed	Initial phase
1	$3n$	0	0
2	$3n + 7m$	$7m\omega/p_{fh}$	$7m\theta_0/p_{fh}$
3	$3n - 7m$	$-7m\omega/p_{fh}$	$-7m\theta_0/p_{fh}$
4	$-(3n-7m)$	$7m\omega/p_{fh}$	$(-\pi + 7m\theta_0)/p_{fh}$

are the number of stator teeth and rotor salient reluctance poles, respectively, n_s is the number of stator teeth between two adjacent PMs, θ_0 and ω are the initial rotor position and the mechanical speed of the rotor, respectively. n and m are positive integer numbers.

It can be seen that because the stator cores and PMs are stationary, a series of stationary spatial harmonic components appear in the airgap flux density waveform of the 12/7 Π-DSPM machine, of which the pole-pair number p_{fh} is equal to $nN_s/(2n_s) = 3n$. Moreover, due to the modulation effect of the rotor salient reluctance poles to the PM magnetic field, other series of spatial harmonic components will show up in the airgap flux density waveform, whose pole-pair numbers p_{fh} are $nN_s/(2n_s) + mN_r = 3n + 7m$ and $|nN_s/(2n_s) - mN_r| = |3n - 7m|$. The spatial harmonic components of the airgap flux density in 12/7 DSPM machines can be classified into four groups, as tabulated in Table 6.1, in which the mechanical rotational speed and the initial phase of each group of harmonics are also summarized.

Based on the presented analysis, the spatial harmonic components with $3n$ pole pairs in Group 1 in Table 6.1 stay stationary relative to the stator, so they will not induce back-EMF in the armature winding. Harmonic components in Group 2, 3, and 4, which are generated by the synchronous modulation of stator teeth and the asynchronous modulation of rotor salient reluctance poles to the PM MMF, rotate together with the rotor. So the back-EMF can be generated in the armature winding, so as the output torque and power when appropriate currents are injected into the windings.

To verify the analytical analysis, two-dimensional (2D) finite element analysis (FEA) has been used to calculate the PM flux density distribution in the 12/7 Π-DSPM machine. As shown in Figure 6.12a, the PM flux density under each stator tooth is unipolar and its amplitude changes with the rotation of rotor. The peaks appear under the 5th and 12th stator teeth when the rotor is at Position A. The peaks move to positions under the 6th and 11th stator teeth when the rotor is at Position B.

Figure 6.12b shows the amplitude of each spatial harmonic is almost independent of the rotor position. In addition, the amplitudes of some harmonics

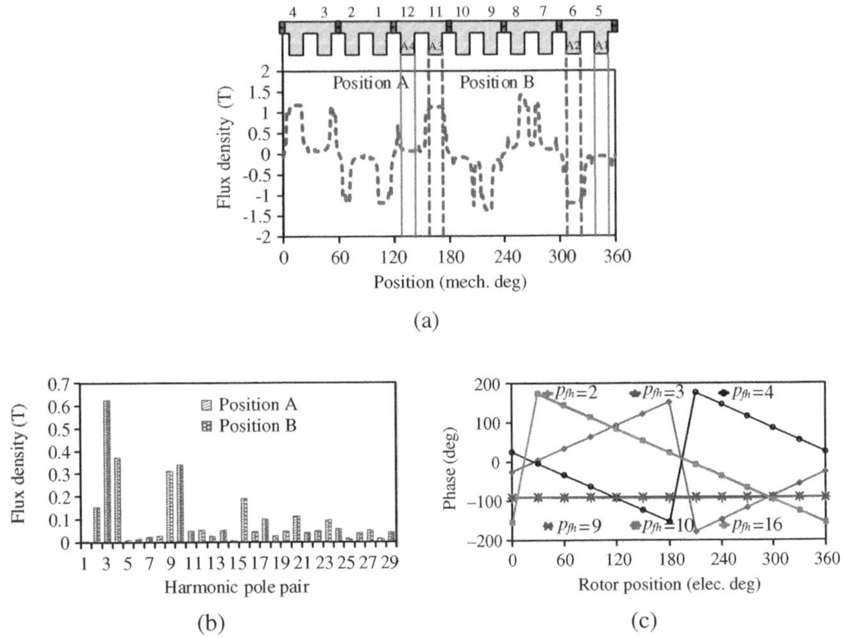

Figure 6.12 No-load airgap flux density distribution of the 12/7 Π-DSPM machine. (a) Radial airgap flux density distributions at Position A and Position B. (b) Amplitude spectrums of waveforms in (a). (c) Phase change of dominating harmonics.

are relatively high, such as the 2nd ($n = 3$, $m = 1$, Group 3), 3rd ($n = 1$, $m = 0$, Group 1), 4th ($n = 1$, $m = 1$, Group 4), 9th ($n = 3$, $m = 0$, Group 1), 10th ($n = 1$, $m = 1$, Group 2), 16th ($n = 3$, $m = 1$, Group 2) components, and so on.

These spatial harmonics with higher amplitudes are termed as the dominating harmonic components. The phase shifts of the above-mentioned six dominating harmonics versus the rotor position is shown in Figure 6.12c. Because the harmonic with three pole pairs is generated directly by the PMs, and the harmonic with nine pole pairs is generated by the synchronous modulation of stator teeth to the 3-pole-pair PM field component, the phases of these two harmonics are constant. The other four harmonic components, which are generated by the asynchronous modulation of rotor salient reluctance poles to the 3-pole-pair and 9-pole-pair harmonics, rotate by one electrical cycle when the rotor rotates by one tooth pitch. The rotational speeds in electrical degrees of these four harmonics are the same as that of the rotor.

On the other hand, the rotational direction of the 2-pole-pair harmonic is opposite to that of the 4-, 10-, and 16-pole-pair harmonics. The phase change of the 2-pole-pair harmonic mirrors that of the 4-pole-pair harmonic about the

Table 6.2 Slot pitch angle based on different dominating harmonics.

Pole-pair number	Slot pitch angle	Pole-pair number	Slot pitch angle
2	$-\pi/3$	10	$5\pi/3$
4	$2\pi/3$	16	$2\pi/3$

horizontal axis. In addition, the phase differences between the 10- and 4-, 16- and 4- pole-pair harmonics are 180°, which are consistent with Table 6.1.

4 Connection of Armature Winding Coils

Tooth-wound concentrated coils are employed in the Π-DSPM machine. According to the slot star diagram, the phase armature winding can be obtained by connecting coils with similar phase angles in the same direction and coils with almost 180° phase shift in the opposition direction. To explain how to connect the armature winding coils, slot pitch angles for dominating harmonics of different pole pairs are calculated first:

$$\alpha = \text{sign}\,(\omega_{fh})\frac{2\pi}{N_s}p_{fh} \tag{6.2}$$

where α is the slot pitch angle in electrical degrees, sign(ω_{fh}) and ω_{fh} are the rotational direction and speed of the magnetic field harmonic, respectively. The corresponding slot pitch angles are tabulated in Table 6.2. The slot pitch angles for the 2-pole-pair and 10-pole-pair spatial harmonics are the same, which are −60°. Therefore, the pattern "A ~ B ~ C ~ (—A) ~ (—B) ~ (—C)" can be employed. The slot pitch angles of the 4-pole-pair and 16-pole-pair harmonics are 120°, suitable for the pattern "A ~ B ~ C ~ A ~ B ~ C." It should be noted that there is an obvious disagreement for different spatial harmonics in terms of the connection pattern.

To visualize the influence of different armature winding coil connections of the 12/7 Π-DSPM machine, the PM flux linkage ψ_{fh} linked by each stator coil due to each dominating harmonic can be calculated as

$$\psi_{fh}(t) = B_{fhj\max} r_{si} l_{stk} N_c \int_{\theta_{w1}}^{\theta_{w2}} \cos[p_{fhj}\theta + \varphi_{fhj}(t)]d\theta \tag{6.3}$$

where $B_{fhj\max}$ and $\varphi_{fhj}(t)$ are the amplitude and phase of the p_{fhj}-pole-pair dominating harmonic, r_{si} is the inner radius of the stator, N_c is the number of turns per coil, l_{stk} is the stack length, θ_{w1} and θ_{w2} are the positions of two coil sides along the airgap circumference in mechanical degrees.

Key dimensions and design specifications of the studied 12/7 Π-DSPM machine are listed in Table 6.3, where the 12/8 E-DSPM machine in [24] is also listed for comparison. The two machines share the same stack length. Due to the different

Table 6.3 Key dimension parameters of the 12/7 Π-DSPM machine.

Parameters	Value	
	Π-DSPM	E-DSPM
Stator inner diameter (mm)	76	75
Rotor outer diameter (mm)	75.1	74.1
Rotor inner diameter (mm)	24	
Airgap length (mm)	0.45	
Stack length (mm)	75	
Stator tooth width (mm)	11.8	9.8
PM width in tangential direction (mm)	4	6
PM width in radial direction (mm)	18.5	
Total PM volume (mm^3)	33 300	
PM remanence (T)	0.98	
Rotor tooth arc (deg)	20.7	20
Rotor tooth height (mm)	11.55	9
Turns in series per phase	280	260
Phase current (Apk)	6	5.2
Area per stator slot (mm^2)	266	206
Current density (A/mm^2)	2.24	2.68

magnetic circuits of the two machines, the stator yoke thickness of the Π-DSPM machine is much smaller than that of the E-DSPM machine.

The PM flux linkage linked by each coil contributed by different dominating harmonics are shown in Figure 6.13, where T_i denotes the ith coil. Figure 6.13a,b depict the flux linkage waveforms due to the 2-pole-pair and 10-pole-pair spatial harmonics linked by each coil, respectively. Figure 6.13c shows their summation. It can be seen that the phase difference between the flux linkages linked by two adjacent coils is $\pi/3$, so the flux linkages linked by Coil T_i and Coil T_{i+7} are the same, and this conclusion is in line with Table 6.2. Moreover, the PM flux linkage linked by each coil produced by the 2-pole-pair component is in phase with those produced by the 10-pole-pair dominating harmonic. Thus, the resultant PM flux linkage due to both components can be obtained by summing them up directly.

Flux linkages linked by each coil contributed by the 4- and 6-pole-pair harmonics are shown in Figure 6.13d. It shows that the phase difference between the flux linkages of two adjacent coils is $2\pi/3$, and the phase difference between flux

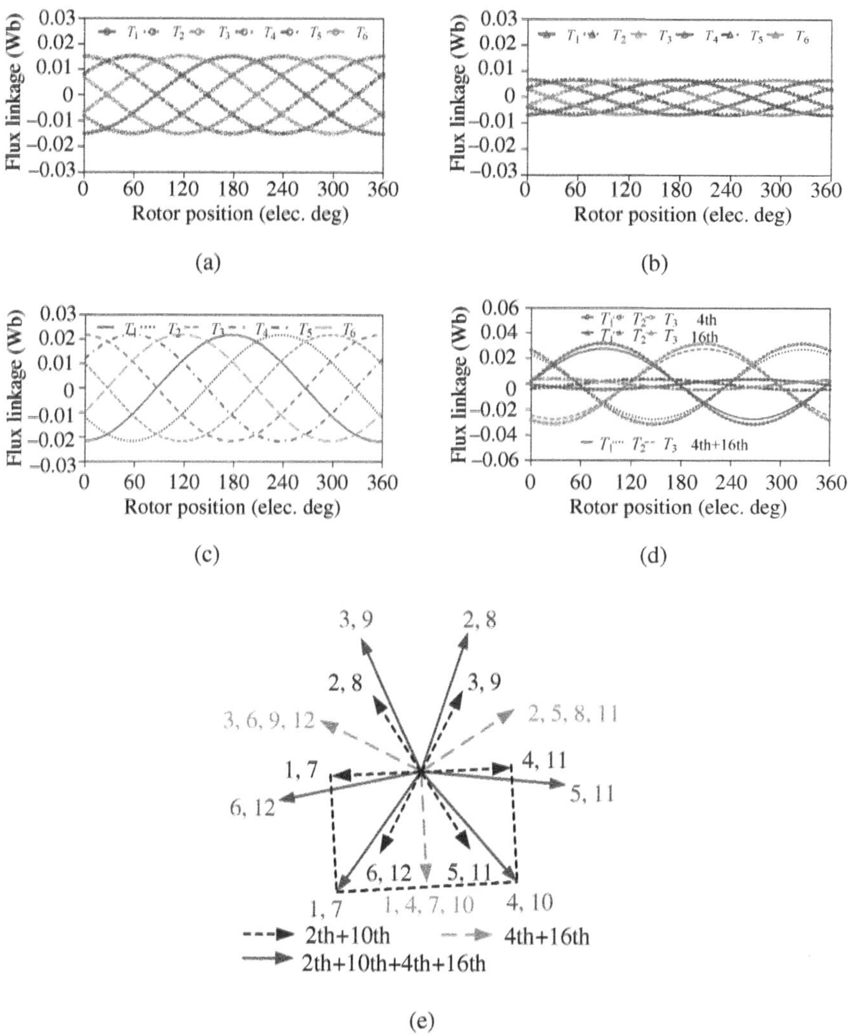

Figure 6.13 PM flux linkage contributed by different dominating harmonics. (a) Flux linkage due to the 2-pole-pair harmonic. (b) Flux linkage due to the 10-pole-pair harmonic. (c) Resultant flux linkage due to the 2- and 10-pole-pair harmonics. (d) Flux linkage due to the 4- and 16-pole-pair harmonics. (e) Slot star diagram considering all dominating harmonics.

linkages contributed by these two dominating harmonics is 180°. Therefore, the flux linkages can be partially canceled in the resultant waveforms.

It is obvious that phase differences of flux linkages in adjacent coils due to these two dominating harmonic families are different. Considering that their

Table 6.4 Armature winding coil connections of the 12/7 Π-DSPM machine.

	Method I	Method II	Method III
Phase A	$\boldsymbol{T_5 + T_{11} - T_6 - T_{12}}$	$T_5 + T_{11} + T_2 + T_8$	$T_5 + T_{11} + T_4 + T_{10}$
Phase B	$\boldsymbol{T_3 + T_9 - T_4 - T_{10}}$	$T_3 + T_9 + T_6 + T_{12}$	$T_3 + T_9 + T_2 + T_8$
Phase C	$\boldsymbol{T_1 + T_7 - T_2 - T_8}$	$T_1 + T_7 + T_4 + T_{10}$	$T_1 + T_7 + T_6 + T_2$

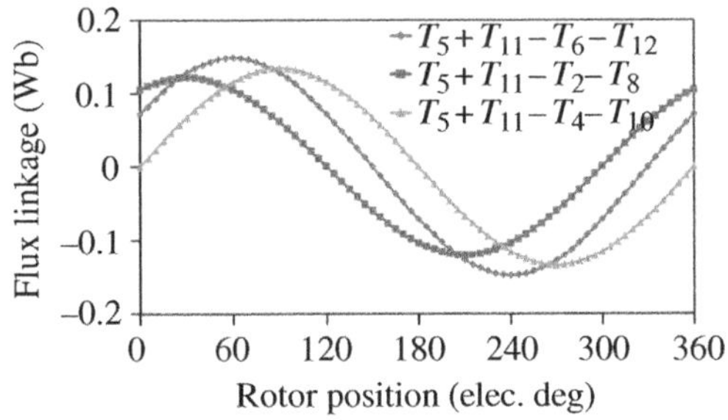

Figure 6.14 Comparison of phase flux linkages due to different coil connections.

frequencies and phase sequences are the same, the total PM flux linkage linked by each coil can be obtained by using the vector addition.

Figure 6.13e shows the slot star diagram considering all the dominating harmonics, based on which, there will be three different connections of armature winding coils. As shown in Table 6.4, for example, the armature winding of phase A can be composed of coil connections: $(T_5 + T_{11} - T_6 - T_{12})$, $(T_5 + T_{11} + T_2 + T_8)$ or $(T_5 + T_{11} + T_4 + T_{10})$. To deliver the highest power output, flux linkages of three coil connection methods are calculated by 2D FEA. As shown in Figure 6.14, the flux linkage waveforms of the three connection methods are all sinusoidal and the $(T_5 + T_{11} - T_6 - T_{12})$ can achieve the maximum flux linkage amplitude.

6.2.2 Performance Analysis and Comparison

Electromagnetic performance of the Π-DSPM machine is calculated based on the GAFMT and FEA to verify the theoretical analysis. Due to the ideal assumptions made in the theoretical analysis, the 3- and 9- pole-pair harmonics calculated by (6.1) are higher than those obtained by 2D FEA, and amplitudes of the 2-, 4-, 10-, and 16-pole-pair components obtained by (6.1) are lower than the 2D FEA results. To focus on the influence of dominating harmonics instead of how to obtain more accurate airgap flux density distributions, this section takes the amplitudes and phase angles of airgap flux density harmonics obtained from 2D FEA

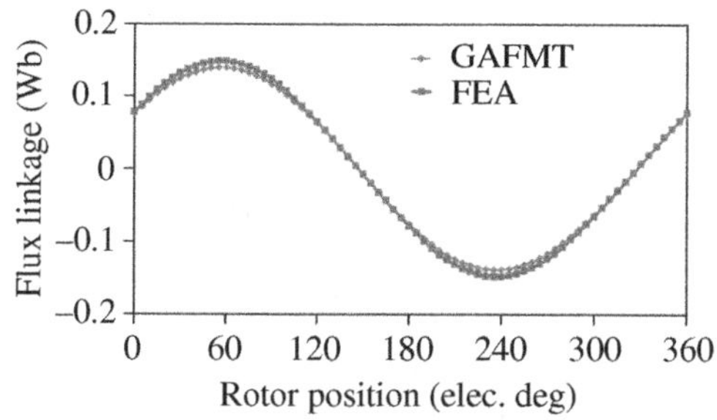

Figure 6.15 PM flux linkage of the 12/7 Π-DSPM machine.

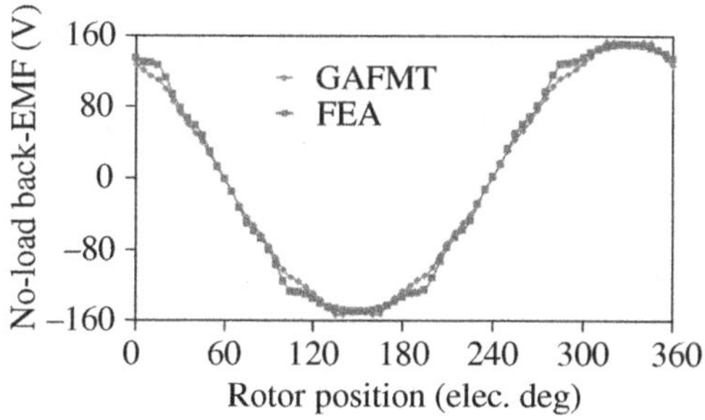

Figure 6.16 No-load back-EMF of the 12/7 Π-DSPM machine at 1500 r/min.

as the starting point to analyze the electromagnetic performance, such as the PM flux linkage, back-EMF, torque, etc.

Figure 6.15 shows the flux linkage waveforms of phase A, where GAFMT denotes the vector sum of flux linkages contributed by the four rotating dominating harmonics calculated using (6.3), and FEA denotes the flux linkage directly from the 2D FEA. The results are very close, so it can be concluded that the total flux linkage linked by each phase winding is mainly contributed by the four dominating harmonics. Thus, the electromagnetic performance prediction and the armature winding design can be simplified to analyzing the four dominating harmonics.

Figure 6.16 compares the no-load back-EMF waveforms from the GAFMT and 2D FEA, showing good agreement. The slight difference is mainly caused by spatial harmonics except the four dominating harmonics.

Figure 6.17 shows the output torque waveform when the rated three-phase currents are applied with $i_d = 0$ control method. It shows that most of the average electromagnetic torque of the Π-DSPM machine is contributed by the four rotating dominating harmonics.

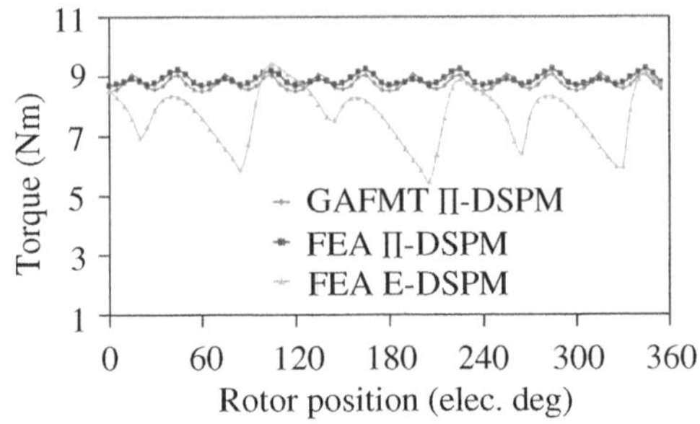

Figure 6.17 Comparison of output torque waveforms.

The torque waveform of the 12/8 E-DSPM machine based on the 120° conduction brushless DC (BLDC) operation is also included in Figure 6.17. It can be seen that the average torque of the Π-DSPM machine is 8.87 Nm, which is 2.22 Nm higher than that of the E-DSPM machine with the same PM volume and similar machine size. In addition, the Π-DSPM machine produces smoother torque waveforms than the E-DSPM machine. Torque ripples of the two machines are 6.27% and 57%, respectively. The torque ripple of the Π-DSPM machine has been reduced greatly due to its balanced magnetic circuit and sinusoidal no-load back-EMF.

6.2.3 Prototype and Experimental Results

Figure 6.18 shows a few photos of the 12/7 Π-DSPM machine prototype. In order to suppress the excessive magnetic flux leakage around PMs sandwiched by stator core modules, an aluminum housing has been used.

Figure 6.19a compares the measured and simulated no-load back-EMFs at 1200 r/min, showing good agreement. The corresponding harmonic spectrums are shown in Figure 6.19b. The fundamental amplitude of the measured no-load back-EMF is 123.4 V, which is 96% of that from the 2D FEA. Since the 2D FEA was adopted, the difference is mainly caused by end effects of PMs and armature windings, and manufacturing imperfections. Additionally, it can be found that both waveforms are sinusoidal and the total harmonic distortions (THDs) of measured and simulated no-load back-EMFs are 8.9% and 5.1%, respectively.

The experimental setup shown in Figure 6.20 was used for testing. The prototype was mounted on the base by two L-shaped supports. A magnetic powder brake was used to produce the load torque. An HBM torque transducer and an encoder were installed to measure the torque and rotor position/speed. The control was implemented on a dSPACE controller board.

Figure 6.21 shows the measured steady-state torque and the corresponding current waveform at 220 r/min and the load torque was 4 Nm. The amplitude

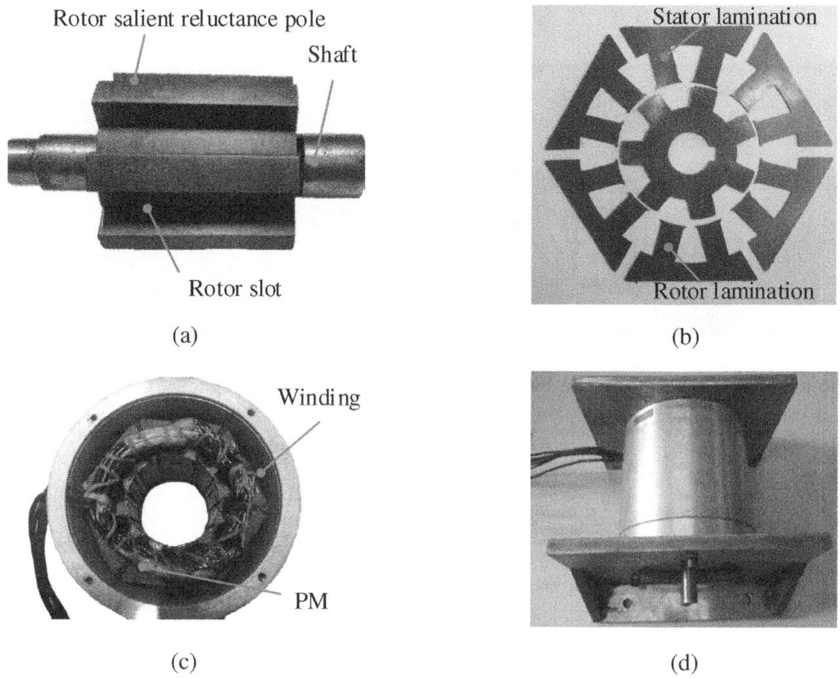

Figure 6.18 Photos of the 12-7 Π-DSPM machine prototype. (a) Rotor, (b) Silicon steel laminations, (c) Stator, (d) Whole machine.

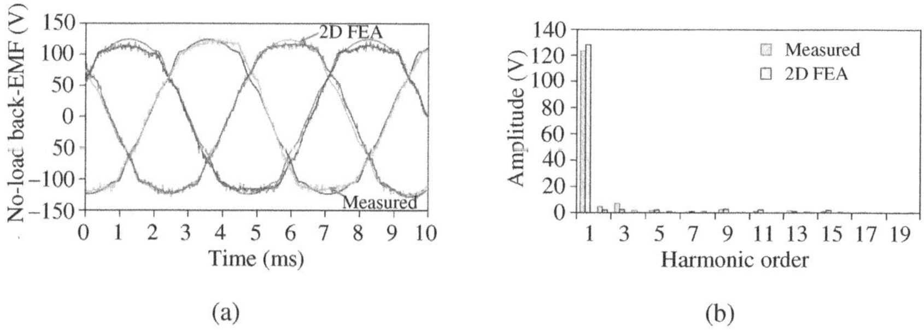

Figure 6.19 No-load EMF waveforms and harmonic analysis. (a) Measured and 2D FEA no-load EMF at 1200 r/min, (b) Amplitude spectrum.

of current was 2.8 A. According to the 2D FEA simulation results, when the amplitude of the armature current reaches 2.8 A, the output torque is 4.14 Nm, which means the measured torque is 96.6% of the FEA-predicted value.

The dynamic response has also been tested. As shown in Figure 6.22a, when the speed reference was changed from 100 to 200 r/min, 300 r/min, and then back to

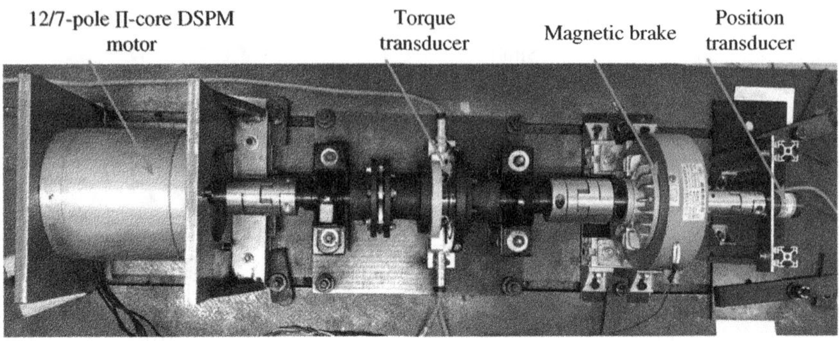

Figure 6.20 Experimental setup.

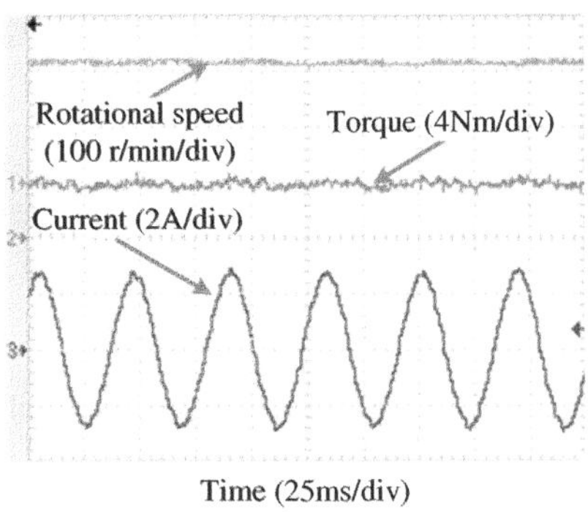

Figure 6.21 Test results of static characteristics.

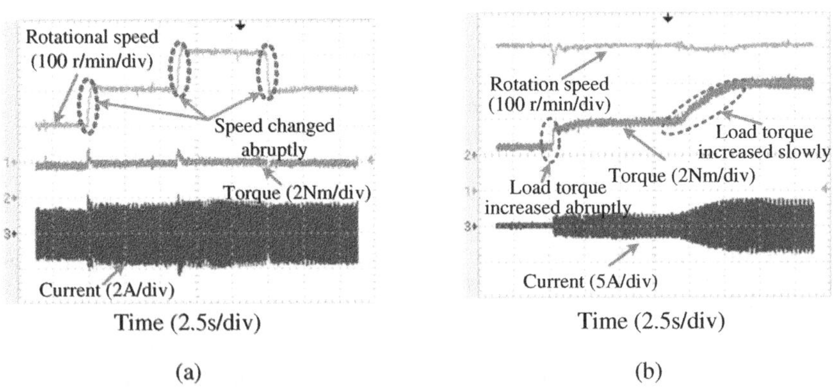

Figure 6.22 Dynamic performance waveforms. (a) With a constant load toque and varied speed references. (b) With a constant speed reference and varied load torques.

200 r/min with a constant load torque of 2 Nm, the output torque stayed almost constant, and the armature winding current rises slightly with the increase of speed due to the increased friction torque.

Then, the dynamic performance when the load torque varied from 0 to 2 Nm and later 4 Nm, were tested when the speed reference was set to 400 r/min. As shown in Figure 6.22b, the armature winding current increased rapidly as an abrupt change of the load torque was applied, showing the fast dynamic response of the control system. When the load torque increased slowly, the armature current followed the linear increase closely.

6.3 Stator-PM Variable Reluctance Resolver

Variable reluctance (VR) resolvers are an ideal option for position sensors in machine control systems. Covering the functions of incremental encoders and hall sensors, the VR resolver system can produce an accurate position signal and locate the initial rotor position [25]. The rotor of a VR resolver is simply an iron block made of silicon steel laminations, which leads to the advantages of uncomplicated structure, low cost, ease of manufacturing and installation, as well as less sensitivity to vibrations and other harsh environments. As such, it is a promising candidate for high- and ultra-high-speed applications, such as electric vehicles/hybrid electric vehicles (EV/HEVs).

In order to reduce the traction system size and improve the power density, the machine speed in the automotive industry is becoming higher and higher. An ultra-high-speed machine seems a competitive potential powertrain solution for future EV/HEVs, but there are still a number of technical challenges, including the rotor position sensing. Even the existing VR resolver is not a perfect solution for such challenges.

Specifically, the VR resolver requires a high-frequency (HF) signal excitation to produce output signals (SIN/COS) for the rotor position measurement based on the modulation effect of rotor teeth. To ensure the accuracy of measurement, the frequency of excitation signal must be much higher than the machine's operating frequency. For example, in an 8-pole PM synchronous machine, if the resolver excitation frequency is 10 kHz, then the rotational speed of the machine itself must be less than 20 kr/min. Otherwise, the position measurement will not be accurate enough to implement the high-performance vector control, which is the so-called high-frequency limitation problem. This issue can be solved by increasing the frequency of the excitation signal, but the cost and risk of the decoding system will increase as well.

The stator-PM machines working by the GAFMT shares some similarities with the VR resolver and is inherently suitable for high-speed applications. As a PM

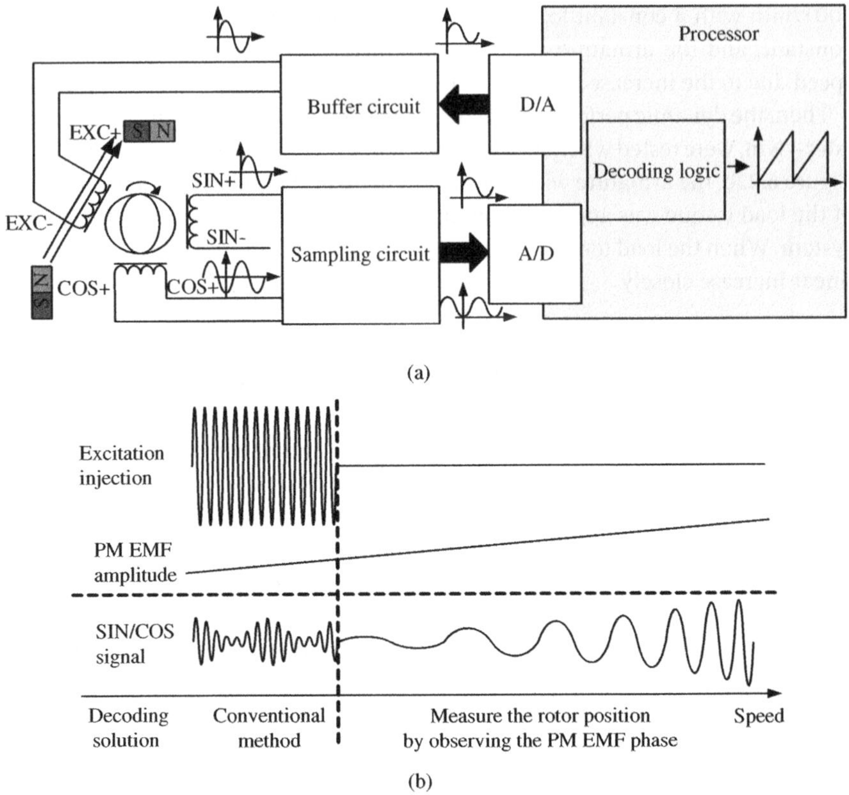

Figure 6.23 Novel stator-PM VR resolver. (a) Schematic diagram, (b) Signal logic diagram.

machine, the no-load back-EMF in synchronization with the rotor position can be induced in windings without any high-frequency signal injection. Therefore, the high-frequency limitation problem can be addressed naturally. Also, since the position sensing does not require strong magnetic field, the inexpensive ferrite magnets can be used to generate the no-load motional EMF. It should be noted that this no-load motional EMF also exists in conventional VR resolvers, but it is usually too weak to be utilized for position measurement.

Based on the GAFMT, a new type of resolver making full use of the field modulation effect is proposed to inherit advantages of stator-PM machines and the conventional VR resolvers [26]. Figure 6.23a shows the schematic diagram of the new stator-PM variable reluctance resolver.

In the low-speed range, the high-frequency signal induces a high-frequency transformer EMF in the output windings termed as SIN/COS, which shares the same operating principle as the conventional VR resolver. As the rotational speed increases, the motional EMF due to PMs increases. Once the motional EMF surpasses a certain threshold value, the decoding system stops the high-frequency signal injection and then the rotor position can be obtained from the PM EMF. The signal logic diagram is illustrated in Figure 6.23b.

6.3.1 Machine Construction and Operating Principle

The stator-PM VR resolver is essentially a special two-phase DSPM generator, of which the two-phase EMF outputs are orthogonal or quasi-orthogonal.

Figure 6.24 shows a 16-slot 4-pole (16/4) stator-PM VR resolver, where four ferrite PMs are mounted on the stator to produce four magnetic poles. Each rotor salient reluctance pole corresponds to one magnetic pole and extends across four stator teeth and covers 360 electrical degrees, namely, each stator tooth is equivalent to a 1/4 electrical cycle.

To obtain two-phase orthogonal EMF outputs with a 90° phase shift with respect to one another, the SIN and COS coils should be wound on the four stator teeth alternately. Coils on the first and third tooth are connected in series to form the SIN winding, and coils on the second and fourth tooth are connected in series to form the COS winding. Moreover, the excitation winding is composed of four coils connected in series on the stator teeth.

It can be found that all the windings and PMs are on the stator in the proposed stator-PM VR resolver, and it can be regarded as a two-phase stator-PM machine equipped with excitation coils and some cheap ferrite magnets to realize the position measurement. Therefore, it is suitable for low-cost applications as well.

1 Excitation Signal Applied Only

In order to explain the operating principle of the stator-PM VR resolver, the output EMF in one electric cycle is analyzed based on Figure 6.24. When the excitation

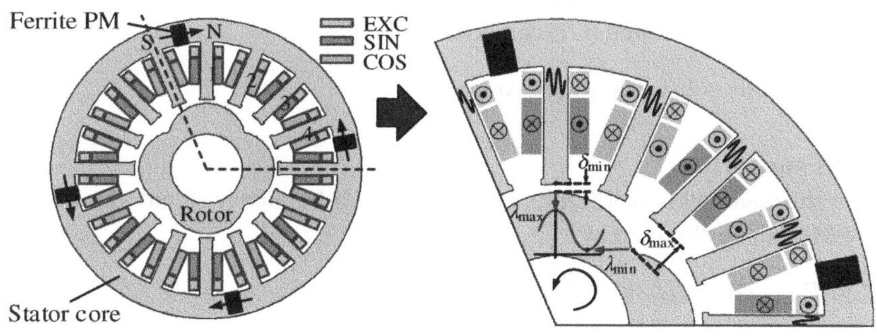

Figure 6.24 Structure of a typical stator-PM VR resolver.

signal works only, the excitation MMF created by each excitation coil is

$$F = F_{EXC} \sin(\omega_{EXC} t) \tag{6.4}$$

The flux flowing through one stator tooth can be expressed as

$$\varphi_{\sin 1}(\theta_r, \theta_0, t) = \pm F_{EXC} \sin(\omega_{EXC} t) \times \left[\lambda_a \cos(\theta_0 - \theta_r) + \lambda_b + \lambda_{La} \sin\left(\theta_0 - \frac{\pi}{4}\right) + \lambda_{Lb}\right] \tag{6.5}$$

where $\lambda_a = (\lambda_{\max} - \lambda_{\min})/2$, $\lambda_b = (\lambda_{\max} + \lambda_{\min})/2$, $\lambda_{La} = (\lambda_{L\max} - \lambda_{L\min})/2$, $\lambda_{Lb} = (\lambda_{L\max} + \lambda_{L\min})/2$. $\lambda_{\max}$ and $\lambda_{\min}$ are the maximum and minimum airgap permeance values of the main magnetic flux path, respectively. $\lambda_{L\max}$ and $\lambda_{L\min}$ are the maximum and minimum permeance values of leakage magnetic flux path, respectively. θ_r is the rotor position, θ_0 is the position angle of the stator tooth. For example, the position angle of tooth 1 and tooth 3 are 0 and π, respectively. F_{EXC} is the amplitude of the excitation MMF and ω_{EXC} is the electrical angular frequency of the excitation signal, whose typical value is 10 kHz. By deducing and summing the fluxes of stator teeth 1, 3, 5…, until 15 corresponding to the eight SIN coils of the resolver, the flux linkage of the SIN winding can be written concisely as:

$$\psi_{\sin\mathrm{HF}}(t) = -8F_{EXC} \sin(\omega_{EXC} t)(\lambda_a \cos\theta_r - 0.7\lambda_{La}) \tag{6.6}$$

Then the output EMF of the SIN winding can be obtained by the time derivative of the flux linkage as

$$E_{\sin\mathrm{HF}}(t) = -\frac{d\psi_{\sin\mathrm{HF}}(t)}{dt} = 8F_{EXC}\lambda_{\mathrm{a}} \times [\omega_{EXC} \cos(\omega_{EXC} t)(\cos\theta_r - 0.7\lambda_{\mathrm{La}}/\lambda_{\mathrm{a}}) - \omega_r \sin(\omega_{EXC} t) \sin\theta_r] \tag{6.7}$$

where $\omega_r = d\theta_r/dt$ is the electrical angular frequency corresponding to the rotor speed. Considering ω_{EXC} is much higher than ω_r at low speeds, the second item in (6.7) can be ignored. Then (6.7) can be simplified to:

$$E_{\sin\mathrm{HF}}(t) = E_{HF} \cos(\omega_{EXC} t)[\cos(\omega_r t) - \lambda_x] \tag{6.8}$$

2 PMs Applied Only

When the stator PMs work only without the high-frequency signal, the no-load PM EMF can be obtained by:

$$E_{\sin\mathrm{PM}}(t) = E_{PM} \sin(\omega_r t) \tag{6.9}$$

where the constant E_{PM} is defined as $E_{PM} = F_{PM}\lambda_a\omega_r$, and F_{PM} is the amplitude of the MMF created by PMs.

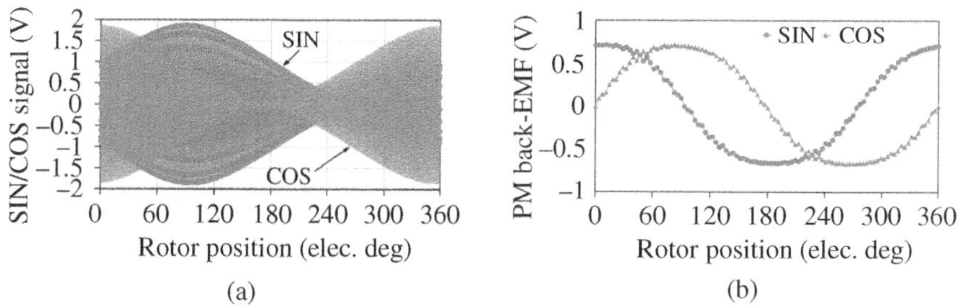

Figure 6.25 EMF waveforms of stator-PM VR resolver. (a) EMF due to high-frequency excitation signal at 500 r/min, (b) PM EMF without high-frequency excitation signal at 10 000 r/min.

Similarly, the transformer EMF and motional EMF due to PMs of the COS winding can be obtained. The resultant EMFs of SIN/COS windings can be written as

$$\begin{cases} E_{\sin}(t) = E_{HF}\cos(\omega_{EXC}t)[\cos((\omega_r t)) - \lambda_x] + E_{PM}\sin(\omega_r t) \\ E_{\cos}(t) = E_{HF}\cos(\omega_{EXC}t)[\sin((\omega_r t)) - \lambda_x] + E_{PM}\cos(\omega_r t) \end{cases} \tag{6.10}$$

When the rotor speed is low, the amplitude of the PM EMF is much lower than that of the transformer EMF induced by the electrical excitation, so the former one can be ignored. The output EMF waveforms at 500 r/min are shown in Figure 6.25a. When the rotor speed is high, the high-frequency excitation signal does not work, so E_{HF} is 0. Consequently, the output EMF is only composed of the PM EMF, whose waveform is shown in Figure 6.25b.

Figure 6.26 shows the prototype of the proposed 16/4 stator-PM VR resolver and its associated decoding circuit. Figure 6.27 shows the decoding circuit diagram,

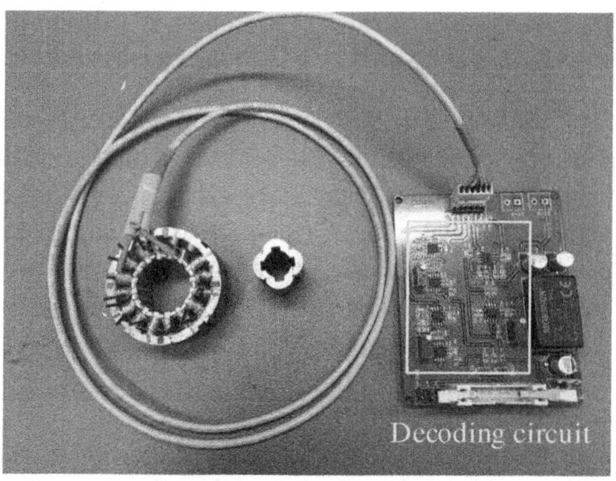

Figure 6.26 Photo of the stator-PM VR resolver and its decoding circuit.

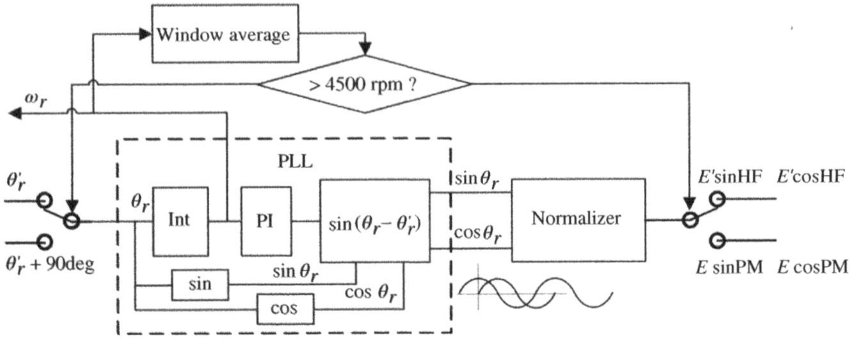

Figure 6.27 Decoding diagram for the entire speed range.

in which E'_{sinHF} and E'_{cosHF} are the demodulated EMF signals. When the rotational speed is lower than the set value (4500 r/min in Figure 6.27), θ'_r, E'_{sinHF}, and E'_{cosHF} are adopted based on the two switches. When the rotational speed is higher than the set value, $\theta'_r + 90°$, E_{sinPM}, and E_{cosPM} are used, in which a 90° phase shift is introduced on the rotor position to compensate the phase difference between the two methods.

Figure 6.28 shows the measured EMF waveforms at different rotational speeds and the rotor position signal based on the decoding EMF. It can be seen that the position signal of the stator-PM VR resolver is consistent with that of conventional VR resolver in low-speed range. When the rotational speed reaches as high as 10 000 r/min, there is an obvious ripple in the position signal from conventional VR resolver, which leads to measurement errors. By contrast, the position signal from the proposed stator-PM VR resolver is more stable and reliable.

6.3.2 Odd-Pole Issue and Solutions Based on GAFMT

The number of PM poles is the same as that of rotor salient reluctance poles in the above-mentioned 16/4 stator-PM VR resolver. When the rotor salient reluctance pole rotates by one PM pole pitch, the EMF experiences a full electrical cycle, but this conclusion is limited within structures whose number of rotor salient reluctance poles is odd. There are also a number of machines with odd pole-pair numbers used in the industry, and resolvers with an odd number of rotor poles are also needed to measure the rotor position for these machines.

1 Problem Formulation

Considering the structure of the proposed stator-PM VR resolver, it is easy to obtain a rotor design with an odd-number of salient poles. The stator PM poles, however, have to be in pairs, which means the number of PM poles needs to be even.

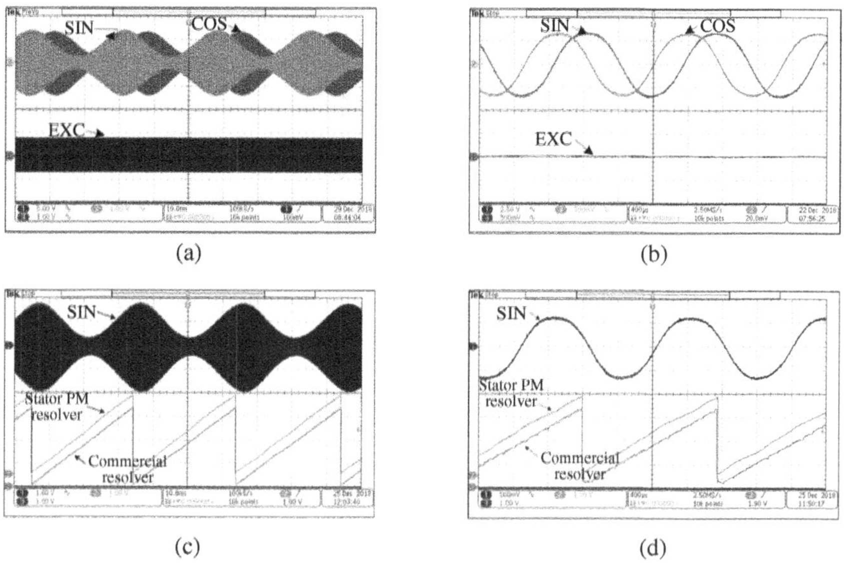

Figure 6.28 Measured waveforms of stator-PM VR resolver prototype. (a) EMF waveform with the HF excitation signal, (b) PM EMF waveform (without HF injection), (c) SIN and position signals with the HF excitation signal at 500 r/min, (d) SIN and position signals with PMs excitation at 10000 r/min.

Here we take the resolver with three rotor salient reluctance poles as an example, as shown in Figure 6.29. According to the design method of even-pole resolvers, each rotor salient reluctance pole corresponds to four stator teeth and each PM pole is expected to cover four stator teeth. However, the polarity of the third PM in the red circle cannot be defined. To solve this problem, the developed GAFMT has

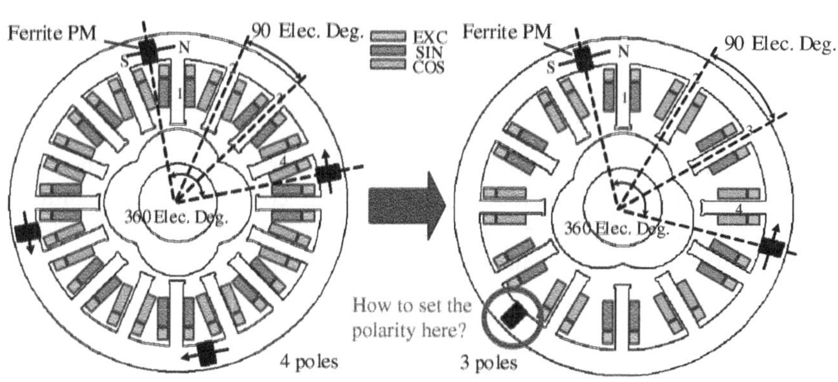

Figure 6.29 Odd-pole issue in the proposed stator-PM VR resolver.

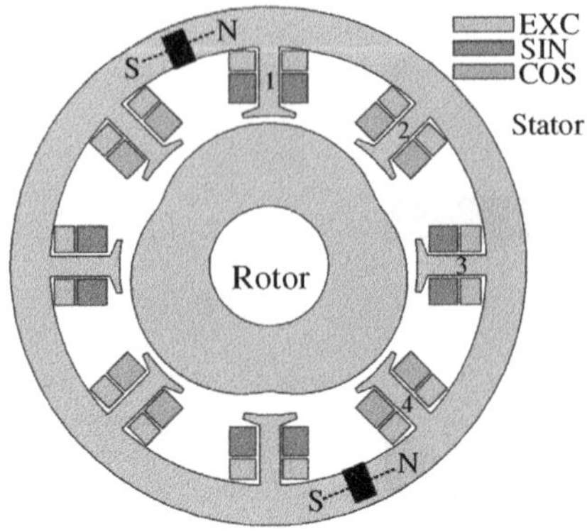

Figure 6.30 Odd-pole structure of stator-PM VR resolver.

been used to match the number of stator PM poles and rotor salient pole numbers [27, 28].

2 Solutions Based on GAFMT

Figure 6.30 shows a stator-PM VR resolver with three rotor salient reluctance poles. Two ferrite PM blocks are mounted on the stator, so a 1-pole-pair PM magnetic field can be created. According to the GAFMT, the stator winding needs to be designed as an 8-slot 4-pole fractional-slot concentrated winding.

The field modulation principle is explained in the high-speed range, where the HF excitation is turned off. The no-load motional EMF is induced by the stationary magnets and moving rotor. Taking the SIN signal as an example, if the leakage flux is ignored, the PM MMF and airgap permeance are shown below. Only the fundamental components are considered for simplicity and clarity.

$$F_{PM}(\theta_{0m}) = F_{PM}\sin(p_{PM}\theta_{0m}) \tag{6.11}$$

$$\lambda(\theta_{0m}, \theta_{rm}) = \frac{\lambda_{max} - \lambda_{min}}{2}\cos[p_r(\theta_{0m} - \theta_{rm})] + \frac{\lambda_{max} + \lambda_{min}}{2} \tag{6.12}$$

where θ_{0m} and θ_{rm} are the mechanical angles of stator tooth and rotor positions, respectively. The system includes magnetic field components with different pole-pair numbers and the result of the magnetic field modulation effect is reflected in the flux linkage. The mathematical formulas are shown as

$$\begin{aligned}\psi_{PM}(\theta_{0m}, \theta_{rm}) = {} & \frac{\lambda_{max} - \lambda_{min}}{2}F_{PM}\cos[p_r(\theta_{0m} - \theta_{rm})]\sin(p_{PM}\theta_{0m}) \\ & + \frac{\lambda_{max} + \lambda_{min}}{2}F_{PM}\sin(p_{PM}\theta_{0m})\end{aligned} \tag{6.13}$$

The motional EMF can be obtained by the time derivative of the flux linkage as

$$\begin{aligned}E_{\sin H}(\theta_{0m},\theta_{rm}) &= -\frac{d\psi_{PM}(\theta_{0m},\theta_{rm})}{dt}\\ &= -F_{PM}\omega_{rm}p_r\frac{\lambda_{\max}-\lambda_{\min}}{2}\sin[p_r(\theta_{0m}-\theta_{rm})]\sin(p_{PM}\theta_{0m})\\ &= -E_{PM}\cos\left[(p_r-p_{PM})\left(\theta_{0m}-\frac{p_r\omega_{rm}}{p_r-p_{PM}}t\right)\right]\\ &\quad + E_{PM}\cos\left[(p_r+p_{PM})\left(\theta_0-\frac{p_r\omega_{rm}}{p_r+p_{PM}}t\right)\right]\end{aligned} \tag{6.14}$$

where ω_{rm} is the angular frequency of rotor. Equation (6.14) describes a typical field modulation case, where the PM MMF is modulated by the rotor salient reluctance poles and two magnetic field components with pole-pair numbers of $(p_r - p_{PM})$ and $(p_r + p_{PM})$ are produced. Considering that the number of rotor salient reluctance poles is 3 ($p_r = 3$) and the PM pole-pair number is 1 ($p_{PM} = 1$), when the principal pole-pair number of the stator winding is 2, the first term of (6.14) can be induced in the winding.

The operating principle of the low-speed range can be explained similarly by simply replacing the PM magnetic field with the high-frequency signal. The transformer EMF of the 8-slot 3-pole (8/3) stator-PM VR resolver can be obtained as

$$\begin{cases} E_{\sin}(t) = E_{EXC}\cos(\omega_{EXC}t)\cos\left(3\omega_r t-\frac{\pi}{4}\right)+E_{PM}\cos\left(3\omega_r t-\frac{\pi}{4}\right)\\ E_{\cos}(t) = E_{EXC}\cos(\omega_{EXC}t)\cos(3\omega_r t)+E_{PM}\cos\left(3\omega_r t-\frac{\pi}{2}\right)\end{cases} \tag{6.15}$$

where $E_{EXC} = \sqrt{2}\omega_{EXC}F_{EXC}(\lambda_{\max} - \lambda_{\min})$ and $E_{PM} = \sqrt{2}\omega_r p_r F_{PM}(\lambda_{\max} - \lambda_{\min})$.

It should be noted that the SIN/COS signals in (6.15) are not orthogonal. The phase difference is not $\pi/2$ but $\pi/4$, which can be verified by the FEA results shown in Figure 6.31.

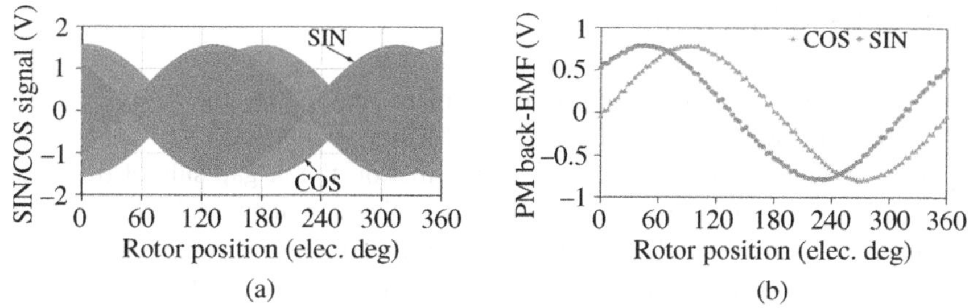

Figure 6.31 FEA results at different rotational speeds. (a) EMF waveform with HF signal injection at 500 r/min, (b) EMF waveform without HF signal injection at 10000 r/min.

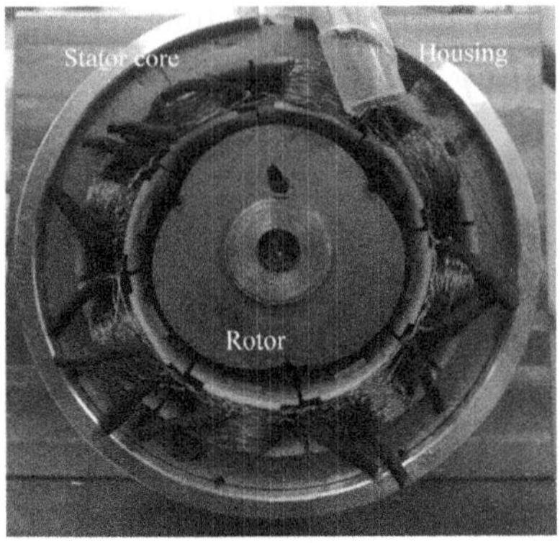

Figure 6.32 Photo of the 8/3 stator-PM VR resolver prototype.

Figure 6.32 shows the studied 8-slot 3-pole (8/3) stator-PM VR resolver. The associated decoding circuit hardware is the same as that shown in Figure 6.27. The only difference is that the SIN/COS EMFs with a $\pi/4$ phase difference are converted into a pair of orthogonal EMFs by a simple mathematical transformation.

Figure 6.33 shows the measured EMF waveforms at different rotational speeds and the rotor position signals based on the decoding of the EMF. It can be seen that the results are consistent basically compared to the commercial VR resolver. The position signal of the prototype has a relatively large error compared with that of the commercial resolver at high speeds. It is caused by the rotor-off-center of the stator-PM resolver prototype at high speeds due to the defects of the experimental platform in the manufacturing process.

The concept of the stator-PM VR resolver can be extended to the field of linear position measurement, leading to the non-contact primary-PM linear VR resolver. See reference [29] for more details.

6.4 FRPM Machine

FRPM machines are one of the three main stator-PM machines [30, 31]. Figure 6.34 illustrates the basic structure of FRPM machines. Two arc PMs with opposite magnetization directions are mounted on the airgap side of the stator tooth tops. According to the modes of magnetization of two adjacent PMs on each stator tooth, FRPM machines can be classified into two types, namely when the neighboring magnets belonging to different stator teeth have different magnetization modes, as shown in Figure 6.34a (called Mode 1), or when they have the same magnetization mode, as shown in Figure 6.34b (called Mode 2).

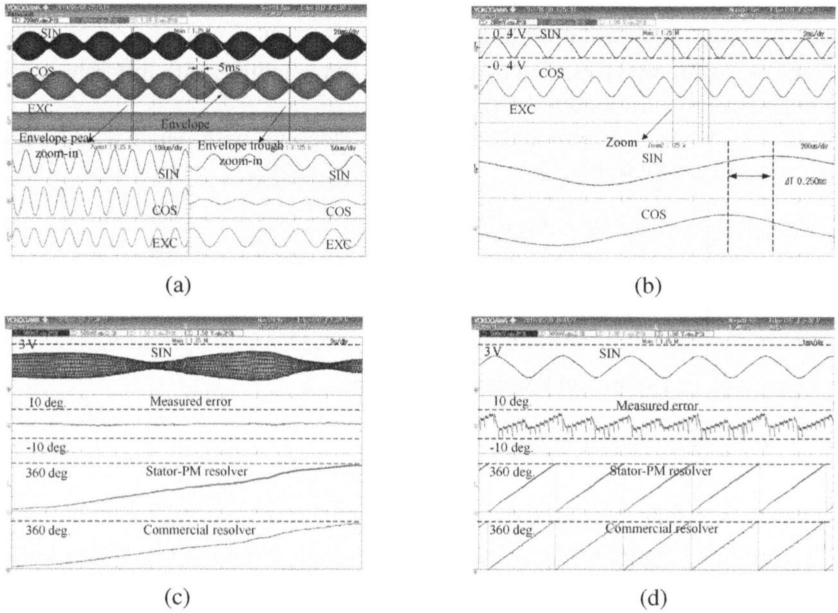

Figure 6.33 Measured waveforms of the 8/3 stator-PM VR resolver. (a) EMF waveform with HF excitation, (b) PM EMF waveform (without HF excitation), (c) SIN and position signals with HF excitation at 3 r/min, (d) SIN and position signals with PM excitation at 10 000 r/min.

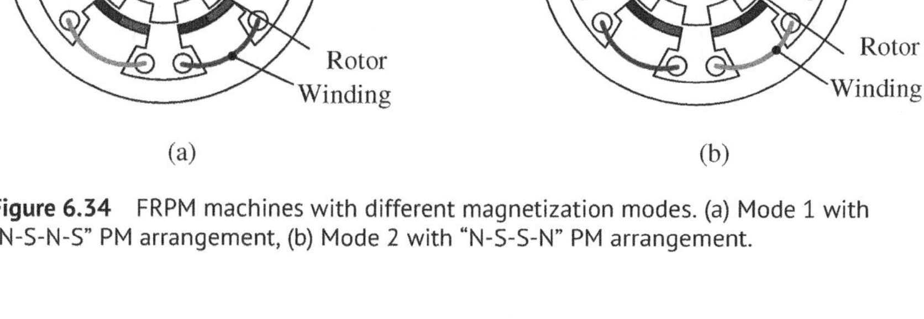

Figure 6.34 FRPM machines with different magnetization modes. (a) Mode 1 with "N-S-N-S" PM arrangement, (b) Mode 2 with "N-S-S-N" PM arrangement.

6.4.1 Operating Principle

The operating principle of FRPM machines can be explained by the GAFMT. The source excitation MMF created by PMs for FRPM machines with different modes of magnetization are shown in Figure 6.35.

The PM MMF distribution of the FRPM machine with Mode 1 shown in Figure 6.35a can be written as

$$F_1 = \begin{cases} 0, & 0 < \theta \le \theta_s/2 \\ F_{pm}, & \theta_s/2 < \theta \le \theta_s/2 + \theta_{pm} \\ -F_{pm}, & \theta_s/2 + \theta_{pm} < \theta \le \tau_s - \theta_s/2 \\ 0, \tau_s & -\theta_s/2 < \theta \le \tau_s. \end{cases} \tag{6.16}$$

Expanding (6.16) into its Fourier series yields

$$F_1 = \sum_{n_1=1}^{\infty} F_{n1} \sin(n_1 N_s \theta), \quad n_1 = 1, 2, 3 \ldots \tag{6.17}$$

where $F_{n1} = \frac{2F_{pm}}{n_1 \pi}[1 - \cos(n_1 N_s \theta_{pm})]$.

According to Figure 6.35b, when Mode 2 is adopted, the PM MMF distribution can be written as

$$F_2 = \begin{cases} 0, & 0 < \theta \le \theta_s/2 \\ -F_{pm}, & \theta_s/2 < \theta \le \theta_s/2 + \theta_{pm} \\ F_{pm}, & \theta_s/2 + \theta_{pm} < \theta \le \tau_s - \theta_s/2 \\ 0, \tau_s & -\theta_s/2 < \theta \le \tau_s + \theta_s/2 \\ F_{pm}, & \tau_s + \theta_s/2 < \theta \le \tau_s + \theta_s/2 + \theta_{pm} \\ -F_{pm}, & \tau_s + \theta_s/2 + \theta_{pm} < \theta \le 2\tau_s - \theta_s/2 \\ 0, 2\tau_s & -\theta_s/2 < \theta \le 2\tau_s \end{cases} \tag{6.18}$$

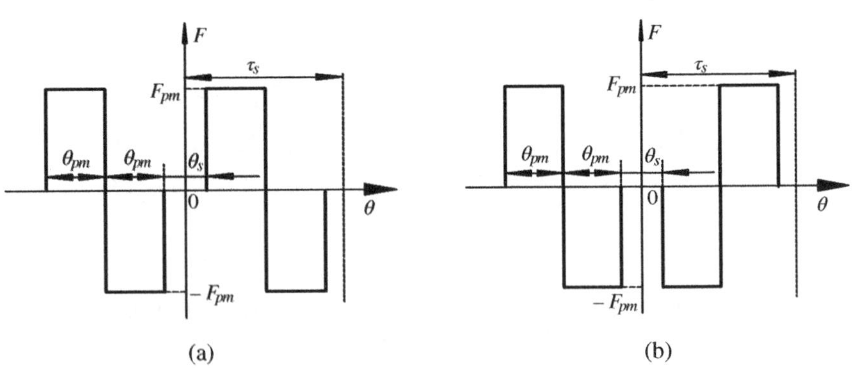

Figure 6.35 PM MMF distributions of two PRPM machines with different magnetization modes. (a) Mode 1. (b) Mode 2.

It can also be written as a Fourier series

$$F_2 = \sum_{n_2=1}^{\infty} F_{n2} \cos\left(\frac{n_2 N_s \theta}{2}\right), \quad n_2 = 1, 3, 5 \ldots \tag{6.19}$$

where $F_{n2} = \frac{4F_{pm}}{n_2\pi} \sin\left(\frac{n_2\pi}{2}\right)\left[1 - \cos\left(\frac{n_2\pi}{2} - \frac{n_2 N_s \theta_s}{2}\right)\right]$.

It can be seen from (6.17) that both odd- and even-order harmonics exist in the machine with Mode 1. However, because $\theta_{pm} = 2\pi/3N_s$ has been used in the initial design, the 3rd, 6th, 9th,… harmonics do not exist. Equation (6.19) shows that only odd-order harmonics exist in the machine with Mode 2. More details are summarized as Table 6.5.

To verify the above analysis, a 6-slot 8-pole (6/8) FRPM machine and a 12-slot 10-pole (12/10) FRPM machine are taken as examples for further analysis. Figure 6.36 shows the PM MMF distributions of the two machines calculated by 2D FEA, which are consistent with Table 6.5. The fundamental harmonic is predominant in the machine with Mode 1. In the machine with Mode 2, both the fundamental and the 3rd harmonics dominates the airgap magnetic field.

By applying the rotor salient reluctance pole modulator, the airgap magnetic permeance shown in Figure 6.37 can be expressed as

$$\Lambda(\theta, \alpha) = \begin{cases} \Lambda_2, 0 < \theta \le \frac{\tau_r - \theta_{rt}}{2} \\ \Lambda_1, \frac{\tau_r - \theta_{rt}}{2} < \theta \le \frac{\tau_r + \theta_{rt}}{2} \\ \Lambda_2, \frac{\tau_r + \theta_{rt}}{2} < \theta \le \frac{2\pi}{N_r} \end{cases} \tag{6.20}$$

Table 6.5 Harmonic orders of PM MMF in FRPM machines with different modes of magnetization.

Harmonic orders (Mode 1)	Harmonic orders (Mode 2)
$n_1 = 1, 2, 4, 5, \ldots, k, k \ne 3i\ (i = 1,2,3\ldots)$	$n_2 = 1, 3, 5, 7, \ldots, 2k-1$

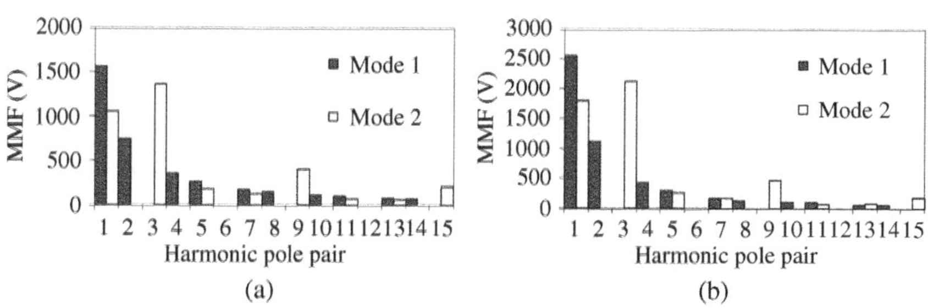

Figure 6.36 PM-MMF distributions of the 6/8 and 12/10 FRPM machines. (a) 6/8 machine, (b) 12/10 machine.

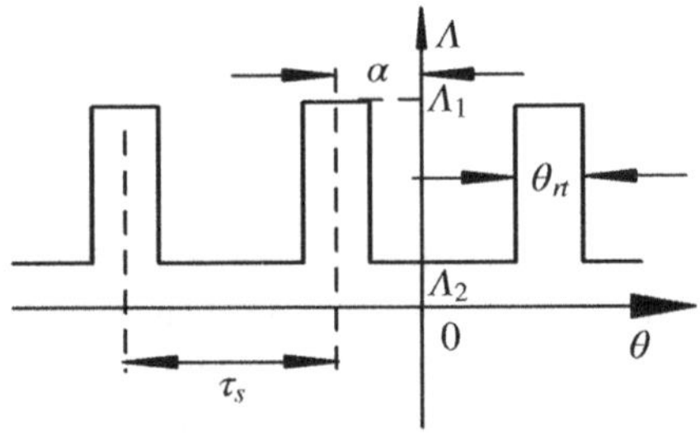

Figure 6.37 Airgap permeance distribution.

Its Fourier Series expansion is

$$\Lambda(\theta, \alpha) = \Lambda_0 + \sum_{m=1}^{\infty} \Lambda_m \cos[mN_r(\theta + \alpha)] \tag{6.21}$$

where

$$\Lambda_0 = \frac{N_r}{2\pi}\theta_{rt}\Lambda_1 \tag{6.22}$$

$$\Lambda_m = \frac{2\Lambda_1}{m\pi} \sin\left(\frac{mN_r\theta_{rt}}{2}\right). \tag{6.23}$$

Rotor salient reluctance poles modulate the PM MMF to produce a spectrum of MMF harmonic components shown in Figure 6.36. The modulation process is discussed as follows.

The no-load airgap flux density distribution can be expressed as:

$$B(\theta, \alpha) = F\Lambda(\theta, \alpha) \tag{6.24}$$

Based on (6.17), (6.21), and (6.24), the expression of the no-load airgap flux density distribution for FRPM machines with Mode 1 can be obtained as:

$$B_1(\theta, \alpha) = \sum_{n_1=1}^{\infty} \Lambda_0 F_{n1} \sin(n_1 N_s \theta) + \sum_{m_1}^{\infty}\sum_{n_1}^{\infty} \Lambda_{m1} F_{n1} \cos(n_1 N_s \theta) \cos[m_1 N_r(\theta + \alpha)] \tag{6.25}$$

It can also be written as

$$B_1(\theta, \alpha) = \sum_{n_1=1}^{\infty} \Lambda_0 F_{n1} \sin(n_1 N_s \theta) + \sum_{m_1}^{\infty}\sum_{n_1}^{\infty} \frac{\Lambda_{m1} F_{n1}}{2} \sin[(n_1 N_s \pm m_1 N_r)\theta \pm m_1 N_r \alpha] \tag{6.26}$$

Similarly, the no-load airgap flux density distribution for FRPM machines with Mode 2 can be obtained as

$$B_2(\theta, \alpha) = \sum_{n_2=1}^{\infty} \Lambda_0 F_{n2} \cos\left(\frac{n_2 N_s \theta}{2}\right) + \sum_{m_2}^{\infty}\sum_{n_2}^{\infty} \frac{\Lambda_{m2} F_{n2}}{2} \cos\left[\left(\frac{n_2 N_s \pm 2m_2 N_r}{2}\right)\theta \pm m_2 N_r \alpha\right] \tag{6.27}$$

Table 6.6 Spatial harmonics of no-load airgap flux density.

Mode 1		Mode 2	
Pole-pair number	**Rotational speed**	**Pole-pair number**	**Rotational speed**
n_1N_s	0	$n_2N_s/2$	0
$n_1N_s + m_1N_r$	$m_1N_r\omega_r/(n_1N_s + m_1N_r)$	$(n_2N_s + 2m_2N_r)/2$	$2m_2N_r\omega_r/(n_2N_s + 2m_2N_r)$
$\|n_1N_s - m_1N_r\|$	$m_1N_r\omega_r/(n_1N_s - m_1N_r)$	$\|(n_2N_s - 2m_2N_r)/2\|$	$2m_2N_r\omega_r/(n_2N_s - 2m_2N_r)$

It can be seen from (6.26) and (6.27) that no matter which magnetization mode is used, there are numerous spatial harmonics with different pole pairs and/or rotational speeds in the no-load airgap flux density waveform, which are summarized in Table 6.6. It should be noted that not all the harmonics contribute to the fundamental component of the no-load back-EMF.

In order to verify the above analysis, the 6/8 and 12/10 machines with two different modes of magnetization methods are studied by 2D FEA. As shown in Table 6.6, due to the modulation behavior of rotor salient reluctance poles, only even-order harmonics ($6n_1$ or $|6n_1 \pm 8\,m_1|$) exist in the airgap flux density distribution for FRPM machines with Mode 1, and only odd-order harmonics ($3n_2$ or $|3n_2 \pm 8\,m_2|$) exist in the airgap flux density distribution for FRPM machines with Mode 2. Only even-order harmonics ($12n_1$, $|12n_1 \pm 10\,m_1|$, $6n_2$, or $|6n_2 \pm 10\,m_2|$) exist in the airgap flux density distribution in the 12/10 FRPM machine, regardless of the mode of magnetization.

Figure 6.38 shows the airgap flux density waveforms and the corresponding harmonic spectrums of the two machines of different magnetization modes obtained by 2D FEA, which are consistent with those shown in Table 6.6.

The PM flux linkage ψ_{PM} of FRPM machines can be obtained by:

$$\psi_{PM} = r_g l_{stk} \int_0^{2\pi} k_w B(\theta, \alpha) N(\theta)\, d\theta \tag{6.28}$$

where r_g is the radius of the airgap, $N(\theta)$ is the winding function, which depends on the winding layout. For clarity, Figure 6.39 shows the winding layouts, i.e. coil connections, and winding functions of two different modes of magnetization of the 6/8 FRPM machine. For Mode 1, the two coils belonging to the same phase are connected in series in the same direction. In the 6/8 FRPM machine with Mode 2, the two coils belonging to the same phase are connected in series in the reverse direction.

Assuming that only the fundamental harmonics are considered, it is easy to find that the following equations should be met,

$$\begin{cases} |n_1N_s \pm N_r| = |jp| \\ \left|\dfrac{n_2N_s \pm 2N_r}{2}\right| = |jp| \end{cases} \tag{6.29}$$

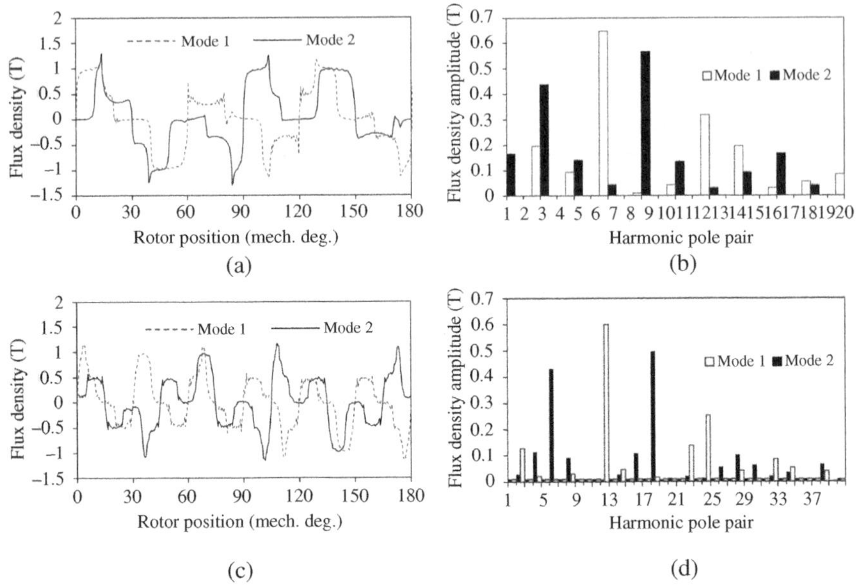

Figure 6.38 PM flux density and harmonic distribution. (a) Flux density (6/8 FRPM machine), (b) Amplitude spectrum (6/8 FRPM machine), (c) Flux density (12/10 FRPM machine), (d) Amplitude spectrum (12/10 FRPM machine).

Otherwise, Equation (6.26) would be zero.

Thus, FRPM machines can be equipped with coil connection 1 or 2, and it depends on the slot-pole combinations. Generally, there are four feasible cases: (i) Mode 1 + coil connection 1, (ii) Mode 2 + coil connection 1, (iii) Mode 1 + coil connection 2, (iv) Mode 2 + coil connection 2. The no-load back-EMF of the machine can be obtained by:

$$e(t) = -\frac{d\psi}{dt} \tag{6.30}$$

a) Machine A: Mode 1 + coil connection 1

Substitute (6.24), (6.26), and (6.27) into (6.30), the no-load EMF is

$$e_1(t) = \begin{cases} -\dfrac{r_g l_{stk} N_w N_r \omega_r}{n_1 N_s + m_1 N_r} \displaystyle\sum_{n_1=1}^{\infty} \sum_{m_1=1}^{\infty} k_{wm1} F_{pmn1} \Lambda_{m1} \cos(m_1 N_r \omega_r t), & n_1 N_s + m_1 N_r = j_1 p_1 \\ \dfrac{r_g l_{stk} N_w N_r \omega_r}{n_1 N_s - m_1 N_r} \displaystyle\sum_{n_1=1}^{\infty} \sum_{m_1=1}^{\infty} k_{wm1} F_{pmn1} \Lambda_{m1} \cos(m_1 N_r \omega_r t), & n_1 N_s - m_1 N_r = j_1 p_1 \\ 0, & \text{Others.} \end{cases} \tag{6.31}$$

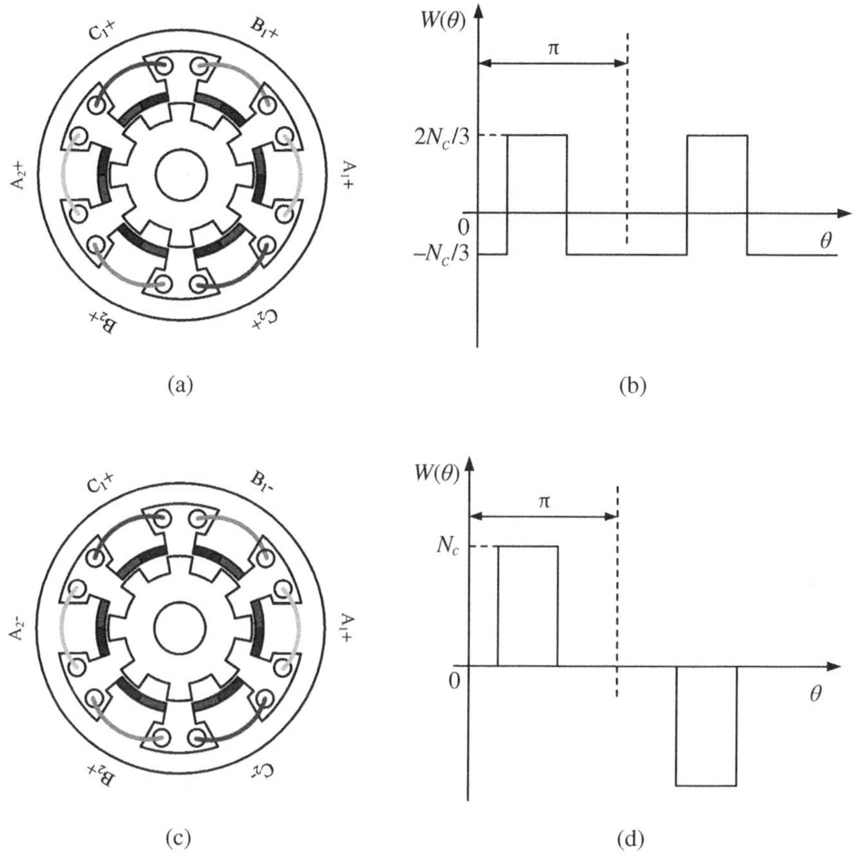

Figure 6.39 Winding layouts of the 6/8 FRPM machine. (a) Coil connection 1 for Mode 1, (b) Winding function of coil connection 1, where Nc is the number of turns per coil. (c) Coil connection 2 for Mode 2, (d) Winding function of coil connection 2.

b) Machine b: Mode 2 + coil connection 1
   Substitute (6.25), (6.26), and (6.27) into (6.30), the no-load EMF is

$$e_2(t) = \begin{cases} \dfrac{2r_g l_{stk} N_w N_r \omega_r}{n_2 N_s \pm 2m_2 N_r} \displaystyle\sum_{n_2=1}^{\infty} \sum_{m_2=1}^{\infty} k_{wm2} F_{pmn2} \Lambda_{m2} \sin(m_2 N_r \omega_r t), & \dfrac{n_2 N_s \pm 2m_2 N_r}{2} = j_2 p_2 \\ 0, & \text{Others.} \end{cases} \quad (6.32)$$

c) Machine c: Mode 1 + coil connection 2
Substitute (6.24), (6.26), and (6.28) into (6.30), the no-load EMF is

$$e_3(t) = \begin{cases} \dfrac{r_g l_{stk} N_w N_r \omega_r}{n_3 N_s \pm m_3 N_r} \sum_{n_3=1}^{\infty} \sum_{m_3=1}^{\infty} k_{wm3} F_{pmn3} \Lambda_{m3} \sin(m_3 N_r \omega_r t), & n_3 N_s \pm m_3 N_r = j_3 p_3 \\ 0, & \text{Others.} \end{cases} \tag{6.33}$$

d) Machine d: Mode 2 + coil connection 2
Substitute (6.24), (6.26), and (6.28) into (6.30), the no-load EMF is

$$e_4(t) = \begin{cases} \dfrac{2 r_g l_{stk} N_w N_r \omega_r}{n_4 N_s + 2 m_4 N_r} \sum_{n_4=1}^{\infty} \sum_{m_4=1}^{\infty} k_{wm4} F_{pmn4} \Lambda_{m4} \cos(m_4 N_r \omega_r t), & \dfrac{n_4 N_s + 2 m_4 N_r}{2} = j_4 p_4 \\ -\dfrac{2 r_g l_{stk} N_w N_r \omega_r}{n_4 N_s - 2 m_4 N_r} \sum_{n_4=1}^{\infty} \sum_{m_4=1}^{\infty} k_{wm4} F_{pmn4} \Lambda_{m4} \cos(m_4 N_r \omega_r t), & \dfrac{n_4 N_s - m_4 N_r}{2} = j_4 p_4 \\ 0, & \text{Others.} \end{cases} \tag{6.34}$$

It can be seen from (6.31)–(6.34) that the fundamental no-load EMF of an arbitrary FRPM machine can always be expressed as

$$E_1 = \frac{r_g l_{stk} N_w N_r \omega_r}{jp} \sum_{n=1}^{\infty} k_{w1} F_{pmn} \Lambda_1 \tag{6.35}$$

6.4.2 Analysis of Open-Circuit Back-EMF Based on GAFMT

As can be seen from (6.31)–(6.34), not all the flux density harmonics listed in Table 6.6 contribute to the fundamental component of the no-load back-EMF. The armature winding works as a magnetic field filter that selects specific airgap flux density harmonics to produce the fundamental component of the no-load back-EMF based on their harmonic pole pairs. See Table 6.7 for more details.

According to (6.29), the preferred coil connections for the 6/8 FRPM machine with Mode 1 and Mode 2 are coil connection 1 and coil connection 2, respectively. The preferred coil connections for the 12/10 FRPM machine with Mode 1 and Mode 2 are coil connection 2 and coil connection 1, respectively. The selected airgap flux density harmonics of these two machines can be obtained according to

Table 6.7 Selected harmonic orders of flux density.

Machine type	Machine a	Machine b	Machine c	Machine d
Selected harmonic flux density	$\lvert n_1 N_s \pm m_1 N_r \rvert$ and $\lvert n_1 N_s \pm m_1 N_r \rvert = j_1 p_1$	$\lvert (n_2 N_s \pm 2 m_2 N_r)/2 \rvert$ and $\lvert (n_2 N_s \pm 2 m_2 N_r)/2 \rvert = j_2 p_2$	$\lvert n_3 N_s \pm m_3 N_r \rvert$ and $\lvert n_3 N_s \pm m_3 N_r \rvert = j_3 p_3$	$\lvert (n_4 N_s \pm 2 m_4 N_r)/2 \rvert$ and $\lvert (n_4 N_s \pm 2 m_4 N_r)/2 \rvert = j_4 p_4$

Table 6.8 Flux density harmonics contributing to the no-load back-EMF.

Machine type		No-load back-EMF time harmonics (Mode 1)					No-load back-EMF time harmonics (Mode 2)				
		1st	2nd	3rd	4th	5th	1st	2nd	3rd	4th	5th
8/6	Harmonic pole pairs	*2,14*	***10***,14	6,***18***	2,10	2,10	***1***,*5*,7 11,13 17,19	***1***,5,***7*** 11,13 17,19	3,9 ***15***	***1***,*5*,7 11,13 17,19	***1***,5,7, 11,13, 17,19
	Back-EMF amplitude (V)	124.0	1.2	0.8	1.3	2.3	239.9	3.4	12.9	2.1	4.4
12/10	Harmonic pole pairs	**2**,14,***22*** 34,38	–	6,18 30	–	2,10 26,38	***4***,20 ***28***	–	12,36	–	***4***,20 28
	Back-EMF amplitude (V)	105.2	0	0	0	8.1	54.5	0	0	0	2.9

Table 6.7, and they are shown in Table 6.8, where numbers in the red italics, blue italics, and underlined green denote flux density spatial harmonics generated by the fundamental, 3rd and higher order MMF harmonics. For example, in the 6/8 machine with Mode 2, the fundamental component of the no-load back-EMF is contributed by the 1-, 5-, 7-, 11-, 13-, 17-, and 19-pole-pair flux density spatial harmonics. The 1- and 5-pole-pair spatial harmonics are generated by the 3rd harmonic and the fundamental PM MMF, respectively. Though the 2nd time harmonic of the no-load back-EMF is contributed by the above flux density harmonics, the 1- and 7-pole-pair airgap flux density harmonics are generated by the 3rd-order and higher order PM MMF harmonics, respectively.

Some conclusions can be drawn from Table 6.8 as follows:

1. Due to the frequency selection capability of armature windings, airgap flux density spatial harmonics in both machines have been filtered more or less when compared with the results shown in Figure 6.38, and the selected ones are listed in Table 6.8.
2. The filtering effect varies with different pole-slot combinations and different modes of magnetization. In the 6/8 machine, more flux density harmonics are filtered for Mode 1 than Mode 2, leading to a reduced no-load back-EMF. The 12/10 machine shows the opposite.
3. In the 12/10 machine, the filtering effect eliminates even-order harmonics of the no-load back-EMF, contributing to the so-called winding complementarity. By contrast, even-order harmonics still persist in the 6/8 machine.
4. The orders of PM MMF harmonics contributing to the no-load back-EMF can be obtained according to Table 6.7. These PM harmonic orders are determined by the modes of magnetization and coil connections. Similarly, taking the 6/8 and 12/10 machine as examples, the fundamental component contributing to the 5-pole-pair MMF harmonics influences the fundamental component of the no-load back-EMF differently for the two machines, as shown in Figure 6.40. It is shown that the no-load back-EMF of the 6/8 machine with Mode 1 is mainly contributed by the 1-pole-pair PM MMF, while with Mode 2, the 1- and 3-pole-pair PM MMF harmonics contribute to the back-EMF. Similar conclusions can also be drawn for the 12/10 machine.

6.5 FSPM Machine with Full-Pitch Windings

Compared with traditional PM synchronous machines, FSPM machines boast the advantages of robust rotor structure suitable for high-speed operation and ease of the PM cooling [32]. In [33], the pole-slot combinations for FSPM machines have been comprehensively studied and the 12-slot 14-pole (12/14) FSPM shows

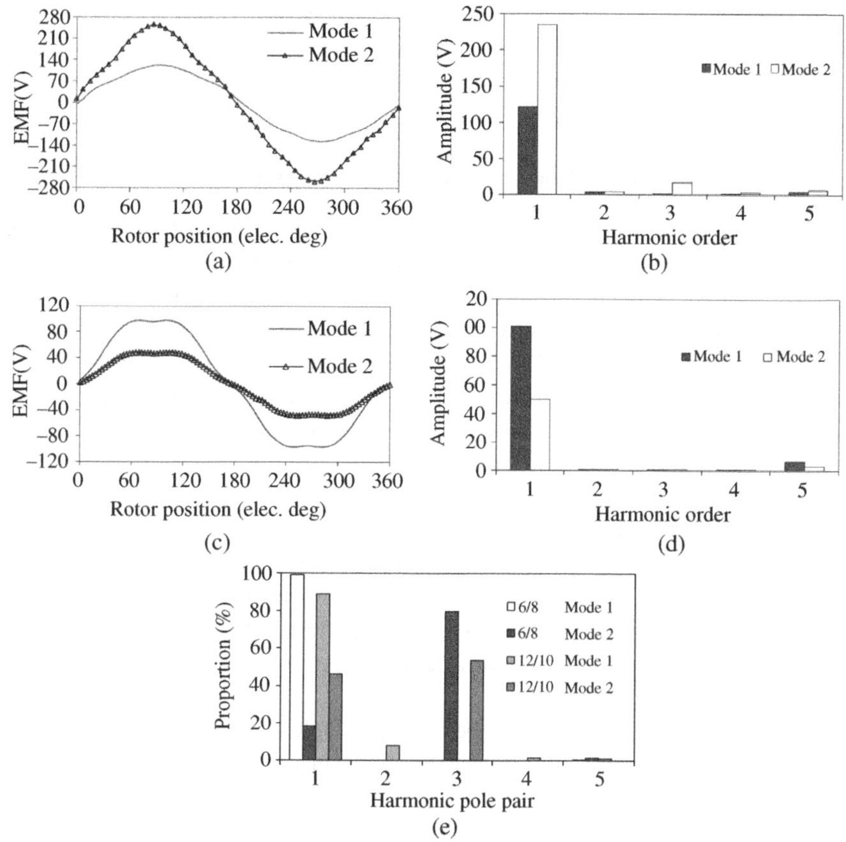

Figure 6.40 No-load back-EMFs and contributing spatial harmonics in the 6/8 and 12/10 machines. (a) No-load back-EMF of the 6/8 machine, (b) Amplitude spectrum of (a), (c) No-load back-EMF of the 12/10 machine, (d) Amplitude spectrum of (c), (e) Contributions of airgap flux density spatial harmonics to the fundamental no-load back-EMF.

higher torque capability. To improve the fault-tolerance, the FSPM machine with E-shaped stator cores has been presented [4] and multi-phase versions have also been proposed to decrease the inverter power rating per phase [34]. The comparative study between optimized E-core and C-core linear FSPM machines shows that the 12/13 C-core machine delivers 18.4% higher thrust force and uses 41% less magnets than the conventional counterpart for the same copper loss [35]. Additionally, in order to realize the online regulation of no-load airgap flux density, the hybrid excitation flux-switching machine [36] and HTS flux-switching machine [37] have been proposed for EVs and power generation applications, respectively. All these flux-switching machines use non-overlapping fractional-slot concentrated windings. Based on the GAFMT, a new 12-slot 7-pole (12/7) FSPM machine with full-pitch windings is proposed and demonstrated to be feasible by changing the filter of the three basic elements [38].

6.5.1 Machine Construction and Operating Principle

The 12/7 FSPM machine with full-pitch windings is shown in Figure 6.41. The stator core is composed of 12 U-shaped laminated modules, between which the PMs are sandwiched. The magnets are magnetized alternately in the tangential direction. The rotor core has seven salient reluctance poles. The key difference compared to the traditional FSPM machine lies in the armature winding [39].

To visualize the operating principle of the machine, 2D FEA models have been used to obtain the PM magnetic field distribution. As shown in Figure 6.42a, when one rotor tooth aligns with stator tooth 1, the flux linkage of phase A reaches the positive maximum. When the rotor rotates by a 1/2 tooth pitch and the middle line of one of the rotor slots aligns with stator tooth 1 as shown in Figure 6.42b, the flux linkage of phase A reaches the negative maximum. In this way, the polarity of phase A flux linkage can be changed. It is shown that the no-load PM field distribution shown in Figure 6.42 has a similar pattern to that of a 2-pole PM synchronous machine, so intuitively the armature winding can be designed to have a principal pole pair number of 1.

In order to further illustrate the operating principle from the perspective of the GAFMT, the no-load airgap flux density distribution and its harmonic analysis are discussed here.

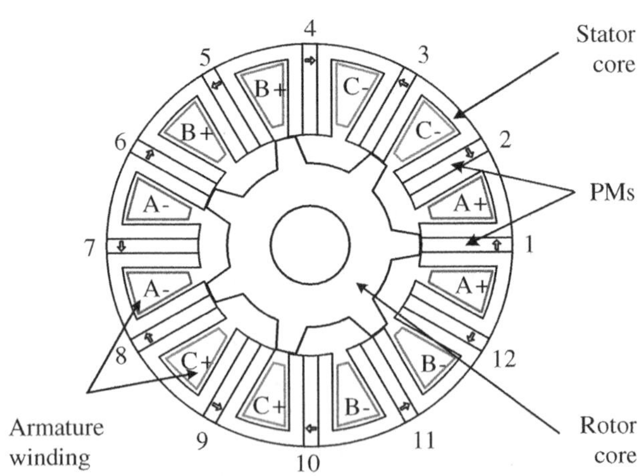

Figure 6.41 Cross-sectional view of the 12/7 FSPM machine with full-pitch windings.

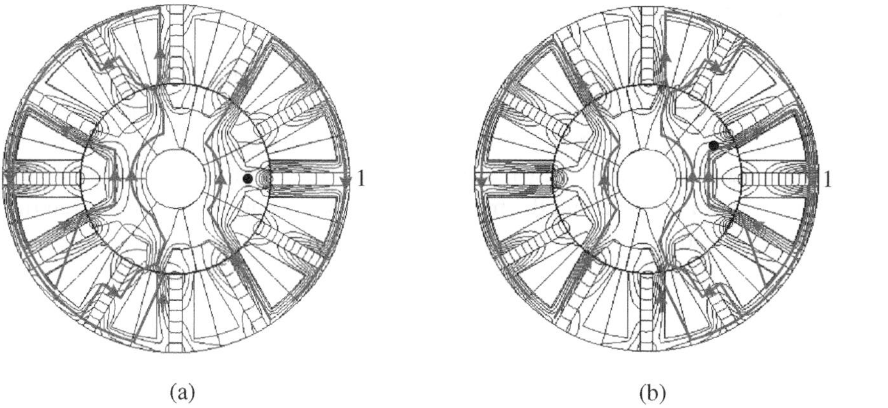

Figure 6.42 Magnetic field distribution. (a) $\theta_e = 0°$, (b) $\theta_e = 180°$.

The no-load airgap flux density waveform and the corresponding harmonic spectrum are shown in Figure 6.43. It can be seen that the flux density waveform contains a series of spatial harmonics, among which amplitudes of the 1-, 6-, 11-, 13-, 18-, and 25-pole-pair components are relatively higher. The 6-pole-pair harmonic is directly produced by the 6-pole-pair PM array. The 18-pole-pair harmonic is produced by the synchronous modulation of the 24 stator teeth to the 6-pole-pair PM magnetic field. Therefore, the two spatial harmonics are stationary. The 1-, 11-, 13-, and 25-pole-pair harmonics are generated by the asynchronous modulation of rotor salient reluctance poles to the 6- and 18-pole-pair stationary harmonics. When the rotor rotates by one rotor pole pitch, the four rotating harmonics will travel by 360 electrical degrees. The mechanical speed relationship between the rotor and these harmonics can be expressed as

$$G_r = \frac{\omega_{fhm}}{\omega_{rm}} = \text{sign}(p_s - N_r)\frac{N_r}{|p_s \pm N_r|} = \pm\frac{N_r}{p_{fh}} \tag{6.36}$$

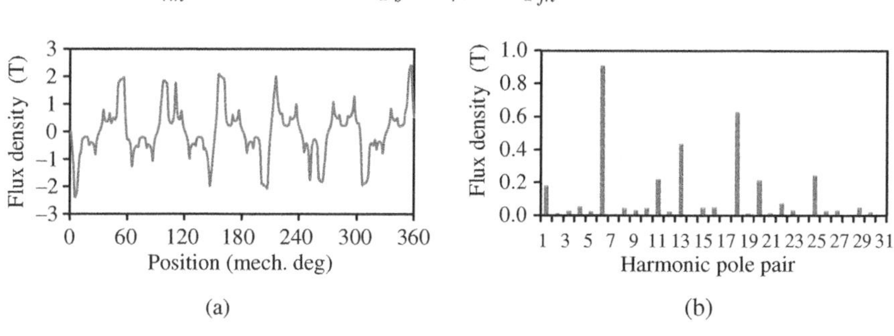

Figure 6.43 No-load airgap flux density of the 12/7 FSPM machine with full-pitch windings. (a) Waveform, (b) Amplitude spectrum of (a).

where G_r is the defined as the magnetic gear ratio, ω_{rm}, and ω_{fhm} are the mechanical rotational speed of the rotor and the airgap magnetic field harmonics, respectively. p_s is the pole-pair number of the stationary airgap magnetic field harmonic. In the 12/7 FSPM machine, $p_s = 6$ or 18. N_r is the number of rotor salient reluctance poles. Sign($p_s - N_r$) shows the rotational direction and p_{fh} is the pole-pair number of rotating airgap flux density harmonics. The 1-pole-pair harmonic has the maximum rotational speed and the minimum pole-pair number, which is seven times the rotor speed.

The slot pitch angles with respective to these four rotating harmonics can be calculated:

$$\alpha = p_{fh} \frac{2\pi}{n_s} \tag{6.37}$$

where n_s is the number of stator teeth, and $n_s = 12$ for the 12/7 FSPM machine. Therefore, the shot pitch angles of the 1-, 11-, 13-, and 25-pole-pair rotating harmonics are $\alpha = 30°$ or $330°$. Considering the rotational directions of the four rotating harmonics, the corresponding slot star diagram can be drawn, as shown in Figure 6.44. It can be seen that though the pole-pair numbers of these harmonics are different, the corresponding slot pitch angles and slot star diagrams are exactly the same. Both the full-pitch winding and fractional-slot concentrated windings can be used in the 12/7 FSPM machine. The winding factor of the full-pitch winding is much higher, contributing to an improved output torque.

6.5.2 Influence of Key Geometric Parameters

The machine performance is sensitive to some key geometric parameters for the FSPM machine with full-pitch windings, as defined in Figure 6.45. In the design

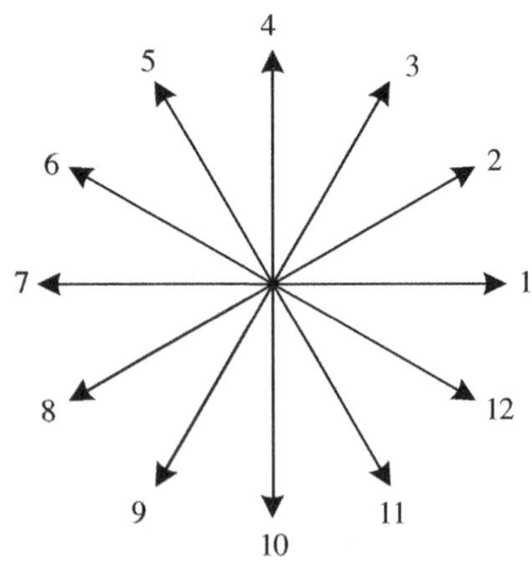

Figure 6.44 Slot star diagram of the 12/7 FSPM machine with full-pitch windings.

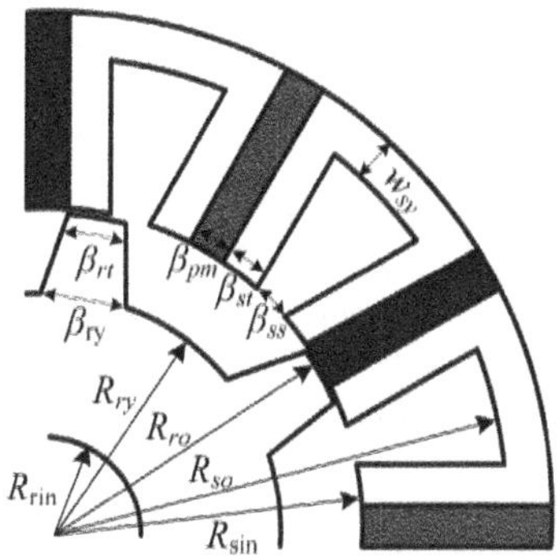

Figure 6.45 Definitions of key geometric parameters of the 12/7 FSPM machine with full-pitch windings.

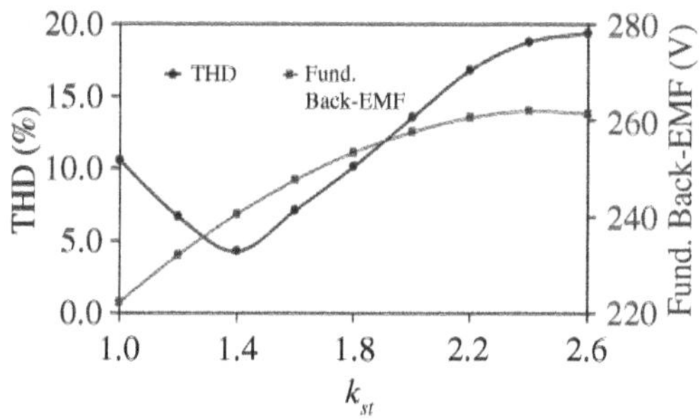

Figure 6.46 No-load back-EMF vs. k_{st}.

process, the stator slot opening arc β_{ss}, the stator tooth pole arc β_{st}, and the PM width arc β_{PM} are kept constant at a quarter of the stator pole pitch arc $\beta_{\tau p}$, i.e. $\beta_{ss} = \beta_{st} = \beta_{PM} = \beta_{\tau p}/4$. Additionally, the stator outer diameter and stack length of machine are fixed at 64 mm and 75 mm, respectively.

Firstly, we define the ratio of the rotor tooth top arc β_{rt} to the 1/4 stator tooth width as:

$$k_{st} = \frac{\beta_{rt}}{\beta_{\tau p}/4} \tag{6.38}$$

The variation of no-load back-EMF, cogging torque, and output torque of the 12/7 FSPM machine are shown in Figures 6.46 and 6.47. When k_{st} equals 2.4 and 2.2, the fundamental component of the no-load back-EMF and average torque reach the maximum, respectively. When k_{st} equals 2.0 and 2.4, the torque ripple defined as the peak-to-peak value of torque, and cogging torque reach the minimum, respectively, in spite of the relatively high THD of no-load back-EMF.

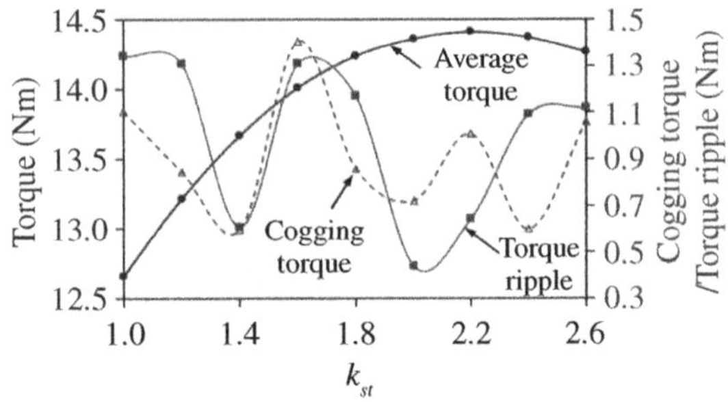

Figure 6.47 Torque performance vs. k_{st}.

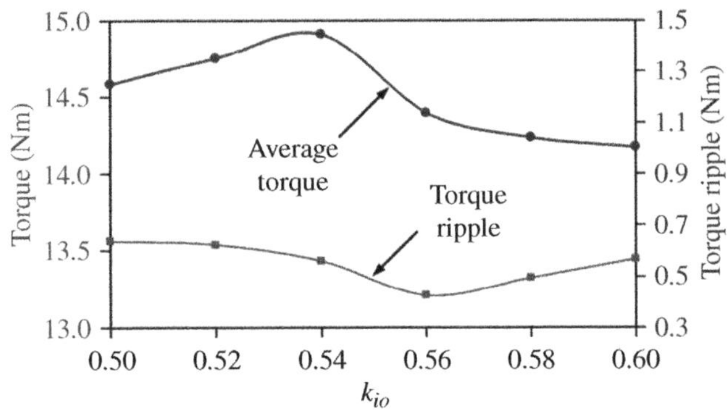

Figure 6.48 Torque performance vs. k_{io}.

Considering all these aspects, $k_{st} = 2.0$ is adopted for better trade-off in this stage. Thus, the output torque is around the maximum value, and the torque ripple reaches the minimum. Meanwhile, the cogging torque, THD, and amplitude of the fundamental component of the no-load back-EMF are all in acceptable ranges.

Secondly, we define the split ratio of machine as the ratio between the inner and outer radius of the stator:

$$k_{io} = R_{sin}/R_{so} \tag{6.39}$$

Figure 6.48 shows the variations of cogging torque, output torque, and torque ripple with k_{io}. It is worth noting that the slot area will change with k_{io}. To conduct a relatively fair comparison, the slot filling factor and the copper loss remain unchanged. Because $\beta_{ss} = \beta_{st} = \beta PM$ during the change of k_{io}, the stator slot area can be calculated based on the value of k_{io}. Then the winding turns per phase and the rated current values can be obtained by using the same wire gauge, as listed in Table 6.9.

Table 6.9 Variations of key parameters versus k_{io}.

k_{io}	S_{slot} (mm^2)	S_{PM} (mm^2)	N (Turns)	I_{rms} (A)	T_{avg} (Nm)	T_{rip} (Nm)	T_{rip}/T_{avg}O (%)	T_{avg}/S_{PM} (Nm/m^2)
0.5	312.34	1608	332	3.51	14.58	0.636	4.36	9.07
0.52	293.37	1605	312	3.62	14.75	0.623	4.22	9.19
0.54	274.25	1597	292	3.74	**14.9**	0.559	3.74	**9.33**
0.56	257.26	1585	272	3.86	14.4	**0.425**	2.95	9.08
0.58	239.27	1567	252	4.01	14.24	0.494	3.47	9.09

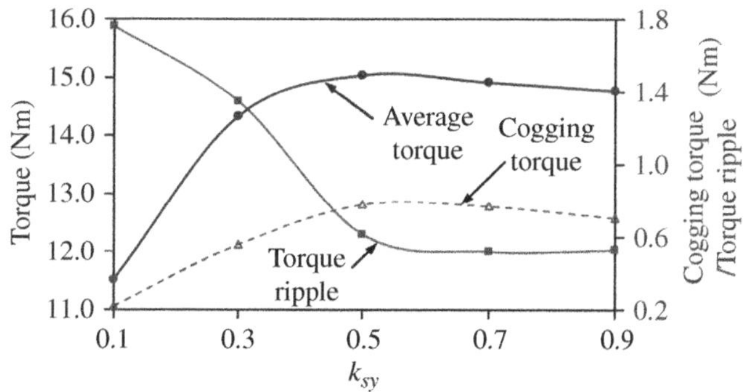

Figure 6.49 Torque performance vs. k_{sy}.

Figure 6.48 shows that the output torque reaches the maximum and the torque ripple is around the minimum when k_{io} is 0.54. The average torque per PM area T_{avg}/S_{PM} has also been calculated. As shown in the last column of Table 6.9, T_{avg}/S_{PM} reaches the maximum when k_{io} is 0.54, which is selected for the design.

The ratio of the rotor root width to the stator pole pitch is defined as:

$$k_{sy} = \beta_{ry}/\beta_{\tau p} \tag{6.40}$$

Figure 6.49 shows the cogging torque, output torque, and torque ripple versus k_{sy}. It can be seen that when k_{sy} equals 0.5, the output torque reaches the maximum, and the cogging torque reaches the maximum at the same time. Finally, $k_{sy} = 0.7$ is chosen for the design.

Figure 6.50 shows the torque characteristic versus the rotor tooth height h_{rt}. It can be observed that the output torque monotonically increases with h_{rt}. When h_{rt} is higher than 9 mm, the increase of average output torque becomes insignificant. Additionally, when $h_{rt} = 9$ mm, the torque ripple and cogging torque are relatively small. Considering different aspects of the torque performance and the mechanical construction, the height of the rotor tooth is set to 9 mm.

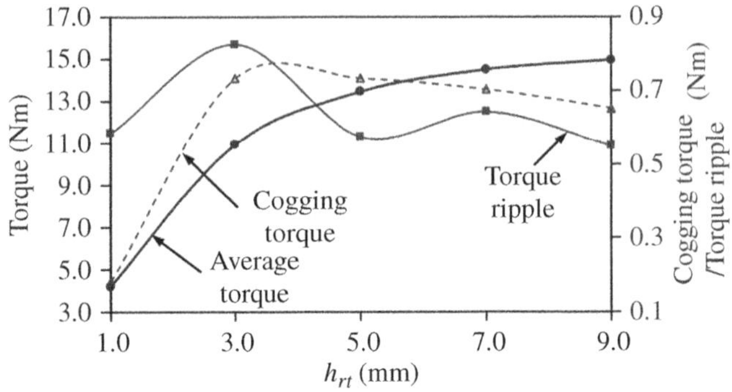

Figure 6.50 Torque performance vs. h_{rt}.

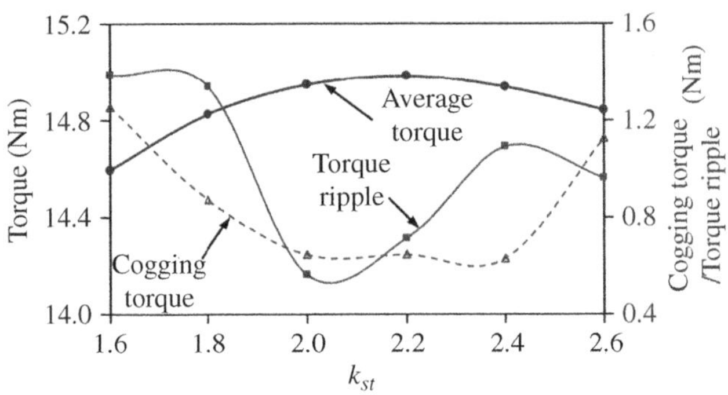

Figure 6.51 Torque performance vs. k_{st}.

Because the cogging torque of the 12/7 FSPM machine is extremely sensitive to k_{st}, and all the other ratios, i.e. k_{io}, k_{sy}, and h_{rt}, have been optimized, it is necessary to study the influence of k_{st} again. After k_{io}, k_{sy}, and h_{rt} are set to their optimal values, the cogging torque, output torque, and torque ripple are shown in Figure 6.51 when k_{st} varies in the range of 1.6–2.6. It can be seen that k_{st} = 2.0 is still the optimal point.

Based on the aforementioned parametric study results, the optimal values of key geometric parameters of the 12/7 FSPM machine are obtained as tabulated in Table 6.10.

6.5.3 Comparative Study

To further validate the merits of the full-pitch winding employed in this machine, the electromagnetic performance of the new FSPM machine with full-pitched windings is studied and compared with that of the fractional-slot

Table 6.10 Optimal values of key geometric parameters of the 12/7 FSPM machine.

Parameters	Values	Parameters	Values
Stator yoke width, w_{sy} (mm)	4.6	Rotor outer radius, R_{ro} (mm)	34.26
Stator slot opening arc, β_{ss} (deg)	7.5	Rotor inner radius, R_{rin} (mm)	11
Stator tooth pole arc, β_{st} (deg)	7.5	Rotor tooth pole arc, β_{rt} (deg)	15
Number of turns per coil, N_{tpc}	73	Rotor tooth root pole arc, β_{ry} (deg)	21
Coils in series per phase, N_{scpp}	4	Airgap length, g (mm)	0.3
PM width pole arc, β_{PM} (deg)	7.5	Rated current, I_{rms} (A)	3.74

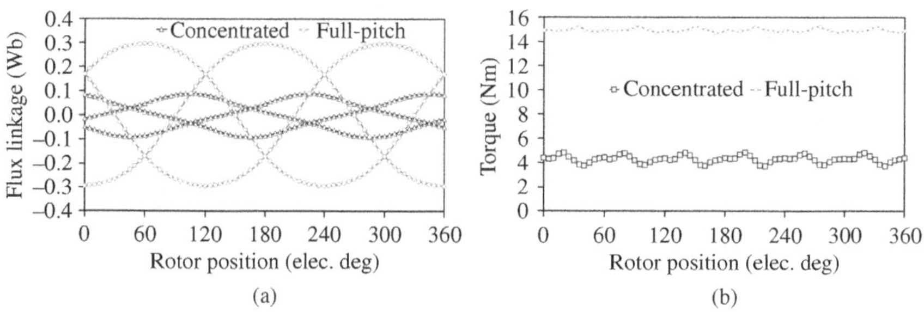

Figure 6.52 Comparison of different winding configurations. (a) Flux linkage waveforms, (b) Torque waveforms.

concentrated winding counterpart. Both 12/7 FSPM machines with different winding configurations were optimized to realize a fair comparison.

Figure 6.52a shows the PM flux linkage waveforms with different winding configurations. It can be seen that the flux linkage amplitude of the 12/7 FSPM machine with full-pitch windings is about 0.3 Wb, which is 3.2 times that with the fractional-slot concentrated windings. Figure 6.52b compares the output torque. The average torques of the FSPM machine with full-pitch windings and fractional-slot concentrated windings are 14.9 Nm and 4.38 Nm, respectively. It can be concluded that the torque output can be increased more than 3.4 times by replacing the concentrated winding with the full-pitch winding, which is in line with the analysis of no-load back-EMF.

6.6 Rotor-PM FSPM Machine

As mentioned in Section 6.5, FSPM machines have advantages of simple rotor structure, ease of PMs cooling, high torque/power density, high efficiency, and so on [40–42]. Because both the PMs and armature windings are placed on the same stator, spatial competition occurs unavoidably. Consequently, the further improvement of torque density and overload capability of this machine type is limited [9], [43]. In addition, armature winding coils wound around PMs increase the risk of demagnetization due to the high temperature [44]. Based on the GAFMT, a new rotor-PM FSPM machine topology has been proposed to mainly address the spatial competition between magnets and armature windings by moving PMs from the stator side to rotor side.

6.6.1 Machine Construction and Operating Principle

The rotor-PM FSPM machine is derived from the FSPM machine, as shown in Figure 6.53, by moving the PMs of a three-phase 12/10 FSPM machine from the stator to the rotor. A rotor module consists of two rotor salient teeth and one PM block, as shown in Figure 6.53b. The whole rotor of the rotor-PM FSPM machine is composed of 10 rotor modules fixed on a non-magnetic support by clamping slots.

Adjacent U-shaped stator core modules of FSPM machines are connected together to obtain 12 stator teeth, termed as armature teeth in this section. The fractional-slot concentrated winding is used and wound on armature teeth, as shown in Figure 6.53c. In order to provide a flowing path for the PM flux and achieve the flux switching action when the rotor rotates, a fault-tolerant tooth is added between two adjacent armature teeth. As a result, a 3-phase 24-slot 10-pole (24/10) rotor-PM FSPM machine is obtained, as illustrated in Figure 6.54.

Fault-tolerant teeth separate adjacent armature winding coils, reducing the mutual inductances between coils and the fault-tolerance is thus greatly improved. It should be mentioned that PMs in the presented rotor-PM FSPM machine example are tangentially magnetized in the same direction, which

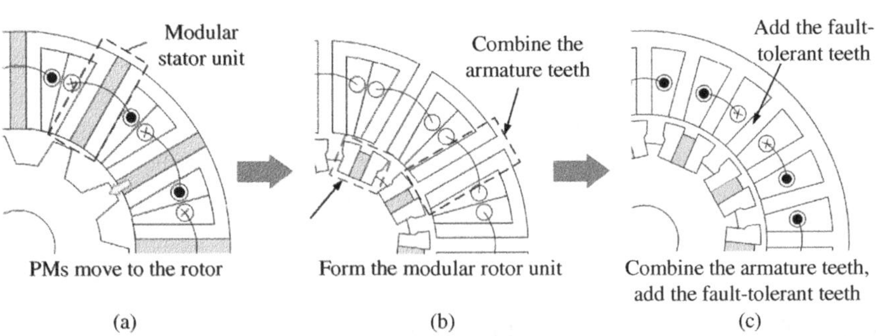

Figure 6.53 Derivation of the rotor-PM FSPM machine from a 12/10 FSPM machine. (a) Moving PMs from stator to rotor, (b) Creating the rotor module, (c) Combining armature teeth and adding fault-tolerant teeth.

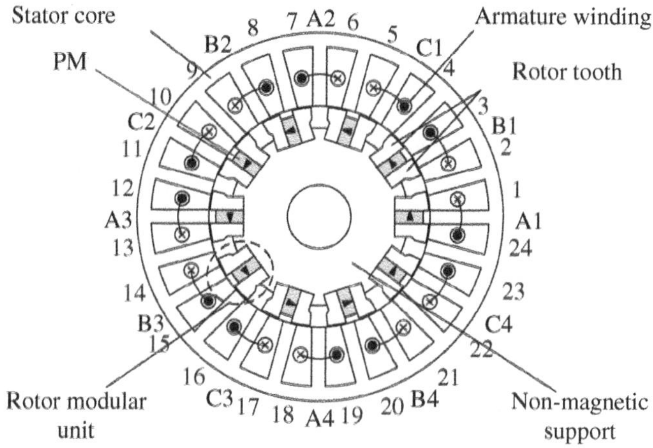

Figure 6.54 Cross-sectional view of the new three-phase 24/10 rotor-PM FSPM machine.

is different from the 12/10 FSPM machine and traditional PM synchronous machines.

The flux switching principle of the rotor-PM FSPM machine is illustrated in Figure 6.55. As shown in Figure 6.55a, when the rotor tooth R1 is aligned with the armature tooth A1, the rotor position is defined as $\theta_r = 0°$. The PM flux passes through armature tooth A1 and enters the rotor tooth through the airgap. At this moment, the amplitude of the PM flux linkage linked by armature coil A1 reaches the positive maximum. When the rotor rotates by 9°. counterclockwise, namely $\theta_r = 9°$, as shown in Figure 6.55b, the rotor PM is aligned with the stator tooth A1, and the effective flux linked by armature winding A1 is 0.

Then when the rotor rotates by 9° counterclockwise again, namely $\theta_r = 18°$, as shown in Figure 6.55c, the rotor tooth R2 is aligned with the armature tooth A1, and the PM flux generated by PMs passes the stator armature teeth from rotor teeth through the airgap and is linked by armature coil A1. At this moment, the amplitude of the PM flux linkage linked by the armature winding A1 reaches the negative maximum. When the rotor position is $\theta_r = 27°$, as shown in Figure 6.55d, the middle line of the gap between adjacent rotor modules is aligned with the stator tooth A1 and the effective flux linked by the armature coil A1 becomes 0.

Because there are 10 rotor modules in the 24/10 rotor-PM FSPM machine, when the rotor continues to rotate by 9° and arrives at $\theta_r = 36°$, the magnetic circuit of the machine repeats that at $\theta_r = 0°$. With the continuous rotation of rotor, the PM flux linkage linked by armature coil A1 will alternate. In one electrical cycle shown in Figure 6.56, the PM flux linkage linked by armature coil A1 changes from the maximum to the minimum and then to the maximum again, and the

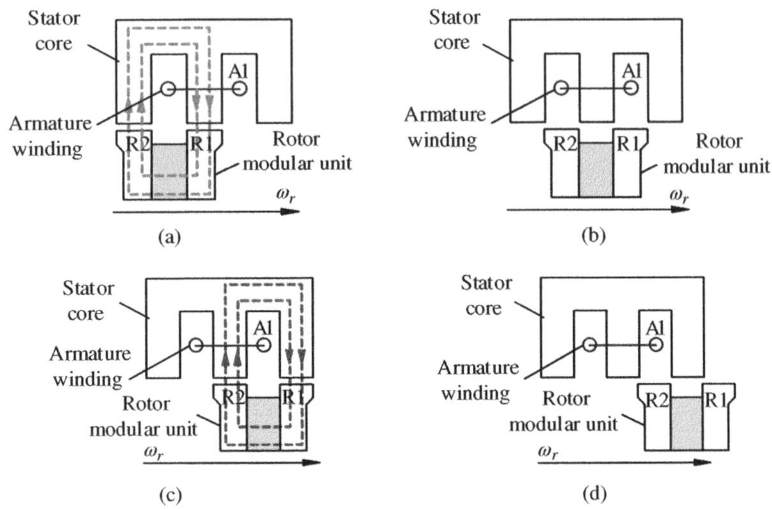

Figure 6.55 Flux switching action of the rotor-PM FSPM machine, (a) $\theta_r = 0°$, (b) $\theta_r = 9°$, (c) $\theta_r = 18°$, (d) $\theta_r = 27°$.

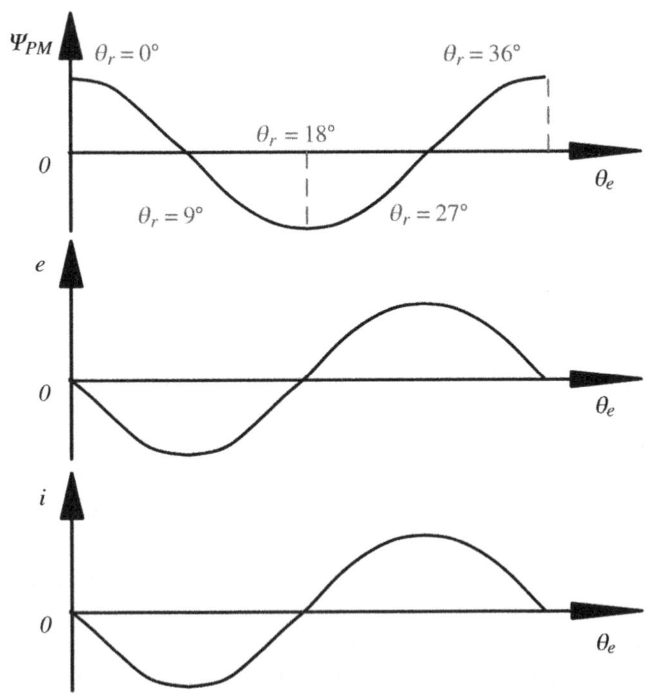

Figure 6.56 PM flux linkage, no-load back-EMF and ideal armature current waveform for the rotor-PM FSPM machine.

polarity changes from positive to negative and then back to positive. Namely, with the rotation of the rotor, the polarity, and amplitude of the PM flux linkage change periodically. Therefore, the rotor-PM FSPM machine shows the "flux-switching action." Considering the fundamental wave only, the ideal PM flux linkage ψ_{PM} and back-EMF e will be sinusoidal. A constant torque can be obtained by applying appropriate three-phase sinusoidal currents to the armature windings in synchronization with the rotor position.

6.6.2 Winding Consistency and Complementarity

Based on the four special rotor positions shown in Figure 6.55, the so-called consistency and complementarity of armature windings in rotor-PM FSPM machines is analyzed. When $\theta_r = 0°$ as shown in Figure 6.57a, the rotor tooth R1 is aligned with armature tooth A1. Assuming that PMs are magnetized clockwise, the PM flux enters rotor tooth R1 from stator armature tooth A1 and the PM flux linkage linked by armature coil A1 reaches the positive maximum. Meanwhile, rotor tooth R3 is aligned with armature tooth A2. Though the PM flux enters armature tooth A2 from rotor tooth R3 at this moment, the PM flux linkage linked by armature coil A2 also reaches the positive maximum considering armature coil A2 and armature coil A1 are of the opposite directions, which is consistent with that of armature coil A1.

When $\theta_r = 9°$, both PM flux linkages linked by armature coils A1 and A2 are 0, but the relative position between these two coils and rotor teeth are completely different. As shown in Figure 6.57b, the armature tooth A1 is aligned with the rotor PM, and this specific position is defined as the first equilibrium position. The armature tooth A2 is aligned with the middle line of the gap between rotor modules. This specific rotor position is defined as the second equilibrium position. Due to differences in the permeability, width of PM blocks, and gap between rotor modules, the magnetic circuits at the two specific positions are different.

When $\theta_r = 18°$, rotor tooth R2 is aligned with armature tooth A1, and rotor tooth R4 is aligned with armature tooth A2, as shown in Figure 6.57c. Both PM linkages linked by armature coils A1 and A2 reach the negative maximum.

When $\theta_r = 27°$, armature tooth A1 is aligned with the middle line of the gap between rotor modules, namely the "second equilibrium position," while armature winding A2 is in line with PMs, which is the "first equilibrium position." Therefore, flux linkages linked by armature coils A1 and A2 are both 0.

It can be seen from the above analysis that the PM flux linkages linked by armature coils A1 and A2 of the 24/10 rotor-PM FSPM machine experience the same change in a whole electrical cycle. This phenomenon is defined as the "winding consistency."

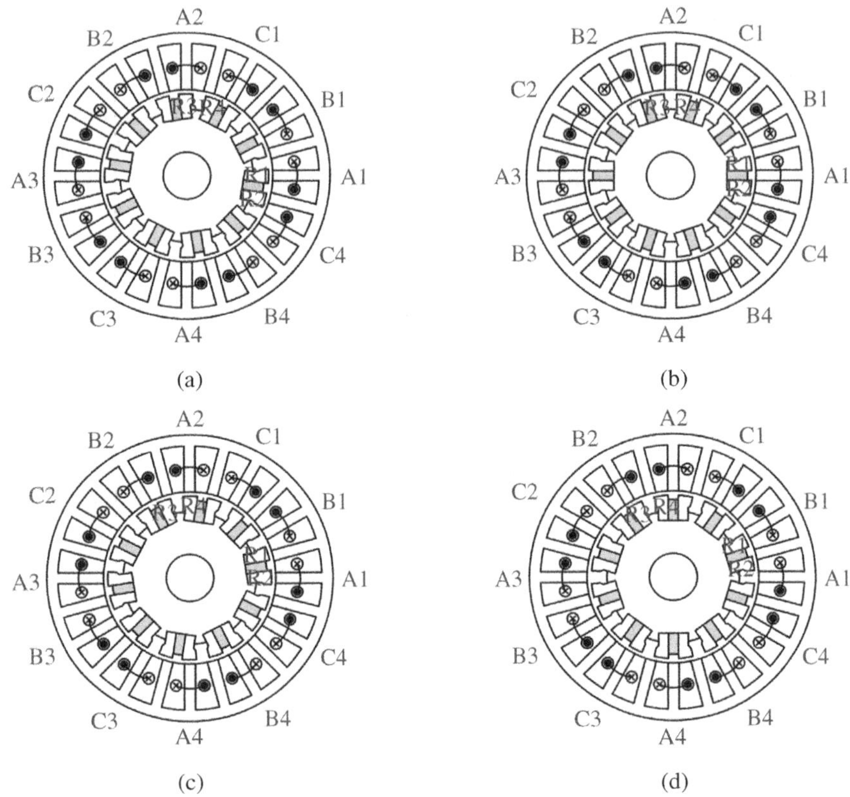

Figure 6.57 Four specific positions of the 24/10 rotor-PM FSPM machine. (a) $\theta_r = 0°$, (b) $\theta_r = 9°$, (c) $\theta_r = 18°$, (d) $\theta_r = 27°$.

On the other hand, the magnetic circuits at the "first equilibrium position" and the "second equilibrium position" are different because the armature tooth is aligned with the PM and gap between adjacent rotor modules, respectively. As a result, PM flux linkages linked by armature coils A1 and A2 are not exactly the same.

Corresponding to the four specific rotor positions, the relative positions between armature coil A1 and the rotor are "positive aligned" → the "first equilibrium position" → "negative aligned" → the "second equilibrium position" → "positive aligned." By contrast, the relative positions of armature coil A2 and the rotor are: "positive aligned" → the "first equilibrium position" → "negative aligned" → the "second equilibrium position" → "positive aligned."

It can be seen that both of them start from the "positive aligned" position, but the rotational directions are opposite. It causes an 180° phase shift in the PM flux linkage and no-load back-EMF waveforms between armature coils A1 and A2. This phenomenon is defined as the "winding complementarity." The "winding consistency" and "winding complementarity" effectively eliminate even-order spatial harmonics, leading to symmetrical PM flux linkage and no-load back-EMF waveforms [45].

6.6.3 Fundamental Electromagnetic Performance

1 No-Load Magnetic Field and Back-EMF

Figure 6.58 shows PM magnetic field distributions of the rotor-PM FSPM machine at the aforementioned four specific rotor positions. It is obvious that the PM

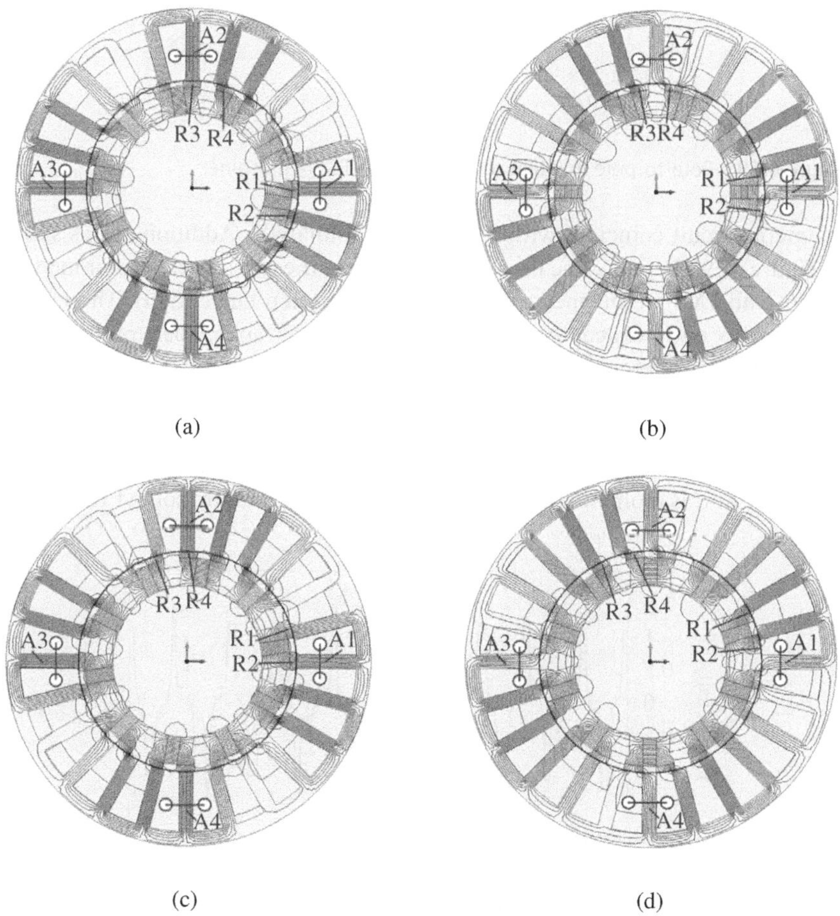

Figure 6.58 No-load PM magnetic field distributions. (a) $\theta_r = 0°$, (b) $\theta_r = 9°$, (c) $\theta_r = 18°$, (d) $\theta_r = 27°$.

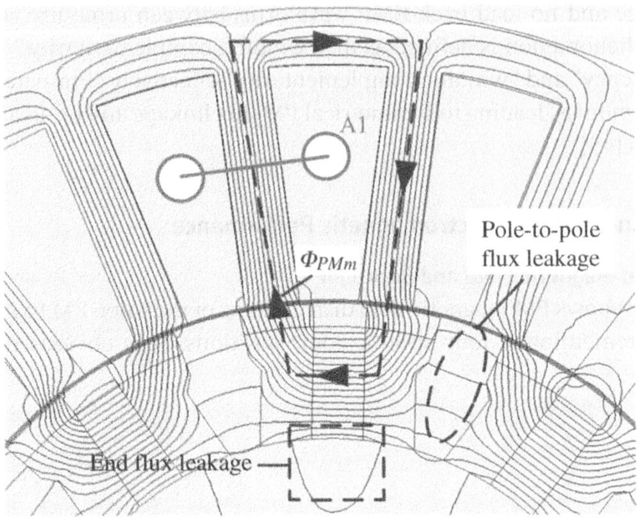

Figure 6.59 Pole-to-pole flux leakage in rotor-PM FSPM machine.

magnetic circuit coincides with the theoretical analysis. Additionally, as shown in Figure 6.59, there is some flux leakage between the two adjacent rotor modules because all PMs are magnetized in the same direction tangentially. Therefore, a leakage flux path, called "pole-to-pole leakage flux branch," occurs throughout the whole circumference of the rotor, resulting in reduced airgap flux density and PM flux linkage.

Figure 6.60 shows the no-load airgap flux density when $\theta_r = 0°$, where the positive maximum occurs at the alignment position of rotor tooth R1 and armature tooth A1. At this moment, the flux linkage linked by armature coil A1 reaches the

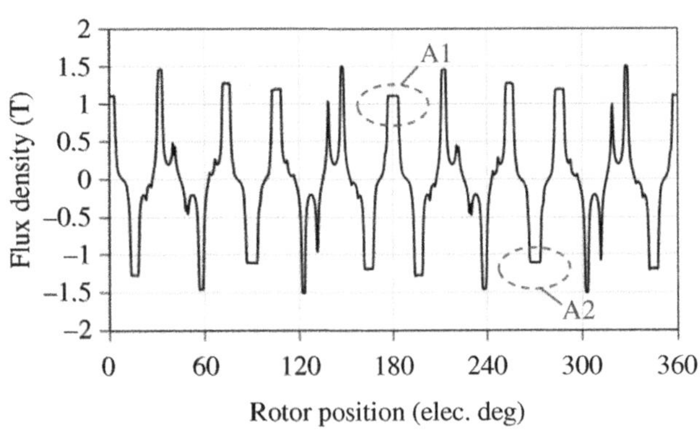

Figure 6.60 No-load airgap flux density when $\theta_r = 0°$.

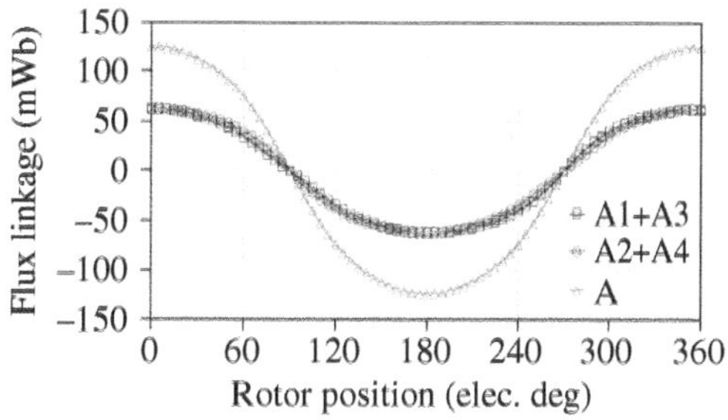

Figure 6.61 Flux linkage waveforms of the 24/10 rotor-PM FSPM machine.

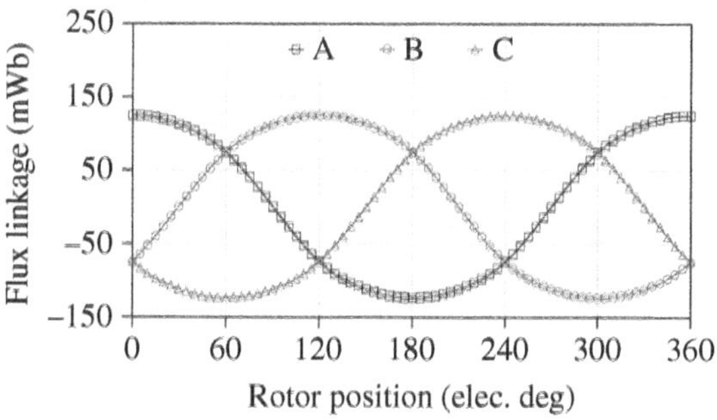

Figure 6.62 Three-phase PM flux linkage of the 24/10 rotor-PM FSPM machine.

maximum. The negative maximum corresponds to the alignment position of rotor tooth R3 and armature tooth A2. In addition, the amplitude of the airgap flux density at the alignment position is 1.1 T, which is much lower than that in the 12/10 FSPM machine. This indicates the magnetic saturation in the main flux path has been alleviated by migrating magnets to the rotor.

The PM flux linkage waveforms of the rotor-PM FSPM machine are shown in Figures 6.61 and 6.62. Due to the geometric and magnetic symmetry of the 24/10 rotor-PM FSPM machine, the flux linkages linked by coils A1 and A3 are completely the same, as well as those of coils A2 and A4.

The flux linkage of phase A winding is analyzed based on the resultant flux linkages obtained from A1 + A3, and A2 + A4. It can be seen that the PM flux linkage of each individual coil is sinusoidal, and waveforms of the two resultant flux linkages obtained from A1 + A3 and A2 + A4 intersect at special positions of $\theta_r = 9°$ and $\theta_r = 27°$, corresponding to the electrical angle of 90° and 270°, respectively in Figure 6.61. There are slight differences at other rotor positions. These observations verify the consistency and complementarity of the windings.

Figure 6.63 shows the harmonic spectrums of PM flux linkages. It can be found that both flux linkages of coils A1 and A2 contain even-order harmonics. The

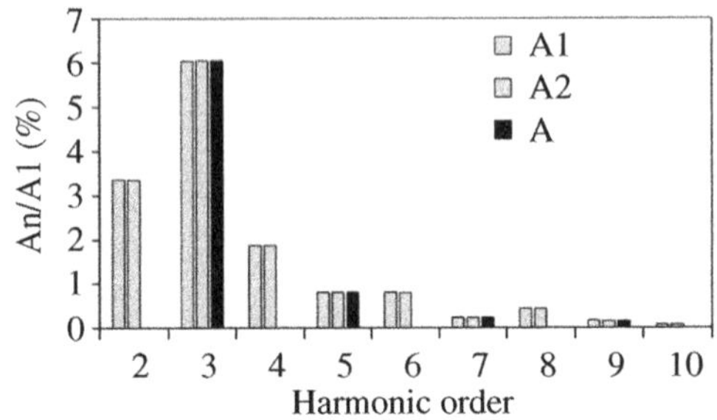

Figure 6.63 PM flux linkage harmonic spectrums of the 24/10 rotor-PM FSPM machine.

Table 6.11 Break-down of PM flux linkage harmonics of the rotor-PM FSPM machine.

Harmonic order	Coil A1	Coil A2	Phase A
2th (%)	3.36	3.35	0.0048
3th (%)	6.03	6.04	6.038
4th (%)	1.88	1.87	0.0018
5th (%)	0.8	0.8	0.805
6th (%)	0.8	0.8	0.0026
7th (%)	0.23	0.23	0.229
THD (%)	7.26	7.27	6.1

THDs for coils A1 and A2 are 7.26% and 7.27%, respectively, as shown in Table 6.11. The even-order harmonics in coil flux linkage waveforms can be canceled thanks to the winding complementarity. There are almost no even-order harmonics in the phase flux linkage, which effectively improves the symmetry of the flux linkage waveform. In addition, the harmonic analysis results show that the 3rd harmonic of the flux linkage is higher, which leads to the higher 3rd harmonic in the no-load back-EMF.

The no-load back-EMF waveforms are shown in Figure 6.64, from which the influence of the winding complementary on the resultant no-load back-EMF is illustrated clearly. The harmonic spectrums of the no-load back-EMF per coil and per phase are shown in Figure 6.65. It can be found that the 3rd harmonic is always the highest one for both the per-coil and per-phase no-load back-EMF. Moreover, amplitudes of the 2nd harmonics of coils A1 and A2 are almost the same but with opposite phases. Therefore, when coils A1 and A2 are connected in series, even-order harmonics are canceled, as shown in Table 6.12.

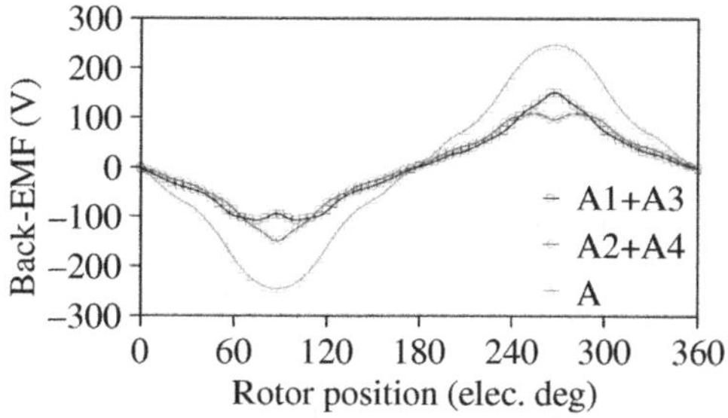

Figure 6.64 Single-phase no-load back-EMF waveform of the 24/10 rotor-PM FSPM machine.

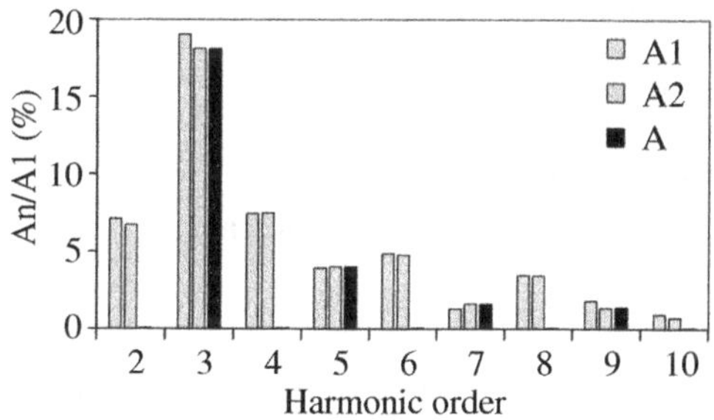

Figure 6.65 No-load back-EMF harmonic spectrums of the 24/10 rotor-PM FSPM machine.

Table 6.12 No-load back-EMF harmonic contents.

Harmonic order	Coil A1 + A3	Coil A2 + A4	Phase A
2th (%)	7.08	6.69	0.0096
3th (%)	18.99	18.08	18.07
4th (%)	7.44	7.46	0.007
5th (%)	3.93	4.01	3.99
6th (%)	4.84	4.74	0.015
7th (%)	1.3	1.58	1.58
THD (%)	22.9	22.06	18.6

Considering harmonics in the PM flux linkages, the three-phase no-load back-EMFs can be expressed as:

$$\begin{cases} e_a(t) = \sum_{n=1}^{5} E_{m(2n-1)} \sin[(2n-1)p_r\omega_r t] \\ e_b(t) = \sum_{n=1}^{5} E_{m(2n-1)} \sin\left[(2n-1)p_r\omega_r t - \frac{2\pi}{3}\right] \\ e_c(t) = \sum_{n=1}^{5} E_{m(2n-1)} \sin\left[(2n-1)p_r\omega_r t + \frac{2\pi}{3}\right] \end{cases} \tag{6.41}$$

where $E_{m(2n-1)}$ is the amplitude of the $(2n-1)$th harmonic, p_r is the rotor pole-pair number, and ω_r is the mechanical angular velocity of the rotor.

The three-phase symmetrical sinusoidal current can be expressed as:

$$\begin{cases} i_a(t) = I_{am} \sin(p_r\omega_r t) \\ i_b(t) = I_{am} \sin\left(p_r\omega_r t - \frac{2\pi}{3}\right) \\ i_c(t) = I_{am} \sin\left(p_r\omega_r t + \frac{2\pi}{3}\right) \end{cases} \tag{6.42}$$

where I_{am} is the amplitude of the phase current.

The electromagnetic torque can be calculated as:

$$T_e = \frac{e_a i_a + e_b i_b + e_c i_c}{\omega_r} \tag{6.43}$$

Substituting (6.41) and (6.42) into (6.43), the electromagnetic torque of the rotor-PM FSPM machine can be expressed as

$$T_e = T_{ef} + T_{e3} + T_{e5} + T_{e7} + T_{e9} \tag{6.44}$$

where

$$\begin{cases} T_{ef} = 1.5\mathrm{E}_{m1} I_{am}/\omega_r \\ T_{e3} = 0 \\ T_{e5} = -1.5E_{m5} I_{am} \cos(6p_r\omega_r t)/\omega_r \\ T_{e7} = 1.5E_{m7} I_{am} \cos(6p_r\omega_r t)/\omega_r \\ T_{e9} = 0 \end{cases} \tag{6.45}$$

The electromagnetic torque expression (6.45) indicates that when the balanced three-phase sinusoidal current are fed, the 3rd harmonic of the no-load back-EMF will neither produce no torque, while the 5th and the 7th harmonics only produce torque ripples.

Considering the fundamental waves only, the three-phase PM flux linkages ψ_{ma}, ψ_{mb}, ψ_{mc} can be written as

$$\begin{cases} \psi_{ma} = \psi_m \cos(\theta_e) \\ \psi_{mb} = \psi_m \cos\left(\theta_e - \frac{2\pi}{3}\right) \\ \psi_{mc} = \psi_m \cos\left(\theta_e + \frac{2\pi}{3}\right) \end{cases} \tag{6.46}$$

where ψ_m is the peak value of the fundamental component of the PM flux linkage, θ_e is the rotor position in electrical radians, and $\theta_e = p_r\theta_r$.

By using the Park transformation, the *d*- and *q*-axis PM flux linkages can be obtained from the three-phase waveforms. Before doing so, positions of the *d*-axis and *q*-axis should be defined. For traditional PM synchronous machines, such as the surface-mounted PM synchronous machine, the *d*-axis is located at the centerline of one of the rotor magnets along the magnetization direction. The *q*-axis can be obtained by rotating the *d*-axis by $\pi/(2p_r)$ mechanical radians, which means the *q*-axis coincides with the centerline between two adjacent PMs [46].

Although PMs in the rotor-PM FSPM machine are also located on the rotor, the arrangement of PMs is different from the traditional PM synchronous machines. Therefore, positions of the *d*- and *q*-axis have to be determined by the flux linkage variations [46].

Figure 6.66 illustrates the definitions of *dq* axes in three-phase rotor-PM FSPM machines. As shown in Figure 6.66a, armature tooth A1 is in line with one of the PMs, namely the rotor, is at the "first equilibrium position." At this moment, the flux linkage of phase A is 0. Therefore, the centerline of the PM is defined as the *q*-axis. Then the *d*-axis is defined at the position lagging the *q*-axis by $\pi/(2p_r)$. When the rotor rotates counterclockwise by $\pi/(2p_r)$, as shown in Figure 6.66b, the *d*-axis coincides with the centerline of armature tooth A1, and the flux linkage of phase A reaches the positive maximum.

Based on the definitions of *dq* axes, the *d*- and *q*-axis flux linkages can be calculated by:

$$\begin{bmatrix} \psi_{md} \\ \psi_{mq} \\ \psi_{m0} \end{bmatrix} = P \begin{bmatrix} \psi_{ma} \\ \psi_{mb} \\ \psi_{mc} \end{bmatrix} \tag{6.47}$$

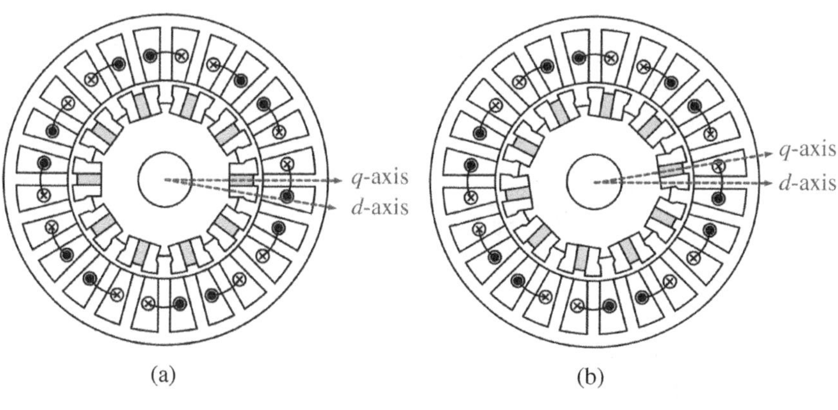

Figure 6.66 Definitions of dq axes in rotor-PM FSPM machines. (a) Position when PM flux linkage equals zero. (b) Position when PM flux linkage reaches the maximum.

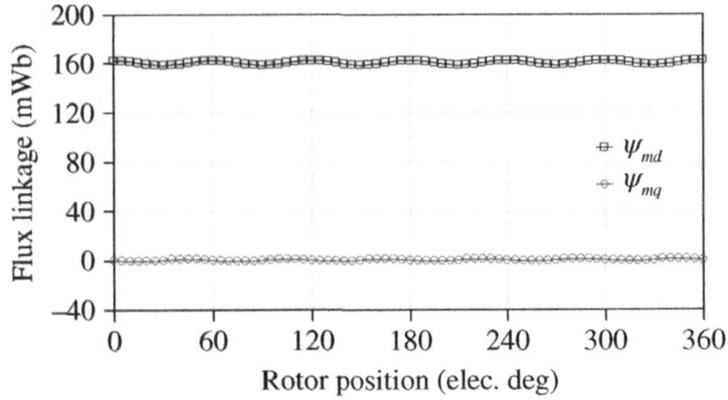

Figure 6.67 PM flux linkage waveforms of the 24/10 rotor-PM FSPM machine in the synchronous reference frame.

where

$$P = \frac{2}{3}\begin{bmatrix} \cos\theta_e & \cos\left(\theta_e - \frac{2\pi}{3}\right) & \cos\left(\theta_e + \frac{2\pi}{3}\right) \\ -\sin\theta_e & -\sin\left(\theta_e - \frac{2\pi}{3}\right) & -\sin\left(\theta_e + \frac{2\pi}{3}\right) \\ 1/2 & 1/2 & 1/2 \end{bmatrix} \tag{6.48}$$

The subscript "0" denotes the zero-axis component. The d- and q-axis flux linkages can be obtained further

$$\begin{cases} \psi_{md} = \psi_m \\ \psi_{mq} = 0 \\ \psi_{m0} = 0 \end{cases} \tag{6.49}$$

It shows that the d-axis component of the PM flux linkage ψ_{md} is a constant and is independent of the rotor position. Moreover, ψ_{md} is equal to the amplitude of the PM flux linkage ψ_m, while both the q-axis and 0-axis components equal zero. Based on the three-phase flux linkage waveforms shown in Figure 6.63, the d- and q-axis flux linkage waveforms are calculated and shown in Figure 6.67.

Similarly, considering the fundamental component only, the no-load back-EMF can be expressed as:

$$\begin{cases} e_{ma}(t) = -E_m \sin(\theta_e) \\ e_{mb}(t) = -E_m \sin\left(\theta_e - \dfrac{2\pi}{3}\right) \\ e_{mc}(t) = -E_m \sin\left(\theta_e + \dfrac{2\pi}{3}\right) \end{cases} \tag{6.50}$$

2 Inductance Characteristics

Accurate inductance values are of great significance for the machine control. When the magnetization of PMs is ignored, the iron core is unsaturated and the

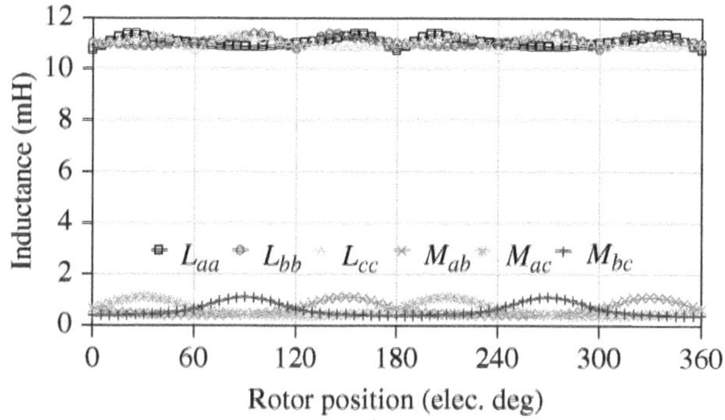

Figure 6.68 Unsaturated self-inductance and mutual inductance waveforms.

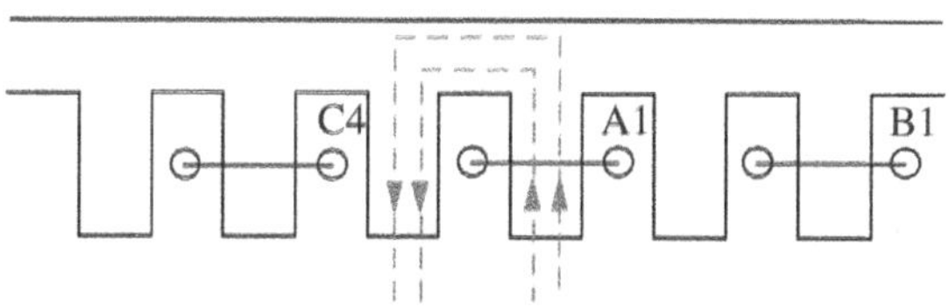

Figure 6.69 Main flux path of the rotor-PM FSPM machine.

three-phase unsaturated self-inductance and mutual inductance waveforms can be obtained, as shown in Figure 6.68. It can be seen that the average value of the mutual inductance is almost zero because coils are separated and isolated by fault-tolerant teeth. The magnetic field isolation effect can be explained by Figure 6.69. When the phase A winding is fed, the armature magnetic field produced by phase A current loops through armature teeth of phase A and two adjacent fault-tolerant teeth. Namely, the armature field of phase A does not pass through adjacent armature teeth belonging to phase B and phase C.

In practice, both the stator and rotor cores are saturated or over-saturated due to the large PM remanence. It is necessary to consider the effect of iron core saturation on inductances. The three-phase saturated self-inductance and mutual inductance waveforms of the rotor-PM FSPM machine are shown in Figure 6.70. It can be seen that the saturated self-inductance is smaller than the unsaturated counterpart due to the reduced magnetic permeability, while the mutual inductance is still almost zero, similar to those shown in Table 6.13.

It is worth noting that there is a noticeable 1st-order harmonic in the self-inductance waveform of each phase, and the phase difference between any two phases is 120 electrical degrees. This observation can be used to build mathematical models of the three-phase rotor-PM FSPM machine considering the magnetic saturation.

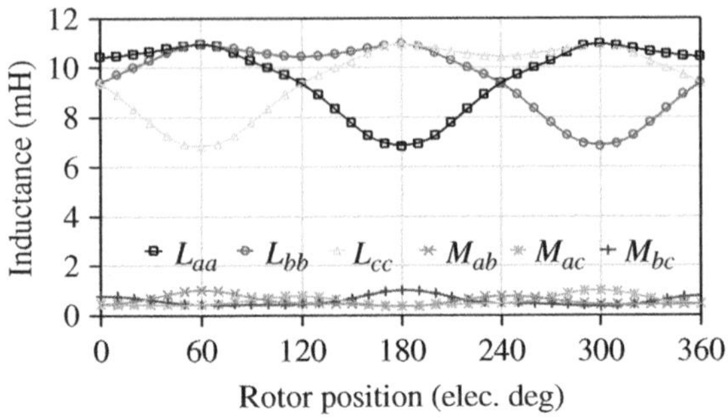

Figure 6.70 Saturated self-inductance and mutual inductance waveforms.

Table 6.13 Inductance characteristics of the rotor-PM FSPM machine.

	Unsaturated inductance	**Saturation inductance**
L_{aa} (mH)	11.1	9.59
L_{bb} (mH)	11.1	9.59
L_{cc} (mH)	11.1	9.59
M_{ab} (mH)	0.61	0.57
M_{ac} (mH)	0.61	0.57
M_{bc} (mH)	0.61	0.57

Based on the three-phase winding inductances, the corresponding *d*- and *q*-axis inductances L_d and L_q can be obtained as

$$L_{dq}(\theta_r) = PL_{abc}(\theta_r)P^{-1} \tag{6.51}$$

where L_{abc} is the three-phase inductance matrix

$$L_{abc}(\theta_r) = \begin{bmatrix} L_{aa} & M_{ab} & M_{ac} \\ M_{ba} & L_{bb} & M_{bc} \\ M_{ca} & M_{cb} & L_{cc} \end{bmatrix} \tag{6.52}$$

The corresponding *dq* axes inductance waveforms are presented in Figures 6.71 and 6.72. It can be found the unsaturated *dq* axes inductances are almost independent of the rotor position, while the saturation inductances have relatively large fluctuations. The average values are tabulated in Table 6.14.

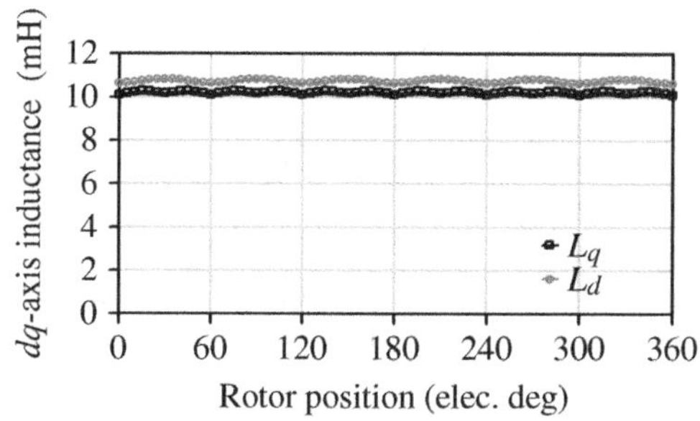

Figure 6.71 Unsaturated dq axes inductance waveforms.

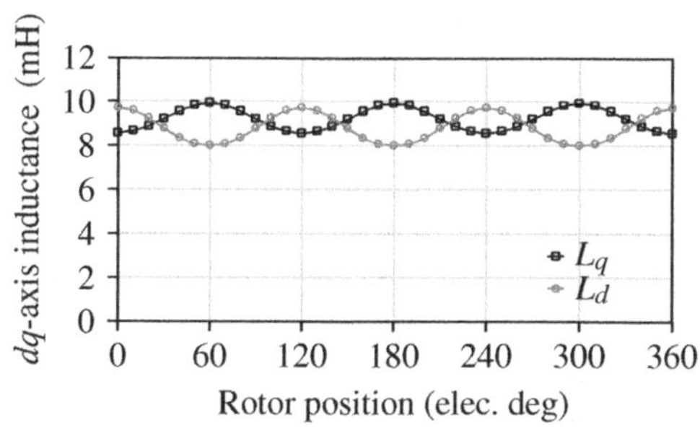

Figure 6.72 Saturated dq axes inductance waveforms.

Table 6.14 Dq axes inductance with and without considering the PM effect.

	Unsaturated	Saturated
Average value of *d*-axis inductance, L_d (mH)	10.7	8.8
Average value of *q*-axis inductance, L_q (mH)	10.2	9.2

3 Electromagnetic Torque

Based on the flux linkage, no-load back-EMF and inductance characteristics shown above, the electromagnetic torque can be written as

$$T_e = T_{PM} + T_r + T_{cog} = \frac{3}{2}p_r[\psi_m i_q + (L_d - L_q)i_d i_q] + T_{cog} \tag{6.53}$$

The electromagnetic torque is mainly composed of three components: T_{PM} is the PM torque generated by the interaction between *q*-axis current i_q and the PM field; T_r is generated by the interaction between *dq* currents i_d and i_q through the *dq* inductances and is typically termed as the reluctance torque; and T_{cog} is the cogging torque.

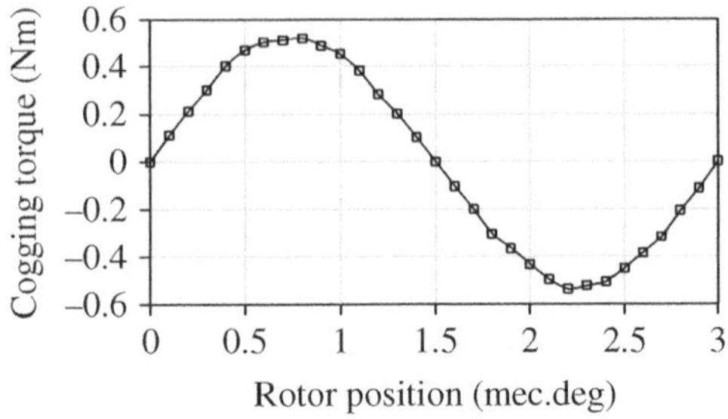

Figure 6.73 Cogging torque of the 24/10 rotor-PM FSPM machine.

The period of the cogging torque can be determined by:

$$C_{cog} = \frac{2\pi}{LCM(p_s, p_r)} \tag{6.54}$$

where $LCM(p_s, p_r)$ is the least common multiple of the stator slot number p_s and the rotor PM pole-pair number p_r. In the 24/10 rotor-PM FSPM machine example, $p_s = 24$, $p_r = 10$, and $LCM(p_s, p_r) = 120$. Thus the cogging torque has a period of $C_{cog} = 3°$. The cogging torque waveform from 2D FEA is shown in Figure 6.73, whose period is consistent with the theoretical analysis.

It has been shown that L_d and L_q of the rotor-PM FSPM machine are almost the same. Thus the reluctance torque component T_r can be ignored, and the $i_d = 0$ control method can benefit the current utilization. Equation (6.53) can be simplified to:

$$T_e = \frac{3}{2} p_r \psi_m I_m + T_{cog} \tag{6.55}$$

where I_m is the amplitude of the phase current.

The electromagnetic torque waveform in one electrical cycle is shown in Figure 6.74. The average electromagnetic torque is 16.05 Nm, and the torque ripple is 11%. In addition, the electromagnetic torque contains the 6th and 12th harmonics, the latter of which is caused by the cogging torque. Without the cogging torque, the torque ripple is significantly reduced to only 5%, so the cogging torque is the main contributor to the torque ripple.

The 12th harmonic of the electromagnetic torque waveform is mainly generated by harmonics in the flux linkage and no-load back-EMF. According to the harmonic analysis results, apart from the highest 3rd harmonic, the three-phase flux linkage and no-load back EMF also contains the 5th and the 7th harmonics with relatively high amplitudes. According to (6.45), when the balanced three-phase sinusoidal currents are fed into the armature winding, the 5th and the 7th harmonics in the no-load back-EMF will interact with the current to produce the 6th torque harmonic.

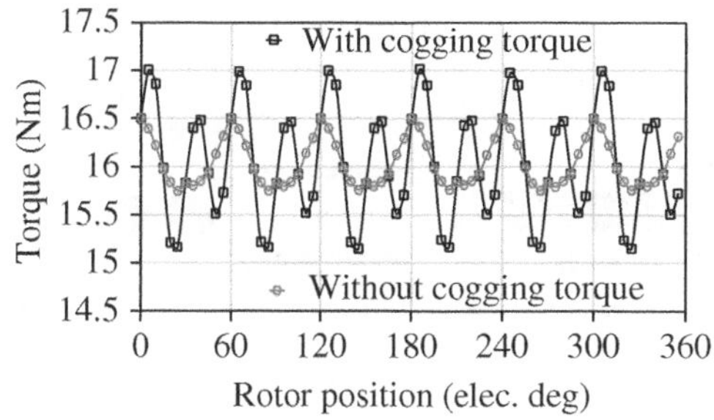

Figure 6.74 Electromagnetic torque of the 24/10 rotor-PM FSPM machine with $I_m = 5.9$ Arms.

6.7 Dual-Rotor Magnetically-Geared Power Split Machine

New energy vehicles have become a focus of development in the automotive industry. The hybrid electric vehicle (HEV), an intermediate step in the transition from the traditional internal combustion engine (ICE) vehicle to the battery-powered EV, has attracted much attention. The ICE can always run with a high efficiency based on the power split system in series-parallel HEVs. As the most commonly used series-parallel hybrid power technology, the planetary gear system by Toyota has monopolized this field for a long time. Nevertheless, the mechanical gear usually suffers from mechanical abrading and backlash, so there is a pressing need for research on an integrated electrified transmission system solution.

Magnetic gear technology has recently become widespread in the electrical engineering field, because this technology can help prevent the mechanical abrading of gears, thereby reducing system maintenance costs and improving reliability. Considering the fact that the kinematic characteristics of coaxial magnetic gears working by the field modulation principle are exactly the same as that of the mechanical planetary gear, the authors proposed a magnetically geared dual-rotor PM machine called a magnetically geared power split machine (MGPSM), which is a derived variation of the FSPM machine based on the GAFMT. It serves as a good candidate to replace the power split device in existing series-parallel HEV power systems.

Figure 6.75 illustrates the hybrid powertrain with input power split based on the MGPSM. The output power of the ICE is split into two parts. One is transmitted to the stator winding of the MGPSM and converted into electric energy stored in the battery, and then transferred out to drive the traction machine when required. The other part is directly transferred from the outer rotor to the inner rotor to drive the vehicle [11], [47–49].

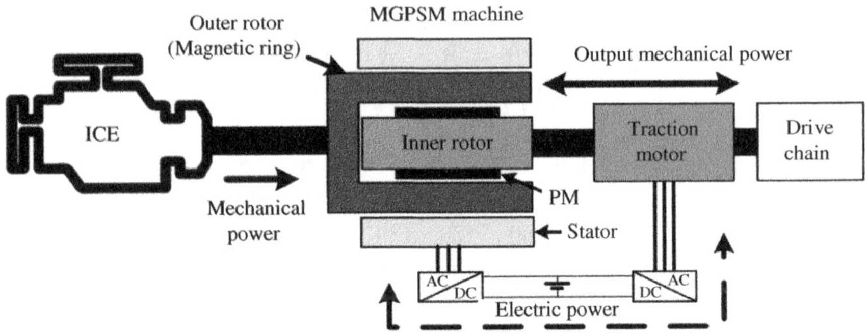

Figure 6.75 Input power split hybrid powertrain based on MGPSM.

6.7.1 Machine Construction and Operating Principle

The basic structure of the MGPSM is shown in Figure 6.76a. PMs are mounted on the inner rotor. The outer rotor is a salient reluctance pole rotor, which serves as the field modulator. A set of three-phase 8-pole distributed winding is placed in the 24 stator slots. In the outer airgap, the magnetic field harmonic with the same pole-pair number as that of the winding can produce the effective flux linkage in the armature winding. According to the GAFMT, the field harmonics will rotate when the outer and/or inner rotor rotates, inducing the no-load back-EMF in the stator winding. Thus, the electromechanical energy conversion can be achieved when appropriate currents are fed into the winding, and the torque can be produced as well.

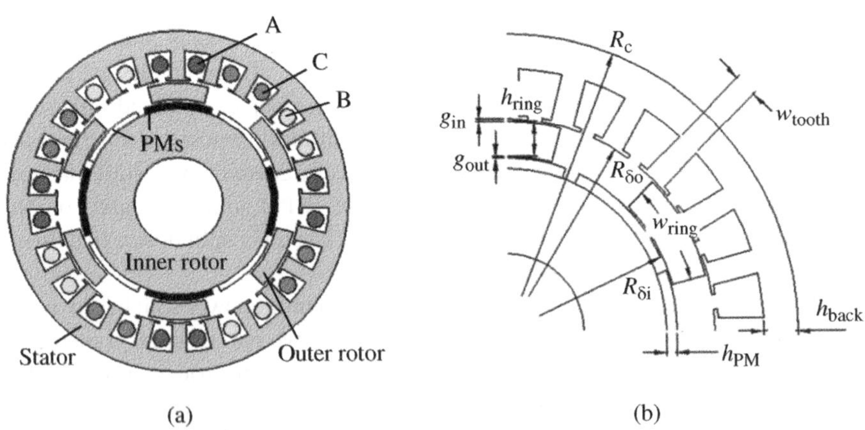

Figure 6.76 MGPSM. (a) Cross-sectional view, (b) Key dimensions.

According to the GAFMT for modulations using salient reluctance poles, the relationships related to pole-pair numbers and rotational speeds can be written as

$$p_w + p_{ir} = p_{or} \tag{6.56}$$

$$n_w p_w + n_{ir} p_{ir} = n_{or} p_{or} \tag{6.57}$$

where n_w, n_{ir}, and n_{or} are the synchronous speed of the armautre field, inner rotor (PM rotor) speed and outer rotor speed, respectively. p_w, p_{ir}, and p_{or} are the pole-pair numbers of the armature field and inner rotor (PM rotor), and number of ferromagnetic blocks of the outer rotor, respectively. The frequency-speed relationship of the armature field can be expressed as

$$n_w = \frac{60f}{p_w} \tag{6.58}$$

Substituting it into (6.57) yields

$$f = \frac{n_{or} p_{or} - n_{ir} p_{ir}}{60} \tag{6.59}$$

where f is the frequency of the armature current. Since the frequency of armature current in the MGPSM is not determined by the speed of a single rotor, the discussion of MGPSM should be relative to the frequency rather than the speed.

It can be found from (6.59) that when the outer rotor speed is very low, the "negative frequency" may appear, which means that the direction of the electric power will change with the speed difference between the inner and outer rotors, even if the inner and outer rotor rotate in the same direction [50].

The positive directions of the inner rotor torque $T_{ir,}$ outer rotor torque T_{or} and the electromagnetic torque T_{em} are defined as the counterclockwise. They are the same as the positive direction of the speed. Considering the torque balance, one can get:

$$T_{ir} + T_{or} + T_{em} = 0 \tag{6.60}$$

Another equation can be obtained by the power balancing:

$$T_{ir} n_{ir} + T_{or} n_{or} + T_{em} n_w = 0 \tag{6.61}$$

Then, substituting (6.57) and (6.60) into (6.61), the torque relationship between the inner and outer rotors can be obtained by eliminating $T_{em} n_w$:

$$T_{ir} n_{ir} + T_{or} n_{or} - (T_{ir} + T_{or}) \frac{n_{or} p_{or} - n_{ir} p_{ir}}{p_w} = 0 \tag{6.62}$$

Considering the relationship of pole-pair numbers shown in (6.56), (6.62) can be expressed further as:

$$\begin{aligned} &T_{ir} n_{ir} p_{or} - T_{ir} n_{ir} p_{ir} + T_{or} n_{or} p_{or} - T_{or} n_{or} p_{ir} \\ &= T_{ir} n_{or} p_{or} - T_{ir} n_{ir} p_{ir} + T_{or} n_{or} p_{or} - T_{or} n_{ir} p_{ir} \end{aligned} \tag{6.63}$$

Then the relationship between the inner rotor torque T_{ir} and the outer rotor torque T_{or} can be obtained:

$$T_{ir} = -\frac{p_{ir}}{p_{or}} T_{or} \tag{6.64}$$

Applying the same operations to (6.61), the relationship between the outer rotor torque T_{or} and the electromagnetic torque T_{em} can be obtained by eliminating $T_{ir}n_{ir}$:

$$T_{em} = -\frac{p_w}{p_{or}} T_{or} \tag{6.65}$$

Applying the same operations to (6.61), the relationship between the inner rotor torque T_{ir} and the electromagnetic torque T_{em} can be obtained to eliminate $T_{or}n_{or}$:

$$T_{em} = \frac{p_w}{p_{ir}} T_{ir} \tag{6.66}$$

It can be seen that the ratios between the electromagnetic torque and the torque of inner and outer rotors are fixed, which is only related to the pole-pair combinations and cannot be changed for a given design. Although the synchronous speed of the armature field can be controlled, it can only change the speeds of the inner rotor and outer rotor of the MGPSM proportionally. Furthermore, the direction of electromagnetic torque is the same as that of the inner rotor, and opposite to that of the outer rotor.

6.7.2 Modes of Operation

Traditional PM synchronous machines can only work as either a generator or a motor at any given time, but the MGPSM can work in multiple modes of operation as long as the power conservation is met, because it has three ports. From this perspective, the MGPSM cannot be simply seen as a generator or a motor. In practice, it often plays the role of a generator and a motor simultaneously.

The so-called power split refers to the transfer and/or conversion of electric power and mechanical power among the three ports of the MGPSM, as depicted in Figure 6.77. Three typical working modes in hybrid power powertrains will be analyzed. First of all, positive directions used in the MGPSM power splitter are defined: power input – negative, power output – positive. The whole system meets the conservation of energy, namely the sum of all the input power and the output power is zero.

1 Working Mode 1

As shown in Figure 6.78, the mechanical power and the electrical power flows into the MGPSM from the outer rotor and the stator winding, respectively, and the inner rotor is producing the mechanical power. Thus, the output power of the

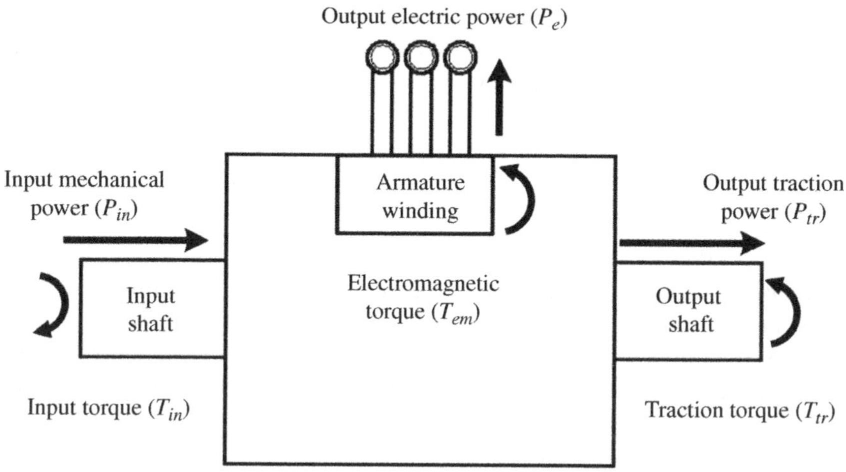

Figure 6.77 Power flow of MGPSM power splitter (Counterclockwise is the positive direction).

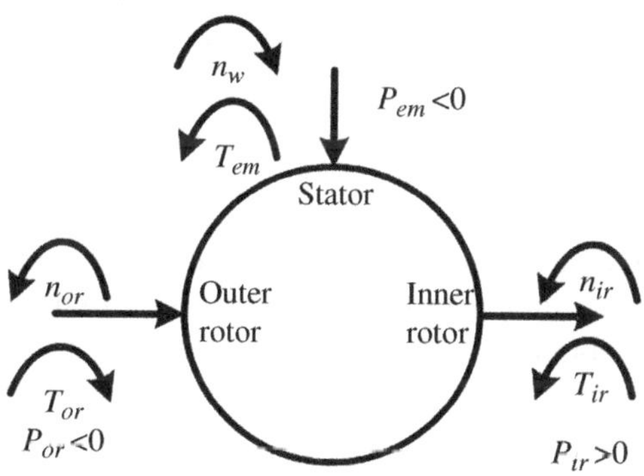

Figure 6.78 Power flowing direction in working mode 1.

inner rotor is equal to the sum of the input mechanical power of the outer rotor and the input electric power of the winding, which can be expressed as

$$p_{ir} = -(P_{or} + P_{em}) > 0 \tag{6.67}$$

assuming that the outer rotor rotates counterclockwise (in the positive direction), and the torque of the outer rotor is negative. Because the torque and the speed directions are opposite, the outer rotor has a negative power and absorbs power from outside. From (6.64) and (6.65), it can be concluded that the electromagnetic torque and inner rotor torque must be positive at this time. The inner rotor outputs power, and the directions of both the speed and torque are positive.

Moreover, because the armature winding absorbs power, the direction of the armature field is negative. According to (6.59), it can be concluded that the operation frequency is negative, namely:

$$f = \frac{n_{or}p_{or} - n_{ir}p_{ir}}{60} < 0 \tag{6.68}$$

Then,

$$n_{ir} > \frac{p_{or}}{p_{ir}} n_{or} \tag{6.69}$$

Equation (6.69) shows that at this moment, the inner rotor is running at a high speed and delivering high power. In the context of an HEV, this working mode achieves the full power output at a high speed. The mechanical power from the ICE and the electrical power from the battery can be used to drive the vehicle simultaneously. Thus, the total output power to the wheels is greater than that from the ICE due to the presence of the MGPSM.

2 Working Mode 2

As shown in Figure 6.79, the mechanical power flows into the outer rotor and the electrical power flows out of the stator winding in this mode of operation. The mechanical power flows out of the inner rotor. The input power of the outer rotor is equal to the sum of the output mechanical power from the inner rotor and the output electrical power from the winding:

$$P_{or} = -(P_{ir} + P_{em}) < 0 \tag{6.70}$$

Assuming that the outer rotor rotates counterclockwise (in the positive direction) at this moment, similar to the derivation for working mode 1, the speed relationship can be obtained:

$$0 < n_{ir} < \frac{p_{or}}{p_{ir}} n_{or} \tag{6.71}$$

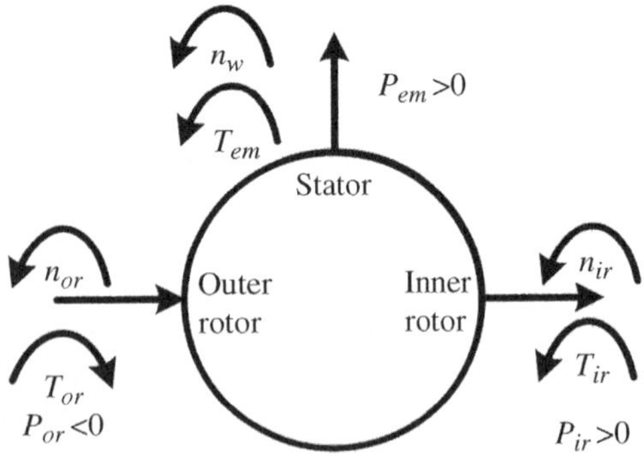

Figure 6.79 Power flowing direction in working mode 2.

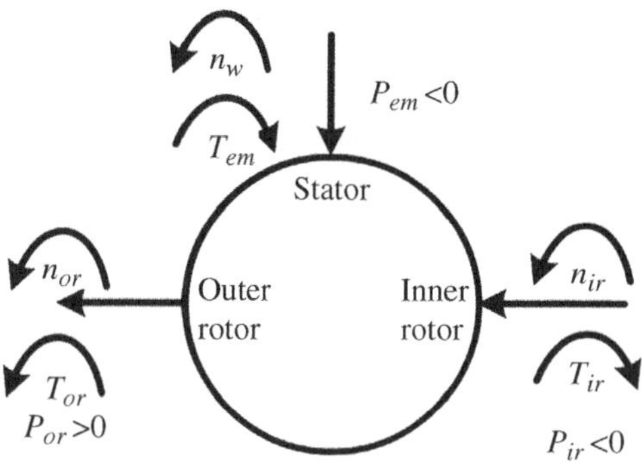

Figure 6.80 Power flowing direction in working mode 3.

This working mode is suitable for normal operation in HEVs. The output mechanical power of the ICE is split into two parts. One part is outputted through the inner rotor in the form of mechanical power to drive the vehicle, and the rest is applied to drive the traction motor or transmitted to the battery through the armature winding in the form of electric power.

3 Working Mode 3

As shown in Figure 6.80, the mechanical power and the electrical power input from the inner rotor and the stator winding, respectively, and the output mechanical power of the MGPSM is used to drive the outer rotor. Then the output power of the outer rotor is equal to the sum of the input mechanical power of the inner rotor and the input electrical power of the winding. Namely,

$$P_{or} = -(P_{ir} + P_{em}) > 0 \tag{6.72}$$

Assuming that the outer rotor rotates counterclockwise (in the positive direction), the torque and speed directions can be obtained in a similar method to the derivation for working mode 1. The power flow directions of Mode 3 and Mode 2 are the opposite. Mode 3 can be used to start the ICE when the underpower occurs during the battery-powered operation of the HEV, and this process usually takes less than one second. The starting process of the ICE in the case of a vehicle stop is a special case of Mode 3 with $n_{ir} = 0$, when the inner rotor is usually locked by the brake mechanism. At this moment, the power required is completely provided by the stator winding.

The above three working modes are summarized in Table 6.15.

6.7.3 Asymmetry in Magnetic Circuits

A reduced-power MGPSM with a simple structure has been designed and prototyped. To meet the requirements of HEVs, the pole-pair number ratio is set to

Table 6.15 Three commonly used MGPSM working modes (outer rotor rotates counterclockwise).

Mode	Power flow direction	Speed relationship	Armature field	Function
1	$P_{ir} = -(P_{or} + P_{em}) > 0$, $P_{or} < 0, P_{em} < 0$	$n_{ir} > \frac{p_{or}}{p_{ir}} n_{or}$	Clockwise	Full power operation
2	$P_{or} = -(P_{ir} + P_{em}) < 0$, $P_{ir} > 0, P_{em} > 0$	$0 < n_{ir} < \frac{p_{or}}{p_{ir}} n_{or}$	Counterclockwise	Power split
3	$P_{or} = -(P_{ir} + P_{em}) > 0$, $P_{ir} \leq 0, P_{em} < 0$	$n_{or} <$ starting speed	Counterclockwise	Start the ICE

$p_{ir}/p_w = 2$. In addition, to reduce the operating frequency of the armature winding, the pole-pair number of the prototype was selected to be as low as possible, and the outer rotor is expected to have an even number of ferromagnetic blocks. Taking into account all of these aspects, a topology with $p_{or} = 6$, $p_{ir} = 4$, and $p_w = 2$ was selected.

Table 6.16 Design parameters and specifications of the reduced-power prototype.

Parameter	Symbol	Value
Rated power	P_{in}	1.1 kW
Outer rotor rating	T_{or} at n_{or}	11 Nm at 1000 r/min
Inner rotor rating	T_{ir} at n_{ir}	7.3 Nm at 750 r/min
Phase voltage	U	≤36 Vrms
Inner airgap radius	$R_{\delta i}$	46.4 mm
Outer airgap radius	$R_{\delta o}$	60.6 mm
Stator outer radius	R_e	85 mm
Stator yoke height	h_{back}	10 mm
Remanence	B_r	1.2 T (NdFeB)
Magnet length	h_{PM}	3 mm
Rated current density	J	5 Arms/mm^2 (natural cooling)
Stack length	l_{stk}	45 mm
Airgap length	g_{in}, g_{out}	0.6 mm
Stator tooth width	w_{tooth}	6.8 mm
Arc of ferromagnetic blocks	w_{ring}	30°
Outer rotor thickness	h_{ring}	10 mm

A 24-slot/8-pole (24/8) stator with a single-layer distributed winding was used, as shown in Figure 6.76. The design parameters are listed in Table 6.16. The installation scheme of the MGPSM in the hybrid powertrain has been shown in Figure 6.75. Assuming that the operating speeds of the ICE and inner rotor are 2500 r/min and 1500 r/min, respectively, and the vehicle speed is about 40 km/h, the frequency of the armature winding current will reach 150 Hz according to (6.59), which is easier to achieve.

The magnetic circuit seen from the phase A winding at the initial rotor position in the MGPSM is not the same as that seen from the phase C winding when the outer rotor rotates by 120 electrical degrees, nor is it the same as that seen from the phase B winding when the outer rotor rotates by 240 electrical degrees. Figure 6.81 illustrates the main magnetic circuits when the inner rotor of MGPSM is fixed and the outer rotor rotates. When the rotors are at the positions shown in Figure 6.81a, the magnetic circuit seen by the phase A winding is the shortest, so the flux linkage of phase A reaches the peak value. When the outer rotor rotates by 120 and 240 electrical degrees, the magnetic circuits seen by phase C and phase B windings are shown in Figure 6.81c,b, respectively. It can be found that the magnetic circuit shown in Figure 6.81c is different from that shown in Figure 6.81a, and mirrors that shown in Figure 6.81b.

To better visualize the influence of the magnetic circuit asymmetry, Figure 6.82 shows the flux linkage and no-load back-EMF waveforms of the MGPSM. Due to the asymmetry of the magnetic circuit, the three-phase flux linkage waveforms in Figure 6.82a are unbalanced. The shape of the phase A flux linkage is different from those of the phase B and phase C. The phase B and phase C flux linkages have similar waveforms, which verifies the magnetic circuit-based analysis.

Compared with flux linkage waveforms, the unbalance of the three-phase no-load back-EMF, which is proportional to the rate of change of the flux linkage with time, is more noticeable, as shown in Figure 6.82b. In particular, the no-load

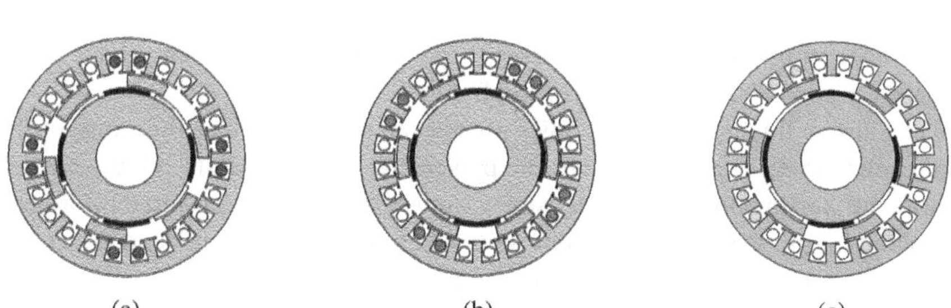

Figure 6.81 Magnetic circuit asymmetry among phases of the MGPSM. (a) Flux linkage of phase A reaches maximum, (b) Flux linkage of phase C reaches maximum, (c) Flux linkage of phase B reaches maximum.

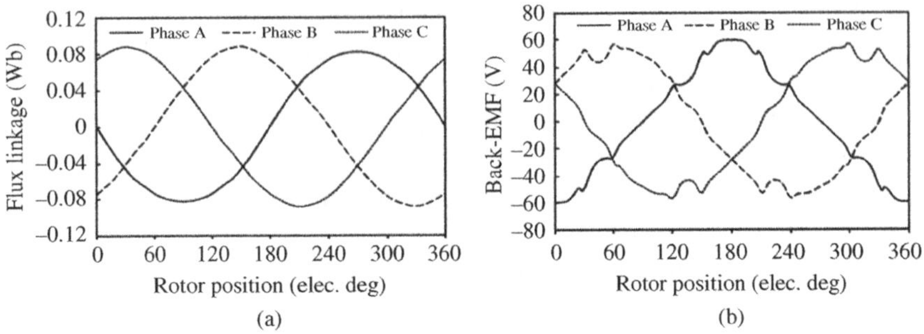

Figure 6.82 PM flux linkage and no-load back-EMF waveforms at $n_{ir} = 0$ r/min and $n_{or} = 1000$ r/min. (a) No-load flux linkage, (b) No-load back-EMF.

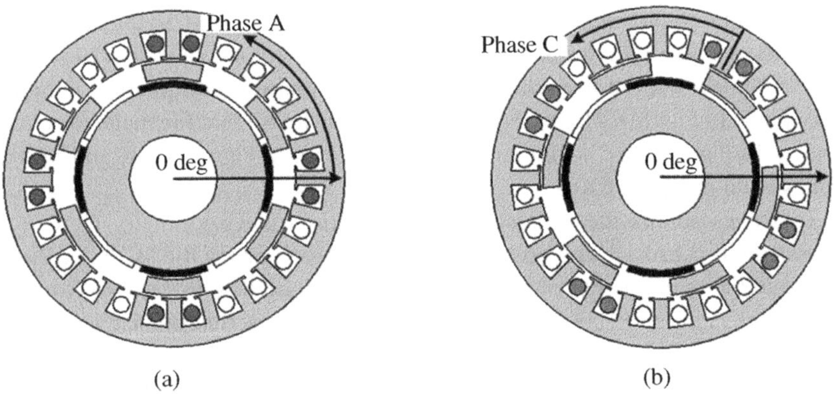

Figure 6.83 Definitions of coordinate systems for phase A and phase C airgap flux density calculation. (a) Starting point of flux density calculation for phase A at the initial rotor position, (b) Starting point of flux density calculation for phase C when the outer rotor rotates by 120 electrical degrees.

back-EMF of phase A is different from those of phase B and C. The phase B and phase C no-load back-EMFs mirror each other.

The airgap flux density harmonics are also analyzed to reveal the cause of the magnetic circuit asymmetry from the perspective of the GAFMT. Figure 6.83a,b show the starting point of flux density waveforms for phase A at initial rotor position and for phase C when the outer rotor rotates by 120 electrical degrees, respectively. The magnetic density waveforms for phase A and phase C at these two moments are shown in Figure 6.84.

It can be seen from Figure 6.84 that the two flux density waveforms are completely different. However, when examining the waveforms from the starting points defined in Figure 6.83, the amplitudes and phase angles of their effective

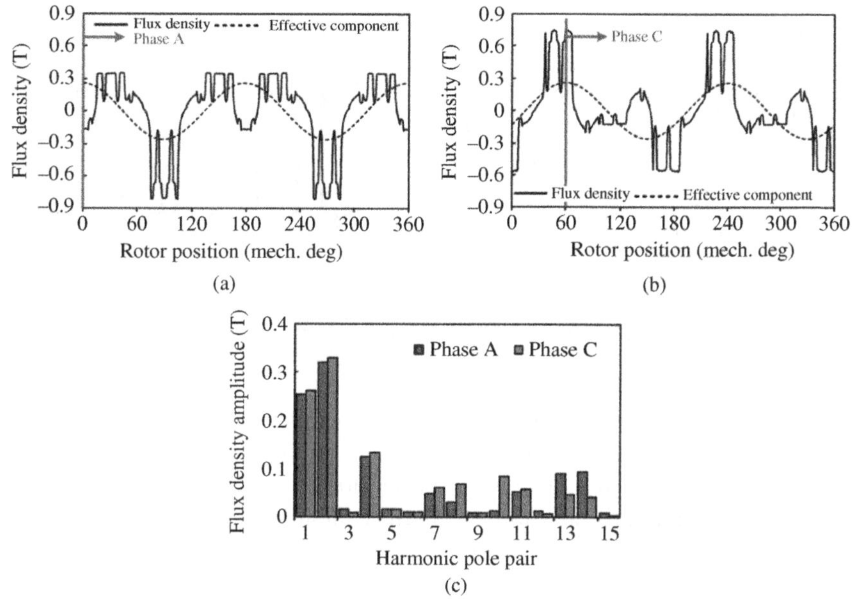

Figure 6.84 Harmonic analysis of airgap flux density waveforms. (a) Flux density for phase A at initial rotor position, (b) Flux density for phase C when the outer rotor rotates by 120 electrical degrees, (c) Amplitude spectrums for phase A and phase C.

flux density components are exactly the same. This explains why the amplitudes of the fundamental components of the no-load back-EMFs per phase are the same.

It can be seen from the amplitude spectrum of the flux density waveforms that the high-order spatial harmonics are totally different. Amplitudes of these harmonics change when the rotor rotates, which is responsible for the observed asymmetry in airgap flux density waveforms, flux linkages, and back-EMFs.

According to the GAFMT, airgap flux density harmonics with different pole-pair numbers are introduced when the PM MMF is modulated by the rotor salient reluctance poles. Among them, the ν pole-pair harmonic can be expressed as:

$$B_\nu(\theta, t) = \sum_{m,n=1,3,5,\ldots}^{\infty} \frac{a_n \lambda_m}{2} \cos\left[\nu\left(\theta - \frac{mp_{or}\omega_{or} - np_{ir}\omega_{ir}}{\nu}t + \frac{mp_{or}\beta_0 - np_{ir}\alpha_0}{\nu}\right)\right] \tag{6.73}$$

where the ν-pole-pair harmonic is contributed by multiple harmonics, which satisfy (6.74) with different m and n.

$$\nu = |mp_{or} - np_{ir}| \tag{6.74}$$

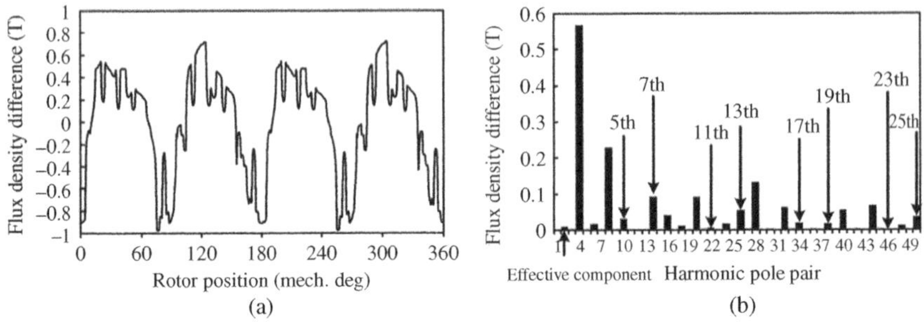

Figure 6.85 The influence of spatial harmonics on airgap flux density distortion. (a) Flux density difference between phase A and phase C, (b) Difference of each harmonic.

Since the PM MMF and the airgap permeance waveform are close to trapezoidal waves, Fourier coefficients for high-order harmonics are much lower than low-order harmonics, m and n are expected to be relatively small. Taking the 14-pole-pair harmonic as an example, all the combinations of m and n satisfying equation (6.75) can contribute to it.

$$|6m - 4n| = 14 \tag{6.75}$$

Among various combinations, the combination with the smallest m and n is considered as the main harmonic component. In order to find out the harmonic components that cause the asymmetry in airgap flux density waveforms, the difference between two airgap flux density waveforms shown in Figure 6.84 is calculated and illustrated in Figure 6.85a.

The waveform difference contains rich spatial harmonic components, whose amplitudes are different in the flux density waveforms for phase A and phase C shown in Figure 6.84. The asymmetry in flux density waveforms is mainly caused by these amplitude-changing harmonics. It should also be noted that not all spatial harmonic components induce no-load back-EMF in the armature winding. Only the $(6k \pm 1)$th harmonic components will affect the back-EMF and output torque, where k is a positive integer. The difference of each harmonic is shown in Figure 6.85b.

6.7.4 Complementary MGPSM and Experimental Validation

In order to suppress the influence of the magnetic circuit asymmetry in MGPSMs, a complementary MGPSM has been proposed by skewing the source MMF created by PMs. It is well known that the rotor skewing is an effective technique to suppress the magnetic circuit asymmetry and associated spatial harmonics. The phase angle of the airgap flux density fundamental component is directly determined by the rotor position in traditional PM synchronous machines, and the rotor skewing will reduce the average electromagnetic torque. By contrast, since the airgap flux density of the MGPSM is determined by both positions of the inner and outer rotors, the magnetic circuit asymmetry can be suppressed based on the rotor skewing but the undesired reduction in the output torque can also be avoided by selecting the skew angles of two rotors properly.

1 Torque Preservation Principle and Design Rules

For the convenience of description, the airgap field modulation process in the MGPSM is described based on a simple magnetic circuit as shown in Figure 6.86.

In Figure 6.86, θ_{ir} and θ_{or} are the angular positions of the inner and outer rotors, respectively, and their changes describe the rotation of the inner and outer rotors. $\mathfrak{R}_{og}$ and $\mathfrak{R}_{ig}$ are the reluctance of the inner airgap and the outer airgap, respectively. Since the inner and outer airgaps of the electrical machine are smooth and uniform, their airgap reluctance is a constant along the circumferential direction. $\mathfrak{R}_r(\theta, \theta_{or})$ refers to the reluctance of the outer rotor area. The outer rotor is composed of ferromagnetic modulation blocks and non-magnetic blocks. The magnetic permeability of modulation blocks is very high, while in non-magnetic blocks, the permeability is close to the air. Therefore, the permeance of the outer rotor changes along the circumference, and changes drastically on the interface between the modulation blocks and the non-magnetic blocks.

It can be seen that the permeance of the outer rotor changes alternately and its value is always greater than zero. Its shape is close to a rectangular wave. The reluctance of each point along the circumference changes with the rotation of the outer rotor, i.e. θ_{or}. $F_{PM}(\theta, \theta_{ir})$ represents the PM MMF created by the magnets on the inner rotor. The PM MMF distribution is an alternating function with symmetrical

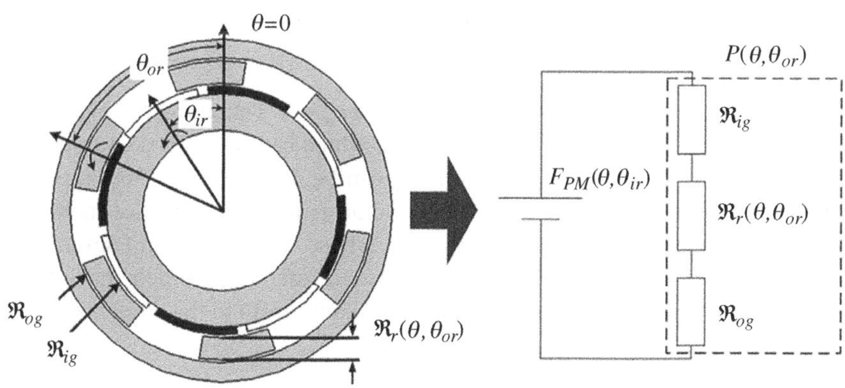

Figure 6.86 Description of MGPSM with no-load operation by simple magnetic circuit method.

positive and negative half cycles. Moreover, the PM MMF waveform is close to the trapezoidal wave. The PM MMF can be expressed as:

$$F_{PM}(\theta, \theta_{ir}) = \sum_{l=1,3,5\ldots}^{\infty} a_l \cos[lp_{ir}(\theta - \theta_{ir} + \theta_{ir0})] \tag{6.76}$$

The permeance coefficient is defined as the permeance per unit area in the magnetic circuit. In the simple magnetic circuit shown in Figure 6.86, the permeance coefficient can be written as:

$$P(\theta, \theta_r) = \frac{1}{\Re_{og} + \Re_{ig} + \Re_r(\theta, \theta_{or})} = c_0 + \sum_{m=1}^{\infty} c_m \cos[mp_{or}(\theta - \theta_{or} + \theta_{or0})] \tag{6.77}$$

Since the flux density is the flux per unit area, the flux density distribution can be obtained by multiplying the MMF distribution $F_{\mathrm{PM}}(\theta, \theta_{ir})$ and the airgap permeance per unit area $P(\theta, \theta_{or})$. Then the airgap flux density can be calculated by:

$$\begin{aligned} B(\theta, \theta_{ir}, \theta_{or}) = F_{PM}(\theta, \theta_{ir})P(\theta, \theta_{or}) = &\sum_{l=1,3,5\ldots}^{\infty} a_l c_0 \cos[lp_{ir}(\theta - \theta_{ir} + \theta_{ir0})] \\ &+ \sum_{l=1,3,5\ldots}^{\infty} \sum_{m=1}^{\infty} \frac{a_l c_m}{2} \\ &\times \cos\left[(mp_{or} + lp_{ir})\left(\theta - \frac{mp_{or}\theta_{or} + lp_{ir}\theta_{ir}}{mp_{or} + lp_{ir}} + \frac{mp_{or}\theta_{or} + lp_{ir}\theta_{ir0}}{mp_{or} + lp_{ir}}\right)\right] \\ &+ \sum_{l=1,3,5\ldots}^{\infty} \sum_{m=1}^{\infty} \frac{a_l c_m}{2} \\ &\times \cos\left[(mp_{or} - lp_{ir})\left(\theta - \frac{mp_{or}\theta_{or} - lp_{ir}\theta_{ir}}{mp_{or} - lp_{ir}} + \frac{mp_{or}\theta_{or} - lp_{ir}\theta_{ir0}}{mp_{or} - lp_{ir}}\right)\right] \end{aligned} \tag{6.78}$$

In the complementary prototype, $p_{or} = 6$ and $p_{ir} = 4$. Therefore, the 2-pole-pair harmonic of the outer airgap magnetic field, which can produce the torque output by interacting with the armature field, can be obtained only from the third term of (6.78). Since both the PM MMF waveform and the permeance coefficient waveform are close to the trapezoidal wave, whose low-order term Fourier coefficients are much larger than high-order ones, a_1 and m_1 are their maximums. When $m = l = 1$, the 2-pole-pair magnetic field component in the outer airgap can be written as:

$$B_{1,1}(\theta, \theta_{ir}, \theta_{or}) = \frac{a_1 c_1}{2} \cos[(p_{or} - p_{ir})\theta - (p_{or}\theta_{or} - p_{ir}\theta_{ir}) + (p_{or}\theta_{or0} - p_{ir}\theta_{ir0})] \tag{6.79}$$

It can be seen that the position of the 2-pole-pair harmonic is determined by both positions of the inner and outer rotors. Moreover, it is very interesting that when the positions of the inner and outer rotors satisfy:

$$p_{or}\theta_{or} = p_{ir}\theta_{ir} \tag{6.80}$$

the position of the 2-pole-pair harmonic will remain unchanged. This is because:

$$\begin{aligned} &B_{1,1}(\theta, \theta_{ir} + \Delta\theta_{ir}, \theta_{or} + \Delta\theta_{or}) \\ &= \frac{a_1 c_1}{2} \cos[(p_{or} - p_{ir})\theta - (p_{or}\theta_{or} + p_{or}\Delta\theta_{or} - p_{ir}\theta_{ir} - p_{ir}\Delta\theta_{ir}) + (p_{or}\theta_{or0} - p_{ir}\theta_{ir0})] \\ &= B_{1,1}(\theta, \theta_{ir}, \theta_{or}) \end{aligned} \tag{6.81}$$

It can be concluded that as long as the rotational angles of the inner and outer rotors satisfy equation (6.80), the amplitude and phase of the outer airgap flux density can be kept unchanged, so that the output torque can be preserved. This finding inspired the authors to appropriately design the rotor skew angle.

2 Solutions for Magnetic Circuit Asymmetry

According to (6.81), if skewing is used for one of the rotors to weaken non-working spatial harmonics, the amplitude and phase of effective components of the air-gap flux density can be kept unchanged by adopting the appropriate relative rotor skewing for the other rotor, so as to preserve the output torque. In the prototype, both the inner and outer rotors are divided into three segments (named M1, M2, and M3 along the axial direction), and each segment rotates by a certain angle relative to the other, as shown in Figure 6.87. Since each segment is displaced by a certain angle relative to the neighboring segment, additional flux leakage will be introduced between adjacent segments. Non-magnetic material has been used to separate these segments and reduce the flux leakage. The three outer rotor segments are shifted by $\Delta\theta_{or0}$ along the circumference in turn and the three inner

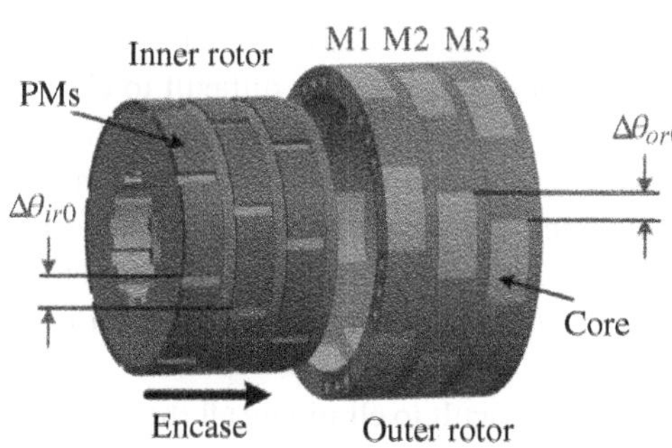

Figure 6.87 Schematic diagram of complementary rotor structure.

rotor segments are shifted by $\Delta\theta_{ir0}$ along the circumference. In order to maintain the same average torque output, the following constraint has to be met

$$p_{or}\Delta\theta_{or} = p_{ir}\Delta\theta_{ir} \tag{6.82}$$

To eliminate the influence of the magnetic circuit asymmetry on the no-load back-EMF and torque ripple, it is necessary to focus on eliminating the $(6k \pm 1)$-th harmonics. In order to eliminate the ν-th harmonic in the MGPSM, the segmented rotor skewing method as shown in Figure 6.87 can be used with $\Delta\theta_{ir0}$ and $\Delta\theta_{or0}$ satisfying:

$$B_\nu(\theta, \theta_{ir}, \theta_{or}) + B_\nu(\theta, \theta_{ir} + \Delta\theta_{ir0}, \theta_{or} + \Delta\theta_{or0}) + B_\nu(\theta, \theta_{ir} + 2\Delta\theta_{ir0}, \theta_{or} + 2\Delta\theta_{or0}) = 0 \tag{6.83}$$

According to (6.78), the ν-th harmonic component can be written as:

$$B_\nu(\theta, \theta_{ir}, \theta_{or}) = \sum_{m,l=1,3,5\ldots}^{\infty} \frac{a_l c_m}{2} \cos[\nu\theta - (mp_{or}\theta_{or} - np_{ir}\theta_{ir}) + (mp_{or}\theta_{or0} - np_{ir}\theta_{ir0})] \tag{6.84}$$

In order to achieve the effect of (6.83), the rotated angle among the three rotor segments is designed as:

$$mp_{or}\Delta\theta_{or0} - lp_{ir}\Delta\theta_{ir0} = k\frac{2\pi}{3}, k \text{ is not an integer multiple of } 3 \tag{6.85}$$

It can be seen that different rotated angles can be obtained when different values of k are selected. Since the rotated arrangement will cause flux leakages between adjacent segments, $\Delta\theta_{ir0}$ and $\Delta\theta_{or0}$ should be as small as possible in order to minimize this leakage. The value of k should be ± 1. Combining (6.82) and (6.85), it is easy to obtain:

$$|m - l|p_{ir}\Delta\theta_{ir0} = |m - l|p_{or}\Delta\theta_{or0} = \frac{2\pi}{3} \tag{6.86}$$

In (6.86), the presence of m and l means that the ν-th harmonic is actually composed of many combined components. It is difficult to eliminate all the components of the ν-th harmonic. Therefore, in practical operations, the shift angle can be designed to eliminate the dominating low-order component. Based on the analysis of the flux density difference waveform in Figure 6.85b, it can be seen that the 7th harmonic has the greatest influence. In order to eliminate this 14-pole-pair air-gap flux density harmonic, the calculated rotation angles for various combinations of m and l are listed in Table 6.17.

As discussed above, it is difficult to eliminate all the components with different combinations of m and l in Table 6.20 at the same time, but $p_{ir}\Delta\theta_{ir0}$ can be designed as $\pi/3$ to eliminate the closest combination ($m = 3$, $l = 1$, and $|m - l| = 2$). Moreover,

Table 6.17 Components and rotated angles of the 7th harmonic.

m	−9	−5	−1	**3**	7	11	15
l	−17	−11	−5	**1**	7	13	19
$\|m-l\|$	8	6	4	**2**	0	2	4
$p_{ir}\Delta\theta_{ir0}$	π/12	π/9	π/6	**π/3**	-	3	π6

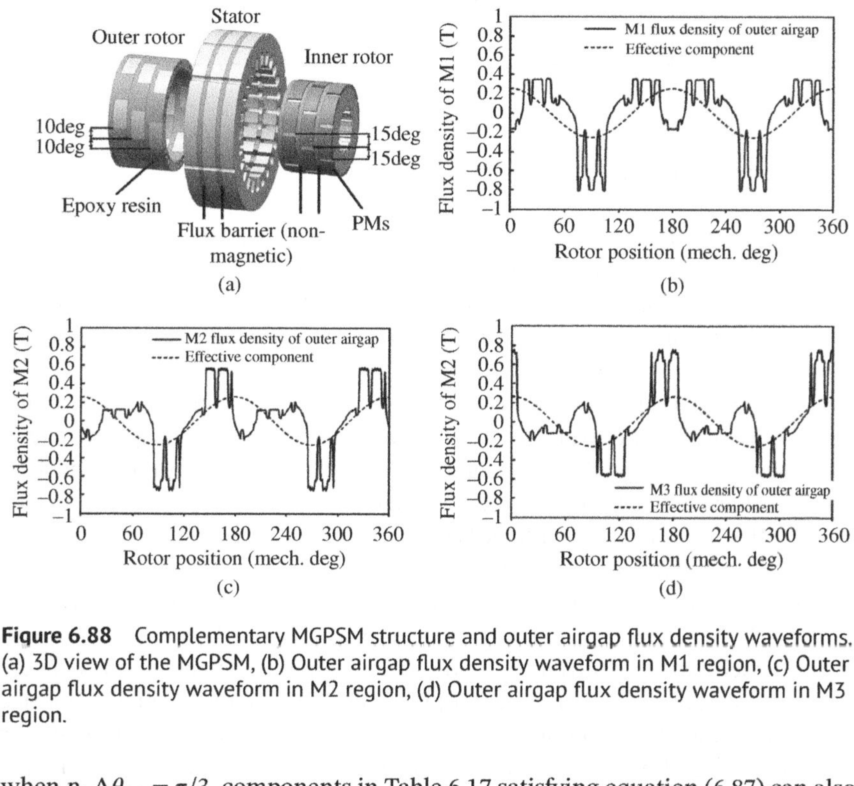

Figure 6.88 Complementary MGPSM structure and outer airgap flux density waveforms. (a) 3D view of the MGPSM, (b) Outer airgap flux density waveform in M1 region, (c) Outer airgap flux density waveform in M2 region, (d) Outer airgap flux density waveform in M3 region.

when $p_{ir}\Delta\theta_{ir0} = \pi/3$, components in Table 6.17 satisfying equation (6.87) can also be eliminated.

$$|m-l|\frac{\pi}{3} = k\frac{2\pi}{3}, \text{k is not an integer multiple of 3} \tag{6.87}$$

Substituting $m = 3$, $l = 1$, $p_{ir} = 4$ and $p_{or} = 6$ into (6.86), it can be obtained that $\Delta\theta_{\text{ir0}}$ and $\Delta\theta_{\text{or0}}$ for eliminating the 14-pole-pair harmonic are 15 and 10 mechanical degrees, respectively. As a result, the MGPSM can be designed as shown in Figure 6.88a, in which the stator is also divided into three segments separated by non-magnetic material to reduce the flux leakage.

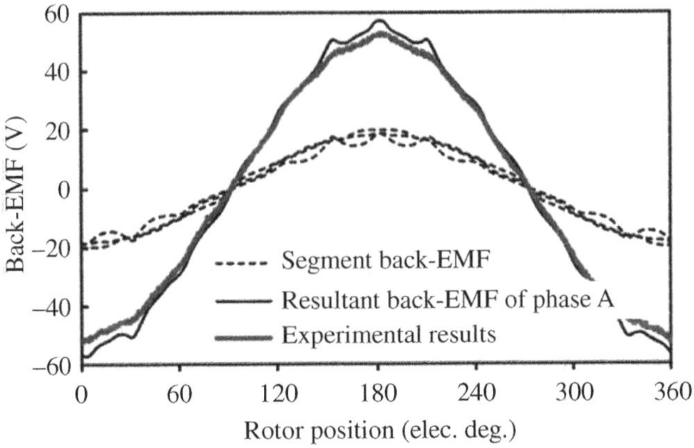

Figure 6.89 Resultant no-load back-EMF of phase A.

Figure 6.88b–d are the outer airgap flux density waveforms of the three segments. It can be seen that although the airgap flux density waveforms of the three modules change greatly, the amplitudes and phase angles of the 2-pole-pair effective harmonic components are the same, which ensures that the output torque will not be reduced by the rotor skewing.

The segmented rotor skewing technique eliminates the asymmetry in flux linkage waveforms and no-load back-EMF waveforms, though the magnetic circuit asymmetry still exists within each rotor segment. In order to verify the effect of the proposed harmonic elimination method, Figure 6.89 shows the resultant no-load back-EMF waveform along with the back-EMF from each segment when the inner rotor is fixed and the outer rotor rotates at 1000 r/min.

Compared with Figure 6.82b, it can be found that the no-load back-EMF waveform has become more sinusoidal at the same speed. This is because the resultant no-load back-EMF is contributed by the three segments together so those undesired spatial harmonics with different phases can be partially or totally canceled by implementing the appropriate rotation angle. Then the balanced three-phase no-load back-EMF waveforms can be obtained. This harmonic cancelation phenomenon is called "complementarity" in this section and the MGPSM showing this feature is termed as the "complementary MGPSM." More possible pole-pair combinations and rotation angle designs can be found in [50].

3 Experimental Verification

A complementary MGPSM design has been prototyped and tested for experimental validation, as shown in Figure 6.90.

Figure 6.91a shows the no-load back-EMF waveforms when both the inner and outer rotors were at the rated speeds. Figure 6.91b shows the no-load back-EMF

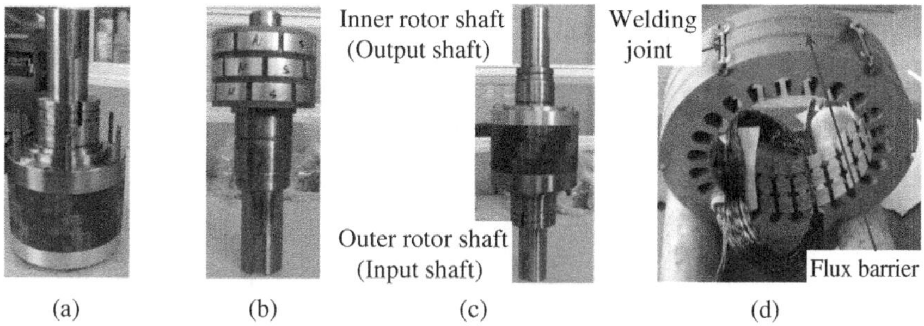

Figure 6.90 Photos of the 1st prototype. (a) Outer rotor, (b) Inner rotor, (c) Assembly of two rotors, (d) Stator.

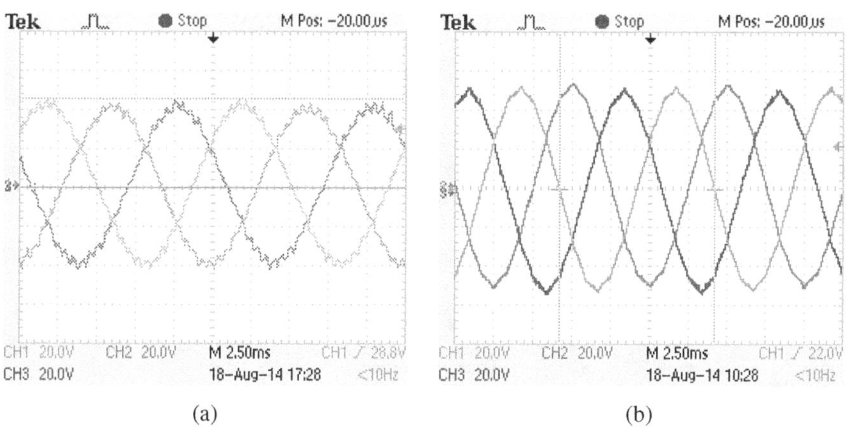

Figure 6.91 No-load back-EMF waveforms (20 V/div). (a) Inner rotor at 1200 r/min, outer rotor at 1600 r/min, (b) Inner rotor at 0 r/min, outer rotor at 1000 r/min.

waveforms when the outer rotor rotated only (n_{or} = 1000 r/min). It can be seen that the no-load back-EMFs are sinusoidal and show good symmetricity.

Figure 6.92 shows the measured cogging torque and no-load back-EMF waveforms when the inner rotor was fixed and the outer rotor rotated by 360 mechanical degrees. In order to reduce the signal interference, the outer rotor speed was set to 50 r/min. The no-load back-EMF waveform was measured at the same time to serve the purpose of counting the cogging torque cycles. It can be seen that there are 12 cycles in the cogging torque waveform in one electrical cycle of the outer rotor, and that the amplitude of the cogging torque is about 0.2 Nm, which is consistent with the simulation results [50].

Figure 6.93a shows the measured average torque of the inner rotor (traction torque) and outer rotor (input torque) versus the current, and the 2D FEA results

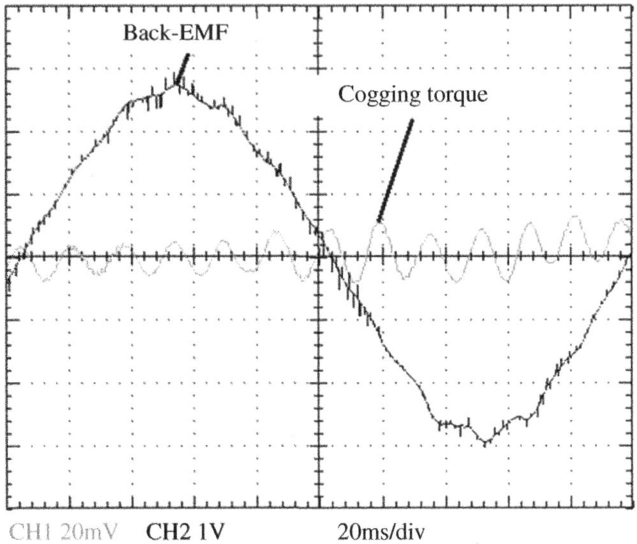

Figure 6.92 Measured cogging torque of the outer rotor in one electrical cycle (cogging torque 0.4 Nm/div, no-load back-EMF 1 V/div).

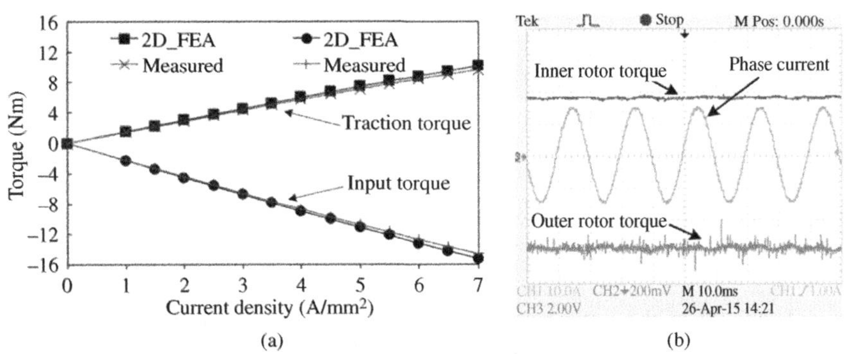

Figure 6.93 Measured steady-state torque. (a) Torque vs. current density, (b) Steady-state torque waveform (current 10 A/div, torque 4 Nm/div).

are also depicted for comparison. It can be seen that the torques change with the increase of the current linearly, which indicates that the MGPSM has good overload capability.

Figure 6.93b shows the torque of the inner and outer rotors and the corresponding current waveforms in the steady state. It can be seen that the current is sinusoidal, and the torque of the inner and outer rotors is very smooth. The measured outer rotor torque is $T_{or} = 10.77$ Nm and inner rotor torque is $T_{ir} = 7.08$ Nm, both

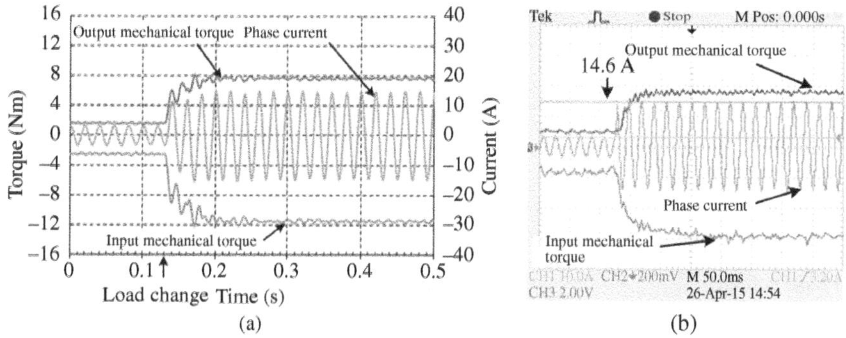

Figure 6.94 Dynamic response. (a) Simulation results, (b) Measured results (current 10 A/div, torque 4 Nm/div).

of which reach 96% of the 2D FEA results (T_{or} = 11.12 Nm and T_{ir} = 7.42 Nm). The difference is mainly caused by the prototyping process and material property errors.

Figure 6.94 shows the dynamic response when the load torque was increased abruptly with the inner rotor fixed and the outer rotor dragged by an induction machine to a constant speed of 500 r/min with $i_d = 0$ control. It can be seen that when i_q increased from the light-load operation to full-load operation abruptly, the torque of the inner and outer rotors changed accordingly. Figure 6.94a shows the MATLAB/Simulink simulation results, and Figure 6.94b depicts the measured waveforms. It can be seen that the simulated and measured results are in good agreement.

With the confidence gained by the first prototype shown in Figure 6.90, the design parameters were analyzed and optimized further to leverage the complementary MGPSM structure. The second complementary MGPSM prototype was manufactured and tested, whose main design parameters are tabulated in Table 6.18. In addition to the differences of dimensions and pole numbers, the main changes are that the spoke-type PM arrangement has been employed for the inner rotor and the two-segmented complementary structure has been adopted, as shown in Figure 6.95 [48, 49].

The second prototype was tested on a hybrid powertrain test platform as shown in Figure 6.96. Testing results of the prototype in different operation modes verify that the complementary MGPSM has similar functions to the planetary gear power splitter used in Toyota Prius hybrid vehicles, namely the power generation, the ICE starting when the vehicle stops and runs, power split, engine speed self-stabilization, torque coordination, etc.

Figure 6.97 shows the measured current and torque waveforms in steady state. To simplify the control, the MGPSM was tested at a relatively low speed. The inner

Table 6.18 Main parameters and design specifications of the prototype.

Parameter	Symbol	value
Rated power	P_{in}	12 kW (51 Nm at 2250 r/min)
Outer rotor rating	T_{or} at n_{or}	51 Nm at 2250 r/min
Inner rotor rating	T_{ir} at n_{ir}	36 Nm at 1500 r/min
Poles configuration	p_{or}, p_{ir}, p_w	10, 7, 3
Phase voltage	U	≤100 Vrms
Stator outer diameter	D_e	230 mm
Magnet remanence	B_r	1.28 T (NdFeB)
Rated current density	J_{max}	14.8 A/mm^2 (peak value, water cooling)
Stack length	L_e	32 mm
Airgap length	g_{in}, g_{out}	0.5 mm
Number of slots	Q	36
Number of slots per pole per phase	q_w	2
Winding factor	K_w	0.966

Figure 6.95 Photos of the 2nd MGPSM Prototype. (a) 3D drawing, (b) Rotor assembly, (c) Prototype with the cooling jacket.

rotor speed was 426 r/min and the outer rotor was fixed. The operating frequency was 50 Hz and $i_d = 0$ control was used. The torque and current were measured under the half-load and full-load operation conditions.

It can be seen that the current waveforms are very sinusoidal, but there are some fluctuations in the torque waveforms. Further inspections showed that the torque pulsation did not increase with the current, so it can be concluded that the torque pulsation is not caused by electromagnetic interactions with the machine. Considering the negligible cogging torque, it can be inferred that the pulsation comes from the dynamometer. In the half-load operation shown in Figure 6.97a,

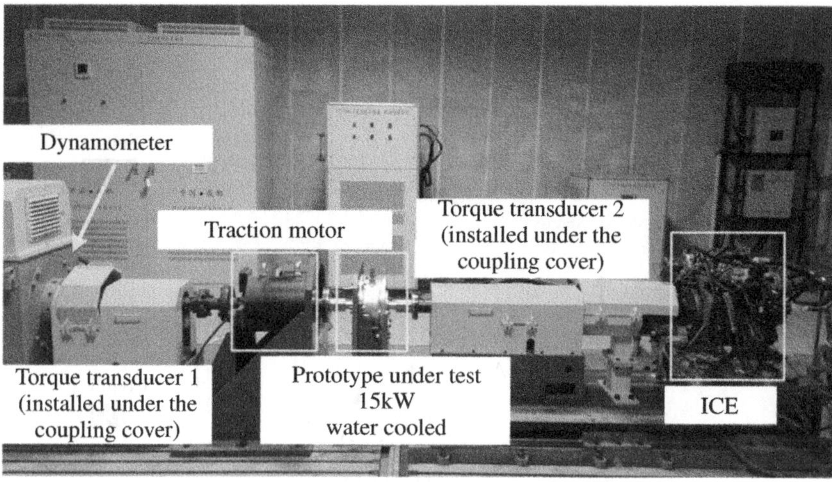

Figure 6.96 Complementary MGPSM prototype test platform.

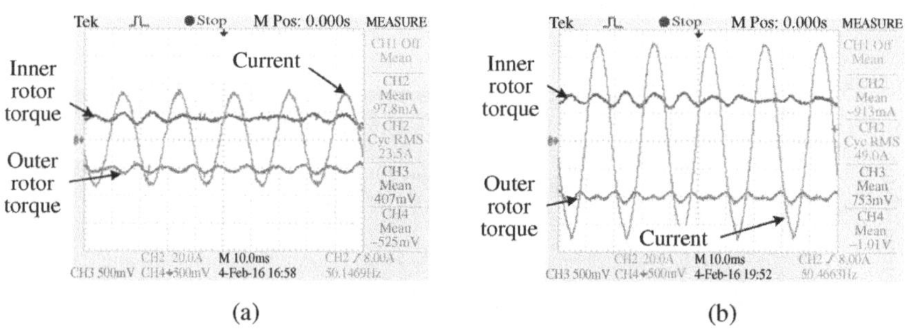

Figure 6.97 Measured torque and current waveforms. (torque 25 Nm/div, current 20 A/div), (a) With half load, (b) With full load.

the phase current is 23.5 Arms. The measured outer rotor torque is 26.25 Nm, and the inner rotor torque is 20.35 Nm. In the full-load operation depicted in Figure 6.97b, the phase current is 49 Arms. The measured outer rotor torque is 50.5 Nm, and the inner rotor torque is 37.65 Nm.

Figure 6.98 shows the torque dynamic response at a constant speed when the outer rotor was fixed and the inner rotor was dragged by the dynamometer to a constant speed of 428 r/min. Correspondingly, the electrical frequency is 50 Hz and the electrical cycle is 20 ms. The current increased step by step every 100 ms from 8 A (light load) until the current reached its rated value. The torques of inner and outer rotors increased accordingly until reaching the rated torque.

Figure 6.99 shows the efficiency map of the full operation range of the outer rotor when the inner rotor speed was fixed at 1500 r/min. The efficiency value represents

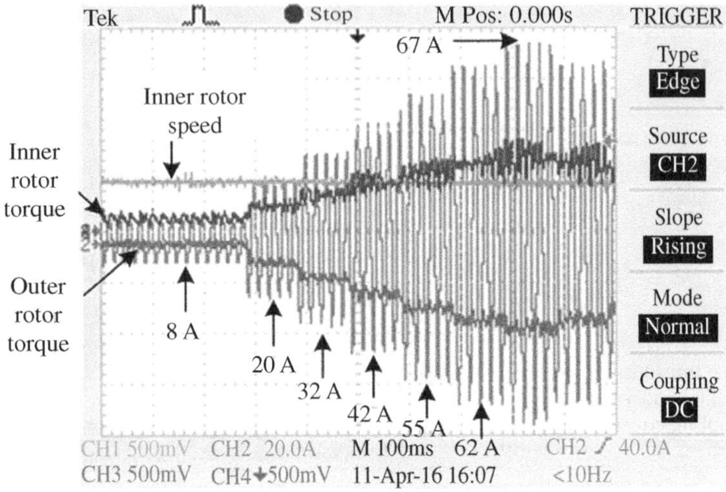

Figure 6.98 Torque response at a constant speed. (Torque 25 Nm/div, speed 500 r/min/div, current 20 A/div).

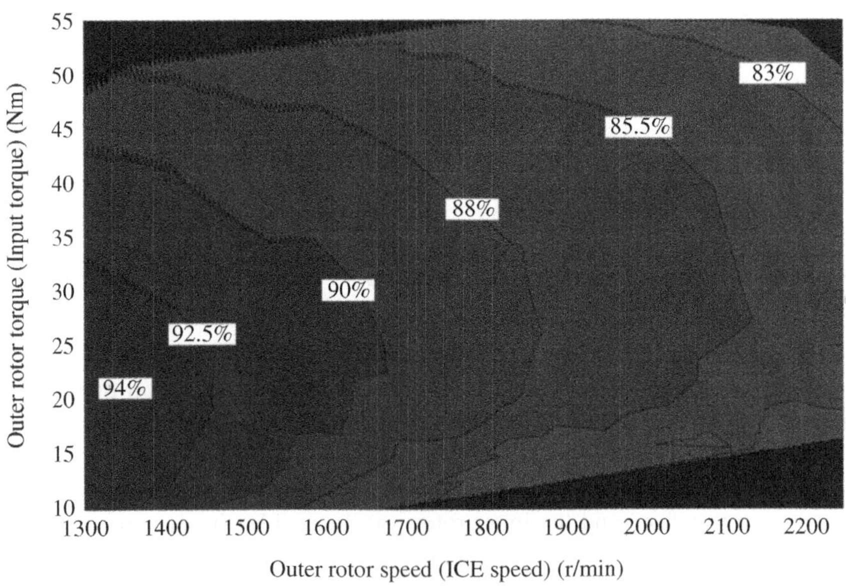

Figure 6.99 Measured efficiency map of the MGPSM (η_{sys}).

the ratio of the output power (divided into mechanical power and electrical power) to the total input power of the entire system. It can be observed that:

1. The lower the speed of ICE is, the higher the efficiency of the system will be. This is because the slip frequency is lower when the ICE speed is close to the inner rotor speed. A low slip frequency means the power conversion completed by the armature winding is taking a small part. The direct power transmission from the outer rotor to the inner rotor is very efficient, and the power loss mainly occurs in the power conversion.
2. In the low-speed region, the torque has a great influence on the efficiency. In contrast, in the high-speed region, the influence is relatively small so the efficiency is relatively stable. This is because the loss in the low-speed region is dominated by the copper loss, which is closely related to the torque, while the loss in the high-speed region is dominated by the iron loss.
3. The operation of the system is close to a parallel hybrid powertrain when the speed is low, and a series hybrid powertrain when the speed is high. It can be seen from the efficiency map that the efficiency of the parallel hybrid mode is higher than that of the series hybrid mode. The speed of the ICE is required to vary with vehicle speed in parallel hybrid powertrains, which reduces the efficiency of the ICE. In general, both the series and the parallel systems have their own merits. The advantage of the hybrid system is to obtain a good trade-off, and the system efficiency can be flexibly adjusted by controlling the slip speed of inner/outer rotors and the armature current. More analysis and testing results of the 2nd complementary MGPSM prototype can be found in [50].

6.8 Stator Field-Excitation HTS Machines

Electrical machines using superconducting (SC) materials are called SC electrical machines. Compared with conventional electrical machines, SC machines have higher specific electric or magnetic loadings, offering the merits of higher power density and higher efficiency, because the DC resistance of SC materials is zero. SC electrical machines have attracted more and more attention in some special application fields, such as large-capacity high-torque-density propulsion machine drives, high-power direct-drive wind energy generators, etc. Since the first low-temperature SC electrical machine was proposed and verified by Aveo-Everett Laboratory of the United States [51] in the 1960s, the research boom of SC electrical machines has expanded globally, and successive SC prototypes have been developed as a result. The most typical cases are the 36.5 MW ship propulsion SC machine developed in the United States [52] and the 3.6 MW direct-drive SC wind energy generator developed by Envision Energy [53].

- Superconducting machine
  - Rotating seal superconducting machines
  - Stationary seal superconducting machines
    - Stationary armature
    - Rotating armature

Figure 6.100 Classification of SC electrical machines.

From the perspective of the operating principle, all existing SC electrical machines are essentially synchronous machines. SC electrical machines are usually classified based on the structure of the cryogenic cooling system into two categories: one with a rotating seal and the other with a stationary seal. A cryogenic transmission coupler has to be employed in a rotating seal SC machine. However, for stationary seal SC machines, whether the SC windings are used as the armature winding or the field winding, the SC windings need to be stationary. Therefore, based on whether the armature winding rotates, stationary seal SC machines can be further divided into two types: the armature rotating type and the armature stationary type. The classification is shown in Figure 6.100 [54].

Regardless of the seal type, SC machines can be divided into two types as well: air core and iron core structures, according to whether the iron core is adopted for the stator and rotor.

The SC wire is usually used as the field winding in rotating seal SC machines to generate a strong magnetic field by feeding DC current. When the machine is running, the SC field winding rotates with the rotor at a synchronous speed, which is relatively stationary with the fundamental armature reaction field in the space. AC loss can almost be made negligible in the SC winding. However, in order to keep superconductivity of the SC field winding on the rotor, the coolant must be delivered to the rotor through a rotating seal cryogenic transmission coupler, which results in a complicated structure and high manufacturing costs. Therefore, an important direction of research has involved how to remove the rotating coupler and achieve a stationary coolant seal.

The rotating armature type of SC machines is a feasible solution for a stationary seal because it locates the SC field winding and the armature winding on the stator and the rotor, respectively. Although a stationary seal of coolant can be achieved with a rotating armature structure, the brush slip ring normally has to be used to transmit the current into the armature winding, which reduces the machine system's reliability, particularly if it has a high capacity.

The other stationary seal solution is the stationary armature type, in which the armature winding made of SC wire is placed on the stator. Cooling is achieved by the stationary seal. However, the armature winding of synchronous machines is typically fed by AC current, which leads to relatively high AC copper loss in the

SC wire in high-current conditions. This both reduces the efficiency and limits the critical current (and therefore the advantages) of the SC wire.

In recent years, one kind of stator SC excitation field modulation (SSE-FM) machine has been proposed by the authors' research group, based on the airgap field modulation theory. The proposed SSE-FM machine not only realizes the stationary seal of coolant but also transmits the DC current to the SC winding, showing significant characteristics and advantages.

6.8.1 Stator Field-Excitation HTS Flux-Switching Machine

Figure 6.101 shows the structure of stator SC excitation flux-switching (SSE-FS) machine [55–57]. Compared with the flux switching PM machine [24], the operating principle of the proposed SSE-FS machine is the same, and the copper armature winding is also adopted. The key difference is that the SC field winding is employed to replace the PMs, and the DC current is fed to generate the excitation field. Both the SC field winding and the armature winding are located on the stator, so that brushless current transmission and a stationary coolant seal can be simultaneously achieved. A prototype was designed and manufactured by the China University of Petroleum. The excitation winding was made of BSCCO-2223 wire, where BSCCO stands for Bismuth strontium calcium copper oxide, and low-cost liquid nitrogen was adopted to keep the temperature below 77 K. A racetrack-shaped SC field winding was designed and and was housed between the inner and outer Dewars, as shown in Figure 6.102.

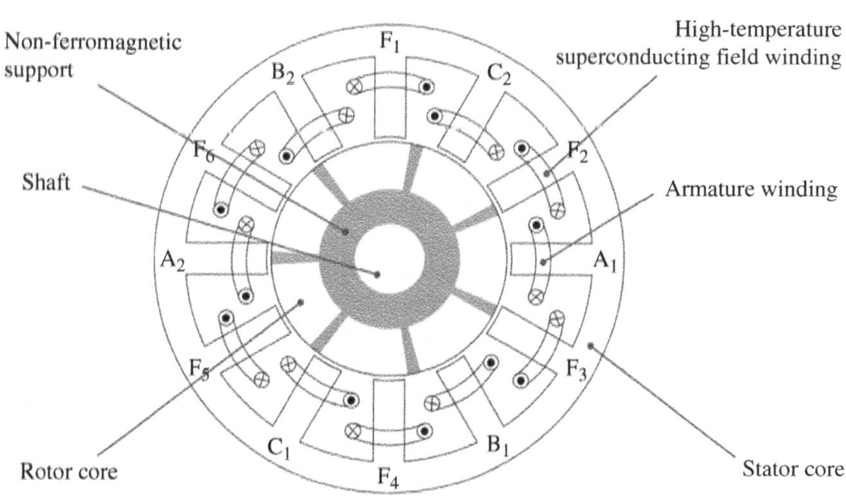

Figure 6.101 Configuration of stator SC excitation flux switching machine.

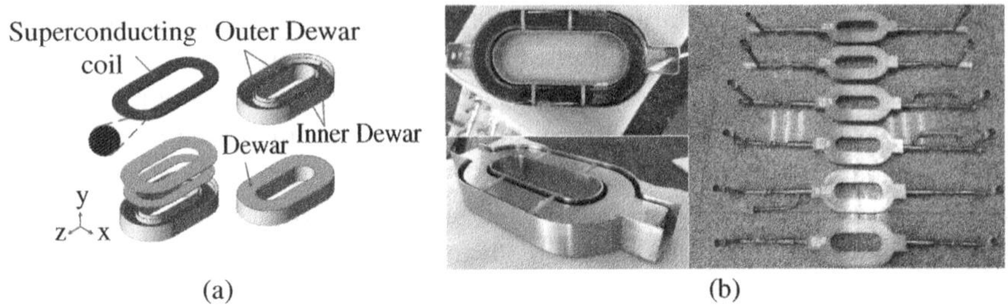

Figure 6.102 SC field winding and Dewar. (a) Schematics, (b) Photos.

Figure 6.103 Prototype and its experimental system. (a) Prototype assembly, (b) Prototype and its cooling device.

Figure 6.103 shows the assembly process of the prototype and the prototype experimental system. A prime machine drives the SSE-FS machine to operate as a generator. The no-load EMF waveforms at different excitation currents are shown in Figure 6.104.

Although the experimental results of the prototype verify the principle and feasibility of the SSE-FS machine, it also reveals some potential problems. First, the Dewar housing the SC field winding occupies a large space in the stator, which leads to difficulty in arranging the armature winding, and also results in a larger stator size. Second, the rotating armature reaction field has a serious impact on the SC field winding, thus limiting the current carrying capacity of the SC winding. Therefore, based on the airgap field modulation theory, a double-stator field modulation SC excitation (DS-FMSE) machine [58] was proposed. The principle and main characteristics are introduced below.

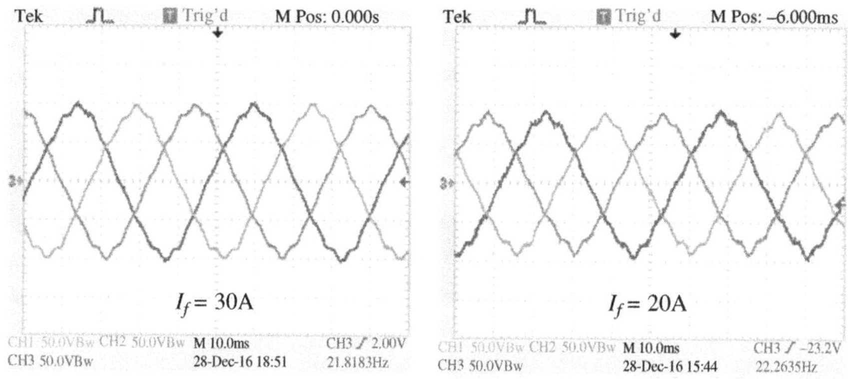

(a) (b)

Figure 6.104 Measured no-load back-EMF of prototype (n = 200 r/min). (a) Field current 30A, (b) Field current 20A.

6.8.2 Double-Stator Field Modulation Superconducting Excitation Machine

1 Machine Construction and Operating Principle

According to the airgap field modulation theory in Chapter 3, the excitation source is in charge of establishing an initial MMF. Then the modulated MMF with multiple harmonics is obtained because of the modulation of the modulator to the initial MMF. Finally, the armature winding selects several harmonic MMF to achieve the energy conversion. Whether the excitation source is PMs or wound field, whether it is located on the stator or rotor, as long as the established initial MMF is the same, the excitation effect will be the same. Therefore, when the SC field winding is moved from the stator, as shown in Figure 6.101, to the inner stator inside the rotor, a DS FMSE machine can be obtained. Thus, only the armature winding is placed on the outer stator (benefiting from the formed double-stator structure), and the SC winding is located on the inner stator. The superconductor can be cooled by the stationary seal and the problem of space competition between the SC field winding and the armature winding is solved.

Figure 6.105 shows the structure of DS-FMSE machine. The inner stator, the rotor, and the outer stator are arranged coaxially. The copper armature winding is placed on the outer stator, and SC magnets composed of SC coil and cryostat are placed on the inner stator. DC current is fed into the SC field winding on the inner stator to generate a stationary excitation MMF. Modulated by the rotor poles, a series of rotating harmonic magnetic fields are generated in the airgap, and the armature winding on the outer stator extracts useful components for energy conversion.

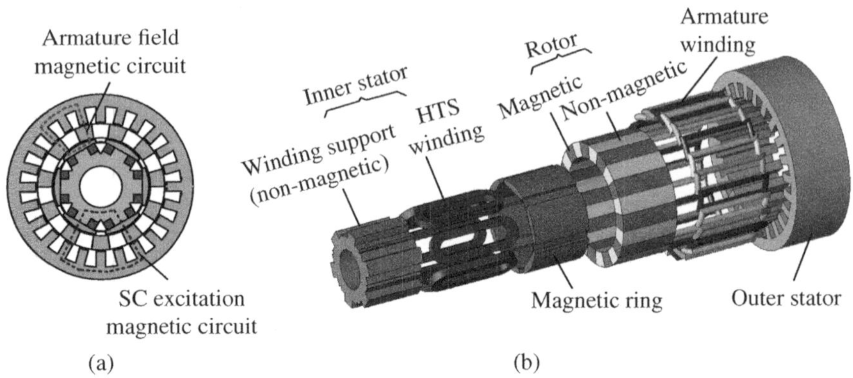

Figure 6.105 Structure of the double-stator field modulation SC excitation machine. (a) Cross-sectional view, (b) 3D view.

2 Influence of Armature Field on Superconducting Winding

In a conventional synchronous SC machine, the fundamental component of armature reaction field rotates synchronously with the rotor in space. There is no relative movement between the armature reaction field and the SC field winding, and no AC loss is generated in the superconductor. However, in the DS-FMSE machine, the armature reaction field will be modulated by the rotor poles to produce a series of harmonic fields, which rotate with respect to the stationary SC field winding. Consequently, AC voltages will be induced in the SC winding, which will cause a ripple of field current and AC loss. The AC loss of the SC winding not only reduces efficiency, but also may cause the superconductor to quench.

To effectively suppress the influence of the armature reaction field on the SC field winding, a magnetic ring is set outside the SC magnets. Based on the reasonable design of the outer stator tooth number, the rotor pole number and the magnetic ring, the armature reaction field can form a closed loop through the outer stator, the rotor poles and the magnetic ring as much as possible so that the coupling between the armature reaction field and the SC winding can be effectively reduced [59]. Thus the path of armature reaction field can be depicted as the dashed line in Figure 6.105a.

To further suppress the coupling effect caused by the armature reaction field not short-circuited by the outer stator, the rotor poles and the magnetic ring, a composite electromagnetic shielding layer is proposed. It is composed of a squirrel cage damping shielding layer arranged on the magnetic ring and a copper shielding layer placed inside the cryostat as shown in Figure 6.106 [60]. In order to verify the effect of the proposed shielding layers, a 10 kW prototype was designed.

The main design parameters are shown in Table 6.19. Figure 6.107 and Figure 6.108 compare the no-load EMF in the SC winding with and without the

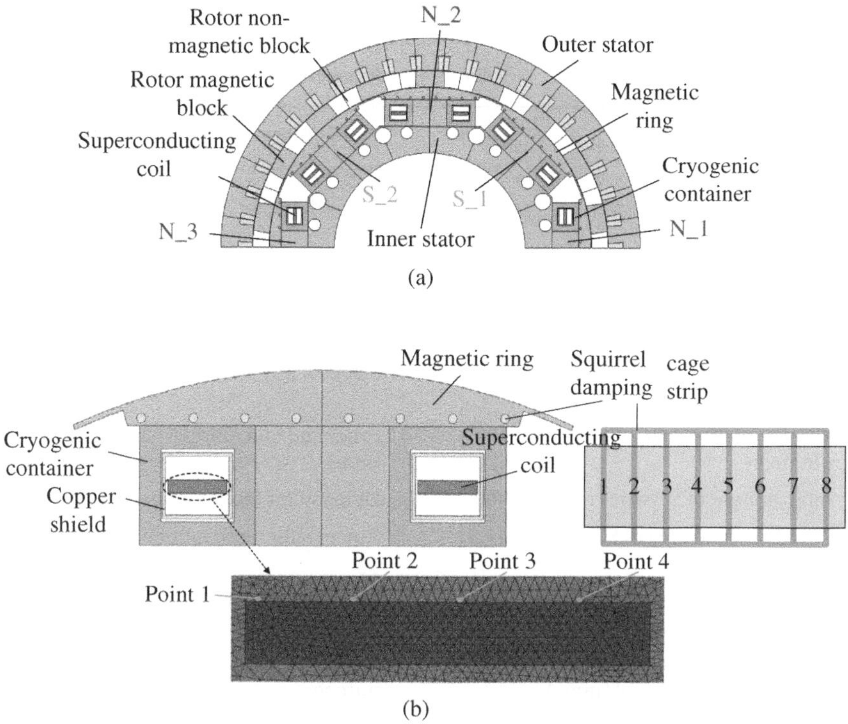

Figure 6.106 Composite d electromagnetic shielding layer. (a) Composite electromagnetic shielding layer, (b) Zoomed-in view of composite electromagnetic shielding layer.

composite electromagnetic shielding layers, in which N_1, S_1, N_2, S_2, and N_3 are the SC winding shown in Figure 6.106a. Figures 6.109 and 6.110 compare the magnetic field levels parallel to and perpendicular to the SC winding. Points 1, 2, 3, and 4 are the magnetic density calculation points shown in Figure 6.106b. It can be seen that the composite electromagnetic shielding layer can effectively reduce the harmonic field around the SC winding and the induced voltage in the SC winding. Figure 6.111 shows the main components and assembly of the 10 kW prototype.

In order to investigate the application prospects in the field of high-power wind energy conversion, a 10 MW DS-FMSE generator was designed according to the technical parameters of a 10 MW offshore wind generator. The basic electromagnetic performances were analyzed and compared with those of available 10 MW synchronous rotating SC excitation generators [14], [61]. The analysis results show that the basic electromagnetic performances of the DS-FMSE generator, such as no-load EMF distortion degree and torque ripple, can meet the requirements of

Table 6.19 Main design parameters of DS-FMSE machine.

Parameter	Value	Parameter	Value
Rated power (kW)	10	Rated voltage (V)	380
Rated speed (r/min)	300	Power factor	1
Rated excitation current of SC winding (A)	40	Outer diameter of outer stator (mm)	760
SC winding material	Bi2223	Rotor outer diameter (mm)	608.8
Working temperature of SC winding (K)	77	Rotor inner diameter (mm)	552.8
Critical current of SC winding (at 77 K, 0 T) (A)	70	Stack length of rotor core (mm)	100
Number of turns per SC winding	100	Inner and outer airgap length (mm)	0.6
Total amount of SC wire (m)	460	Outer stator tooth number	42
Cross-sectional area of SC coil (mm^2)	13.5×8.4	Inner stator tooth number	8
Outer Dewar cross-sectional area of cryogenic vessel (mm^2)	49×45	Rotor pole number	18
Inner Dewar cross-sectional area of cryogenic vessel (mm^2)	40×37	Armature winding pole-pair number	14
Number of pole pairs of SC field winding	4		

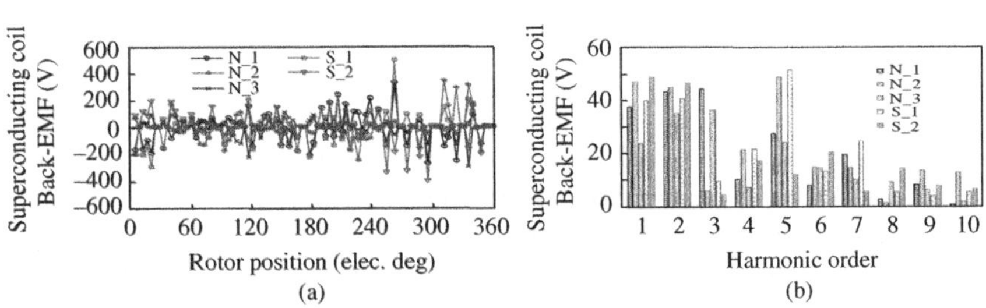

Figure 6.107 No-load back-EMF of SC coil in rated working conditions without the composite electromagnetic shielding layer. (a) Waveforms, (b) Amplitude spectrums.

offshore wind energy conversion, and the torque density is equivalent to that of the synchronous rotating SC excitation generator, simultaneously. The DS-FMSE generator can successfully overcome the drawbacks of current transmission and coolant seal. Generally, the DS-FMSE generator has obvious advantages in terms of the reliability and the cost in whole service life. More details can be found in [14].

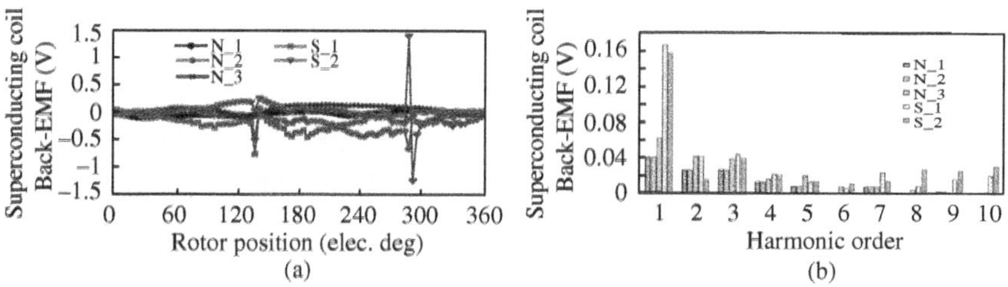

Figure 6.108 No-load back-EMF of SC coil in rated working conditions with the composite electromagnetic shielding layer. (a) Waveforms, (b) Amplitude spectrums.

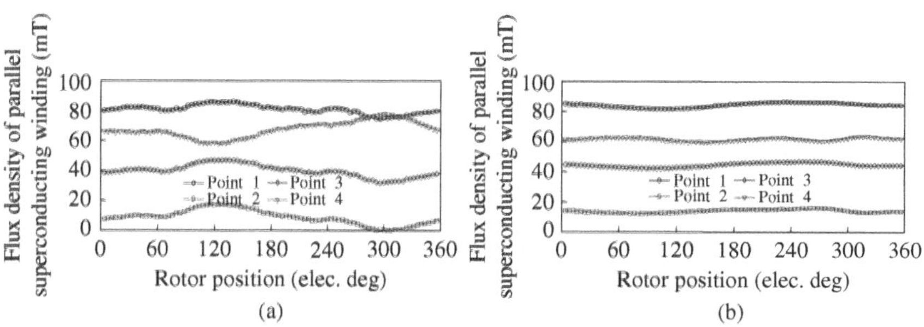

Figure 6.109 Flux density in the direction parallel to the SC winding in rated condition. (a) Without (b) With the composite electromagnetic shielding layer.

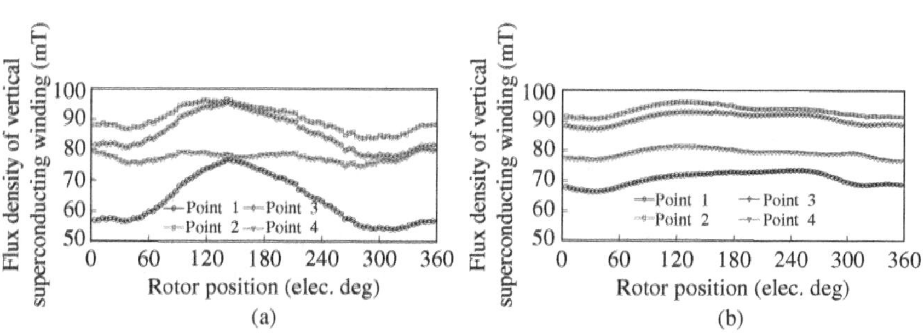

Figure 6.110 Flux density in the direction perpendicular to the SC winding in rated condition. (a) Without (b) With the composite electromagnetic shielding layer.

6.8.3 Technical Challenges and Outlook of Field Modulation HTS Machines

Thanks to the breakthroughs in cryogenic system seals, current transmission, and torque production, the field modulation SC machine shows significant advantages over traditional SC machines. In order to further promote the development of

(a) (b) (c) (d)

Figure 6.111 10 kW DS-FMSE machine prototype. (a) Inner stator with SC magnets, (b) Cup-shaped rotor, (c) Outer stator, (d) Prototype assembly.

the field modulation SC machine technology, efforts and collaborations in the following aspects are in great need.

1. Machine topology and design method.
   The field modulation SC machine has many unique challenges, such as the influence of abundant field harmonics on the performance of SC materials, the maximum achievable airgap flux density due to the use of ferromagnetic materials, large machine size due to the cryogenic container carrying SC materials, etc. Therefore, the machine topology and design method for existing field modulation machines cannot be applied directly. The suitable topologies and design methods can be evaluated based on the developed theory.
2. Accurate modeling, analysis, and multi-objective optimization.
   SC machines generally target high-power applications. Using FEA in the initial design stage is time-consuming, even for the minimal partial geometry model. The fast and accurate modeling of field modulation SC machines (considering the high magnetic field, strong coupling, and heavy magnetic saturation) can effectively shorten the development cycle of SC field modulation machines, facilitating the systematic multi-objective optimization in the future [62].
3. Harmonic suppression and loss calculation.
   In general, there are rich field harmonics in the airgap of field modulation SC machines. These field harmonics will increase core losses, reduce the power factor and efficiency, as well as deteriorate the torque and voltage quality. Moreover, the field harmonics also induce AC loss in SC magnets, increasing the cryogenic cooling system rating and the risk of quench. Therefore, it is necessary to enhance effective working harmonics, and suppress useless components by studying the harmonic production mechanism in depth. The complex magnetic field and heavy saturation of ferromagnetic materials should be considered in evaluating core losses. AC losses in superconductors under complex

magnetic field environments should also be studied to provide guidance for the design of cryogenic cooling systems and improvement of operating efficiency.

4. Low-temperature system design and thermal management.
   The cryogenic cooling system is one of the most crucial parts for SC electric machines to ensure the safe operation of SC magnets. It is expected to have simple structures and a stable internal temperature during normal operation. It is important to study the thermal management technology to effectively control the internal temperature and avoid the local quench of SC magnets.
5. Excitation system design and control strategy.
   SC windings typically have high currents, low voltages, low resistances, and high inductances. A small voltage change will vary the current significantly, leading to an unstable excitation field. Therefore, a constant current power supply with a low voltage and high current is the key for the excitation system. In addition, the excitation current should be regulated to be neither too fast (to avoid excessive AC losses) nor too slow (to meet the dynamic performance requirements).

Generally, the advantages of SC machines over traditional machines have been recognized all over the world, especially in applications having demanding requirements on the size and weight of electric machines, such as ship electric propulsion and direct-drive wind power generation. Due to the technical merits of low-temperature system seal and high current/torque transmission capability, the field modulation SC machine shows great promise.

6.9 Brushless Doubly-Fed Reluctance Machine with an Asymmetrical Composite Modulator

The brushless doubly-fed reluctance machine (BDFRM) is a family of multi-port reluctance machines with two AC electrical ports and a common mechanical port. It is also a category of magnetic field modulated machines performing asynchronous modulation behavior. Based on the two accessible electrical ports, the BDFRM can be endowed with a greater degree of freedom for speed and power control, inherent fault-tolerant capability and high reliability. Most recently, new topological structures of BDFRM with composite modulators were essentially designed to enhance the magnetic field conversion effects based on the dual asynchronous modulation behaviors [63, 64].

The conventional symmetrical composite rotor, featuring multi-layer flux barrier and short-circuited coils, or the composite rotor featuring salient pole reluctance and short-circuited coils, improves the torque characters, increases the magnetic cross-coupling capability, enhances the effective harmonic component and suppresses the unproductive harmonic components of the BDFRM. However, the short-circuited coil modulator in the composite rotor causes spatial phase shift of the modulated airgap magnetic field harmonics, that is the harmonics modulated by the two modulators have a certain phase difference in space. The phase difference of the principal and summation modulated harmonic respectively reduces the primitive and summation magnetic field conversion factor and affects the BDFRM electromagnetic performances such as modulated harmonic distribution and output torque capability. Therefore, there is still potential for continued improvement of the magnetic field modulation capability for the composite modulator and torque characters of BDFRMs.

The objective of this section is to investigate the modulation behavior of the composite modulator from the perspective of the magnetic field conversion mechanism based on the GAFMT. The modulated harmonic phase shift phenomenon, particularly for the principal modulated harmonic, caused by the short-circuited coil modulator is firstly verified in detail. A modified 4/2 BDFRM with an asymmetrical composite modulator is proposed and the torque character comparison between the BDFRMs with asymmetrical and symmetrical composite modulators is presented [65].

6.9.1 Phase Shift Phenomenon of Modulated Harmonics

The topologies and magnetic field distributions at no-load of the 4/2 BDFRM are illustrated in Figure 6.112, and the major common design parameters are listed in Table 6.20.

Compared to the BDFRM with salient rotor shown in Figure 6.112a whose ratio of pole-arc to pole-pitch is 0.48, the composite rotor introduced in Figure 6.112b

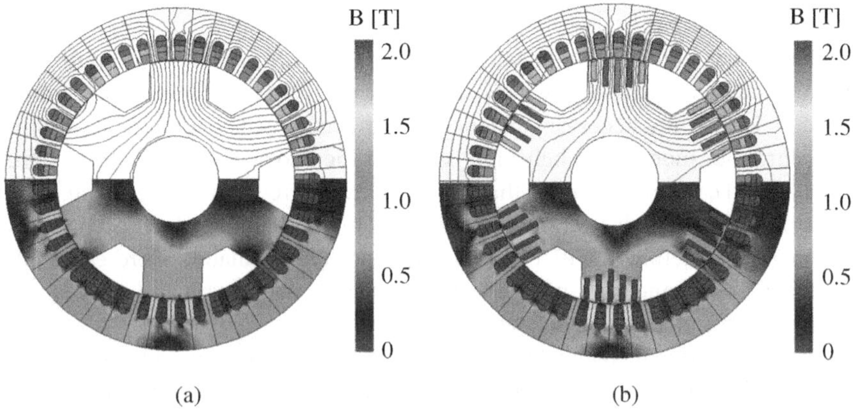

Figure 6.112 Topology and no-load flux lines/density distribution of the 4/2 BDFRM. (a) Salient reluctance pole modulator, (b) Composite modulator.

Table 6.20 Main parameters of the studied BDFRM with a composite rotor.

Quantity	Value		Quantity	Value
Stator slot number	45	Control winding	Pole pair	2
Outer stator radius	91 mm		Coil pitch	10
Stator inner radius	63.5 mm		Turns in series per phase	255
Rotor inner radius	23 mm		Winding factor	0.9406
Airgap length	0.35 mm	Power winding	Pole pair	4
Stack length	90 mm		Coil pitch	5
Number of rotor poles	6		Turns in series per phase	510
Number of loops	3		Winding factor	0.9406

has six salient poles with five flux ducts per pole, and the copper bars are inserted into the central position of each flux duct. Bars with the same color are connected together, while the bars in the other end of machine are short-circuited and connected to form the three-layer nest-loop cage. Therefore, the composite rotor can be termed as a combination of the salient pole reluctance and short-circuited coil modulators.

To understand the modulated harmonic distribution of the BDFRM, the modulation operator and modulated MMF are introduced first. The rotor salient poles in the BDFRM utilize alternative variation of magnetic reluctance to perform the asynchronous modulation on the source MMF, the modulation operator $M(N_{RT}, \varepsilon_{RT})[.]$ can be defined as:

$$M(N_{RT}, \varepsilon_{RT})[f(\phi, t)] = \begin{cases} f(\phi, t), \phi \in C^R \\ 0, \quad \phi \in [0, 2\pi] - C^R \end{cases} \tag{6.88}$$

where C^R is the discontinuous intervals occupied by salient rotor poles in mechanical radians. N_{RT} is the number of rotor poles, and ε_{RT} denotes the ratio of pole-arc to pole-pitch of rotor. It is assumed that the source magnetizing MMF $f(\phi,t)$ is a unit cosine function and all harmonic items are neglected for simplification, then the modulated MMF can be derived in Fourier series as:

$$\begin{aligned} M(N_{RT}, \varepsilon_{RT})[f(\phi, t)] &= C_p \cos(p\phi - \omega_s t) \\ &\quad - \sum_{k=lN_{RT}-p}^{\infty} C_{sum} \sin\left[k\phi + \omega_s t + (k+p)\frac{\pi}{N_{RT}}\right] \\ &\quad + \sum_{k=lN_{RT}+p}^{\infty} C_{dif} \sin\left[k\phi - \omega_s t + (k-p)\frac{\pi}{N_{RT}}\right], \ l \in Z^+ \end{aligned} \tag{6.89}$$

where ω_s is the electrical angular frequency of stator winding current. ϕ is the mechanical angle along the circumference in modulator-fixed reference frame. The subscripts *sum* and *dif* represent the summation and differential modulation process, respectively. Equation (6.89) shows that the modulated magnetic field harmonics appear in pairs, and the summation and differential modulated harmonics are equivalent for the same l, indicating weak effect on the reduction of high-order unproductive harmonics. In addition, the asynchronous modulation of salient pole reluctance modulator shifts the spectrum lines of harmonic components, while remains the spatial phase distributions unchanged.

The short-circuited coil structure such as the nest-loop or series ring is generally utilized to achieve qualified magnetic field conversion capability for the magnetic field modulated machine. The modulated magnetizing MMF equals to the summation of source MMF and all the extra components, and can be characterized by proposing modulator $M(N_{SC},\gamma)[.]$ as:

$$M(N_{SC},\gamma)[f(\phi,t)] = f(\phi,t) + \sum_{j=1}^{N_{SC}}\sum_{x=1}^{S} W_{jx} i_{jx},\quad \phi \in [0, 2\pi] \tag{6.90}$$

where N_{SC} is the number of short-circuited coil group. γ is the coil span factor of a specific loop. i_{jx} and W_{jx} are the current flowing and winding function of x-th loop ($x = 1, 2, \ldots, S$) in j-th nest ($j = 1, 2, \ldots, N_{SC}$), respectively. S is the number of loops within a single nest. Then the modulated MMF can be rewritten in Fourier series as:

$$\begin{aligned} M(N_{SC},\gamma)[f(\phi,t)] &= C_p\cos(p\phi - \omega_s t - \varphi) \\ &\quad - \sum_{k=lN_{SC}-p}^{\infty} C_{sum}\sin\left[k\phi + \omega_s t + (k+p)\frac{\pi}{N_{SC}} - \varphi\right] \\ &\quad + \sum_{k=lN_{SC}+p}^{\infty} C_{dif}\sin\left[k\phi - \omega_s t + (k-p)\frac{\pi}{N_{SC}} + \varphi\right], l \in Z^+ \end{aligned} \tag{6.91}$$

where φ is the impedance angle, which can be expressed as:

$$\varphi = \arctan(\omega_s L_x/R_x), L_x = 2\pi\mu_0 r_g l_{stk}\gamma_x/gN_{SC} + L_\sigma \tag{6.92}$$

where R_x, L_x, and L_σ are the resistance, inductance and leakage inductance of the x-th loop in j-th nest, respectively. There will be three different components with p, lN_{SC}-p and lN_{SC}+p pole pairs in the modulated MMF corresponding to the source p-pole-pair MMF. The p-pole-pair modulated harmonic component and the primitive magnetic field conversion factor offer high amplitude. Furthermore, the amplitudes of summation and differential modulated harmonics depend on rotor structure (N_{SC}, γ, S, R_x, L_x, and L_σ) and winding principal pole pair p.

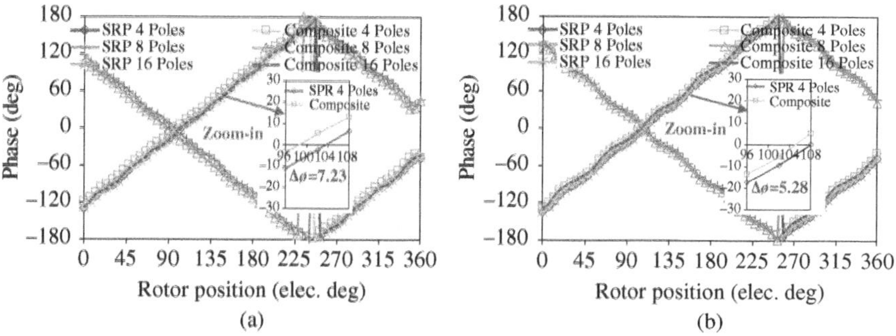

Figure 6.113 Spatial phase distribution. (a) Control winding excited and power winding load, (b) Control winding excited and power winding open-circuited.

The principal 2-pole-pair and differential modulated 8-pole-pair harmonics are both forward-rotating, while the summation modulated 4-pole-pair harmonic is reverse-rotating when observing from the modulator-fixed reference frame ϕ as illustrated in (6.90), (6.91) and Figure 6.113. In addition, the harmonics modulated by the salient pole reluctance modulator and composite modulator have a certain phase difference in space caused by the short-circuited coil structure in the composite rotor, which is defined as $\Delta\phi$ electric degree in the modulator-fixed reference frame. The harmonics modulated by the composite modulator are lagged to those modulated by the salient pole reluctance modulator, i.e. they are leftward shift in space caused by the copper nests in composite modulator as expressed in (6.90), (6.91), and Figure 6.113. The spatial phase distribution fluctuates significantly at no-load condition as given in Figure 6.113b, and the phase shift $\Delta\phi$ at load condition is larger than that at no-load, which is respectively 7.23° and 5.28° electric degrees. It should be noted that, the $\Delta\phi$ is obtained by eliminating individual points with high errors and averaging over a period in space. Moreover, the phase shift $\Delta\phi$ of p-pole-pair modulated harmonic is same with that of the summation modulated harmonics as predicted in (6.91), and the phase shift of the differential modulated harmonic is $-\Delta\phi$.

The spatial phase distribution of p-pole-pair modulated harmonic established by the symmetrical composite modulator is not consistent with that modulated by the salient pole reluctance modulator as shown in Figure 6.114. Thus the modulated MMF or magnetic field conversion factors of the two modulators cannot be directly added up, and the resultant MMF (the red line in Figure 6.114) does not reach the maximum. The phase shift $\Delta\phi$ for the p-pole-pair modulated harmonic is reduced based on an asymmetrical composite modulator configuration introduced in next section to further improve magnetic coupling capability of the composite modulator.

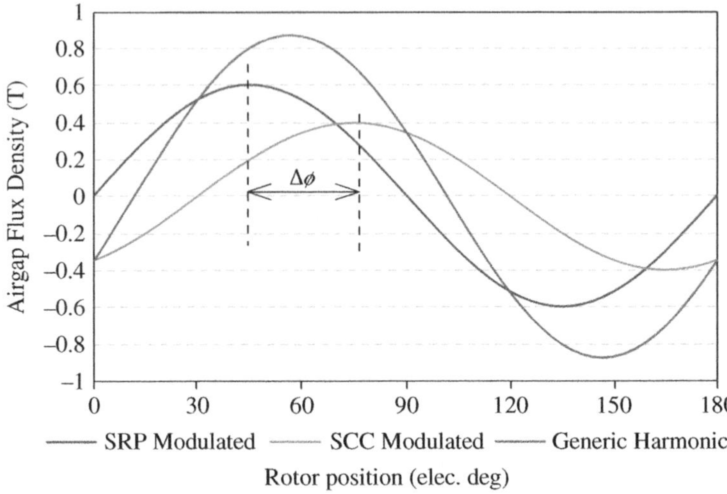

Figure 6.114 Schematic of modulated p-pole-pair harmonic superposition for the symmetrical composite modulator.

6.9.2 Asymmetrical Composite Modulator

To enhance the primitive conversion factor and keep the effective summation modulation conversion factor simultaneous with high amplitude, the phase shift $\Delta\phi$ of p-pole-pair modulated harmonic in space should be reduced as far as possible. Thus, an asymmetrical composite modulator configuration is proposed, as shown in Figure 6.115a. The short-circuited coil modulator features asymmetry distribution by offset a specific design of assembling angle δ along the magnetic field rotation direction, and $\delta = \Delta\phi/N_{RT/SC}$ for the purpose of reducing the phase shift $\Delta\phi$, improving the primitive conversion factor and comprehensively enhance the output electromagnetic performances.

Firstly, the control winding (CW) excited and power winding (PW) load condition is the most typical excitation condition for BDFRM, and the phase shift phenomena under this kind of condition exhibited in Figure 6.113a is taken into account. Secondly, since the phase shift $\Delta\phi$ does nothing with the excited current value under unsaturated condition while it is linearly related to the excited current frequency, $\Delta\phi$ should be obtained by averaging over a frequency range to simply replace its trend. Thirdly, $\Delta\phi$ is purposely designed as 7.2 electrical degrees and δ is 1.2 mechanical degrees for the 4/2 pole pair BDFRM. Then the spatial phase distribution of p-pole-pair modulated harmonic established by the modified asymmetrical composite modulator is consistent to that modulated by the salient pole reluctance modulator, i.e. the p-pole-pair spatial harmonic stays in phase, as displayed in Figure 6.115b. The resultant MMF as denoted by the red line in Figure 6.115b, and the modulation capability reach the theoretical maximum.

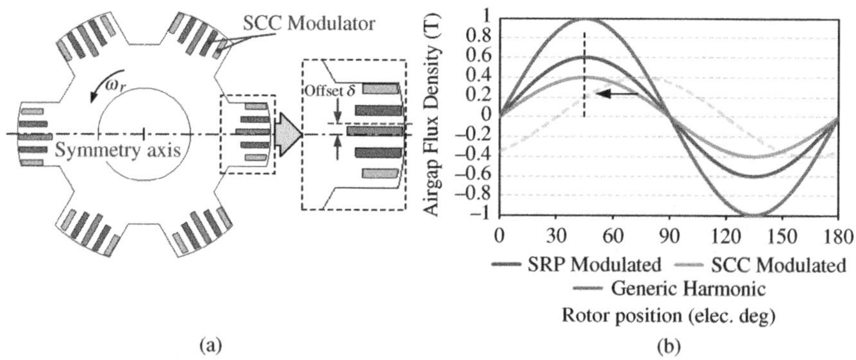

Figure 6.115 Torque improvement of the 4/2 BDFRM. (a) Topology diagram: along the rotation direction, (b) Schematic of modulated p-pole-pair harmonic superposition for the asymmetrical composite modulator.

Torque waveforms of the BDFRM featuring an asymmetrical composite modulator have no significant fluctuation in each excitation condition as illustrated in Figure 6.116. Taking the CW excited and PW load condition for an example, the asymmetrical composite modulator offers roughly a 9.83% growth in average output torque compared to the original symmetrical counterpart shown in Figure 6.112b, and together with a 20.26% reduction in torque ripple at the same time.

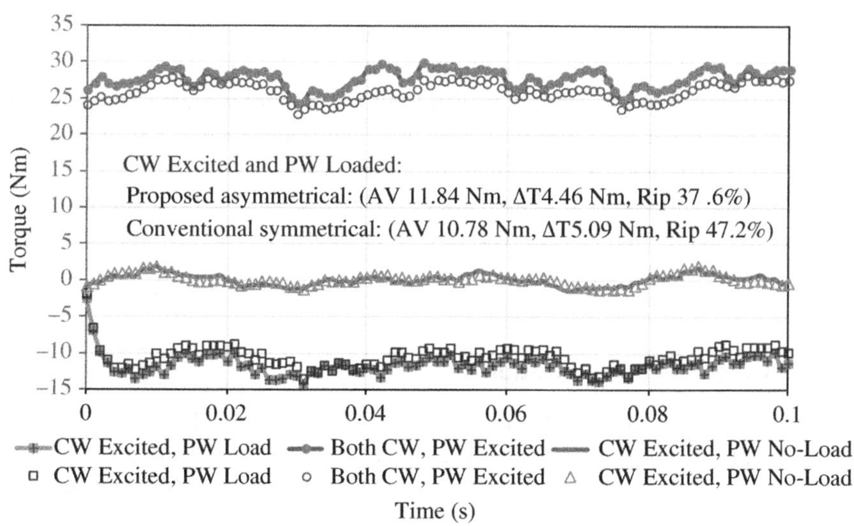

Figure 6.116 Torque waveforms comparison of the 4/2 BDFRM with symmetrical and asymmetrical composite modulators. Solid lines with markers denote asymmetric modulators, and the others are for symmetric modulators.

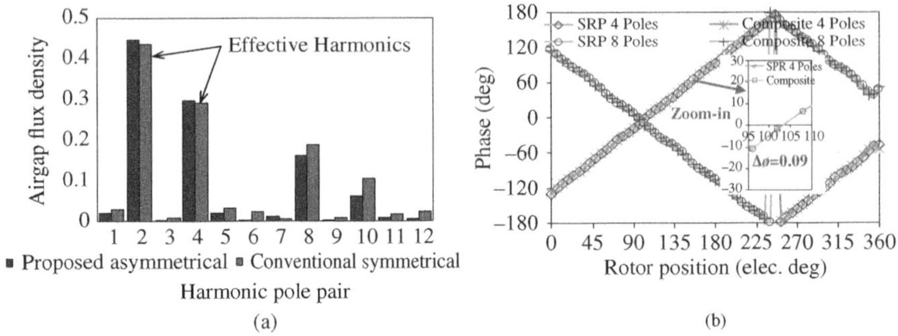

Figure 6.117 Modulated harmonic distributions. (a) Harmonic spectrum comparison between symmetrical and asymmetrical composite modulators, (b) Spatial phase shift distribution of the 4/2 BDFRM with an asymmetrical composite modulator.

The effective summation modulated 4-pole-pair ($N_{RT/SC}-p$) and the unproductive differential modulated 8 ($N_{RT/SC}+p$), 10 ($2N_{RT/SC}-p$)-pole-pair harmonics of the modified asymmetrical composite modulator are respectively further improved and reduced compared with the traditional symmetrical composite modulator as given in Figure 6.117a. The enhancement of principal 2-pole-pair harmonic amplitude is around 4.6%, while the contributed torque component increases by about 23.7%, which is attributed to the small phase difference between the radial and tangential harmonics from the perspective of Maxwell Stress Tensor method. The amplitude of effective summation modulated 4-pole-pair harmonic is almost unchanged (both the asymmetrical and symmetrical composite modulators offer around 0.3 T). Moreover, the amplitude of the unproductive differential modulated 8-pole-pair harmonic is 0.18 T, which is slightly decreased compared with the symmetrical composite modulator. The phase shift $\Delta\phi$ of p-pole-pair harmonic is significantly reduced to 0.09 electrical degrees under CW excited and PW load condition as demonstrated in Figure 6.107b, then the primitive conversion factor, the contributed torque component and the overall output torque characters are improved.

6.9.3 Experimental Verification

In order to verify the effectiveness of the above theoretical analysis, the 4/2 BDFRM prototype machines with different composite modulators, namely the symmetrical and asymmetrical composite modulator as illustrated in Figure 6.118a,b are manufactured and tested. Figure 6.118c shows the stator structure with two standard short-pitch distributed windings and Figure 6.118d illustrates the composite rotor assembling.

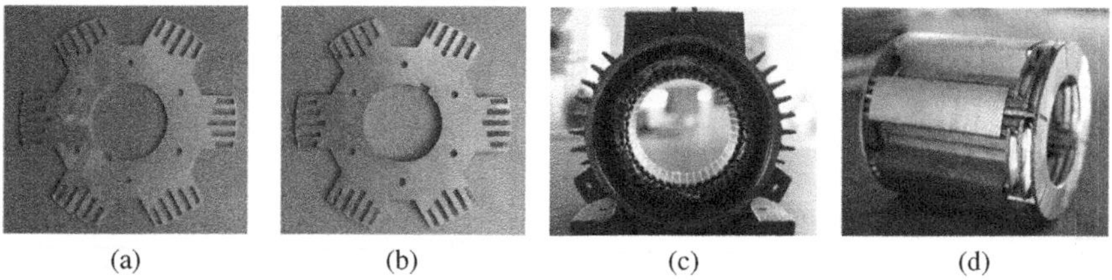

Figure 6.118 Prototype machines. (a) Symmetrical modulator, (b) Asymmetrical modulator, (c) Stator, (d) Assembled rotor.

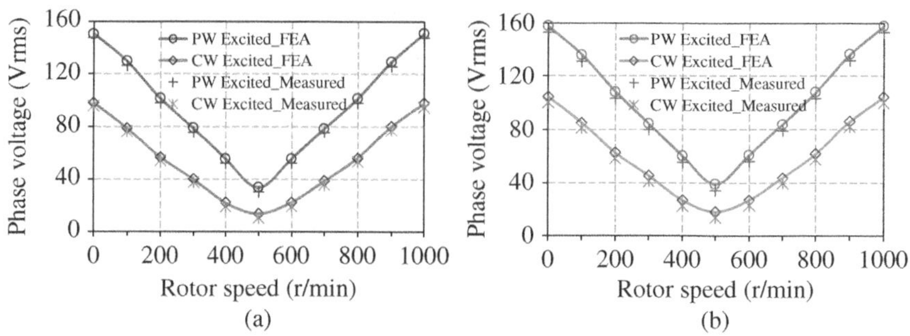

Figure 6.119 Measured cross-coupling characteristics (excitation phase voltage: 220 Vrms, frequency 50 Hz). (a) BDFRM with a symmetrical composite modulator, (b) BDFRM with an asymmetrical composite modulator.

The cross-coupling capability was tested in voltage control mode (excitation phase voltage: 220 Vrms, frequency 50 Hz) as exhibited in Figure 6.119. It is shown that the cross-coupling capability is linearly related to the rotation speed and gets nearly lost at around the synchronous speed of 500 r/min (the rotor pole pair number $N_{RT/SC}$ is 6), matching well with the FEA simulated values.

Figure 6.120 compares the FEA simulated inductances with the experiments, and the results agree well with each other. It is evident that both the self-inductances of the two stator windings are nearly independent of rotor position.

In addition, the coupling factor C^{r}_{pc} between PW and CW is a pure alternating component and repeat one complete cycle for π electrical radians, whose fundamental cycle depends on the rotor pole pair number $N_{RT/SC}$. Further analysis shows that relative ratio of phase self-inductances and coupling factor between PW and CW are in good accordance respectively with the theoretical prediction magnetic field conversion factors.

Figure 6.121 illustrates the measured torque performances under two different excitation conditions. The tested average torques are litter smaller than those

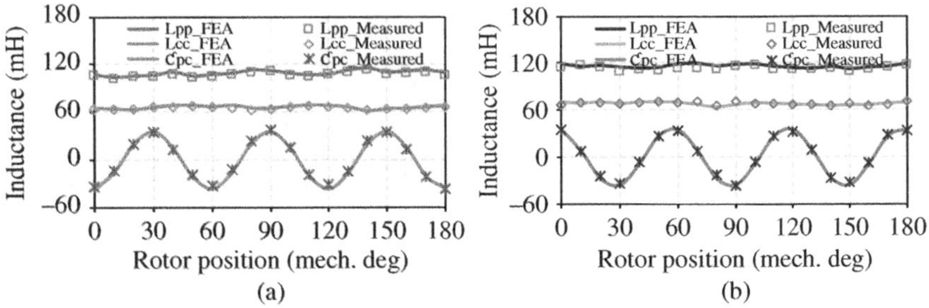

Figure 6.120 Inductance characteristics. (a) BDFRM with a symmetrical composite modulator, (b) BDFRM with an asymmetrical composite modulator.

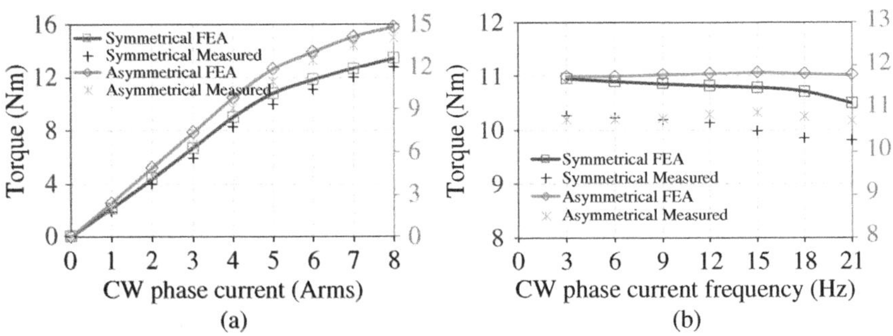

Figure 6.121 Torque performance. (a) Torque versus current, (b) Torque versus CW current frequency.

of the FEA simulated as shown in Figure 6.121a, since the end effect and manufacture tolerance are not considered in 2-D FEA simulation. The overload capabilities of the two BDFRMs are similar and qualified. Moreover, the torque characters remain almost unchanged with an excited current frequency in a given speed/frequency range. The cross coupling ability along with the inductance characters and output torques of the 4/2 BDFRM with an asymmetrical composite modulator are always higher than those of a machine with a symmetrical composite modulator. All the FEA simulated results and experiments indicate the proposed asymmetrical composite modulator is effective.

References

1 Wu, D., Shi, J.T., Zhu, Z.Q., and Liu, X. (2014). Electromagnetic performance of novel synchronous machines with permanent magnets in stator yoke. *IEEE Trans. Magn.* 50 (9), Article# 8102009.

2 Zhang, Y., Hua, W., Cheng, M. et al. (2012). Comprehensive comparison of novel stator surface-mounted PM electrical machines. In: *Proceedings of International Conference on Electrical Machines (ICEM)*, 589–594. Marseille, France.

3 Shi, J.T., Zhu, Z.Q., Wu, D. et al. (2014). Comparative study of novel synchronous machines having permanent magnets in stator poles. In: *Proceedings of International Conference on Electrical Machines (ICEM)*, 429–435. Berlin, Germany.

4 Chen, J.T., Zhu, Z.Q., and Iwasaki, S. (2011). A novel E-core switched-flux PM brushless AC machine. *IEEE Trans. Ind. Appl.* 47 (3): 1273–1282.

5 Zhang, J., Cheng, M., and Chen, Z. (2008). Investigation of a new stator interior permanent magnet machine. *IET Electr. Power Appl.* 2 (2): 77–87.

6 Xu, W., Zhu, J.G., Zhang, Y. et al. (2011). New axial laminated-structure flux-switching permanent magnet electrical machine with 6/7 poles. *IEEE Trans. Magn.* 47 (10): 2823–2826.

7 Owen, R.L., Zhu, Z.Q., Thomas, A.S. et al. (2010). Alternate poles wound flux-switching permanent magnet brushless AC machines. *IEEE Trans. Ind. Appl.* 46 (2): 1406–1415.

8 Du, Y., Xiao, F., Hua, W. et al. (2016). Comparison of flux-switching PM motors with different winding configurations using magnetic gearing principle. *IEEE Trans. Magn.* 52 (5): Article# 8201908.

9 Hua, W., Su, P., Zhang, G., and Cheng, M. (2015). A novel rotor-permanent magnet flux-switching machine. In: *Proceedings of International Conference on Ecological Vehicles and Renewable Energies (EVER)*, 1–10. Monte, Monaco.

10 Wu, Z.Z. and Zhu, Z.Q. (2016). Analysis of magnetic gearing effect in partitioned stator switched flux PM machines. *IEEE Trans. Energy Convers.* 31 (4): 1239–1249.

11 Sun, L., Cheng, M., and Jia, H. (2015). Analysis of a novel magnetic-geared dual-rotor motor with complementary structure. *IEEE Trans. Ind. Electron.* 62 (11): 6737–6747.

12 Cheng, M., Han, P., Buja, G., and Jovanović, M.G. (2017). Emerging multiport electrical machines and systems: past developments, current challenges and future prospects. *IEEE Trans. Ind. Electron.* 65 (7): 5422–5435.

13 Hua, W., Zhang, G., and Cheng, M. (2015). Flux-regulation theories and principles of hybrid-excited flux-switching machines. *IEEE Trans. Ind. Electron.* 62 (9): 5359–5369.

14 Zhu, X. and Cheng, M. (2019). Design and analysis of 10 MW class HTS exciting double stator direct-drive wind generator with stationary seal. *IEEE Access*, vol. 7, 51129–51139.

15 Cheng, M., Chau, K.T., and Chan, C.C. (2001). Static characteristics of a new doubly salient permanent magnet motor. *IEEE Trans. Energy Convers.* 16 (1): 20–25.

16 Chau, K.T., Sun, Q., Fan, Y., and Cheng, M. (2005). Torque ripple minimization of doubly salient permanent-magnet machines. *IEEE Trans. Energy Convers.* 20 (2): 352–358.

17 Yang, W., Xu, W., and Xiao, X. (2012). Torque ripple reduction strategy of model based predictive torque control for doubly salient permanent magnet synchronous machines. In: *Proceedings of IEEE Energy Conversion Congress and Exposition, ECCE (ECCE)*, 113–120. Raleigh, NC, USA.

18 Cui, W., Gong, Y., Jiang, J., and Zhang, Y. (2011). Optimized doubly salient memory motors with symmetric features using transposition design methods. In: *Proceedings of International Conference on Power Engineering, Energy and Electrical Drives*, 1–7. Torremolinos, Málaga, Spain.

19 Gong, Y., Chau, K.T., Jiang, J.Z. et al. (2009). Design of doubly salient permanent magnet motors with minimum torque ripple. *IEEE Trans. Magn.* 45 (10): 4704–4707.

20 Cao, R., Cheng, M., Mi, C. et al. (2011). A linear doubly salient permanent-magnet motor with modular and complementary structure. *IEEE Trans. Magn.* 47 (12): 4809–4821.

21 Liu, X. and Zhu, Z.Q. (2013). Electromagnetic performance of novel variable flux reluctance machines with DC-field coil in stator. *IEEE Trans. Magn.* 49 (6): 3020–3028.

22 Du, Y., Zhang, C., Zhu, X. et al. (2019). Principle and analysis of doubly salient PM motor with Π-shaped stator iron core segments. *IEEE Trans. Ind. Electron.* 66 (3): 1962–1972.

23 Heller, B. and Hamata, V. (1977). *Harmonic Field Effects in Induction Machines*. Amsterdam: Elsevier.

24 Cheng, M. and Hua, W. (2021). *Stator Permanent Magnet Brushless Machines-Theory, Design and Control*. Beijing: Science Press.

25 Cheng, M. (2022). *Small & Special Electrical Machines and Systems*, 3e. Beijing, China: China Electric Power Press.

26 Sun, L., Taylor, J., Callegaro, A.D., and Emadi, A. (2020). Stator-PM-based variable reluctance resolver with advantage of motional back-EMF. *IEEE Trans. Ind. Electron.* 67 (11): 9790–9801.

27 Cheng, M., Han, P., and Hua, W. (2017). General airgap field modulation theory for electrical machines. *IEEE Trans. Ind. Electron.* 64 (8): 6063–6074.

28 Sun, L., Luo, Z., Wang, K. et al. (2021). A stator-PM resolver with field modulation principle. *IEEE Trans. Energy Convers.* 36 (1): 159–172.

29 Sun, L., Taylor, J., Guo, X. et al. (2021). A linear position measurement scheme for long-distance and high-speed applications. *IEEE Trans. Ind. Electron.* 68 (5): 4435–4447.

30 Hua, W., Su, P., Shi, M. et al. (2015). The influence of magnetizations on bipolar stator surface-mounted permanent magnet machines. *IEEE Trans. Magn.* 51 (3): Article# 8201904.

31 Zhu, X., Hua, W., Wang, W., and Huang, W.T. (2019). Analysis of back-EMF in flux-reversal permanent magnet machines by airgap field modulation theory. *IEEE Trans. Ind. Electron.* 66 (5): 3344–3355.

32 Cheng, M., Hua, W., Zhang, J., and Zhao, W. (2011). Overview of stator-permanent magnet brushless machines. *IEEE Trans. Ind. Appl.* 58 (11): 5087–5101.

33 Chen, J.T. and Zhu, Z.Q. (2010). Winding configurations and optimal stator and rotor pole combination of flux-switching PM brushless AC machines. *IEEE Trans. Energy Convers.* 25 (2): 293–302.

34 Xue, X., Zhao, W., Zhu, J. et al. (2013). Design of five-phase modular flux-switching permanent-magnet machines for high reliability applications. *IEEE Trans. Magn.* 49 (7): 3941–3944.

35 Min, W., Chen, J.T., Zhu, Z.Q. et al. (2011). Optimization and comparison of novel E-Core and C-Core linear switched flux PM machines. *IEEE Trans. Magn.* 47 (8): 2134–2141.

36 Zhang, G., Hua, W., Cheng, M. et al. (2015). Investigation of an improved hybrid-excitation flux-switching brushless machine for HEV/EV applications. *IEEE Trans. Ind. Appl.* 51 (5): 3791–3799.

37 Wang, Y., Chen, M., Ching, T.W., and Chau, K.T. (2015). Design and analysis of a new HTS axial-field flux-switching machine. *IEEE Trans. Applied Supercond.* 25 (3): Article# 5200905.

38 Du, Y., Zou, C., Zhu, X. et al. (2016). A full-pitched flux-switching permanent-magnet motor. *IEEE Trans. Applied Supercond.* 26 (4): Article# 0604505.

39 Wu, Z.Z. and Zhu, Z.Q. (2015). Analysis of airgap field modulation and magnetic gearing effects in switched flux permanent magnet machines. *IEEE Trans. Magn.* 51 (5): Article# 8105012.

40 Hua, W., Su, P., Zhao, G., and Zhang, G. (2015). Flux-switching machines. In: *Wiley Encyclopedia of Electrical and Electronics Engineering*. USA: John Wiley & Sons.

41 Thomas, A.S., Zhu, Z.Q., and Jewell, G.W. (2011). Comparison of flux switching and surface mounted permanent magnet generators for high-speed applications. *IET Electr. Syst. Transport.* 1 (3): 111–116.

42 Sulaiman, E., Kosaka, T., and Matsui, N. (2011). High power density design of 6-slot–8-pole hybrid excitation flux switching machine for hybrid electric vehicles. *IEEE Trans. Magn.* 47 (10): 4453–4456.

43 Awah, C.C., Zhu, Z.Q., Wu, Z.Z. et al. (2016). Comparison of partitioned stator switched flux permanent magnet machines having single- or double-layer windings. *IEEE Trans. Magn.* 52 (1): Article# 9500310.

44 Cai, X., Cheng, M., Zhu, S. et al. (2016). Thermal modeling of flux-switching permanent-magnet machines considering anisotropic conductivity and thermal contact resistance. *IEEE Trans. Ind. Electron.* 63 (6): 3355–3365.

45 Hua, W. (2007). Analysis, Design and Control of New Flux-Switching Permanent Magnet Machines. Nanjing, China: Ph.D. Thesis of Southeast University.

46 Tang, R. (1997). *Theory and Design of Modern Permanent Magnet Machines.* Beijing: China Machine Press.

47 Sun, L., Cheng, M., Zhang, J., and Song, L. (2016). Analysis and control of complementary magnetic-geared dual-rotor motor. *IEEE Trans. Ind. Electron.* 63 (11): 6715–6725.

48 Sun, L., Cheng, M., Wen, H., and Song, L. (2017). Motion control and performance evaluation of a magnetic-geared dual-rotor motor in hybrid powertrain. *IEEE Trans. Ind. Electron.* 64 (3): 1863–1872.

49 Sun, L., Cheng, M., and Tong, M. (2017). Key issues in design and manufacture of magnetic-geared dual-rotor motor for hybrid vehicles. *IEEE Trans. Energy Convers.* 32 (4): 1492–1501.

50 Sun, L. (2016). Analysis, Design and Control of The Magnetic-Geared Power Split Machine. Nanjing, China: Ph.D. Thesis of Southeast University.

51 Stekly, Z., Woodson, H.H., Hatch, A.M. et al. (1966). A study of alternators with superconducting field windings: II–experiment. *IEEE Trans. Power Apparat. Syst.* PAS-85 (3): 274–280.

52 Gamble, B., Snitchler, G., and MacDonald, T. (2011). Full power test of a 36.5 MW HTS propulsion motor. *IEEE Trans. Applied Supercond.* 21 (3): 1083–1088.

53 Song, X., Bührer, C., Brutsaert, P. et al. (2019). Designing and basic experimental validation of the world's first MW-class direct-drive superconducting wind turbine generator. *IEEE Trans. Energy Convers.* 34 (4): 2218–2225.

54 Wang, Y. (2020). Development status of rotating superconducting motor. *Electr. Mach. Contr. Appl.* 47 (2): 1–8.

55 Wang, Y., Wang, C., Feng, Q., and Li, X. (2017). Fabrication and experiment of racetrack HTS magnet for stator field-excitation HTS machine. *IEEE Trans. Applied Supercond.* 27 (4): Article# 5201605.

56 Wang, Y., Feng, Q., Li, X., and Ma, W. (2018). Design, analysis, and experimental test of a segmented-rotor high-temperature superconducting flux-switching generator with stationary seal. *IEEE Trans. Ind. Electron.* 65 (11): 9047–9055.

57 Wang, Y. and Ma, Y. (2020). Electromagnetic force calculation of field-excitation coil in high temperature superconducting flux-switching machine. *Electr. Mach. Contr. Appl.* 24 (7): 82–89.

58 Cheng, M., Wang, Y., Hua, W., and Wang, Z. (2019). Dual-stator superconductive exciting field modulating machine, European Patent No. 3293870 (PCT/CN2016/073416), 4.10.

59 Zhu, X., Cheng, M., Li, X., and Wang, Y. (2020). Topology analysis, design, and comparison of high temperature superconducting double stator machine with stationary seal. *IEEE Trans. Applied Supercond.* 30 (1): Article# 5200110.

60 Cheng, M., Zhu, X., Wang, Y. et al. (2020). Effect and inhibition method of armature-reaction field on superconducting coil in field-modulation superconducting electrical machine. *IEEE Trans. Energy Convers.* 35 (1): 279–291.

61 Cheng, M., Ning, X., Zhu, X., and Wang, Y. (2020). Selection of excitation operating points of 10 MW HTS exciting double stator direct-drive wind generators having single and double polarity inner stator. *Politeknik Dergisi (Journal of Polytechnic)* 23 (2): 537–545.

62 Zhu, X., Cheng, M., Wang, Y. et al. (2022). Mathematical analysis model of double-stator field modulation HTS machine based on general airgap field modulation theory. *IEEE Trans. Energy Convers.* 37 (1): 475–486.

63 Zhang, F., Song, Y., Yu, S. et al. (2019). Influence of slot combination on performance of brushless doubly fed generator with hybrid rotor. *IEEE Trans. Mag.* 55 (11): Article# 6700406.

64 Zhang, F., Wang, H., Jia, G. et al. (2018). Effects of design parameters on performance of brushless electrically excited synchronous reluctance generator. *IEEE Trans. Ind. Electron.* 65 (11): 9179–9189.

65 Wen, H., Cheng, M., and Zhang, G. (2022). Principle and performance of a new brushless doubly–fed reluctance machine with asymmetrical composite modulator. *IEEE Trans. Ind. Electron.* 69 (12): 12086–12095.

7

Other Applications of General Airgap Field Modulation Theory

7.1 Introduction

Chapters 4–6 present three main applications of the general airgap field modulation theory (GAFMT), that is, analysis of magnetic field modulation behaviors, calculation of machine performance, and guiding the innovation of machine topologies. Since the developed GAFMT focuses on the spatial distribution and temporal variation of magnetic field harmonics within the airgap of electrical machines, various aspects related to the magnetic field harmonics in electrical machines will be relevant to the magnetic field modulation phenomenon and can be analyzed with the aid of the developed GAFMT.

To showcase more possible applications of the GAFMT, the brushless doubly-fed machine is taken as an example to analyze the distribution of radial forces, laying a solid foundation for the analysis of electromagnetic vibration and noise in electrical machines [1]. Then, suspension windings for bearingless homopolar and consequent pole PM machines are studied to show the effectiveness and superiority of the GAFMT in the design and analysis of pole-changing windings [2]. By combining the GAFMT and 2D finite element analysis (FEA), new methods for the stray load loss calculation in squirrel-cage induction machines (IMs) and the core loss calculation in flux-switching PM (FSPM) machines with pulse width modulation (PWM) inverter supplies are presented [3].

Apart from these applications, the GAFMT can also be used to assist optimization of electrical machines. Two examples are included. The first example studies the influence of rotor salient reluctance pole structures on the torque performance and general guidelines for design optimization of electrical machines with salient reluctance poles are concluded [4]. The second example shows how to incorporate the GAFMT in systematic large-scale design optimization of electrical machines by examining the working harmonics emerging from the airgap field modulation process [5].

General Airgap Field Modulation Theory for Electrical Machines: Principles and Practice,
First Edition. Ming Cheng, Peng Han, Yi Du, and Honghui Wen.

Companion website: www.wiley.com/go/genairgapfieldmodulationtheory

7.2 Analysis of Radial Forces in Brushless Doubly-fed Machines

7.2.1 Electromagnetic Vibration and Noise in Electrical Machines

The machine noise consists of electromagnetic noise, aerodynamic noise, and mechanical noise. Electromagnetic noise is caused by the vibration in the iron core and its associated mechanical components caused by the alternating electromagnetic force generated by the airgap magnetic field harmonics. Due to the effect of the airgap flux density, there are two electromagnetic forces in the stator tooth, namely, the radial and tangential components. The radial component contributes to the vibration and deformation of the stator core which is the main reason for the electromagnetic noise, whereas the tangential component as a secondary source of electromagnetic vibration produces electromagnetic torque, causing bending of the teeth and local deformation. Therefore, analysis of electromagnetic vibration and noise for the machine is equivalent to analyzing the distribution of radial electromagnetic force in the airgap and its variation with time, as well as the radial inherent vibration characteristics of the stator. When the airgap harmonic order of the electromagnetic force with appropriate speed coincides with a certain structural mode of the stator, the resonance is produced, resulting in significant vibration and electromagnetic noise.

7.2.2 Analysis of Radial Forces

The radial force density in an electrical machine is closely related to the airgap flux density. Generally, the tangential component of the airgap flux density is much smaller than the radial one, so the radial force density $\sigma_r(\phi,t)$ can be simplified to

$$\sigma_r(\phi, t) = \frac{B_r^2(\phi, t)}{2\mu_0} \tag{7.1}$$

where $B_r(\phi,t)$ is the radial component of airgap flux density and μ_0 is the permeability of vacuum. ϕ is the mechanical angular position along airgap peripheral and t is time.

The resultant radial force $\vec{F}(t)$ acting on the rotor is calculated by

$$\vec{F}(t) = F_x(t) + jF_y(t) = r_g l_{stk} \int_0^{2\pi} \sigma_r(\cos\phi + j\sin\phi)d\phi \tag{7.2}$$

where r_g is the radius of airgap and l_{stk} is the stack length. "x" and "y" denote the components of x-axis and y-axis, respectively.

For traditional AC rotating electrical machines with symmetrical windings and uniform airgap whose slot number per pole per phase q is an integer, the resultant radial force is always zero. The spectrum of the radial airgap flux density is discrete.

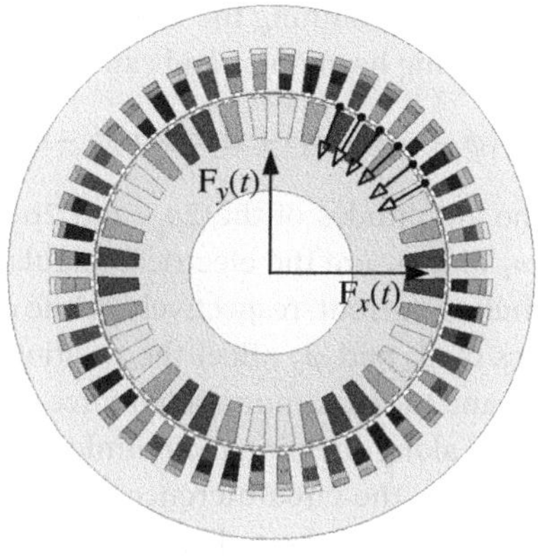

Figure 7.1 Radial force distribution on the top of the stator teeth for brushless doubly-fed induction machine (BDFIM) (local) and resultant vector of radial force on the rotor.

The minimum pole-pair difference between two adjacent harmonics is two. The distribution of the radial force density in the airgap and the resultant vector of the radial force for the rotor are shown in Figure 7.1.

7.2.3 Calculation of Radial Forces

There are many spatial harmonics in BDFMS, with different pole pairs rotating in the airgap at different mechanical speeds. A well-designed BDFM has two principal working harmonics produced by the power winding and control winding, whose numbers of pole pairs are p_1 and p_2 ($p_1 \neq p_2$). The rotor can be short-circuited coils, simple-salient-pole reluctance or a multi-layer magnetic barrier structure, and contains $N_r = (p_1 + p_2)$ repeating segments. For clarity, the pole pair combination is denoted in the format of (p_1-N_r-p_2). The analytical expression of the airgap flux density can be easily obtained by the GAFMT which is established in Chapter 3, and thus, providing the possibility for the analysis of the radial force for those machines with a variety of useful harmonics such as the BDFIMs. This is the advantage that cannot be offered by the traditional electrical machine analysis theory.

According to the GAFMT in Chapter 3, it is known that there are various flux density harmonics in the airgap of BDFMs due to the modulation of the short-circuited coils, simple-salient-pole reluctance or multi-layer magnetic barriers. Detailed expressions for the airgap flux densities of three typical BDFMs are given in Section 4.4. Two kinds of airgap flux density harmonics of the three BDFMs are the dominant components, which have the largest amplitudes and contribute to the average torque, differing from the conventional AC machines.

To obtain a general conclusion, assuming the resultant radial flux density $B_r(\phi,t)$ only consists of two dominating harmonics and can be expressed as

$$B_r(\phi, t) = B_1 \cos(p_1\phi - \omega_1 t + \varphi_1) + B_2 \cos(p_2\phi - \omega_2 t + \varphi_2) \tag{7.3}$$

where B_1 and B_2 are the amplitudes of the $2p_1$- and $2p_2$-pole airgap field components, respectively. ω_1 and ω_2 are the electrical angular velocity of the power winding and control winding current, respectively, while φ_1 and φ_2 are the initial phase angles. The values of φ_1 and φ_2 affect the position and time of the maximum airgap flux density and the maximum radial force density, but not the mode of radial force distribution along the airgap circumference. The variation of the radial force distribution along the circumference of the airgap in one electrical period is the so-called the mode of the radial force. Substituting (7.3) into (7.2) and simplifying the equation can obtain

$$\mathrm{F}_x(t) + j\mathrm{F}_y(t) = \begin{cases} \dfrac{\pi B_1 B_2 r_g l_{\mathrm{stk}}}{2\mu_0} e^{\pm j[(\omega_1-\omega_2)t-(\varphi_1-\varphi_2)]}, & p_1 - p_2 = \pm 1, \\ 0, & p_1 - p_2 \neq \pm 1. \end{cases} \tag{7.4}$$

Equation (7.4) indicates that the resultant radial force is non-zero only if the pole pairs of the two field components are differed by 1. When the harmonics of the airgap flux density need to be considered, the detailed expression given in Section 4.4 can be used for the analysis instead of (7.3).

Tables 7.1 and 7.2 show the radial force patterns of different pole-pair combinations in one electrical period where φ_1 and φ_2 are zero. It can be concluded that different pole pair combinations offer various radial forces when the frequencies of two windings are different. For a BDFM, the radial force mode should change when the two stator windings are switched as shown the pole-pairs combinations (2-3-1) and (1-3-2) in Table 7.1. From Tables 7.1 and 7.2, it can be summarized that the number of the peak waves in the radial force equals to the repeatable units N_r of the rotor.

Integrating the radial force distribution in Tables 7.1 and 7.2 along the circumference of the airgap, the corresponding resultant radial forces acting on the rotor are summarized in Figure 7.2. Corresponding to Tables 7.1 and 7.2, the supply frequencies of the two stator windings are $f_1 = 50$ Hz and $f_2 = 25$ Hz, respectively. It can be observed from Figure 7.2a–f that if the pole-pair of two stator windings p_1 and p_2 are differed by 1, the resultant radial force acting on the rotor is not zero. F_x and F_y vary with the sinusoidal (cosinusoidal) distributions, that is, the resultant radial force has a constant amplitude. The radial force wave rotates at a frequency of $|\omega_1 - \omega_2|$ and the rotating direction is closely related to the pole-pair combinations and the supply frequencies. When the difference of the pole-pairs of the two stator windings is larger than 1, shown in Figure 7.2d,e, the resultant radial force

Table 7.1 Radial magnetic force patterns along circumference of airgap of BDFMs under pole-pair combinations of (2-3-1), (1-3-2) and (3-5-2) ($\omega_1 = 50 \times 2\pi$ Hz, $\omega_2 = 25 \times 2\pi$ Hz, T = 1/(50−25) = 40 ms, $\varphi_1 = \varphi_2 = 0$).

Time	(2-3-1)	(1-3-2)	(3-5-2)
0T/5 = 0 ms	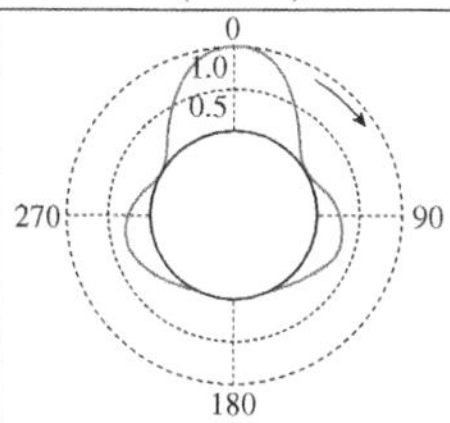	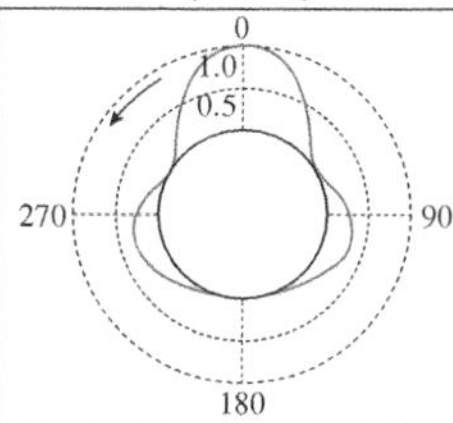	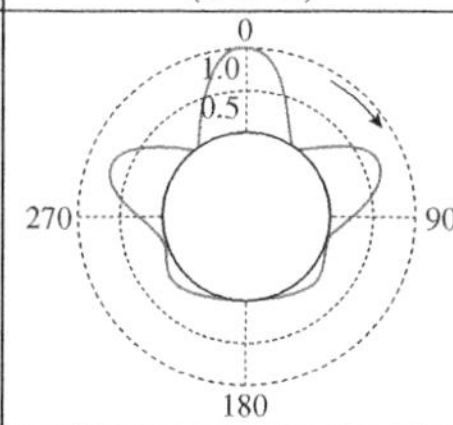
1T/5 = 8 ms	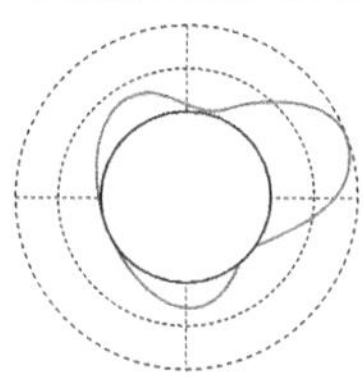	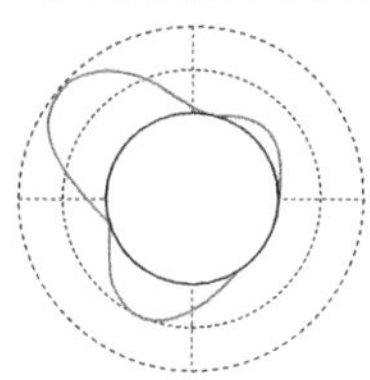	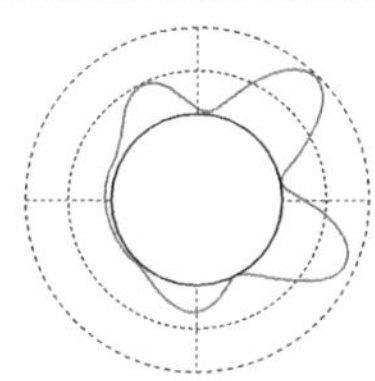
2T/5 = 16 ms	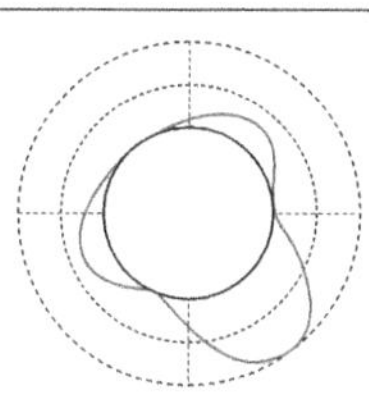	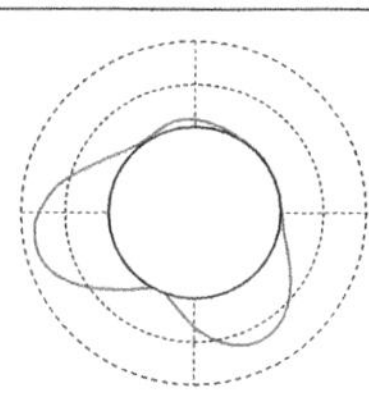	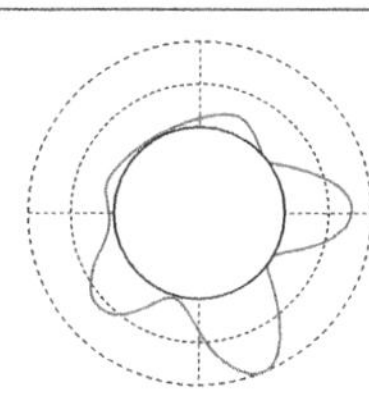
3T/5 = 24 ms	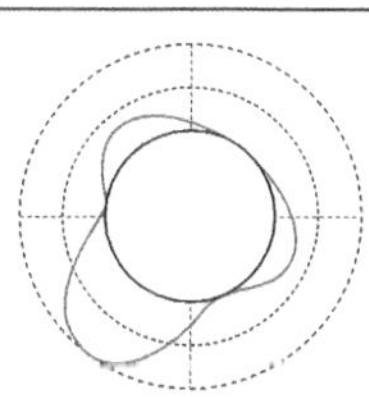	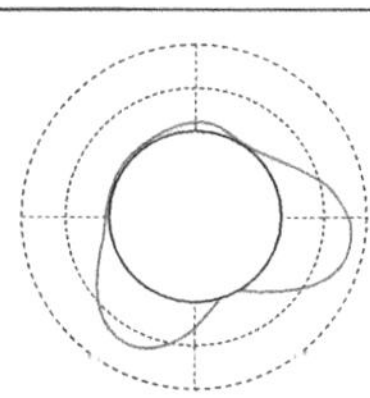	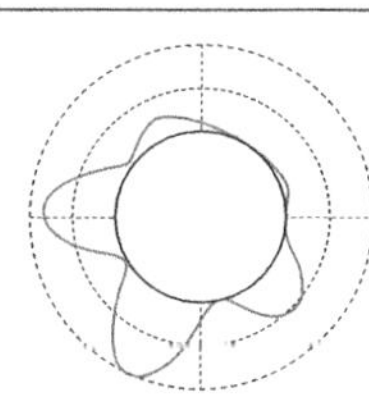
4T/5 = 32 ms	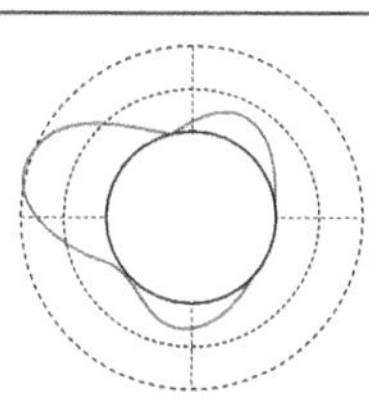	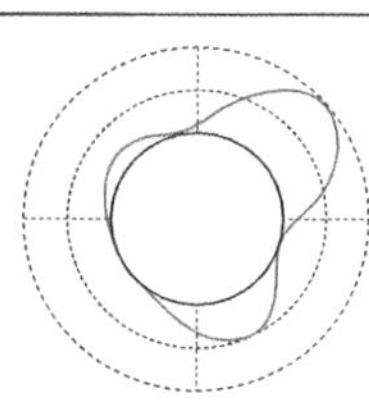	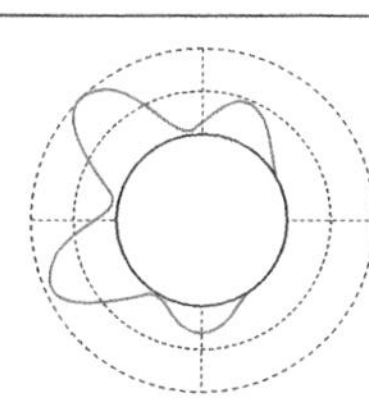
5T/5 = 40 ms	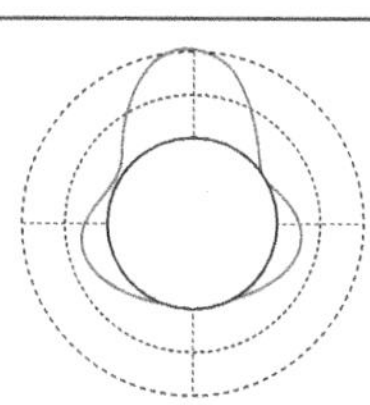	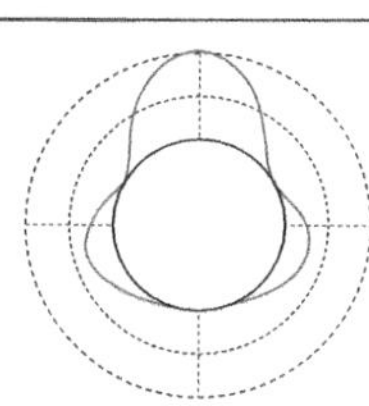	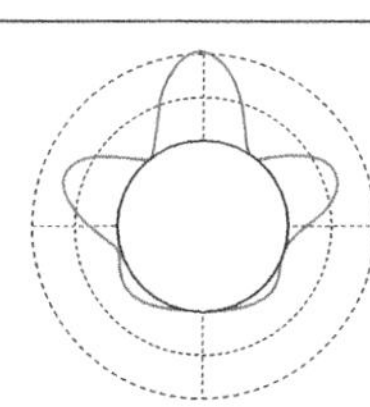

Table 7.2 Radial magnetic force patterns along circumference of airgap of BDFMs under pole-pair combinations of (4-5-1), (4-6-2) and (4-7-3) ($\omega_1 = 50 \times 2\pi$ Hz, $f_2 = 25 \times 2\pi$ Hz, $T = 1/(50-25) = 40$ ms, $\varphi_1 = \varphi_2 = 0$).

Time	(4-5-1)	(4-6-2)	(4-7-3)
0T/5 = 0 ms	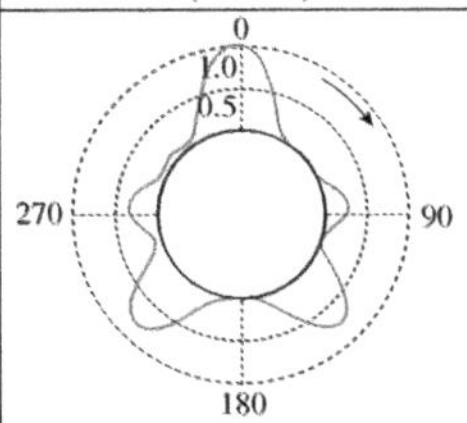	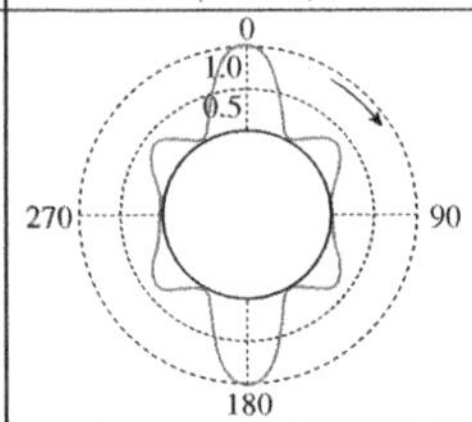	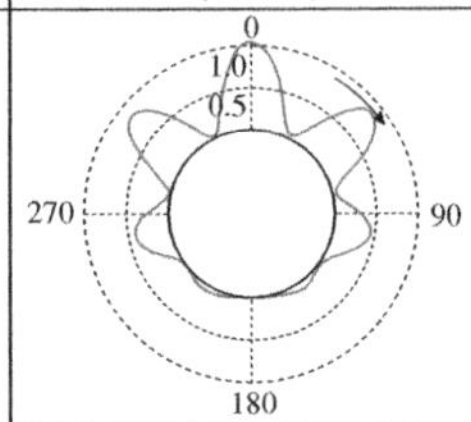
1T/5 = 8 ms	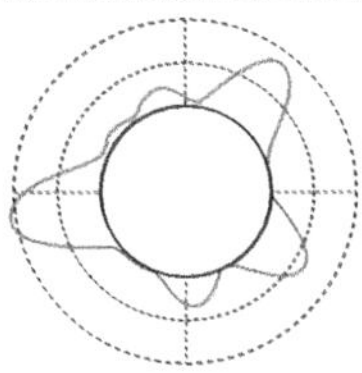	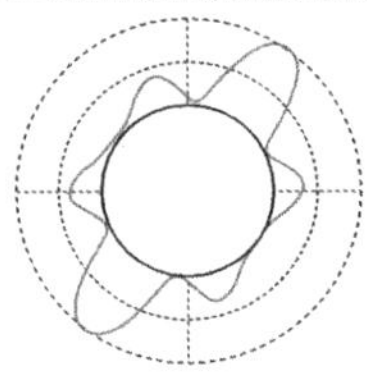	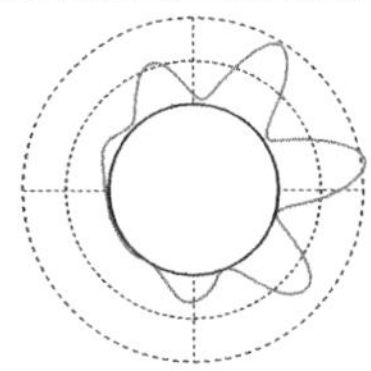
2T/5 = 16 ms	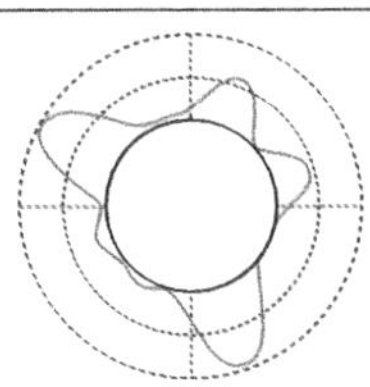	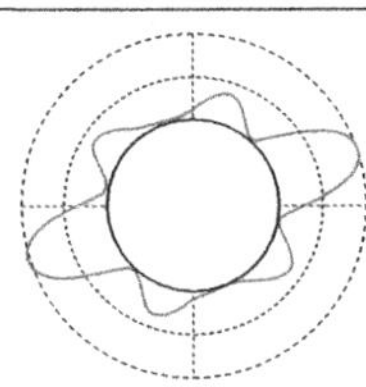	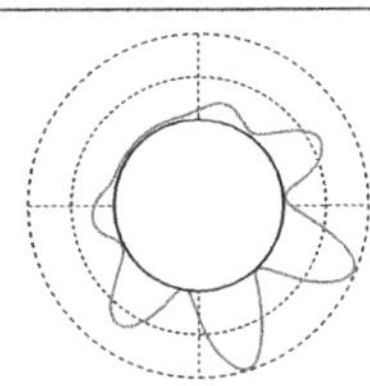
3T/5 = 24 ms	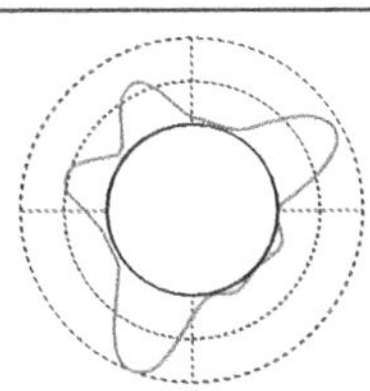	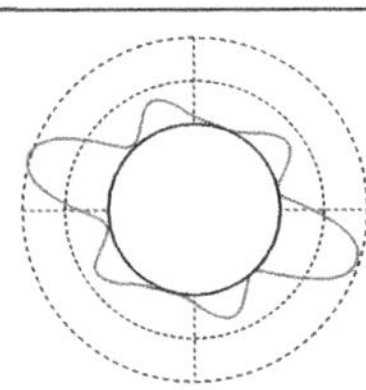	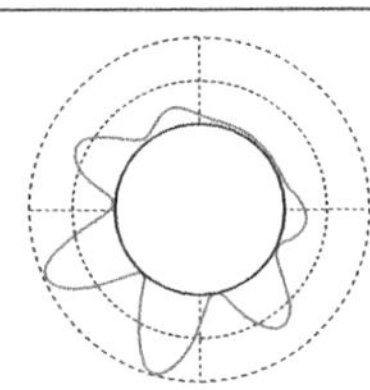
4T/5 = 32 ms	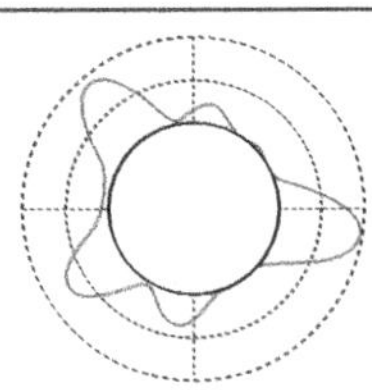	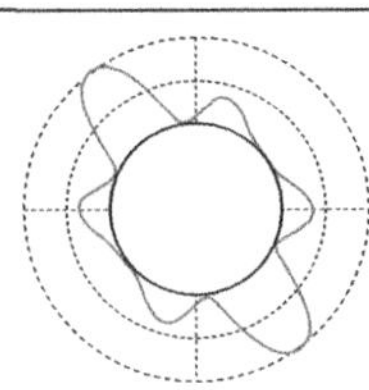	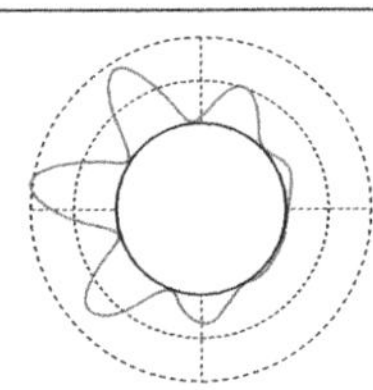
5T/5 = 40 ms	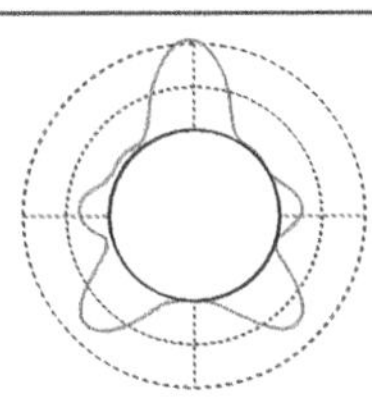	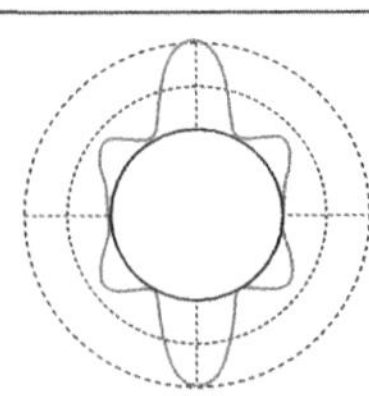	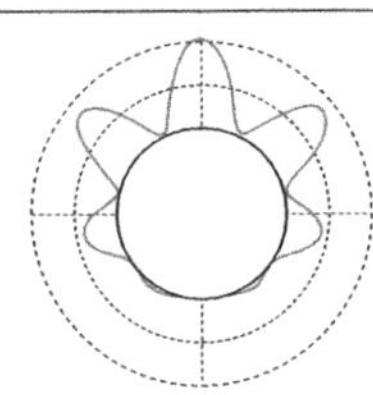

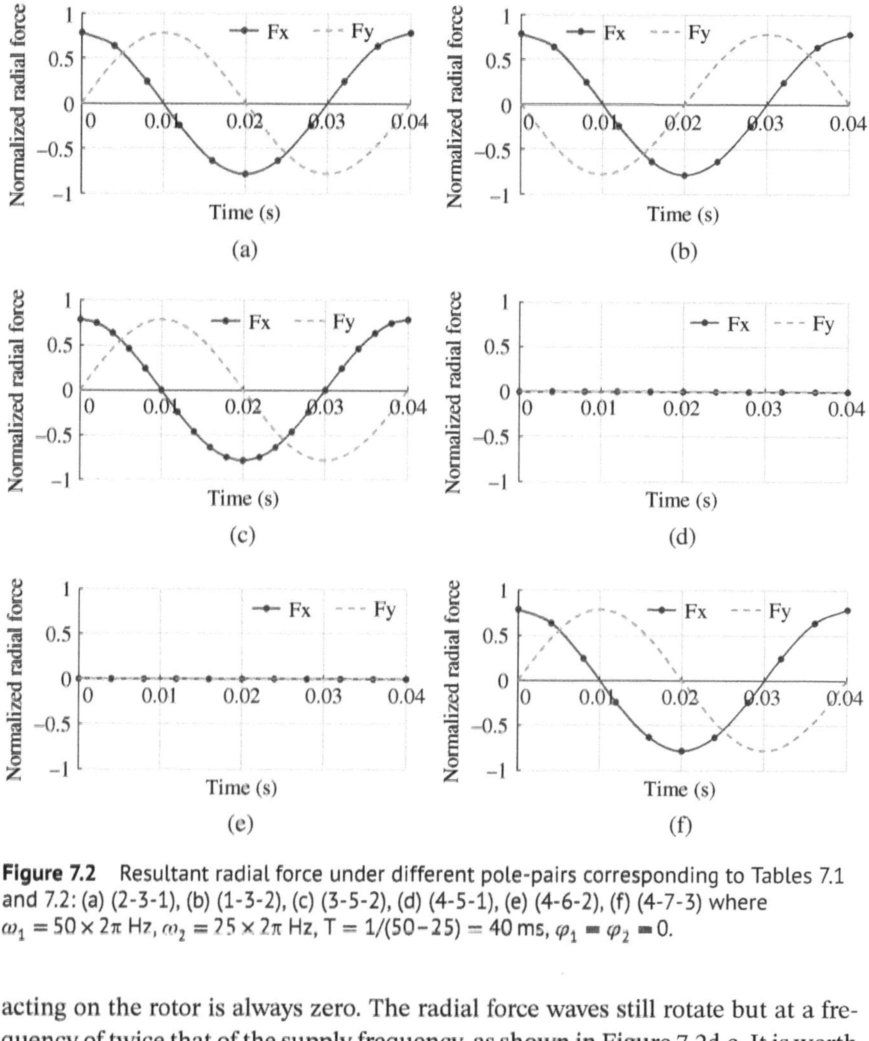

Figure 7.2 Resultant radial force under different pole-pairs corresponding to Tables 7.1 and 7.2: (a) (2-3-1), (b) (1-3-2), (c) (3-5-2), (d) (4-5-1), (e) (4-6-2), (f) (4-7-3) where $\omega_1 = 50 \times 2\pi$ Hz, $\omega_2 = 25 \times 2\pi$ Hz, $T = 1/(50-25) = 40$ ms, $\varphi_1 = \varphi_2 = 0$.

acting on the rotor is always zero. The radial force waves still rotate but at a frequency of twice that of the supply frequency, as shown in Figure 7.2d,e. It is worth noting that, though the radial force patterns of the pole pair combination (4-5-1) are not diametrically symmetrical, the resultant radial force is always zero.

For investigating the influence of the supply frequencies on the radial force mode and the resultant radial force, the radial force distributions with different pole-pairs during one electrical period are given in Tables 7.3 and 7.4 where the frequencies of the two stator windings are 50 Hz. The radial force waves still rotate but at a frequency of twice that of the supply frequency. When the supply frequencies of the two stator windings are exactly the same, and the two stator windings are switched, the resultant radial force modes are completely the same as shown

Table 7.3 Radial electromagnetic force patterns along circumference of airgap of BDFMs under pole-pair combinations of (2-3-1), (1-3-2) and (3-5-2) ($\omega_1 = 50 \times 2\pi$ Hz, $\omega_2 = 50 \times 2\pi$ Hz, T = 1/2 × 1/50 Hz = 10 ms, $\varphi_1 = \varphi_2 = 0$).

Time	(2-3-1)	(1-3-2)	(3-5-2)
0T/5 = 0 ms	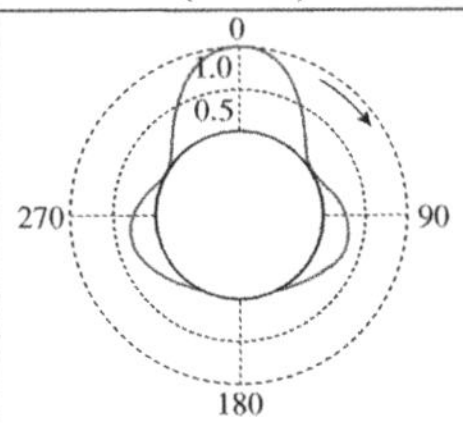	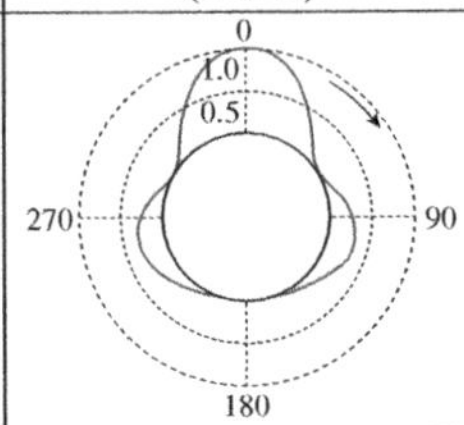	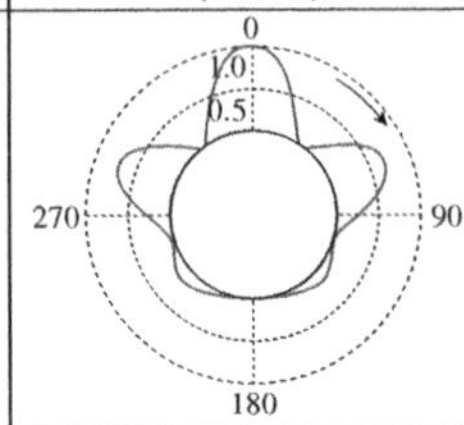
1T/5 = 2 ms	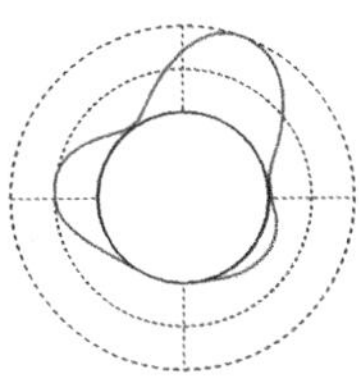	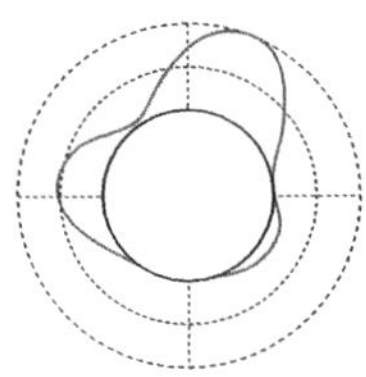	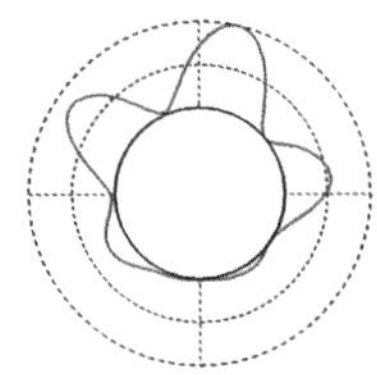
2T/5 = 4 ms	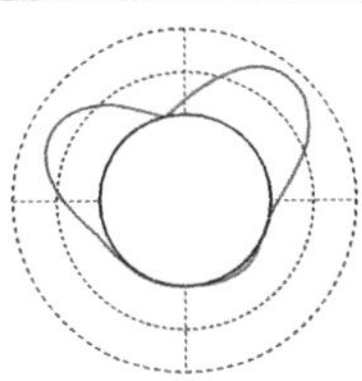	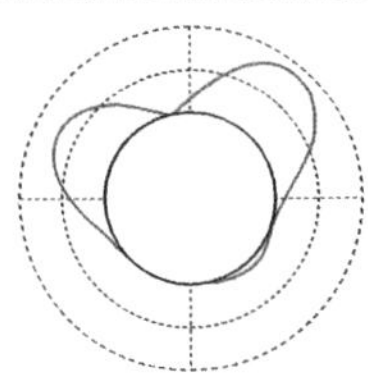	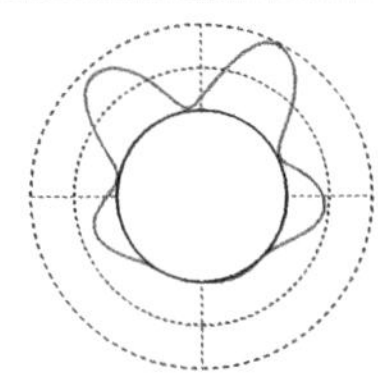
3T/5 = 6 ms	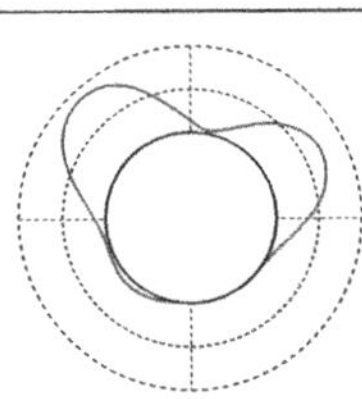	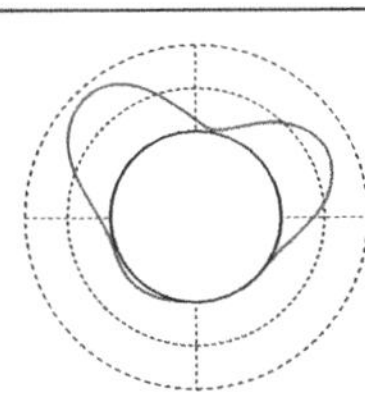	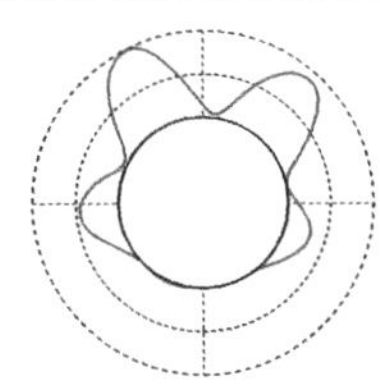
4T/5 = 8 ms	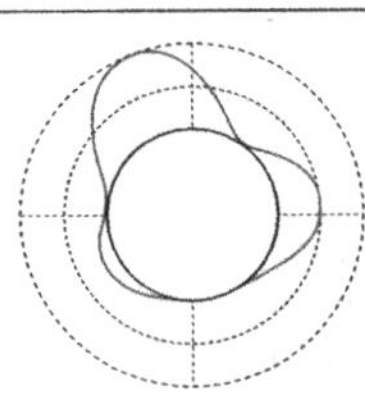	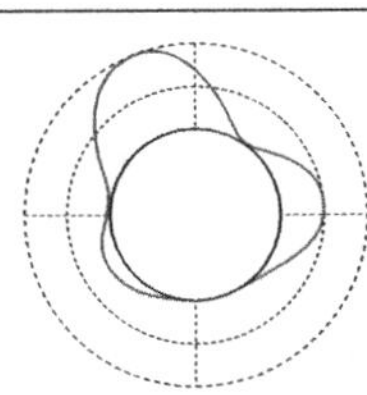	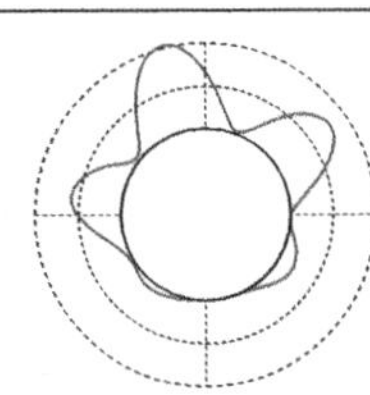
5T/5 = 10 ms	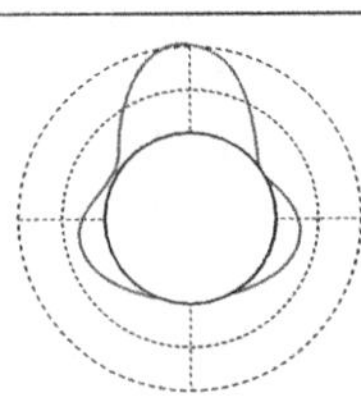	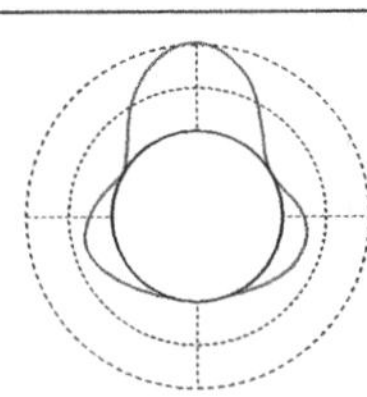	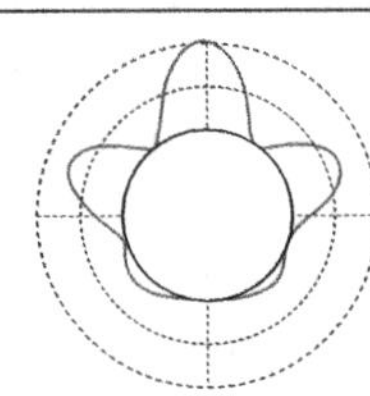

Table 7.4 Radial electromagnetic force patterns along circumference of airgap of BDFMs under pole-pair combinations of (4-5-1), (4-6-2) and (4-7-3) ($\omega_1 = 50 \times 2\pi$ Hz, $\omega_2 = 50 \times 2\pi$ Hz, T = 1/2 × 1/50 Hz = 10 ms, $\varphi_1 = \varphi_2 = 0$).

Time	(4-5-1)	(4-6-2)	(4-7-3)
0T/5 = 0 ms	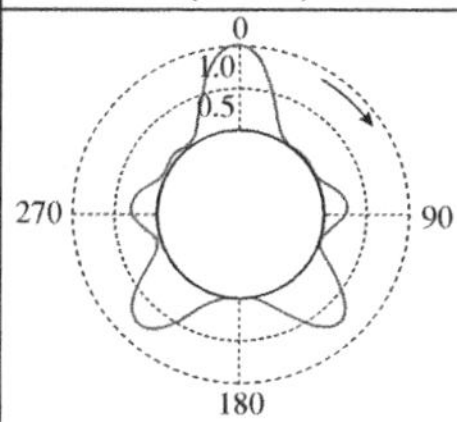	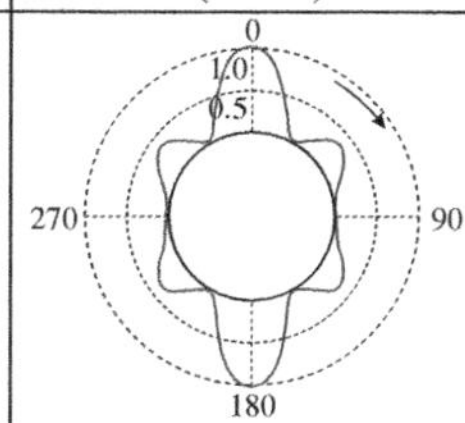	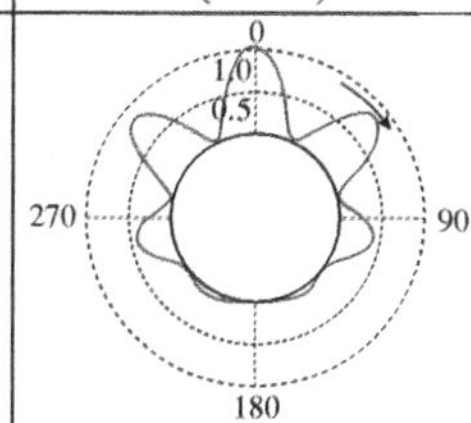
1T/5 = 2 ms	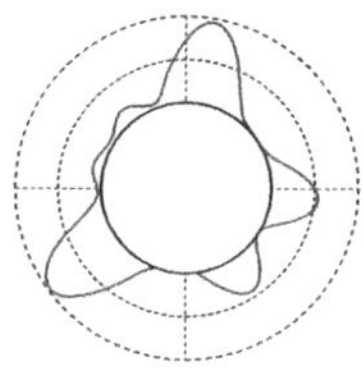	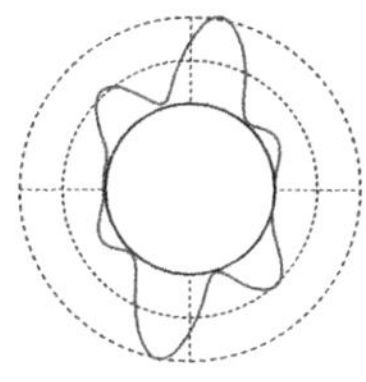	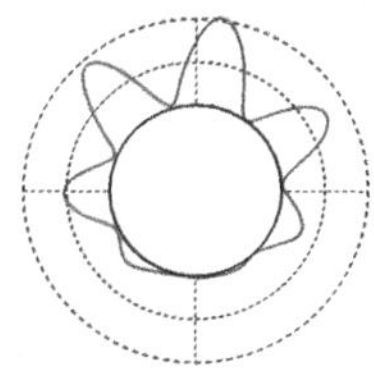
2T/5 = 4 ms	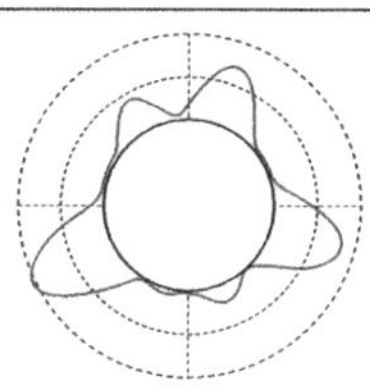	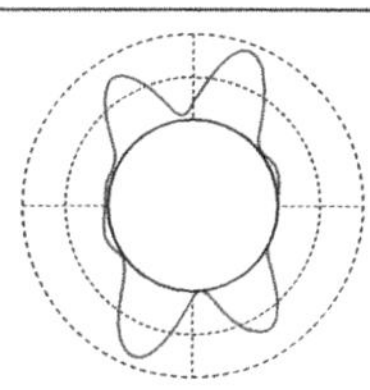	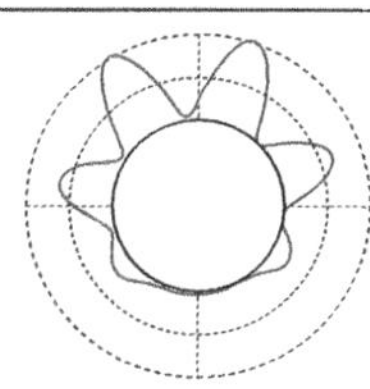
3T/5 = 6 ms	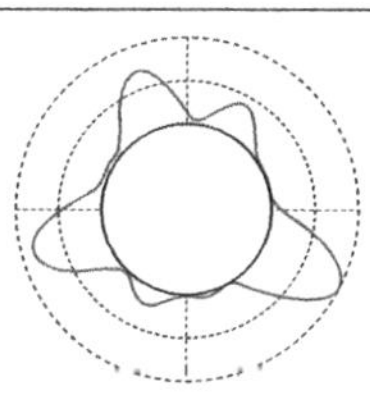	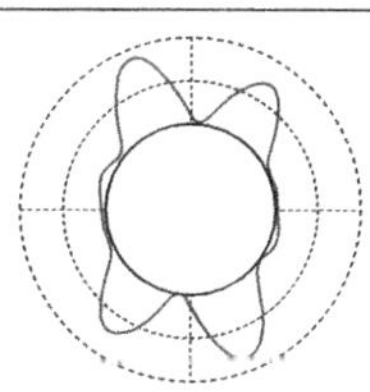	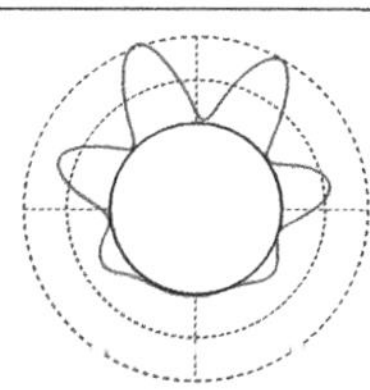
4T/5 = 8 ms	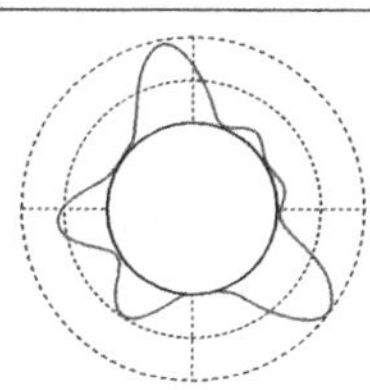	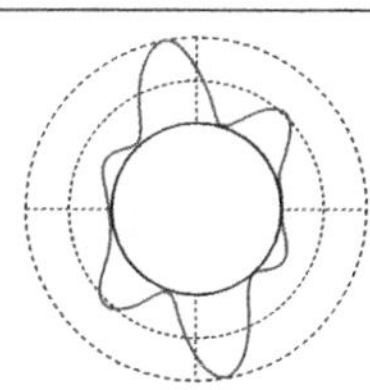	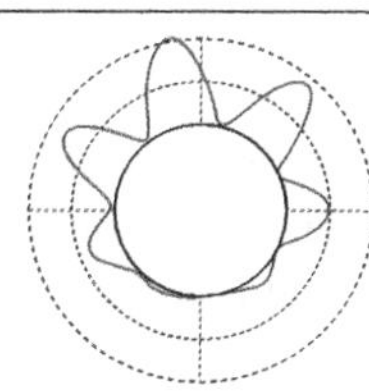
5T/5 = 10 ms	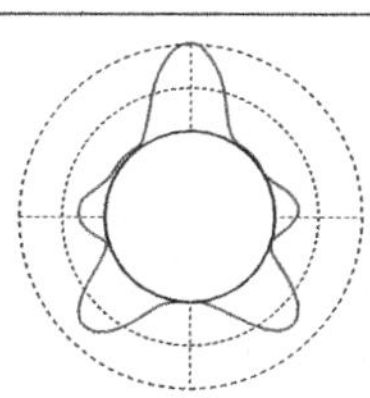	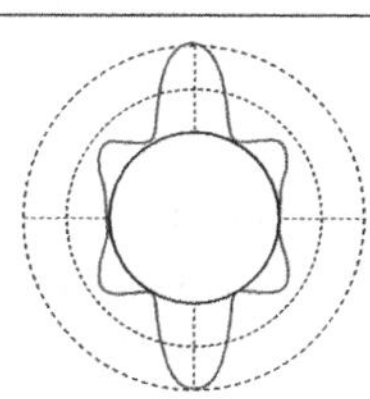	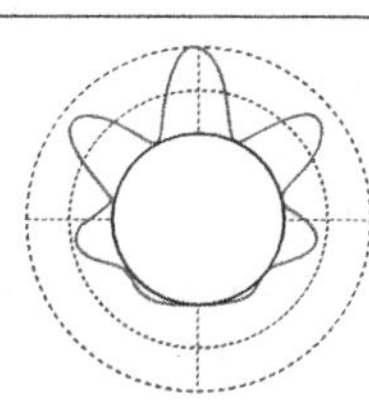

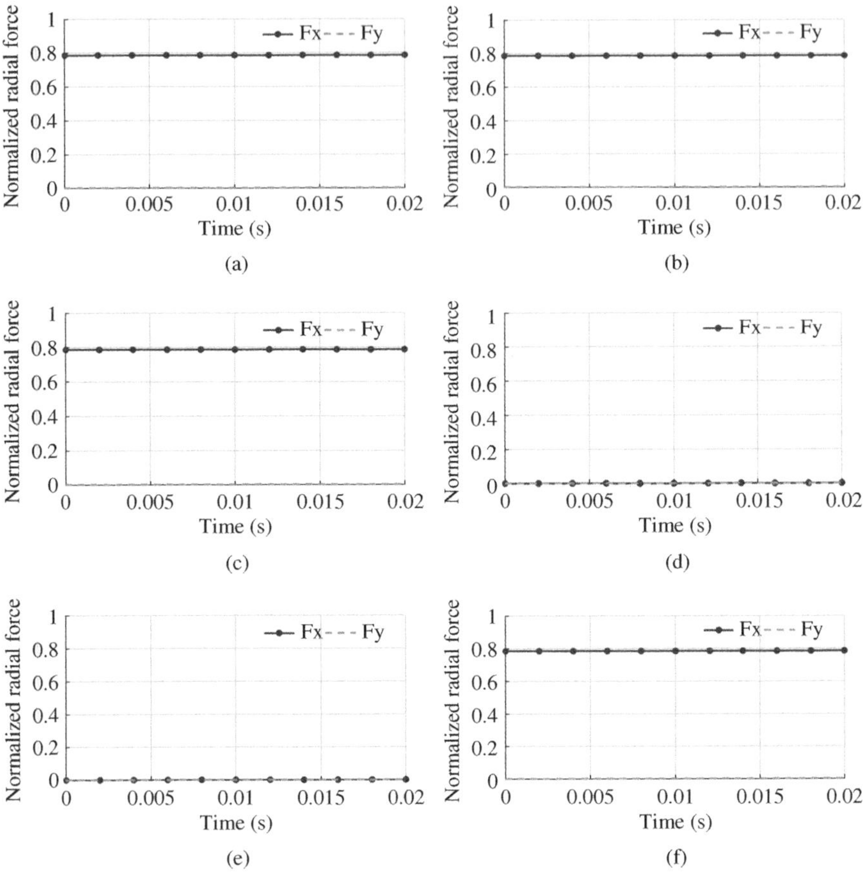

Figure 7.3 Resultant radial force under different pole-pair corresponding to Tables 7.3 and 7.4: (a) (2-3-1), (b) (1-3-2), (c) (3-5-2), (d) (4-5-1), (e) (4-6-2), (f) (4-7-3) where $\omega_1 = 50 \times 2\pi$ Hz, $\omega_2 = 50 \times 2\pi$ Hz, T = 1/2 × 1/50 Hz = 10 ms, $\varphi_1 = \varphi_2 = 0$.

the pole-pair combinations of (2-3-1) and (1-3-2) in Table 7.3. For any pole-pair combination, the number of the peak waves in the radial force still equals to the repeatable units N_r of the rotor.

From Figure 7.3, the supply frequencies of the two stator windings are the same so that the resultant radial force becomes a constant vector instead of rotation which is in accordance with the prediction of (7.4).

To find out the contribution of each airgap flux density pairs to the non-zero resultant radial force, the airgap flux density distribution at a certain instant of time is extracted from the FEA results and the associated amplitude spectrum is calculated, as shown in Figures 7.4 (unsaturated) and 7.5 (saturated). The radial

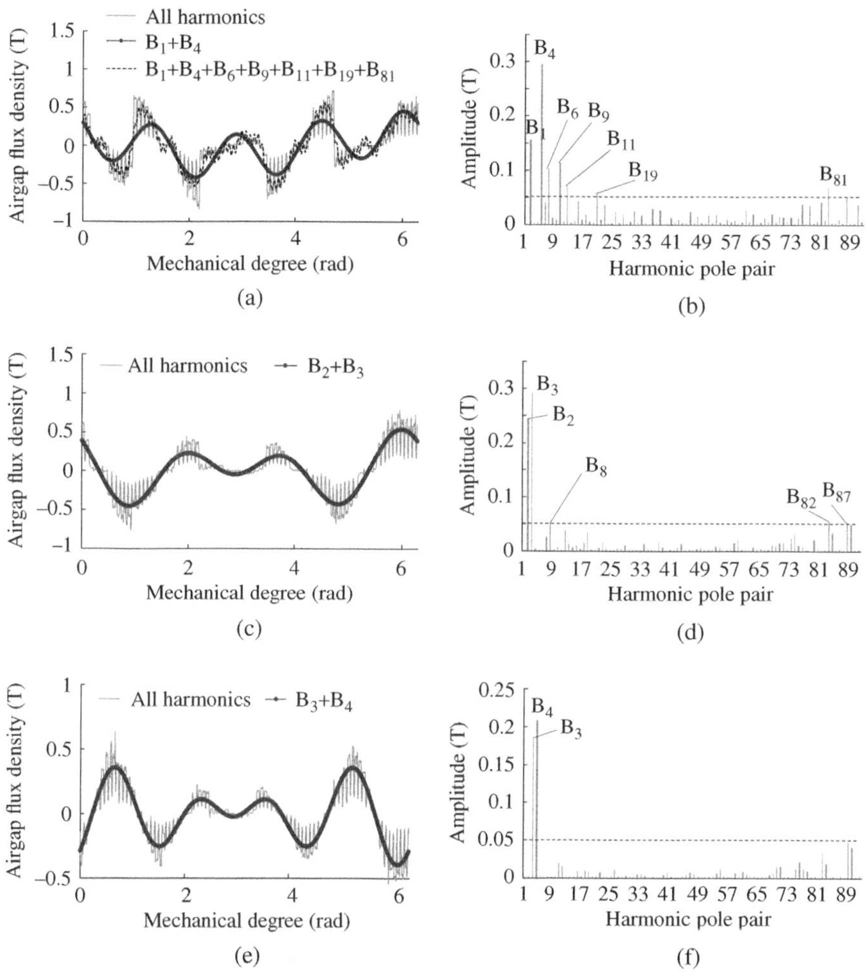

Figure 7.4 FEA-predicted radial airgap flux density waveforms and associated amplitude spectrums of different pole-pair combinations under unsaturated condition, (a) airgap flux density waveform of (4-5-1), (b) amplitude spectrum of (4-5-1), (c) airgap flux density waveform of (3-5-2), (d) amplitude spectrum of (3-5-2), (e) airgap flux density waveform of (4-7-3), (f) amplitude spectrum of (4-7-3).

force comparison is tabulated in Tables 7.5 (unsaturated) and 7.6 (saturated). In the steady state, the amplitudes of main airgap magnetic field harmonic components are nearly constant and irrespective of the rotor position. Therefore, the airgap flux density waveform and the associated spectrum at an instant of time are enough for the radial force analysis.

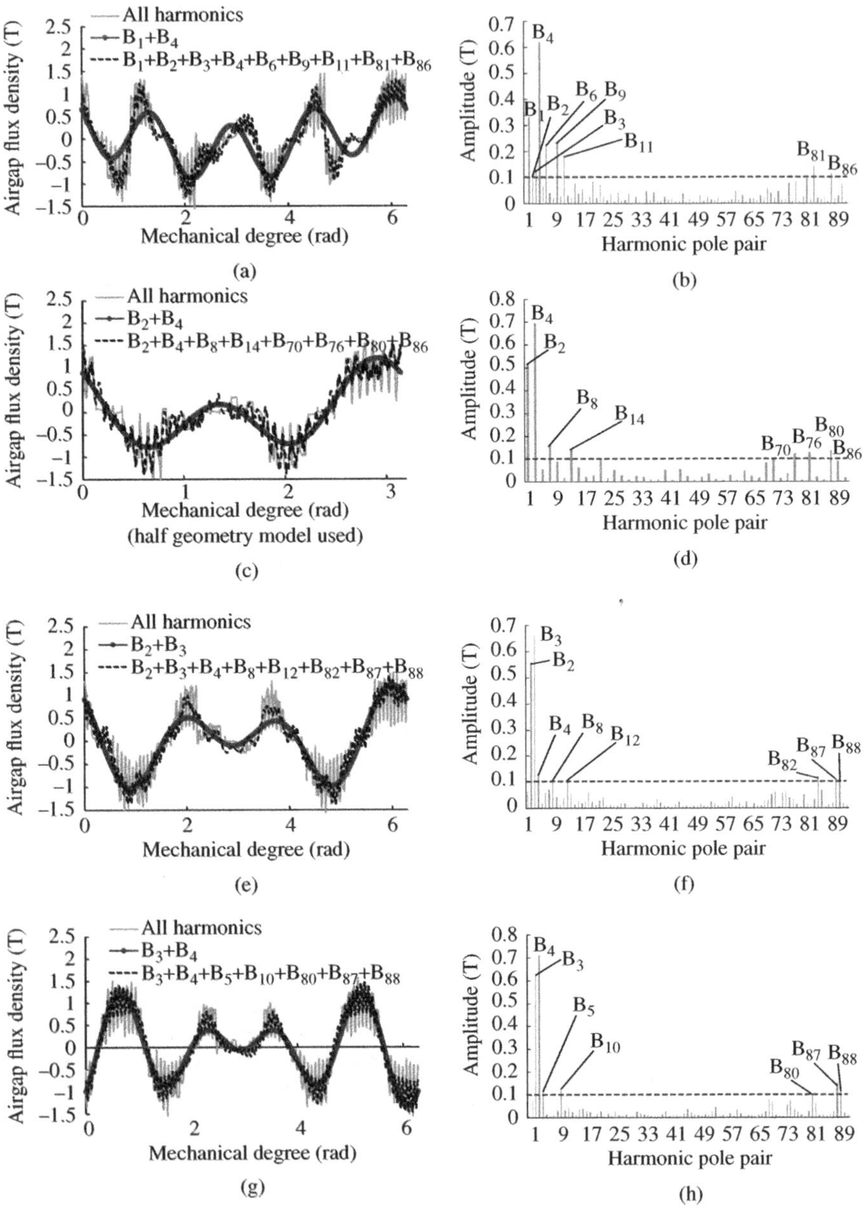

Figure 7.5 FEA-predicted radial airgap flux density waveforms and associated amplitude spectrums of different pole-pair combinations under saturated condition, (a) flux density waveform of (4-5-1), (b) amplitude spectrum of (4-5-1), (c) flux density waveform of (4-6-2), (d) amplitude spectrum of (4-6-2), (e) flux density waveform of (3-5-2), (f) amplitude spectrum of (3-5-2), (g) flux density waveform of (4-7-3), (h) amplitude spectrum of (4-7-3).

Table 7.5 Radial forces of different pole-pair combinations under unsaturated condition.

Pole-pair combinations	Calculated magnetic field harmonics	Resultant radial force vectors (N)	Amplitudes (N)
(4-5-1)	$(B_1,B_4,B_6,B_9,B_{11},B_{19},B_{81})$	0	0
	All harmonics	(2228.92 − j2786.17)	3568.02 (FEA result = 3595.03)
(3-5-2)	(B_2,B_3)	(13 113.38 − j2799.03)	13 408.78
	All harmonics	(15 168.47 − j3023.37)	15 466.84 (FEA result = 15 490)
(4-7-3)	(B_3,B_4)	(7089.37 − j1531.78)	7252.96
	All harmonics	(8239.69 − j1563.9)	8386.79 (FEA result = 8355)

Table 7.6 Radial forces of different pole pair combinations under saturated condition.

Pole-pair combinations	Calculated magnetic field harmonics	Resultant radial force vectors (N)	Amplitudes (N)
(4-5-1)	(B_1,B_2)	(6334.73 + j1560.69)	6524.16
	(B_2,B_3)	(607.44 − j2588.18)	2588.18
	(B_3,B_4)	(13 441.50 + j7113.80)	15 207.89
	(B_1,B_2,B_3,B_4)	(20 383.68 + j6158.61)	21 293.73
	All harmonics	(32 309.2 + j15272.27)	35 736.91 (FEA result = 35 807.8)
(3-5-2)	(B_2,B_3)	(67 609.66 − j13686.72)	68 981.10
	(B_3,B_4)	(−15 817.16 + j2767.25)	16 057.40
	(B_{87},B_{88})	(2069.80 − j386.24)	2105.53
	$(B_2,B_3,B_4,B_{87},B_{88})$	(53 862.30 − j11305.71)	55 036.05
	All harmonics	(46 963.65 − j10320.89)	48 084.36 (FEA result = 47 891.0)
(4-6-2)	$(B_2,B_4,B_8,B_{14},B_{70},B_{76},B_{80},B_{86})$	0	0
	All harmonics	0	0 (FEA result = 0)
(4-7-3)	(B_3,B_4)	(82 365.80 − j17526.85)	84 209.95
	(B_4,B_5)	(−15 521.80 − j3140.39)	15 836.30
	(B_{87},B_{88})	(3212.47 − j554.44)	3259.97
	$(B_3,B_4,B_5,B_{87},B_{88})$	(70 056.47 − j14940.90)	71 631.97
	All harmonics	(63 355.16 − j12210.9)	64 521.18 (FEA result = 63 683.8)

It can be found from the FEA results that the unbalanced magnetic pulls (UMPs) are mainly contributed by the work harmonics for the combinations (3-5-2) and (4-7-3). However, the UMP in the combination (4-5-1) is mainly contributed by useless harmonics. This is because in the combination (4-5-1), the pole-pair number of the two working harmonics differs by 3, so UMP cannot be generated. It can be concluded that the assumption for radial force analysis is not reasonable that only the two work harmonics are considered and all other spatial harmonics originating from winding layouts, the saturation effect and slotting effect are negligible. This also highlights the importance of using GAFMT to analyze the effect of various flux density harmonics in the airgap on the radial force distribution.

The comparison of Tables 7.5 and 7.6 shows that the saturation effect introduces extra low-order harmonics aside from the dominating harmonics. As the amplitude spectrum in the low-order part becomes denser, there will be more flux density harmonic pairs with the difference of 1 in the pole-pair number. As a consequence, the difference between the resultant radial force calculated by all the airgap flux density harmonics and that predicted from the two work harmonics becomes larger. This indicates that, under saturated conditions, useless harmonics account for a large fraction of the radial force and cannot be neglected anymore.

7.2.4 Pole-Pair Combinations Without UMP

To avoid unbalanced magnetic pull, the radial force distribution of the brushless doubly-fed machine should be radially symmetrical,

$$\sigma_r(\phi, t) = \sigma_r(\phi + \pi, t) \tag{7.5}$$

Taking into account all airgap flux density harmonics introduced by the modulation behavior of short-circuited coils, simple-salient-pole reluctance or multi-layer magnetic barriers, (7.5) can be rewritten as

$$\begin{aligned} &\frac{1}{2\mu_0}\left\{\frac{\mu_0}{g_e}M(N_r)[F_1(\phi,t)+F_2(\phi,t)]\right\}^2 \\ &\quad=\frac{1}{2\mu_0}\left\{\frac{\mu_0}{g_e}M(N_r)[F_1(\phi+\pi,t)+F_2(\phi+\pi,t)]\right\}^2 \end{aligned} \tag{7.6}$$

where $M(N_r)[\cdot]$ is the modulation operator proposed in Chapter 3. g_e is the effective airgap length. $F_1(\phi, t)$ and $F_2(\phi, t)$ are the magnetizing MMF produced by the power winding and control winding, respectively. (7.6) is further simplified and can be obtained by

$$p_1 \pm p_2 = N_r = 2k, (p_1, p_2 \in Z^+, k \in Z) \tag{7.7}$$

The saturation of iron core will introduce the additional third harmonic, which imposes more constraints on the available pole-pair combinations

$$\begin{cases} p_1 \pm p_2 = N_{MB} = 2k, & (p_1, p_2 \in Z^+, k \in Z) \\ 3p_1 \pm p_2 = N_{MB} = 2k, & (p_1, p_2 \in Z^+, k \in Z) \\ p_1 \pm 3p_2 = N_{MB} = 2k, & (p_1, p_2 \in Z^+, k \in Z) \\ 3p_1 \pm 3p_2 = N_{MB} = 2k, & (p_1, p_2 \in Z^+, k \in Z) \end{cases} \tag{7.8}$$

By simplifying (7.8), one can get the same equation as (7.7) because the last three equations will be naturally met if the pole-pair combination satisfies the first equation in (7.8). In summary, when considering the iron core saturation, the unbalance magnetic pull can be completely eliminated only when the rotor of the brushless doubly-fed machine contains an even number of repeating units.

7.3 Design of Suspension Windings for Bearingless Homopolar and Consequent Pole PM Machines

The application of the GAFMT allows the rapid analysis and design [2] of two-pole suspension winding for bearingless homopolar and consequent pole PM machines in [6]. The application requires a set of stator windings capable of generating two different main pole pairs at the same time, which coincides with the multiple spatial harmonics of the GAFMT. Therefore, the modulation analysis of the source magnetizing MMF is considered, which in turn determines the connection of the coils in the suspension winding.

For the investigated torque winding ($Q = 36$, $p = 4$, $m = 3$), the number of slots per pole per phase $q = Q/(2mp) = 3/2$, pole pitch $\tau = Q/(2p) = 9/2$, and coil span $y = 4$ or 5, which needs to be an integer. The conductor distribution of the torque winding is illustrated in Figure 7.6a, where the cross sign means the current flows into the page and the dot sign means the current flows out of the page. The corresponding conductor distribution and linear current density distribution are shown in Figure 7.6b. The conductor distribution is a train of Dirac functions (arrows in Figure 7.6), which denote the locations and current directions of slot conductors. The turn function and winding function are shown in Figure 7.6c,d, respectively. The turn function $T(\phi,\varphi_0)$ is the integral of the conductor distribution $D(\phi)$ from an arbitrary angular position φ_0. By removing the DC component of $T(\phi,\varphi_0)$, the winding function $W(\phi)$ is achieved, which is independent of φ_0. According to the GAFMT, a set of winding can be described by its winding function. The winding factors of all kinds of harmonics can be obtained by the Fourier transforms of the winding function.

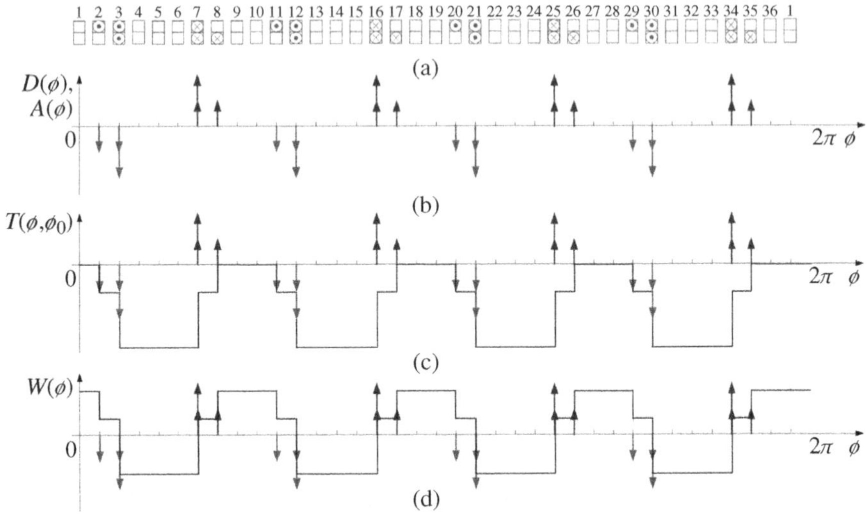

Figure 7.6 The torque winding ($Q = 36, p = 4, m = 3$) for bearingless homopolar and consequent pole PM machines, (a) conductor distribution, (b) distribution functions of conductor and linear current density, (c) turn function, (d) winding function.

7.3.1 Design Principle of Pole-Changing Windings

The conductor distribution function of the torque winding can be summarized by

$$C_t(\phi) = \sum_j R_j \delta(\phi_j), \left(\sum_j R_j = 0 \right) \tag{7.9}$$

where R_j is the number of conductors for the j-th slot and contains the current direction. For example, $R = -5$ represents that there are five conductors in a slot, while the current of conductors flows out of the page. The distribution function of the linear current density A, the distribution function of magnetizing MMF F, the turn function T and the winding function W are, respectively, expressed by:

$$A_t(\phi) = i \cdot C_t(\phi) \tag{7.10}$$

$$F_t(\phi) = \int_{\phi_0}^{\phi} A_t(\phi)d\phi - \frac{1}{2\pi}\int_0^{2\pi}\left[\int_{\phi_0}^{\phi} A_t(\phi)d\phi\right] d\phi = i \cdot W_t(\phi) \tag{7.11}$$

$$T_t(\phi, \phi_0) = \int_{\phi_0}^{\phi} C_t(\phi)d\phi \tag{7.12}$$

$$W_t(\phi) = T_t(\phi, \phi_0) - \int_0^{2\pi} T_t(\phi, \phi_0)d\phi \tag{7.13}$$

Since the magnetizing MMF created by the torque winding has p_t pole pairs, the dominating fundamental component with p_t pole pairs of the conductor distribution function can be written as:

$$C_t(\phi) = C\sin(p_t\phi + \varphi_d) \tag{7.14}$$

The source magnetizing MMF established by the winding can be modulated by the short-circuited coils, simple-salient-pole reluctance or multi-barrier rotors based on the GAFMT described in Chapter 3, generating new magnetizing MMF harmonics with different pole-pairs. The use of variable-pole winding enables the source magnetizing MMF change which can be regarded as the result of the simple-salient-pole reluctance modulation. Further, when the modulation operator defined in Chapter 3 for salient reluctance poles is applied to the conductor distribution function $C_t(\phi)$ where $\lambda(x,\theta_r)$ is $M\sin(p_m\phi + \varphi_m)$, the conductor distribution will be:

$$\begin{aligned} M(p_m)[C_t(\phi)] &= MC\sin(p_m\phi + \varphi_m)\sin(p_t\phi + \varphi_d) \\ &= \left(\frac{MC}{2}\right)\{\cos[(p_m - p_t)\phi + (\varphi_m - \varphi_d)] \\ &\quad - \cos[(p_m + p_t)\phi + (\varphi_m + \varphi_d)]\} \end{aligned} \tag{7.15}$$

The magnetizing MMF created by the modulated conductor distribution will be:

$$\begin{aligned} F_s(\phi) &= i\cdot\left\{\int_{\phi_0}^{\phi} M(\phi)C_t(\phi)d\phi - \frac{1}{2\pi}\int_0^{2\pi}\left[\int_{\phi_0}^{\phi} M(\phi)C_t(\phi)d\phi\right]d\phi\right\} \\ &= i\cdot\left(\frac{MC}{2}\right)\left(\frac{1}{p_m - p_t}\right)\{\sin[(p_m - p_t)\phi + (\varphi_m - \varphi_d)]\} \\ &\quad - i\cdot\left(\frac{MC}{2}\right)\left(\frac{1}{p_m + p_t}\right)\{\sin[(p_m + p_t)\phi + (\varphi_m + \varphi_d)]\} \end{aligned} \tag{7.16}$$

It can be seen from (7.16) that the magnetizing MMF produced by the modulated conductor distribution has two different pole pairs: $(p_m - p_t)$ and $(p_m + p_t)$, and the component with $(p_m - p_t)$-pole pairs has a larger amplitude than that of the component with $(p_m + p_t)$-pole pairs. Therefore, if we want a set of the winding with 4-pole pairs to induce the magnetizing MMF with 1-pole pair by changing the current directions of existing conductors, the modulation wave should have $(4+1)$-pole pairs, i.e. $p_m = 5$. After the number of pole pairs of the modulation waves is determined, we can use this modulation wave to determine the current direction in which conductors should be reversed to generate the required two-pole pairs magnetizing MMF and the winding functions of new conductor distributions.

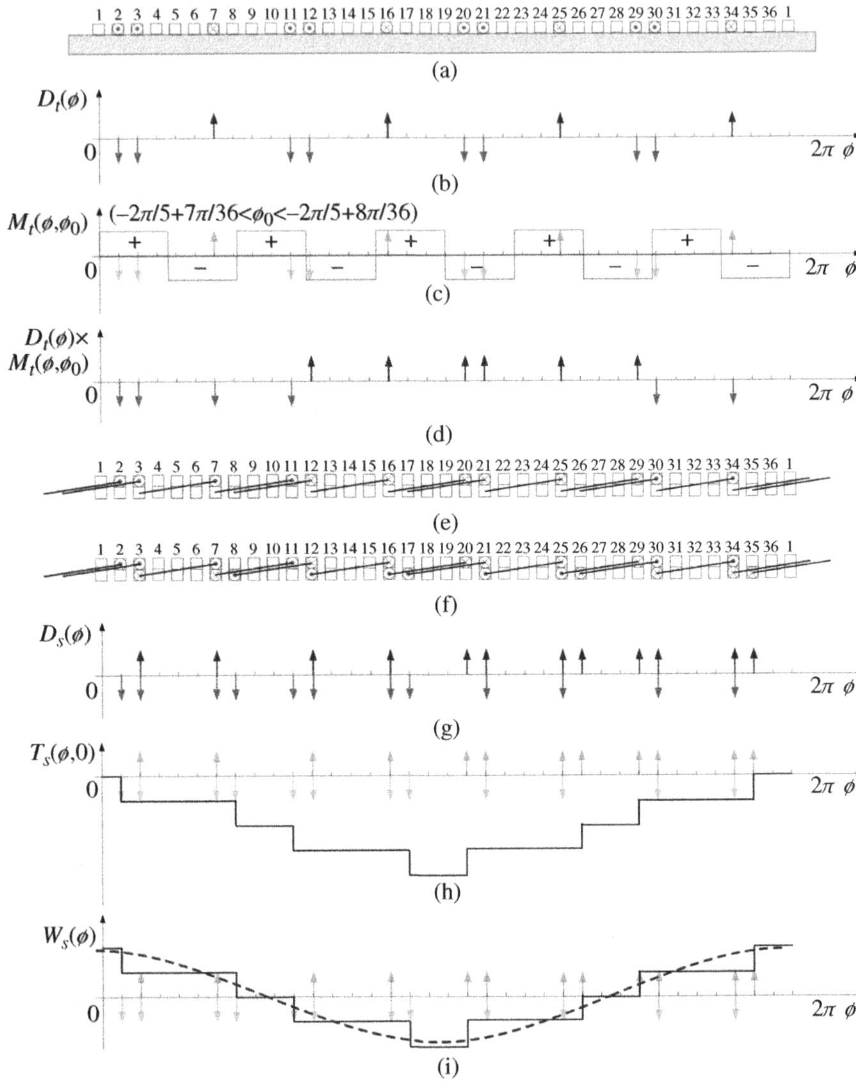

Figure 7.7 The derivation of the conductor distribution of the suspension winding I from existing torque winding ($Q=36$, $p=4$, $m=3$, $y=4$), (a) top-layer conductors of the torque winding, (b) conductor distribution of the top-layer torque winding, (c) modulation wave, (d) conductor distribution of the top-layer suspension winding, (e) top-layer conductors of the suspension winding, (f) conductors connection of the suspension winding, (g) conductor distribution of the suspension winding (including top-layer and bottom-layer), (h) turn function of the suspension winding, (i) winding function of the suspension winding.

7.3.2 Solution 1: Coil Span y = 4

Let us look at the top-layer conductors and the corresponding conductor distribution, which are shown in Figure 7.7a,b, respectively. If the initial position of the modulation wave falls into the interval $(-2\pi/5+7\pi/36, -2\pi/5+8\pi/36)$, as shown in Figure 7.7c, the modulated conductor distribution of top-layer conductors will be that shown in Figure 7.7d. For the case where the winding has a coil span of $y=4$, the current directions in bottom-layer conductors can be determined accordingly. It can be observed here that different initial positions of the modulation wave can lead to different suspension winding layouts.

Another possibility for $y=4$ is shown in Figure 7.8. To distinguish two possible suspension windings, they are named here suspension winding I and suspension winding II, respectively. Once the direction of the current in the top and bottom conductors is determined, the turn function and the winding function of the suspension winding can be obtained. The green dashed lines in Figures 7.7 and 7.8 indicate the fundamental components of the magnetizing MMFs established by the suspension windings. By calculating the Fourier transforms of the turn function and the winding function of the suspension winding, all winding factors of harmonics can be obtained, as shown in Figures 7.9 and 7.10.

The winding factors of harmonics of suspension windings I and II are tabulated in Table 7.7. The GAFMT facilitates graphical design and analysis of the complex windings, and this whole process can be easily done by the signal analysis and transforms by programing.

7.3.3 Solution 2: Coil Span y = 5

Similarly, the case with coil span $y=5$ can be analyzed and another two possible designs, namely, suspension windings III and IV, can be obtained as shown in Figures 7.11 and 7.12. The Fourier transforms are given in Figures 7.13 and 7.14, whereas the winding factors of harmonics are tabulated in Table 7.7.

7.4 Loss Calculation

The developed GAFMT is very useful in analyzing and calculating airgap magnetic field harmonic related losses, such as the machine core loss, eddy current loss in PMs, AC winding losses, stray load loss, etc. This section includes one example for the calculation of stray load loss in squirrel-cage IMs and another example for core loss estimation of FSPM machines supplied by PWM inverters. These two examples well demonstrate the effectiveness and superiority of incorporating the developed GAFMT in harmonic-related loss calculation.

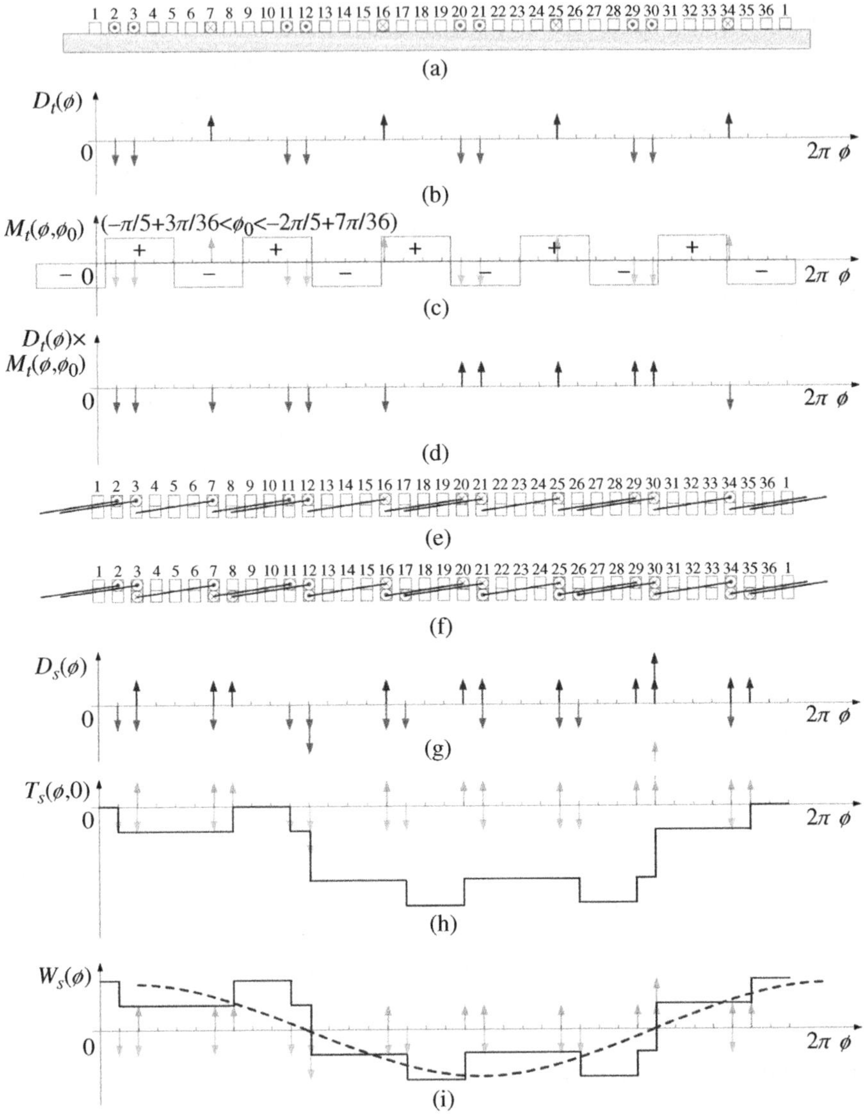

Figure 7.8 The derivation of the conductors distribution of the suspension winding II from existing torque winding ($Q = 36$, $p = 4$, $m = 3$, $y = 4$), (a) top-layer conductors of the torque winding, (b) conductor distribution of the top-layer torque winding, (c) modulation wave, (d) conductor distribution of the top-layer suspension winding, (e) top-layer conductors of the suspension winding, (f) conductors of the suspension winding, (g) conductor distribution of the suspension winding (including top-layer and bottom-layer), (h) turn function of the suspension winding, (i) winding function of the suspension winding.

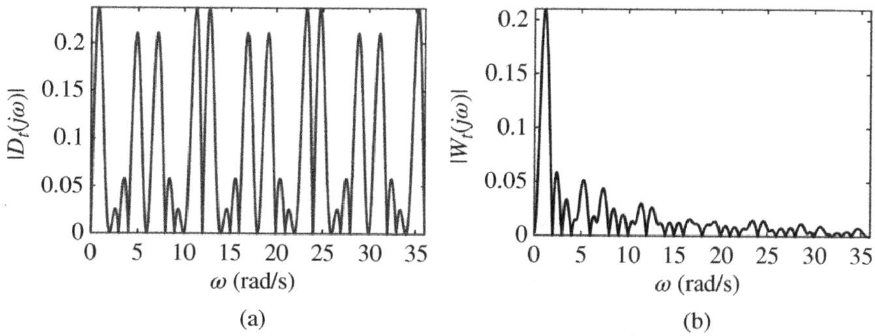

Figure 7.9 Fourier transforms of the suspension winding I (Fourier coefficients are values at $\omega = 1, 2, 3, 4, 5, \ldots$), (a) normalized Fourier transform of the conductor distribution of the suspension winding $D_t(\alpha)$, (b) normalized Fourier transform of the winding function of the suspension winding $W_t(\alpha)$.

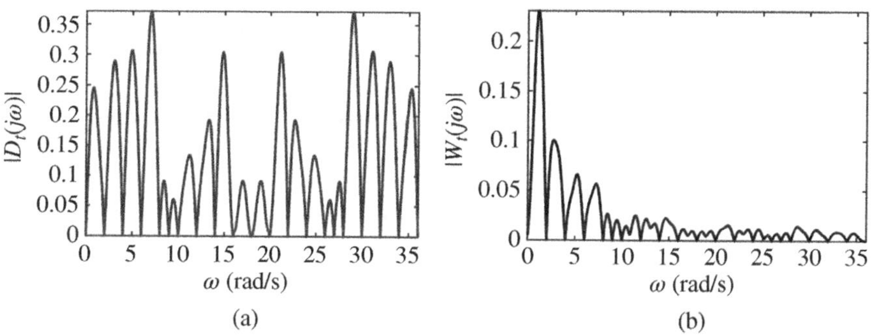

Figure 7.10 Fourier transforms of the suspension winding II (Fourier coefficients are values at $\omega = 1, 2, 3, 4, 5, \ldots$), (a) normalized Fourier transform of the conductor distribution of the suspension winding $D_t(\alpha)$, (b) normalized Fourier transform of the winding function of the suspension winding $W_t(\alpha)$.

Table 7.7 Winding factors of available winding designs.

Cases	**Case 1:coil span $y = 4$**		**Case 2:coil span $y = 5$**	
Pole-pairs of harmonics	**Suspension winding I**	**Suspension winding II**	**Suspension winding III**	**Suspension winding IV**
1	0.2041	0.2250	0.2522	0.2780
3	0	0.2887	0	0.3220
5	0.2041	0.3046	0.1698	0.2534
7	0.2041	0.3711	0.0277	0.0503
9	0	0	0.5000	0.3727
11	0.2041	0.1262	0.3164	0.1956

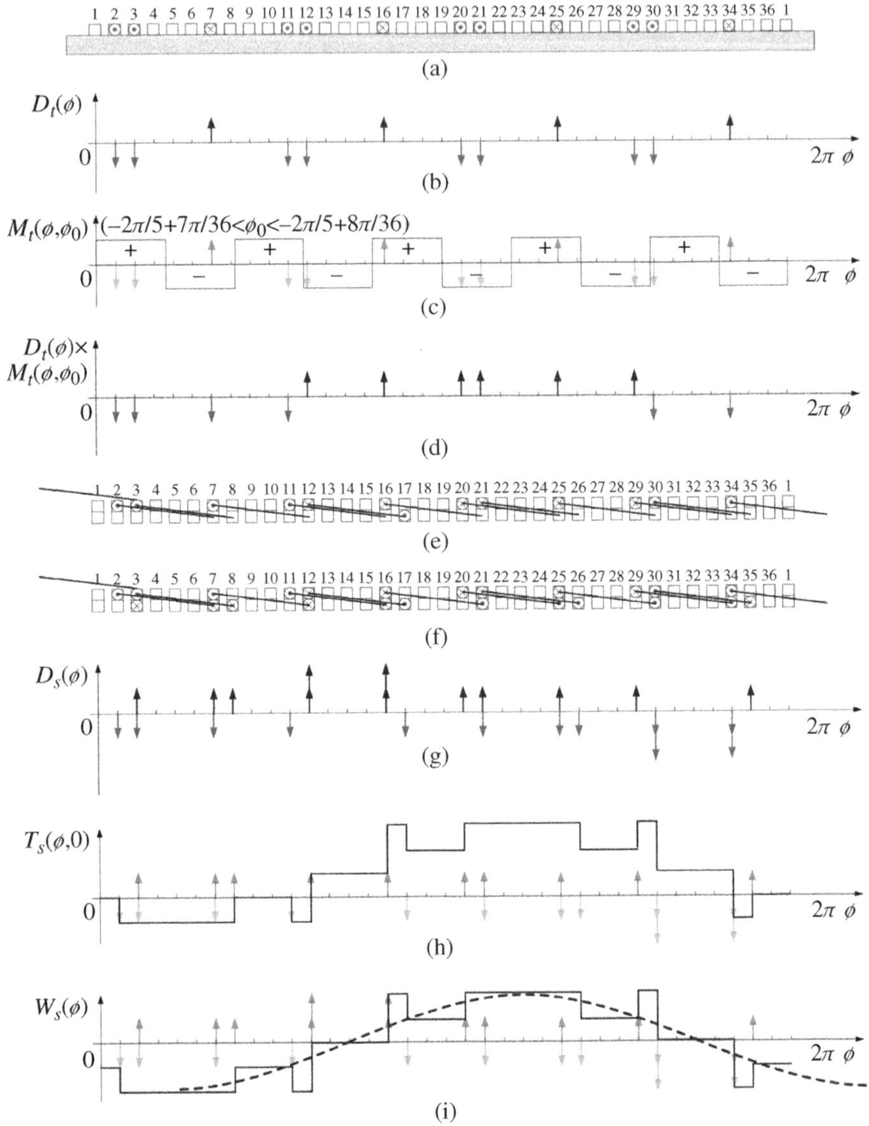

Figure 7.11 The derivation of the conductors distribution of the suspension winding III from existing torque winding ($Q = 36$, $p = 4$, $m = 3$, $y = 5$), (a) top-layer conductors of the torque winding, (b) conductor distribution of the top-layer torque winding, (c) modulation wave, (d) conductor distribution of the top-layer suspension winding, (e) top-layer conductors of the suspension winding, (f) conductors of the suspension winding, (g) conductor distribution of the suspension winding (including top-layer and bottom-layer), (h) turn function of the suspension winding, (i) winding function of the suspension winding.

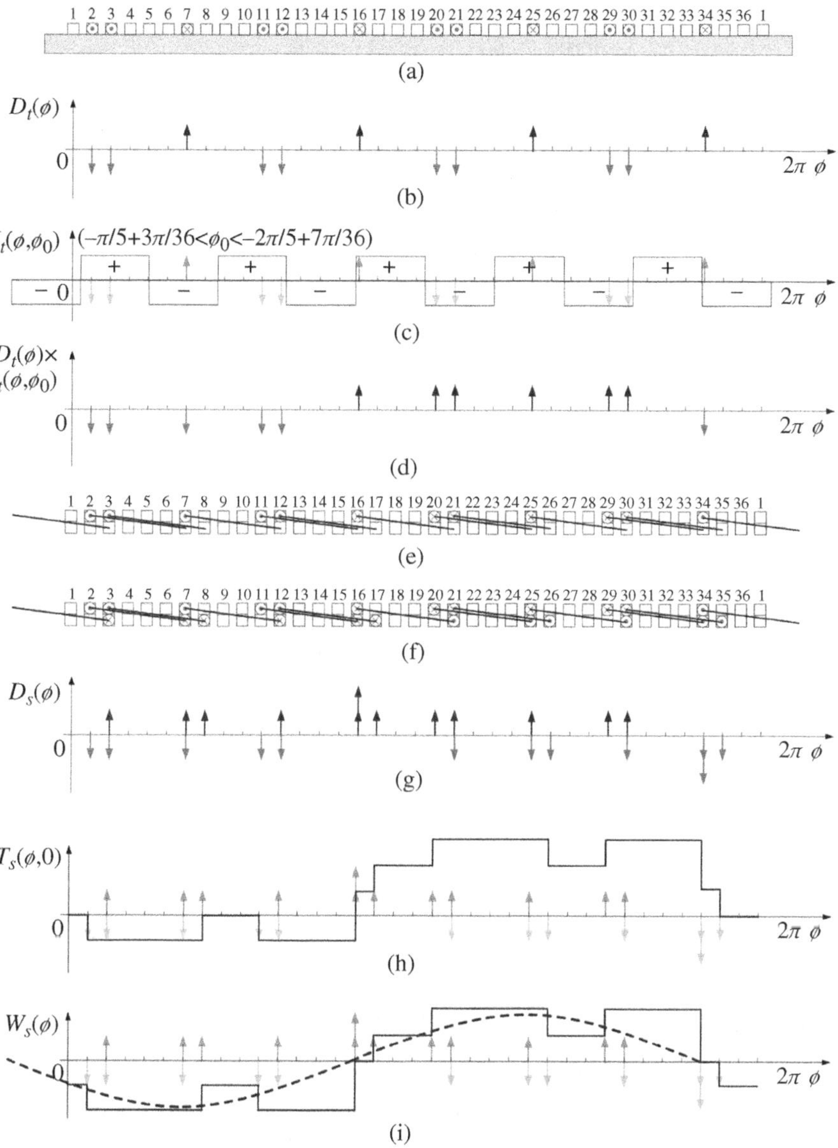

Figure 7.12 The derivation of the conductors distribution of the suspension winding IV from existing torque winding ($Q = 36$, $p = 4$, $m = 3$, $y = 5$), (a) top-layer conductors of the torque winding, (b) conductor distribution of the top-layer torque winding, (c) modulation wave, (d) conductor distribution of the top-layer suspension winding, (e) top-layer conductors of the suspension winding, (f) conductors of the suspension winding, (g) conductor distribution of the suspension winding (including top-layer and bottom-layer), (h) turn function of the suspension winding, (i) winding function of the suspension winding.

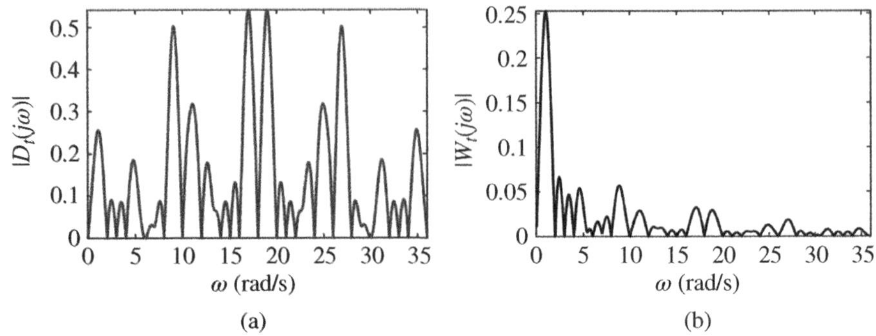

Figure 7.13 Fourier transforms of the suspension winding III (Fourier coefficients are values at $\omega = 1, 2, 3, 4, 5, \ldots$), (a) normalized Fourier transform of the conductor distribution of the suspension winding $D_t(\alpha)$, (b) normalized Fourier transform of the winding function of the suspension winding $W_t(\alpha)$.

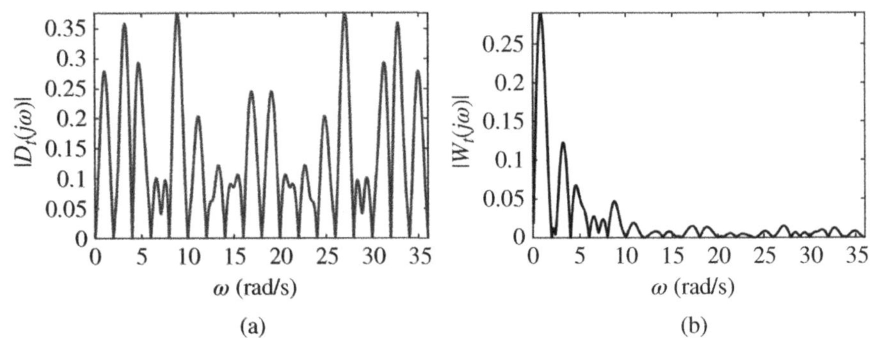

Figure 7.14 Fourier transforms of the suspension winding IV (Fourier coefficients are values at $\omega = 1, 2, 3, 4, 5, \ldots$), (a) normalized Fourier transform of the conductor distribution of the suspension winding $D_t(\alpha)$, (b) normalized Fourier transform of the winding function of the suspension winding $W_t(\alpha)$.

7.4.1 Stray Load Loss Calculation for IMs

Main loss components in squirrel-cage IMs include core losses, copper losses (stator winding loss + rotor cage loss), stray load loss and friction-windage loss. The stray load loss is defined as the difference between the total power loss of the machine on load and losses determined by the segregation method, which include the friction-windage loss, stator core loss deduced from the no-load test, stator winding loss, and fundamental winding loss of the rotor under loaded condition [7, 8].

The stray load loss is usually roughly estimated in practice but represents a key component for the efficiency evaluation. Limited by testing facilities and experimental methods, the stray load loss under the rated load operation is typically considered as a certain percentage of the total input power in standards, such as IEC 60034-2-1 [9]. Alternatively, it may also be assumed as a fixed percentage of the machine's output power varying from 0.9% to 1.8%, depending upon the machine's power rating [10]. Those methods based on empirical values do not consider the influence of motor electromagnetic parameters [11] and operating conditions on stray load loss [12].

Currently, the commonly used methods for stray load loss calculation include the three-dimensional (3D) FEA [13], the empirical relationship method [14], the equivalent circuit method [15], and the analytical approach [16]. However, all these methods have some inherent limitations. New methods that can improve the calculation accuracy and address the following two challenges are in great need. Firstly, it should not only be independent of the prototype but also benefit the machine design. Secondly, it should be able to account for the influence of operating conditions, pole-slot combinations, and airgap magnetic flux density harmonics introduced by slots.

The GAFMT provides a new angle to calculate the stray load loss in IMs by using two-dimensional (2D) FEA [3]. Figure 7.15 illustrates the proposed stray load loss calculation workflow. Firstly, the amplitudes and rotational speeds of main spatial harmonics in the airgap magnetic flux density distribution are obtained from the GAFMT [17], and the relationships between these spatial harmonics and losses are analyzed. These harmonics are then synthesized by the equivalent current layers in the airgap and substituted into 2D FEA loss models.

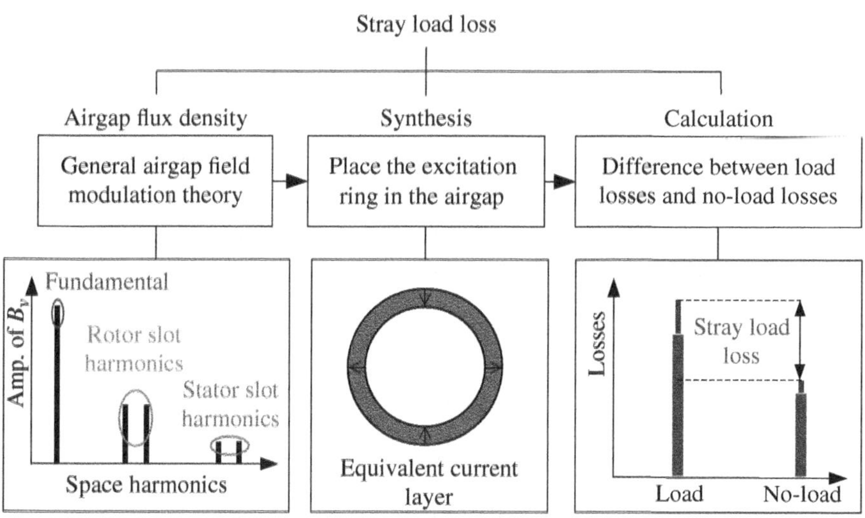

Figure 7.15 Stray load loss calculation.

1 Analysis of Airgap Magnetic Field Harmonics due to Separate Modulators

The airgap magnetic flux density spatial harmonics of IMs are considered as the leading cause of stray load loss [15]. It is therefore important to appropriately account for the airgap magnetic field harmonics into stray load loss calculation. A squirrel-cage IM shown in Figure 7.16 is taken as an example to show a detailed analysis. The key parameters and specifications are tabulated in Table 7.8. The lamination material is N23-50 and the rotor has a cast aluminum cage.

The source magnetizing MMF created by the three-phase stator winding can be expressed as:

$$F_f = \sum_{m=1}^{3} W_m(\phi) i_m = \left(\frac{3}{2}\right)\left(\frac{4}{\pi}\right)\frac{N_t k_w}{2}\sqrt{2}I\cos(p_f\phi - \omega t) \tag{7.17}$$

$$k_w = \frac{\sin\left(qp_f\frac{\alpha}{2}\right)}{q\sin\left(p_f\frac{\alpha}{2}\right)}\cos\left(\frac{p_f\beta}{2}\right) \tag{7.18}$$

where $W_m(\phi)$ and i_m are the winding function and the instantaneous current of the m-phase stator winding, respectively, and ϕ is the mechanical angle along the circumference in the modulator-fixed reference frame. I is the root-mean-square value of the input current. The source magnetizing MMF distribution has a

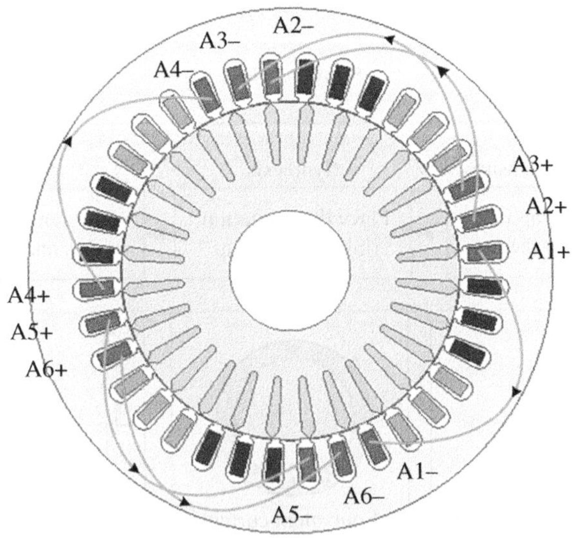

Figure 7.16 Cross-sectional view and winding layout of the squirrel-cage IM under study. Only the winding pattern of Phase A is shown for clarity.

Table 7.8 Key parameters and specifications of the squirrel-cage IM under study.

Symbol	Description	Value
N_{st}	Number of stator teeth	36
N_{rt}	Number of rotor teeth	28
N_t	Turns in series per pole per phase	141
p_f	Number of stator pole pairs	2
q	Number of slots per pole per phase	3
α	Slot pitch angle (deg)	10
β	Short pitch angle (deg)	10
g	Airgap length (mm)	0.4
l_{stk}	Stack length (mm)	115
b_{0s}	Stator slot opening width (mm)	3.5
τ_s	Stator pole pitch (mm)	11.87
b_{0r}	Rotor slot opening width (mm)	1
τ_r	Rotor pole pitch (mm)	15.17
I_n	Rated phase current (Arms)	6.81
ω_r	Rated rotor angular speed (rad/s)	150.8
ω	Angular speed of source MMF (rad/s)	314.16

sinusoidal waveform when three-phase balanced currents are applied to the stator windings.

To determine the mechanical rotational speed of the p_f-pole-pair magnetizing MMF, it is necessary to find those points along the inner circumference that have equal MMF values at the same instant. Then this condition needs to be met:

$$\cos(p_f\phi - \omega t) = const. \tag{7.19}$$

The derivative to the time is taken for both sides of (7.19), and the mechanical rotational angular speed of p_f-pole-pair magnetizing MMF in the airgap is:

$$\omega_{p_f} = \frac{\omega}{p_f} \tag{7.20}$$

Since the airgap is almost uniform, the magnetic flux density distribution around the airgap can be expressed by:

$$B_f = \frac{3\sqrt{2}}{\pi}\frac{\mu_0 N_t I k_w}{g}\cos(p_f\phi - \omega t) \tag{7.21}$$

where μ_0 is the vacuum permeability. The airgap magnetic flux density produced by the source magnetizing MMF has a sinusoidal distribution as well.

Equation (7.21) assumes that the lamination material has an infinite magnetic permeability while the magnetic material has some saturation characteristic. When the magnetic field increases, the stator and rotor teeth of the machine will become saturated, and the magnetizing MMF drop in the stator and rotor teeth will not be zero anymore, resulting in a decrease of the magnetizing MMF in the airgap. To consider the influence of the finite permeability of the magnetic material on the airgap flux density, (7.21) can be modified to:

$$B_f = B_{fm} \cos(p_f \varphi - \omega t) \tag{7.22}$$

$$B_{fm} = \lambda_\delta \frac{3\sqrt{2}}{\pi} \frac{\mu_0 N_t I k_w}{g} \tag{7.23}$$

where λ_δ is the coefficient representing the attenuation of airgap magnetic flux density due to the influence of the finite permeability and B_{fm} is the amplitude of airgap magnetic flux density.

Figure 7.17 illustrates the influence of the magnetic saturation on the airgap magnetic flux density amplitude. $B_{\mu\infty}$ represents the airgap magnetic flux density amplitude when the magnetic material has an infinite permeability and $B_{\mu D23\_50}$ is the airgap magnetic flux density amplitude when the magnetic material is D23_50. It is shown that with the increase of the stator magnetizing current, the airgap flux density amplitude will not increase proportionally due to the saturation in the magnetic core.

The squirrel cage is a type of short-circuited winding comprising multiple interconnected loops. It is installed on the rotor side and rotates asynchronously with respect to the source magnetizing MMF. Currents will be induced in the rotor cage,

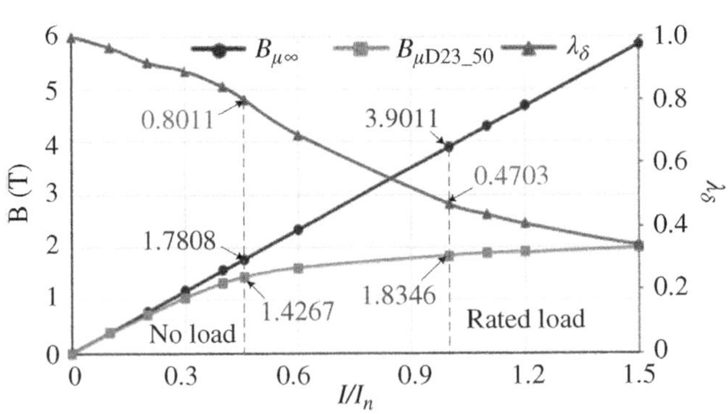

Figure 7.17 Influence of lamination material permeability on the airgap magnetic flux density amplitude when the magnetizing MMF is kept constant.

and an extra magnetizing MMF distribution is further established in the airgap to modify the time-varying source magnetizing MMF distribution, whose modulation process has been illustrated in Chapter 3. The eventual magnetizing MMF modulated by the rotor short-circuited coils is the superposition of the source magnetizing MMF and the extra magnetizing MMF.

Given that the rotor contains N_{SC} independent short-circuited coils and the coil span is γ, the modulated magnetizing MMF of rotor short-circuited coils considering the effect of resistance R and leakage inductance L can be described as:

$$\begin{aligned} M_{SC}(N_{SC},\gamma)[f(\phi,t)] &= \cos(p_f\phi-\omega t)-(1-C_{p_f})\sin(p_f\phi-\omega t+\varphi) \\ &\quad -\sum_{k=lN_{SC}-p_f}^{\infty} C_{sum}\sin\left[k\phi+(\omega-lN_{SC}\omega_r)t+(p_f+k)\frac{\pi}{N_{SC}}-\varphi\right] \\ &\quad +\sum_{k=lN_{SC}+p_f}^{\infty} C_{dif}\sin\left[k\phi-(\omega+lN_{SC}\omega_r)t+(k-p_f)\frac{\pi}{N_{SC}}+\varphi\right] \end{aligned} \tag{7.24}$$

where $\varphi = \arctan(\omega_s L/R)$ represents phase angle offset after the source magnetizing MMF is modulated by the short-circuited coils and $\omega_s = \omega - p_f\omega_r$ is the rotor SCC slip angular velocity. C_{pf}, C_{sum}, and C_{dif} represent the magnetic field conversion factors, as expressed:

$$C_{p_f} = 1 - K_{SC}\left[\frac{\sin\left(\gamma\frac{p_f\pi}{N_{SC}}\right)}{\left(\gamma\frac{p_f\pi}{N_{SC}}\right)}\right]^2 \tag{7.25}$$

$$C_{sum} = K_{SC}\frac{\sin\left(\gamma\frac{p_f\pi}{N_{SC}}\right)}{\left(\gamma\frac{p_f\pi}{N_{SC}}\right)}\frac{\sin\left(\gamma\frac{k\pi}{N_{SC}}\right)}{\left(\gamma\frac{k\pi}{N_{SC}}\right)},\quad k = lN_{SC}-p_f \tag{7.26}$$

$$C_{dif} = K_{SC}\frac{\sin\left(\gamma\frac{p_f\pi}{N_{SC}}\right)}{\left(\gamma\frac{p_f\pi}{N_{SC}}\right)}\frac{\sin\left(\gamma\frac{k\pi}{N_{SC}}\right)}{\left(\gamma\frac{k\pi}{N_{SC}}\right)},\quad k = lN_{SC}+p_f \tag{7.27}$$

$$K_{SC} = 2\pi N_{SC}\frac{\omega_s}{\sqrt{R^2+(\omega_s L)^2}}\frac{\mu_0 r_g l_{stk}}{g}\left(\frac{\gamma}{N_{SC}}\right)^2 \tag{7.28}$$

where r_g is the radius of the airgap.

From (7.24), it can be found that the main harmonics of the modulated magnetizing MMF contain p_f and $N_{rt} \pm p_f$ pole pair harmonics. The modulated magnetizing MMF in (7.24) is obtained in the short-circuited coil modulator coordinate system, and the mechanical rotational angular speed of short-circuited

Table 7.9 Magnetic field conversion factors of the rotor cage in the studied squirrel-cage IM.

Conditions	Methods	C_{pf}	C_{sum}	C_{dif}
Load	Theory	0.9275	0.0044	0.0038
	FEA	0.925	0.0076	0.0056
	Abs. difference	0.0025	0.0032	0.0018
	Percent difference (%)	0.27	72.73	47.37
No load	Theory	1	0	0
	FEA	0.9926	0.0049	0.003
	Abs. difference	0.0074	0.0049	0.003
	Percent difference (%)	0.74	—	—

coils relative to the stator is ω_r. Therefore, the mechanical rotational angular speed of the p_f-pole-pair magnetic flux density in the airgap is ω/p_f, which is equal to that of the source magnetizing MMF. The mechanical rotational angular speeds of $(N_{SC} \pm p_f)$-pole-pair harmonic magnetic flux densities relative to the stator are $\pm(\omega \pm N_{SC}\omega_r)/(N_{SC} \pm p_f)$, respectively. The positive or negative signs denote that the corresponding airgap magnetic flux density harmonics modulated by short-circuited coils rotate in the same or opposite direction of the rotor, respectively.

Table 7.9 shows the results of the magnetic field conversion factors obtained from the GAFMT and 2D FEA method under no-load and loaded operation conditions. For squirrel-cage IMs, $\gamma = 1$, N_{SC} is much larger than p_f, so the values of C_{sum} and C_{dif} are very low. Besides, since C_{sum} and C_{dif} are very low in theory, losses generated by spatial harmonics corresponding to magnetic conversion factors C_{sum} and C_{dif} have little effect on the calculated stray load loss.

The stator and rotor teeth are salient reluctance poles, and they use the variation of reluctance to modulate the source magnetizing MMF, as pointed out in Chapter 3. The modulated magnetizing MMF by rotor teeth in the Fourier series form can be expressed as:

$$\begin{aligned} M(N_{rt}, \varepsilon_r)[\cos(p_f\phi - \omega_s t)] &= C_{p_f,p_f,r}\cos(p_f\phi - \omega_s t) \\ &+ \sum_{k=lN_{rt}-p_f}^{\infty} C_{p_f,k,r}^{sum}\cos\left[k\phi + \omega_s t + \frac{(p_f+k)\pi}{N_{rt}}\right] \\ &+ \sum_{k=lN_{rt}+p_f}^{\infty} C_{p_f,k,r}^{dif}\cos\left[k\phi - \omega_s t + \frac{(k-p_f)\pi}{N_{rt}}\right], \\ &\times l \in Z^+ \end{aligned} \tag{7.29}$$

where $C_{P_f,P_f,r}$, $C^{dif}_{p_f,k,r}$, and $C^{sum}_{p_f,k,r}$ represent the magnetic field conversion factors. The $C_{P_f,P_f,r}$ is the factor converting the p_f- to p_f-pole-pair magnetizing MMF, and $C^{dif}_{p_f,k,r}$, and $C^{sum}_{p_f,k,r}$ are factors converting the p_f- to $(lN_{rt} \pm p_f)$-pole-pair magnetizing MMF component, respectively. Three magnetic field conversion factors can be obtained by:

$$C_{P_f,P_f,r} = \varepsilon_r \tag{7.30}$$

$$C^{sum}_{p_f,k,r} = \frac{1}{\pi}\left[\frac{\sin(\varepsilon_r l\pi)}{l}\right], l = 1, 2, \dots, \infty \tag{7.31}$$

$$C^{dif}_{p_f,k,r} = \frac{1}{\pi}\left[\frac{\sin(\varepsilon_r l\pi)}{l}\right], l = 1, 2, \dots, \infty \tag{7.32}$$

where ε_r is the rotor salient pole arc coefficient and is equal to $(\tau_r - b_{0r})/\tau_r$. Equations (7.30)–(7.32) indicate that the magnetic field conversion factors of the rotor salient reluctance pole modulator are only the function of the rotor salient pole arc coefficient.

After the source magnetizing MMF is modulated by rotor teeth, the most significant harmonics are of $(N_{rt} \pm p_f)$ pole pairs, besides the p_f-pole-pair magnetizing MMF. The modulated magnetizing MMF in (7.29) is obtained in the rotor speed coordinate system, and the mechanical rotational angular speed of the rotor is ω_r. The mechanical angular speed of the modulated p_f -pole-pair magnetizing MMF in the airgap is ω/p_f. The mechanical rotational angular speeds of $(N_{rt} \pm p_f)$-pole-pair harmonics relative to the stator are $\pm(\omega \pm N_{rt}\omega_r)/(N_{rt} \pm p_f)$, respectively. The positive or negative signs denote that the corresponding modulated airgap magnetic flux density spatial harmonics rotate in the same or opposite direction of the rotor, respectively.

Similarly, the magnetic field conversion factors of the stator salient reluctance poles can also be calculated. Table 7.10 lists the magnetic field conversion factors of the rotor and stator salient reluctance poles and gives the absolute errors between the results obtained from the GAFMT and 2D FEA method.

2 Synthesis of Airgap Magnetic Field Harmonics

The eventual airgap MMF turns out to be the superposition of magnetizing MMF modulated by p_f-pole-pair spatial harmonic and all the additional ones established by the $(N_{rt} \pm p_f)$- and $(N_{st} \pm p_f)$-pole-pair spatial harmonics. Based on (7.22), (7.24), and (7.29), the eventual airgap MMF F_g can be written as:

$$F_g = C_{pf}C_{pf,pf,r}C_{pf,pf,s}F_{fm}\cos(p_f\phi - \omega t)$$
$$+ \left(C_{dif} + C^{dif}_{pf,k,r}\right)F_{fm}\cos[(N_{rt} - p_f)\phi + (\omega - N_{rt}\omega_r)t + \phi_1]$$

$$+\left(C_{sum}+C^{sum}_{pf,k,r}\right)F_{fm}\cos[(N_{rt}+p_f)\phi-(\omega+N_{rt}\omega_r)t+\phi_2]$$
$$+C^{dif}_{pf,k,s}F_{fm}\cos[(N_{st}-p_f)\phi+\omega t]+C^{sum}_{pf,k,s}F_{fm}\cos[(N_{st}+p_f)\phi-\omega t] \tag{7.33}$$

$$F_{fm}=\lambda_\delta\frac{3\sqrt{2}}{\pi}N_t I k_w \tag{7.34}$$

where φ_1' and φ_2' represent the phase shifts of the modulated magnetic field.

Then, the eventual airgap magnetic flux density can be expressed as:

$$B_g=\frac{\mu_0}{g}\begin{Bmatrix} C_{pf}C_{pf,pf,r}C_{pf,pf,s}F_{fm}\cos(p_f\phi-\omega t) \\ +\left(C_{dif}+C^{dif}_{pf,k,r}\right)F_{fm}\cos[(N_{rt}-p_f)\phi+(\omega-N_{rt}\omega_r)t+\phi_1] \\ +\left(C_{sum}+C^{sum}_{pf,k,r}\right)F_{fm}\cos[(N_{rt}+p_f)\phi-(\omega+N_{rt}\omega_r)t+\phi_2] \\ +C^{dif}_{pf,k,s}F_{fm}\cos[(N_{st}-p_f)\phi+\omega t]+C^{sum}_{pf,k,s}F_{fm}\cos[(N_{st}+p_f)\phi-\omega t] \end{Bmatrix} \tag{7.35}$$

To verify the accuracy of the GAFMT, the amplitudes of airgap magnetic flux density spatial harmonics obtained from the GAFMT and 2D FEA method under no-load and rated load operation conditions are compared in Table 7.11. For the p_f-pole-pair spatial harmonic, the absolute difference between the amplitudes obtained from the GAFMT and 2D FEA is 0.0105, and the percentage difference is 1.45%. The maximum absolute difference (0.077) between the two methods occurs in the $(N_{st}+p_f)$-pole-pair harmonic modulated by the stator salient reluctance poles under both no-load and loaded conditions.

Table 7.10 Magnetic field conversion factors of the rotor and stator salient reluctance poles (teeth) in the studied squirrel-cage IM.

Modulator	Methods	$C_{P_f,P_f,r}$	$C^{dif}_{P_f,k,r}$	$C^{sum}_{P_f,k,r}$
Rotor	Theory	0.9330	0.0665	0.0665
	FEA	0.9555	0.0543	0.0672
	Abs. difference	0.0225	0.0122	0.0007
	Percent difference (%)	2.41	18.35	1.05
Modulator	Methods	$C_{P_f,P_f,r}$	$C^{dif}_{P_f,k,r}$	$C^{sum}_{P_f,k,r}$
Stator	Theory	0.533	0.2545	0.2545
	FEA	0.5347	0.2954	0.1563
	Abs. difference	0.0017	0.0409	0.0982
	Percent difference (%)	0.32	16.07	38.59

Note: $C_{P_f,P_f,r}$, $C^{dif}_{P_f,k,r}$, and $C^{sum}_{P_f,k,r}$ represent the magnetic field conversion factors of the stator salient reluctance poles.

Table 7.11 Airgap magnetic flux density breakdown.

Cond.	Pole pair	Amp. theory (T)	Amp. FEA (T)	Abs. diff. (T)	Percentage diff. (%)	Relative to stator speed (r/min)	Relative to rotor speed (r/min)
No-load	p_f	0.7340	0.7235	0.0105	1.43	1500	0
	$N_{rt}-p_f$	0.0520	0.0308	0.0212	40.77	1500	0
	$N_{rt}+p_f$	0.0562	0.0480	0.0082	14.59	1500	0
	$N_{st}-p_f$	0.2111	0.2357	0.0246	11.65	−88.24	1588.24
	$N_{st}+p_f$	0.2111	0.1341	0.077	36.48	78.947	1421.05
Loaded	p_f	0.848	0.8644	0.0164	1.93	1500	60
	$N_{rt}-p_f$	0.0616	0.0809	0.0193	31.33	1435.3	−4.7
	$N_{rt}+p_f$	0.0697	0.1194	0.0497	71.31	1444	4
	$N_{st}-p_f$	0.2500	0.2703	0.0203	8.12	−88.24	1528.24
	$N_{st}+p_f$	0.2500	0.1730	0.077	30.8	78.947	1361.05

From Table 7.11, the relationship between losses and airgap magnetic flux density can be obtained and the following conclusions can be drawn:

1) When the motor is under no-load or loaded operation condition, $(N_{rt} \pm p_f)$-pole-pair harmonics in the airgap magnetic flux density are almost stationary relative to the rotor, which means that these harmonics do not induce losses in the rotor. On the contrary, under the two operation conditions, $(N_{st} \pm p_f)$-pole-pair harmonics move relative to the rotor, and losses will be induced in the rotor. The p_f-pole-pair magnetic flux density rotates relative to the rotor only under load operation condition.
2) Whether it is under no-load or loaded operation condition, the p_f-, $(N_{rt} \pm p_f)$-, and $(N_{st} \pm p_f)$-pole-pair harmonics in the airgap magnetic flux density rotate relative to the stator, and these harmonics will induce losses in the stator.

3 Loss Calculation Based on Airgap Flux Density Harmonics

Core losses in electrical machines are closely related to the magnetic flux density distribution in stator and rotor cores. As we have seen, analysis based on the GAFMT shows that the airgap magnetic flux density mainly contains p_f-, $(N_{rt} \pm p_f)$- and $(N_{st} \pm p_f)$-pole-pair spatial harmonics, which contribute most to the stray load loss. To consider the influence of each flux density spatial harmonic on the stray load loss separately, thin current layers that can generate the same airgap flux density distributions are applied to the stator and rotor surface in 2D FEA models. It is assumed that fringing and leakage flux at the two ends of the

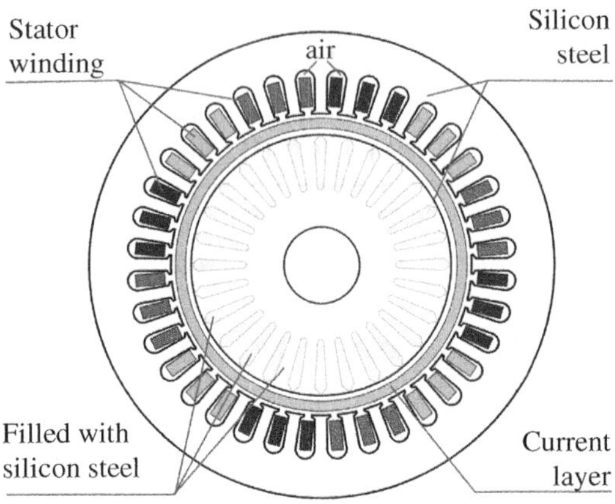

Figure 7.18 2D FEA model for stator core loss calculation. Rotor slots are filled with the same lamination material used for machine cores.

lamination stack is negligible and there is no axial component for flux density vectors in the airgap and machine cores. In addition, time harmonics are not considered due to the pure sinusoidal voltage supply.

The main influence of the p_f-pole-pair spatial harmonic on the stray load loss is introduced by the stator core losses under no-load and loaded operation conditions. The corresponding 2D FEA model for stator core loss calculation is shown in Figure 7.18. The stator is slotted but the rotor surface is covered by a smooth thin current layer and the lamination material is applied to areas occupied by rotor bars.

Based on (7.33) and (7.34), the amplitude of the magnetizing MMF corresponding to the p_f-pole-pair spatial harmonic in the airgap obtained by the GAFMT is:

$$F_{p_f} = \lambda_\delta \left(\frac{3\sqrt{2}}{\pi}\right) N_t I k_w C_{p_f} C_{p_f,p_f,s} C_{p_f,p_f,r} \tag{7.36}$$

The current density distribution j_{pf} in the thin current layer takes the form of:

$$j_{p_f} = J_{p_f} \cos\left(\omega t - \frac{\pi}{\tau_{p_f}} \phi\right) \tag{7.37}$$

where J_{pf} is the amplitude of the current density in A/m and τ_{pf} is the pole pitch corresponding to the p_f-pole-pair airgap flux density spatial harmonic, and x represents the arc length along the airgap circumference in meters.

The pole pitch of the p_f-pole-pair spatial harmonic is:

$$\tau_{p_f} = \frac{2\pi r_g}{2p_f} \tag{7.38}$$

The magnetizing MMF f^e_{pf} corresponding to p_f-pole-pair spatial harmonic created by the equivalent current layer is:

$$f^e_{pf} = F^e_{pf} \cos\left(\omega t - \frac{\pi}{\tau_{p_f}}\phi - \frac{\pi}{2}\right) \tag{7.39}$$

$$F^e_{pf} = C_{pf,pf,s}\lambda_\delta \frac{\tau_{pf}}{\pi} J_v \tag{7.40}$$

where F^e_{pf} is the magnetizing MMF amplitude.

To produce the same magnetizing MMF, the equivalent current density amplitude corresponding to the p_f-pole-pair spatial harmonic is:

$$J_{p_f} = p_f \frac{3\sqrt{2}}{\pi r_g} N_t I k_w C_{p_f} C_{p_f,p_f,r} \tag{7.41}$$

With this current density applied to the thin current layer in 2D FEA models, the core losses produced by the p_f-pole-pair spatial harmonic in the stator can be determined.

Similarly, the main contribution of $(N_{rt} \pm p_f)$-pole-pair spatial harmonics to the stray load loss is to produce core losses in the stator, so the calculation domain is the stator core and the 2D FEA model is the same as that shown in Figure 7.18.

Based on (7.33) and (7.34), the amplitudes of the magnetizing MMFs corresponding to the $(N_{rt} \pm p_f)$-pole-pair spatial harmonics in the airgap obtained by the GAFMT are:

$$F_{N_{rt}+p_f} = \lambda_\delta \left(\frac{3\sqrt{2}}{\pi}\right) N_t I k_w \left(C_{sum} + C^{sum}_{p_f,p_f,r}\right) \tag{7.42}$$

$$F_{N_{rt}-p_f} = \lambda_\delta \left(\frac{3\sqrt{2}}{\pi}\right) N_t I k_w \left(C_{dif} + C^{dif}_{p_f,p_f,r}\right) \tag{7.43}$$

In corresponding 2D FEA models, thin equivalent current layers are added, and the current densities $j_{N_{rt}\pm p_f}$ are:

$$j_{N_{rt}\pm p_f} = J_{N_{rt}\pm p_f} \cos\left[(\omega - N_{rt}\omega_r)t \mp \frac{\pi}{\tau_{N_{rt}\pm p_f}}\phi\right] \tag{7.44}$$

where $J_{N_{rt}\pm p_f}$ and $\tau_{N_{rt}\pm p_f}$ are the current density amplitudes and the pole pitches corresponding to the $(N_{rt} \pm p_f)$-pole-pair spatial harmonics in the airgap, respectively.

The pole pitches of $(N_{rt} \pm p_f)$-pole-pair airgap flux density spatial harmonics are:

$$\tau_{N_{rt}\pm p_f} = \frac{2\pi r_g}{2(N_{rt} \pm p_f)} \tag{7.45}$$

The magnetizing MMF amplitudes $F^e_{N_{rt}\pm p_f}$ created by the current layers corresponding to $(N_{rt} \pm p_f)$-pole-pair space harmonics are:

$$F^e_{N_{rt}\pm p_f} = C_{p_f,p_f,s}\lambda_\delta \frac{\tau_{N_{rt}\pm p_f}}{\pi} J_{N_{rt}\pm p_f} \tag{7.46}$$

Therefore, the equivalent current density amplitudes corresponding to the $(N_{rt} \pm p_f)$-pole-pair spatial harmonics are:

$$J_{N_{rt}+p_f} = \frac{3\sqrt{2}}{\tau_{N_{rt}\pm p_f}} N_t I k_w \left(\frac{C_{sum} + C^{sum}_{p_f,p_f,r}}{C_{p_f,p_f,s}} \right) \tag{7.47}$$

$$J_{N_{rt}-p_f} = \frac{3\sqrt{2}}{\tau_{N_{rt}\pm p_f}} N_t I k_w \left| \frac{C_{dif} + C^{dif}_{p_f,p_f,r}}{C_{p_f,p_f,s}} \right| \tag{7.48}$$

By contrast, the $(N_{st} \pm p_f)$-pole-pair spatial harmonics' main contribution to the stray load loss is to produce core losses and rotor cage loss, so the calculation domain is the rotor. In the corresponding 2D FEA model, the rotor is slotted and the stator slots are filled with the same magnetic material used for cores, as shown in Figure 7.19.

Based on (7.33) and (7.34), the amplitudes of the magnetizing MMFs corresponding to the $(N_{st} \pm p_f)$ -pole-pair space harmonics in the airgap obtained by the GAFMT are:

$$F_{N_{st}+p_f} = \lambda_\delta \left(\frac{3\sqrt{2}}{\pi} \right) N_t I k_w C^{sum}_{p_f,p_f,s} \tag{7.49}$$

$$F_{N_{st}-p_f} = \lambda_\delta \left(\frac{3\sqrt{2}}{\pi} \right) N_t I k_w C^{dif}_{p_f,p_f,s} \tag{7.50}$$

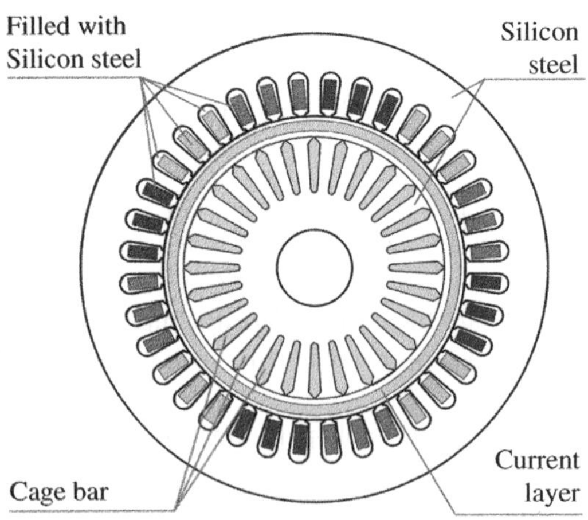

Figure 7.19 2D FEA model for rotor core loss calculation. Stator slots are filled with the same lamination material used for machine cores.

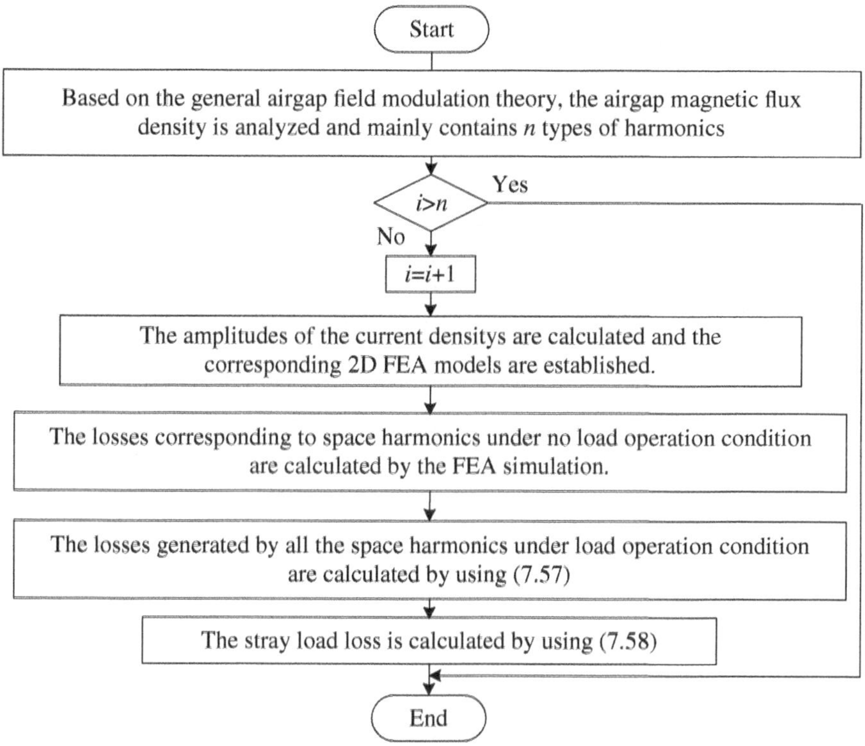

Figure 7.20 Calculation process for stray load loss.

In the corresponding 2D FEA models, thin equivalent current layers are also added, and the current densities $j_{N_{st}\pm p_f}$ are:

$$j_{N_{st}\pm p_f} = J_{N_{st}\pm p_f} \cos\left(\omega t \mp \frac{\pi}{\tau_{N_{st}\pm p_f}} \phi\right) \tag{7.51}$$

where $J_{N_{st}\pm p_f}$ and $\tau_{N_{st}\pm p_f}$ are the current density amplitudes and pole pitches corresponding to the $(N_{st} \pm p_f)$- pole-pair spatial harmonics in the airgap, respectively.

The pole pitches of $(N_{st} \pm p_f)$-pole-pair spatial harmonics are:

$$\tau_{N_{st}\pm p_f} = \frac{2\pi r_g}{2(N_{st} \pm p_f)} \tag{7.52}$$

The magnetizing MMF amplitudes $F^e_{N_{st}\pm p_f}$ induced by the current layers corresponding to $(N_{st} \pm p_f)$-pole pair spatial harmonics obtained by the GAFMT are:

$$F^e_{N_{st}\pm p_f} = C_{p_f,p_f,r} \lambda_\delta \frac{\tau_{N_{st}\pm p_f}}{\pi} J_{N_{st}\pm p_f} \tag{7.53}$$

Therefore, the equivalent current density amplitudes corresponding to the $(N_{st} \pm p_f)$-pole-pair spatial harmonic are:

$$J_{N_{st}+p_f} = \frac{3\sqrt{2}}{\tau_{N_{st}\pm p_f}} N_t I k_w \left(\frac{C^{sum}_{p_f,p_f,s}}{C_{p_f,p_f,r}} \right) \tag{7.54}$$

$$J_{N_{st}-p_f} = \frac{3\sqrt{2}}{\tau_{N_{st}\pm p_f}} N_t I k_w \left(\frac{C^{dif}_{p_f,p_f,s}}{C_{p_f,p_f,r}} \right) \tag{7.55}$$

The rotor core losses and rotor cage loss produced by the $(N_{st} \pm p_f)$-pole-pair spatial harmonics can be determined by post-processing the field results from 2D FEA models.

To reuse FEA models built for no-load operation (and thus reducing the modeling and calculation time), a coefficient λ_v denoting the amplitude ratio of the v-th flux density harmonic between no-load and loaded operation conditions is defined. Since core losses are only related to the square of magnetic flux density when the difference in rotational speeds of each flux density spatial harmonic between no-load and loaded operation is neglected, losses generated by the v-th flux density spatial harmonic under loaded operation can be readily obtained by:

$$\lambda_v = \frac{B_{v1}}{B_{v0}} \tag{7.56}$$

$$P_{v1} = \lambda_v^2 P_{v0} \tag{7.57}$$

where B_{v0} and B_{v1} are the amplitudes of the v-th magnetic flux density under no-load and loaded operation conditions, respectively. P_{v0} is the loss produced by the v-th flux density spatial harmonic under no-load operation obtained from the 2D FEA model. For p_f- and $(N_{rt} \pm p_f)$-pole-pair spatial harmonics, P_{v0} and P_{v1} only represent core losses, while for spatial harmonics with $(N_{st} \pm p_f)$ pole pairs, P_{v0} and P_{v1} represent the sum of core losses and the rotor cage loss.

The stray load loss is regarded as the difference of the total core losses, eddy current end region losses, and harmonic rotor cage losses between the loaded and no-load conditions [13], i.e. P_{sll} can be defined as:

$$P_{sll} = \sum P_{v1} - \sum P_{v0} \tag{7.58}$$

The entire calculation process for the stray load loss is illustrated in Figure 7.20.

Table 7.12 shows losses corresponding to the main airgap flux density spatial harmonics obtained from the proposed method combining the GAFMT with 2D FEA models. P_{ir} represents the core losses produced by each harmonic of the airgap magnetic flux density, and p_{cage} represents the rotor cage loss induced by each airgap magnetic flux density harmonic. According to (7.58), the stray load loss contributed by main airgap flux density spatial harmonics from the GAFMT is 43.15 W.

Table 7.12 Losses produced by main flux density harmonics.

Condition	Loss	p_f	$N_{rt}-p_f$	$N_{rt}+p_f$	$N_{st}-p_f$	$N_{st}+p_f$
Load	P_{ir} (W)	108.7	3.16	2.99	12.46	11.25
	P_{cage} (W)	—	—	—	1.2	3.8
No load	P_{ir} (W)	76.7	2.04	1.93	9.30	7.44
	P_{cage} (W)	—	—	—	0.7	2.3

4 Experimental Validation

To verify the accuracy of the proposed model, both no-load and rated load testing were conducted on the studied squirrel-cage IM using the experimental setup shown in Figure 7.21. The tested IM has a skewed cast aluminum cage. Its rated power is 5.5 kW at the rated speed 1440 r/min. Thermal couplers were buried in stator slots to monitor the operating temperature.

The testing equipment and procedures met the requirements specified in Standard IEEE 112-B. The measured stator winding temperature was very stable, and it is thus possible to perform several measurements for each load torque.

The tested IM was supplied by a 15 kVA three-phase variac. Electrical quantities at the winding terminals were measured by a power analyzer, and the load torque was measured by a torque transducer with a high accuracy of 0.2%.

Table 7.13 shows the measured results. U_1, U_2, and U_3 are the measured line voltages. I_1, I_2, and I_3 are the measured line currents. P_{in} is the measured input electrical power. n and T are the measured rotational speed and torque, respectively.

The core loss and mechanical loss under no-load operation can be separated graphically as shown in Figure 7.22, where

$$P_0' = P_{in0} - P_{Cu0} \tag{7.59}$$

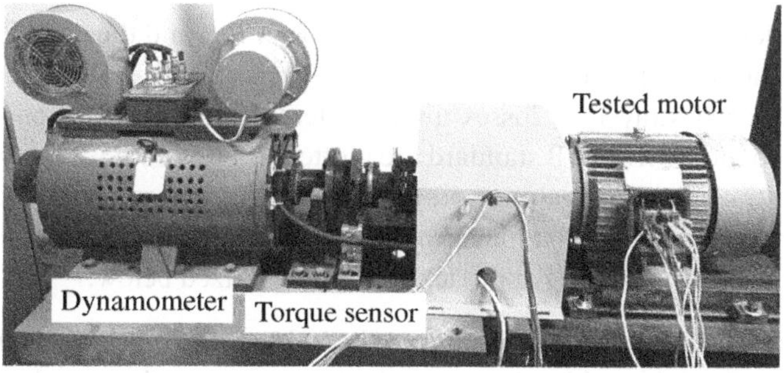

Figure 7.21 Experiment setup.

Table 7.13 Measured results of no-load and rated load tests.

Symbol	Unit	No load	Load
U_1		379.31	380.06
U_2	Vrms	380.00	381.93
U_3		380.04	381.81
I_1		5.390	11.386
I_2	Arms	5.489	11.805
I_3		5.319	11.605
P_{in}	W	348.1	6317
n	r/min	1500	1440
T	Nm	0	36.4

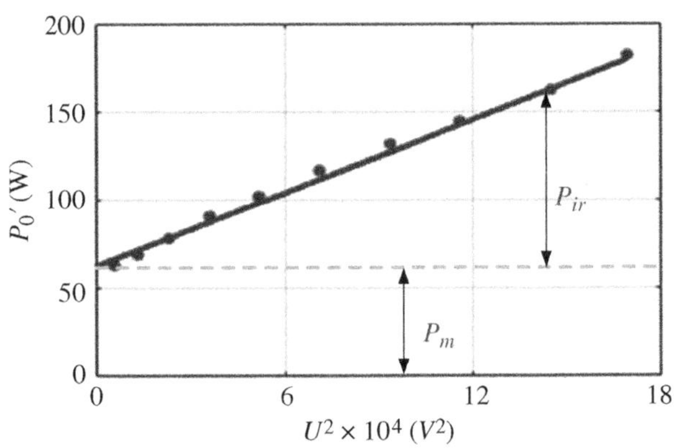

Figure 7.22 Total loss from no-load test vs. voltage squared for loss separation.

where P'_0 denotes the difference of the input power P_{in0} and the stator copper loss P_{cu0} under no-load operation.

The measured stray load losses under different loaded operation conditions according to the IEEE 112-B standard are plotted in Figure 7.23, showing that the measured stray load loss is almost proportional to the square of the torque.

The presented new method for calculating stray load losses in squirrel-cage IMs has advantages over existing approaches, as summarized below:

1) The proposed method can account for the influence of machine operating conditions, pole-slot combinations, and slot harmonics on the stray load loss.

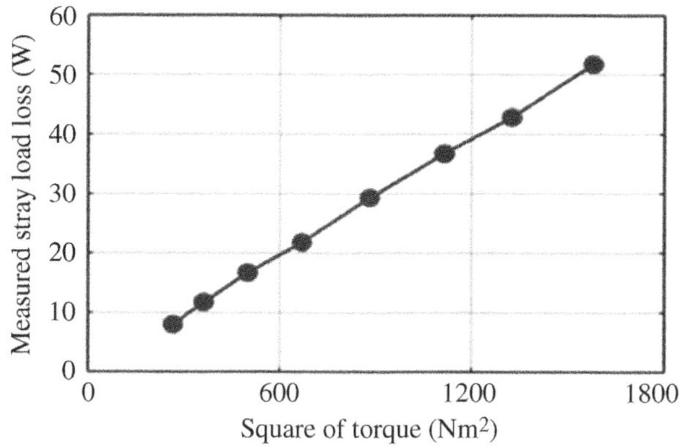

Figure 7.23 Measured stray load loss vs. torque squared. The rated load operation produces 44.32 W stray load losses in comparison to the predicted 43.15 W from the combined GAFMT and 2D FEA approach, showing good agreement.

2) The contribution of each spatial harmonic component to the stray load loss is known, providing exclusive insights into the loss breakdown and benefiting the machine design.

7.4.2 Computationally Efficient Core Loss Calculation for FSPM Machines Supplied by PWM Inverters

PWM inverters have been widely used in variable speed machine drive systems. High-order time harmonics introduced by PWM inverters, nevertheless, unavoidably increase the machine core losses, eddy current loss in magnets, and AC winding losses. The calculation of core losses considering the influence of PWM inverters plays a vital role for the accurate prediction of efficiency and operating temperature [18–20].

The most straightforward method to predict core loss under the PWM inverter supply is to feed the measured voltage or current waveforms to FEA [21, 22] and then the flux density variation in all elements can be obtained for iron loss prediction. This method requires fine time steps to distinguish carrier harmonics and thus a long computational time, so is not suitable for the machine design stage.

It is difficult to build an accurate analytical model to predict core losses under PWM inverter supplies considering that the modulation index, carrier frequency, and modulation techniques will influence the results [23, 24]. Therefore, engineering approaches for fast PWM-induced core loss estimation were mainly developed [25–27], showing that there exists a direct correlation between PWM voltages and induced core losses. For example, in [26, 27], the carrier harmonic core loss was calculated by the linear frequency-domain FEA with differential permeability

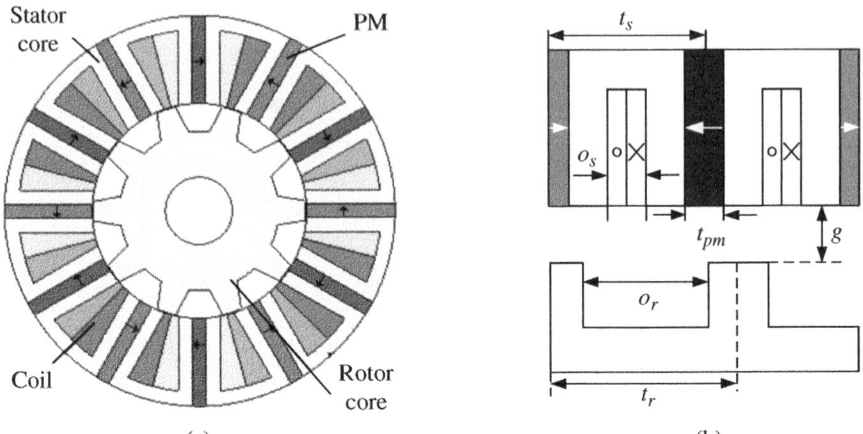

Figure 7.24 Cross-sectional view of the FSPM machine under study. (a) Full cross section, (b) Zoomed-in view of one stator pole pitch. Arrows in PMs denote the magnetization direction. Circles and crosses denote directions of currents in windings.

distribution obtained from the time-domain FEA. This method did not consider the DC-biased magnetic inductions, which are common in field modulation PM (FMPM) machines and will increase the hysteresis loss [18, 28].

The GAFMT developed in Chapter 3 provides additional insights into spatial harmonics in electrical machines and their contributions to core losses, as exemplified in Section 7.4.1 for stray load loss calculation in squirrel-cage IMs [3]. This section takes one step further to incorporate the influence of PWM inverter supply on core losses and a typical field modulation machine topology that has been proposed for electric vehicle and wind turbine applications – the FSPM machine [29, 30], is used for this study. The cross-sectional view and main parameters of the machine are shown in Figure 7.24 and Table 7.14, respectively.

1 Source Magnetizing MMF

In FSPM machines, the source magnetizing MMF is created by PMs on the stator. Stator and rotor salient reluctance poles play the roles of modulators. The source magnetizing MMF is firstly modulated by stator and rotor salient reluctance poles, and the modulated MMF in the airgap contains abundant spatial harmonics. The armature winding selects effective spatial harmonics to generate the electromotive force. When the FSPM machine is operated as a generator or motor, currents flow into armature windings, thereby generating the armature reaction MMF. In short, the working MMF in the airgap is a synthesis of the source magnetizing MMF and armature reaction MMF.

To study the source magnetizing MMF distribution in the FSPM machine, both the stator and rotor cores are assumed to be unslotted, as shown in Figure 7.25, and the airgap length is still g. The source magnetizing MMF generated by PMs is considered as a periodic rectangular wave distribution along the circumference of the airgap, and the period is $2\pi/p_f$. The existence of α in Figure 7.26 is because the relative permeability of PM is 1.05, which can be regarded as a nonmagnetic material. In Figure 7.26, the vertical coordinate is coincident with the center line of a magnet, where α can be expressed as:

$$\alpha = \frac{t_{pm}}{2R_{si}} \tag{7.60}$$

Table 7.14 Geometric parameters of the studied FSPM machine.

Parameter	Value
Stator slot opening width o_s (mm)	4.607
Rotor slot opening width o_r (mm)	12.1
Stator width t_s (mm)	18.4
Rotor width t_r (mm)	21.9
PM width t_{pm} (mm)	4.6
Stator outer radius R_{so} (mm)	64
Stator inner radius R_{si} (mm)	35.2
Rotor inner radius R_{ri} (mm)	11
PM pole pair p_f	6
Stator slot number N_{st}	12
Rotor slot number N_{rt}	10
Phase resistance (Ω)	1.436
Rotor angular velocity ω_r (rad/s)	942.48
Airgap length g (mm)	0.3
Stack length l_{stk} (mm)	75

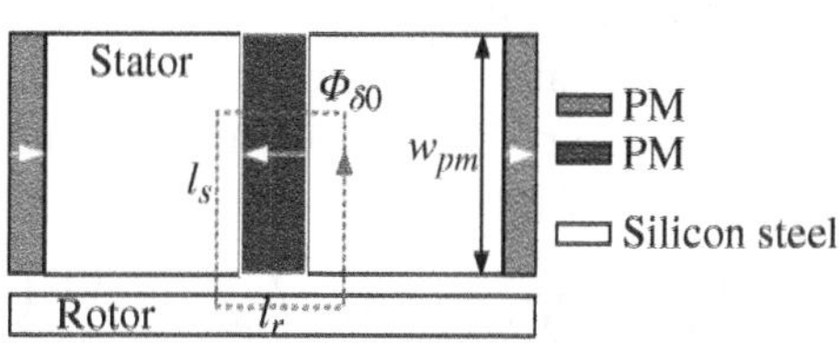

Figure 7.25 Simplified model for the source magnetizing MMF.

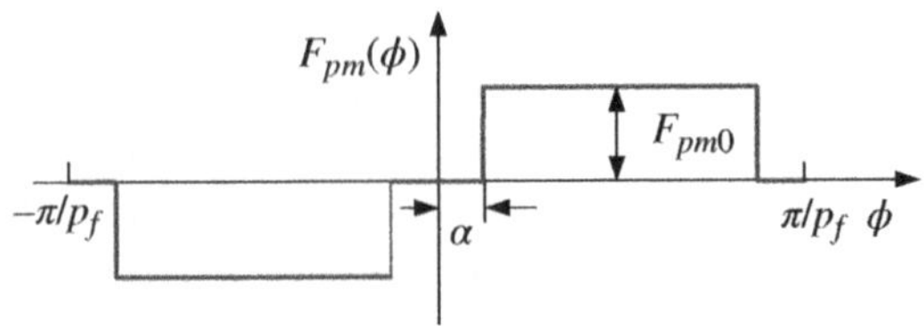

Figure 7.26 Source magnetizing MMF distribution.

The Fourier series expansion of the source magnetizing MMF shown in Figure 7.26 can be written as:

$$F_{pm}(\phi) = \sum_{\nu=2k+1}^{\infty} f_{pm\nu} \sin(\nu p_f \phi) \tag{7.61}$$

$$f_{pm\nu} = \frac{4}{\pi} \frac{F_{pm0}}{\nu} \sin\left(\frac{\nu\pi}{2}\right) \cos\left(\frac{\nu\pi}{2} - \nu\alpha N_{st}\right) \tag{7.62}$$

where k is a natural number. F_{pm0} is the source magnetizing MMF rectangular wave amplitude shown as in Figure 7.26. Figure 7.27 shows the detailed calculation process for F_{pm0}, where the saturation of stator and rotor cores is considered. B_r is the PM remanence. A_m (=$w_{pm}L$) is the PM area and w_{pm} is the PM length. σ_0 is the magnetic flux leakage coefficient, which can be accurately obtained by FEA [31]. A_t (=($t_s - t_{pm}$)$L/2$) is the stator silicon steel sheet area. μ_0 and μ_{pm} are the vacuum permeability and the PM permeability, respectively. μ_s and μ_r are the stator and rotor permeability, which can be obtained from the magnetization B-H curve of the silicon steel sheet. l_{st} and l_{rt} are the equivalent stator and rotor core lengths in the magnetic circuit.

In the GAMFT, the airgap is assumed to be uniform. Hence, the source airgap magnetic flux density induced by the source magnetizing MMF can be written as:

$$B_{pm}(\phi) = \frac{\mu_0}{g} F_{pm}(\phi) = \sum_{\nu=2k+1}^{\infty} b_{pm\nu} \sin(\nu p_f \phi) \tag{7.63}$$

$$b_{pm\nu} = \frac{\mu_0}{g} f_{pm\nu} = \frac{\mu_0}{g} \left(\frac{4}{\pi}\right) \frac{F_{PM0}}{\nu} \sin\left(\frac{\nu\pi}{2}\right) \cos\left(\frac{\nu\pi}{2} - \nu\alpha N_{st}\right) \tag{7.64}$$

where $b_{pm\nu}$ is the amplitude of the ν-pole-pair spatial harmonic in the airgap magnetic flux density corresponding to the source magnetizing MMF.

2 Modulation of Salient Reluctance Poles

In FSPM machines, both the stator and rotor salient reluctance poles can be regarded as effective modulators. The modulation operator of stator salient reluctance poles can be expressed as:

$$M(p_s, \varepsilon_s)[f(x)] = \left\{ m_{s0} - \sum_{i=1}^{\infty} m_{si} \cos\left[ip_s\left(\phi + \frac{\pi}{N_{st}}\right)\right]\right\} f(x) \tag{7.65}$$

$$m_{s0} = \frac{1}{k_{cs}} = \frac{1}{\frac{t_s}{t_s - g\frac{\varepsilon_s^2}{5+\varepsilon_s}}} \tag{7.66}$$

$$m_{si} = \frac{4\beta_s}{i\pi}\left[\frac{1}{2} + \frac{(i\varepsilon_s)^2}{\frac{25}{32} - 2(i\varepsilon_s)^2}\right]\sin\left(\frac{8}{5}\pi i\varepsilon_s\right) \tag{7.67}$$

where $\varepsilon_s = o_s/t_s$ represents the slot-opening-to-pole-pitch ratio, o_s the stator slot opening width, and t_s the stator pole pitch. Moreover, p_s is the number of the stator salient reluctance poles. i is a positive integer. If the stator core is not slotted, the airgap flux density should be a constant, and its value is B_{max}. When the stator is slotted, the airgap magnetic flux density is shown by the blue line in Figure 7.28.

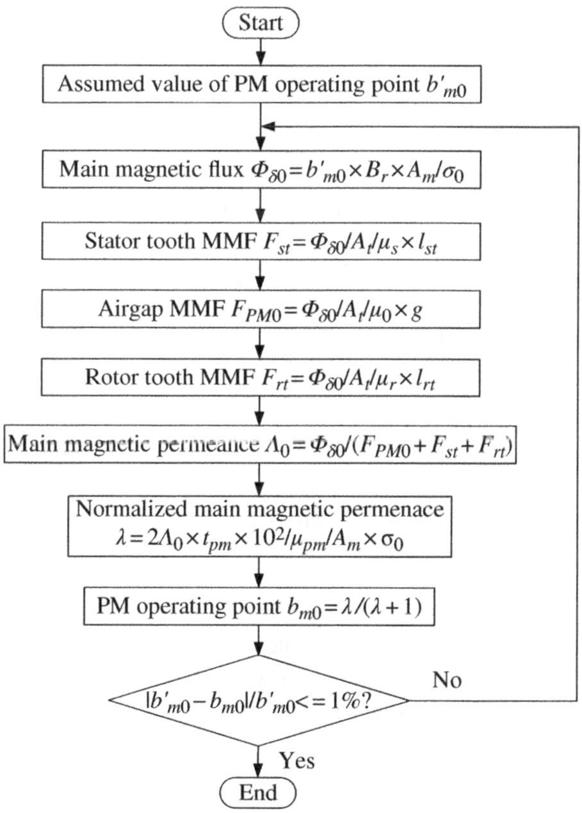

Figure 7.27 Calculation process for PM operating point.

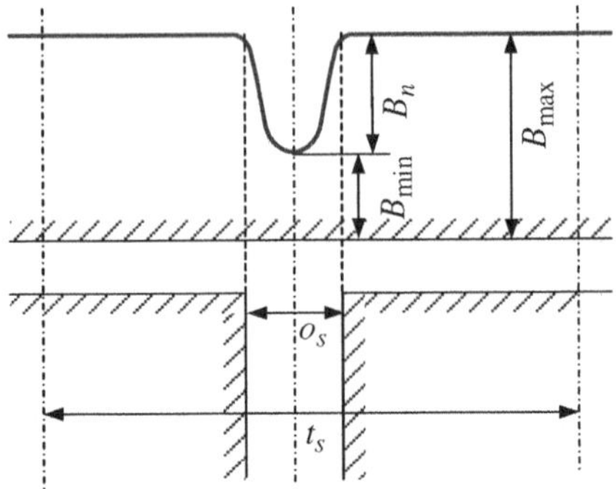

Figure 7.28 The radial airgap flux density distribution when the stator/rotor core is slotted.

Defining β_s as the reduction of airgap magnetic flux density when the stator/rotor core is slotted, and its expression is [32]:

$$\beta_s = \frac{B_n}{2B_{max}} \tag{7.68}$$

Similarly, the modulation operator of rotor salient reluctance poles can be readily obtained by replacing the subscript "s" with "r" in (7.65), (7.66), and (7.67):

$$M(p_r, \varepsilon_r)[f(x)] = \left\{ m_{r0} - \sum_{l=1}^{\infty} m_{rl} \cos[lp_r(\theta - \omega_r t)] \right\} f(x) \tag{7.69}$$

$$m_{r0} = \frac{1}{k_{cr}} = \frac{1}{\dfrac{t_r}{t_r - g\dfrac{\varepsilon_r^2}{5+\varepsilon_r}}} \tag{7.70}$$

$$m_{rl} = \frac{4\beta_r}{l\pi}\left[\frac{1}{2} + \frac{(l\varepsilon_r)^2}{\dfrac{25}{32} - 2(l\varepsilon_r)^2}\right]\sin\left(\frac{8}{5}\pi l\varepsilon_r\right) \tag{7.71}$$

where $\varepsilon_r = o_r/t_r$ represents the slot-opening-to-pole-pitch ratio, o_r the rotor slot opening width and t_r the rotor salient pole pitch. β_r is defined as the reduction of airgap magnetic flux density when the rotor core is slotted. p_r is the number of the rotor salient reluctance poles. l is a positive integer.

3 No-load Airgap Flux Density Distribution

For no-load operation, stator windings are left open. The source magnetizing MMF created by PMs is modulated by stator salient reluctance poles synchronously and then by rotor salient reluctance poles asynchronously. The no-load airgap flux density distribution is:

$$\begin{aligned} B_{pm\_s}(\phi) &= M(N_{rt},\varepsilon_r)[M(N_{st},\varepsilon_s)[B_{pm}(\phi)]] \\ &= M(N_{rt},\varepsilon_r)\left[\sum_{v=2k+1}^{\infty} C_{pmv}\sin(vp_f\phi)\right] \\ &= \sum_{v=2k+1}^{\infty} C_{pm\_r}\sin(vp_f\phi) \\ &\quad - \sum_{v=2k+1}^{\infty}\sum_{l=1}^{\infty} C_{pm\_r}^{sum}\sin[(vp_f + lN_{rt})\phi - lN_{rt}\omega_r t] \\ &\quad - \sum_{v=2k+1}^{\infty}\sum_{l=1}^{\infty} C_{pm\_r}^{dif}\sin[(vp_f - lN_{rt})\phi + lN_{rt}\omega_r t] \end{aligned} \tag{7.72}$$

where

$$C_{pmv} = \frac{2p_f}{\pi}\int_0^{\frac{\pi}{p_f}} M_s(N_{st},\varepsilon_s)[B_{pm}(\phi)]\sin(vp_f\phi)d\phi \tag{7.73}$$

$$C_{pm\_r} = C_{pmv}m_{r0} \tag{7.74}$$

$$C_{pm\_r}^{sum} = \frac{C_{pmv}m_{rl}}{2} \tag{7.75}$$

$$C_{pm\_r}^{dif} = \frac{C_{pmv}m_{rl}}{2} \tag{7.76}$$

Equation (7.72) shows that stator salient reluctance poles do not change harmonic orders contained in the source magnetizing MMF but alter their amplitudes. By contrast, rotor salient reluctance poles change not only harmonic orders but also amplitudes of MMF harmonics, which can be characterized by three magnetic field conversion factors C_{pm_r}, $C_{pm\_r}^{sum}$, and $C_{pm\_r}^{dif}$.

Figure 7.29a shows the no-load airgap magnetic flux density distributions from the GAFMT and 2D FEA. Figure 7.29b shows the corresponding amplitude spectrums. Good agreement has been achieved between the GAFMT and 2D FEA results.

Table 7.15 lists angular velocities of the no-load airgap flux density spatial harmonics relative to the stator/rotor and orders of corresponding time harmonics. Spatial harmonics with vp_f pole pairs associated with PMs in the no-load airgap flux density waveform are stationary relative to the stator, leading to the DC magnetic bias in the stator core. In contrast, they rotate relative to the rotor, and the corresponding vth magnetic flux density is induced in the rotor. Spatial harmonics with $|vp_f \pm lN_{rt}|$ pole pairs produced by the salient reluctance pole modulators rotate relative to the stator and rotor, and the lth and vth flux density are induced

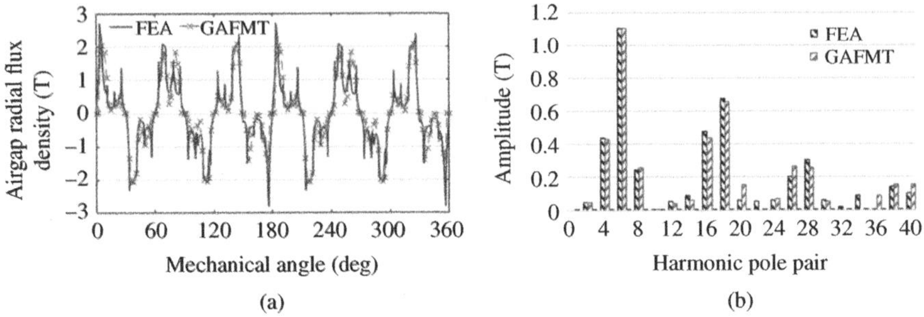

Figure 7.29 No-load airgap flux density distribution. (a) Waveforms, (b) Amplitude spectrums. Good agreement between the GAFMT and 2D FEA results has been shown.

Table 7.15 Open-circuit PM field harmonics in the airgap.

Pole-pair	Speed relative to stator, $N_{rt}\omega_r$	Time harmonic order in stator	Speed relative to rotor, $p_f\omega_r$	Time harmonic order in rotor
νp_f	0	0 (DC bias)	$-\nu/\nu p_f$	ν
$\nu p_f + lN_{rt}$	$+l/(\nu p_f + lN_{rt})$	l	$-\nu/(\nu p_f + lN_{rt})$	ν
$\nu p_f - lN_{rt} > 0$	$-l/\lvert\nu p_f - lN_{rt}\rvert$	l	$-\nu/\lvert\nu p_f - lN_{rt}\rvert$	ν
$lN_{rt} - \nu p_f > 0$	$+l/\lvert\nu p_f - lN_{rt}\rvert$	l	$+\nu/\lvert\nu p_f - lN_{rt}\rvert$	ν

in the stator and rotor, respectively. Signs "+" and "−" in Table 7.15 denote the rotational direction of spatial harmonics relative to the rotor, and the "+" sign means the same direction relative to the rotor.

4 Armature-Reaction MMF

In analyzing the armature-reaction MMF, the coercivity of PMs is set to 0, and the relative permeability of PMs is 1.05. In addition, the stator and rotor are slotted.

The three-phase sinusoidal currents are fed into the armature winding:

$$\begin{cases} i_a = I_{\max} \cos(N_{rt}\omega_r t + \varphi_i) \\ i_b = I_{\max} \cos\left(N_{rt}\omega_r t + \varphi_i - \dfrac{2\pi}{3}\right) \\ i_c = I_{\max} \cos\left(N_{rt}\omega_r t + \varphi_i - \dfrac{4\pi}{3}\right) \end{cases} \tag{7.77}$$

where $N_{rt}\omega_r$ is the electrical angular frequency and φ_i the initial phase of the Phase-A current.

The winding function of Phase-A with N_t turns is:

$$W_{sA}(\phi) = \sum_{u=1}^{\infty} C_{swu} \cos(uN_t\phi) \tag{7.78}$$

where

$$C_{swu} = \frac{2}{\pi}\frac{N_t}{u}\sin\left(\frac{u\pi}{3}\right) \tag{7.79}$$

The resultant armature-reaction MMF can then be expressed as:

$$\begin{aligned}
F_{ABC}(\phi, t) &= W_{sA}(\phi)i_a + W_{sB}(\phi)i_b + W_{sC}(\phi)i_c \\
&= W_{sA}(\phi)i_a + W_{sA}\left(\phi - \frac{2\pi}{N_{st}}\right)i_b + W_{sA}\left(\phi + \frac{2\pi}{N_{st}}\right)i_c \\
&= \frac{3}{2}I_{\max}\sum_{u=3h-1}^{\infty} C_{swu}\cos(uN_{st}\phi + N_{rt}\omega_r t + \varphi_i) \\
&\quad + \frac{3}{2}I_{\max}\sum_{u=3h-2}^{\infty} C_{swu}\cos(uN_{st}\phi - N_{rt}\omega_r t - \varphi_i)
\end{aligned} \tag{7.80}$$

Considering the modulation behavior of stator salient reluctance poles, the modulated armature-reaction MMF is:

$$\begin{aligned}
M(2N_{st}, \varepsilon_s)[F_{ABC}(\phi, t)] &= \sum_{u=3h-1}^{\infty} C_{swu,s}\cos(uN_{st}\phi + N_{rt}\omega_r t + \varphi_i) \\
&\quad + \sum_{u=3h-2}^{\infty} C_{swu,s}\cos(uN_{st}\phi - N_{rt}\omega_r t - \varphi_i)
\end{aligned} \tag{7.81}$$

where

$$C_{swu,s}\Big|_{u\neq 3h} = \frac{N_t}{\pi}\int_{-\frac{\pi}{N_{st}}}^{\frac{\pi}{N_{st}}} M(2N_{st}, \varepsilon_s)F_{ABC}(\phi, t)\cos(uN_{st}\phi)d\phi \tag{7.82}$$

Equation (7.82) indicates that the armature MMF modulated by stator salient reluctance poles do not introduce new harmonic orders and only the amplitude of each space harmonic changes, which can be characterized by coefficients $C_{swu,s}$.

Considering the modulation behavior of rotor salient reluctance poles further, the modulated armature MMF is:

$$\begin{aligned}
M(N_{rt}, \varepsilon_r)[M(2N_{st}, \varepsilon_s)F_{ABC}(\phi, t)] &= \sum_{u=3h-1}^{\infty} C_{p_r}^{4u}\cos(uN_{st}\phi + N_{rt}\omega_r t + \varphi_i) \\
&\quad + \sum_{l=1}^{\infty}\sum_{u=3h-1}^{\infty} C_{sum}^{4u}\cos[(uN_{st} + lN_{rt})\phi - (l-1)N_{rt}\omega_r t + \varphi_i] \\
&\quad + \sum_{l=1}^{\infty}\sum_{u=3h-1}^{\infty} C_{dif}^{4u}\cos[(uN_{st} - lN_{rt})\phi + (l+1)N_{rt}\omega_r t + \varphi_i] \\
&\quad + \sum_{u=3h-2}^{\infty} C_{p_r}^{4u}\cos(uN_{st}\phi - N_{rt}\omega_r t - \varphi_i) \\
&\quad + \sum_{l=1}^{\infty}\sum_{u=3h-2}^{\infty} C_{sum}^{4u}\cos[(uN_{st} + lN_{rt})\phi - (l+1)N_{rt}\omega_r t - \varphi_i] \\
&\quad + \sum_{l=1}^{\infty}\sum_{u=3h-2}^{\infty} C_{dif}^{4u}\cos[(uN_{st} - lN_{rt})\phi + (l-1)N_{rt}\omega_r t - \varphi_i]
\end{aligned} \tag{7.83}$$

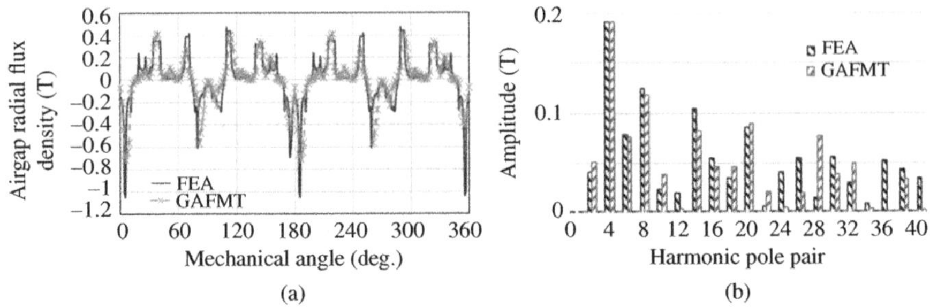

Figure 7.30 Armature-reaction airgap flux density distributions, (a) Waveforms, (b) Amplitude spectrums. Good agreement has been achieved between the GAFMT and 2D FEA results.

where

$$\begin{cases} C_{p_r}^{4u} = C_{swu,s} m_{r0} \\ C_{sum}^{4u} = C_{dif}^{4u} = C_{swu,s} m_{rl} \end{cases} \tag{7.84}$$

Figure 7.30a shows the armature-reaction airgap flux density distributions from the GAFMT and 2D FEA. Figure 7.30b shows the corresponding amplitude spectrums, from which we can tell that the results obtained by GAFMT are very close to those obtained from the 2D FEA.

Table 7.16 Armature field harmonics in the airgap, where $m = 2(h+1) - 1$ and $n = 2(h-1) - 1$.

u	Pole pair		Speed relative to the stator, $N_{rt}\omega_r$	Speed relative to the rotor, $p_f\omega_r$
$3h-1$	uN_t	$mp_f - N_{rt}$	$-1/(mp_f - N_{rt})$	$-m/(mp_f - N_{rt})$
	$uN_t + N_{rt}$	mp_f	static	$-m/mp_f$
	$uN_t + lN_{rt}$ $(l>1)$	$mp_f + (l-1)N_{rt}$	$+(l-1)/[mp_f + (l-1)N_{rt}]$	$-m/[mp_f + (l-1)N_{rt}]$
	$uN_t - lN_{rt} > 0$	$mp_f - (l+1)N_{rt}$	$-(l+1)/[mp_f - (l+1)N_{rt}]$	$-m/[mp_f - (l+1)N_{rt}]$
	$lN_{rt} - uN_t > 0$	$(l+1)N_{rt} - mp_f$	$+(l+1)/[(l+1)N_{rt} - mp_f]$	$+m/[(l+1)N_{rt} - mp_f]$
$3h-2$	uN_t	$np_f + N_{rt}$	$+1/[np_f + N_{rt}]$	$-n/[np_f + N_{rt}]$
	$uN_t + lN_{rt}$	$np_f + (l+1)N_{rt}$	$+(l+1)/[np_f + (l+1)N_{rt}]$	$-n/[np_f + (l+1)N_{rt}]$
	$\lvert uN_t - N_{rt} \rvert$	np_f	static	$-n/np_f$
	$uN_t - lN_{rt} > 0$ $(l>1)$	$np_f - (l-1)N_{rt}$	$-(l-1)/[np_f - (l+1)N_{rt}]$	$-n/[np_f - (l+1)N_{rt}]$
	$lN_{rt} - uN_t > 0$ $(l>1)$	$(l-1)N_{rt} - np_f$	$+(l-1)/[(l+1)N_{rt} - np_f]$	$+n/[(l+1)N_{rt} - np_f]$

Table 7.16 lists rotational speeds of the armature-reaction airgap flux density spatial harmonics relative to the stator and the rotor, respectively. Comparing Table 7.16 with Table 7.15, we can see that spatial harmonics with the same number of pole pairs in the no-load and armature reaction airgap flux density waveforms share the same rotational speeds relative to the stator and rotor.

In addition to the fundamental wave, currents from the PWM inverter also contains high-order time harmonics, which can be expressed as:

$$\begin{cases} i_a = \sum_{\kappa} I_{\max} \cos(\kappa N_{rt}\omega_r t + \varphi_i) \\ i_b = I_{\max} \cos\left[\kappa\left(N_{rt}\omega_r t - \frac{2\pi}{3}\right) + \varphi_i\right], \quad \kappa \neq 3j \\ i_c = I_{\max} \cos\left[\kappa\left(N_{rt}\omega_r t - \frac{4\pi}{3}\right) + \varphi_i\right] \end{cases} \tag{7.85}$$

where j is a positive integer.

The armature reaction MMF without considering the slotting effect can be expressed as:

$$\begin{aligned} F_{ABC\kappa}(\phi, t) &= W_{sA}(\phi)i_a + W_{sB}(\phi)i_b + W_{sC}(\phi)i_c \\ &= W_{sA}(\phi)i_a + W_{sA}\left(\phi - \frac{2\pi}{N_{st}}\right)i_b + W_{sA}\left(\phi + \frac{2\pi}{N_{st}}\right)i_c \\ &= \frac{3}{2}I_{\max}\sum_{\kappa}\sum_{u+\kappa=3h}^{\infty} C_{swu}\cos(uN_{st}\phi + \kappa N_{rt}\omega_r t + \varphi_\kappa) \\ &\quad + \frac{3}{2}I_{\max}\sum_{\kappa}\sum_{u+\kappa\neq 3h}^{\infty} C_{swu}\cos(uN_{st}\phi - \kappa N_{rt}\omega_r t - \varphi_\kappa), \\ &\quad u \neq 3i, i \in Z^+ \end{aligned} \tag{7.86}$$

The armature reaction MMF is firstly synchronously modulated by stator salient reluctance poles, and the modulated MMF in Fourier series form is:

$$\begin{aligned} M(2N_{st}, \varepsilon_s)[F_{ABC\kappa}(\phi, t)] &= \sum_{\kappa}\sum_{u+\kappa=3h}^{\infty} C_{swu,s,\kappa}\cos(uN_{st}\phi + \kappa N_{rt}\omega_r t + \varphi_\kappa) \\ &\quad + \sum_{\kappa}\sum_{u+\kappa\neq 3h}^{\infty} C_{swu,s,\kappa}\cos(uN_{st}\phi - \kappa N_{rt}\omega_r t - \varphi_\kappa) \end{aligned} \tag{7.87}$$

where

$$C_{swu,s,\kappa}\Big|_{u\neq 3h} = \frac{N_t}{\pi}\int_{-\frac{\pi}{N_{st}}}^{\frac{\pi}{N_{st}}} M(2N_{st}, \varepsilon_s)F_{ABC\kappa}(\phi, t)\cos(uN_{st}\phi)d\phi \tag{7.88}$$

where $C_{swu,s,\kappa}$ is the magnetic field conversion factor considering the modulation effect of stator salient reluctance poles.

Considering the slotting effects of the stator and rotor cores, the armature reaction MMF created by currents from the PWM inverter can be expressed as:

$$
\begin{aligned}
&M(N_{rt}, \varepsilon_r)[M(2N_{st}, \varepsilon_s)F_{ABC\kappa}(\phi, t)] \\
&= \sum_{\kappa} \sum_{u+\kappa=3h}^{\infty} C_{P_r}^{4u} \cos(uN_{st}\phi + \kappa N_{rt}\omega_r t + \varphi_\kappa) \\
&+ \sum_{l=1}^{\infty} \sum_{\kappa} \sum_{u+\kappa=3h}^{\infty} C_{sum}^{4u} \cos[(uN_{st} + lN_{rt})\phi - (l - \kappa)N_{rt}\omega_r t + \varphi_\kappa] \\
&+ \sum_{l=1}^{\infty} \sum_{\kappa} \sum_{u+\kappa=3h}^{\infty} C_{dif}^{4u} \cos[(uN_{st} - lN_{rt})\phi + (l + \kappa)N_{rt}\omega_r t + \varphi_\kappa] \\
&+ \sum_{\kappa} \sum_{u+\kappa\neq 3h}^{\infty} C_{P_r}^{4u} \cos(uN_{st}\phi - \kappa N_{rt}\omega_r t - \varphi_\kappa) \\
&+ \sum_{l=1}^{\infty} \sum_{\kappa} \sum_{u+\kappa\neq 3h}^{\infty} C_{sum}^{4u} \cos[(uN_{st} + lN_{rt})\phi - (l + \kappa)N_{rt}\omega_r t - \varphi_\kappa] \\
&+ \sum_{l=1}^{\infty} \sum_{\kappa} \sum_{u+\kappa\neq 3h}^{\infty} C_{dif}^{4u} \cos[(uN_{st} - lN_{rt})\phi + (l - \kappa)N_{rt}\omega_r t - \varphi_\kappa]
\end{aligned}
\tag{7.89}
$$

Equation (7.89) shows that spatial harmonics with uN_t pole pairs are introduced by stator armature currents, and their rotational speeds relative to the stator are $\kappa N_{rt}\omega_r/uN_t$.

Tables 7.17 and 7.18 show the comparison of airgap flux density spatial harmonics produced by armature currents with κ-order temporal harmonics. Compared with Table 7.16, it can be found that when currents with rich high-order time harmonics are fed into the FSPM machine, no new spatial harmonic orders will be introduced but the rotational speed of each spatial harmonic will be changed.

5 Determination of Equivalent Currents

Subsections 7.4.2.1–7.4.2.4 analyze the amplitudes and rotational speeds of airgap flux density spatial harmonics based on the GAFMT. These harmonics are represented by equivalent current layers in the airgap and substituted into 2D FEA loss models.

The calculation model of stator core losses is shown in Figure 7.31, where the rotor is regarded as a unit modulator. That is, a cylindrical rotor is used. The coercivity of PMs is set to 0 and the stator windings are open-circuited. The only

Table 7.17 Airgap magnetic field harmonics introduced by high-order current harmonics with $\kappa = 3j - 1$.

u	**Pole-pair**		**Speed relative to the stator, $N_{rt}\omega_r$**
$3h-1$	uN_t	$mp_f - N_{rt}$	$\kappa/(mp_f - N_{rt})$
	$uN_t + lN_{rt}$	$mp_f + (l-1)N_{rt}$	$(l+\kappa)/[mp_f + (l-1)N_{rt}]$
	$uN_t - lN_{rt}, l = \kappa$	$mp_f - (l+1)N_{rt}$	stationary
	$\lvert lN_{rt} - uN_t\rvert, l \neq \kappa$	$\lvert(l+1)N_{rt} - mp_f\rvert$	$(l-\kappa)/[(l+1)N_{rt} - mp_f]$
$3h-2$	uN_t	$np_f + N_{rt}$	$\kappa/(np_f + N_{rt})$
	$uN_t + lN_{rt}, l = \kappa$	$np_f + (l+1)N_{rt}$	Stationary
	$uN_t + lN_{rt}, l \neq \kappa$	$np_f + (l+1)N_{rt}$	$(l-\kappa)/[np_f + (l+1)N_{rt}]$
	$\lvert uN_t - lN_{rt}\rvert$	$\lvert np_f - (l-1)N_{rt}\rvert$	$(l+\kappa)/[np_f - (l-1)N_{rt}]$

Table 7.18 Airgap magnetic field harmonics introduced by high-order current harmonics with $\kappa = 3j - 2$.

u	**Pole-pair**		**Speed relative to the stator, $N_{rt}\omega_r$**
$3h-1$	uN_t	$mp_f - N_{rt}$	$\kappa/(mp_f - N_{rt})$
	$uN_t + lN_{rt}, l = \kappa$	$mp_f + (l-1)N_{rt}$	Stationary
	$uN_t + lN_{rt}\ l \neq \kappa$	$mp_f + (l-1)N_{rt}$	$(l-\kappa)/[mp_f + (l-1)N_{rt}]$
	$\lvert uN_t - lN_{rt}\rvert$	$\lvert mp_f - (l+1)N_{rt}\rvert$	$(l+\kappa)/[mp_f - (l+1)N_{rt}]$
$3h-2$	uN_t	$np_f + N_{rt}$	$\kappa/(np_f + N_{rt})$
	$uN_t + lN_{rt}$	$np_f + (l+1)N_{rt}$	$(l+\kappa)/[np_f + (l+1)N_{rt}]$
	$\lvert uN_t - lN_{rt}\rvert, l = \kappa$	$\lvert np_f - (l-1)N_{rt}\rvert$	Stationary
	$\lvert uN_t - 1N_{rt}\rvert, l \neq \kappa$	$\lvert np_f - (l-1)N_{rt}\rvert$	$(l-\kappa)/[np_f - (l-1)N_{rt}]$

excitation is the equivalent current layer used to represent each airgap flux density spatial harmonic under the PWM supply.

The equivalent current layer is split into a high number of segments, e.g. N_p, whose distance is $2\pi R_{si}/N_p$, as shown in Figure 7.31. The current in each segment corresponds to the integral of the current density distribution over an arc length $2\pi R_{si}/N_p$.

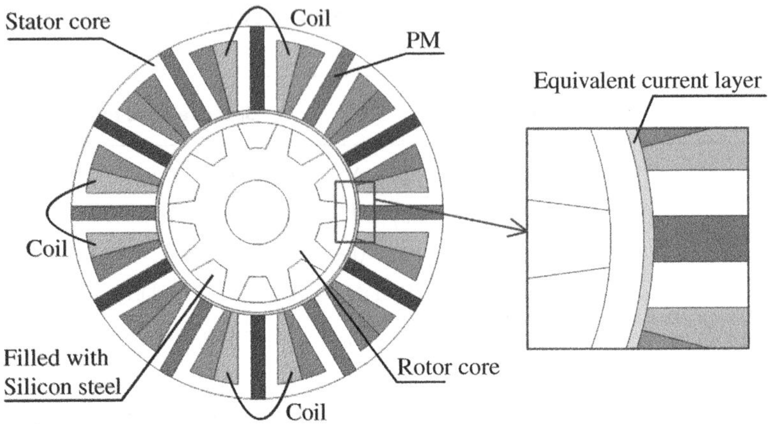

Figure 7.31 Stator core loss calculation model.

To solve out the current density in the equivalent current layer, taking spatial harmonics with $(uN_t + lN_{rt})$ pole pairs in (7.89) as an example, we assume the magnetic flux density $B'_v(\theta, t)$ to be:

$$\begin{aligned} B'_v(\phi, t) &= C^{4u}_{sum} \cos[(uN_{st} + lN_{rt})\phi - (l - \kappa)N_{rt}\omega_r t + \varphi_i] \\ &= B'_v \cos(v\phi + \eta N_{rt}\omega_r t + \varphi_v) \end{aligned} \tag{7.90}$$

where

$$\begin{cases} v = uN_{st} + lN_{rt}, & u = 3h - 1 \\ \eta = 1 - \kappa, & \kappa = 3j - 2 \\ \varphi_v = \varphi_i \end{cases} \tag{7.91}$$

If the stator and rotor cores are not slotted, the current density can be obtained by [33]:

$$j_v(\phi, t) = \frac{g}{\mu_0 R_{si}} \frac{\partial}{\partial \phi} B'_v(\phi, t) \tag{7.92}$$

where $j_v(\phi,t)$ is the current density corresponding to the spatial harmonic with the v pole pairs, B'_v the flux density amplitude, R_{si} the stator inner radius, g the airgap length, and μ_0 the vacuum permeability.

According to the GAFMT, due to the stator slotting effect, the MMF created by the current density distribution $j_v(\phi,t)$ and modulated by stator salient reluctance poles contains the vth spatial harmonic and a series of other field harmonics related to the number of stator slots. The corresponding airgap magnetic flux density can be written as:

$$\begin{aligned} B''_v(\phi, t) &= M(2N_{st}, \varepsilon_s)\left[B'_v(\phi, t)\right] \\ &= B'_v m_{s0} \cos(v\phi + \eta N_{rt}\omega_r t + \varphi_v) \end{aligned}$$

$$-\sum_{i=1}^{\infty}\frac{m_{si}B'_v}{2}\cos[(2iN_{st}+v)\phi+\eta N_{rt}\omega_r t+\varphi_v]$$
$$-\sum_{i=1}^{\infty}\frac{m_{si}B'_v}{2}\cos[(2iN_{st}-v)\phi-\eta N_{rt}\omega_r t-\varphi_v] \tag{7.93}$$

The stator slotting effect can be accounted for by the equivalent MMF of the equivalent current layer. In this case, the current density distribution can be expressed as:

$$\begin{aligned} j_v(\phi,t) = &\frac{g}{\mu_0 R_{si}}\frac{\partial}{\partial\phi}\left[\frac{B'_v}{m_{s0}}\cos(v\phi+\eta N_{rt}\omega_r t+\varphi_v)\right] \\ &+\frac{g}{\mu_0 R_{si}}\frac{\partial}{\partial\phi}\left\{\sum_{i=1}^{\infty}\left(\frac{m_{si}B'_v}{2m_{s0}^2}\right)\cos[(2iN_{st}+v)\phi+\eta N_{rt}\omega_r t+\varphi_v]\right\} \\ &+\frac{g}{\mu_0 R_{si}}\frac{\partial}{\partial\phi}\left\{\sum_{i=1}^{\infty}\left(\frac{m_{si}B'_v}{2m_{s0}^2}\right)\cos[(2iN_{st}-v)\phi-\eta N_{rt}\omega_r t-\varphi_v]\right\} \end{aligned} \tag{7.94}$$

To consider the influence of the DC magnetic bias on stator core losses, the current density needs to be modified to:

$$\begin{aligned} j_v(\phi,t) = &\frac{g}{\mu_0 R_{si}}\frac{\partial}{\partial\phi}\left[\frac{B'_v}{m_{s0}}\cos(v\phi+\eta N_{rt}\omega_r t+\varphi_v)\right] \\ &+\frac{g}{\mu_0 R_{si}}\frac{\partial}{\partial\phi}\left\{\sum_{i=1}^{\infty}\left(\frac{m_{si}B'_v}{2m_{s0}^2}\right)\cos[(2iN_{st}+v)\phi+\eta N_{rt}\omega_r t+\varphi_v]\right\} \\ &+\frac{g}{\mu_0 R_{si}}\frac{\partial}{\partial\phi}\left\{\sum_{i=1}^{\infty}\left(\frac{m_{si}B'_v}{2m_{s0}^2}\right)\cos[(2iN_{st}-v)\phi-\eta N_{rt}\omega_r t-\varphi_v]\right\} \\ &+\frac{g}{\mu_0 R_{si}}\frac{\partial}{\partial\phi}\left[\frac{B'_6}{m_{s0}}\cos(6\phi+\varphi_6)+\frac{B'_{18}}{m_{s0}}\cos(18\phi+\varphi_{18}) \right. \\ &\left. +\frac{B'_6 m_{s1}}{2m_{s0}}\cos(6\phi-\varphi_6)\right] \end{aligned} \tag{7.95}$$

where B'_6 and φ_6 are the amplitude and initial phase of the 6th spatial harmonic, respectively. B'_{18} and φ_{18} are the amplitude and initial phase of the 18th spatial harmonic, respectively. The amplitudes of spatial harmonics with vp_f pole pairs ($v > 3$) are negligible.

The calculation model of rotor core losses is shown in Figure 7.32, where the stator is regarded as a unit modulator. That is, the stator bore is cylindrical. On the stator side, there is no PMs or armature currents. The rotor maintains the salient reluctance pole structure. The only excitation is the equivalent current layer.

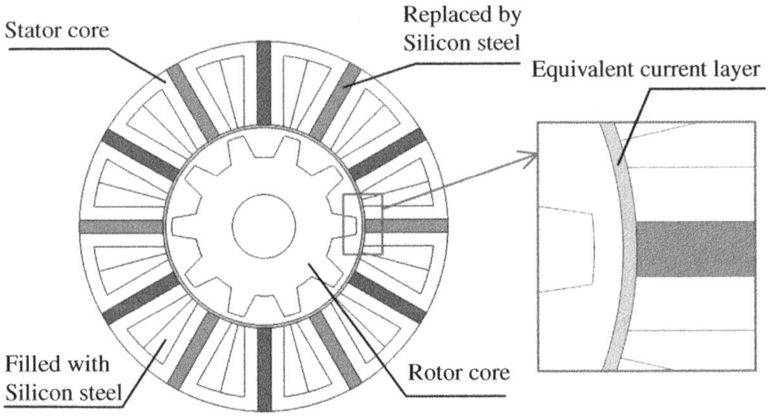

Figure 7.32 Rotor core loss calculation model.

The current density distribution corresponding to the νth spatial harmonic $B'_{\nu}(\theta, t)$ in (7.90) in the rotor core loss calculation model can be expressed by:

$$\begin{aligned} j_{\nu}(\phi, t) = & \frac{g}{\mu_0 R_{si}} \frac{\partial}{\partial \phi} \left[\frac{B'_{\nu}}{m_{r0}} \cos(\nu\phi + \eta N_{rt}\omega_r t + \varphi_{\nu}) \right] \\ & + \frac{g}{\mu_0 R_{si}} \frac{\partial}{\partial \phi} \left\{ \sum_{i=1}^{\infty} \left(\frac{m_{ri} B'_{\nu}}{2m_{r0}^2} \right) \cos[(iN_{rt} + \nu)\phi + \eta N_{rt}\omega_r t + \varphi_{\nu}] \right\} \\ & + \frac{g}{\mu_0 R_{si}} \frac{\partial}{\partial \phi} \left\{ \sum_{i=1}^{\infty} \left(\frac{m_{ri} B'_{\nu}}{2m_{r0}^2} \right) \cos[(iN_{rt} - \nu)\phi - \eta N_{rt}\omega_r t - \varphi_{\nu}] \right\} \end{aligned} \tag{7.96}$$

For spatial harmonics with low rotational speeds, the core loss calculation equations are:

$$P = K_e f^2 B_m^2 + \varepsilon(\Delta B) K_h f B_m^2 \tag{7.97}$$

$$\varepsilon(\Delta B) = 1 + K_{dc} \Delta B^{\beta} + K_l \Delta B^2 \tag{7.98}$$

where K_e and K_h are the eddy current loss and hysteresis loss coefficients [34] and are only suitable to the frequency range from 50 to 400 Hz. ΔB is the DC magnetic bias. K_{dc} and K_l are the coefficients relative to the DC magnetic bias, which can be obtained by experimental measurements [34]. These core loss coefficients are temperature-dependent. The ambient temperature 20 °C is used to simplify the example calculation in this study. B_m is the flux density amplitude.

Amplitudes of spatial harmonics with high rotational speeds are usually as low as a few mT, the core loss coefficients in (7.97) are no longer valid for calculating the carrier-harmonic-induced core losses. Core losses P_{high} induced by the spatial harmonics with high rotational speeds (frequency range from 2.5 to 40 kHz) can be obtained by [18]:

$$P_{high} = k_{e0}f^2B_m^2k_{fe}(f, \mu_d) + k_{hm}(\Delta B, B_m)fB_m^2k_{fh}(f, \mu_d) \tag{7.99}$$

where

$$k_{e0} = \frac{\sigma h^2 \pi^2}{6} \tag{7.100}$$

$$k_{fe}(f, \mu_d) = \frac{3\Delta}{h}\frac{\sinh(h/\Delta) - \sin(h/\Delta)}{\cosh(h/\Delta) - \cos(h/\Delta)} \tag{7.101}$$

$$\Delta = \frac{1}{\sqrt{\pi f \mu_d \sigma}} \tag{7.102}$$

$$k_{fh}(f, \mu_d) = \frac{h}{2\Delta}\frac{\sinh(h/\Delta) + \sin(h/\Delta)}{\cosh(h/\Delta) - \cos(h/\Delta)} \tag{7.103}$$

where $k_{hm}(\Delta B, B_m)$ is the hysteresis iron loss coefficient, which can be obtained by fitting the experiment data [18]. σ and h are the electrical conductivity and thickness of the silicon steel sheets. μ_d is the differential permeability and f the frequency.

6 Core Loss Calculation Under Different Source Types

Figure 7.33 illustrates the detailed workflow for calculating the core losses with the PWM inverter supply. First, the winding currents supplied by the PWM inverter i_{abc} is measured or simulated, and the fundamental wave together with PWM harmonics are extracted by FFT. Then the amplitudes and rotational speeds of airgap flux density spatial harmonics in the airgap magnetic flux density are obtained by the GAFMT.

The distribution of the differential permeability μ_d in stator and rotor cores is obtained from 2D magnetic transient (time-stepping) FEA excited by the fundamental current with the amplitude I_a and phase φ_a. The differential permeability data is substituted into core loss models for each spatial harmonic to account for the magnetic saturation in stator and rotor cores, which is reasonable because the amplitudes of the high-order time harmonics are so small that they have little impact on differential permeability [26].

Then the amplitude and frequency of the current density in the equivalent current layer for the νth spatial harmonic is calculated by (7.95)–(7.96) and substituted into the core loss models illustrated in Figures 7.31 and 7.32. The magnetic flux density variation in each mesh element is then extracted and substituted into (7.97) and (7.99) to calculate core losses.

Both the sinusoidal current source and the PWM inverter supply are considered for comparison. The influence of different control strategies on core losses is also studied, using the sinusoidal PWM (SPWM) and space vector PWM (SVPWM) as

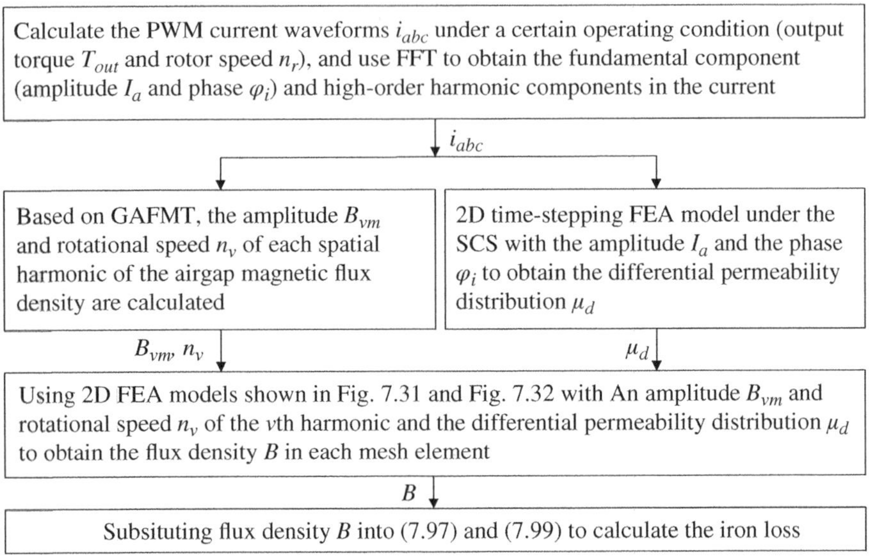

Figure 7.33 Workflow for core loss calculation with the current excitation from PWM voltage source inverters.

examples. The FSPM machine delivers 10 Nm at 900 r/min for both the sinusoidal current source and PWM inverter supply scenarios.

The amplitude and initial phase of the voltage waveform are 188 V and 265.26°, respectively. The corresponding current is 3.66 A and the frequency f_0 is 150 Hz. Table 7.19 compares core loss results from the presented new method and the traditional FEA under a sinusoidal current power supply with different sampling number N in one electrical period. In the presented method, core losses produced by main spatial harmonics can be separated, among which the fourth spatial harmonic produces the most core losses. Ideally, the larger the sampling number

Table 7.19 Core loss breakdown with the sinusoidal current source excitation.

	Core loss contributed by each spatial harmonic from the presented method (W)								Core loss from FEA (W)
N	**4th**	**6th**	**8th**	**16th**	**18th**	**26th**	**28th**	**Total**	
120	37.9	12.2	9.67	15.4	3.37	4.76	2.17	85.5	86.7
60	37.9	12.2	9.67	15.4	3.37	4.76	2.16	85.5	86.7
30	37.9	12.1	9.66	15.4	3.36	4.75	2.16	85.3	86.6
15	37.8	12.0	9.64	15.4	3.34	4.74	2.14	85.1	86.3
7	36.4	11.5	9.27	14.7	3.19	4.56	2.05	81.7	83.1

per electrical cycle is, the more accurate the calculation results will be. It can be seen from Table 7.19 that the increase of core losses with more than 15 sampling points per electrical cycle is marginal. To minimize the computation time, therefore, all the FEA models for core loss calculations in the presented method use 15 sampling points within one electrical cycle.

When the FSPM machine is operated at 900 r/min with an output of 10 Nm, the time harmonic amplitudes in the PWM voltage waveform can be obtained by FFT with the SPWM and SVPWM control strategies, as shown in Figure 7.34. The amplitudes of the fundamental for the SPWM and SVPWM voltages are 188.1 and 188.2 V, respectively. Figure 7.35 shows the corresponding amplitude spectrum. The carrier frequency f_c ($f_c = N_c f_0$, N_c is the frequency modulation ratio) of the two control strategies are 10.05 kHz.

In the current waveform obtained from SPWM, besides the fundamental wave, the most significant harmonic orders are $(N_c \pm 2)f_0$ and $(2N_c \pm 1)f_0$, while in the

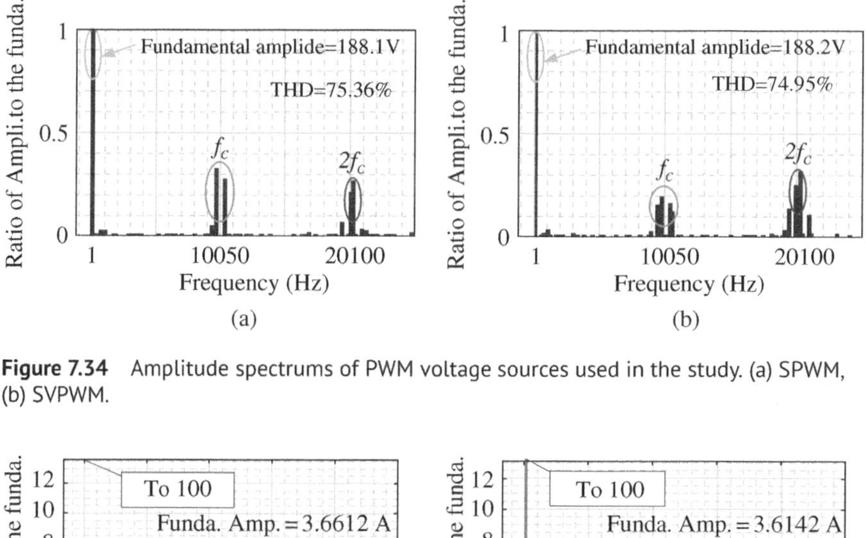

Figure 7.34 Amplitude spectrums of PWM voltage sources used in the study. (a) SPWM, (b) SVPWM.

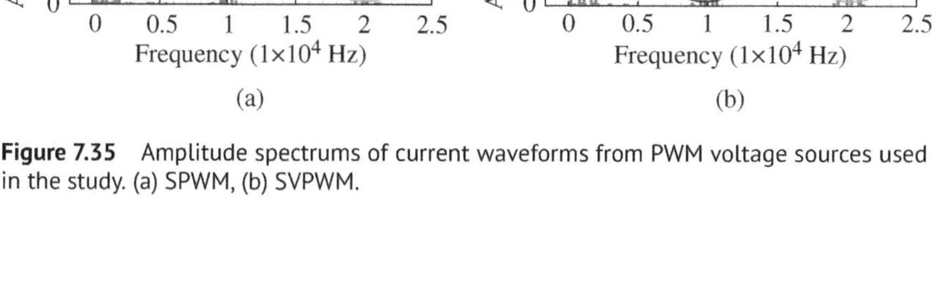

Figure 7.35 Amplitude spectrums of current waveforms from PWM voltage sources used in the study. (a) SPWM, (b) SVPWM.

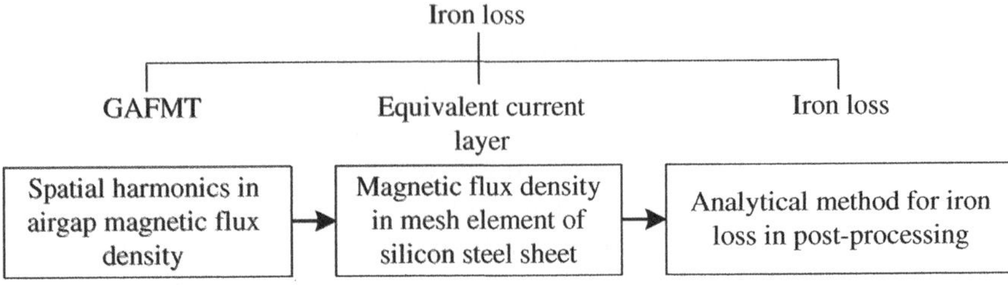

Figure 7.36 Core loss calculation process based on the GAFMT.

current waveform from SVPWM, the dominating harmonic orders are $(N_c \pm 2)f_0$, $(N_c \pm 4)f_0$, and $(2N_c \pm 1)f_0$.

After PWM harmonics are obtained, three steps are required to calculate the core losses based on the GAFMT, as shown in Figure 7.36.

Step 1: By substituting the currents shown in Figures 7.35a,b into (7.85)–(7.89), the amplitude and rotational speed of each airgap flux density spatial harmonic produced by each PWM time harmonic can be calculated. In addition, to consider the magnetic saturation of stator and rotor cores, the differential permeability is calculated based on the fundamental current.

Step 2: The corresponding amplitude and frequency of the equivalent current density are obtained using (7.95) and (7.96).

Step 3: Substituting the amplitude and frequency of the equivalent current density into models shown in Figures 7.31 and 7.32, the magnetic flux density variation in silicon steel sheets can be obtained. Core losses can be calculated by (7.97) and (7.99).

Figure 7.37 shows the amplitude distribution of airgap magnetic flux density under the SPWM and SVPWM control strategies. In Figure 7.37, the horizontal

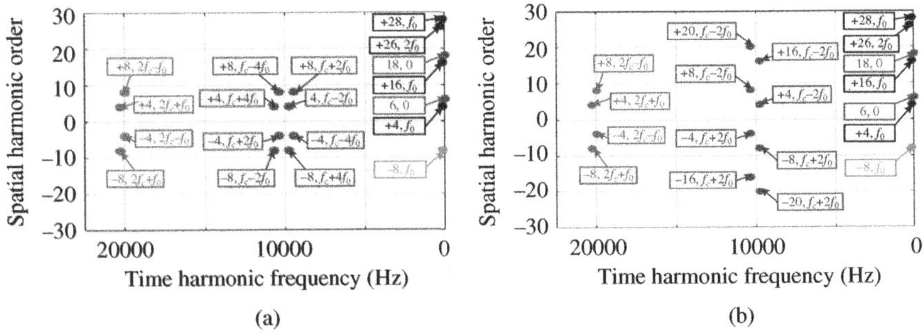

Figure 7.37 Amplitude distribution of the airgap magnetic flux density. (a) SPWM, (b) SVPWM.

Table 7.20 Core loss breakdown with PWM voltage source excitations.

Harmonic pole pair	Time harmonic order	SPWM			SVPWM		
		Flux density amp. (T)	Loss stator (W)	Loss rotor (W)	Flux density amp. (T)	Loss stator (W)	Loss rotor (W)
4	f_0	0.444	33.4	4.46	0.443	33.43	4.46
	f_c-4f_0	0	0	0	6×10^{-4}	0.23	0.11
	f_c+4f_0	0	0	0	4×10^{-4}	0.12	0.07
	f_c-2f_0	9×10^{-4}	0.76	0.54	7×10^{-4}	0.28	0.15
	f_c+2f_0	7×10^{-4}	0.45	0.28	6×10^{-4}	0.19	0.08
	$2f_c-f_0$	3×10^{-4}	0.15	0.08	3×10^{-4}	0.32	0.12
	$2f_c+f_0$	4×10^{-4}	0.25	0.13	4×10^{-4}	0.43	0.18
6	0	1.095	0	12.2	1.104	0	12.2
8	f_0	0.245	8.04	1.63	0.241	8.04	1.63
	f_c-4f_0	0	0	0	5×10^{-4}	0.16	0.08
	f_c+4f_0	0	0	0	3×10^{-4}	0.09	0.05
	f_c-2f_0	6×10^{-4}	0.42	0.38	5×10^{-4}	0.23	0.12
	f_c+2f_0	5×10^{-4}	0.23	0.16	5×10^{-4}	0.16	0.05
	$2f_c-f_0$	2×10^{-4}	0.12	0.05	3×10^{-4}	0.28	0.09
	$2f_c+f_0$	3×10^{-4}	0.19	0.08	3×10^{-4}	0.32	0.14
16	f_0	0.4711	9.13	6.29	0.464	9.13	6.29
	f_c-2f_0	3×10^{-4}	0.13	0.06	0	0	0
	f_c+2f_0	2×10^{-4}	0.08	0.03	0	0	0
18	0	0.684	0	3.37	0.688	0	3.37
20	f_c-2f_0	3×10^{-4}	0.11	0.04	0	0	0
	f_c+2f_0	2×10^{-4}	0.06	0.03	0	0	0
26	$2f_0$	0.197	3.55	1.21	0.197	3.55	1.21
28	f_0	0.314	0.03	2.14	0.317	0.03	2.14

axis represents the magnetic density alternating frequency relative to the stator side, and the vertical axis represents the spatial harmonic order in the airgap magnetic field. The 6th and 18th spatial harmonics are stationary relative to the stator. In addition, signs "+" and "−" indicate that the harmonic rotates in the same or opposite direction as the rotor in the airgap, respectively.

Table 7.20 lists the amplitudes of main harmonics in the airgap flux density waveform and the corresponding stator and rotor core losses. The stator core loss induced by the fourth spatial harmonic under the SPWM control strategies accounts for 61.31% of the total stator core loss, of which the PWM harmonics account for 2.82%. The stator core loss generated by the 16th spatial harmonic with the SPWM control strategies accounts for 16.36% of the total stator core loss, of which the PWM harmonics account for 0.37%.

The 6th and 18th spatial harmonics related to the PM pole pairs do not contribute to the stator core losses. The sixth spatial harmonic contributes the most to the rotor core loss. The 4th, 8th, 16th, and 28th spatial harmonics of the airgap flux density induce the fundamental flux density on the stator. The contribution of each spatial harmonic to the stator core loss is calculated separately and then superimposed to get the total core loss. The total core losses with the SPWM and SVPWM control strategies are 90.26 and 89.53 W, respectively.

The following conclusions can be drawn from Figure 7.37 and Table 7.20:

1) When the armature reaction MMF created by the PWM time harmonics is modulated by the stator/rotor salient reluctance poles, the airgap magnetic field mainly contains spatial harmonics with uN_t pole pairs. Compared with the airgap magnetic field generated by the fundamental wave current, the spatial harmonic orders are unchanged but the corresponding rotational speeds are different.
2) In traditional core loss calculation methods, only the total core loss generated by each PWM harmonic in the stator/rotor core can be calculated [37]. The presented new method can not only calculate the total core loss generated by each PWM harmonic in the stator/rotor, but also indicate which spatial harmonic contributes the most to the core loss. For example, with the SPWM control strategy, the total core loss generated by the PWM harmonic with the frequency of f_c-$2f_0$ on the stator and rotor is 2.44 W, in which the fourth spatial harmonic accounts for $(0.76 + 0.54)/2.44 = 53.3\%$.
3) The stator iron loss is mainly produced by spatial harmonics with uN_t pole pairs in the airgap magnetic field. These spatial harmonics are related to the rotor salient reluctance poles. Therefore, to reduce the total core loss and improve the machine efficiency, it is necessary to reduce the amplitudes of modulation harmonics with uN_t pole pairs, which provides theoretical guidance for the machine design.

Table 7.21 shows the result comparison between the traditional FEA method and the presented core loss method based on the GAFMT under the PWM inverter supply. In the traditional FEA method, the measured or simulated current waveforms are applied to FEA models directly. The fundamental electrical frequency is 150 Hz. To fully capture the switching effect of PWM inverters, the time step needs to be set to an extremely small value and the number of sampling points in one fundamental electrical cycle in the tradition method should be $1/150/4\text{e-}6 = 1667$,

Table 7.21 Comparison of core loss calculation methods.

Items	Traditional FEA	Presented method
Core loss with SPWM voltage source excitation (W)	91.58	90.26
Core loss with SVPWM source excitation (W)	91.06	89.53
FEA calculation time steps	1667	540
Total calculation time (hour)	8	5

where 150 Hz is the fundamental frequency and 4e-6s is the sampling period. It took eight hours in total to calculate the core losses of the stator and rotor with the SPWM and SPWM control strategies on a powerful workstation with Intel Xeon Gold 6164 CPU at 3.2 GHz and 256 GB RAM.

For the presented new method, there are 36 FEA core loss calculation models to run, where only 15 sampling points per electrical cycle per FEA model are needed. It took five hours in total to complete the whole calculation using the same hardware.

7 Experimental Validation

The experimental setup shown in Figure 7.38 is used for validation. The torque transducer measured and recorded the output torque. The power analyzer was used to measure the input electrical power and the oscilloscope recorded the input

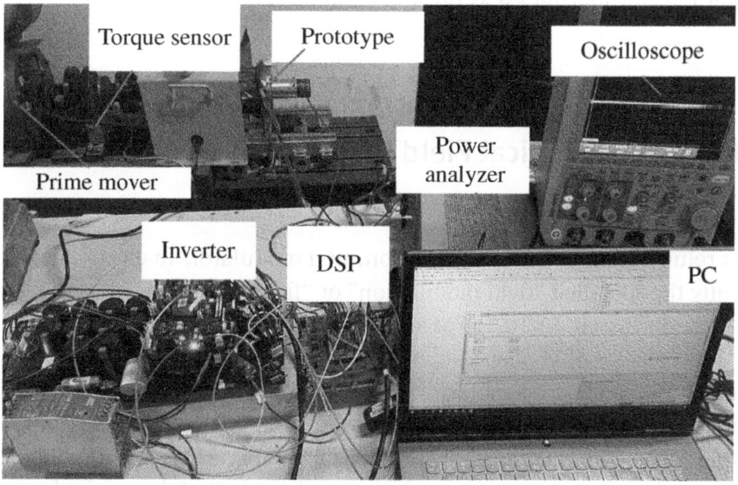

Figure 7.38 Experimental setup.

Table 7.22 Comparison between the calculated and measured losses under different operating conditions.

Operating condition	Calculated loss (W)					Measured total loss (W)	Percent difference (%)
	Core	copper	PM	Mech	Total		
10 Nm at 1200 r/min with SPWM	146.07	31.57	26.27	41.73	245.6	252.7	2.81
10 Nm at 1200 r/min with SVPWM	145.46	31.04	26.19	41.73	244.4	250.3	2.36
5 Nm at 900 r/min with SPWM	87.16	8.24	6.47	31.3	133.2	134.9	1.26
5 Nm at 900 r/min with SVPWM	86.77	8.06	6.35	31.3	132.5	133.1	0.45
10 Nm at 900 r/min with SPWM	90.26	31.3	12.6	31.3	165.5	167.5	1.19
10 Nm at 900 r/min with SVPWM	89.53	30.1	12.2	31.3	163.1	165.2	1.27

current waveforms. The closed-loop control system with the SPWM and SVPWM modulation strategies were used to drive the FSPM machine. Analytical models in [35] were used to estimate the copper loss and mechanical losses. The analytical model in [36] was used in the calculation of the PM eddy current loss.

Table 7.22 lists the comparison between the calculated and measured total losses with different speeds, loads, and modulation strategies. It has been shown that the influence of manufacturing on core losses can be neglected when the machine operating frequency is lower than 200 Hz [38].

Overall, the maximum error is with 3% for the six simulation and testing scenarios, validating the effectiveness of the presented GAFMT-based core loss calculation method considering PWM inverter time harmonics.

7.5 Optimization of Salient Reluctance Pole Modulators for Typical Field Modulation Electrical Machines

Salient reluctance poles are the most common modulators in electrical machines, especially the so-called "field modulation" or "field modulated" types. Taking the previously discussed 12/10 FSPM machine as an example, the stator core with open slots consists of U-shaped lamination modules between which the circumferentially magnetized magnets are placed. The source magnetizing MMF is synchronously modulated by stator teeth (i.e. salient reluctance poles) to produce principal MMF components with $(2k-1)p_f$ pole pairs, where the amplitude of the fundamental harmonic component becomes about half of the original one, while the amplitudes of other odd harmonics increase significantly.

The synchronous modulation behavior will not introduce new harmonic components, so the stator salient reluctance poles can be designed purposely to further improve the mechanical robustness and/or reduce the flux leakage. By contrast, rotor salient reluctance poles usually perform asynchronous modulation and will shift the positions of spectrum lines, changing the principal pole pair number (PPPN) to produce additional working harmonic components. It is these additional effective harmonics that play an instrumental role in optimizing the electromagnetic performance of electrical machines with salient reluctance pole modulators.

The torque production mechanism and magnetic gearing effect of representative magnetic field modulated machines have been studied based on the simple MMF permeance model in [39–41]. Generally, the synchronously modulated MMF by stator salient reluctance poles plays the role of magnets in a magnetic gear, and rotor salient reluctance poles are equivalent iron pieces. Rotor geometries have been optimized to maximize the torque performance in [42]. New design methods applying asymmetric rotor shapes or skew patterns have also been proposed to reduce the torque ripple in [43, 44]. These studies rely heavily on complicated FEA, and only provide the overall output characteristics, lacking enough insights into the airgap magnetic field harmonics to be useful for machine design and systematic performance optimization.

This subsection takes two typical field modulated machines as example to show how we can enhance the airgap field harmonic distribution by improving the rotor salient reluctance pole structures used in asynchronous modulation behaviors from the perspective of the GAFMT.

The torque production mechanisms of two typical field modulated machines, namely the magnetically-geared machine (MGM) and the flux-reversal permanent magnet (FRPM) machine are discussed in detail, whose main parameters are tabulated in Table 7.23. A few common rotor salient reluctance pole structures, such as the upright/inverted trapezoidal, are introduced, based on which general design and optimization guidelines for the two field modulated machines are concluded. The theoretical analysis is validated by 2-D FEA and experimental measurements.

7.5.1 Typical Salient Reluctance Poles

Four rotor salient reluctance pole structures together with the shapes of corresponding modulation functions are presented in Figure 7.39. This section focuses on improving the rotor geometry only; all the other geometric parameters are neglected. For example, the slot depth and magnet length in the studied MGM and FRPM machine are kept unchanged in the whole process. In addition, the following assumptions commonly used in analytical analysis of electrical machines are made for simplicity and clarity:

Table 7.23 Main parameters of the studied MGM and FRPM machines.

Symbol	Quantity	MGM	FRPM
N_{st}/N_{rt}	Pole pair combination	12/10, 12/14, 12/16	6/8, 12/10, 12/14
R_{s_out}	Stator outer radius (mm)	62 mm	64 mm
R_{s_in}	Stator inner radius (mm)	41 mm	35.2 mm
R_{r_out}	Rotor outer radius (mm)	40.5 mm	33 mm
$R_{\_shaft}$	Shaft radius (mm)	15 mm	11 mm
$S_{\_r}$	Rotor slot depth (mm)	5 mm	9 mm
G	Airgap length (mm)	0.35 mm	0.35 mm
l_{stk}	Stack length (mm)	70 mm	75 mm
B_r	Magnet remnance (T)	1.26 T	1.26 T
μ_r	Relatively permeance	1.2	1.2
h_m	PM length (mm)	3.16 mm	1.84 mm
	Modulator type	Salient reluctance pole	

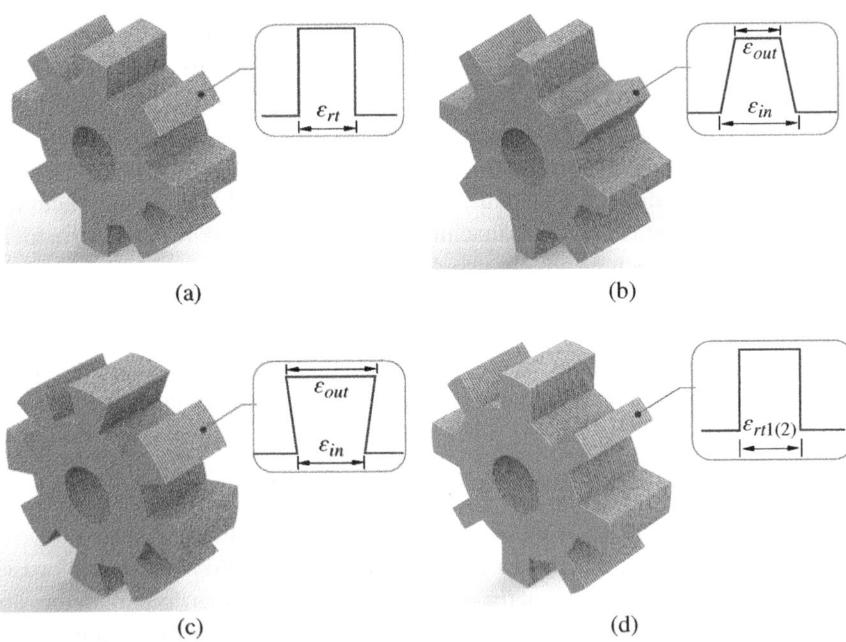

Figure 7.39 Rotor salient reluctance pole structures under study and shapes of their modulation functions. (a) Type 1 – parallel rotor pole structure with equal pole width, (b) Type 2 – Trapezoidal rotor pole structure, (c) Type 3 – Inverted Trapezoidal rotor pole structure, (d) Type 4 – parallel rotor pole structure with unequal pole width.

- The airgap MMF in intervals occupied by rotor slots is zero.
- The lamination material has an infinite magnetic permeability.
- Flux leakage is neglected.
- The influence of finite axial length is neglected.

1 Type 1: Parallel Rotor Pole Structure with Equal Pole Width

As shown in Figure 7.39a, Type 1 is the most common parallel rotor pole (tooth) structure, which serves as the benchmark design in this study. ε_{rt} denotes the pole-arc-to-pole-pitch ratio. The rotor salient reluctance pole modulation function in its Fourier series form is expressed as:

$$R(\phi) = \frac{a_0}{2} + \sum_{l=1}^{\infty} a_n \cos(lN_{rt}\phi) \tag{7.104}$$

where N_{rt} is the number of rotor salient reluctance poles. ϕ is the mechanical angle along the circumference in the modulator-fixed reference frame, i.e. the rotor-speed reference frame in this case.

$$a_0 = 2\varepsilon_{rt} \tag{7.105}$$

$$a_n = \frac{2}{l\pi}\sin(l\varepsilon_{rt}\pi) \tag{7.106}$$

According to Chapter 3, the modulation operator of the rotor salient reluctance pole modulator $M(N_{rt}, \varepsilon_{rt})[.]$ can be defined as:

$$M(N_{rt}, \varepsilon_{rt})[f(\phi, t)] = \begin{cases} f(\phi, t)R(\phi), & \phi \in C^R \\ 0, & \phi \in [0, 2\pi] - C^R. \end{cases} \tag{7.107}$$

where C^R is the union of intervals occupied by rotor salient reluctance poles in mechanical radians. It is assumed here that the source magnetizing MMF is a unit cosine function, and all harmonic items are neglected for simplicity. The normalized resultant MMF modulated by rotor reluctance poles asynchronously can be generally expressed as:

$$\begin{aligned} M(N_{rt}, \varepsilon_{rt})[f(\phi, t)] &= \sum_{v=1} C_p \cos(vp_f\phi) \\ &+ \sum_{v=1}\sum_{l=1} C_{sum} \cos[(vp_f - lN_{rt})\phi + lN_{rt}\omega_r t] \\ &+ \sum_{v=1}\sum_{l=1} C_{dif} \cos[(vp_f + lN_{rt})\phi - lN_{rt}\omega_r t] \end{aligned} \tag{7.108}$$

where the corresponding coefficients can be derived as:

$$C_p = C_{vp,vp} = \frac{a_0}{2} = \varepsilon_{rt} \tag{7.109}$$

$$C_{sum} = C_{vp,vp-lN_{RT}} = \frac{a_n}{2} = \frac{2}{l\pi}\sin(l\varepsilon_{rt}\pi) \tag{7.110}$$

$$C_{dif} = C_{vp,vp+lN_{RT}} = \frac{a_n}{2} = \frac{2}{l\pi}\sin(l\varepsilon_{rt}\pi) \tag{7.111}$$

where ω_r is the mechanical angular velocity of rotor. l and v are arbitrary positive integers. Subscripts "*sum*" and "*dif*" denote the sum and differential modulation type, respectively.

Magnetic field conversion factors C_p, C_{sum}, and C_{dif} are used to characterize interactions between vp_f-pole-pair, $(lN_{rt} - vp_f)$-pole-pair and $(lN_{rt} + vp_f)$-pole-pair harmonics, respectively. Rotor salient reluctance poles alter the relative amplitudes of harmonic components and positions of spectrum lines. Their modulation capability on the source magnetizing MMF distribution can be characterized by magnetic field conversion factors, which depend on the topological parameter ε_{rt} but is independent of pole combinations. The parameter ε_{rt} can be purposely adjusted to improve the electromagnetic performance.

2 Type 2: Trapezoidal Rotor Pole Structure

The ratio ε_{out} is smaller than ε_{in} for Type 2. The modulated MMF can be rewritten in Fourier series form like (7.108) but with different magnetic field conversion factors:

$$C_p = C_{vp,vp} = \frac{a_0}{2} = \frac{\varepsilon_{in} + \varepsilon_{out}}{2} \tag{7.112}$$

$$\begin{aligned} C_{sum} = C_{dif} = \frac{a_n}{2} &= \frac{\sin(l\varepsilon_{out}\pi)}{l\pi} + \frac{\varepsilon_{in}}{l\pi(\varepsilon_{in} - \varepsilon_{out})}[\sin(l\varepsilon_{in}\pi) - \sin(l\varepsilon_{out}\pi)] \\ &- \frac{1}{l\pi(\varepsilon_{in} - \varepsilon_{out})} \\ &\times \left[\varepsilon_{in}\sin(l\varepsilon_{in}\pi) - \varepsilon_{out}\sin(l\varepsilon_{out}\pi) + \frac{1}{l\pi}\cos(l\varepsilon_{in}\pi) - \frac{1}{l\pi}\cos(l\varepsilon_{out}\pi)\right] \end{aligned} \tag{7.113}$$

3 Type 3: Inverted Trapezoidal Rotor Pole Structure

Different from the trapezoidal structure, the ratio ε_{out} of inverted trapezoidal rotor pole structures is greater than ε_{in}, whose magnetic field conversion factors are:

$$C_p = C_{vp,vp} = \frac{a_0}{2} = \frac{\varepsilon_{in} + \varepsilon_{out}}{2} \tag{7.114}$$

$$\begin{aligned} C_{sum} = C_{dif} = \frac{a_n}{2} &= \frac{\sin(l\varepsilon_{out}\pi)}{l\pi} + \frac{\varepsilon_{in}}{l\varepsilon_{out}\pi}[\sin(l\varepsilon_{out}\pi) - \sin(l\varepsilon_{in}\pi)] \\ &- \frac{1}{l\pi(\varepsilon_{out} - \varepsilon_{in})} \\ &\times \left[\varepsilon_{out}\sin(l\varepsilon_{out}\pi) - \varepsilon_{in}\sin(l\varepsilon_{in}\pi) + \frac{1}{l\pi}\cos(l\varepsilon_{out}\pi) - \frac{1}{l\pi}\cos(l\varepsilon_{in}\pi)\right] \end{aligned} \tag{7.115}$$

4 Type 4: Parallel Rotor Pole Structure with Unequal Pole Width

Analogous to the PWM with different duty cycles, ratios ε_{rt1} and ε_{rt2} in Type 4 are not equal. Specifically, it is designed to have $\varepsilon_{rt1} > \varepsilon_{rt2}$, where the salient reluctance pole with ε_{rt1} is defined as the primary pole and that with ε_{rt2} is the auxiliary pole. The primary and auxiliary poles are equally spaced along the circumference. The magnetic field conversion factors can be derived as:

$$C_p = \frac{a_0}{2} = \varepsilon_{rt1} + \varepsilon_{rt2} \tag{7.116}$$

$$C_{sum} = C_{dif} = \frac{a_n}{2} = \left(\frac{1}{l\pi}\right)[\sin(l\varepsilon_{rt1}\pi) + (-1)^l \sin(l\varepsilon_{rt2}\pi)] \tag{7.117}$$

Magnetic field conversion factors with typical parameters ε_{rt}, ε_{in}, ε_{out} are tabulated in Table 7.24. A few conclusions can be drawn as:

1) Compared with Type 1, Type 2 will increase the conversion factor for sum modulation C_{sum} and thus enhance working harmonics from the modulation, while Type 3 will reduce amplitudes of the modulated harmonics. The fundamental harmonics of Type 2 and Type 3 will maintain high amplitudes when parameters ε_{in} and ε_{out} are properly designed.
2) The fundamental harmonic is improved by Type 4 as expected, while the low-order ($l = 1, 2$) modulated harmonics are significantly decreased.
3) For the MGM, combining Type 2 and Type 3 is a good choice due to the different locations of the magnetizing MMF (at rotor) and armature MMF (at stator).
4) For the FRPM machine, combining Type 2 and Type 4 is a good option to accommodate both the fundamental and other working harmonics introduced by the modulation behaviors.

7.5.2 Optimization for Magnetically-Geared PM Machine

As illustrated in Figure 7.40a, the source magnetizing MMF distribution along the airgap of the MGM created by embedded PMs is a rectangular waveform,

Table 7.24 Magnetic field conversion factors of different rotor salient reluctance pole structures and their influence on harmonic amplitudes.

Configuration	ε_1	ε_2	C_p	C_{sum}	Fundamental harmonic	Modulated harmonic
Type 1	0.5		**0.472**	**0.309**	Baseline design	
Type 2	0.35	0.65	**0.424**	**0.478**	−	+
Type 3	0.58	0.42	**0.481**	**0.295**	−	−
Type 4	0.60	0.40	**0.875**	**0.142**	+	−

"+" and "−"denote that the harmonic amplitude is increased and decreased, respectively.

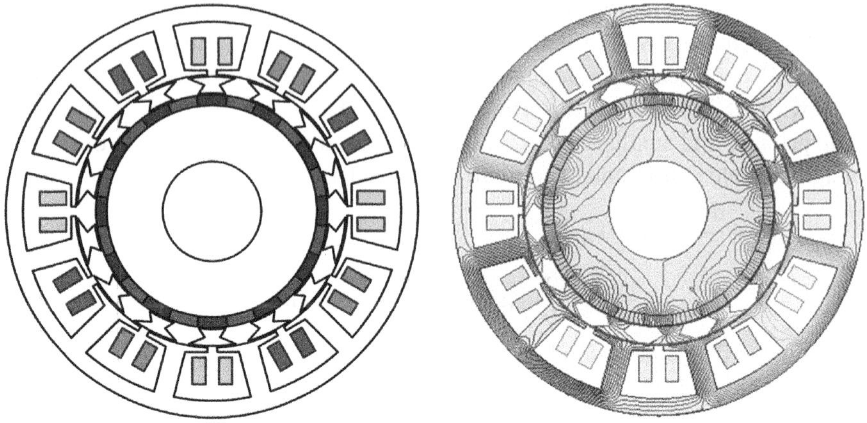

Figure 7.40 Cross-sectional view and flux lines distribution of the MGM under study. Both Type 2 and Type 3 are used with parameters $\varepsilon_{out} = 0.78$, $\varepsilon_{mid} = 0.27$, $\varepsilon_{in} = 0.51$. (a) Cross-sectional view, (b) Flux lines under loaded condition.

which is asynchronously modulated by the rotor salient reluctance poles. The rotor has $N_{rt} = 14$ salient poles to produce principal harmonic components with $(N_{rt} \pm p_f) = 14 \pm 12 = +2/(+26)$ pole pairs. The armature winding located on the stator extracts the fundamental component with $N_{rt} - p_f = 2$ pole pairs to produce flux linkage.

For the studied MGM, the number of rotor salient reluctance poles is close to the PM pole pairs. Rotor salient reluctance poles provide flowing paths for the modulated magnetizing magnetic field, causing low amplitudes of effective harmonic components in the outer airgap. Therefore, the torque is mainly produced in the inner airgap. The rotor salient pole structure shown in Figure 7.40 is a combination of Type 2 and Type 3. Flux lines are illustrated in Figure 7.40b.

The radial and tangential components of the fundamental wave (12th) in the modulated magnetizing and armature magnetic fields show the opposite trends when ε_{out} and ε_{in} change, as illustrated in Figure 7.41. When the pole-arc-to-pole-pitch ratio of middle rotor salient poles ε_{mid} and ε_{in} are fixed, the torque ripple and peak-to-peak values of the cogging torque are almost independent of the change of ε_{out}, as shown in Figure 7.42.

The torque ripple and peak-to-peak values of the cogging torque will increase faintly with the increase of ε_{mid}. On the contrary, the average torque decreases as ε_{mid} increases due to the relatively low amplitude for the p_f–pole–pair B_{tm} component and the torque item $B_{ra}B_{tm}$.

The comparison of torque performance with different rotor salient pole structures is illustrated in Figure 7.42 with a fixed ε_{mid}. In short, the torque

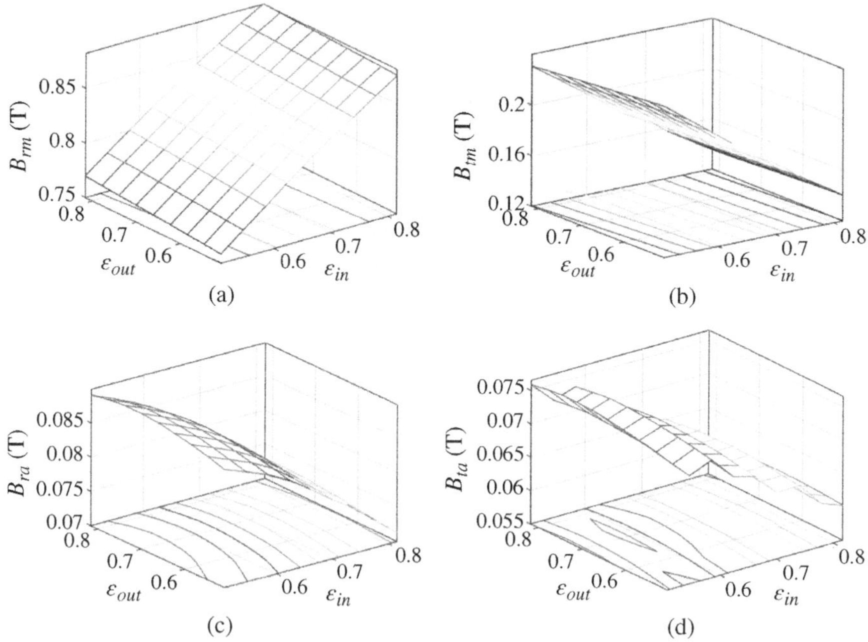

Figure 7.41 Variations of the fundamental harmonic (12th) amplitude vs. ε_{out} and ε_{in} when $\varepsilon_{mid} = 0.27$. (a) B_{rm}, (b) B_{tm}, (c) B_{ra}, (d) B_{ta}.

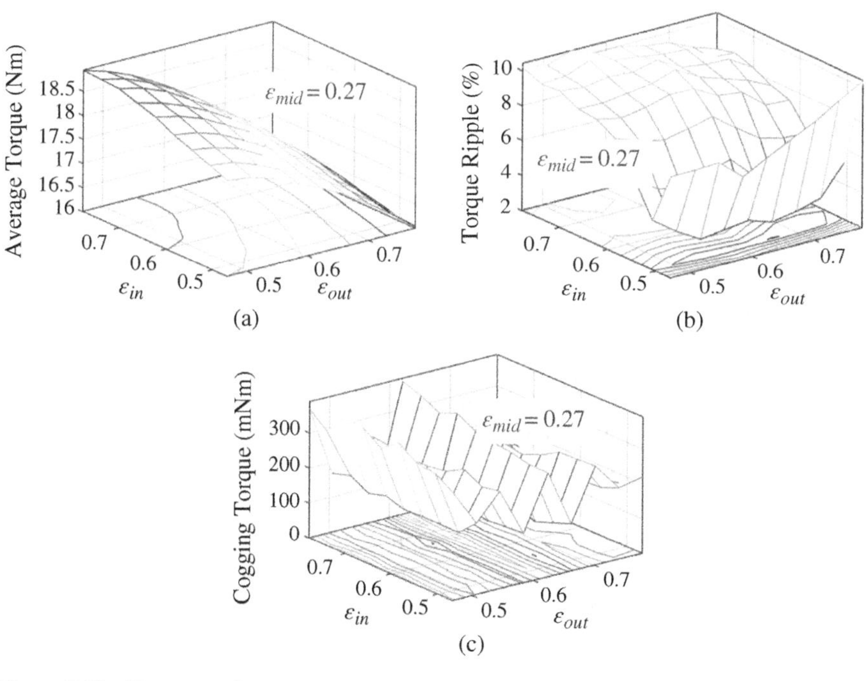

Figure 7.42 Torque performance vs. ε_{out} and ε_{in} when $\varepsilon_{mid} = 0.27$. (a) Average torque, (b) Torque ripple. (c) Peak to peak cogging torque.

Table 7.25 Comparison of working airgap magnetic field harmonics produced by modulation.

Harmonic component		Modified rotor configuration		
		Type 1	Type 3	Type (2 + 3)
12th	B_{rm}	**0.776**	0.751	0.763
	B_{tm}	0.226	0.274	**0.288**
	B_{ra}	0.087	0.090	**0.091**
	B_{ta}	0.075	0.077	**0.080**

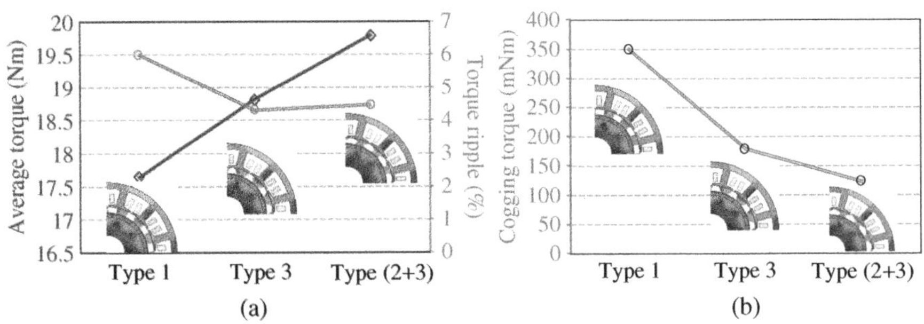

Figure 7.43 Comparison of torque performance for different rotor salient pole types. (a) Torque under loaded operation, (b) Cogging torque under no-load operation.

production of the MGM analyzed in this subsection is due to the interaction of the fundamental magnetizing field harmonic with p_f pole pairs and the modulated armature field harmonic with $(N_{rt} - p_a)$ pole pairs, and the optimal ε_{out} should be larger than ε_{in}.

The 12-pole-pair harmonic component of the modulated armature magnetic field is enhanced by the combined Type 2 and Type 3 rotor pole structure, as tabulated in Table 7.25. As shown in Figure 7.43, the improved rotor configuration can increase the torque output by 12.16% while keeping the torque ripple and cogging torque low. The results indicate that Type (2 + 3) is a good choice for the MGM to improve the torque performance.

The above-mentioned theoretical analysis shows that when ε_{out} is about 2.85–3.0 times ε_{mid}, the inverted Type 2 structure will improve the mutual magnetic conversion factor C_{sum} and then increase the modulated armature magnetic field harmonic with $(N_{rt} - p_a)$ pole pairs. Moreover, when ε_{in} is about 1.8–2.0 times ε_{mid}, the Type 3 structure will increase the amplitude of the fundamental magnetizing magnetic field harmonic with (p_f) pole pairs and suppress the modulated harmonic with $(N_{rt} - p_f)$ pole pairs for the MGM under study.

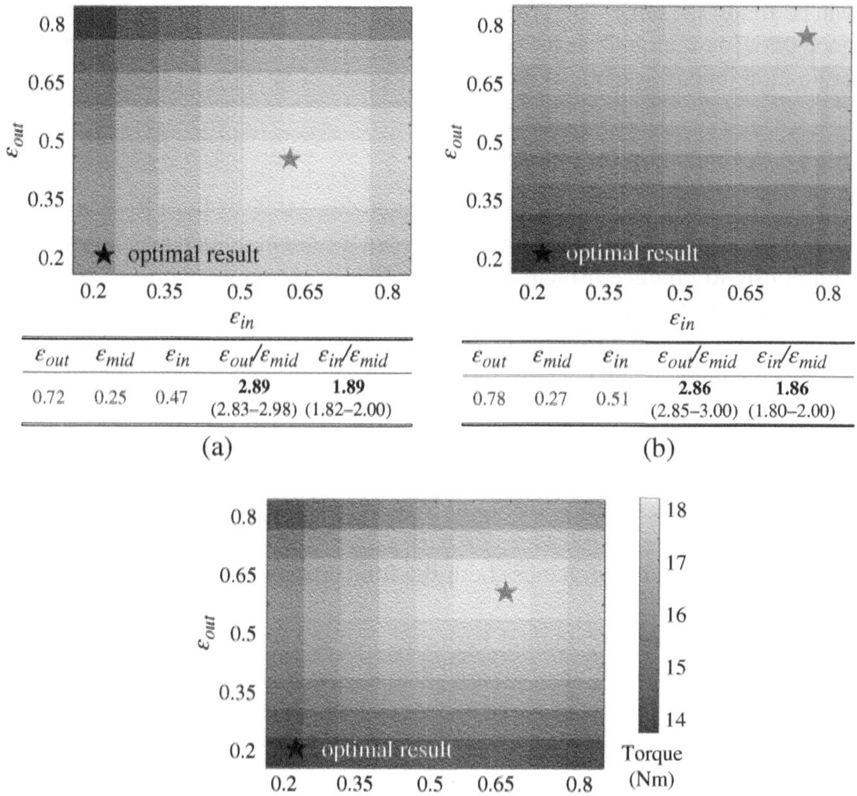

ε_{out}	ε_{mid}	ε_{in}	$\varepsilon_{out}/\varepsilon_{mid}$	$\varepsilon_{in}/\varepsilon_{mid}$
0.72	0.25	0.47	**2.89** (2.83–2.98)	**1.89** (1.82–2.00)

(a)

ε_{out}	ε_{mid}	ε_{in}	$\varepsilon_{out}/\varepsilon_{mid}$	$\varepsilon_{in}/\varepsilon_{mid}$
0.78	0.27	0.51	**2.86** (2.85–3.00)	**1.86** (1.80–2.00)

(b)

ε_{out}	ε_{mid}	ε_{in}	$\varepsilon_{out}/\varepsilon_{mid}$	$\varepsilon_{in}/\varepsilon_{mid}$
0.64	0.22	0.44	**2.90** (2.84–2.95)	**2.00** (1.80–2.05)

(c)

Figure 7.44 The optimum combination of ε_{out} and ε_{in} for MGMs with different PM pole pairs (compared with Type 1). (a) 8 PM pole pairs, (b) 12 PM pole pairs, (c) 14 PM pole pairs.

The modified rotor salient pole structures for the MGM with 8 and 14 PM pole pairs are actually both the combination of Type (2 + 3) structures. The average torque vs. ε_{out} and ε_{in} for the three MGMs are presented in Figure 7.44. Taking the MGM with 8 PM pole pairs as an example, ε_{out} should be about 2.83–2.98 times ε_{mid}, and ε_{in} should be around 1.82–2.0 times ε_{mid}. Similar conclusions can also be drawn for the MGM with 14 pole pairs. In short, the optimal combinations of ε_{out}, ε_{mid}, and ε_{in} will be different for different pole/slot combinations but the modified rotor salient pole structure and optimal ranges of $\varepsilon_{out}/\varepsilon_{mid}$ and $\varepsilon_{in}/\varepsilon_{mid}$

will be almost the same. This study provides generalized guidelines to help achieve good initial designs and has the potential to reduce the searching and performance evaluation time in systematic design optimization of MGMs.

7.5.3 Optimization for FRPM Machine

The rectangularly distributed source magnetizing MMF of the FRPM machine shown in Figure 7.45a is modulated by rotor salient reluctance poles asynchronously to produce principal magnetic field components with $(N_{rt} \pm p_f) = 8 \pm 6 = +2/(+14)$ and $(N_{rt} \pm 2p_f) = 8 \pm 12 = -4/(+20)$ pole pairs. The latter are caused by the second harmonic in the source magnetizing MMF.

The fundamental and effective modulated harmonic components contribute to the average torque in FRPM machines. The trapezoidal structure will increase the modulated harmonics (p to $N_{rt} - p$) and the parallel rotor salient pole structure with unequal widths will enhance the fundamental harmonic component. The modified rotor pole for the studied FRPM machine is a combination of Type 2 and Type 4 structures ($\varepsilon_{out1} = 0.44$, $\varepsilon_{in1} = 0.53$, $\varepsilon_{out2} = 0.31$, $\varepsilon_{in2} = 0.44$), as illustrated in Figure 7.45b, which enhances both the fundamental and effective modulated harmonics.

Variations of the fundamental harmonic and the torque performance versus different parameter combinations are illustrated in Figures 7.46 and 7.47, respectively. It is shown that the performance changes little within a certain range of $\varepsilon_{out1}/\varepsilon_{out2}$. The electromagnetic torque shows a dramatic increase and then

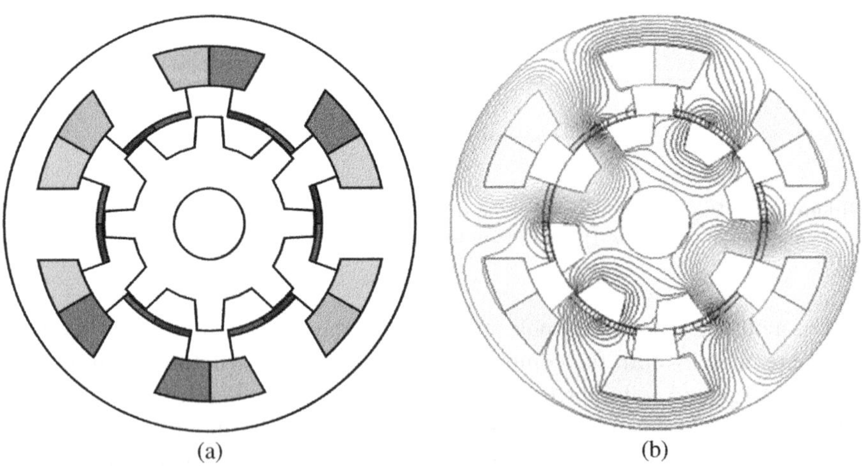

Figure 7.45 Cross-sectional view and flux lines of the 6/8 FRPM machine with a combined rotor structure, i.e. Type (2 + 4): $\varepsilon_{out1} = 0.44$, $\varepsilon_{in1} = 0.53$, $\varepsilon_{out2} = 0.31$, $\varepsilon_{in2} = 0.44$. (a) Cross-sectional view, (b) Flux lines under loaded condition.

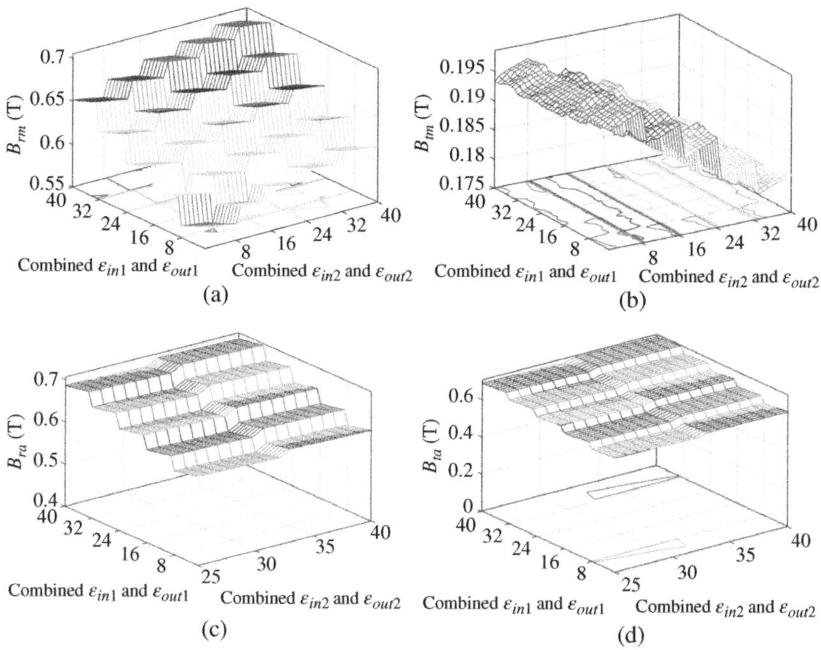

Figure 7.46 Modulated harmonic distributions (the fundamental harmonic: 6th). (a) B_{rm}, (b) B_{tm}, (c) B_{ra}, (d) B_{ta}.

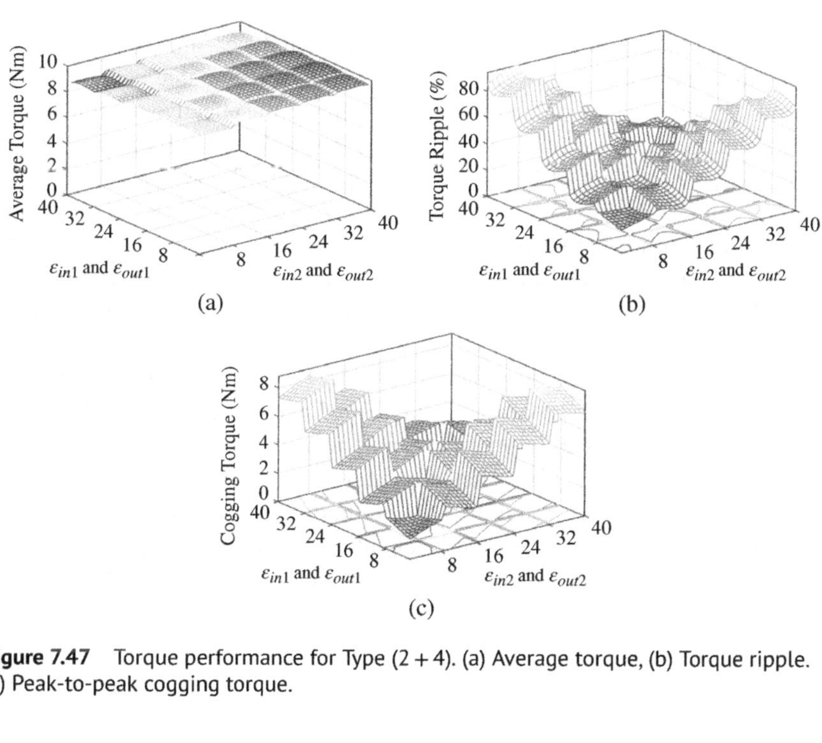

Figure 7.47 Torque performance for Type (2 + 4). (a) Average torque, (b) Torque ripple. (c) Peak-to-peak cogging torque.

Table 7.26 Comparison of airgap magnetic field harmonics between Type (2 + 4) and Type 2.

Harmonic component	Main contribution			Little contribution		
	2	6	12	4	10	14
B_{rm}	+	+	+	−	+	−
B_{tm}	+	−	−	−	+	−
B_{ra}	+	+	+	+	+	+
B_{ta}	+	+	+	−	+	+

decreases with the increase of ε_{out2}. It also increases gradually with ε_{out1}. The auxiliary rotor pole reshapes the periodicity of the rotor modulation function and introduces other harmonics. These spatial harmonics, which have high amplitudes when the difference between ε_{out1} and ε_{out2} is large, increase the torque ripple and cogging torque.

The torque ripple and cogging torque show similar distribution and stay almost constant within a certain range of $\varepsilon_{out1}/\varepsilon_{out2}$. All the results indicate that $\varepsilon_{out1(2)}$ is the key parameter affecting the airgap magnetic field modulation effect and machine performance. Amplitudes of the 2nd, 6th, and 12th B_{rm}, along with all the corresponding harmonic components of $B_{r(t)a}$ have been selectively enhanced, as tabulated in Table 7.26.

In short, the torque enhancement of the combined Type 2 and Type 4 structure compared to Type 2 is due to two aspects: the more reasonable combination of ε_{out} and ε_{in}, where ε_{out} is smaller than ε_{in} to enhance the effective modulated harmonics and the increased fundamental harmonics – 6th and 12th B_{rm}, and 2nd B_{ra} and B_{ta}. These fundamental harmonics are improved by the Type 4 structure with auxiliary rotor salient poles.

The above mentioned analysis confirms that when ε_{in1} is about 1.18–1.55 times ε_{out1}, Type 2 structure will improve the mutual conversion factor and enhance the modulated magnetizing and armature magnetic field harmonics. Additionally, when ε_{in2} is about 0.70–0.85 times ε_{in1} and ε_{out2} is about 0.67–0.88 times ε_{out1}, i.e. when adjacent rotor salient poles act as auxiliary poles, amplitudes of the fundamental magnetizing and armature magnetic field harmonics will be increased.

Similarly, the modified rotor salient pole structure and optimal ranges of $\varepsilon_{in1}/\varepsilon_{out1}$, $\varepsilon_{in2}/\varepsilon_{in1}$ and $\varepsilon_{out2}/\varepsilon_{out1}$ remain almost unchanged though the pole-slot combinations are different, as illustrated in Figure 7.48. Taking the 12/10 FRPM machine as an example, ε_{in1} should be about 1.19–1.58 times ε_{out1}, which corresponds to a Type 2 structure. Moreover, ε_{in2} should be around 0.70–0.88 times ε_{in1}

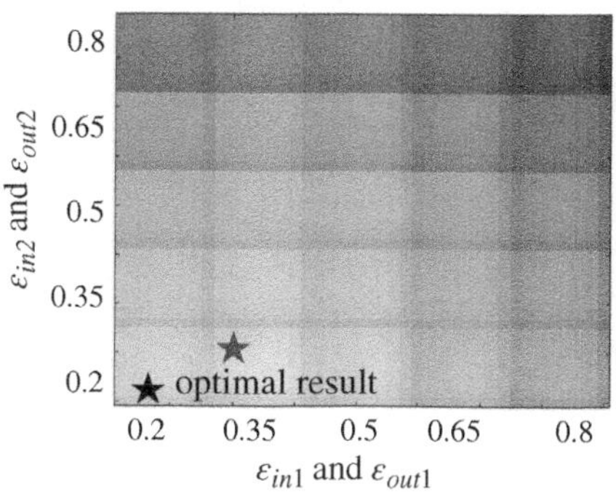

Primary pole		Auxiliary pole	
ε_{out1}	ε_{in1}	ε_{out2}	ε_{in2}
0.44	0.53	0.31	0.44

$\varepsilon_{in1}/\varepsilon_{out1}$	$\varepsilon_{in2}/\varepsilon_{in1}$	$\varepsilon_{out2}/\varepsilon_{out1}$
1.20	**2.89**	**1.89**
(1.18–1.55)	(2.83–2.98)	(1.82–2.00)

(a)

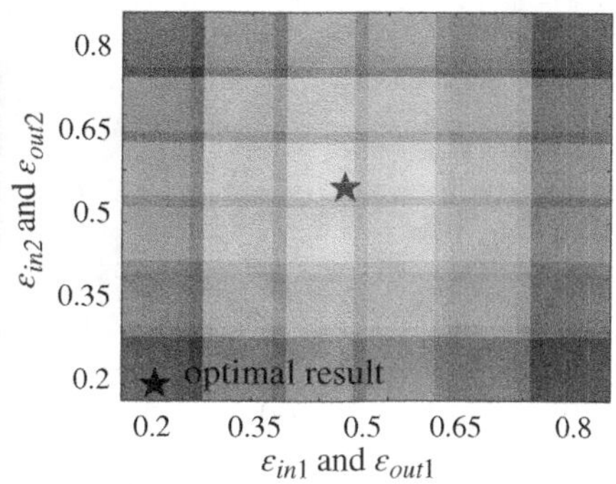

Primary pole		Auxiliary pole	
ε_{out1}	ε_{in1}	ε_{out2}	ε_{in2}
0.36	0.50	0.28	0.42

$\varepsilon_{in1}/\varepsilon_{out1}$	$\varepsilon_{in2}/\varepsilon_{in1}$	$\varepsilon_{out2}/\varepsilon_{out1}$
1.38	**0.83**	**0.83**
(1.19–1.58)	(0.70–0.88)	(0.67–0.88)

(b)

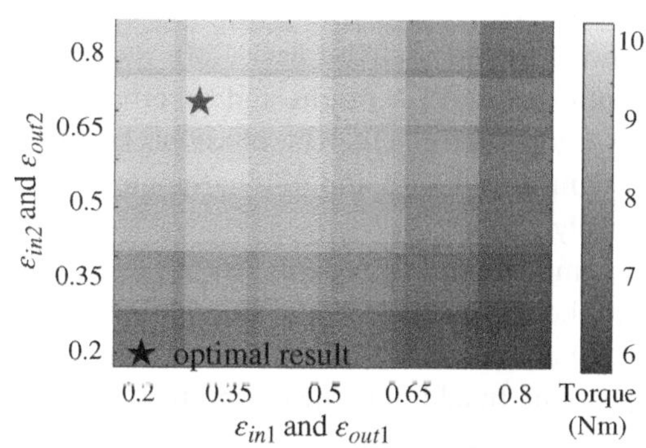

Primary pole		Auxiliary pole	
ε_{out1}	ε_{in1}	ε_{out2}	ε_{in2}
0.43	0.62	0.31	0.51

$\varepsilon_{in1}/\varepsilon_{out1}$	$\varepsilon_{in2}/\varepsilon_{in1}$	$\varepsilon_{out2}/\varepsilon_{out1}$
1.45	**0.81**	**0.73**
(1.19–1.58)	(0.70–0.88)	(0.67–0.88)

(c)

Figure 7.48 The optimum solution for FRPM machines with different pole-slot combinations (compared with Type 1). (a) The 6/8 FRPM machine, (b) The 12/10 FRPM machine, (c) The 12/14 FRPM machine.

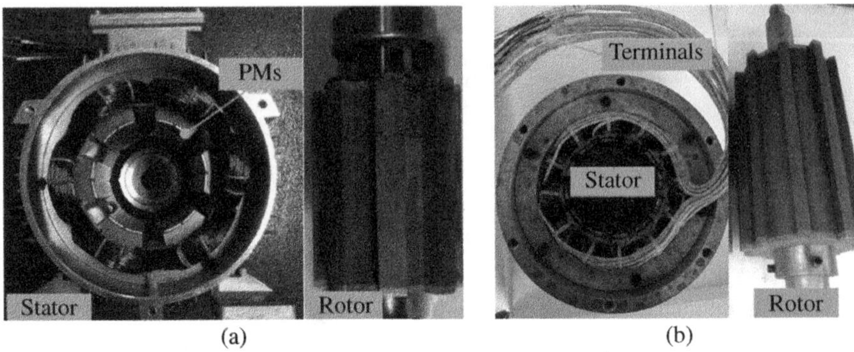

Figure 7.49 FRPM porotypes with different pole-slot combinations. (a) The 6/8 FRPM machine, (b) The 12/10 FRPM machine.

and ε_{out2} should be about 0.67–0.88 times ε_{out1}. Similar conclusions can also be drawn for the 12/14 FRPM machine.

To validate the above-mentioned theoretical analysis, FRPM machines with different pole-slot combinations, namely 6/8 and 12/10, were prototyped and tested, as illustrated in Figure 7.49.

Figure 7.50a,b shows the open-circuit back-EMF waveforms and their amplitude spectrums. Both the 2D FEA results and experimental measurements are provided, showing a good agreement. The phase back-EMFs of the two FRPM machines contain significant fourth and fifth harmonic components corresponding to lN_{rt}, respectively.

Figures 7.50c,d compares the FEA-predicted inductances with experimental measurements, and the results agree well with each other. It is evident that both the self and mutual inductances of the two machines are constant and independent of the rotor position. In addition, the low mutual inductance indicates weak coupling between phases and the inherent fault tolerance.

The measured average torque values are slightly lower than those from 2D FEA, as shown in Figure 7.51a, because the 3D effects, such as magnet end effect and end windings, and manufacture tolerances were not considered in FEA simulations. Moreover, the torque increase of the 12/10 FRPM machine becomes saturated when the phase current density is larger than 10 A/mm^2. The torque of the 6/8 FRPM machine increases linearly with the phase current density, showing better overload capability.

Cogging torque waveforms from 2D FEA and measurements are illustrated in Figure 7.51b. The period of the cogging torque is related to the least common multiplier of N_{st} and N_{rt}. For the 6/8 FRPM machine, the period is 360°/LCM(6, 8) = 360°/24 = 15° (mechanical).

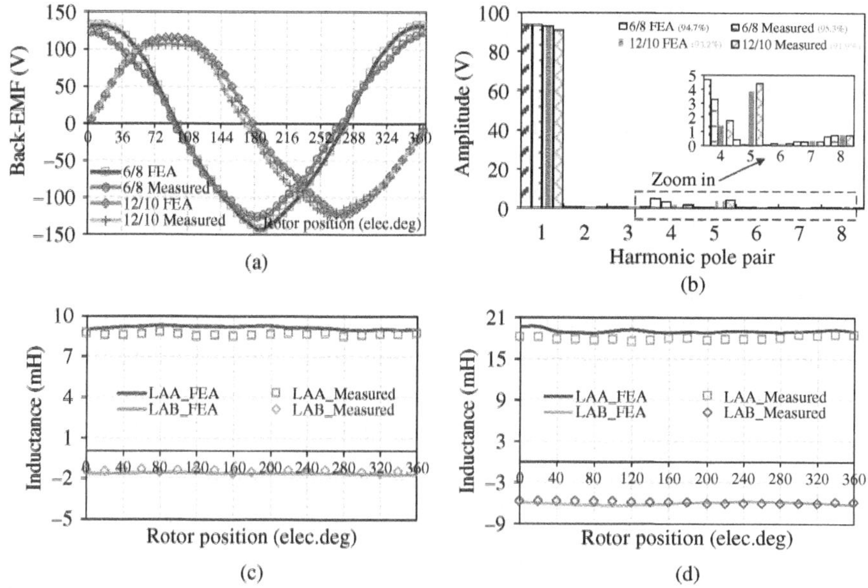

Figure 7.50 Open-circuit back-EMF and inductance characteristics of the 6/8 (at 1500 r/min) and 12/10 (at 1000 r/min) FRPM machine. (a) Open-circuit back-EMF waveforms, (b) Amplitude spectrums, (c) Inductance waveforms of the 6/8 FRPM machine, (d) Inductance waveforms of the 12/10 FRPM machine.

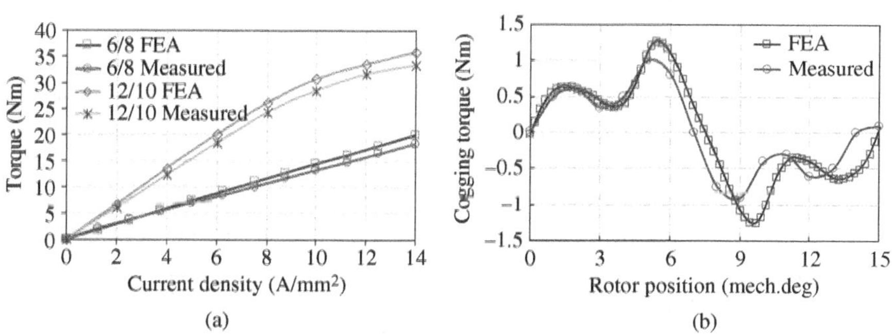

Figure 7.51 Torque performance. (a) Average torque vs. current density, (b) Cogging torque waveform of the 6/8 FRPM machine.

7.5.4 General Guidelines

This section studies the possibility of improving the torque performance of electrical machines with salient reluctance pole structures from the perspective of modifying rotor pole structures based on the GAFMT. Effective modulated magnetic field harmonics for the MGMs and FRPM machines are changed on purpose. A few conclusions can be drawn from the study and serve as the generalized design guidelines as listed blew:

(1) The MGM relies on a single working harmonic to produce torque like the traditional PM synchronous machine, while the torque of FRPM machines is produced by multiple spatial harmonics.
(2) The optimal rotor salient pole structure for the MGM is a combination Type 2 and Type 3 because the magnetizing and armature MMFs are placed on different machine parts, while the combination of Type 2 and Type 4 is more suitable for FRPM machines.
(3) To improve the magnetic field conversion factor C_{sum} and increase the modulated armature field harmonic with $(N_{rt} - p_a)$ pole pairs, it is recommended to keep ε_{out} of the MGM about 2.85–3.00 times ε_{mid}. ε_{in} should be about 1.80–2.00 times ε_{mid} to increase the amplitude of the fundamental magnetizing field harmonic and suppress the modulated harmonic with $(N_{rt} - p_f)$ pole pairs.
(4) ε_{IN1} of the FRPM machine should be about 1.18–1.55 times ε_{out1} to improve the magnetic field conversion factor C_{sum} and enhance the modulated magnetizing/armature field harmonics. Additionally, for auxiliary salient poles, ε_{IN2} ought to be about 0.70–0.85 times ε_{in1} and ε_{out2} should be about 0.67–0.88 times ε_{out1} to achieve the torque enhancement.

7.6 Airgap-Harmonic-Oriented Design Optimization Methodology

FMPM machines are representative machine topologies working by the principle of the magnetic field modulation and some of them have been considered as promising candidates for direct-drive applications.

Currently, the existing different flux modulation techniques in FMPM motors now trend to form a unified theory in general [17]. From the perspective of the GAFMT, more and more attention has been paid to the redesigns of motor topology and winding distribution, as well as investigation and application of flux-modulated theory in various new PM motor types [45–47].

The existing studies on FMPM machines mainly focus on the three key elements – PMs (i.e. MMF sources), modulators and armature windings. For better understanding, the relationship between the three key elements and the machine's performance is illustrated in Figure 7.52. The variation of each element will lead to a change of machine performance, such as the torque

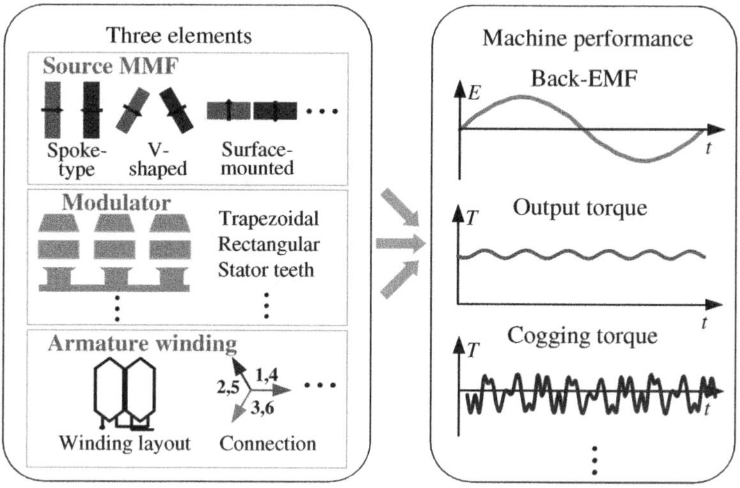

Figure 7.52 Relationship between the three chained key elements and the performance of FMPM machines.

enhancement, reduction in torque ripple, improvement in back-EMF, etc. It means that by sophisticated design of the three key elements, the FMPM machine can obtain higher electromagnetic performance. Several challenges arise from design practices, mainly including the difference in elements' sensitivity to the performance, trade-offs between key design elements and geometric constraints. How to effectively evaluate the influence of the three key elements on the machine performance, and to approach optimization designs efficiently has therefore become a topic of great interest.

The GAFMT provides a new perspective to systematically optimize electrical machines by examining the airgap magnetic flux density harmonics (i.e. the airgap-harmonic-orientated design method). The main idea of this method is to achieve optimal designs by using the airgap flux density harmonics as the bridge between the three key elements and the machine performance. In the design optimization process, airgap field harmonics not only serve as design variables for the torque performance but are also considered as optimization objectives for adjusting geometric parameters. That is, the airgap harmonics' role is twofold, and both the torque performance and geometric parameters can be investigated and evaluated effectively. An outer-rotor field modulation machine with V-shaped PMs, termed as the V-FMPM machine, as illustrated in Figure 7.53, is taken as an example to show the GAFMT-based analysis and optimization workflow.

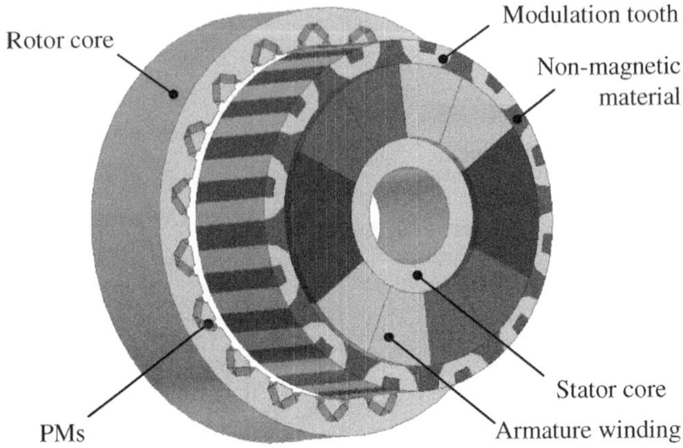

Figure 7.53 3D geometry of the V-FMPM machine under study.

7.6.1 Airgap-Harmonic-Oriented Design Optimization Concept

1 Airgap Magnetic Field Harmonic Analysis

To incorporate airgap field harmonics into the optimization process, the relationships between field harmonics and the performance index of interest, such as the average output torque or torque ripple, should be developed first. Generally, for field-modulation PM machines, by the Fourier series expansion, the radial and tangential components of PM MMFs – $F_{pmr}(\phi,t)$ and $F_{pmt}(\phi,t)$, can be written as:

$$\begin{cases} F_{pmr}(\phi, t) = \sum\limits_{i=1,3,5....}^{\infty} A_i \cos ip_r(\omega_r t + \phi + \phi_{r0}) \\ F_{pmt}(\phi, t) = \sum\limits_{i=1,3,5....}^{\infty} C_i \cos ip_r(\omega_r t + \phi + \phi_{t0}) \end{cases} \tag{7.118}$$

where p_r is the PM pole pairs and ω_r is the mechanical angular velocity. i is a positive odd integer. A_i and C_i denote the amplitudes of radial and tangential components of the ith harmonic in MMF generated by PMs, respectively. Besides, ϕ_{r0} and ϕ_{t0} represent the initial phases of the ith radial and tangential harmonic components. ϕ is the position angle along the airgap circumference.

The Fourier series expansion of the stator-side airgap permeance function is:

$$\Lambda_s(\phi) = \sum_{j=0,\pm1,\pm2......}^{\infty} P_j \cos(jN_s\phi) \tag{7.119}$$

where N_s is the number of stator teeth, j a natural number and P_j the amplitude of the jth harmonic in the PM magnetic field.

Based on (7.118) and (7.119), the radial and tangential components of the modulated airgap flux density $B_r(\phi,t)$ and $B_t(\phi,t)$ can be obtained. When the effective area is unit, $B_r(\phi,t)$ and $B_t(\phi,t)$ can be expressed as:

$$\begin{cases} B_{pmr}(\phi,t) = F_{pmr}(\phi,t)\Lambda_s(\phi) \\ \quad = \frac{1}{2}\sum_{i=1,3,5\dots}^{\infty}\sum_{j=0,\pm1,\pm2\dots}^{\infty} A_i P_j \cos[ip_r\omega_r t + (jN_s + ip_r)\phi + ip_r\phi_{r0}] \\ \quad + \frac{1}{2}\sum_{i=1,3,5\dots}^{\infty}\sum_{j=0,\pm1,\pm2\dots}^{\infty} A_i P_j \cos[ip_r\omega_r t + (ip_r - jN_s)\phi + ip_r\phi_{r0}] \\ B_{pmt}(\phi,t) = F_{pmt}(\phi,t)\Lambda_s(\phi) \\ \quad = \frac{1}{2}\sum_{i=1,3,5\dots}^{\infty}\sum_{j=0,\pm1,\pm2\dots}^{\infty} C_i P_j \cos[ip_r\omega_r t + (jN_s + ip_r)\phi + ip_r\phi_{t0}] \\ \quad + \frac{1}{2}\sum_{i=1,3,5\dots}^{\infty}\sum_{j=0,\pm1,\pm2\dots}^{\infty} C_i P_j \cos[ip_r\omega_r t + (ip_r - jN_s)\phi + ip_r\phi_{t0}] \end{cases} \tag{7.120}$$

It can be observed from (7.120) that apart from the fundamental harmonic with p_r pole pairs, there are abundant other spatial harmonics in the radial and tangential airgap flux density waveforms, which is quite different from conventional PM synchronous machines.

The radial and tangential components of the armature MMF – $F_{awr}(\phi,t)$ and $F_{awt}(\phi,t)$ can be expressed as:

$$\begin{cases} F_{awr}(\phi,t) = f_{Ar}(\phi,t) + f_{Br}(\phi,t) + f_{Cr}(\phi,t) \\ \quad = \sum_{m}^{\infty} F_{mr}\cos(\omega t + m\phi + \phi_{r0m}) + \sum_{n}^{\infty} F_{nr}\cos(\omega t - n\phi + \phi_{r0n}) \\ F_{awt}(\phi,t) = f_{At}(\phi,t) + f_{Bt}(\phi,t) + f_{Ct}(\phi,t) \\ \quad = \sum_{m}^{\infty} F_{mt}\cos(\omega t + m\phi + \phi_{t0m}) + \sum_{n}^{\infty} F_{nt}\cos(\omega t - n\phi + \phi_{t0n}) \end{cases} \tag{7.121}$$

where ω is the electrical angular velocity, and $\omega = p_r\omega_r$. ϕ_{r0n} and ϕ_{t0n} are initial phases of the n-th harmonic in the armature MMF. The Fourier series expansion of the rotor-side airgap permeance function is:

$$\Lambda_r = \sum_{k=0,\pm1,\pm2\dots}^{\infty} P_k \cos[kp_r(\omega_r t + \phi)] \tag{7.122}$$

where, P_k is the amplitude of the k-th harmonic of the rotor-side permeance.

The radial and tangential components of the airgap flux density – $B_{awr}(\phi,t)$ and $B_{awt}(\phi,t)$, generated by the armature winding can be obtained by (7.121) and (7.122), which can be written as:

$$\begin{cases} B_{awr}(\phi,t) = F_{awr}(\phi,t)\Lambda_r(\phi,t) \\ \quad = \frac{1}{2}\sum_k^{\infty}\sum_m^{\infty} F_m P_k\{\cos[(1+k)\omega t + (m+kp_r)\phi + \phi_{r0}] \\ \quad + \cos[(1-k)\omega t + (m-kp_r)\phi + \phi_{r0}]\} \\ \quad + \frac{1}{2}\sum_k^{\infty}\sum_n^{\infty} F_n P_k\{\cos[(1+k)\omega t + (-n+kp_r)\phi + \phi_{r0}] \\ \quad + \cos[(1-k)\omega t + (-n-kp_r)\phi + \phi_{r0}]\} \\ B_{awt}(\phi,t) = F_{awt}(\phi,t)\Lambda_r(\phi,t) \\ \quad = \frac{1}{2}\sum_k^{\infty}\sum_m^{\infty} F_m P_k\{\cos[(1+k)\omega t + (m+kp_r)\phi + \phi_{t0}] \\ \quad + \cos[(1-k)\omega t + (m-kp_r)\phi + \phi_{t0}]\} \\ \quad + \frac{1}{2}\sum_k^{\infty}\sum_n^{\infty} F_n P_k\{\cos[(1+k)\omega t + (-n+kp_r)\phi + \phi_{t0}] \\ \quad + \cos[(1-k)\omega t + (-n-kp_r)\phi + \phi_{t0}]\} \end{cases} \tag{7.123}$$

Equation (7.123) indicates that there are also rich spatial harmonics in the radial and tangential components of the airgap flux density generated by the armature MMF. According to (7.120) and (7.123), airgap field harmonics in FMPM machines can be expressed as:

$$\begin{cases} B_r(\phi,t) = \sum_\nu B_{r\nu}\cos[\nu\phi - \phi_{r\nu}(t)] \\ B_t(\phi,t) = \sum_\nu B_{t\nu}\cos[\nu\phi - \phi_{t\nu}(t)] \end{cases} \tag{7.124}$$

where $B_{r\nu}$ and $B_{t\nu}$ are Fourier coefficients of the νth harmonic in the radial component B_r and tangential component B_t. $\phi_{r\nu}$ and $\phi_{t\nu}$ are phases of the νth airgap field harmonic. The electromagnetic torque based on Maxwell Stress Tensor can be expressed as [40]

$$T_{out}(t) = \frac{r_g^2 l_{stk}}{\mu_0}\int_0^{2\pi} B_r B_t d\phi = \sum_q T_{outq}(t) \tag{7.125}$$

where r_g is the radius of the airgap, l_{stk} the machine stack length, and μ_0 the vacuum permeability. B_r and B_t are the radial and tangential components of the on-load airgap magnetic flux density at the radius r_g. T_{outq} is the electromagnetic torque generated by the q-th flux density harmonic pair.

Combining (7.124) with (7.125), $T_{outq}(t)$ can be expressed as:

$$T_{outq}(t) = \frac{\pi r_g^2 l_{stk}}{\mu_0} B_{rq}B_{tq}\cos[\phi_{rq}(t) - \phi_{tq}(t)] \tag{7.126}$$

Meanwhile, the torque ripple generated by the qth flux density harmonic pair can be expressed as:

$$T_{rq} = \frac{\max(T_{outq}(t)) - \min(T_{outq}(t))}{\text{mean}(T_{outq}(t))} \tag{7.127}$$

According to (7.126) and (7.127), the amplitude of the h-th torque component can be calculated by:

$$T_{out}(h) = \sum_{h} T_{out\text{-}h} = \frac{\pi r_g^2 l_{stk}}{\mu_0} \sum_{o} B_{rh} B_{th} \cos \varphi_h \tag{7.128}$$

where h is the order of harmonics which have a great influence on the output torque of the FMPM machine, and φ_h the phase difference between the radial and tangential components of the h-th airgap field harmonic.

Similarly, the torque ripple can be derived and expressed as:

$$\begin{aligned} T_r(l,t) &= \frac{\max(T_{out}(t)) - \min(T_{out}(t))}{\text{mean}(T_{out}(t))} \\ &= \frac{1}{\sum_{l} B_{rl} B_{tl} \cos \varphi_l} \left\{ \max\left(\sum_{l} B_{rl} B_{tl} \cos[\phi_{rl}(t) - \phi_{tl}(t)] \right) \right. \\ &\qquad \left. - \min\left(\sum_{l} B_{rl} B_{tl} \cos[\phi_{rl}(t) - \phi_{tl}(t)] \right) \right\} \end{aligned} \tag{7.129}$$

where l is the order of airgap field harmonics having a great influence on the torque ripple.

2 Airgap-Harmonic-Oriented Design Optimization Method

It can be observed from (7.128) and (7.129) that airgap field harmonics are directly related to the average output torque and torque ripple, which can be leveraged to

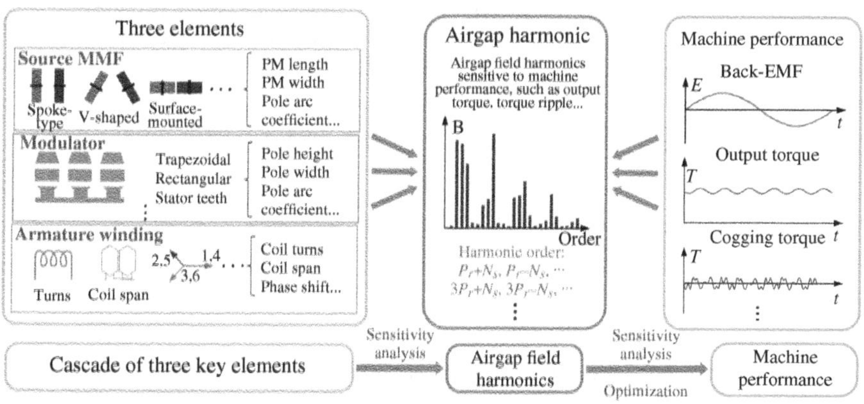

Figure 7.54 Workflow of the airgap-harmonic-orientated design optimization method.

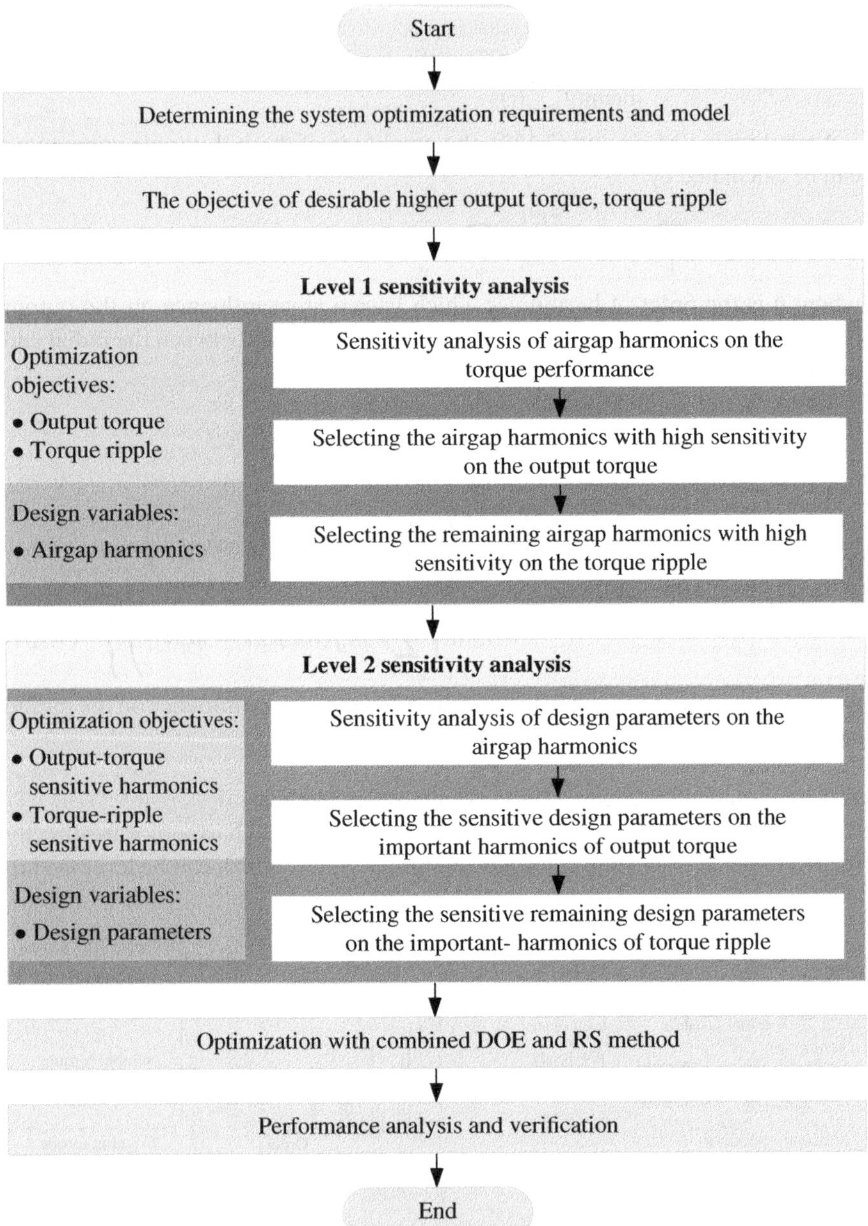

Figure 7.55 Implementation of the airgap-harmonic-oriented design optimization.

optimize FMPM machines. The workflow for the new harmonic-orientated design method is illustrated in Figure 7.54. Airgap magnetic field harmonics serve as not only design variables for the machine performance, but also an objective for geometric parameters. A two-level sensitivity analysis is used to identify key sensitive airgap flux density harmonics and leading design parameters, and contribute to improving the optimization efficiency. With the assistance of the optimization algorithm, field harmonics having a positive effect on the machine performance can be effectively enhanced while those having adverse effects will be suppressed. The key to implement the airgap-harmonic-oriented method is to establish the correlation among the airgap magnetic field harmonics, geometric parameters related to the three elements and performance objectives.

The detailed implementation of the airgap-harmonic-oriented design method is shown in Figure 7.55. First, the design requirements, design variables, and optimization objectives are determined. Then sensitivity analysis is performed to identify airgap field harmonics and design parameters sensitive to the objectives. This step consists of two levels. Level 1 is the sensitivity analysis of airgap field harmonics on the output torque and torque ripple. Level 2 is the sensitivity analysis of design parameters on sensitive airgap field harmonics. Based on the selected airgap field harmonics and design parameters, the optimization is performed using the combined design of experiment (DOE) method and response surface (RS) method to search the optimal designs.

7.6.2 Sensitivity Analysis

Geometric parameters of the machine under study are illustrated in Figure 7.56. Armature winding employs fractional-slot concentrated windings, and dummy

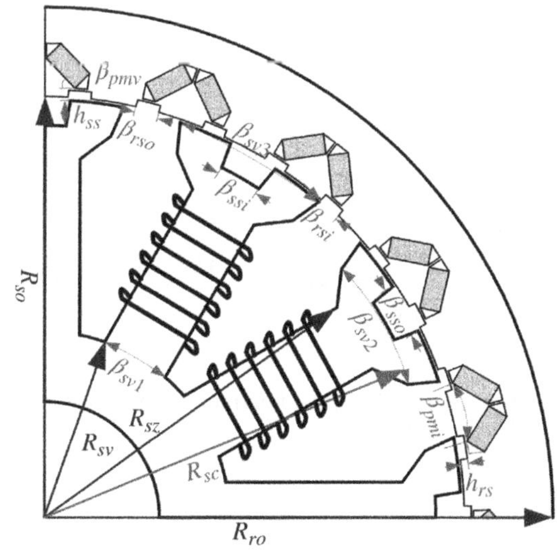

Figure 7.56 Visualization of geometric parameters.

Table 7.27 Design specification and basic parameters.

Parameter	Value
Rated current I_{max} (A)	25
Rated voltage (V)	90
Rated speed n (r/min)	300
Rated torque T (Nm)	$72 \leq T \leq 85$
Torque ripple (peak-to-peak) T_r (Nm)	$1 \leq T_r \leq 5$
Number of stator slots P_s	12
Number of rotor poles P_r	38
Rotor outer radius R_{ro} (mm)	110
Stack length l_{stk} (mm)	60
Airgap length g (mm)	0.7

slots are used on each main stator tooth. The design specifications and key initial parameters are listed in Table 7.27.

1 Sensitivity Analysis of Airgap Harmonics on Torque Performance

To evaluate the sensitivity of airgap field harmonic on the torque performance, the contribution of each field harmonic on the output torque and torque ripple is calculated independently based on (7.128) and (7.129). Figure 7.57a shows that the 19th and 43rd field harmonics from the modulated fundamental MMF are more sensitive to the output than other harmonics, which means the output torque in

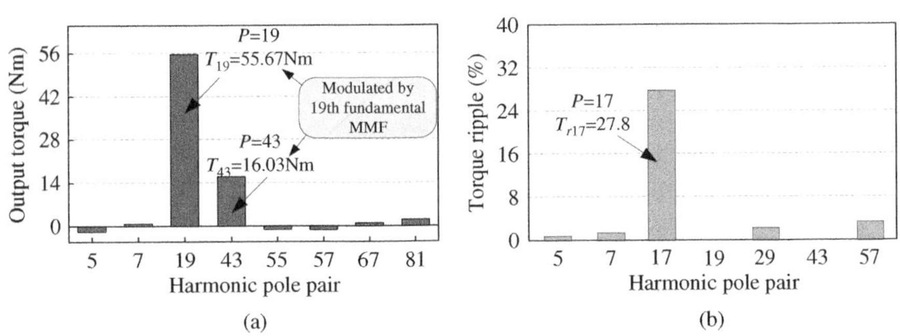

Figure 7.57 The contribution of airgap field harmonics to the torque performance. (a) Output torque. (b) Torque ripple.

the studied V-FMPM machine is mainly produced by the fundamental MMF, and the harmonic MMF contributes little to the output torque.

Similarly, the contribution of each airgap field harmonic to the torque ripple is also calculated and shown in Figure 7.57b. It is shown that the ripple component generated by the 17th airgap field harmonic has the largest amplitude. Namely, the 17th harmonic dominates the torque ripple performance.

2 Sensitivity Analysis of Design Parameters on Airgap Harmonics

Twelve geometric parameters are selected as design parameters to control the machine geometry. β_{pmi} and β_{pmv} are the inclination angle and opening angle of the V-shaped PMs, which are parameters related to the source MMF, i.e. PMs. In addition, some design parameters related to the modulator are selected, where h_{rs}, β_{rsi}, and β_{rso} are the height, internal and external angle of dummy slot on the rotor, respectively. h_{ss}, β_{ssi}, and β_{sso} are the height, internal and external angle of dummy slot on the stator, respectively. β_{sv1} and β_{sv3} are the stator top and bottom tooth angle. R_{sc} and β_{sv2} are the radius and angle of the inner arc in main teeth.

To identify the influence of each design parameter on the airgap field harmonics selected as optimization objectives effectively, the sensitivity index $S(x)$ is introduced as [48]

$$S(x) = \text{mean}[S(X_i)] = \text{mean}\left[\frac{\left(\dfrac{F(x_0 \pm \Delta X_i) - F(X_0)}{F(X_0)}\right)}{\left(\dfrac{\pm\Delta(X_i)}{X_0}\right)}\right] \tag{7.130}$$

where X is the design parameter and X_0 is the initial value of X that is usually defined by adding 10% or 20% to its initial value; $F(X)$ is the optimization objective under the value of X. mean($S(X_i)$) is the average value of $S(X_i)$.

The sensitivity of selected design parameters on the leading airgap field harmonics is evaluated based on (7.130) and the corresponding results are tabulated in Table 7.28. It is shown that each design parameter usually shows different sensitivities on various optimization objectives. Design parameters having high sensitivities are highlighted in Table 7.28.

It can be observed that the opening angle of the V-shaped PM β_{pmi} and the stator bottom tooth angle β_{sv3} have a dominant influence on the 19th harmonic. For the 43rd harmonic, the inclination angle of the V-shaped PMs β_{pmv} and the internal angle of dummy slot on the stator β_{ssi} have larger influence than other parameters. Design parameters with higher sensitivity on the 17th harmonic are the external angle of dummy slot on rotor β_{ssi} and the height of dummy slot on stator h_{ss}.

Table 7.28 Sensitivity indexes of design variables.

Items	Selected objectives		
	19th	43rd	17th
β_{pmi}	**0.3976**	0.3965	−0.3796
β_{pmv}	0.2395	**0.5945**	−0.0161
β_{rsi}	−0.0060	−0.0311	0.1686
β_{rso}	−0.0085	−0.0078	**1.9099**
β_{sv1}	0.0141	0.0069	−0.4348
β_{sv2}	0.0962	0.0845	−0.3952
β_{sv3}	**0.5188**	0.1879	2.9699
β_{ssi}	−0.0433	**2.0714**	2.8863
β_{sso}	−0.1431	−0.2075	−0.2435
h_{rs}	−0.0464	0.3095	0.0345
R_{sc}	0.0150	0.0136	−0.2063
h_{ss}	−0.0360	−0.3761	**2.4287**

To better visualize the sensitivity of design variables on optimization objectives, the data in Table 7.28 are further presented by radar charts, as shown in Figure 7.58. Red points represent positive data for the harmonic sensitivity and green points denote negative ones. In addition, the closer the point is to the radar center, the more sensitive the parameter will be.

7.6.3 Multi-Objective Optimization

The initial values and ranges of the selected design parameters listed in Table 7.29 are used in the optimization. Relationships between design parameters and the torque performance related to the 17th, 19th, and 43rd harmonics are constructed by the RS method [49]. For better understanding, three reprehensive surfaces are visualized in Figure 7.59, where dots denote the actual data points.

It can be seen from Figure 7.59a that the output torque produced by the 19th harmonic increases with the increase of β_{pmi}. It increases first and then decreases as β_{sv3} increases. Based on the trend showing in the RS, β_{pmi} and β_{sv3} can be selected set to 4.2° and 23.5° respectively to achieve the relatively large output torque produced by the 19th harmonic. Similarly, β_{pmv} and β_{ssi} are preferable to be 57.0° and 3.2°, respectively, to achieve a relatively large torque produced by the 43rd harmonic. To obtain a low torque ripple, 3.2° and 5.5 mm have been selected for

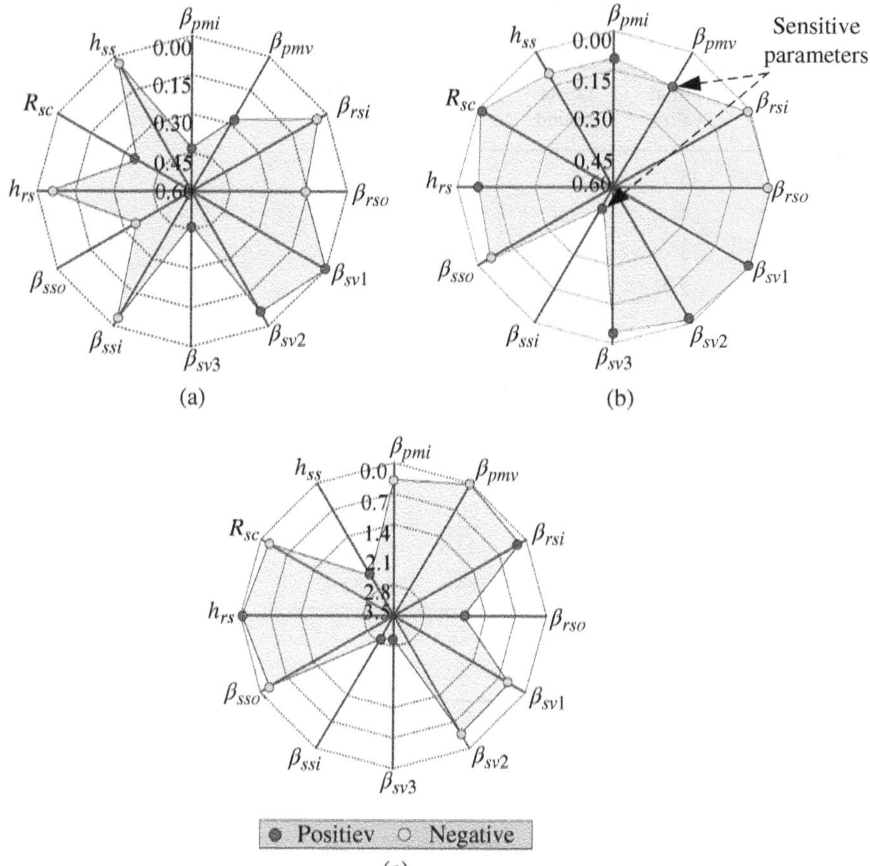

Figure 7.58 Sensitivity analysis for the selected airgap filed harmonics. (a) 19th harmonic. (b) 43rd harmonic. (c) 17th harmonic.

β_{rso} and h_{ss}, respectively. The optimal parameter values are listed in Table 7.29 as well.

Performance comparison between the initial design and optimal design of the V-FMPM machine is tabulated in Table 7.30. To clearly show the optimization results, the output torque and torque ripple contributed by various airgap field harmonics are compared in Figure 7.60. Figure 7.60a shows that the torque produced by the 19th airgap field harmonic in the optimal design increases by 16.7% compared with the initial design, which is from 55.6 to 64.9 Nm. The torque component produced by the 43rd airgap field harmonic reduces slightly from 16.0 to 15.3 Nm. The total torque produced by the 19th and 43rd airgap field harmonic is

Table 7.29 Initial values, ranges and optimal values of design variables.

Variable	Initial value	Lower bound	Upper bound	Optimal value
β_{pmv} (deg)	45	54	57	57
β_{pmi} (deg)	4	3.9	4.2	4.2
β_{rsi} (deg)	3	2.8	3.2	3
β_{rso} (deg)	2.9	2.8	3.4	3.2
h_{rs} (mm)	3	2	7	6.5
β_{ssi} (deg)	6.2	3	7	3.8
β_{sso} (deg)	7	8.5	9.3	9
h_{ss} (mm)	5.5	5.05	6.55	5.5
β_{sv1} (deg)	21.8	21	22	21.8
β_{sv2} (deg)	19.8	19	20	19.8
β_{sv3} (deg)	23	21.5	23.5	23.5
R_{sc} (mm)	83.4	83	84	83.4

Table 7.30 Performance comparison.

Performance	Initial model	Optimal model
Total output torque (Nm)	74.5	83.5
Torque contributed by 19th harmonic (Nm)	55.6	64.9
Torque contributed by 43rd harmonic (Nm)	16.0	15.3
Total torque ripple T_r/T_{out} (%)	3.9	1.7
Torque ripple produced by the 17th harmonic T_{r17}/T_{out17} (%)	27.8	0.8

80.2 Nm, which almost equals the rated torque of 83.5 Nm. This means the output torque in the V-FMPM machine is mainly contributed by the fundamental MMF, and the harmonic MMF has a very limited contribution to the output torque.

Moreover, the torque ripple is reduced from 3.9% to 1.7%, which is mainly due to the significant reduction of the torque ripple caused by the 17th harmonic, which is consistent with the analysis shown in Figure 7.57b.

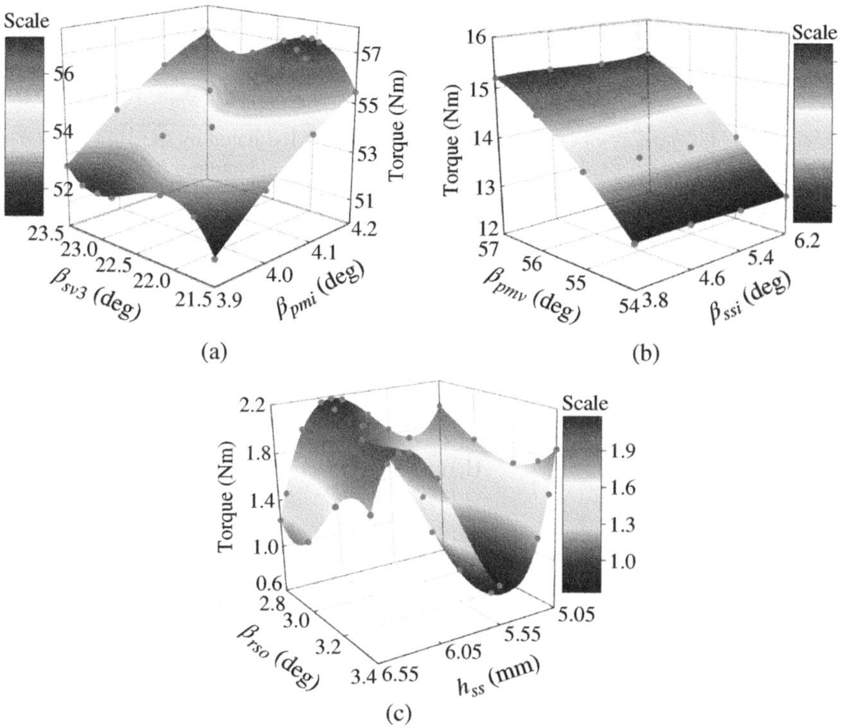

Figure 7.59 Representative surface responses. (a) Output torque produced by the 19th harmonic vs. β_{sv3} and β_{pmi}. (b) Output torque produced by the 43rd harmonic vs. β_{pmv} and β_{ssi}. (c) Output torque produced by the 17th harmonic vs. β_{rso} and h_{ss}.

7.6.4 Optimization Results and Experimental Validation

To verify the effectiveness of the airgap-harmonic-oriented design optimization methodology, the performance evaluation of the V-FMPM machine is evaluated in this subsection. The flux lines and flux density distributions of the initial and optimal designs are compared in Figure 7.61. The optimal design has thinner but wider magnets to provide more PM flux.

The no-load back-EMFs of the V-FMPM machine are compared at 300 r/min, as shown in Figure 7.62. The no-load back-EMF has been improved by 15.3% compared to the initial design and the optimal design exhibits a lower THD for the no-load back-EMF.

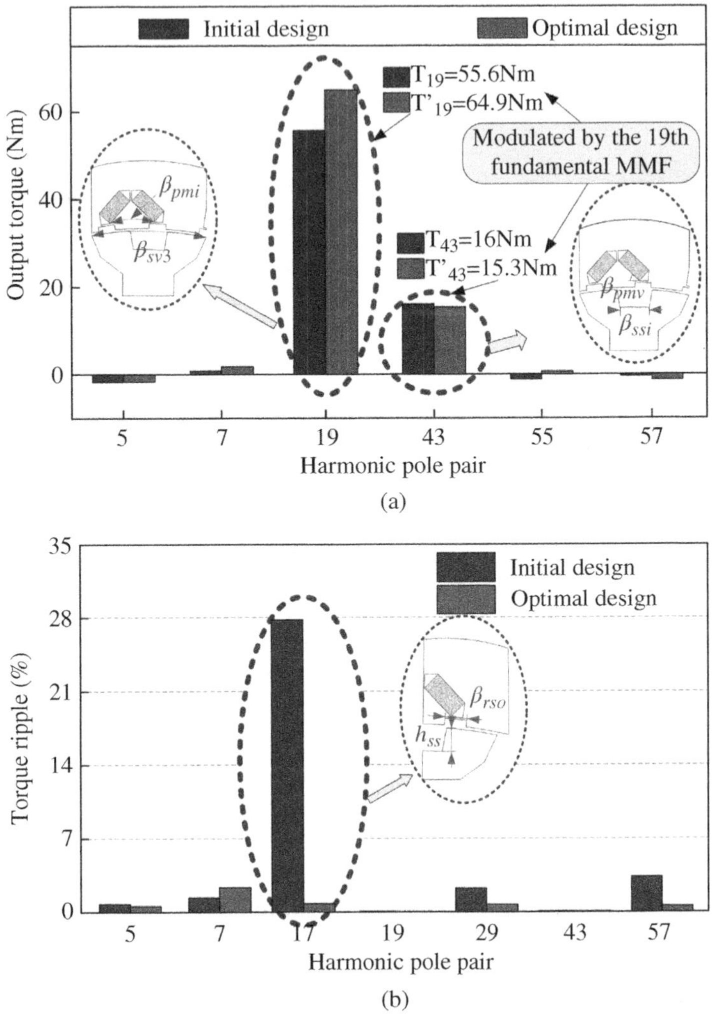

Figure 7.60 Comparisons of torque performance contributions in the initial and optimal motors. (a) Output torque. (b) Torque ripple.

The torque performance is compared in Figure 7.63. The peak-to-peak value of the cogging torque reduces from 1.03 to 0.73 Nm after optimization, as shown in Figure 7.63a. The output torque of the optimal design increases by 12% reaching an average value of 83.5 Nm, which is mainly due to the optimized 19th and 43rd harmonics. This result is consistent with the analysis shown in Figure 7.60a. With the reduction of the 17th harmonic, the torque ripple has been decreased effectively.

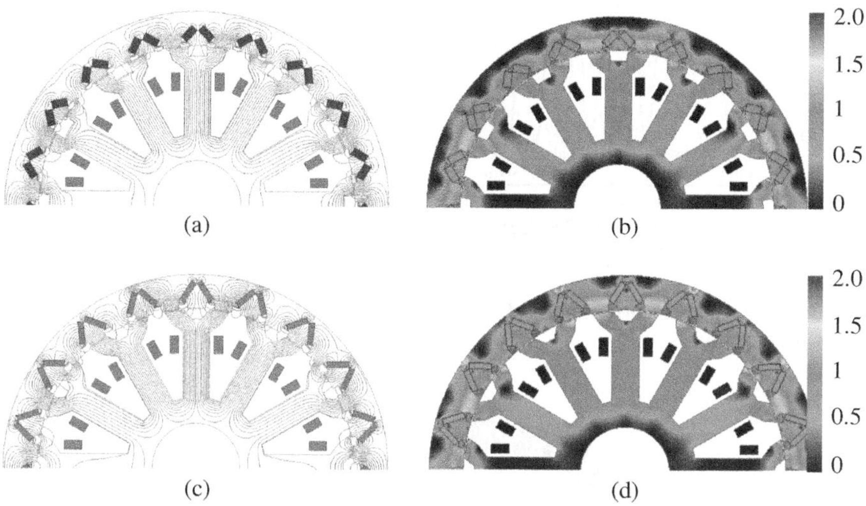

Figure 7.61 Flux lines and flux density distributions of the V-FMPM machine. (a) Flux lines of the initial design. (b) Flux density distribution of the initial design. (c) Flux lines of the optimal design. (d) Flux density distribution of the optimal design.

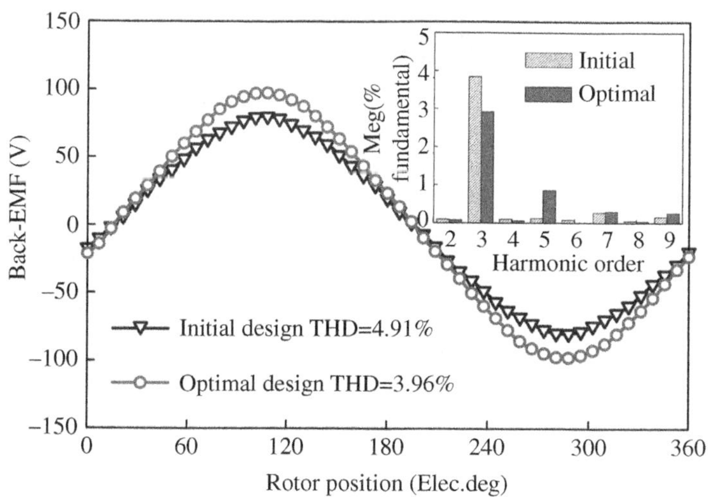

Figure 7.62 No-load back-EMF waveforms.

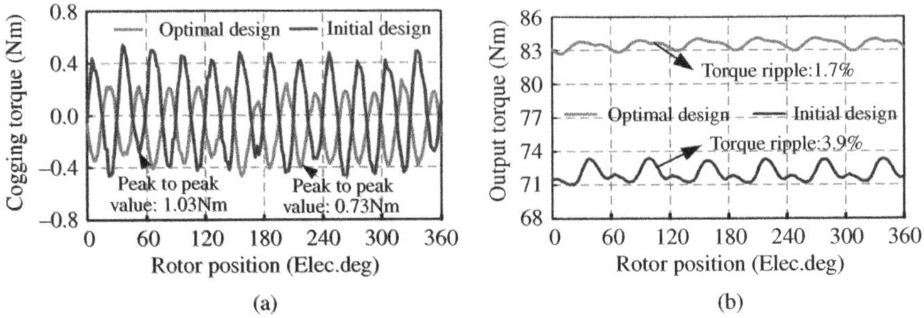

Figure 7.63 Torque performance of the initial and optimal designs. (a) Cogging torque. (b) Output torque.

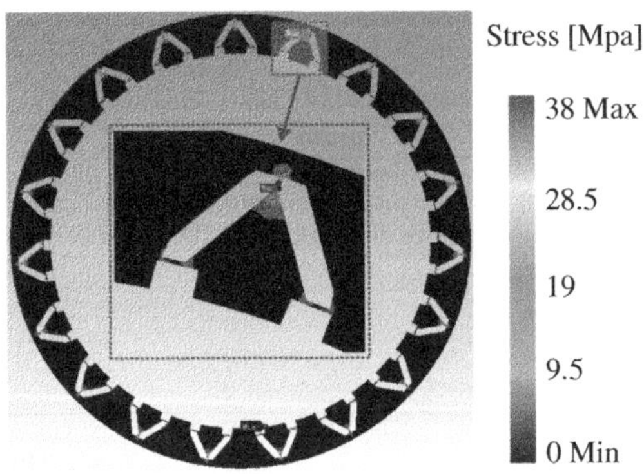

Figure 7.64 Stress distribution of the V-FMPM machine at 5000 r/min.

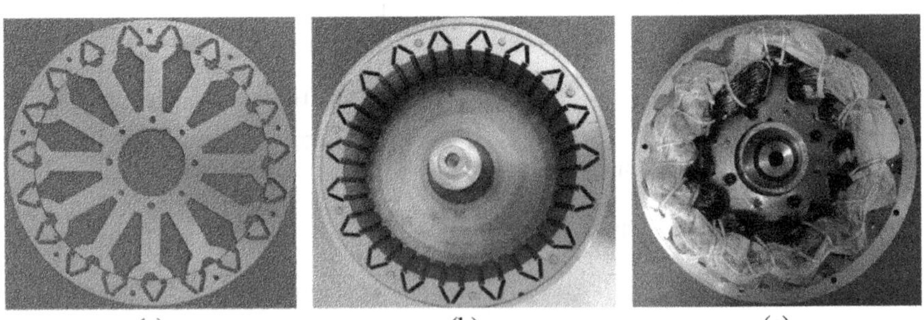

Figure 7.65 Parts of the V-FMPM machine prototype. (a) Stator and rotor lamination sheets. (b) Rotor assembly. (c) Stator assembly.

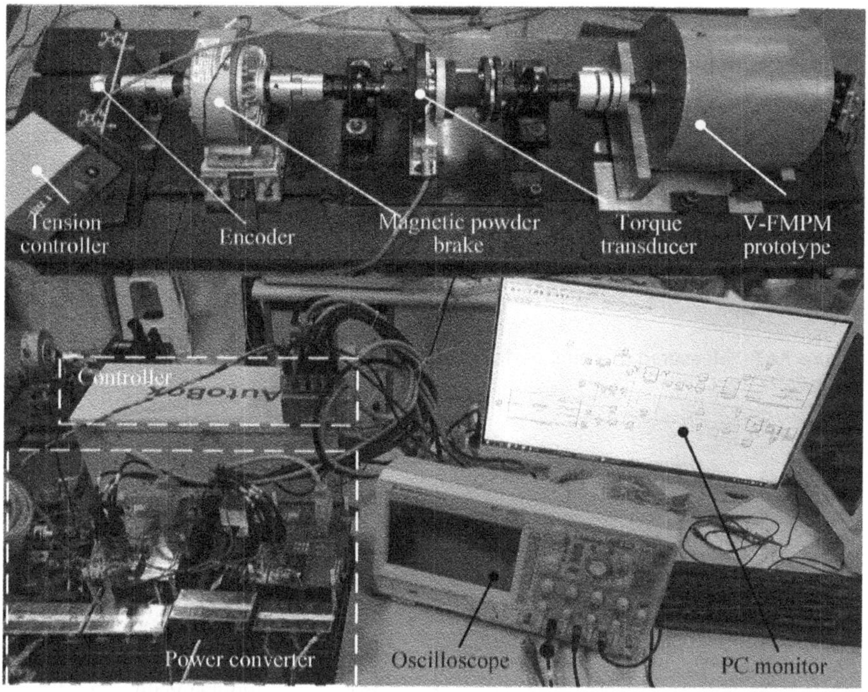

Figure 7.66 Experimental setup for testing the V-FMPM machine prototype.

The stress analysis was conducted to check the mechanical feasibility of the optimized design as shown in Figure 7.64. The maximum stress occurs at the top of V-shaped PMs, where the specific stress value reaches 38 MPa, which is within the acceptable range of the used lamination material.

A 2.5 kW machine prototype was manufactured for experimental testing, as shown in Figure 7.65. The experimental setup is depicted in Figure 7.66.

The measured no-load back-EMF at 300 r/min is sinusoidal and its amplitude reaches 94 V as shown in Figure 7.67. The amplitude spectrum and THD of the measured waveform are shown in Figure 7.67b. The THD of the measured back-EMF is 4.17% while the simulated value is 3.96%, showing good agreement.

The measured torque at different phase currents is shown in Figure 7.68 together with the simulation results. The machine delivers 80 Nm in the measurement under the rated operation with a rated current of 25 A. the simulated rated output torque is 83.5 Nm. The slight difference between the simulation and test results may be due to the neglecting of 3D effects in the 2D FEA model, difference in material properties, and manufacturing tolerances.

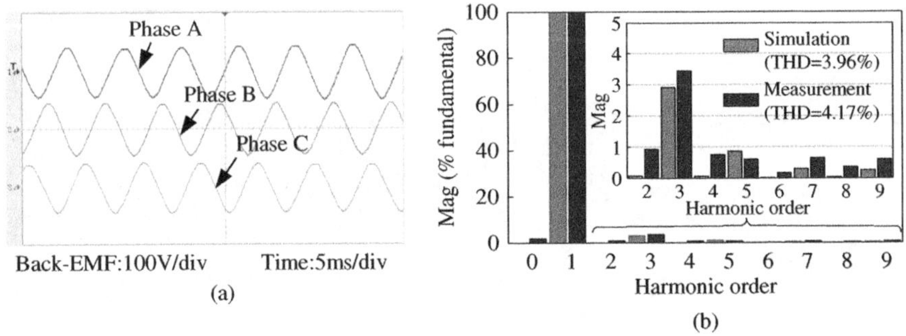

Figure 7.67 Measured machine performance. (a) Back-EMF waveforms. (b) THD of the back-EMF.

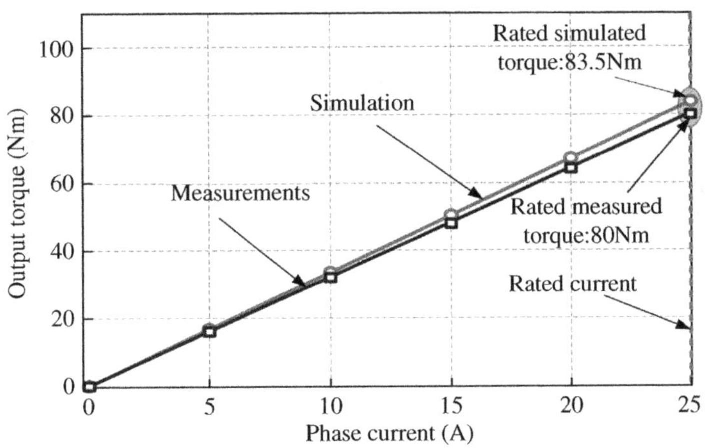

Figure 7.68 Torque versus phase current.

The transient performance was also tested for the V-FMPM machine prototype at low speeds. The starting performance is shown in Figure 7.69a. When the torque command was set to 8 Nm, the rotor speed stepped up to 100 r/min quickly and the phase current increased to 2 A from zero.

Figure 7.69b shows that when the speed was changed from 50 to100 r/min and from 100 to 120 r/min, the motor offers a stable output torque 6 Nm. When the load torque changed from 13 to 19 Nm abruptly, the rotor speed was almost undisturbed, as shown in Figure 7.69c.

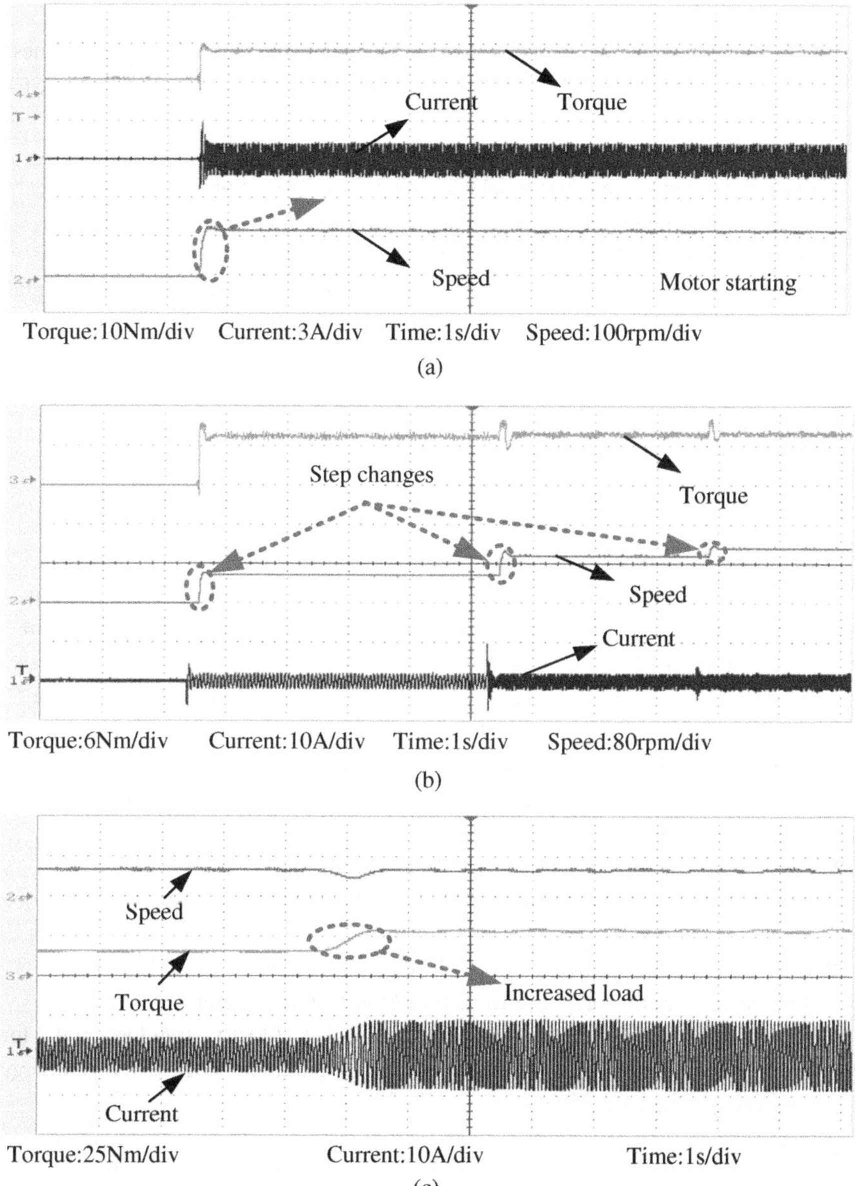

Figure 7.69 Transient waveforms. (a) Staring performance. (b) Transient variable-speed response. (c) Transient variable-load response.

References

1 Han, P. and Cheng, M. (2019). Radial forces in brushless doubly-fed machines. In: *Proceedings of the IEEE Energy Conversion Congress and Exposition (ECCE)*. Baltimore, USA.

2 Han, P. and Cheng, M. (2019). Synthesis of airgap magnetic field modulation phenomena in electric machines. In: *Proceedings of the IEEE Energy Conversion Congress and Exposition (ECCE)*. Baltimore, USA.

3 Wang, J., Cheng, M., Wen, H. et al. (2021). Stray load loss calculation for induction motor by combination of general airgap field modulation theory and 2D FEA. *IEEE Trans. Energy Convers.* 36 (3): 2524–2533.

4 Wen, H., Cheng, M., Zhou, J. et al. (2022). Optimization of rotor salient pole reluctance for typical field modulated electric machines. *IEEE J. Emerging Sel. Top. Power Electron.* 10 (2): 1847–1859.

5 Zhu, X., Jiang, M., Xiang, Z. et al. (2020). Design and optimization of a flux-modulated permanent magnet motor based on an airgap-harmonic-orientated design methodology. *IEEE Trans. Ind. Electron.* 67 (7): 5337–5348.

6 Severson, E., Nilssen, R., Undeland, T., and Mohan, N. (2015). Dual-purpose no-voltage winding design for the bearingless AC homopolar and consequent pole motors. *IEEE Trans. Ind. Appl.* 51 (4): 2884–2895.

7 Alger, P.L., Angst, G., and Davies, E.J. (1957). Stray load losses in polyphase induction machines. *Trans. AIEE* 78: 349–357.

8 Jimoh, A.A., Findlay, R.D., and Poloujadoff, M. (1985). Stray losses in induction machines: part I, definition, origin and measurement. *IEEE Trans. Power Apparat. Syst.* PAS-104 (6): 1500–1505.

9 IEC Std 60034-2-1. (2007).Rotating Electrical Machines – Part 2-1: Standard Methods for Determining Losses and Efficiency from Tests (Excluding Machines for traction vehicles).

10 IEEE Std 112-2017 (Revision of IEEE Std 112-2004). (2017). IEEE Standard Test Procedure for Polyphase Induction Motors and Generators.

11 Kumar, R., Kumar, P., Kanekawa, T., and Oishi, K. (2020). Stray loss model for induction motors with using equivalent circuit parameters. *IEEE Trans. Energy Convers.* 35 (2): 1036–1045.

12 Yamazaki, K. and Haruishi, Y. (2004). Stray load loss analysis of induction motor-comparison of measurement due to IEEE standard 112 and direct calculation by finite-element method. *IEEE Trans. Ind. Appl.* 40 (2): 543–549.

13 Cheaytani, J., Benabou, A., Tounzi, A., and Dessoude, M. (2017). Stray load losses analysis of cage induction motor using 3-D finite-element method with external circuit coupling. *IEEE Trans. Magn.* 53 (6): Article# 8202104.

14 Bašić, M., Vukadinović, D., and Polić, M. (2018). Stray load and iron losses in small induction machines under variable operating frequency and flux: a simple estimation method. *IEEE Trans. Energy Convers.* 33 (2): 869–876.

15 Yamazaki, K., Suzuki, A., Ohto, M. et al. (2011). Equivalent circuit modeling of induction motors considering stray load loss and harmonic torques using finite element method. *IEEE Trans. Magn.* 47 (5): 986–989.

16 Boldea, I. (2020). *Induction Machines Handbook: Steady State Modeling and Performance*, 3e. Boca Raton, USA: CRC Press.

17 Cheng, M., Han, P., and Hua, W. (2017). General airgap field modulation theory for electrical machines. *IEEE Trans. Ind. Electron.* 64 (8): 6063–6074.

18 Zhu, S. and Shi, B. Modeling of PWM-induced iron losses with frequency-domain methods and low-frequency parameters. *IEEE Trans. Ind. Electron.* 69 (3): 2402–2413.

19 Sun, X. and Cheng, M. (2013). Thermal analysis and cooling system design of dual mechanical port machine for wind power application. *IEEE Trans. Ind. Electron.* 60 (5): 1724–1733.

20 Chen, X., Wang, J., Sen, B. et al. (2015). A high-fidelity and computationally efficient model for interior permanent-magnet machines considering the magnetic saturation, spatial harmonics, and iron loss effect. *IEEE Trans. Ind. Electron.* 62 (7): 4044–4055.

21 Yamazaki, K. and Seto, Y. (2006). Iron loss analysis of interior permanent-magnet synchronous motors-variation of main loss factors due to driving condition. *IEEE Trans. Ind. Appl.* 42 (4): 1045–1052.

22 Xue, S., Feng, J., Guo, S. et al. (2017). Iron loss model for electrical machine fed by low switching frequency inverter. *IEEE Trans. Magn.* 53 (11): 2801004.

23 Boglietti, A., Ferraris, P., Lazzari, M., and Profumo, F. (1993). Effects of different modulation index on the iron losses in soft magnetic materials supplied by PWM inverter. *IEEE Trans. Magn.* 29 (6): 3234–3236.

24 Boglietti, A., Ferraris, P., Lazzari, M., and Pastorelli, M. (1995). Change of the iron losses with the switching supply frequency in soft magnetic materials supplied by PWM inverter. *IEEE Trans. Magn.* 31 (6): 4250–4252.

25 Boglietti, A., Cavagnino, A., Lazzari, M., and Pastorelli, M. (2003). Predicting iron losses in soft magnetic materials with arbitrary voltage supply: an engineering approach. *IEEE Trans. Magn.* 39 (2): 981–989.

26 Yamazaki, K., Iida, K., and Terai, Y. (2020). Fast estimation of harmonic losses caused by inverter carrier in interior permanent-magnet synchronous motors by using combination of time- and frequency-domain finite-element analyses. *IEEE Trans. Magn.* 56 (1): Article# 7504704.

27 Zhu, S., Dong, J., Li, Y., and Hua, W. (2020). Fast calculation of carrier harmonic iron losses caused by pulse width modulation in interior permanent magnet synchronous motors. *IET Electr. Power Appl.* 14 (7): 1163–1176.

28 Cheng, M., Wang, J., Zhu, S., and Wang, W. (2018). Loss calculation and thermal analysis for nine-phase flux switching permanent magnet machine. *IEEE Trans. Energy Convers.* 33 (4): 2133–2142.

29 Hua, W., Zhang, G., and Cheng, M. (2015). Investigation and design of a high power flux-switching permanent magnet machine for hybrid electric vehicles. *IEEE Trans. Magn.* 51 (3): 8201805.

30 Cheng, M. and Zhu, Y. (2014). The state of the art of wind energy conversion systems and technologies: a review. *Energy Convers. Manag.* 88: 332–347.

31 Gao, P., Gu, Y., Shah, S. et al. (2019). Calculation and analysis of flux leakage coefficient of interior permanent magnet synchronous motors with fractional slot concentrated windings. *IEEE Trans. Appl. Supercond.* 29 (2): Article# 0602004.

32 Heller, B. and Hamata, V. (1977). *Harmonics Field Effects in Induction Machines*. New York, NY, USA: Elsevier Scientific Pub.

33 Tessarolo, A., Ciriani, C., Elloumi, N. et al. (2021). Fast computation method for stator winding skin-effect additional losses in synchronous machines with open slots and arbitrary rotor geometry. *IEEE Trans. Energy Convers.* 36 (2): 1156–1168.

34 Wang, J., Hu, Y., Cheng, M. et al. (2020). Bidirectional coupling model of electromagnetic field and thermal field applied to the thermal analysis of the FSPM machine. *Energies* 13 (12): 3079.

35 Gieras, J. (2002). *Permanent Magnet Motor Technology: Design and Applications*. London: CRC Press.

36 Wang, J., Cheng, M., and Li, B. (2018). An accurate and fast PM eddy current loss calculation method for flux switching permanent magnet machine. In: *Proceedings of the IEEE Student Conference on Electric Machines and Systems*, 1–6. HuZhou, China.

37 Zhu, S., Cheng, M., Dong, J., and Du, J. (2014). Core loss analysis and calculation of stator permanent-magnet machine considering DC-biased magnetic induction. *IEEE Trans. Ind. Electron.* 61 (10): 5203–5212.

38 Boubaker, N., Matt, D., Enrici, P. et al. (2019). Measurements of iron loss in PMSM stator cores based on CoFe and SiFe lamination sheets and stemmed from different manufacturing processes. *IEEE Trans. Magn.* 55 (1): Article# 8100309.

39 McFarland, J.D., Johns, T.M., and EL-Refaie, A.M. (2015). Analysis of the torque production mechanism for flux-switching permanent–magnet machines. *IEEE Trans. Ind. Appl.* 51 (4): 3041–3049.

40 Wu, Z.Z. and Zhu, Z.Q. (2015). Analysis of air-gap field modulation and magnetic gearing effects in switched flux permanent magnet machines. *IEEE Trans. Magn.* 51 (5): 8105012.

41 Huang, L., Feng, J., Guo, S. et al. (2017). Analysis of torque production in variable flux reluctance machines. *IEEE Trans. Energy Convers.* 32 (4): 1297–1308.

42 Zhang, Z., Tang, X., Wang, D. et al. (2018). Novel rotor design for single-phase flux switching motor. *IEEE Trans. Energy Convers.* 33 (1): 354–361.

43 Jung, Y., Lim, M., Yoon, M. et al. (2018). Torque ripple reduction of IPMSM applying asymmetric rotor shape under certain load condition. *IEEE Trans. Energy Convers.* 33 (1): 333–340.

44 Jiang, J., Bilgin, B., Yang, Y. et al. (2016). Rotor skew pattern design and optimization for cogging torque reduction. *IET Electr. Syst. Transport.* 6 (2): 126–135.

45 Zhu, Z.Q. and Liu, Y. (2018). Analysis of air-gap field modulation and magnetic gearing effect in fractional-slot concentrated-winding permanent magnet synchronous machines. *IEEE Trans. Ind. Electron.* 65 (5): 3688–3698.

46 Xiang, Z., Zhu, X., Quan, L. et al. (2016). Multilevel design optimization and operation of a brushless double mechanical port flux-switching permanent-magnet motor. *IEEE Trans. Ind. Electron.* 63 (10): 6042–6054.

47 Du, Y., Xiao, F., Hua, W. et al. (2016). Comparison of flux-switching PM motors with different winding configurations using magnetic gearing principle. *IEEE Trans. Magn.* 52 (5): Article# 8201908.

48 Lei, G., Liu, C., Zhu, J., and Guo, Y. (2015). Techniques for multilevel design optimization of permanent magnet motors. *IEEE Energy Convers.* 30 (4): 1574–1584.

49 Zhu, X., Huang, J., Quan, L. et al. (2019). Comprehensive sensitivity analysis and multi-objective optimization research of permanent magnet flux-intensifying motors. *IEEE Trans. Ind. Electron.* 66 (4): 2613–2627.

Appendix A

Derivation of Modulation Operators

This appendix shows how to derive the modulation operators of three typical modulators, i.e. the short-circuited coil, salient reluctance pole, and multilayer flux barrier. The modulation operator is defined as a mathematical mapping from the source magnetomotive force (MMF) distribution function $f(x, t)$ to the resultant airgap MMF distribution function, where $f(x, t)$ is an arbitrary function of spatial variable x and temporal variable t, and $f(x, t) = f(x + 2\pi, t)$.

A.1 Derivation of Modulation Operator for Short-circuited Coils

The presence of short-circuited rotor windings introduces additional MMFs, i.e. the rotor MMF, into the air gap. The rotor MMF depends on the rotor winding currents and winding functions. Taking a cylindrical rotor with N_{SC} equally spaced short-circuited coil group as an example, the resultant MMF in the air gap can be obtained by superimposing the rotor winding MMF onto the stator winding MMF:

$$F(\phi, t) = F_s(\phi, t) + F_r(\phi, t) = F_s(\phi, t) + \sum_{j=1}^{N_{SC}} W_j i_j(t), \quad \phi \in [0, 2\pi]. \qquad \text{(A.1)}$$

where W_j and $i_j(t)$ are the winding function and current of the j-th short-circuited coil group. The current flowing in the short-circuited coil group $i_j(t)$ is dependent on the source MMF $f(x, t)$, the speed of the source MMF relative to the rotor, and the structure of rotor short-circuited windings. Therefore, we can define the mathematical mapping from the source MMF distribution (function) to the resultant airgap MMF distribution as follows and denote it by a modulation operator $M(N_{SC})[.]$:

$$M(N_{SC})[f(x, t)] = f(x, t) + \sum_{j=1}^{N_{SC}} W_j i_j(t), \quad x \in [0, 2\pi]. \qquad \text{(A.2)}$$

General Airgap Field Modulation Theory for Electrical Machines: Principles and Practice,
First Edition. Ming Cheng, Peng Han, Yi Du, and Honghui Wen.

Companion website: www.wiley.com/go/genairgapfieldmodulationtheory

To determine the currents in the rotor short-circuited windings, one needs to look at the voltage equations, which can be expressed in matrix form as:

$$\mathbf{0} = [\boldsymbol{r}_r]_{S\times S}[\boldsymbol{i}_r]_{S\times 1} + \frac{d[\boldsymbol{\psi}_r]_{S\times 1}}{dt} \tag{A.3}$$

where $[\boldsymbol{r}_r]_{S\times S}$ is the resistance matrix, $[\boldsymbol{i}_r]_{S\times 1}$ is the rotor current vector, and $[\boldsymbol{\psi}_r]_{S\times 1}$ is the rotor flux linkage vector. The rotor flux linkage consists of two main components, one contributed by the mutual coupling between the rotor short-circuited coils and stator windings, and the other by the rotor winding itself:

$$[\boldsymbol{\psi}_r]_{S\times 1} = [\boldsymbol{\psi}_{rs}]_{S\times 1} + ([\boldsymbol{L}_r]_{S\times S} - [\boldsymbol{L}_{r-r}]_{S\times S})[\boldsymbol{i}_r]_{S\times 1} \tag{A.4}$$

where $[\boldsymbol{\psi}_{rs}]_{S\times 1}$ is the flux linkage linked by a rotor short-circuited coil group resulting from the airgap magnetic field created by the stator winding. $[\boldsymbol{L}_r]_{S\times S}$ is the group inductance matrix, which comprises the self- and mutual inductances of the loops within one coil group. $[\boldsymbol{L}_{r-r}]_{S\times S}$ is the group-group mutual inductance matrix, which comprises the mutual inductance terms between loops of different groups.

The currents solved from (A.3) and (A.4) can then be used as inputs for (A.2) to get the resultant airgap MMF taking into account the influence of short-circuited coils. The derivation process is illustrated in Figure A.1, where at each position along the airgap peripheral, the resultant airgap MMF is obtained by adding the additional MMF introduced by rotor short-circuited coils h_2 to the source MMF h_1. Rotor currents needs to be determined beforehand to get the rotor winding MMF distribution.

A.2 Derivation of Modulation Operator for Salient Reluctance Poles

For the sake of generality, we consider a machine with salient reluctance pole structures on both the stator and rotor sides. In such cases, the airgap magnetic flux density distribution $B(\phi, t)$ is typically obtained by multiplying the source MMF distribution by the airgap permeance:

$$B(\phi, t) = F(\phi, t)\Lambda(\phi, t) \tag{A.5}$$

where $F(\phi, t)$ is the source MMF distribution and $\Lambda(\phi, t)$ the airgap permeance.

By multiplying g/μ_0 to both sides of (A.5), we can get:

$$F_g(\phi, t) = \frac{g}{\mu_0}B(\phi, t) = F_s(\phi, t)\frac{g}{\mu_0}\Lambda(\phi, t) = F_s(\phi, t)\lambda(\phi, t) \tag{A.6}$$

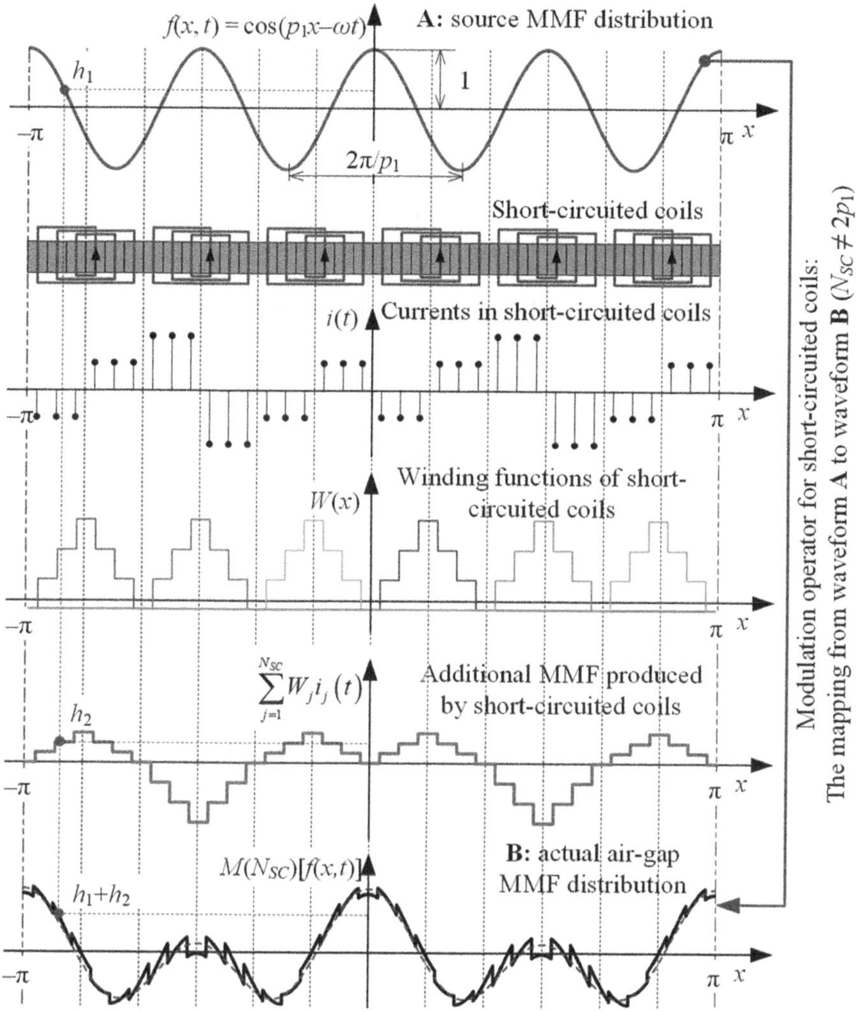

Figure A.1 Graphical derivation of the modulation operator for short-circuited coils.

where $\lambda(\phi, t)$ is the normalized airgap permeability function, and it can be expressed as:

$$\lambda(\phi, \theta_r) = \frac{g}{g(\phi, \theta_r)} = \frac{g}{\frac{1}{g_s^{-1}(\phi, \theta_r)} + \frac{1}{g_r^{-1}(\phi, \theta_r)} - g} \tag{A.7}$$

where $g(\phi, \theta_r)$ is the actual airgap length. $g_s^{-1}(\phi, \theta_r)$ is the inverse airgap function based on the assumption that the rotor is cylindrical and only the stator salient reluctance poles (teeth) are considered. Similarly, $g_r^{-1}(\phi, \theta_r)$ is the inverse airgap function assuming that the stator is cylindrical and only the rotor salient reluctance poles are taken into account.

When the stator bore is cylindrical, the inverse airgap function for the stator side will turn into a constant $1/g$. Similarly, when the rotor is cylindrical, the inverse airgap function for the rotor side will be a constant $1/g$. They can be approximated by "simply" trigonometric functions for the simplicity of analytical derivations or extracted from finite element analysis (FEA) for better accuracy, such as the FE-based approach to identifying the permeance functions.

Therefore, we can define the mathematical mapping from the source MMF distribution (function) to the resultant airgap MMF distribution as follows and denote it by a modulation operator $M(N_{SP})[.]$:

$$M(N_{SP})[f(x,t)] = f(x,t)\lambda(x,\theta_r) \tag{A.8}$$

The derivation process is illustrated in Figure A.2, where at each position along the airgap peripheral, the resultant airgap MMF is obtained by multiplying the source MMF h_1 by a factor h_2.

A.3 Derivation of Modulation Operator for Multilayer Flux Barriers

The stator and rotor cores are assumed to have infinite permeability, and all the flux lines passing through the rotor will be confined in flux guides and parallel to each other. Considering an arbitrary closed surface Γ enclosing a single flux guide, as shown in Figure A.3, the Gauss's Theorem can be applied as:

$$\begin{aligned}
&\int_0^{l_{stk}} \int_G r_g \left[\mu_0 \frac{F(\phi,t)}{g(\phi)}\right] d\phi dl = 0, \quad r_g \in (R_r, R_s) \\
&G = (\phi_0 - \Delta\phi, \phi_0 + \Delta\phi) \bigcup (\phi - \Delta\phi, \phi + \Delta\phi), \phi_0 = (2i+1) \times \frac{2\pi}{N_{MB}} - \phi, \\
&i = 0, 1, \ldots, N_{MB} - 1.
\end{aligned} \tag{A.9}$$

where $F(\phi, t)$ and $g(\phi)$ are the actual airgap MMF distribution function and airgap length, respectively. $\Delta\phi$ is half the width of a single flux guide.

As $\Delta\phi$ approaches 0, (A.9) can be simplified to:

$$\frac{F(\phi,t)}{g(\phi)} + \frac{F(\phi_0,t)}{g(\phi_0)} = 0 \tag{A.10}$$

$$F(\phi,t) = \int_{\phi_0}^{\phi} A(x,t)dx + F(\phi_0,t) \tag{A.11}$$

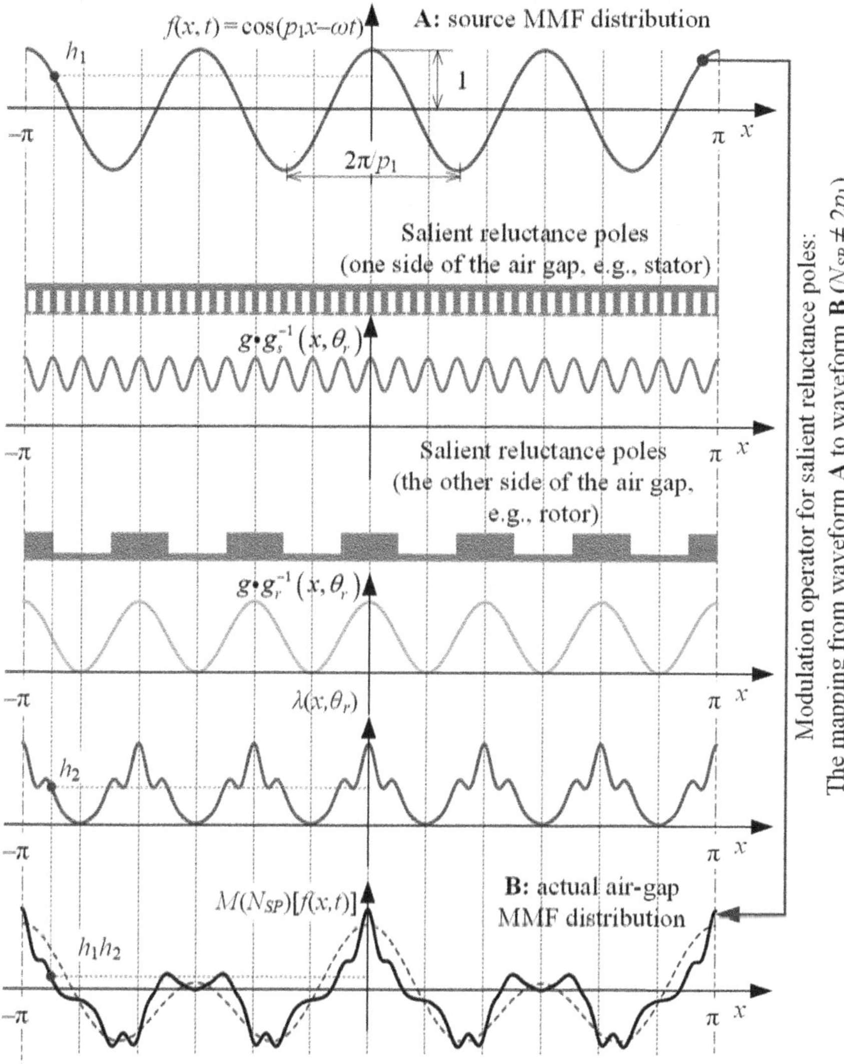

Figure A.2 Graphical derivation of the modulation operator for salient reluctance poles.

where $A(\phi, t)$ is the linear current density distribution along the airgap peripheral. Substituting (A.9) and (A.10) into (A.11) yields:

$$F(\phi, t) = \left\{ \left[1 + \frac{g(\phi_0)}{g(\phi)} \right]^{-1} \int_{\phi_0}^{\phi} A(x, t)dx \right\} i(t)$$

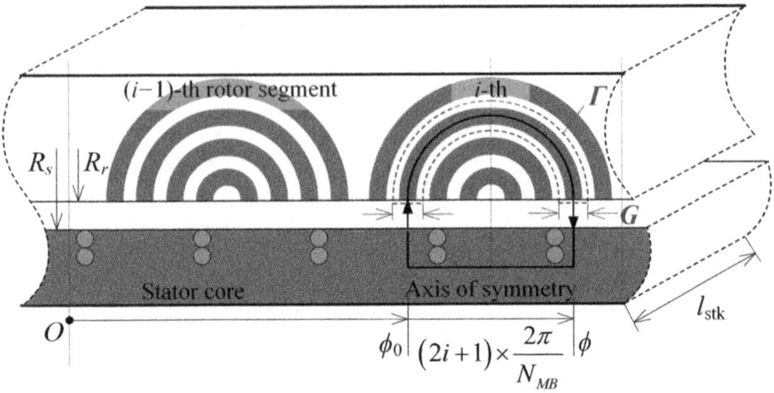

Figure A.3 Derivation of the multilayer flux barrier modulation operator.

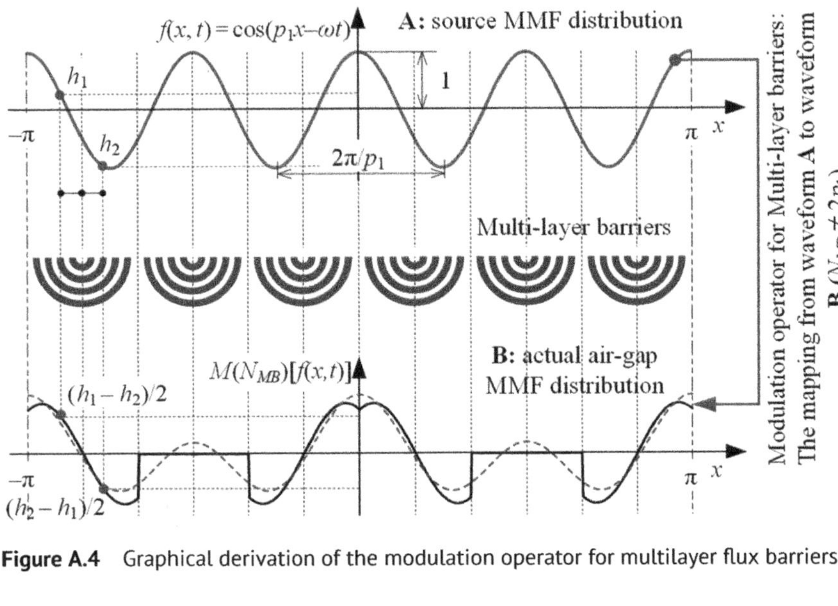

Figure A.4 Graphical derivation of the modulation operator for multilayer flux barriers.

$$= \left[1+\frac{g(\phi_0)}{g(\phi)}\right]^{-1}\{[T(\phi)-\overline{T(\phi)}]-[T(\phi_0)-\overline{T(\phi)}]\}i(t)$$

$$= \left[1+\frac{g(\phi_0)}{g(\phi)}\right]^{-1}[W(\phi)-W(\phi_0)]i(t) \tag{A.12}$$

where $\overline{T(\phi)}$ is the average value of the turn function $T(\phi)$. A mathematical operator $M(N_{MB})[\bullet]$, termed as modulation operator, can now be defined to describe the modification to the source magnetizing MMF distribution:

$$M(N_{MB})[F_s(\phi,t)] = \left\{1+\frac{g(\phi_0)}{g(\phi)}\right\}^{-1}[F_s(\phi,t)-F_s(\phi_0,t)],$$
$$\phi \in \left(i\times\frac{2\pi}{N_{MB}},(i+1)\times\frac{2\pi}{N_{MB}}\right), i=0,1,2,\ldots,N_{MB}-1. \tag{A.13}$$

where $f(\phi)$ is an arbitrary function of ϕ with a period of 2π. The MMF distribution considering the influence of the multi-barrier rotor can then be written as:

$$F(\phi,t) = M(N_{MB})[W(\phi)i(t)] = \left[1+\frac{g(\phi_0)}{g(\phi)}\right]^{-1}$$
$$[W(\phi)-W(\phi_0)]i(t) = M(N_{MB})[W(\phi)]i(t) \tag{A.14}$$

Equation (A.14) indicates that the modulated source MMF can be readily obtained by multiplying the modulated winding function with the corresponding winding current. Generally, the airgap is uniform, and the modulation operator can be further simplified to:

$$M(N_{MB})[F_s(\phi,t)] = \frac{1}{2}\left\{F_s(\phi,t)-F_s\left((2i+1)\times\frac{2\pi}{N_{MB}}-\phi,t\right)\right\},$$
$$\phi \in \left(i\times\frac{2\pi}{N_{MB}},(i+1)\times\frac{2\pi}{N_{MB}}\right) \tag{A.15}$$

where ϕ denotes the angular position in the modulator reference frame. For example, in the brushless doubly-fed reluctance machine (BDFRM) with a multi-barrier rotor, the modulator reference frame means the rotor-speed reference frame.

Therefore, we can define the mathematical mapping from the source MMF distribution (function) to the resultant airgap MMF distribution as follows and denote it by a modulation operator $M(N_{MB})[.]$:

$$M(N_{MB})[f(x,t)] = \frac{1}{2}\left\{f(x,t)-f\left((2i+1)\times\frac{2\pi}{N_{MB}}-x,t\right)\right\},$$
$$x \in \left(i\times\frac{2\pi}{N_{MB}},(i+1)\times\frac{2\pi}{N_{MB}}\right) \tag{A.16}$$

The derivation process is illustrated in Figure A.4, where at each position along the airgap peripheral, the resultant airgap MMF is obtained by the difference between the source MMF values at two points symmetric to the centerline of the flux barrier group. On the left-hand-side of the centerline, it is $(h_1 - h_2)/2$, and on the other side, it is $(h_2 - h_1)/2$.

Appendix B

Magnetic Force of Current-Carrying Conductors in Airgap and in Slots

This appendix shows how magnetic force is produced when a current-carrying conductor is placed in the airgap and in the slot.

When the armature is smooth, the tangential magnetic force on the armature will concentrate on the current-carrying conductors placed on the surface of the armature, and the tangential force and torque can be calculated readily by $F = Bli$.

When the armature core is slotted and conductors are placed inside the slots, the case will be totally different. More details on this topic can be found in [1–3]. Because the permeability of the magnetic core is much larger than the air or vacuum, the flux density in the slot will be much lower. As a result, the magnetic force on the in-slot conductors will reduce dramatically and the tangential magnetic force and torque will be mainly produced on the slot walls.

Figure B.1a illustrates the airgap magnetic field and field inside the slot produced by the rotor under no-load condition. Figure B.1b shows the magnetic field created by the in-slot conductor alone. Assuming the permeability of the magnetic core is constant, the resultant magnetic field can be obtained by superimposing the fields shown in Figure B.1a,b, as illustrated in Figure B.1c.

It is shown that the magnetic field distribution created by the rotor under no-load condition and the magnetic field solely created by the in-slot conductor are symmetrical about the centerline of the slot. As a result, the magnetic force on the left slot wall and that on the right slot wall are balanced and the resultant force/torque is zero.

Under loaded condition, the magnetic field in the airgap and inside the slot become asymmetrical. In the example presented in Figure B.1c, the magnetic force on the left slot wall is enhanced while that on the right slot wall is reduced in accordance with the resultant magnetic fields. As a result, a magnetic force/torque pointing to the right is produced on the stator.

To quantify this force, we assume B_{s0} is the magnetic flux density on the slot wall under no-load condition and B_{si} is the magnetic flux density on the left and right

General Airgap Field Modulation Theory for Electrical Machines: Principles and Practice, First Edition. Ming Cheng, Peng Han, Yi Du, and Honghui Wen.

Companion website: www.wiley.com/go/genairgapfieldmodulationtheory

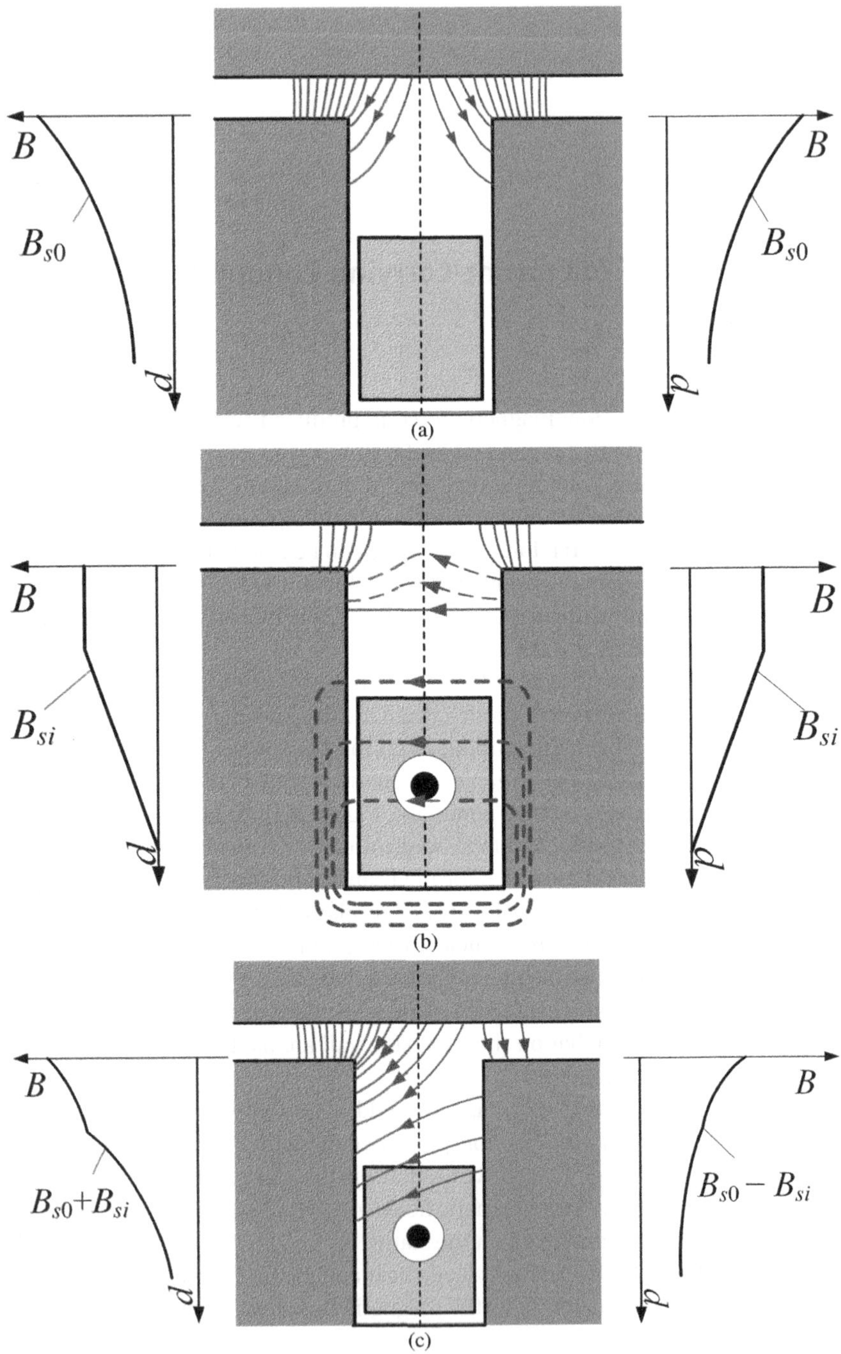

Figure B.1 Flux density distributions in the airgap and along the armature slot walls, (a) under no-load condition, (b) when excited by the in-slot conductor only, (c) under loaded condition.

slot walls created solely by the in-slot conductor. Then the magnetic flux density on the left slot wall under loaded condition is $B_{s1} = B_{s0} + B_{si}$ and that on the right slot wall is $B_{s2} = B_{s0} - B_{si}$. The resultant surface magnetic stress on the left and right slot walls is:

$$\sigma \approx \frac{1}{2\mu_0}\left(B_{s1}^2 - B_{s2}^2\right) = \frac{1}{2\mu_0}(B_{s1} + B_{s2})(B_{s1} - B_{s2})$$
$$= \frac{1}{2\mu_0}(2B_{s0})(2B_{si}) = \frac{2}{\mu_0}B_{s0}B_{si} \tag{B.1}$$

The resultant magnetic force is:

$$F = \int_0^h \sigma l_{stk} dy = \frac{2}{\mu_0} l_{stk} \int_0^h B_{s0}B_{si} dy \tag{B.2}$$

where h is the slot depth and l_{stk} is the stack length. This force is perpendicular to the slot wall but for the armature, it is tangential and responsible for the electromagnetic torque.

Typically, it is very difficult to get the analytical expressions for B_{s0} and B_{si} for an arbitrary slot geometry. For simplicity, here we show a very special case where a current-carry conductor is placed at the bottom of a rectangular slot and slot depth is relatively large. In this case, we can get B_{s0} and B_{si} based on the Schwarz–Christoffel transformation, and then use them to calculate the resultant magnetic force. The following useful conclusions can be drawn:

- When the slot has an infinite depth, the resultant magnetic force on the left and right slot walls is equal to the tangential force acted on the current-carrying conductor placed on the smooth armature surface.
- The resultant force on the slot wall from slot opening to depth d is a function of slot depth as well as slot opening width. Figure B.2 shows that more than 90% magnetic force is produced on the slot wall from slot opening to depth $\leq 2.5g$, where g is the airgap length.
- When current-carrying conductors are placed in the slot, the tangential magnetic force on conductors reduce dramatically due to the shielding effect of teeth. For example, when $b_s/g = 4$, the tangential magnetic force on the in-slot conductor will be reduced to approximately 10% of the value when the conductor is placed in the airgap when $d = 2g$.

When the armature is slotted, the sum of the resultant tangential magnetic force on the armature tooth walls and the tangential force on in-slot conductors is always the same as the tangential force on the current-carrying conductors when they are placed on the smooth armature surface. Therefore, when only the total electromagnetic torque is needed without knowing the torque breakdown between the two, we can always assume that current-carrying conductors are placed on the smooth armature surface and can use $F = Bli$ to calculate the total electromagnetic torque, which is the basis for the developed general airgap field modulation theory.

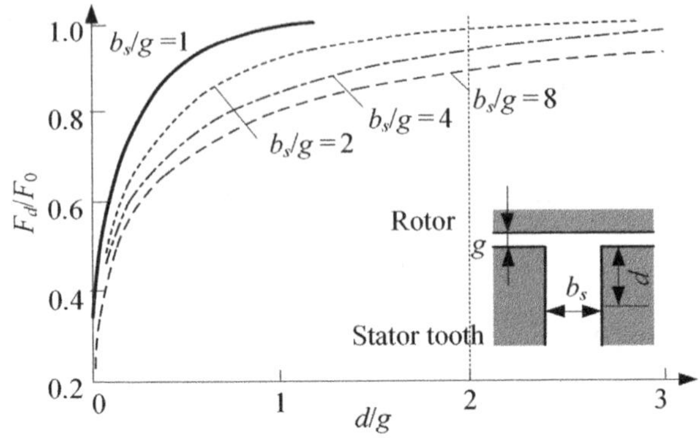

Figure B.2 F_d versus d/g and b_s/g, where b_s is the slot opening width and g the airgap length.

It should be also noted that the armature surface has to be a smooth cylindrical surface when this method is used, so we treat all the influence of the short-circuited coils, salient reluctance poles and multilayer flux barriers on the source magnetomotive force (MMF) as the modulation effect and convert every machine airgap into a uniform one with smooth cylindrical surfaces.

References

1 Williams, F.C. and Mamak, R.S. (1961). Electromagnetic forces in slotted structures. In: *The IEE Monograph*, vol. 456 U, 11–17.
2 Carter, G.W. (1965). Distribution of mechanical forces in magnetized material. *Proc. IEE* 112 (9): 1771–1777.
3 Tang, Y. (2011). *Electric Machinery*, 4e. Beijing: China Machine Press.

Appendix C

Methods for Force and Torque Calculation

This appendix shows two methods commonly used for force and torque calculation in electrical machines, i.e. the Maxwell stress tensor method and virtual work method.

C.1 Maxwell Stress Tensor Method

Current carrying conductors in a magnetic field are subject to magnetic forces. The local force vector is:

$$d\vec{F} = \vec{J} \times \vec{B} \tag{C.1}$$

where $\vec{J}$ is the current density vector, either input and therefore known, or computed in an eddy current region, and $\vec{B}$ is the local value of flux density.

Equation (C.1) is useful for calculating the magnetic force on conductors directly and it can also be used to find the global force on ferromagnetic objects by replacing the ferromagnetic object with an equivalent current distribution, as shown in the derivation of Maxwell stress tensors shown below. In general, it would involve both the surface and volume current distribution.

The Maxwell stress formula has been well documented in [1] by using the magnetic force Eq. (C.1) and the fictitious equivalent current distributions to replace the ferromagnetic material. A volume υ of ferromagnetic material, can be replaced by a distribution of surface current $\vec{J}_s$ and volume current $\vec{J}_\upsilon$ defined by

$$\vec{J}_s = \frac{M_t}{\mu_0}\vec{a}_t \tag{C.2}$$

$$\vec{J}_\upsilon = \frac{\nabla \times M}{\mu_0} \tag{C.3}$$

General Airgap Field Modulation Theory for Electrical Machines: Principles and Practice,
First Edition. Ming Cheng, Peng Han, Yi Du, and Honghui Wen.

Companion website: www.wiley.com/go/genairgapfieldmodulationtheory

where $\vec{a}_t$ is the unit vector tangential to the surface and the subscript t refers to the tangential component of the vector. The magnetization vector is defined as

$$\vec{M} = \frac{\vec{B}}{\mu_0} - \vec{H} \tag{C.4}$$

Applying these formulas, we can replace the ferromagnetic material by volume and surface currents in free space. The idea also applies to the modeling of permanent magnets in finite element analysis by replacing them with equivalent currents. Then the force density can be expressed as

$$\vec{f}_v = \vec{J} \times \vec{B} \tag{C.5}$$

Using the Ampere's law $\nabla \times \vec{H} = \vec{J}$ and the constitutive relationship $\vec{B} = \mu_0 \vec{H}$, (C.5) becomes

$$\vec{f}_v = \left(\nabla \times \frac{\vec{B}}{\mu_0} \right) \times \vec{B} \tag{C.6}$$

By expanding (C.6), we can get the x component of the force density as

$$f_x = \frac{1}{\mu_0} \left(B_z \frac{\partial B_x}{\partial z} - B_z \frac{\partial B_x}{\partial x} + B_y \frac{\partial B_x}{\partial y} - B_y \frac{\partial B_y}{\partial x} \right) \tag{C.7}$$

By adding and subtracting a new term $\frac{1}{\mu_0} B_x \frac{\partial B_x}{\partial x}$ to the right-hand-side of (C.7), we can get

$$f_x = \frac{1}{\mu_0} \left(B_x \frac{\partial B_x}{\partial x} + B_z \frac{\partial B_x}{\partial z} - B_z \frac{\partial B_x}{\partial x} - B_x \frac{\partial B_x}{\partial x} + B_y \frac{\partial B_x}{\partial y} - B_y \frac{\partial B_y}{\partial x} \right) \tag{C.8}$$

Considering the relationship that

$$\frac{\partial B^2}{\partial r} = 2B \frac{\partial B}{\partial r} \tag{C.9}$$

Equation (C.8) becomes

$$\begin{aligned} f_x &= \frac{1}{\mu_0} \left[\frac{1}{2} \frac{\partial B_x^2}{\partial x} + B_z \frac{\partial B_x}{\partial z} + B_y \frac{\partial B_x}{\partial y} - \frac{1}{2} \frac{\partial}{\partial x} \left(B_x^2 + B_y^2 + B_z^2 \right) \right] \\ &= \frac{1}{\mu_0} \left[\frac{\partial}{\partial x} \left(B_x^2 - \frac{1}{2} |B|^2 \right) + \frac{\partial}{\partial y} (B_x B_y) + \frac{\partial}{\partial z} (B_x B_z) - B_x \nabla \cdot \mathrm{B} \right] \end{aligned} \tag{C.10}$$

Considering the Gauss's law $\nabla \cdot \vec{B} = 0$, we can get

$$f_x = \frac{1}{\mu_0} \left[\frac{\partial}{\partial x} \left(B_x^2 - \frac{1}{2} |B|^2 \right) + \frac{\partial}{\partial y} (B_x B_y) + \frac{\partial}{\partial z} (B_x B_z) \right] \tag{C.11}$$

Equation (C.11) is in the form of the divergence of a vector. If we assume the vector is $\vec{T}$, then

$$f_x = \nabla \cdot \vec{T}_x = \frac{\partial T_{xx}}{\partial x} + \frac{\partial T_{xy}}{\partial y} + \frac{\partial T_{xz}}{\partial z} \tag{C.12}$$

where the x, y, and z components of $\vec{T}_x$ are defined as

$$\begin{cases} T_{xx} = \dfrac{1}{\mu_0}\left(B_x^2 - \dfrac{1}{2}|B|^2\right) \\ T_{xy} = \dfrac{1}{\mu_0}B_x B_y \\ T_{xz} = \dfrac{1}{\mu_0}B_x B_z \end{cases} \tag{C.13}$$

Similarly, the y and z components of the force density can be obtained by replacing x with y and z, respectively. The three vectors $\vec{T}_x$, $\vec{T}_y$, and $\vec{T}_z$ can be combined into a tensor T

$$\mathrm{T} = \frac{1}{\mu_0}\begin{pmatrix} B_x^2 - \frac{1}{2}|B|^2 & B_x B_y & B_x B_z \\ B_y B_x & B_y^2 - \frac{1}{2}|B|^2 & B_y B_z \\ B_z B_x & B_z B_y & B_z^2 - \frac{1}{2}|B|^2 \end{pmatrix} \tag{C.14}$$

The volume force density can now be written as the divergence of T as

$$\vec{f}_v = \nabla \cdot \mathrm{T} \tag{C.15}$$

The total force can now be obtained by integrating the volume force density over the whole volume

$$\vec{F} = \int_v (\nabla \cdot \mathrm{T}) dv \tag{C.16}$$

Using the vector divergence theorem, we can transform the volume integral shown in (C.16) into a surface integral and get

$$\vec{F} = \iint_S \mathrm{T} dS \tag{C.17}$$

For two-dimensional geometries, the surface of integration is a line. The unit tangential and normal vectors to the line are

$$\begin{cases} \vec{a}_t = s_x \vec{a}_x + s_y \vec{a}_y \\ \vec{a}_n = -s_y \vec{a}_x + s_x \vec{a}_y \end{cases} \tag{C.18}$$

The incremental force is

$$\begin{aligned} d\vec{F} &= \mathrm{T} \cdot ds = \mathrm{T} \cdot \vec{a}_n dl \\ &= \frac{1}{\mu_0}\begin{pmatrix} B_x^2 - \frac{1}{2}|B|^2 & B_x B_y \\ B_y B_x & B_y^2 - \frac{1}{2}|B|^2 \end{pmatrix}(-s_y \vec{a}_x + s_x \vec{a}_y) dl \end{aligned} \tag{C.19}$$

where $ds = \vec{a}_n dl$ is the incremental integration path, where dl is a differential length along the integration path. The tangential component of the incremental force is

$$\begin{aligned} dF_t &= d\vec{F} \cdot \vec{a}_t \\ &= s_x dF_x + s_y dF_y \\ &= s_x \frac{dl}{\mu_0}\left[-s_y\left(B_x^2 - \frac{1}{2}|B|^2\right) + s_x B_x B_y\right] + s_x \frac{dl}{\mu_0}\left[-s_y B_y B_x + s_x\left(B_y^2 - \frac{1}{2}|B|^2\right)\right] \\ &= \frac{dl}{\mu_0}\left[B_x B_y\left(s_x^2 - s_y^2\right) + s_x s_y\left(B_y^2 - B_x^2\right)\right] \end{aligned} \tag{C.20}$$

The normal component of the incremental force is

$$\begin{aligned} dF_n &= d\vec{F} \cdot \vec{a}_n \\ &= -s_y dF_x + s_x dF_y \\ &= -s_y \frac{dl}{\mu_0}\left[-s_y\left(B_x^2 - \frac{1}{2}|B|^2\right) + s_x B_x B_y\right] + s_x \frac{dl}{\mu_0}\left[-s_y B_y B_x + s_x\left(B_y^2 - \frac{1}{2}|B|^2\right)\right] \\ &= \frac{dl}{\mu_0}\left[B_x^2 s_y^2 + B_y^2 s_x^2 - \frac{1}{2}|B|^2\left(s_x^2 + s_y^2\right) - 2B_x B_y s_x s_y\right] \\ &= \frac{dl}{\mu_0}\left[B_x^2 s_y^2 + B_y^2 s_x^2 - \frac{1}{2}B_x^2\left(s_x^2 + s_y^2\right) - \frac{1}{2}B_y^2\left(s_x^2 + s_y^2\right) - 2B_x B_y s_x s_y\right] \\ &= \frac{dl}{2\mu_0}\left[2B_x^2 s_y^2 + 2B_y^2 s_x^2 - B_x^2 s_x^2 - B_x^2 s_y^2 - B_y^2 s_x^2 - B_y^2 s_y^2 - 4B_x B_y s_x s_y\right] \\ &= \frac{dl}{2\mu_0}\left[\left(B_x^2 s_y^2 + B_y^2 s_x^2 - 2B_x B_y s_x s_y\right) - \left(B_x^2 s_x^2 + B_y^2 s_y^2 + 2B_x B_y s_x s_y\right)\right] \end{aligned} \tag{C.21}$$

The tangential and normal components of the flux density vector are

$$\begin{cases} B_t = \vec{B} \cdot \vec{a}_t = B_x s_x + B_y s_y \\ B_n = \vec{B} \cdot \vec{a}_n = -B_x s_y + B_y s_x \end{cases} \tag{C.22}$$

Then

$$\begin{aligned} B_n B_t &= -B_x^2 s_x s_y + B_x B_y s_x^2 - B_x B_y s_y^2 + B_y^2 s_x s_y \\ &= B_x B_y\left(s_x^2 - s_y^2\right) + s_x s_y\left(B_y^2 - B_x^2\right) \end{aligned} \tag{C.23}$$

and

$$B_n^2 - B_t^2 = \left(B_x^2 s_y^2 - 2B_x B_y s_x s_y + B_y^2 s_x^2\right) - \left(B_x^2 s_x^2 + 2B_x B_y s_x s_y + B_y^2 s_y^2\right) \tag{C.24}$$

It is clear that the tangential and normal components of the incremental force are

$$\begin{cases} dF_t = \dfrac{B_n B_t}{\mu_0} dl \\ dF_n = \dfrac{B_n^2 - B_t^2}{2\mu_0} dl \end{cases} \tag{C.25}$$

The force densities are

$$\begin{cases} f_t = \dfrac{B_n B_t}{\mu_0} \\ f_n = \dfrac{B_n^2 - B_t^2}{2\mu_0} \end{cases} \tag{C.26}$$

which are the familiar expressions for Maxwell stresses.

For radial-flux rotating electrical machines, the radial (normal) and tangential components of airgap flux density can be expressed as

$$\begin{cases} B_t(\phi, t) = \sum_k B_{tk} \cos[k\phi - \theta_{tk}(t)] \\ B_n(\phi, t) = \sum_k B_{nk} \cos[k\phi - \theta_{nk}(t)] \end{cases} \tag{C.27}$$

where B_{tk} and B_{nk} are the Fourier coefficients of k-pole-pair components from B_t and B_n, respectively, while θ_{tk} and θ_{nk} are the corresponding phases.

The electromagnetic torque calculated by the Maxwell stress tensor method based on the obtained airgap flux density waveforms can be expressed as

$$T_{em}(t) = \int_0^{2\pi r_g} r_g l_{stk} dF_t = \frac{r_g^2 l_{stk}}{\mu_0} \int_0^{2\pi} B_n(\phi, t) B_t(\phi, t) d\phi \tag{C.28}$$

where r_g is the airgap radius, μ_0 is the vacuum permeability, and l_{stk} is the stack length. Considering the torque contributions of each spatial harmonic pole pair, the torque equation can be further written as

$$T_{em}(t) = \sum_k T_{emk}(t) = \sum_k \frac{\pi r_g^2 l_{stk}}{\mu_0} B_{tk} B_{nk} \cos[\theta_{tk}(t) - \theta_{nk}(t)] \tag{C.29}$$

Equation (C.29) indicates that only the interaction between the radial (normal) and tangential airgap flux density components of the same harmonic pole pairs can contribute to the electromagnetic torque. Equation (C.29) has been used in Chapter 3 and Chapter 5, and Chapter 7 to decompose the torque into contributions produced by airgap flux density harmonic pairs (B_{tk}, B_{nk}) of different pole pairs.

C.2 Principle of Virtual Work

The virtual work method has also been widely used to calculate the electromagnetic torque or force within electrical machines using magnetic field coupling to accomplish the electromechanical energy conversion. Both the magnetic stored energy and co-energy can be used for the derivation.

C.2.1 Torque Derived from Magnetic Stored Energy and Virtual Displacement

For the sake of generality, we assume the studied electrical machine has m electrical ports, i.e. windings connected to the power supply, and one mechanical port, i.e. the shaft. By neglecting the core losses and considering the winding losses using resistors at the winding terminals, the electrical machine can be treated as an ideal loss-free magnetic energy storage system, as illustrated in Figure C.1.

The magnetic stored energy W_m is a state function, determined uniquely by the values of the independent state variables ψ and θ, where ψ is the flux linkage, and θ the angular displacement. This can be shown explicitly by the following equation

$$dW_m(\psi_1, \ldots, \psi_k, \ldots, \psi_m, \theta) = \sum_{k=1}^{m} i_k d\psi_k - T_{em} d\theta \tag{C.30}$$

The total differential of W_m with respect to the two state variables ψ and θ is

$$dW_m(\psi_1, \ldots, \psi_k, \ldots, \psi_m, \theta) = \sum_{k=1}^{m} \left.\frac{\partial W_m}{\partial \psi_k}\right|_{\psi_1, \ldots, \psi_{k-1}, \psi_{k+1}, \ldots, \psi_m, \theta} d\psi_k + \left.\frac{\partial W_m}{\partial \theta}\right|_{\psi_1, \ldots, \psi_k, \ldots, \psi_m} d\theta$$

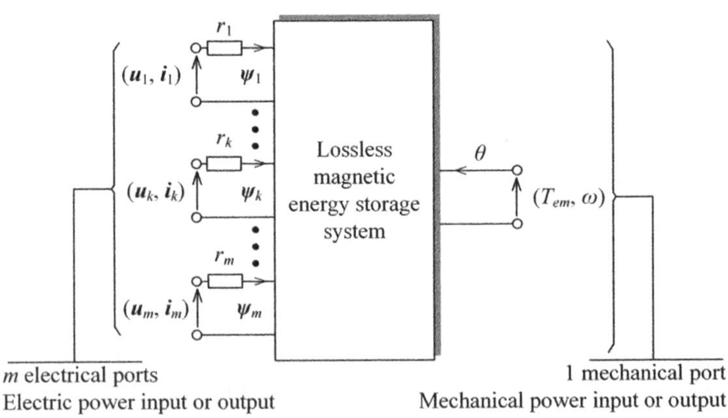

Figure C.1 Representation of an electrical machine with m electrical ports and one mechanical port as a lossless magnetic energy storage system.

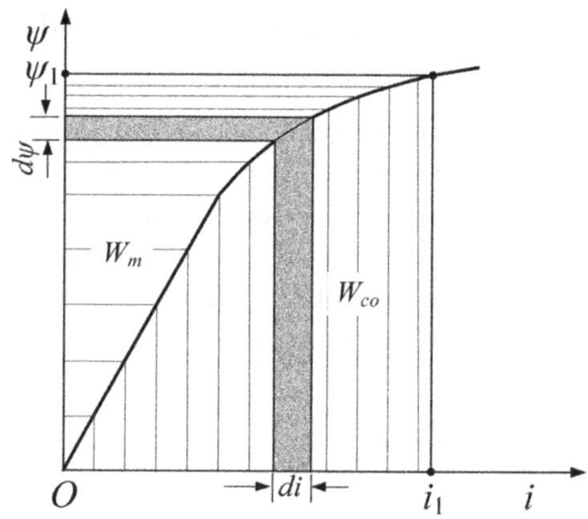

Figure C.2 Magnetic stored energy and co-energy when $m = 1$.

(C.31)

Since ψ and θ are independent variables, (C.30) and (C.31) must be equal for all values of $d\psi$ and $d\theta$, so

$$i_k = \left.\frac{\partial W_m(\psi_1, \dots, \psi_k, \dots, \psi_m, \theta)}{\partial \psi_k}\right|_{\psi_1,\dots,\psi_{k-1},\psi_{k+1},\dots,\psi_m,\theta} \tag{C.32}$$

where the partial derivative is taken while holding θ constant and

$$T_{em} = -\left.\frac{\partial W_m(\psi_1, \dots, \psi_k, \dots, \psi_m, \theta)}{\partial \theta}\right|_{\psi_1,\dots,\psi_k,\dots,\psi_m} \tag{C.33}$$

In this case, ψ is held constant while taking the partial derivative.

For a system with a rotating mechanical displacement, the magnetic stored energy can be expressed as

$$\begin{aligned} W_m(\psi_1, \dots, \psi_k, \dots, \psi_m, \theta) = &\int_0^{\psi_1} i\left(\psi_1', \dots, 0, \dots, 0, \theta\right) d\psi_1' \\ &+ \int_0^{\psi_2} i\left(\psi_1, \psi_2', \dots, 0, \dots, 0, \theta\right) d\psi_2' \\ &+ \dots + \int_0^{\psi_k} i\left(\psi_1, \psi_2, \dots, \psi_{k-1}, \psi_k', \dots, 0, \theta\right) d\psi_k' \\ &+ \dots + \int_0^{\psi_m} i\left(\psi_1, \psi_2, \dots, \psi_k, \dots, \psi_m', \theta\right) d\psi_m' \end{aligned} \tag{C.34}$$

It can also be expressed in terms of the energy density of magnetic field integrated over the volume V of the magnetic field. In this case

$$W_m(\psi, \theta) = \int_V \left(\int_0^B \vec{H} \cdot d\vec{B}\right) dV \tag{C.35}$$

C.2.2 Torque Derived from Co-energy and Virtual Displacement

A mathematical manipulation of (C.30) can be used to define a new state function, known as the co-energy, from which the force can be obtained directly as a function of the current, as shown in Figure C.2. The selection of magnetic stored energy or co-energy as the state function is purely a matter of convenience. They both give the same result, but one or the other may be simpler analytically, depending on the desired result and the characteristics of the system being analyzed.

The co-energy W_{co} is defined as a function of i and θ such that

$$W_{co}(i,\theta) = \sum_{k=1}^{m} i_k\psi_k - W_m(\psi_1,\dots,\psi_k,\dots,\psi_m,\theta) \tag{C.36}$$

Taking the differential of both side of (C.34), we can get

$$\begin{aligned} dW_{co}(i_1,\dots,i_k,\dots,i_m,\theta) &= d\left(\sum_{k=1}^{m} i_k\psi_k\right) - dW_m(\psi_1,\dots,\psi_k,\dots,\psi_m,\theta) \\ &= \left(\sum_{k=1}^{m}\psi_k di_k + \sum_{k=1}^{m} i_k d\psi_k\right) - \left(\sum_{k=1}^{m} i_k d\psi_k - T_{em}d\theta\right) \\ &= \sum_{k=1}^{m}\psi_k di_k + T_{em}d\theta \end{aligned} \tag{C.37}$$

Equation (C.35) indicates that $W_{co}(i,\theta)$ is a state function of the two independent variables i and θ. Thus, its differential can be expressed as

$$dW_{co}(i_1,\dots,i_k,\dots,i_m,\theta) = \sum_{k=1}^{m} \left.\frac{\partial W_{co}}{\partial i_k}\right|_{i_1,\dots,i_{k-1},i_{k+1},\dots,i_m,\theta} di_k + \left.\frac{\partial W_{co}}{\partial \theta}\right|_{i_1,\dots,i_k,\dots,i_m} d\theta \tag{C.38}$$

Equations (C.37) and (C.38) must be equal for all values of di and $d\theta$, thus

$$\psi_k = \left.\frac{\partial W_{co}(i_1,\dots,i_k,\dots,i_m,\theta)}{\partial i_k}\right|_{i_1,\dots,i_{k-1},i_{k+1},i_m,\theta} \tag{C.39}$$

$$T_{em} = \left.\frac{\partial W_{co}(i_1,\dots,i_k,\dots,i_m,\theta)}{\partial \theta}\right|_{i_1,\dots,i_k,\dots,i_m} \tag{C.40}$$

Equation (C.40) gives the mechanical torque directly in terms of i and θ. The partial derivative in (C.38) is taken while holding i constant, thus W_{co} must be a known function of i and θ. For any given systems, (C.33) and (C.40) will give the same result. The choice as to which to use to calculate the torque depends on the user preference or convenience.

For a system with a rotating mechanical displacement, the co-energy can be expressed in terms of the current and angular displacement θ as

$$W_{co}(i_1, \dots, i_k, \dots, i_m, \theta) = \int_0^{i_1} \psi\left(i'_1, \dots, 0, \dots 0, \theta\right) di'_1 + \int_0^{i_2} \psi\left(i_1, i'_2, \dots, 0, \dots 0, \theta\right) di'_2 + \dots + \int_0^{i_k} \psi\left(i_1, \dots, i_{k-1}, i'_k, \dots, 0, \theta\right) di'_k + \dots + \int_0^{i_m} \psi\left(i_1, \dots, i_k, \dots, i'_m, \theta\right) di'_m \quad \text{(C.41)}$$

Equation (C.41) is expressed in circuit-theory terms. In field-theory terms, the co-energy can be expressed as

$$W_{co}(i, \theta) = \int_V \left(\int_{H_c}^{H} \vec{B} \cdot d\vec{H} \right) dV \quad \text{(C.42)}$$

where H_c is the coercivity. For soft magnetic materials, it is typically assumed $B = 0$ when $H = 0$. For hard magnetic materials, for instance, permanent magnets, $B = 0$ when $H = H_c$.

In most cases, further detailed analytical representations of co-energy are difficult to obtain due to the complexity of analyzed geometries and/or magnetic saturation of materials. In such cases, numerical techniques are typically used to evaluate the system energy using (C.34) or (C.35), or the co-energy using (C.41) or (C.42). More details on this topic can be found in [2, 3].

References

1 Salon, S.J. (1995). *Finite Element Analysis of Electrical Machines*, 1e. New York: Springer Science + Business Media, LLC.

2 Fitzgerald, A.E., Kingsley, C., and Umans, S.D. (2003). *Electric Machinery*, 6e. New York: McGraw-Hill Companies, Inc.

3 Tang, Y. (2011). *Electric Machinery*, 4e. Beijing: China Machine Press.

Index

General Airgap Field Modulation Theory for Electrical Machines: Principles and Practice,
First Edition. Ming Cheng, Peng Han, Yi Du, and Honghui Wen.

Companion website: www.wiley.com/go/genairgapfieldmodulationtheory

s

t

u

v

w

Printed and bound by CPI Group (UK) Ltd, Croydon, CR0 4YY

17/09/2026

14964255-0007